HALOGENATED HYDROCARBONS

Solubility–Miscibility with Water

HALOGENATED HYDROCARBONS

Solubility— Miscibility with Water

A. L. Horvath
Imperial Chemical Industries Limited
Runcorn, England

MARCEL DEKKER, INC.   New York and Basel

Library of Congress Cataloging in Publication Data

Horvath, A. L.
    Halogenated hydrocarbons.

    Bibliography: p.
    Includes index.
    1. Halocarbons--Solubility. I. Title.
QD305.H15H67   547'.0204542   81-17424
ISBN 0-8247-1166-1   AACR2

MARCEL DEKKER, INC.

270 Madison Avenue, New York, New York 10016

Current printing (last digit):
10 9 8 7 6 5 4 3 2 1

PRINTED IN THE UNITED STATES OF AMERICA

Preface

For many years the lack of a unifying treatment of solubility and miscibility
between halogenated hydrocarbons and water that is both comprehensive and systematic
has unquestionably delayed more rapid advances in the environmental, medical, and
related sciences. Therefore, it seemed desirable to survey the entire field from a
unified viewpoint and to present it in a consistent format. An abundant amount of
original material as well as numerous references to the literature have been judi-
ciously compiled in this book. The author's purpose is to promote a basic understanding of the mutual solubility concept underlying the selection, behavior, and
computation of solubilities in addition to the data presented in tables, graphs, and
equations. The author points out some of the regularities existing between solubility
and physical properties of solute and solvent, which are useful in the estimation of
solubilities of compounds that have not been experimentally measured.

The book has nine chapters, systematically organized, dealing with experimental
techniques, theoretical treatment of solubility, sources of information, methods of
collection, discrepancies, critical evaluation, and reliability of the solubility
data. In general, the examples given have been chosen to provide simple illustrations of the methods described for application, particularly for solubility data
from experiments and estimation or prediction methods.

The book will be particularly valuable to chemists and chemical engineers in
the field, but it will also serve as a reference source for many scientists in various disciplines, as well as for students, who will find it helpful in gaining a
better knowledge of the solubility of substances and in understanding the inter-
relationships between the various properties of solutes and solvents. It also
provides an extensive compilation of solubility data. The text is directed especially
to researchers concerned with waste disposal, pollution of rivers and seas, and
carcinogenic effects on animals and human beings; it is interdisciplinary in scope
and should be of interest to physicists, physical chemists, chemical engineers,
biologists, medical research workers, and others who are interested in solubility.

I am indebted to many manufacturers of chemicals for their readiness to provide
brochures, booklets, pamphlets, leaflets, and often unpublished data for this book:

> Allied Chemical Corporation
> British Petroleum Chemicals (U.K.) Limited

Dow Chemical Company

du Pont de Nemours and Company, Inc.

Ethyl Corporation

Farbwerke Hoechst AG

Imperial Chemical Industries Limited

Pennwalt Corporation

Phillips Petroleum Company

Pittsburgh Plate Glass Industries, Inc.

Rhone Progil

Shell Chemical Company

Union Carbide Corporation

Special thanks are due du Pont de Nemours and Company, Inc., without whose co-operation this book would not have been possible.

I am grateful to the following bodies for permission to reproduce tables and figures of which they hold the copyright: Academic Press, Fig. 2.77; American Association for the Advances of Science, Fig. 2.29; American Chemical Society, Figs. 2.72 and 2.88, and Tables 2.57, 2.84, 2.85, 2.89, 2.95, 2.104, 2.108, and 2.110; American Institute of Chemical Engineers, Figs. 2.83 and 2.84 and Table 2.103; Canadian Society for Chemical Engineering, Fig. 2.84 and Table 2.56; Gulf Publishing Company, Fig. 2.81; National Research Council of Canada, Table 2.80; Pion Limited, Fig. 2.78; and Springer Verlag, Figs. 2.89 and 2.90.

Thanks are also due a number of colleagues in Imperial Chemical Industries Limited for helpful discussions, especially Drs. D. M. Taylor, G. E. Edward, and P. Rathbone for their welcome suggestions.

I am also grateful to Mr. E. J. Ellis for linguistic help, criticisms, and improvement of the manuscript.

Finally, I want to thank my dear wife Joan, who helped prepare the manuscript, checked the text, and contributed greatly to the whole enterprise.

A. L. Horvath

Contents

Introduction

I. 1 Need for Solubility Data

Users of solubility data prefer to take the information from compiled books which are readily available in most technical libraries. The advantages are apparent considering that the data are scattered in various journals in different languages, and the difficulty of access to many original articles cited in Chemical Abstracts, Referativnyi Zhurnal, Chemisches Zentralblatt, Nuclear Science Abstracts, Citation Index, and so on. However, most of the compiled books on solubility data are at least a decade old, whereas the halogenated hydrocarbons have been developed, marketed, and applied in the commercial sence primarily during the last 10 to 20 years. Therefore, the books available on solubility data contain only a small part of the published results.

The most frequently used books on solubility data are: Stephen and Stephen (1963), Linke (1958-1965), Seidell and Linke (1952), Landolt-Börnstein (1962, 1976a, 1980a), Beilstein (1958-1964), and Kafarov (1961, 1967-1968). In addition to the very limited number of solubility data on halogenated hydrocarbons in these books, there is no selection or recommendation associated with the values presented, particularly if several sets were available, except by Linke (1958-1965) and Seidell and Linke (1952), who usually add some comments on the data presented (although for halogenated hydrocarbons and water systems, no comments are given). Most of the data presented were reproduced from the original sources without further comments. Evaluation and critical examination of the data require considerable effort and experience (Stockmayer, 1978). In a subsequent chapter the selection procedure is discussed in detail and the reasons for discrepancies are examined. Riddick and Bunger (1970) presented some more recent solubility data, however, without any comments on or description of their methods of selection or recommendation. Most of these values are only single solubility data at 20 or 25°C for a few halogenated hydrocarbons without compounds of mixed halogenated hydrocarbons (several kinds of halogen atom(s) in the molecule).

The major users of solubility data are design engineers, who deal with miscellaneous problems in the chemical industries for which the solubilities of various substances, in both the gaseous and liquid phases, are required for design calculations of distillation, extraction, scrubber, disposal, and other types of equipment.

Here, the main emphasis is directed toward the solubility and miscibility between halogenated hydrocarbons and water, and their applications for various objectives. The halogenated hydrocarbons have a very wide spectrum of application and thus easily come into contact with water by mixing or condensation. In a brief review of the miscellaneous aspects of solubility and miscibility between halogenated hydrocarbons and water, the author (Horvath, 1972b) outlined the needs and various uses of solubility data in industry, research, medicine, cosmetics, geothermal power cycle, and so on. In view of the wide application of solubility data, it is apparent that a compilation on the subject is needed. Furthermore, there is no comprehensive work on the theoretical and practical aspects of solubility and miscibility between halogenated hydrocarbons and water. This is a rapidly developing field and future progress will depend entirely on the availability and retrieval of the results achieved by earlier investigators (Rosenzweig, 1975).

Halogenated hydrocarbons are manufactured in large quantities. The world production has been estimated by McConnell et al. (1975) and Pearson and McConnell (1975) of chlorinated solvents in 1973 (see Table I.1). The 10 largest manufacturers are listed in Table I.2. A more recent survey, "Solvents -- The Neglected Parameters," is that of Ball (1977).

One of the largest groups of halogenated hydrocarbons is that of the refrigerants (Freon, Carrence, Genetron, Isotron, Kulene, Ucon, Arcton, Isceon, C. F. Electro, Halon, Aclar, Frigedohn, Edifren, Col-flon, Daiflon, Fresane, Eskimon, Flurion, Flugene, Frigen, Algofrene, Fluorino, etc.) or range of chlorofluorohydrocarbons (see Fig. I.1)(ASHRAE, 1959; Stacey et al., 1963; Banks, 1979; Sanders, 1979). Two additional areas of halocarbon applications are those of household and personal products (e.g., deodorants, perfumes, waxes, polishes, shaving lathers, medicines, anesthetics, biologicals, hair sprays, starches, coatings, fire suppressants)(Gann, 1975; Ford, 1975). In other words, hydrocarbons serve as aerosol propellants, refrigerants, solvents, plastic foam blowing agents, insecticides, weed killers, coatings, finishes, water tracers (Schultz, 1979), lubricants, and dielectrics. A good review of the aerosols is that of Shepherd (1961). Hamilton (1963) has also discussed the use and application of halogenated hydrocarbons. Industrial and commercial applications of methylchloroform and trichloroethylene have been reviewed by Aviado et al. (1976). Other sources of information on the use and application of halocarbons are: Dow Chemical Co. (1966, 1972, 1973), Ethyl Corp. (1958a, 1964, 1972), Hellstrom et al. (1976), Key (1973), Norman and Sherwell (1976), Sherwood et al. (1975), Smith (1973), Stacey et al. (1963), Banks (1979), Sanders (1979), Andelman (1978), Bean et al. (1974), Rosenzweig (1975), Schultz (1979), Hayes and Thompson (1977), and Ball (1977).

Applications and boiling-point relationships are given in Fig. I.1.

During the manufacturing process of halogenated hydrocarbons they may come into contact with process water; or a leak in a refrigeration system (ASHRAE, 1959) may

Table I.1 Estimated Production of Chlorinated Solvents in 1973.

Solvent	Production in 1973 (10^3 tons yr^{-1})	U. S. consumption in 1974 (10^3 tons yr^{-1}) (Ball, 1977)
$CHCl=CCl_2$	1,010	175
$CCl_2=CCl_2$	1,050	332
CH_3-CCl_3	480	236
CH_2Cl_2	400	236
CCl_3F	485	-
CCl_2F_2	570	-
$CH_2=CHCl$	10,500	-
CH_2Cl-CH_2Cl	19,500	-
CCl_4	1,000	-
$CHCl_3$	245	132
CH_3Cl	350	-
C_6H_5Cl	-	172
$1,2-C_6H_4Cl_2$	-	32

Source: McConnell et al., 1975; Pearson & McConnell, 1975.

Table I.2 Major Manufacturers of Halogenated Hydrocarbons.

Manufacturer	Location
Allied Chemical Corporation	U. S. A.
Dow Chemical Company	U. S. A.
du Pont de Nemours and Company	U. S. A.
Farbwerke Hoechst	West Germany
Hooker Chemical Company	U. S. A.
Imperial Chemical Industries Limited	England
Minnesota Mining and Manufacturing Company	U. S. A.
Montecatini	Italy
Pennsalt Chemical Corporation	U. S. A.
Union Carbide Corporation	U. S. A.

Source: Hamilton, 1963.

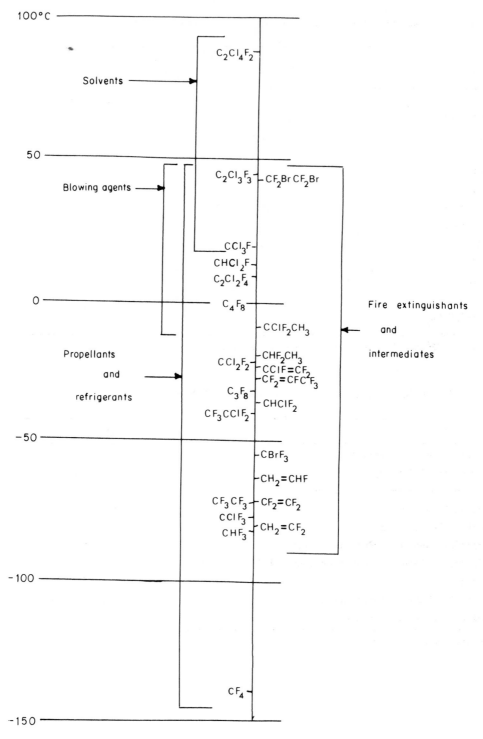

Fig. I.1 Application of commercial halocarbons and their boiling points.

cause some spillage and inevitable mixing with water. As the disposal of wastewater into rivers is now carefully controlled by local authorities, information on the solubility of halogenated hydrocarbons in water are needed so that the expected concentrations of such impurities in the wastewater can be determined.

During the last few years more and more attention has been given to the potential hazards of halocarbons (Pegler and Muir, 1975; National Technical Information Service, 1977). The emphasis is mainly on two aspects: destruction of ozone in the statosphere and the carcinogenic effect of halocarbons on the human population (Anonymous, 1976f; Ferstandig, 1978; Borchardt, 1976; Davis, 1975; Goodman, 1974; Gurnham, 1965; Hay, 1978; Howe, 1975; Lillian et al., 1975; Liptak, 1974; Liss and Slater, 1974; Lovelock, 1975; Lovelock et al., 1973; Molina and Rowland, 1974; Wilkniss, 1973; Wofsy, 1975; McConnell et al., 1975; Norman and Sherwell, 1976; Pearson and McConnell, 1975; Rademacher, 1976; Ricci, 1976; Rowland, 1974; Saffiotti and Cooper, 1976; Scribner, 1976; Sherwood et al., 1975; Singh et al., 1976; Simmonds et al., 1974; Su and Goldberg, 1973; Sugden and West, 1980; Young and Heesen, 1974). An annotated bibliography of water pollution and population problems has been presented in a report by Moore and Moore (1976).

The presence of such common solvents as chloroform and chlorobromomethane in drinking water has been recognized only since 1974 (Andelman, 1978; Environmental Protection Agency, 1978). The presence of halogenated hydrocarbons in water was not reported at all in the 2nd edition of "Water Quality Criteria," (McKee and Wolf, 1963) or in "International Standards for Drinking-Water" (World Health Organization, 1971).

Concentrations of chlorinated hydrocarbons in foodstuffs, human tissue, and marine organisms have also been determined (McConnell et al., 1975; Pearson and McConnell, 1975; Dickson and Riley, 1976).

Photochemical reactions carried out at the National Bureau of Standards indicated that chlorofluorocarbons can form chlorine atoms by reacting with ultraviolet light in the stratosphere. The main emphasis was on two important chlorofluorocarbons: refrigerants 11 ($CFCl_3$) and 12 (CF_2Cl_2). The reasearch showed that up to two atoms of chlorine can be formed from each molecule (Anonymous, 1975f). The stratospheric reactivity of halocarbons by sunlight irradiations was simulated in the laboratory (Lillian et al., 1975). Further excellent reports have been published by Rowland (1974), Rowland and Molina (1975, 1976), and Rowland et al. (1975). The diffusion of chlorofluorocarbons up to the stratosphere has been established by a balloonborne measurement over New Mexico. The principal components found were $CFCl_3$ (R 11) and CF_2Cl_2 (R 12) in various concentrations up to 160 parts per trillion at an altitude of 18.5 km (Anonymous, 1975g).

Good reviews of halocarbons and their effects on stratospheric ozone have been published by Gutowsky (1976), Molina and Rowland (1974), Rowland and Molina (1975, 1976), Anonymous (1976f), Andelman (1978), Gribbin (1979), Sugden and West (1980), and Pegler and Muir (1975).

Table I.3 Concentrations of Halomethanes in the Atlantic.

Halomethane	Temperature (°C)	Concentration in sea (10^{-9} ml gas/ml water, NTP)
CH_3Cl	45	7.2
CH_3Br	45	2.0
CH_3I	45	1.3
CCl_3F	45	0.076
CCl_4	45	0.6

Source: Lovelock, 1975; Lovelock et al., 1973.

Table I.4 Chlorinated Hydrocarbons in Northeast Atlantic Surface Water.

Chlorinated hydrocarbon	Temperature (°C)	Concentration in surface water (ng liter^{-1})
$CHCl_3$	21	8
CCl_4	21	0.14
$CHCl=CCl_2$	21	7.0
$CCl_2=CCl_2$	21	0.5

Source: Murray & Riley, 1973.

Because of the presence of chlorofluorocarbons in the atmosphere, through their solubilities in surface waters, their concentrations will increase in addition to the contents originating from wastewater disposal. Although the solubilities of chlorofluorocarbons are small in water, the large surface area and volume of oceans and rivers will dissolve some fraction of the atmospheric gases, and at the same time some part will evaporate from the surface area (Dilling et al., 1975; Mackay and Meinonen, 1975; Mackay and Wolkoff, 1973; Ryckman et al., 1967; Stumm, 1967; Stumm and Morgan, 1970; Swinnerton and Lamentagne, 1974). Some part of the atmospheric gases will dissolve in rainwater. The impurities in rainwater have been studied by Kölle (1974), particularly for halogenated hydrocarbons. A detailed discussion of the solubility in and the removal processes by the oceans is that of Gutowsky (1976). The concentrations of halomethanes in sea and surface waters have been published by Lovelock (1975) and Lovelock et al. (1973). The numerical values are given in Table I.3 at 45°C. The concentrations of chlorinated hydrocarbons in Northeast Atlantic surface water have been reported by Murray and Riley (1973) at 21°C. The average values are given in Table I.4. Further study was reported by

Table I.5 Average Concentrations of Chlorocarbons in Liverpool Bay Seawater.

Chlorocarbon	Average concentration (ppb by mass)
Trichloroethylene	0.3
Perchloroethylene	0.12
1,1,1-Trichloroethane	0.25
Carbon tetrachloride	0.25
Chloroform	0.1
Hexachlorobutadiene	0.004

Source: Pearson & McConnell, 1975.

Hammer et al. (1978). The concentrations of chlorocarbons in Liverpool Bay sewater have been compiled by Pearson and McConnell (1975) (see Table I.5).

Since the oceans constitute about two-thirds of the earth's surface, the indication is that about 0.07% of the tropospheric R 11 and 0.03% of the R 12 could be dissolved in the oceans at equilibrium. These figures, based upon solubility in water at 288 K, are 4.4×10^{18} molecules cc^{-1} atm^{-1} for R 11 and 1.6×10^{18} molecules cc^{-1} atm^{-1} for R 12. The percentage of other halocarbons in the oceans can be evaluated from their relative solubilities in water. A further study regarding the oceans as a sink of chlorofluoromethanes is that of Junge (1976) in connection with solubility measurements. Other investigators on the subject include Bidleman and Olney (1974), Petrakis and Weiss (1980), Dilling et al. (1975), Andelman (1978), and Hammer et al. (1978).

Ever since it has been suspected that halocarbons can cause cancer, intensive studies have been under way for the establishment of water contaminants throughout the United States. (Anonymous, 1975c; Dowty, Carlisle, and Laseter, 1975; Dowly, Carlisle, Laseter, et al., 1975; Dagani, 1981). The survey of the Environmental Protection Agency showed that drinking water supplies for 79 cities contained halogenated hydrocarbons in the following ranges:

Chloroform .... 0.1 - 311 ppb
Carbon tetrachloride .... 2.0 - 3 ppb
$CHBrCl_2$ .... 0.3 - 116 ppb
$CHBr_2Cl$ .... 0.4 - 100 ppb
$CHBr_3$ .... 0.8 - 92 ppb
CH_2Cl-CH_2Cl .... 0.2 - 6 ppb
$CHCl=CH_2$ .... 1.2 - 5.6 ppb

Large amounts of CCl_4 have been found in the Ohio River (Marx, 1974, 1977). The compounds listed above were found among 700 specific organic chemicals in both

groundwater and surface water systems. In addition to the seven chemicals listed
above, o-dichlorobenzene was also found. In England, the "Annual Report of the Water
Pollution Research Laboratory of the Department of the Environment in 1976" (Anonymous,
1976j) did not report any survey similar to that by the Environmental Protection
Agency. Zürcher and Giger (1976) have reported the presence of more than 70 substances in the River Glatt (Switzerland), particularly CCl_4, C_6H_6, $CCl_2=CCl_2$, and
1,4-dibromobenzene.

Regarding drinking water regulations, the Environmental Protection Agency issued
proposals for interim standards (Environmental Protection Agency, 1975, 1978; Environmental Studies Board, 1973). The water quality parameters were also discussed during
the symposium co-sponsored by the Canada Centre for Inland Waters and the Chemical
Institute of Canada (Anonymous, 1975j). A good account on the various physiological
response to various concentrations of halocarbons in humans is that compiled by
Patty (1962). However, agreement about pollutant thresholds, or levels safe for human
health, is still far from complete (Murray, 1975). Maximum permissible concentrations
in reservoir waters are well-accepted concept in the U. S. S. R. A list of the reported
values for halogenated hydrocarbons has been compiled by Horvath (1976a) (see Chap. 5).
The only proposed Maximum Contaminant Level -- 0.10 mg $liter^{-1}$ for total trihalomethanes -- was reported in the U. S. A. (Environmental Protection Agency, 1978) in
1978. In some cases the levels ranged as high as 0.695 and 0.784 mg $liter^{-1}$ in
terminal water samples in the U. S. A.

The presence of chlorofluorocarbons in drinking water, food, and mediciene was
of concern as early as 1973 (Crossland and Brodine, 1973; Ember, 1975). Investigation
has been widened to establish whether halocarbons are carcinogens (Howe, 1975).
Purchase et al. (1976) proposed six short-term tests for detecting organic chemical
carcinogens. They used 120 compounds in their investigations, of which half were
carcinogens. The relationships among carcinogens, drinking water, and health have
been studied by Bierenbaum (1976); see also Maugh (1978).

The National Cancer Institute (Anonymous, 1975a, 1976c) reported that trichloroethylene induces tumors in mice (liver cancer). The chief application of trichloroethylene is in vapor degreasing of components in metal fabrication plants (Key, 1973;
Silverman, 1974).

Another widely used chlorinated hydrocarbon -- chloroform -- has been found to
be carcinogenic by the National Cancer Institute (Anonymous, 1976a). Chloroform is
widely used for making fluorocarbons in pharmaceutical preparations, dyes, pesticides,
dry cleaning, and solvents. In 1976, the output of five American producers was about
260 million pounds of chloroform. As a result of early findings, it is expected that
chloroform will be prohibited for further use in drug, cosmetic, and food packaging
products (Anonymous, 1976d). Toxicological studies are in progress (Munson et al.,
1977). The source or origin of chloroform in drinking water has been studied by

several investigators (Anonymous, 1977e; Zogorski et al., 1978; Larson & Rockwell, 1978; Andelman, 1978; Maier & Mäckle, 1976; Harms & Looyenga, 1977; Kaiser & Lawrence, 1977; Rook, 1974; Glaze & Henderson, 1975; Kajino, 1977; Environmental Protection Agency, 1978). The chlorination of water showed the formation of chloroform. The analysis of drinking water requires extra care, because the results may be in error owing to delays in the analysis (Fritz, 1976, 1977; see also Hammarstrand, 1976, and Pfaender et al., 1978).

The results of a chronic toxicological study by Dow Chemical U. S. A. indicate that hexachlorobutadiene causes multiple toxicological effects, such as kidney cancer (Anonymous, 1976k).

A 70-week experiment on 6-week-old Syrian golden hamsters, fed ad libitum for life a diet containing 200 ppm of hexachlorobenzene, indicated carcinogenic activity in various tissues. Regarding the carcinogenicity of hexachlorobenzene in relation to humans, it is proposed to obtain epidemiological information for confirmation (Cabral et al., 1977).

Meanwhile, a toxicological study of methyl chloride revealed no evidence of cancer in test animals (Anonymous, 1977a). Methylene chloride is used in aerosols, paint strippers, vapor degreasers, adhesives, and coatings.

Similarly, good news was reported for 1,1,1-trichloroethane and perchloroethylene. An interim report by Dow Chemical summarizing a 24-month inhalation study on rats indicated that no cancer was found (Anonymous, 1975e). These two solvents might become substitutes for trichloroethylene, which is a potential carcinogen (Anonymous, 1975a; National Institute for Occupational Safety and Health, 1976).

A list of dangerous chemicals published by Saffiotti and Cooper (1976) includes hexachlorobenzene and 1,2-dibromoethylene. The National Cancer Institute has stated that 1,2-dibromoethane is carcinogenic (Anonymous, 1978c). A monograph by the American Chemical Society (Searle, 1976) compiles an excellent collection of chemical carcinogens. A list has also been published by Steere (1975). The toxicity of various organic compounds in drinking water has been studied by Tardiff and Deinzer (1973). The list of compounds identified from finished water contains many halogenated hydrocarbons.

Because of the great risk posed by various chemicals in our environment, particular concern is devoted to those that might cause cancer (Ember, 1975; Malone, 1975; Ricci, 1976; Rademarker, 1976; Maugh, 1978). At the 18th annual meeting of the American Cancer Society, several delegates urged the establishment of a national clearinghouse for environmental carcinogenesis (Anonymous, 1976c). The evaluation and testing of chemicals are of concern because there are tens of thousands of chemicals in large-scale commercial use, of which only 6000 to 7000 have been tested in any form. It has been shown that about 1000 of these cause cancer in animals. It is established that about 500 to 1000 new substances enter largescale production each year. The

U. S. National Cancer Institute hopes that the clearinghouse committee will be able
to make recommendation for testing (Anonymous, 1976b). The information gathered
should be made accessible to scientists (Delfino, 1976).

The Environmental Protection Agency has set out new standardized carcinogen
guidelines. The cancer risk assessment will include information on exposure patterns,
metabolic characteristics of chemicals, experimental carcinogenesis studies, epidemiological studies, and dose-response relationships (Anonymous, 1976e; Rich, 1976;
Anonymous, 1977d). A short-term screening test for carcinogens has been proposed by
Bridges (1976). The Occupational Safety and Health Administration has proposed the
following classification of substances:

1. Confirmed carcinogens

2. Suspected carcinogens

3. Noncarcinogens

The classification is based on the results of epidemiological or animal studies or
both (Anonymous, 1977b). The proposed classification would establish three uniform
health standards for carcinogens in our environment. Safety standards for research
involving chemical carcinogens were reported by Steere (1975).

Do chemicals cause cancers? According to Peto (1980) and Epstein and Swatz (1981)
a cancer requires exposure to some carcinogenic agent (e.g., vinyl chloride). Consequently, the recently increased manufacture of organic chemicals, has become the
main hazard of cancer. However, the phenomenon of human cancer is a problem in biology
which is not yet understood. The mechanisms have not been worked out and the causes
of cancer are obscure (Cairns, 1981). There are many kinds of cancer and no single
cause. The proper interpretation of the epidemiological data is often disputed. Many
arguments with all its inconclusiveness dominate the published literature (Anonymous,
1981; Garner and Hertzog, 1981; Whelan, 1981).

To carry out carcinogenic studies, it is essential to know the solubility of
halogenated hydrocarbons in water, blood, and tissues (Horvath, 1976a; Howe, 1975;
Wallnofer et al., 1973).

As a consequence of the research results reported on the potential hazards of
chlorofluorocarbons, a government timetable was announced in the U. S. A. for phasing
out some aerosol propellants. The timetable, which stretched over nearly 2 years,
began on October 15, 1978. Since April 15, 1979, no aerosol products containing
chlorofluorocarbons could be sold in interstate commerce (Anonymous, 1977c, 1978a).

Several manufacturers are very concerned about the production of carbon tetrachloride (about 1.03 billion pounds in 1974), which is the main source for R 11 and
R 12 production (Anderson, 1975). A ban on chlorofluorocarbon aerosols will reduce
the demand for CCl_4 and could trigger plant closings. The search continues for alternative refrigerants, solvents, aerosols, and so on (Bathe, 1976; Birks and Leck, 1976;
Cotton, 1976; Cottrell, 1976; Huttenlocker, 1976; Megaw, 1976; Ward, 1976).

Regarding the 1972 clean water law, 10 big chemical firms have found it difficult to improve wastewater quality significantly. Massive further spending will not reduce pollutants significantly beyond 1977 (Gibney, 1975a, b, 1976).

In addition to the general aspects regarding the need for solubility data as described above, there are a few more specific problems outlined in the published literature, which are discussed next.

I. 2 Specific Problems

I. 2-1 Refrigerants and Corrosion

Halogenated hydrocarbons are used in household refrigerating systems, medium-temperature commercial equipment, airconditioning equipment, and low-temperature ice cream and frozen food cabinets. In these applications there is a risk of leakage, resulting in contact with water. Water entering the refrigeration system will exist in a separate phase and it may freeze and clog in the system. There are several ways in which water can enter a refrigeration system: the system may not be dry; the charging pipes may be damp; the oil added may contain water; or in replacing a part, moist air can enter the system. The formation of ice crystals in refrigerants is inevitable, with subsequent blocking of expansion valves and capillary orifices. Another problem of hydrate formation may arise as the consequence of small traces of water in the system, which is not desirable. The presence of water will also cause a risk of corrosion (Gladis, 1960; Chistyakov, 1965). The water present in a refrigeration system can quickly cause corrosion, with subsequent failure of moving parts and formation of solids that plug valves and tubing and reduce the efficiency of the heat-transfer surfaces. For example, 25 to 50 ppm of water in CF_2Cl_2 can cause corrosion and the formation of ice crystals. Depending on the solubility of water in the refrigerant, freeze-up is likely with refrigerants in which water has greater solubility (e.g., R 23 is a better refrigerant than is R 13). Therefore, solubility and miscibility data are important for design engineers (Ackroyd, 1978; Allied Chemical Corp., 1960b, 1963b; Chinworth and Katz, 1947; Davies et al., 1949; du Pont, 1950; Elsey and Flowers, 1949; Komedera, 1966; Parmelee, 1953-1954; Wittstruck et al., 1961; Petrak, 1975; ASHRAE, 1959; Chistyakov, 1965; Chistyakov and Balezin, 1961; Chistyakov and Shapurova, 1964).

I. 2-2 Propellants

Halogenated hydrocarbons are nonflammable, chemically inert, and virtually non-toxic and therefore are also used as aerosol propellants. For example, refrigerants R 142b and R 152a are miscible in water-ethanol systems and their solubilities in water often produce low-stability or quick-breaking foams. Their solubility in water makes it easier to solubilize or emulsify the propellant in water-based systems. In the proper formulation the propellants can also produce quick-breaking or unstable foams in aqueous solutions. Solubility data are also needed for the use of these substances as propellants for aqueous aerosol mixtures (Allied Chemical Corp., 1960a, 1963a; Parmelee, 1953-1954; Banks, 1979; Sanders, 1979; Rhone Progil, 1972).

I. 2-3 Solvents

In studying the extraction of water and acids by halogenated hydrocarbons, it has been found that monomeric water predominates in most solvents, but that dimer and trimers are also present. This is discussed further in Chap. 2. The solvent is comple- tely miscible with the organic substance but practically immiscible with water. Such cases include the extraction of nicotine from aqueous solutions with trichloroethylene (Reill et al., 1941) and the extraction of acids with methylene dichloride (Sabinin et al., 1970). Considerable industrial effort is expended to determine the partition coefficients in systems in which one component is a halogenated hydrocarbon. Many gases are widely used in the dehydration of delicate substances or to extract them from aqueous solutions. Ethylene dichloride is used as the principal solvent in water- soluble paint removers (BP Chemicals (U.K.) Ltd., 1968). In the case of the separation of HDO and H_2O in the presence of a separating agent by distillation, the relative volatilities can be obtained from solubility data (Eidinoff, 1955). Solubility data are also required to understand the action of hydrotropic solutions in industrial applications (Booth and Everson, 1948, 1949). These solvents are also used for solvent dewaxing of lubricating oils, kerosenes, naphthalenes, and diesel oils (Allied Chemical Corp., 1960a). For further investigations of solute - solvent interactions, the sol- ubility of water in the solvents studied must be known. The wide application of chlorinated solvents is indicated by the large number of common, brand, and trade names by various manufacturers (e.g., trichloroethylenen, see Table I.6) (Högfeldt and Fredlund, 1970; Mannheimer, 1956; Luft, 1957; Ohnishi and Tanabe, 1971; Koradia and Kiovsky, 1977; Banks, 1979; Kinetic Chemicals, 1956; Kirk-Othmer, 1969).

I. 2-4 Anesthetics

Anesthesia occurs when a certain amount of anesthetic agent (e.g., carbon tetra- fluoride, chloroform, halothane, trichloroethylenen) is present in the body. Therefore, it is important to know the factors that influence the uptake and distribution of gases and to understand the differences in behavior found in aqueous solutions. Through solubility measurements it is possible to elucidate the side action of general anesthetics and the rise in decompression sickness. It must be emphasized that several of the earlier solubility determinations did not appreciate the need for equilibration between the liquid and vapor phases, and therefore their results were not reliable. During clinical investigations in connection with anesthetic effects, it is apparent that the solubility of an agent in water, blood, and body tissues is the primary de- terminant. The solubility relationships of drugs and their metabolites were studied by Goldsmith (1974), among others. During the conversion of drugs to metabolities, these experiments showed the general manner in which metabolism changes solubility and hence the penetration of the material into lipid tissue. A study that examined

Table I.6 Common, Brand, and Trade Names for Trichloroethylene

Acetylene trichloride	TRI
Algylen	Triad
Blacosolv	Triasol
Chlorylen	Trichloran
Circosolv	Trichloren
Dow-Tri	Trichloroethene
Ethinyl trichloride	Triclene
Fleck-Flip	Tri-Clene
Gemalgene	Trielene
Lanadin	Trielin
Lethurin	Trilene
Nialk	Triline
Perm-A-Clor	Trimar
Petzinol	Vestrol
Philex	Vitran
TCE	Westrosol
Threthylen	
Trethylene	
Trial	

various anesthetics was reported by Robbins (1946; see also Ashton et al., 1968; Banks, 1979; Larson et al., 1962a, b; Papper and Kitz, 1963; Robbins, 1946; Saidman et al., 1966; Secher, 1971; Wright and Schaffer, 1932; Cohen et al., 1975; Cowles, 1970; Feingold, 1976; Steward et al., 1973, 1974). The disease potential of anesthetic gases has been reviewed by Ferstanding (1978).

I. 2-5 Halogenated Substances in Wastewater

Despite the low solubilities of halogenated hydrocarbons in water, roughly inversely proportional to the molecular weight of the compound, ranging from about 0.05 wt % for heavier compounds to approximately 1 wt % for lighter compounds, industries are discharging large amounts of water containing expensive dissolved compounds, which affects the economics of the processes in which these compounds are used (Nathan, 1978). For example, the waste products of vinyl chloride production contains about 80 substances, mostly chlorinated paraffins (Jensen et al., 1975). Furthermore, wastewater saturated with solvents is not allowed in sewers, because it has a toxic effect on the fuana found in water supplies. Some wastewater plants use chlorination of secondary municipal effluents; although the dosage of the terminal chlorination is moderate, about 10 mg liter^{-1}, the formation of organochlorine com-

Table I. 7 Chemical Indicators of Industrial Contamination

Industrial Chemicals

Benzene	Ethylbenzene
Bis (2-chloroethyl) ether	Hexachlorobenzene
Bis (2-chloroisopropyl) ether	Hexachlorobutadiene
Bromobenzene	Hexachlorocyclopentadiene
4-Bromophenylphenyl ether	Hexachloroethane
Carbon tetrachloride	Nitrobenzene
Chlorinated naphthalenes	Phthalate esters
Chlorobenzene	Polychlorinated biphenyls
Chlorophenols	Propylbenzene
2-Chlorovinyl ether	Styrene
Dichlorobenzenes	1,1,2,2-Tetrachloroethane
Dichlorodifluoromethane	Tetrachloroethylene
1,1-Dichloroethane	Toluene
1,2-Dichloroethane	Trichlorobenzene
1,1-Dichloroethene	1,1,1-Trichloroethane
1,2-Dichloroethene	1,1,2-Trichloroethane
Dichloromethane	Trichloroethylene
1,2-Dichloropropane	Trichlorofluoromethane
1,3-Dichloropropane	Vinyl chloride
Dinitrotoluene	Xylenes

Pesticides

Aldrin	Heptachlor
Atrazine	Heptachlor epoxide
Chlordane	Kepane
DDD, DDE, DDT	Lindane and hexachlorocyclohexanes
Dieldrin	Pentachlorophenol
Endrin	Toxaphene

Polynuclear Aromatic Hydrocarbons

3,4-Benzofluoranthene	Benzo(a) pyrene
1,12-Benzoperylene	Fluoranthene
11,12-Benzofluoranthene	Indenol (1,2,3-cd) pyrene

Source: Environmental Protection Agency, 1978.

pounds is of interest (Glaze and Henderson, 1975). A list of chlorinated organics found in wastewater effluent given by Glaze and Henderson (1975) contains 38 substances. The chemicals listed by the Envirenmental Protection Agency (1978) are given in Table I.7. Regarding microbial cometabolism and the degradation of organic substances in nature, see R. S. Horvath (1972), Lu and Metcalf (1975), and Richardson and Miller (1960). Halogenated hydrocarbons are also present as intermediates for the production of other substances. Wastewaters containing these substances are often formed in these processes and it is therefore important to know the maximum concentrations of the substances in water at various temperatures. Halogenated hydrocarbon compounds exhibit high toxicity to aquatic life. In accessing the effects of spills and aqueous plant effluents containing halogenated hydrocarbons, it is important to know the solubilities in water of individual components of a mixture. From these data the toxicity can be estimated. The disposal of waste material with halogen contents creates another problem. Most of the disposal takes place in effective incineration systems. The waste gases and liquids discharged at various points in the process are absorbed in water to produce dilute hydrochloric acid (in the case of substances with chlorine contents, e.g., polyvinyl chloride), which is then neutralized. However, the combustion gases contain some unburn organic substances which will be absorbed in the wastewater. The solubility of these halogenated compounds in water is important for estimating their concentration and making sure that the toxicity level allowed is not exceeded (Baranaev et al., 1954; Gurnham, 1965; Jensen et al., 1975; Johnston et al., 1976; Nathan, 1978; Hay, 1978; Nex and Swezey, 1954; Kisarov, 1962; Leinonen and Mackay, 1973; Santolari, 1973; Seluzhitskii, 1967; Earhart et al., 1977; Ciaccio, 1971-1973; Dilling et al., 1975). Good reviews of water pollution control have been compiled by Sitting (1969), Richardson and Miller (1960), Robertson et al. (1980), Verschueren (1977), Matthews (1975, 1978), and Wiley (1976). The removal of organic substances from river water has been reported by O'Connor et al. (1977).

I. 2-6 Hydrating Agents and Demineralization

Gas hydrates are a type of clathrate compound, which is a special type of inclusion substance. Clathrate structures involving an aqueous host are referred to as gas hydrates or aqueous clathrates. The monocage gas hydrates appear to be icelike, and many are stable at temperatures above $0^{o}C$. In the hydration process for demineralizing seawater, sewage, and other kinds of impure waters, one of the problems is to select the most economical hydrating agent. The value of the agent chosen depends on its stability and solubility in biological and other aqueous-organic systems. Another important requirement of this process is the prevention of loss of the hydrating agent which might occur by leakage, dissolution in the product stream, or hydrolysis. The purification procedure involves cycling the hydrate and decomposition in the system. There are several prominent hydrating agents (e.g., refrigerants R 21, R 31, R 142b, CH_3Br, CH_3CClF_2) whose economic potential must be evaluated on the basis

of their solubility in water at various temperatures. Hydrate formation is also relevant to the investigation of crystallographic structures and the change of small
molecule gas hydrates with temperature as a result of varying partial occupancy by
the hydrate former within the hydrate water lattice. Some clarification of the nature
of aqueous solutions can be obtained by examining the distribution of the solute between the vapor and liquid phases saturated with the crystalline hydrate. Hydrate
formation has been suggested by Paulin to be a possible mechanism to explain the
anesthetic properties of xenon (Briggs and Barduhn, 1963; Banks, 1979; Glew and
Moelwyn-Hughes, 1953; Glew, 1960; Huang et al., 1965; Kass et al., 1965; Vlahakis et
at., 1969; Wittstruck et al., 1961).

I. 2-7 Hydrolysis

To study the reaction rate of hydrolysis of chlorinated hydrocarbons, solubility
data are needed as a function of temperature. The hydrolysis of dissolved compounds
in water at elevated temperatures is of interest because of the formation of HCl and
its corrosive action on equipment. The solubility of the gas is an approximate measure
of the forces of interaction between methyl halide and water molecules. To obtain
detailed knowledge of the reaction mechanism, the effect of pressure was studied on
the hydrolysis of chloroform, dichlorodifluoromethane, and 3-chloro-3-methylbutyne.
In addition to this, some interesting conclusions have been drawn on the progress of
bound formation and breakage and concomitant compression or decompression in the
solvent surrounding the reacting molecules (le Noble, 1965; le Noble and Duffy, 1964).
In particular, as partially and fully chlorinated hydrocarbons undergo hydrolysis in
the presence of water, mainly at higher temperatures, acid formation increases, which
causes corrosion and damage to equipment. The reaction of chlorinated hydrocarbons
with water is comparatively slow when all water present is dissolved in the solvent.
However, hydrolysis is inevitable, and therefore, those compounds that readily form
acid in the presence of water must be used either in the absence of water or in
specially designed acid-resistant equipment (Allied Chemical Corp., 1960b; Baranaev
et al., 1954; Boggs and Buck, 1958; Boggs and Mosher, 1960; Carlisle and Levine,
1932a, b; McGovern, 1943; Ohnishi and Tanabe, 1971; Simonov et al., 1970). A good
review has been published by Mabey and Mill (1978).

I. 2-8 Mass-Transfer Design

Information on the solubility and miscibility of halogenated hydrocarbons in
water is required by design and process engineers to estimate the water content of
process streams at saturation, to estimate the buildup of a water phase in a frac-
tionating tower, to design absorption and other gas-liquid mass-transfer equipment
for many chemical processes, and to calculate the fractionation of water from halogenated hydrocarbon stream by distillation. Many industrial processes could be further
developed or improved by knowing the very small amount of water in a variety of substances (Hayduk and Castaneda, 1973; Hayduk and Chang, 1970; Hayduk and Laudie, 1973;

Hoot et al., 1957; Karlsson, 1972; Kass et al., 1965; Nathan, 1978; Earhart et al., 1977).

I. 2-9 Separating Minerals

Data on the solubility of halogenated hydrocarbons in water are needed for an evaluation of the effectiveness of continuous processes for separating minerals followed by the removal of the organic liquids from the processed ores by scrubbing with water. Because of the low solubility, the heavy liquid will be prevented from dilution with water carried in by wet ores. However, the presence of water is not desirable in organic liquids because it affects the separation of the minerals, increases the corrosion rate on the equipment, and decreases the stability of the organic compound. Water is usually used to scrub the organic compound from the ore, so the solubility of the organic in water may be an important factor. Most halogenated hydrocarbons hydrolyse slowly in the presence of water, but some stabilizers will retard this action. Furthermore, solubility and miscibility data will be required for the examination of the quality of the effluent streams from the aqueous recovery process (Gooch et al., 1972; O'Connell, 1963).

I. 2-10 Salting-Out Agents

Water-insoluble organic compounds can be used for to study to the salting-out effect of various salts in water. For example, the geochemical pathways could be altered by salting-in or salting-out processes. Solubility data on halogenated hydrocarbons in water are needed to evaluate the effectiveness of salting-out and salting-in processes (Gross, 1929a; Sutton and Calder, 1974, 1975).

I. 2-12 Bactericidal and Algicidal Agents

Some of the chlorinated and brominated compounds of hydrocarbons find application as bactericidal and algicidal agents, generally added to the water in a water-treating system at the rate of 2 to 50 ppm. The accumulated algae or bacteria grown in water-cooling and disposal systems have been treated successfully with chlorinated hydrocarbons. Methyl bromide has also been used as a fumigant of larvae of the oriental fruit fly. Solubility data in water and juices are necessary to determine whether the fruit will absorb or hinder penetration of the gas. Furthermore, solubility data will indicate how to control their toxic limits (Haight, 1951; Banks, 1979; Harnden, 1947; Harvey, 1975; R. S. Horvath, 1972; Walsh and Mitchell, 1974).

I. 2-12 Hydroscopicity and Chemical Reactions

Traces of water in organic substances affect their reactions; therefore, solubility data are needed. Most of the organic solutes are hygroscopic to some extent and so contain moisture as an impurity resulting from their preparation or handling. Chemical or physical procedures might be applied for the drying or removal of water. Homogeneous and heterogeneous catalytic reactions are especially water-sensitive, and a small moisture content may have serious consequences, such as the reduction of yield or an increase in the consumption of catalysts. In some polymerization reactions, the presence of even minute traces of water may modify or even rapidly deactivate and destroy the catalyst. For this and other reasons, a knowledge of the equilibrium water content may provide one of several criteria for selection of the most suitable solvent for carrying out reactions. During recent years, solution chemistry has become one of the most rapidly developing areas of separation methods. Water-free solvents are one of the fundamental requirements for successful operations. The worth of the solvent chosen depends on its solubility in water (Luft, 1957; Ohnishi and Tanabe, 1971; Riddick and Toope, 1955; Riddick and Bunger, 1970; Treger et al., 1964).

I. 2-13 Study of Liquid Structure

To understand aqueous solubility, the study and examination of the water structure are required. However, water has unique structural characteristics as a solvent (see Sec. 2.6), and attention has been directed to various phenomena, such as the formation of clathrates in aqueous solutions, iceberg formation, hydrophobic bonding, and structure breaking. The unique structural characteristics of water in the liquid phase can be approached through study of the solubility behavior of various substances in water at various temperatures and pressures. To understand the structure of water and aqueous solutions, numerous investigations have been carried out on the solubilities and miscibilities of water in combination with other substances (Andrews and Keefer, 1950, 1951; Getzen and Ward, 1971; Gingold, 1973; Chen, 1971; Pierotti and Liabastre, 1972; Moelwyn-Hughes, 1971; Egelstaff, 1967; Pryde, 1966; Rice, 1965; Barker, 1963; Eyring and Thou, 1969; Egelstaff, 1973; Franks, 1972-1975: Vol. 1; Eisenberg and Kauzmann, 1969). A detailed discussion of the effect of water structure on solubility is presented in Chap. 2.

From a wide variety of applications of solubility data between halogenated hydrocarbons and water, it is apparent that sources of information will be scattered throughout the published literature, from marine biology to perfumery. In addition to the difficulty of compiling solubility data from the published literature, there are numerous manufacturers' bulletins containing unpublished data. As these bulletins are normally available only to customers of the product, it is extremely difficult to acquire all data sheets or pamphlets that contain usable data. Several textbooks and handbooks give data without referring to the source of information; in these cases a compiler must make a subjective evaluation of their merit. The main source of solubility data included in this book was Chemical Abstracts; however, Physical Abstracts, Nuclear Science Abstract, Biological Abstracts, Bioresearch Index, Citation Index, Chemisches Zentralblatt, and Referativnyi Zhurnal were also searched. Most of the solubility data published prior to the 1960s were compiled in multivolume handbooks, but since then no comprehensive work has been published, particularly on the halogenated hydrocarbons and water systems.

The great need for solubility data has been recognized by large international organizations such as IUPAC and CODATA, and in a communication in Chemistry and Industry (Anonymous, 1974), it was reported that a subcommission on solubility data was recuiting compilers and evaluators for a solubility data project. Such a project was later set up by the Commission on Equilibrium Data of the Analytical Chemistry Division of the International Union and Applied Chemistry (IUPAC) (see CODATA Bull. 21, 1976, and Kertes et al., (1977). The project, which involves the compilation and critical evaluation of solubility data in all physical systems, is being coordinated by A. S. Kertes of the Institute of Chemistry of the Hebrew University in Jerusalem, Israel. The results will be published by Pergamon Press in three parts: Solubility of Gases in Liquids, Solubility of Liquids in Liquids, and Solubility of Solids in Liquids. For further detail see Chap. 3 and Table 3.2.

The American Chemical Society continues to compile solubility data and their publication can be expected in the near future (CODATA, 1969).

Another major difficulty in publishing solubility data relates to the heterogenity or absence of solubility units in the various sources. Most of the older literature sources fall into this category of obscure data. Information relating to the total or partial pressure of systems is often inadequate or omitted entirely, which makes interpretation of the data very difficult. The solubility data are often not presented in an easily understood manner, and it may be necessary to read the full article carefully, particularly the experimental part (if available) for an understanding of the solubility results. The main problem involves the numerous ways in which the data are expressed, such as in terms of the Bunsen coefficient, Kunen coefficient, Ostwald coefficient, Henry's law constant, weight percent, mole fraction, mole ratio and others. An obvious difficulty arises in the conversion of one solubility value into other units when the density of the pure solvent or solution is requiried but is not easily available. For example, to calculate the mole fraction from a Bunsen coefficient, the density and molecular weight of the solvent and the mole volume of the gas are needed for the conditions specified. A detailed discussion of units and conversions between various expressions of solubility values is provided in Chap. 9.

Because of the lack of solubility data on halogenated hydrocarbon and water systems, more and more experimental measurements have been carried out during the last decade. In other cases, older data were found unreliable and were remeasured. Some experimental determinations are difficult, time consuming, and expensive, and a good reason is required to justify the expenditure, particularly when the accuracy of solubility values is not critical.

Recent techniques are more reliable, because automatic equipment is now available which provides easier and faster determination procedures. Solubility theory is more advanced today than it was, for example 20 years ago, and the new instruments, together with modern solubility theory, provide a more solid basis for experiments than was available in the past.

This compilation has several objectives:

1. To review the latest theoretical and practical aspects of solubility for halogenated hydrocarbon – water systems

2. To update the compilations already available on solubility data

3. To critically examine and evaluate available data

4. To select and recommend the most reliable data

5. To compile and present the selected values in uniform and practical units

6. To apply reliable estimation, prediction, and correlation methods for some substances that have not been reported previously

This compilation covers halogenated hydrocarbons consisting of carbon(s), hydrogen(s), chlorine(s), fluorine(s), bromine(s), and iodine(s) atoms up to six carbon atoms. These substances are those required most frequently for the purposes outlined above. There are over 50 million halogenated hydrocarbons containing up to six carbon atoms (see Chap. 4). The precise water solubilities of these compounds will not be know in the near future. Therefore, the estimation methods discussed will be useful for the time being.

The available solubility data are tabulated and plotted in Chaps. 7 and 8. The presentation is such that they can be understood without reading the text or referring to other chapter(s) of the book or a list of abreviations. There are also comments and discussions of data presented, to indicate their reliability and accuracy.

HALOGENATED HYDROCARBONS

Solubility—Miscibility with Water

Chapter 1

Experimental Methods and Analytical Techniques

Numerous methods of solubility measurement are described in the literature. The techniques have been improved from time to time as a new instrumentation or control setup become available. Consequently, in recent years the apparatus used is reliable and the solubility data are more accurate than previous measurements.

The methods and setups used are well reviewed in the technical literature (Battino and Clever, 1966; Clever and Battino, 1975; Criss and Salomon, 1973; Czolbe, 1975; Drozdov et al., 1969; Kertes et al., 1975; Mader et al., 1963; Mader and Grady, 1971; Markham and Kobe, 1941; Mitchell and Smith, 1977; Ostwald, 1894; Riddick and Toops, 1955; Riddick and Bunger, 1970; Simonov et al., 1970; Tranchant, 1968; Zimmerman, 1952). Therefore, only a short discussion is given here. Further details can be found in the references cited.

In principle, the various methods applicable to the measurement of solubility can be related to the physical state of the solute and solvent. The physical state of a substance naturally depends upon the condition, that is, the temperature and pressure of the system considered. Three phases -- solid, liquid, and gas -- are of great interest to scientists and engineers. The plasma state is beyond the scope of this book.

The principles of solubility measurements or determinations presented by reviewers are given in Tables 1.1 to 1.3.

In this chapter the emphasis is directed toward experimental techniques used in the following three types of systems at the equilibrium condition:

1. solid-liquid
2. liquid-liquid
3. gas-liquid

In general, solubility measurements are well-established methods. To achieve good results, three aspects require extra attention and care during the experiments:

1. Use of purified solutes and solvents including the degassing of the solvents
2. Making sure that equilibrium exists between the solute and solvent
3. Use of a reliable method for determining the solute concentration in a solvent

1

Table 1.1 Principle of Methods for the Determination of Gas Solubility in Liquids

A. Manometric-volumetric methods

    1. Standards for comparison: solubility of O_2 in H_2O at $25^{\circ}C$ and 1 atm

    2. Degassing or removal of gas from the solvent

    3. Control and effect of temperature on solubility measurement

    4. Apparatus and procedure of Cook and Hanson (1957)

    5. Saturation methods

B. Mass spectrometric methods

C. Gas chromatographic methods

D. Chemical Methods

E. Miscellaneous methods

Source: Battino and Clever, 1966.

    Chemicals purchased through major distributors are properly labeled and the purity or impurity of compounds is clearly indicated on the container. It is possible to obtain highly purified solvents and solutes from several suppliers,* so that sometimes the difficult purification method can be avoided. However, when a specific substance is not available commercially in high purity, careful purification becomes necessary. A great number of purification methods are described by Janz and Tomkins (1972), Perrin et al. (1980), and Staveley (1971). In addition to these, it is worthwhile to search in Chemical Abstracts for the best method available. Under the heading of the substance, it is not difficult to locate references to the available methods. Most of later investigations refer to patents; these are usually available in the applicable patent office and can be purchased.

    How important the purity of the solvent is has been demonstrated by Gokcen and Chang (1977) for the solubility of gaseous oxygen in liquid dinitrogen tetroxide, N_2O_4, under pressure. Before the solubility determination, N_2O_4 was purified by double distillation. The result showed that the solubility of O_2 was proportional to $(p_{O_2})^{\frac{1}{2}}$, that is, O_2 dissolved as a monoatomic molecule:

$$\tfrac{1}{2} O_2(gas) \quad \rightarrow \quad O \text{ dissolved in } N_2O_4(liquid)$$

This means that Henry's law is not obeyed. The reason for this curious phenomenon was that a small amount of dissolved N_2O_3 remained present in the solvent $N_2O_4(l)$, with the consequence that

$$\tfrac{1}{2} O_2(gas) + N_2O_3(in\ N_2O_4) \quad \rightarrow \quad N_2O_4(liquid)$$

---

*Standard samples of organic compounds certified with respect to purity are available from national laboratories (Cali, 1976; e.g., National Bureau of Standards, U.S.A.; National Physical Laboratory, England). Commercial distributors include British Drug House, Fluorochem Limited, and Riedel-de-Haän AG, Hannover.

Table 1.2 Various Methods for the Determination of Solubility

---

A. Solubility of solids in liquids

    1. General or gravimetric methods

    2. Rapid solubility measurements

        a. Based upon the rate of solution

        b. Residue-volume method

    3. Determination of solubility in volatile solvents

        a. Synthetic method

    4. Slightly soluble solutes in liquids

        a. Using an interferometer

        b. Ultraviolet spectrometry

        c. Conductometric determination

        d. Radioactive indicators

        e. Film-balance technique (e.g., interfacial tension)

    5. Measurement of solubility at high temperature

        a. Pressure-bomb method

        b. Using small tubes placed in an air thermostat

    6. Determination of solubility in liquefied gases

        a. Condensing the solvent on a known solute sample

        b. Weight measurement

        c. Chemical analysis

B. Solubility of liquids in liquids

    1. General methods

        a. Analytical determination

        b. Synthetic process (e.g., cloud point)

        c. Using a dye

        d. Thermostatic method involving the phase-rule principle

    2. Rapid methods

        a. Dropwise addition of solute to the solvent

        b. Usage of minute, fagged crystals as indicator

    3. Slightly soluble liquids in liquids

        a. Gas chromatography

        b. Karl Fischer titrimetry

        c. Infrared spectrometry

        d. Interferometry

        e. Turbidity titration

C. Solubility of gases in liquids

    1. Rapid methods

        a. Dymond and Hildebrand setup

        b. Loprest apparatus employing manometric measurements

Table 1.2 (Continued)

---

2. Gas saturation apparatus
   a. Modified Ostwald apparatus
   b. Cook and Hansen method
3. Gas extraction methods
   a. Manometric apparatus of Van Slyke
   b. Modified Van Slyke design
   c. Baldwin-Daniel apparatus for viscous liquids
4. Special methods
   a. Modified pressure vessels of Newitt and Weale
   b. Bott and Schulz process
   c. Doremus method

Source: Mader and Grady, 1971.

This reaction took place at ambient temperature. However, in addition to the double distillation, the solvent was vacuum-degasified and redistilled before the solubility was measured. The result indicated that in the absence of N_2O_3 in the solvent, oxygen dissolved in N_2O_4 according to Henry's law, similarly to the solubility of $N_2(g)$ in the same solvent. The double distillation of N_2O_4 was capable of removing a trace amount of N_2O_3 from the N_2O_4 solvent, and this trace quantity caused the misleading results.

It is essential, particularly in the solubility measurements of gases in liquids, to make sure that the solvent is properly degassed before contact is established between the solute gas and the liquid solvent. The various degassing methods have been reviewed by Battino and Clever (1966) and Clever and Battino (1975). The degassing of the solvent can be considered as part of the purification, because the objective is the same as that for purification -- to get rid of impurities in the samples.

Regarding the time required for the equilibrium condition between the solute and solvent, it depends upon the type of system -- gas-liquid, liquid-liquid, or solid-liquid -- and the sort of apparatus used for the measurement. Special care is required for the determination of the solubility of a solid in a liquid, where the particle size has a great influence on the time necessary to achieve the equilibrium condition (Dundon and Mack, 1923; Hulett, 1901; Pedersen and Brown, 1976a, b; Fürer and Geiger, 1977). Essential for a rapid equilibrium condition are several factors, including the type of stirring, shaking, or mixing for the establishment of the largest contact surfaces between the solute and solvent phases; and the nature of the contact (e.g., bubbling the gas through the liquid). Whether equilibrium has or has not been accomplished can be checked: for example, by the results of two successive solubility measurements carried out during a ceratin period in the same sample. If the solubility values show a tendency to increase, it is apparent that the system

Table 1.3 Classification of Methods Used for the Measurements of Solubility

---

A. Solubility of solids in liquids

    1. Saturation methods based upon the percolation principle, circulating saturators, continuous stirring with a siphon for sampling

    2. Cloud point, cloud detection, and volume measurement

    3. Chemical and instrumental analysis; acidimetry, argonometry, iodometry, spectrophotometry, colorimetry, conductivity, surface tension, density, refractive index, polarity, tracer techniques, evaporation

B. Solubility of liquids in liquids

    1. Measuring volume changes upon equilibration

    2. Cloud point, or appearance and disappearance of a fine cloud

    3. Miscellaneous methods; ultraviolet spectrometry, Karl Fischer titration, tracer analysis using tritiated water, dispersion of a water insoluble dye in water, acid-base titration, vapor-oressure measurements, gas chromatography

C. Solubility of gases in liquids

    1. Manometric-volumetric methods

        a. Degassing or removal of gases from the solvent

        b. Calibration of all the components (filling)

        c. Mixing while a complete equilibration is reached

        d. Manometric reading yields the volume of the solute gas

        e. Solubility at high pressure in solubility bomb

        f. Solubility at high temperature

    2. Chemical-analytical methods

        a. Specific chemical reactions, titrimetry, gravimetry, combustion

        b. Gas chromatography fitted with a dehydrating agent

        c. Mass spectrometry for relative solubility or solvent/solute ratio

    3. Miscellaneous methods

        a. Radioactive tracer method

        b. Static and dynamic equilibration

        c. Microgasometry

Source: Kertes et al., 1975.

was not in equilibrium when the first sample was taken. In the case of gas solubility in liquids, the pressure of the gas indicates the state of the equilibrium. No change in the gas pressure gives the equilibrium condition during an interval of time.

    After equilibration, the solute concentration in the solvent can be measured directly or indirectly. Most solubility results of a gas-liquid solubility are obtained by direct measurements (e.g., by manometric-volumetric methods). The concentrations might also be determined by titration, chemical analysis, gas-liquid chromatography (Petrov et al., 1970), mass spectrometry, interferometry (O'Brien, 1972), spectrofluorometry (Diamond, 1959), ultraviolet absorption (Schwarz, 1977; Schwarz and Wasik, 1977), or microparticle dispersion (Fürer and Geiger, 1977). In addition to these methods, there are some more-unconventional analytical procedures [e.g., radioactive analysis (Clark, 1972; Atach and Schneider, 1950), microgasometry, use of dyes]. The accuracy of the method used is an essential part of a successful solubility determination. Different results originate from the various analytical methods, and how much they contribute to the discrepancies between the reported data will be discussed in Chapters 5 and 6. A particularly good illustration of the problem is the comparison of the determination of water content in nonpolar solvents using gas-liquid chromatography, Karl Fischer titration, and CaH_2 method (Seller, 1971). Further comparisons have been reported by Högfeldt and Bolander (1963).

Hunter and Honaker (1979) compared four conventional methods (titration with water, gas chromatography, refractive index, Karl Fischer titration) with the tritium tracer technique for the determination of water solubility in organic liquids. The authors listed the following advantages of the tritium method; rapid, simple, no visual detection, no calibration, can be used for impure liquids. Consequently, the use of a tritium tracer is a simple and easy method for determining water in organic liquids.

The dynamic coupled-column liquid chromatographic technique was used by May (1980) to obtain aqueous solubility data on 11 aromatic hydrocarbons at 25°C. The precision of replicate solubility measurements was better than ±3%.

Otson et al., (1979) compared the dynamic head space, solvent (hexane) extraction, and static head space techniques for the determination of trihalomethanes ($CHCl_3$, $CHBrCl_2$, $CHBr_2Cl$, $CHBr_3$) in water. The investigators found that the dynamic head space technique was the most sensitive, the solvent extraction technique gave comparable precision while the static head space technique showed relatively poor precision and sensitivity. Consequently, the solvent extraction technique is recommended for routine determination of trihalomethanes in water.

According to Korenman and Aref'eva (1978) both the pH-metric and the adsorption on active carbon methods for determining the solubilities of liquid hydrocarbons in water are somewhat laborious or complicated. Therefore the authors proposed a new method, including the addition of a bright color (dithizone, Sudan dye, carbon black, or phenolphthalein) to pure water and shaken vigorously. Small portions of liquid hydrocarbons whose solubility is being determined are added from a microburet, with vigorous shaking after each addition. After saturation, a very small excess of the solute causes flotation of the indicator, which is easily observed from the appearance of a red (Sudan), dark green (dithizone), black (carbon black), or white (phenolphthalein) film on the inner surface of the cylinder above the liquid level.

The headspace gas chromatography in conjunction with electron capture detection was compared with the purge-trap procedure for determining the concentration of chlorinated hydrocarbons (chloroform, carbon tetrachloride, trichloroethylene, tetrachloroethylene) in drinking, natural, and industrial waters by Dietz and Singley (1979).

In a Russian patent Aref'eva et al. (1979) described the determination of the solubility of liquid and solid organic substances in water, using diphenylthio-carbazone. The saturation point was determined by measuring the optical density of the filtered solution.

McAuliffe (1980) used the multiple gas-phase equilibration method to measure the solubility of hydrocarbons in water. The method is based on the partition of the organic solutes favorably from water to a gas phase, followed by gas chromatographic analysis of the gas phases. Two or more successive gas-phase equilibrations give all the necessary data to calculate the concentrations in the aqueous phase.

A good review was reported on the various techniques and problems in analysing halogenated hydrocarbons in water by the Panel on Low Molecular Weight Halogenated Hydrocarbons (Andelman, 1978). The report concludes that electron capture-gas chromatography and the gas chromatography-mass spectrometry are currently the most practical, versatile, and economical methods available.

A collection of papers selected from those presented at the International Symposium on the Analysis of Hydrocarbons and Halogenated Hydrocarbons in the Aquatic Environment deal with all relevant aspects of analytical techniques for determining organic solutes in water (Afghan and Mackay, 1980).

The moisture content in refrigerant 22 and in a mixture of R 22/R 12 was determined by Diamond (1959) and Rhodes (1947) using infrared spectrophotometry up to a maximum of 65 ppm by weight. The accuracy of the instrument in the operating range used was ±1.4 ppm. The results were compared with the water content of R 22 using the P_2O_5 method. Good agreement has been reported.

Analytical data for water in organic solvents have been presented by Mitchell and Smith (1977). The water content was determined by the tert-butyl o-vanadate method for comparison with the Karl Fischer titration. The tert-butyl o-vanadate method is applicable to the anlysis of water in hydrocarbons, alcohols, alkyl halides, esters, and tertiary amines.

The magnesium nitride method also provided good results for samples of benzene, ethanol, ether, and acetone containing from 0.004 to 10% water.

The reaction of chemicals with water is a commonly used analytical technique. The determination of water in the concentration range 0.002% was studied by Luke (1971) using the following reaction in the liquids:

$$K_2Cr_2O_7 + H_2O \xrightarrow{\text{organic phase}} K_2CrO_4 + H_2CrO_4$$

The transmittance of the filtrate is determined by a spectrophotometry at 370 nm. A large number of solvents -- alcohols, hydrocarbons, halogenated hydrocarbons -- were tested and the results were good in comparison with the Karl Fischer method (see Table 1.4).

Schatzberg (1963) also compared various procedures for the determination of water in nonpolar solvents. The diverse apparatus and methods for the measurement of the solubility of water in organic liquids are presented in Table 1.5.

The difference between the Karl Fischer and nuclear magnetic resonance (MNR) methods for the determination of water content in solvents were studied by Ho and Kohler (1974). Their finds are presented in Table 1.6.

A comparison of the solubilities of oxygen in water as given by various investigators is that of Battino and Clever (1966). Typical setups to measure the solubility of gases in liquid are shown in Table 1.7, and for the solubility of halogenated gaseous hydrocarbons in water in Table 1.8. Some of the methods and apparatus for the determination of solid organic substances in liquids are summarized in Table 1.9. Typical descriptive methods for liquid-liquid solubility measurements are given in Table 1.10.

The concentration of various halogenated hydrocarbons in water samples changes markedly over an 8-day period (Fritz, 1976), and therefore the analysis of drinking water may be in error because the samples are not analyzed quickly enough. For example, the chloroform concentration of drinking water increases rapidly for 3 or 4 days and then decreases markedly. The concentration of $CHBrCl_2$ increases and levels off, $CHBr_2Cl$ increases, and $CHBr_3$ increases steadily throughout the storage period (Fritz, 1977; Larson and Rockwell, 1978). The analytical problem is formidable because of the extremely low concentrations of most halogenated hydrocarbons in drinking water. For example, the concentrations of chloroform and chlorobromohalocarbons are in the range 1-300 µg liter^{-1}. For further reports, see Hammarstrand (1976), Pfaender et al. (1978), Zogorski et al. (1978), and Afghan and Mackay (1980).

The separation of water impurities is carried out using gas chromatography. However, prior to this separation, it is necessary to concentrate the trace components. In his review, Fritz (1977) discussed the following methods:

1. Resin sorption
2. Solvent extraction with n-pentane
3. Water-gas distribution
4. Gas stripping
5. Direct injection
6. Resin sorption/pyridine desorption
7. Resin sorption/thermal desorption

Vapor-phase extraction was used by Mackay et al. (1975) for the determination of low concentration of hydrocarbons in water.

Table 1.4 Determination of Water by Reaction with Dichromate

Solvent	Water content (wt %)	
	Dichromate method	Karl Fischer method
Benzene	0.013	0.014
Toluene	0.031	0.030
Xylene	0.02	0.02
Ethylether	0.002	0.003
Ethyl acetate	0.018	0.019
Isopropanol	0.021	0.022
n-Octanol	0.07	0.065
Trichloroethylene	0.004	0.006
1,2-Dichloroethane	0.007	0.007
Tetrahydrofuran	0.02	0.02
Acetonitrile	0.06	0.06
2-Methyl-2-pentanone	0.011	0.012
1,4-Dioxane	0.09	0.09
Pyridine	0.06	0.07
Acetone	0.30	0.30
Cyclohexanone	0.21	0.23

Source: Luke, 1971.

The methods for the determination of water in various materials have been comprehensively reviewed by Mitchell and Smith (1977) under the overall title Aquametry. In Part I the main emphasis is directed to techniques such as chemical, gravimetric, thermal, separation, ultraviolet and visible spectrometric, infrared, nuclear magnetic resonance, radiochemical, physical, and miscellaneous instrumental methods. Part II deals with various electrical measurements, and Part III is devoted to the Karl Fischer technique. In this comprehensive treatise of the various determination methods the reader will find a detailed treatment of all aspects of water determinations. Comments, comparisons, and illustrations are provided for an easier understanding of and selection between the available techniques. The main features of Part I of Aquametry are summarized in Table 1.11.

Probably the three most common determination techniques used today are Karl Fischer titration, gas-liquid chromatography, and the use of indicator paper for water determination in solvents. The first two procedures mentioned are very accurate and reliable. Indicator papers are very quick and cheap to use and provide a good approximation of the water present. Therefore, it will be worthwhile to describe these three methods in some detail.

Table 1.5 Apparatus and Methods for the Measurement of the Solubility of Water in Organic Liquids

References	Solvent	Temperature (°C)	Analytical method
Caddock & Davies (1959)	C_6H_6, toluene	20	Tritiated water and radioactive tracer method
Carlisle (1932), Carlisle & Levine (1932)	CH_2Cl_2, C_2HCl_3	-50 to 25	Cloud point; cloud point versus percent moisture (Alexejeff)
Clifford (1921a, b)	$CHCl_3$, CCl_4, CS_2, C_6H_6, etc.	5-50	Calcium chloride method
Eidinoff (1955)	29 organic liquids	25	Titration by Karl Fischer reagent
Englin et al. (1965)	Hydrocarbons	10-50	CaH_2 method
Evans (1936)	Organic compounds	20	Hill's method based on phase rule compared with cloud point
Filippov & Furman (1952)	C_6H_5Cl, $C_6H_5C_2H_5$	18-50	Alexejeff's cloud-point method
Glasoe & Schultz (1972)	CCl_4, toluene, C_6H_{12}	15-45	Karl Fischer titration using a "dead-stop" end point
Goldman (1969, 1974)	C_6H_6, C_6H_5Cl, $o-C_6H_4Cl_2$	24-45	Titration with Karl Fischer reagent
Högfeldt & Bolander (1963)	Aromatic hydrocarbons	25	Karl Fischer titration has been selected
Högfeldt & Fredlund (1970)	C_6H_6, toluene, CCl_4, C_6H_{14}, etc.	25	Titration with Karl Fischer reagent
Hunter & Honaker (1979)	Organic liquids	25	Tritiated water and radioactive tracer method
Jones & Monk (1963)	C_6H_6, C_6H_5Cl, C_6H_5Br, C_6H_5I, $C_6H_4Cl_2$	25-35	Tritiated water and radioactive tracer method
Joris & Taylor (1948)	C_6H_6	10-26	Tritiated water and radioactive tracer method
Karlsson (1972), Karlsson & Karrman (1971)	CH_3OH, hydrates, etc.	25	Karl Fischer titration and colorimetry
Karyakin et al. (1960)	CCl_4	-10 to 60	Infrared spectroscopy

Reference	Substance	Temperature	Method
Klemenc & Löw (1930)	o-, m-, p-$C_6H_4Cl_2$	20-60	Gravimetry
Kobayashi & Katz (1953)	C_2H_6, CO_2, SF_6, etc.	0-171	Volumetric analysis
Leland et al. (1955)	1-Butene	38-149	Increased weight of $MgClO_4 \cdot 3H_2O$ tubes
Lyle & Smith (1975)	1,4-Dioxane, toluene	25	Gas-liquid chromatography with katharometric bridge
Mannheimer (1956)	$(CH_3)_2O$	25	Increased weight of NaOH or $CaCl_2$ drying tubes
Ödberg & Högfeldt (1969), Ödberg et al. (1972)	$CHCl_3$, C_6H_6, $C_2H_4Cl_2$, etc.	25	Titration with Karl Fischer reagent
Polak & Lu (1973)	Hydrocarbons	0, 25	Titration with Karl Fischer reagent
Reynolds & Harris (1969)	$CHCl_3$, C_6H_6, toluene, etc.	25	Direct-injection enthalpymetry
Rhodes (1947)	CCl_2F_2	25	Infrared spectrophotometry
Rigby & Prausnitz (1968)	N_2, Ar, CH_4	25-100	Volumetric and pressure measurement
Rosenbaum & Walton (1930)	C_6H_6, CCl_4, toluene	10-60	CaH_2 method
Rotariu et al. (1952)	C_7F_{16}	25, 50	Titration with Karl Fischer reagent
Sellers (1971)	C_6H_{14}, C_6H_6, $C_2H_4Cl_2$, etc.	25	CaH_2 method compared with gas-liquid chromafography
Simonov et al. (1970)	CCl_4, C_2Cl_4, C_4Cl_6, C_3Cl_6	10-80	Alexejeff's cloud-point method
Staverman (1941b)	CCl_4, $CHCl_3$, CH_2Cl_2, $C_2H_4Cl_2$	0-35	Titration with Karl Fischer reagent
Utschick et al. (1976)	i-Butanol	25	Measuring the refractive index in a refractometer
Wing (1956), Wing & Johnston (1957)	Aromatic halides	25	Tritiated water and radioactive analysis: CaH_2 method

Table 1.6 Water Contents of Solvents

Solvent	Water content (wt %)	
	Nuclear Magnetic Resonance Method	Karl Fischer Method
Methyl chloride	0.01	0.01
Chloroform	0.00	0.01
Acetone	0.59	0.55
Acetonitrile	0.04	0.02
Dimethyl sulfoxide	0.06	0.05
Dioxane	0.05	0.04
Ethyl acetate	0.03	0.02
Nitromethane	0.10	0.11
Tetrahydrofuran	0.28	0.30

Source: Ho and Kohler, 1974.

1.1 TITRATION WITH KARL FISCHER REAGENT

It has become apparent from several comparative experimental investigations that titration with Karl Fischer reagent is a very accurate method for the determination of water in organic substances. The technique involves the volumetric determination of water in organic liquids. A sample containing water is titrated with a methyl alcohol solution of iodine, sulfur dioxide, and pyridine. At the appearance of free iodine, the solution becomes brown in color, indicating overtitration. The reaction is

$$H_2O + SO_2 + CH_3OH + I_2 = CH_3HSO_4 + 2HI$$

Because of the difficulty in observing the change in color from yellow to brown, two platinum electrodes are dipped into the solution and the other ends are connected to a "dead-stop" end-point instrument, which consists of an ammeter, a voltmeter, and a Wheatstone bridge. A small potential difference upon the electrodes is provided by batteries and the resistance. During the titration the ammeter shows no change until the end point, when it indicates a sudden change. Typical specifications for the instrument are: potential, 10 to 20 mV; ammeter, 0 to 100 µA; battery, 1.5 V; and variable resistor, 150 ohms.

The titration buret is an all-glass apparatus connected to the bottle holding the Karl Fischer reagent. The reagent corresponds to 2 to 3 mg of water per cubic centimeter. The air is passed through drying tubes before entering the setup and coming into contact with the Karl Fischer reagent. Standardization of the reagent is necessary from time to time. Usually, absolute methanol is used for this purpose. When the end point is indicated, sodium tartrate is added to the solution and the titration is repeated. Several measurements are recommended to achieve accurate reproductions.

Table 1.7 Typical Setups to Measure the Solubility of Gases in Liquids

Reference	Solute Gas	Solvent Liquid	Temperature (°C)	Pressure	Analytical method
American Society for Testing and Materials (Anonymous, 1973)	Gas	Liquid	–	–	Gas extraction and volume measurement
Baldwin & Daniel (1952)	N_2, O_2, air	Oils	20	1 atm	Volumetric gas analysis
Ben-Naim & Baer (1963)	Ar	H_2O	3-28	1 atm	Volumetric gas analysis
Burrows & Preece (1953)	He	Oils	20-80	1 atm	Volumetric gas analysis
Claussen & Polglase (1952)	CH_4, C_2H_6, etc	H_2O	0-40	1 atm	Volumetric gas analysis
Cook & Hanson (1957)	H_2	$n\text{-}C_7H_{16}$	25, 35	1 atm	Volumetric gas analysis
Cox & Head (1962)	CO_2	H_2O	20-30	1 atm	Volumetric gas analysis
Douglas (1964)	O_2, Ar, N_2	H_2O	5-30	1 atm	Microgasometric analysis
Dymond & Hildebrand (1967)	CO_2, CF_4, N_2	C_6H_{12}	15-35	1 atm	Volumetric gas analysis
Farkas (1965)	C_6H_{12}	H_2O	25	1 atm	Volumetric gas analysis
Hayduk & Cheng (1970)	C_2H_4	Hydrocarbons	15-40	1 atm	Volumetric gas analysis
Karasz & Helsey (1958)	He, Ne	Ar	84-88 K	10-160 mm Hg	Volumetric gas analysis
Kobatake & Hildebrand (1961)	He, SF_6, CH_4, etc.	CS_2, C_7F_{16}, etc.	5-30	1 atm	Volumetric gas analysis
Kogan et al. (1963)	Cl_2	CCl_4	-20 to 71	1 atm	Volumetric gas analysis
Koonce & Kobayashi (1964)	CH_4	n-Decane	-20 to 40°F	200-1000 psi	Volumetric gas analysis
Kritchevsky & Iliinskaya (1945)	H_2, N_2, CO_2, etc.	H_2O	0-50	1 atm	Volumetric gas analysis

Table 1.7 (Continued)

Reference	Solute	Solvent	Temperature (°C)	Pressure	Analytical method
Lamb & Shair (1976)	SF_6, Freons	H_2O, CS_2, alcohols	10-50	1 atm	GLC with electron-capture detector
Lannung (1930)	He, Ne, Ar	H_2O, C_6H_6, etc.	15-37	1 atm	Volumetric gas analysis
Liabastre (1974)	Hydrocarbons, etc.	H_2O, D_2O	5-45	1 atm	GLC with flame-ionization detector
Loprest (1957)	H_2, CO_2	H_2O, $n\text{-}C_7H_{16}$	20-60	1 atm	Volumetric gas analysis
McDaniel (1911)	CH_4, C_2H_6	C_6H_6, alcohols	25	1 atm	Volumetric gas analysis
Michels et al. (1936)	CH_4	H_2O	25-150	Up to 200 atm	Volumetric gas analysis
Morrison & Billett (1948)	CH_4, C_2H_6, O_2, N_2	H_2O	25	1 atm	Volumetric gas analysis
Moudgil et al. (1974)	CH_4, C_4H_{10}	H_2O	25	1 atm	Volumetric gas analysis
Novak & Conway (1973)	Ar, O_2	H_2O	25	1 atm	Volumetric gas analysis
O'Brien & Hyslop (1975)	Ar, H_2, N_2, O_2	H_2O	1-60	Up to 10 Pa	Refractive index with interferometer
Patsatsiya & Krestov (1970)	Ar	H_2O	25-35	1 atm	Volumetric gas analysis
Pray et al. (1952)	H_2, O_2, N_2, He	H_2O	24-316	Up to 500 psi	Volumetric gas analysis
Slobodin et al. (1964)	N_2	Acetone, CCl_4, etc.	-60 to 60	1 atm	Volumetric gas analysis
Strakhov et al. (1975)	Ne	H_2O	5-72	1 atm	Volumetric gas analysis
Wang & Cheung (1962)	N_2, Ar, H_2	C_2H_6, C_3H_8	-196 to -185	1-20 atm	Transient technique
Wheatland & Smith (1955)	O_2	H_2O	0	1 atm	Gasometric and titrimetric method

Table 1.8 Apparatus and Methods for the Measurement of the Solubility of Gaseous Halogenated Hydrocarbons in Water

Reference	Gaseous solutes	Temperature (°C)	Pressure	Experimental error (%)	Analytical method
Ashton et al. (1968)	CF_4, SF_6, NF_3	0-50	600-1200 mm Hg	1	Volume and pressure measurement
Boggs & Buck (1958)	CH_3Cl, CH_2FCl, CHF_2Cl	10-80	1 atm	1	Volume and pressure measurement
Carey et al. (1966)	$CHCl_2F$, CH_2ClF, CH_3Br, CH_3CClF_2	10-20	0.5-2 atm	1.8-3	Volume and weight measurement
du Pont (1961b)	R 11, R 21, R 112, R 113, R 114	27-75	Sat. press.	±5	Volume and density measurement
Hammer et al. (1978)	CCl_3F	16	1 atm	5	GLC and electron-capture detector
Hayduk & Laudie (1974b)	$CH_2=CHCl$	0-75	1-6 atm	2	Volume and pressure measurement
Maharajh (1973)	CF_4	25	1 atm	1	Gas chromatography
Mamedaliev & Musachanly (1940)	CH_3Cl	20	Sat. press.	-	Volume and pressure measurement
Mizutani & Yamashita (1950)	$CH_2=CHCl$	3-50	1 atm	2	Volume and pressure measurement
Parmelee (1953, 1954)	CCl_2F_2, $CClF_3$, CF_4, $CHClF_2$, CHF_3, C_2F_5Cl	25-75	15-300 psia	5	Volume and pressure measurement
Swain & Thornton (1962)	CH_3F, CH_3Cl, CH_3Br, CH_3I	30-40	Up to 1 atm	-	Volume and pressure measurement
Volokhonovich et al. (1966)	$CF_2=CF_2$	0-70	150-600 mm Hg	-	Volume and pressure measurement

Table 1.9 Apparatus and Methods for the Determination of the Solubility of Organic Solids in Water

References	Solute	Solvent	Temperature ($^{\circ}$C)	Analytical method
Aquan-Yuen et al. (1979)	Halogenated hydrocarbons	H_2O	25	GLC with electron-capture detector
Biggar & Riggs (1974)	Organochloride insecticides	H_2O	15-45	GLC with electron-capture detector
Brooker & Ellison (1974)	p-Chlorobenzoic acid, etc.	H_2O	25	Turbidity measurement by absorption
Dexter & Pavlou (1978)	Polychlorinated biphenyls	H_2O	11.5	GLC with electron-capture detector
Dundon & Mack (1923)	$CaSO_4 \cdot 2H_2O$	H_2O	25	Electrical conductivity measurement
Eganhouse & Calder (1976)	Aromatic hydrocarbons	H_2O	25	Extraction and flame-ionization detector
Hancock & Laws (1955)	Benzene hexachloride	H_2O	25	Extraction and measurement of optical density
Kurihara et al. (1973)	Hexachlorocyclohexane	H_2O	28	GLC and radioactive analysis
Mackay & Shiu (1977)	Polynuclear aromatic hydrocarbons	H_2O	25	Extraction and spectrophotofluoro-metry
May (1980)	Polycyclic aromatic hydrocarbons	H_2O	5-30	Dynamic-coupled column liquid chromatographic technique
May et al. (1978a, b)	Polynuclear aromatic hydrocarbons	H_2O	5-31	Dynamic-coupled column liquid chromatographic technique
Schoor (1974, 1975)	Polychlorinated biphenyls	H_2O	20	GLC with electron-capture detector
Sutton & Calder (1974, 1975)	Paraffins up to C_{26}	H_2O	25	Extraction and GLC with ionization detector
Wauchope (1970), Wauchope & Getzen (1972)	Aromatic hydrocarbons	H_2O	0-75	Extraction and spectrophotometric method
Wiese & Griffin (1978)	Polychlorinated biphenyls	H_2O	16.5	GLC with electron-capture detector

Table 1.10 Apparatus and Methods for the Measurement of the Solubility of Liquids (not H_2O) in Liquids

References	Solute	Solvent	Temperature (°C)	Analytical method
Aref'eva et al. (1979)	CCl_4, $CHCl_3$, etc.	H_2O	25	Optical density via refractive index
Arnold et al. (1958)	C_6H_6	H_2O	0-70	Ultraviolet spectrophotometry for optical density
Bohon & Claussen (1951)	Aromatic hydrocarbons	H_2O	0-40	Ultraviolet spectrophotometry for optical density
Bradley et al. (1973)	C_6H_6, toluene	H_2O	25-60	Ultraviolet spectrophotometry for optical density
Brooker & Ellison (1974)	o-Xylene, etc.	H_2O	25	Turbidimetry and filtration
Brown & Wasik (1974)	C_6H_6, toluene, ethylbenzene	H_2O	5-20	GLC
Carless & Swarbrick (1964)	Benzaldehyde	H_2O	25	Optical density via refractive index
Charykov & Tal'nikova (1973)	Organic liquids	H_2O	25	Saturated point indicated by pH discontinuity
Chey & Calder (1972)	C_6H_6, toluene, C_6H_5Cl	H_2O	21	Extraction with isopropyl ether and GLC
Dietz & Singley (1979)	Chlorinated hydrocarbons	H_2O	4	Headspace gas chromatography
Evans (1936)	Organic compounds	H_2O	20	Hill's method and compared to cloud point of Alexejeff
Filonenko & Korol (1976)	C_2-C_5 alcohols, C_6H_6	Squalee	20-50	Comparison of static and GLC methods
Franks et al. (1963)	C_6H_6	H_2O	17-63	Ultraviolet spectrophotometry for optical density
Glew & Robertson (1956)	Cumene	H_2O	25-80	Ultraviolet spectrophotometry for optical density
Gooch (1971), Gooch et al. (1972)	$1,1,2,2$-$C_2H_2Br_4$	H_2O	1-97	X-ray spectrophotometry and GLC
Gross (1929a, b)	$C_2H_4Cl_2$, $C_3H_6Cl_2$, CCl_4	H_2O	25	Refractive index by interferometer
Gross & Saylor (1931), Gross et al. (1933)	Halogenated hydrocarbons	H_2O	15, 30	Refractive index by Zeiss interferometer
Haight (1951)	CH_3Br	H_2O	10-32	Analysis by weight

Table 1.10 (Continued)

References	Solute	Solvent	Temperature (°C)	Analytical method
Hayashi & Sasaki (1956)	C_6H_6, alcohols	H_2O	20-30	Titration with Tween-80 surfactant
Karger, Castells, et al. (1971), Karger, Chatterjee, et al. (1971), Karger, Sewell, et al. (1971)	CH_2Cl_2, $CHCl_3$, CCl_4, etc.	H_2O	12.5	GLC
Kisarov (1962)	C_6H_5Cl	H_2O	30-90	Measuring pressure and volume
Kobayashi & Katz (1953)	C_2H_6, CO_2, SF_6, etc.	H_2O	32-340	Volumetric and pressure measurement
Korenman & Aref'eva (1978)	Hydrocarbons	H_2O	25	Indicator dye
Leland et al. (1955)	1-Butene	H_2O	38-149	Volumetric and pressure measurement
Liabastre (1974)	Hydrocarbons	H_2O	5-45	GLC with flame-ionization detector
McAuliffe (1980)	Hydrocarbons	H_2O	25	GLC
Mannheimer (1956)	$(CH_3)_2O$	H_2O	24	Weight measurement of the solvent and gas
Meleshchenko (1960b)	$C_6H_3Cl_3$	H_2O	19	Titrimetry ($Na_2S_2O_3$)
Mitchell et al. (1964)	Benzaldehyde	H_2O	15, 25	GLC and gravimetry
Nelson & de Ligny (1968a)	Hydrocarbons	H_2O	4-55	GLC with flame-ionization detector
Otson et al. (1979)	Trihalomethanes	H_2O	25	GLC with electron-capture detector
Paull & Hewett (1976)	Halothane	Olive oil	25	Measuring the partial pressure
Pierotti & Liabastre (1972)	C_6H_6, toluene, etc.	H_2O	5-50	GLC with flame-ionization detector
Polak & Lu (1973)	Hydrocarbons	H_2O	0, 25	GLC with flame-ionization detector
Rauws et al. (1973)	CCl_3F, CCl_2F_2, $C_2Cl_3F_3$, etc.	Body fluids	20	GLC with electron-capture detector
Rex (1906)	Halocarbons	H_2O	0-30	Gravimetric determination
Saylor et al. (1952)	C_6H_6, C_6H_5Cl, $C_6H_5NO_2$	H_2O	30	Zeiss combination liquid-gas interferometer
Schwarz (1980)	Halocarbons	H_2O	23.5	Elution chromatography
Selenka & Bauer (1978)	Halocarbons	H_2O	25	Microcoulometry and GLC

Sobodka & Khan (1931)	Sparingly soluble solutes	H_2O	20	Sudan dye and microburet measurement
Stadnik & El'tekov (1975)	C_6H_6, toluene	H_2O	30	Spectrophotometry
Treger at al. (1964)	Allyl chloride	H_2O	25–70	Colorimetric analysis
Utschick et al. (1976)	i-Butanol	H_2O	25	Refractive index
Vesala (1974)	24 nonelectrolytes	H_2O	25	Titrimetry, spectrophotometry, and radiochemistry
Wasik & Brown (1973)	Hydrocarbons	H_2O	0	GLC
Korenman (1975)	Hydrocarbons	H_2O	20	Indicator dye

Table 1.11 Determination of Water in Various Substances
_____

A. Chemical methods

   Acid chlorides, acid anhydrides, lead tetraacetate, Mg_3N_2, CaC_2

B. Gravimetric methods

   Drying in oven and desiccation, thermogravimetry, absorption, condensation

C. Thermal methods

   Thermal conductivity, heat of reaction, differential thermal analysis

D. Separation methods

   Distillation, extraction, centrifugation, gas and paper chromatography

E. Visible, ultraviolet, and infrared spectrometric methods

   Visible, ultraviolet, and vacuum spectroscopy, infrared spectrophotometry

F. Nuclear magnetic resonance method

   Proton resonance and nuclear magnetic resonance

G. Radiochemical method

   Exchange reactions, neotron scattering, β- and γ-ray counting

H. Physical methods

   Turbidity, density, refractive index, displacement, vapor pressure, dew point,
      hygrometry, cryoscopy, sonic technique, piezoelectric sorption hygrometry,
      chemical reaction-gasometry

I. Miscellaneous instrumental methods

   Mass spectrometry, glow discharge, microwave, x-ray method
_____

Source: Mitchell and Smith, 1977.

1.2 GAS-LIQUID CHROMATOGRAPHY

Since the introduction of gas-liquid chromatography (GLC) during the 1930s, it has proved to be the most powerful analytical tool yet available for use in laboratories. By injecting gaseous or liquid mixtures into the apparatus, the packed column separates the components and they appear on the recorder according to their retention time. The relative concentrations are represented by peak areas.

In principle there are two type of columns: polar and nonpolar. The nonpolar column separates the substances according to their normal boiling points; meanwhile, the polar column separates the compounds according to their polarity.

There are two often-used improvements in the GLC apparatus: the flame-ionization and the electron-capture detectors. Both devices considerably improve the performance and accuracy of the columns. The great advantage of flame-ionization gas chromatograph is its high sensitivity; it detects organic compounds in a concentration of 0.1 ppm. The peak areas are often conveniently measured by an electronic digital integrator, which converts microvolt-seconds into frequencies, and the output is directly proportional to the peak area. The chromatographic column is usually made of stainless steel packed with 80- to 100-mesh Chromosorb W. The column is calibrated daily using prepared standard solutions.

The electron-capture detector contains tritium foil, which produces a current. The operation potential is about 90 V. The stainless column is packed with 80- to 100-mesh Poropak Q or Chromosorb W. The carrier gas is N_2 or He delivered at a rate of 50 ml min^{-1}. The radioactive foil becomes contaminated rapidly if the solvents pass through the detector. Usually, the peaks are integrated using a digital integrator similar to the one described above. The detection range is about 10^{-15} mol.

1.3 INDICATOR PAPERS FOR WATER CONTENT OF REFRIGERANTS

Häntzschel and Kitter (1969) have described the use of indicator papers for the determination of water concentration in various refrigerants. The principle of indicator papers is based upon the color changes that occur upon the formation of hydrates of cobaltous bromide:

$CoBr_2$ green (dry refrigerant)
$CoBr_2 \cdot H_2O$ blue
$CoBr_2 \cdot 2H_2O$ purple
$CoBr_2 \cdot 6H_2O$ pink (wet color)

There are three types of indicator paper; A, B, and C. At $20^{\circ}C$, the recommended paper (A) indicates between $3\frac{1}{4}$ and 50 ppm of water in refrigerant 22, corresponding to a 3 to 5% mois content. Paper (B) gives the color indication for R 12 between 15 and 75 ppm of water with 20% relative humidity. In the case of 45% relative humidity, paper (C) is recommended.

The following table gives the color indications for R 12 at $20^{\circ}C$ for various water contents:

Water content in R 12 (ppm)	Type of paper	Color
1	A	Blue
	B	Blue
5	A	Red
	B	Light blue
10	A	Pink
	B	Light brown-white
20	A	Pink
	B	Pink
30	A	Pink
	B	Pink

Apparently, these papers give only an approximate value for the water content, but have the advantage of a rapid result.

Line and Hoftiezer (1956) used paper containing a uniform distribution of 0.03 to 0.14 g of anhydrous cobaltous bromide per cubic centimeter of cellulosic material. The water content of several halogenated hydrocarbons was determined at 15.5, 26.6, and 37.7°C (see Table 1.12). The limitation of the water content in the sample is 5000 ppm. The indicator is based on a visible color change. The blue color in liquids indicates that the water content is 25 ppm or below. The pink color shows that the water content is above 55 ppm. The method is simple and inexpensive, but is not accurate enough for the scientific determination of water content in organic liquids.

Table 1.12 Solubility of Water in Halogenated Hydrocarbons

Solvent	Solubility of water (ppm by weight)		
	15.5°C	26.6°C	37.7°C
CCl_2F_2	58	98	165
CCl_3F	70	113	168
$CHClF_2$	970	1350	1800
CH_3Cl	1880	2550	3500

Source: Line and Hoftiezer, 1956.

Chapter **2**
Theoretical Consideration
Of Solubility

2.1 Intermolecular Forces

To understand, evaluate, and interpret the various phenomena occurring between molecules in a solution, a knowledge of intermolecular forces is necessary. The chief
object of this chapter is to summarize the main principles of intermolecular forces,
emphasizing the latest developments and particularly their relevance to aqueous
solutions of nonpolar molecules. The reader who is interested in a comprehensive
treatise of the subject is advised to study one or several of the following compilations: Ben-Naim (1971a, b, 1974), Carra et al. (1980), Certain and Bruch (1972), Chu
(1967), Coetzee and Ritchie (1969-1976: Vol. 1), Dainton (1965), Ewing (1975), Fisher
(1964), Good and Elbing (1970), Gorodyskii et al. (1975), Guggenheim (1952, 1966),
Hadzi and Thompson (1959), Hirschfelder (1967), Hirschfelder et al. (1964), Hobza
and Zahradnik (1974a, b, 1975, 1976), Kavanau (1964), Kihara (1970, 1978), Kollman
and Allen (1972), London (1937), Margenau and Kester (1969), Miller (1966), Moelwyn-
Hughes (1964, 1971), Neff and McQuarrie (1973), Pimentel and McClellan (1960), Pitzer
(1959), Prigogine et al. (1957), Prigogine and Defay (1967), Ratajczak and Orville-
Thomas (1980), Reed and Gubbins (1973), Rice and Gray (1965), Rowlinson (1959, 1970),
Scatchard (1976), Schuster et al. (1976), Shimanouchi (1970), Shinoda (1978),
Simonov et al. (1974), Temperley et al. (1968), Temperley and Trevena (1978),
Teresawa et al. (1975), Tompkins (1978), Troitskaya and Tyulin (1976), Urban and
Hobza (1975a, b), Vinogradov and Linnell (1971), and Zellhoefer et al. (1938).

The study of intermolecular interactions is based upon either quantum theory or
experimental results (Hirschfelder, 1967). There are three types of intermolecular
forces:

1. Short-range forces, $r \leqslant \delta$ (chemical energies)
2. Intermediate-range forces, $0 < r \leqslant r_0$ (residual valence interaction, hydrogen
   bonding, and association interaction)
3. Long-range or van der Waals forces, $r \geqslant r_0$ (dispersion energies or London forces,
   polarization energies, direct electrostatic energies, and induction forces)

Applying these concepts to attractive (molecules are far apart) and repulsive molecules are close together) forces between like and unlike molecules of various shapes,
the distance between the molecular centers will vary with the force in question.
Another way to interpret this phenomenon is by use of the potential energy of

interaction rather than the force of interaction. The relationship between the intermolecular forces $F(r)$ and intermolecular potentials (r) is expressed

$$F(r) = - \frac{d\varphi}{dr}$$

That is, the force between molecules is the gradient of the potential, or

$$\varphi(r) = \int_{r}^{\infty} F(r) \, dr$$

where $F(r)$ = force of interaction

$\varphi(r)$ = potential energy of interaction

r = distance between molecular centers

In the case of the short-range repulsion forces between molecules, the potential
energy of interaction is explained by the quantum theory of electron spin and can be
calculated from the quantum mechanical relationships. The potential energy of
interaction increases as the distance becomes shorter and shorter between the two
molecules. A schematic representation is given in Fig. 2.1. If the distance decreases between the two molecules (i.e., $r \to 0$), the potential energy reaches the
Coulomb repulsion energy between two positively charged nuclei. Meanwhile, when the
distance increases, the potential energy becomes negative, passes through a minimum,
and starts to increase again until at $r = \infty$ it becomes zero.

2.1-1 Short-Range Forces

Short-range atomic and molecular repulsions have the same origin as chemical
bonds. These forces are often called valence, overlap, or exchange forces. Although
the physical origin of the short-range forces is clear, the calculations are very
cumbersome or not accurate enough for practical use. Therefore, most often the
theoretical calculations are combined with some experimental results. The theoretical calculations are based upon the Hellmann-Feynman electrostatic theorem and the
Pauli exclusion principle. The force acting between nuclei is the classical Coulomb
interactions, and the Pauli exclusion principle requires a large overlap of charge
clouds, causing some distortion in the electronic clouds of the two atoms.

One must distinguish between the interaction of two atoms with regard to their
closed or incomplete electronic shells. In the first case, the net effect is a
repulsion between the atoms; in the second case the electron distribution between
the nuclei leads to the formation of a chemical bond.

According to molecular-beam scattering experiments, the repulsion pair potential
(φ_{repuls}) is a linear function of r for $r > 0.5 \text{Å}$ in a semilogarithmic plot. This
has been demonstrated by Abrahamson (1963) using a quantum-theory for the argon-
argon pair:

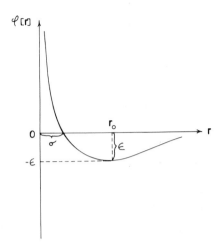

Fig. 2.1 Schematic illustration of the potential energy of interaction.

$$\varphi|_{repuls} = Ae^{ar}$$

where A and a are constants characteristic of the pair of molecules. This exponential expression represents well the results of the scattering experiments.

The quantum theory of electron spin and the Paulin exclusion principle are widely used to explain the short-range intermolecular forces and to predict them, particularly in the case where the interaction takes place between atoms with incomplete electron shells. A typical example is the interaction between two ground-state hydrogen atoms with parallel electronic spins.

Salem (1961) presented a theoretical analysis for the additivity nature of molecular interactions of short-range repulsive forces between closed-shell poly-atomic molecules whose electron clouds overlap slightly. The main contribution to the repulsive forces originates from a reduction of the combined electron density in the overlap region. He demonstrated the additive forces for He-He and also for H_2-H_2 systems. It has been established that the He-He repulsive potential is well represented by Amdur's empirical expression

$$\varphi'_{repuls} = 196.0e^{-4.21r}$$

in the range from 0.5 to 2.3 Å, and does not exceed 20% (Hirschfelder, 1967).

2.1-2 Intermediate-Range Forces

At present, there are several methods proposed for the calculation of intermediate-range intermolecular forces; however, all are subject to criticism. Since we lack a reliable theoretical procedure for the calculation of the intermediate-range forces, the most common practice is to add the long-range forces to the short-range forces.

In the case of intermediate-range forces, the atomic orbitals overlap and it is not possible to associate an electron of the electronic cloud with particular atoms. While at large separations the molecular system is described by the atomic orbitals, at intermediate separations the molecular system should be expressed by the molecular orbitals. That is, the system can be considered as a diatomic molecule. Regarding the calculation of the interaction energy, the Born-Oppenheimer approximation is in good agreement with experimental results. However, greater accuracy can be achieved in some cases by, for example, adiabatic calculation (Hirschfelder, 1967), depending on the range of the potential-energy curve in question. For a consistent calculation the symmetrized wave function is used, which proceeds at least as far as a second-order perturbation calculation. The generated new energy terms or second-order exchange terms are significant only at intermediate separations. These terms were used by Eisenschitz and London (1930) and by Margenau and Kester (1969). The unperturbed wave functions and the second-order perturbation theory provide a reliable semiempirical method, but the algebra behind it is rather complicated (Dalgarno and Lynn, 1956). A good review on the various aspects and present understanding of intermediate-range perturbation theory, including motivation, mathematical complications, a survey of exchange perturbation theories, and applications of exchange perturbation theories is given by Certain and Bruch (1972).

2.1-3 Long-Range or van der Waals Forces

Among the theories of short-, intermediate-, and long-range intermolecular forces, the latter has been most highly developed (Longuet-Higgins and Salem, 1961). At long-range separation the overlap between the electronic clouds of the two molecules is negligible (no electron exchange between the molecules), and consequently the mutual distortion of their electronic charge clouds is small. Therefore, the long-range forces can be calculated or correlated with the physical properties of the individual molecules (or atoms). These forces make the most significant contribution to an understanding of solubility theory, so a more comprehensive outline is adopted here for their description.

Long-range intermolecular forces can be divided into three types:

1. The electrostatic interactions between molecules with charges or electric multipole moments
2. Polarization or induction forces, which arises from the polarization of one molecule by the electric field of the other
3. Dispersion forces between neutral nonpolar molecules, arising from correlated fluctuations in the electron distributions of the two molecules.

The interaction between two molecules whose electrons are kept separated (long-range intermolecular forces) can be calculated by using perturbation theory. The perturbation to the Hamiltonian of noninteracting molecules is the Coulombic interaction between the charges of molecules a and b. The energy can be expressed in ascending powers of the perturbation, but the perturbation energy is usually expressed in inverse powers of the distance between the molecules. The electrostatic energy is the first-order perturbed energy and originates from the interaction of the permanent electric multiple moments of the molecules. Both the induction and dispersion energies belong to the second-order perturbed energies. The interaction between the permanent and induced electric multipole moments generates the induction energy, which can be expressed in terms of the field gradients produced by the multiple moments of one molecule and the static electric dipole and other polarizabilities of the other molecule. The dispersion energy is independent of the permanent electric multipole of either molecule. It depends upon the separation distances to the power of −6 for large r and can be expressed approximately in terms of the static polarizability of the molecules.

The charge and multiple moments (dipole, quadrupole, etc.) of the molecules will determine the electrostatic contribution to the intermolecular potential energy of the system. The electrostatic forces are locally additive and the Coulombic law of electrostatic interaction provides the formulas for the different types of interaction between two molecules. By the incorporation of angular dependence in the Coulombic formulas, the final equations become more complicated. If one takes a statistical average of the potential energy between two dipoles (Keeson alignment energy), the assumption is that the intermolecular distance (r) does not change appreciably as the molecule undergoes a rotation. These averaged potential functions are temperature (T)-dependent because the expressions include the Boltzmann weighting factor (k):

$$\varphi_{ab}(\text{electrostatic}) = - \frac{2\,\mu_a^2\,\mu_b^2}{3\,r^6\,kT}$$

The interaction between a charged and a neutral molecule induces a dipole moment (μ) in the neutral molecule. A dipole moment will also form when two neutral molecules are subjected to an electric field. The interaction energy will be proportional to the polarizability of the molecule (α_b) and the reciprocal of the fourth power of the intermolecular distance (r). The potential energy of interaction forms a similar relationship when averaged over the angles:

$$\varphi_{ab}(\text{induction}) = - \frac{\mu_a^2\,\alpha_b}{r^6}$$

The dipole moment and polarizability of halogenated hydrocarbons are given in the Appendix. Methods for the calculation of polarizability are given by Denbigh (1940), Atoji (1956), Bogaard and Orr (1975), and Landolt-Börnstein (1951). When both polar and nonpolar molecules are subjected to an electric field, the main potential energy due to induction is expressed by the Debye formula:

$$\varphi_{ab}(\text{induction}) = - \frac{\mu_a^2 \alpha_b + \mu_b^2 \alpha_a}{r^6}$$

Usually, the induction energy is greater when it is caused by permanent dipoles, quadrupoles, and so on.

The nonpolar molecules show a temporary dipole moment that changes rapidly with time; however, during this time it induces dipoles in the surrounding molecules. At any instant the electron in molecule a causes an instantaneous dipole moment in molecule b. The interactions between dipoles a and b produce an energy of attraction between the two molecules. This interaction was expressed by London (1937) as

$$\varphi_{ab}(\text{dispersion}) = - \frac{3 \alpha_a \alpha_b}{2 r^6} \left(\frac{h v_a \, h v_b}{h v_a + h v_b}\right)$$

where $h v$ is approximately equal to the ionization potential (I) (Jaffe and Orchin, 1962):

$$\varphi_{ab}(\text{dispersion}) = - \frac{3 \alpha_a \alpha_b}{2 r^6} \left(\frac{I_a I_b}{I_a + I_b}\right)$$

For the same molecules

$$\varphi_{ab}(\text{dispersion}) = - \frac{3 \alpha_a^2 I_a}{4 r^6}$$

The iónization potentials of halogenated hydrocarbons are given in the Appendix. Several methods are compared for the calculation of ionization potentials by Hayashi and Nakajima (1975, 1976). Good reviews of ionization potentials have been compiled by Streitwieser (1963) and Gibson (1977).

The foregoing expressions show that the dispersion energy is independent of temperature. However, London's formulas consider only the first term of a series:

$$\varphi_{ab}(\text{dispersion}) = - \frac{C_6}{r^6} - \frac{C_8}{r^8} - \frac{C_{10}}{r^{10}} - \cdots$$

There are several proposals for the calculation of the dispersion coefficients (C_6, C_8, C_{10}, etc.). One of the recently reported articles (Amos and Yoffe, 1976) applies the Frost model wave function.

Klein et al. (1974) has presented the (m, 6, 8) intermolecular potential functions. In the expression

$$\varphi(r) = \frac{A}{r^m} - \frac{B}{r^6} - \frac{C}{r^8}$$

where m is an index of repulsion, the coefficient ranges between 9 and 18, and the exponents 6 and 8 are indices of attraction. Furthermore, A represents the coefficient for repulsion, and B and C are the coefficients for attraction. This modification or extension of the (m, 6) function provides much more accurate calculations for collision integrals, diffusion, thermal conductivity, thermal diffusion, and viscosity.

The three types of long-range intermolecular forces - - electrostatic, polarization (or induction), and dispersion - - calculated for several halogenated hydrocarbon-water systems at 25°C are presented in Table 2.1. When the dipole moment of the halogenated hydrocarbon is zero, the electrostatic forces are unimportant. The dispersion forces are very similar in all systems (see Table 2.1). However, the expression used for the calculation of dispersion forces is valid only for spherical molecules. For long molecules there is a much more complicated equation (see Hirschfelder et al., 1964).

A simple but illustrative representation of the intermolecular separation is the assumption of the molecules as rigid spheres. This spherical model gives a good description for most nonpolar molecules. Their intermolecular potential energy energy function is given by the Lennard-Jones (12,6) potential:

$$\varphi(r) = 4\,\epsilon\left[\left(\frac{\sigma}{r}\right)^{12} - \left(\frac{\sigma}{r}\right)^{6}\right]$$

where ϵ and σ are characteristic constants of the molecules and have dimensions of energy and length, respectively. Numerical values of these constants for halogenated hydrocarbons have been reported by Nikul'shin and Petriman (1976) and Mourits and Rummens (1977) (see Table 2.7). If the two molecules are far from each other, then the sixthpower term is dominant, whereas at short distances the twelfthpower term provides the repulsive forces and is dominant. The potential energy becomes zero at $r = \sigma$, which is the closest distance between two molecules when the collision takes place with zero initial relative kinetic energy.

In addition to the Lennard-Jones (12,6) function, it is necessary to include an angle-dependent term to describe the intermolecular potential energy for polar molecules. The Stockmayer potential (see later) is widely used for simple polar molecules with dipole-dipole interaction. This describes well the interaction between molecules for which dipole-quadrupole and higher multiple interactions are negligible. Apparently, neither the Lennard-Jones (12,6) nor the Stockmayer potential functions can describe interactions between long molecules, free radicals, molecules in excited states, and ions.

Hirschfelder et al. (1964) classified the various intermolecular potentials into two groups:

Table 2.1 Intermolecular Forces Between Halogenated Hydrocarbons and Water at 25°C

Substance	Dipole moment (debyes)	Polarizability $(cm^3 \times 10^{25})$	Ionization potential (eV)	Intermolecular forces $(erg\ cm^6 \times 10^{60})$		
				Dipole	Induction	Dispersion
CH_3Cl	1.87	45.6	11.3	192.0	21.0	104.0
CH_3Br	1.81	55.5	10.5	180.0	24.0	122.0
CH_2Cl_2	1.60	64.8	11.3	140.0	26.0	148.0
$CHCl_3$	1.01	82.3	11.5	55.9	29.5	189.0
CCl_4	0.0	105.0	11.5	0.0	35.5	241.0
C_2H_5Cl	2.05	64.0	11.0	230.0	28.3	143.0
C_6H_5Cl	1.69	122.0	9.07	157.0	46.0	247.0
$p-C_6H_4Cl_2$	0.0	145.0	8.95	0.0	49.0	289.0
$m-C_6H_4Cl_2$	1.72	142.0	9.12	162.0	52.9	288.0
$o-C_6H_4Cl_2$	2.50	142.0	9.06	343.0	57.9	285.0

1. Angle-independent potentials
    a.   Rigid impenetrable spheres
    b.   Point centers of repulsion
    c.   The square well
    d.   The Sutherland model
    e.   Lennard-Jones (12,6) potential
    f.   Buckingham potential
    g.   Buckingham corner potential
    h.   Modified Buckingham (6-exp.) potential

2. Angle-dependent potentials
    a.   Rigid ellipsoids of revolution
    b.   Kihara spherocylindrical molecules
    c.   Rigid spheres containing a point dipole (Keesom)
    d.   Stockmayer potential

A detailed discussion of these models with illustrations and comments is given by Hirschfelder et al. (1964); therefore, only a brief summary is given here, with some graphical illustrations.

Fig. 2.2 shows the rigid impenetrable spheres of diameter σ. This is a very simple model and gives only an approximation of the strong short-range repulsive forces. The repulsive potential has only two values

$$\varphi(r) = \begin{cases} 0 & \text{when } r > \sigma \\ \infty & \text{when } r < \sigma \end{cases}$$

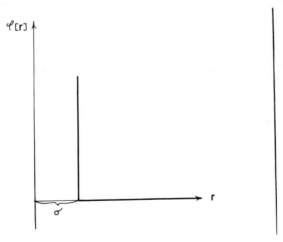

Fig. 2.2 Intermolecular potential of rigid impenetrable spheres.

Another very simple model is the point center of repulsion, which also gives an approximate value only (see Fig. 2.3). The repulsive potential is represented by an inverse-power law

$$\varphi(r) = \frac{A}{r^x}$$

where x is usually between 8 and 16.

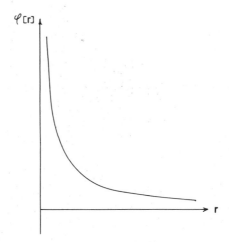

Fig. 2.3 Intermolecular potential of point centers of repulsion.

The square-well potential-energy model (see Fig. 2.4) involves hard, attracting spheres, and it has three adjustable parameters: r, σ, and ε, where

r = intermolecular distance between molecular centers

σ = collision diameter

ε = depth of the well

R = reduced well width

The square-well potential is zero everywhere except at r = σ and r = R σ. In other words,

$$\varphi(r) = \begin{cases} \infty & \text{when } r < \sigma \\ -\varepsilon & \text{when } \sigma < r < R\,\sigma \\ 0 & \text{when } r > R\,\sigma \end{cases}$$

These relationships are well illustrated in Fig. 2.4. The model is an improvement over the hard-sphere theory and is particularly useful for liquids. The potential

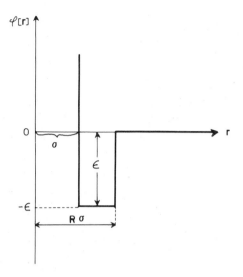

Fig. 2.4 Square-well potential-energy model.

parameters σ, R, and ε can be obtained from second-virial-coefficient data.

   The Sutherland model, a combination of the London dispersion and hard-sphere models, is shown in Fig. 2.5. This is a special case of the (6,n) potential when n = ∞ for   r ⩽ σ. Furthermore,

$$\varphi(r) = \begin{cases} \infty & \text{when } r < \sigma \\ -\epsilon\left(\dfrac{\sigma}{r}\right)^6 & \text{when } r > \sigma \end{cases}$$

Like the square-well potential, the Sutherland model is a considerable improvement over the hard-sphere theory.

   The Lennard-Jones (12,6) potential model

$$\varphi(r) = 4\,\epsilon\left[\left(\frac{\sigma}{r}\right)^{12} - \left(\frac{\sigma}{r}\right)^6\right]$$

is a modification of Mie's equation,

$$\varphi(r) = \frac{(n^n/m^m)^{(1/(n-m))}}{n-m}\left[\left(\frac{\sigma}{r}\right)^n - \left(\frac{\sigma}{r}\right)^m\right]$$

where n and m are positive constants and n >m. The first term $(\sigma/r)^{12}$ represents the repulsive energy, and the second term $(\sigma/r)^6$ gives an account of the attractive energy. The graphical illustration of this potential is shown in Fig. 2.1. There are

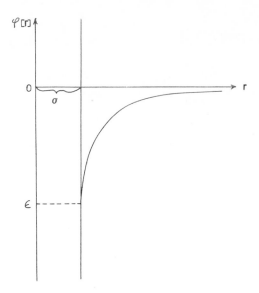

Fig. 2.5 Sutherland model of intermolecular potential.

two parameters: σ represents the distance between the two molecules when the potential energy becomes zero and ε is the maximum energy of attraction or the depth of
the potential well at $r = 2^{1/6} \sigma$. The distance parameter can also be expressed by
$\sigma = 0.5^{1/6} r_o$.

The Lennard-Jones (12,6) potential is very widely used. The potential barriers
of halogenated hydrocarbons were calculated by Heublein et al. (1970). The parameters
σ and ε are most often derived from viscosity measurements (Bae and Reed, 1967;
Casparian and Cole, 1974; Olbregts and Walgraeve, 1976; Nikul'shin and Petriman,
1976). The numerical values for a few halogenated hydrocarbons are presented in
Table 2.7 (see also Clifford et al., 1979 and Wen and Muccitelli, 1979). The parameters can also be calculated from second virial coefficients (Hirschfelder et al.,
1964; Heintz and Lichtenthaler, 1976). Tee et al. (1966) developed a corresponding-
state correlation for the parameters σ and ε/k, with good results. The relationships
between the parameters and viscosity, thermal conductivity, and self-diffusion are
given by Stiel and Thodos (1962). The energy parameter is most often expressed as
the ratio ε/k, where k is Boltzmann's constant. The unit for ε/k is absolute temperature (degrees Kelvin) and the unit for σ is angstroms. A simple, but less accurate method for the calculation of the parameters σ and ε/k is based on the critical properties of the substances (Renon et al., 1967; Stiel and Thodos, 1962;
Salsburg and Kirkwood, 1953; Gotoh, 1972; Rao and Majumdar, 1975; Pesuit, 1978;
Carra et al., 1980):

$$(2/3) \pi N_A \sigma^3 = (2/4) v_{cr}$$

and
$$\epsilon/k = 0.74 \, T_{cr}$$

but for noble gases (Nelson and de Ligny, 1968):

$$\epsilon/k = 0.77 \, T_{cr}$$

where T_{cr} = critical temperature, K
v_{cr} = critical volume, cm^3 g-mole^{-1}
N_A = Avogadro's constant
σ = distance parameter, cm

These relations are based upon the theory of kinetic energy. In a more recent publication Carra et al. (1980) correlated the two parameters (σ and ϵ/k) mentioned for 24 nonpolar substances by the equations:

$$\sigma = -0.35 + 0.93 \, v_{cr}^{1/3}$$

and

$$\epsilon/k = -4.1 + 1.076 \, T_{cr}$$

The simple relations

$$\sigma = 0.841 \, v_{cr}^{1/3}$$

and

$$\epsilon/k = 0.77 \, T_{cr}$$

where used by Afshar and Saxena (1980) for refrigerants R 152a and R 142b. Further approximations have been applied to calculate σ and ϵ/k from the molar volume at the boiling point and from the Boyle temperature, respectively. The parameters were derived from the equation of state of liquids (Gotoh, 1971, 1972), from the velocity of sound (Aziz, 1974), and from vapor pressure and thermodynamic perturbation theory (Goldman, 1976). Troitskaya and Tyulin (1976) correlated the force constant with the length of the chemical bond for diatomic molecules. Boublik (1972) and Boublik and Aim (1972) obtained the hard-sphere diameter from the latent heat of vaporization. Rodriguez (1978) found the parameters from the pressure-volume-temperature behavior of mixtures. For nonpolar molecules the correlation of Tee et al. (1966) has been used by Tiepel and Gubbins (1972) to obtain both σ and ϵ/k; however, for polar molecules a method using corresponding-states analysis suggested by Bae and Reed (1971) was satisfactory. A summary of methods for determining σ and ϵ/k has been presented by Svehla (1962). In a recent publication Mourits and Rummens (1977) presented a critical evaluation of Lennard-Jones potential parameters and some correlation methods. The arithmetical and geometrical means of σ and ϵ,

$$\sigma_{ab} = (\sigma_a + \sigma_b)/2$$

and

$$\varepsilon_{ab} = (\varepsilon_a \, \varepsilon_b)^{\frac{1}{2}}$$

provide the convenient way to calculate the parameters for interactions between unlike molecules. These approximations have been proposed on the basis of the London's theory and are used quite widely for mixtures (e.g., for vapor-liquid equilibria) (Pesuit, 1978).

Searching for a universal solvent polarity scale, Gurikov (1980) pointed out that there are considerable discrepancies between the force constants determined from transport properties and thermodynamic properties in the gas phase. Furthermore, there is no justification for believing that the intermolecular interaction in the liquid phase is described by the same pair potential as in the gas phase. This is explained by the collective effects in the condensed state, the non-sphericity of the molecular shape, and the presence of electric multipoles in the molecules. The (12,6) pair potential is usually used to describe the interaction between the species in simple liquids.

In the Buckingham potential model,

$$\varphi(r) = \frac{A}{e^{B\,r}} - \frac{C}{r^6} - \frac{D}{r^8}$$

the repulsive potential has an exponential term as a function of r in addition to the inverse-power terms. This model has been used successfully for the calculation of intermolecular forces in brominated benzene crystals (Burgos and Bonadeo, 1977). It is more difficult to handle, however, because it represents a more realistic model than the Lennard-Jones (12,6) potential. Fig. 2.6 shows the model. The potential becomes infinite when r approaches zero. Like the Buckingham corner potential, the expression has four coefficients. If r is greater than the r_m value at the energy minimum, the Buckingham corner potential is

$$\varphi(r) = \frac{A}{e^{(13.5\,r/r_m)}} - \frac{C}{r^6} - \frac{D}{r^8}$$

This expression eliminates the unrealistic situation at the origin found with the original Buckingham potential.

A further improvement of the simple Buckingham potential is the modified Buckingham or 6-exp. potential model:

$$\varphi(r) = \begin{cases} \dfrac{\varepsilon}{1 - 6/\alpha} \left[(6/\alpha) \, e^{\alpha(1 - r/r_{min})} - (r_{min}/r)^6 \right] & \text{for } r \geqslant r_{max} \\ \infty & \text{for } r \leqslant r_{max} \end{cases}$$

where α is the steepness of the exponential repulsion and r_{min} the value of r for

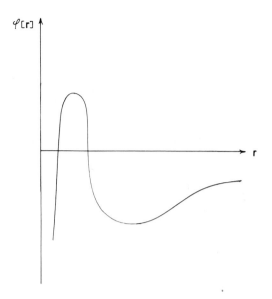

Fig. 2.6 Buckingham potential model.

the energy minimum. The model is shown schematically in Fig. 2.7. The parameters of
the 6-exp. potential are given by Nikul'shin and Petriman (1976) and Heublein et al.
(1970) for halogenated hydrocarbons (see Table 2.7). The mixing rule for unlike molecules is obtained from the geometrical mean of the third parameter (Pesuit, 1978)

$$\alpha_{ab} = (\alpha_a \alpha_b)^{\frac{1}{2}}$$

The model of the rigid ellipsoids of revolution belongs to the angle-dependent
potentials. The flat and elongated molecules are represented by oblate and prolate
ellipsoids respectively. This model is used for the investigation of nonspherical
fields on various properties.

Another model is the Kihara model, representing spherical-rigid cores (Kihara,
1970; Good and Elbing, 1970). The model is illustrated in Fig. 2.8 and described by
the following expressions:

$$\varphi(r) = \begin{cases} \varepsilon\left[(r_{min}/r)^{12} - 2\,(r_{min}/r)^6\right] & \text{when } r > d \\ \infty & \text{when } r < d \end{cases}$$

where ε is the depth of the potential minimum, r_{min} the internuclear distance
corresponding to this minimum, and d the rigid-core diameter. The shape of the
potential energy curve is influenced by the core size. When the core diameter (d)

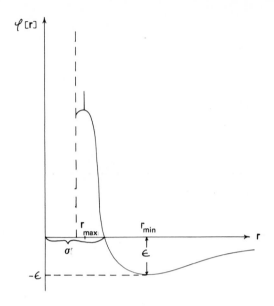

Fig. 2.7 Modified Buckingham or 6-exp. potential model.

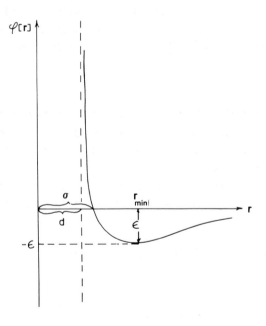

Fig. 2.8 Kihara spherical-rigid-cores model.

equals zero, the Kihara potential becomes the Lennard-Jones (12,6) model. The Kihara
model gives a good representation of the attraction during interactions. If the
parameters are obtained by fitting, for example, the second virial coefficient of a
gas, the derived values from the Lennard-Jones fit need not bear any relationship
to the parameters of the Kihara fit. The difference between the two models is that
the intermolecular distance is between the molecular centers in the Lennard-Jones
potential, whereas the Kihara model takes the minimum distance between the surfaces
of the two molecule cores. The Kihara potential is the most successful potential
available at present, except at very high temperatures. The Kihara potential values
of halogenated hydrocarbons have been reported by Nikul'shin and Petriman (1976)
(see Table 2.7).

The intermolecular potential function of a real gas has been determined by
choosing the Lennard-Jones potential and the Kihara potential and simultaneously
fitting the parameters to experimental results of the isothermal throttling coefficient and the second virial coefficient (Bier et al., 1975). In a more recent paper,
Speedy (1977) presented an equation of state that provides a more accurate result
compared with the previous theoretical predictions for hard-sphere fluids.

The spherical-rigid-core Kihara potential model can be divided into two parts:
a rigid core and an outer soft spherical repulsion region. The rigid-core diameter
is denoted by d (see Fig. 2.8). In the case of polyatomic molecules, there is a
uniform distribution of the repulsion and attraction forces about the center of
mass of each molecule. The unlike-pair empirical parameters used in the Kihara
model can be expressed by geometrical and arithmetical means;

$$d_{ab} = (d_a + d_b)/2$$

The corresponding-states correlations were developed by Tee et al. (1966) for the
prediction of force constants with good accuracy.

The Keesom model belongs to the angle-dependent potentials and represents the
rigid spheres containing a point dipole. The angular dependence of the interaction
between molecules a and b is very complicated. In this model the intermolecular distance r and the angles θ_a, θ_b, ϕ_a, and ϕ_b serve to define the orientations of the
two molecules. The average interaction between two dipoles is called the Keesom
alignment energy, which is negligible relative to the London dispersion forces.

The fourth angular-dependent potential is the Stockmayer model (Malek and Stiel,
1972):

$$\varphi(r) = 4\,\varepsilon\left[(\sigma/r)^{12} - (\sigma/r)^6\right] - (\mu_a\,\mu_b)/r^3 \; g(\theta_a,\theta_b,\phi_b - \phi_a)$$

where

$$g(\theta_a,\theta_b,\phi_b - \phi_a) = 2\cos\theta_a\cos\theta_b - \sin\theta_a\sin\theta_b\cos\phi_{ab}$$

The Stockmayer (12,6,3) potentials for refrigerants have been reported by Nikul'shin
and Petriman (1976) (see Table 2.7). The Stockmayer potential is an extension of the
Lennard-Jones (12,6) potential, with the addition of dipole-dipole interactions and
angular-dependent parameters. The Stockmayer potential functions are shown in Fig.
2.9a and b for dipoles lined up in attractive and repulsive potentials, respectively.
This potential describes well the interaction between those polar molecules whose
higher multipole interactions are negligible. Whereas the Lennard-Jones model is
applicable for the nonpolar molecules, the Stockmayer potential function is used for
polar and nonassociated molecules (Yen and McKetta, 1962). However, Belzile et al.
(1976) used this potential function for CHF_3, which shows molecular association. The
preceding expression is sometimes described by a modification (Halkiadakis and Bowrey,
1975):

$$\varphi(r) = 4\ \varepsilon \left[(\sigma/r)^{12} - (\sigma/r)^6 - \delta\ (\sigma/r)^3 \right]$$

where $\delta = \mu^2/(4\sigma^3)$. The term $\delta\ (\sigma/r)^3$ is a function of the relative orientation of
the molecules. It has been assumed that the relative orientation of the dipoles

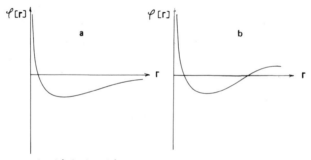

Fig. 2.9 Stockmayer potential functions.

remained constant during the actual collision and that all orientations were equi-
probable. The molecular parameter has been calculated for polar and nonpolar molecules by

$$\sigma = 0.687\ (v_{cr}/N_A)^{1/3}$$

and

$$\varepsilon = 0.404\ k\ T_{cr}\ z_{cr}^{-0.53}$$

However, Malek and Stiel (1972) presented other expressions which generate more accurate values:

$$\sigma\ (P_{cr}/T_{cr})^{1/3} = 2.3454 + 0.2972\ \omega - 40.4271\ x + 61.7675\ \omega\ x - 15.8248\ x^2$$
$$- 0.9706\ \omega^2$$

and

$$\varepsilon/(k\ T_{cr}) = 0.8082 - 0.4504\ \omega + 27.3866\ x - 48.9402\ \omega\ x - 48.7293\ x^2$$
$$+ 0.8784\ \omega^2$$

where the acentric factor (ω) is

$$\omega = -\log_{10}(P^o/P_{cr})\Big]_{T_r = 0.7}^{-1.0}$$

and

$$x = \log_{10}(P^o/P_{cr})\Big]_{T_r = 0.6}^{1.70\ \omega + 1.522}$$

Several correlations have been proposed for the calculation of the acentric factor (ω) from physical properties. For example, Deak (1973) proposed the temperature (T, K) and liquid molar volume (v, cm^3 g-mole^{-1}) as parameters,

$$\omega = A + B\ T + C\ T^2 + (D + E\ T + F\ T^2)\ v + (G + H\ T + I\ T^2)\ v^2$$

where A, B, C, D, E, F, G, H, and I are coefficients of the equation.

An empirical equation relating the reduced vapor pressure, the reduced temperature, and the acentric factor for numerous polar and nonpolar organic liquids was reported by Nath et al. (1976).

The parameters of the Stockmayer potential for halogenated hydrocarbons have been calculated by Nikul'shin and Petriman (1976) from viscosity data.

A critical evaluation of the Lennard-Jones and Stockmayer potential parameters and of some correlation methods has been given by Mourits and Rummens (1977).

Liabastre (1974) investigated the values of σ and ε/k parameters from the solubility of small organic molecules in water using the modified form of the scaled particle theory. The solute size σ of the dissolved molecule in water has been derived from the expression

$$(1 + \alpha_p\ T)\ \Delta G_s - \Delta H_s = (1 + \alpha_p\ T)\ \bar{G}_c - \bar{H}_c + (1 + \alpha_p\ T)\left[R\ T\ \ln(R\ T/V_{soln})\right]$$
$$- R\ T\ (\alpha_p\ T - 1)$$

where α_p = coefficient of thermal expansion

ΔG_s = free energy of solution

ΔH_s = enthalpy of solution

$\bar{G}_c$ = partial molar free energy of cavity formation

$\bar{H}_c$ = partial molar enthalpy of cavity formation

The molecular sizes (σ) of several halogenated hydrocarbons calculated from this equation are listed in Table 2.2. By plotting σ versus the polarizability α, we obtained Fig. 2.10. The fitted data produced the following polynominal expression for σ:

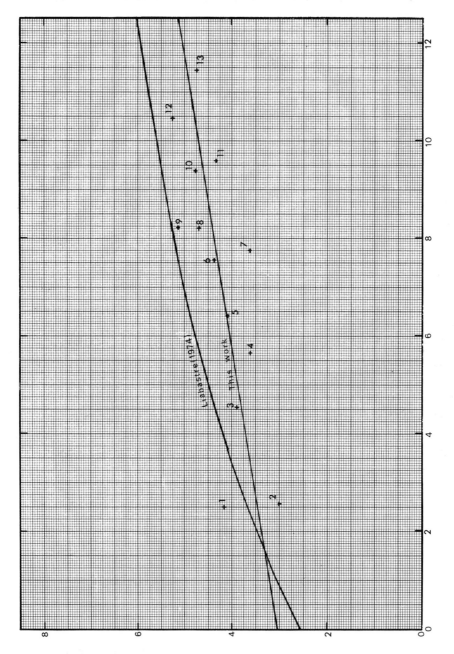

Fig. 2.10 Molecular size (σ, Å) verus the polarizability (α x 10^{24} cc molecule^{-1}). 1, CF$_4$; 2, CH$_3$F; 3, CH$_3$Cl; 4, CH$_3$Br; 5, C$_2$H$_5$Cl; 6, C$_2$H$_5$Br; 7, CH$_3$I; 8, C$_3$H$_7$Cl; 9, CHCl$_3$; 10, C$_3$H$_7$Br; 11, C$_2$H$_5$I; 12, CCl$_4$; 13, C$_3$H$_7$I.

Table 2.2 Molecular Sizes of Halogenated Hydrocarbons

Substance	σ (Å)	Substance	σ (Å)
CCl_4	5.25	CH_3-CHF_2	4.76
CF_4	4.18	$CF_3-CF_2-CF_3$	6.00
$CHCl_3$	5.12	$CH_3-CH_2-CH_2Br$	4.77
CH_3Br	3.64	$CH_3-CH_2-CH_2Cl$	4.68
CH_3Cl	3.90	$CH_3-CH_2-CH_2I$	4.77
CH_3F	3.04	$c-C_4F_8$	5.82
CH_3I	3.62	$CH_3-CH_2-CH_2-CH_2Br$	5.22
CF_3-CF_3	5.56	$CH_3-CH_2-CH_2-CH_2Cl$	5.08
CH_3-CH_2Br	4.36	$CH_3-CH_2-CH_2-CH_2I$	5.22
CH_3-CH_2Cl	4.10	$CH_3-(CH_2)_3-CH_2Br$	5.62
CH_3-CH_2I	4.35	$CH_3-(CH_2)_3-CH_2Cl$	5.50
CH_3-CClF_2	4.87	$CH_3-(CH_2)_3-CH_2I$	5.62

Source: Afshar and Saxena, 1980; Liabastre, 1974; Muccitelli, 1978.

$$\sigma = 2.5505 + 0.54433 \ \alpha - 0.0404 \ \alpha^2 + 0.0022335 \ \alpha^3 - 0.63468 \times 10^{-4} \ \alpha^4$$
$$+ 0.7003 \times 10^{-6} \ \alpha^5$$

From this equation we can derive the molecular size if the polarizability of the molecule is known.

The relationships among intermolecular force constants σ and ε of the Lennard-Jones (12,6) potentials, polarizability (α), and critical constants were discussed by Shanmugasundaram and Thiyagarajan (1980). The polarizability of a large number of halogenated hydrocarbons is listed in Table 2.3. For further values see Gross (1931). The molecular diameter can also be correlated with cube roots of molal volumes of gases at their boiling points (Hildebrand, 1969).

Generally speaking, the molecular interaction varies with the type of compound in question: nonpolar, polar, associated, and so on, species (Barfield and Johnston, 1973; Becker et al., 1972). The nonpolar substances are characterized by the short- and long-range forces, particularly dispersion and direct electrostatic energies. Typical nonpolar molecules include He, Ne, Ar, Kr, Xe, CH_4, CCl_4, and O_2. In the polar molecules, contrary to the nonpolar substances, the positive and negative charges are not coincident, forming a dipole moment. The intermolecular forces belong to the short- and long-range regions. Some typical polar molecules are CH_3Cl, C_6H_5Cl, and acetone. The associated compounds constitute the intermediate-range interaction: H_2O, HF, and others.

Potential-energy curves for aqueous solutions of nonpolar molecules are very sparse in the literature. The CH_4 - H_2O pair interaction for three mutual orienta-

Table 2.3 Polarizability and Dipole Moment of Halogenated Hydrocarbons

Substance	Polarizability (cc molecule^{-1} x 10^{24})	Dipole moment (debyes)
CF_4	2.52	0.0
CH_3F	2.592	1.83
C_2F_6	6.03	0.0
C_2H_5F	4.43	1.96
C_3H_8	8.41	–
C_3H_7F	5.987	–
$n-C_4H_9F$	6.318	–
$n-C_5H_{11}F$	9.950	–
CH_3Cl	4.567	1.92
C_2H_5Cl	6.405	1.96
C_3H_7Cl	8.244	1.97
$n-C_4H_9Cl$	10.08	1.90
$n-C_5H_{11}Cl$	11.94	1.94
CH_3Br	5.705	1.80
C_2H_5Br	7.545	1.90
C_3H_7Br	9.385	1.93
$n-C_4H_9Br$	11.22	2.15
$n-C_5H_{11}Br$	13.03	2.19
CH_3I	7.722	1.48
C_2H_5I	9.604	1.78
C_3H_7I	11.44	1.84
$n-C_4H_9I$	13.28	2.08
$n-C_5H_{11}I$	15.12	1.85
C_6H_5F	10.28	1.42
C_6H_5Cl	12.36	1.51
C_6H_5Br	13.46	1.50
C_6H_5I	17.49	1.29
CCl_4	10.49	0.0
C_6F_6	14.13	0.33
$n-C_7F_{16}$	14.57	–
CCl_2H-CCl_2H	12.14	1.71
$CCl_2F-CClF_2$	10.37	0.77
CH_2Cl_2	6.48	1.62
$CHCl_3$	8.23	1.06
$p-C_6H_4Cl_2$	14.47	0.0
$m-C_6H_4Cl_2$	14.23	1.35
$o-C_6H_4Cl_2$	14.17	2.20

Table 2.3 (Continued)

$c-C_4F_8$	10.8	–
$CFCl_3$	8.25	0.46
CF_2Cl_2	6.25	0.50
CF_3Cl	4.57	0.50
$o-C_6H_4Br_2$	16.60	2.05
$o-C_6H_4BrCl$	15.60	2.20

When substances have no published data on polarizability, it can be calculated from the Lorentz-Lorentz equation,

$$\alpha = \left(\frac{n^2 - 1}{n^2 + 1}\right) \frac{M}{P} \left(\frac{3}{4\pi}\right) \frac{1}{N}$$

or from the bond polarizabilities (Denbigh, 1940; Atoji, 1956).

The polarizability of a solvent can also be calculated from the refractive index and the radius (r) of the spherical cavity (Onsager, 1936):

$$\alpha = \frac{n^2 - 1}{n^2 + 1} r^3$$

For chain molecules with critical volumes greater than 90 cc g-mole^{-1}, the polarizabilities are calculated from the expression (Pesuit, 1978)

$$\alpha = 0.345(V_{cr} - 22.4) \times 10^{-25} \text{ cc}$$

Source: Landolt-Börnstein, 1951; Liabastre, 1974; Martin, 1979; Minkin et al., 1970; Muccitelli, 1978.

tions is given by Zahradnik et al. (1975, 1976) and Goldman and Krishnan (1976) (see Fig. 2.11). The enthalpy of cavity formation was calculated from the expression

$$\Delta H_c = \frac{\alpha_a}{\beta_a} T V_b = \frac{2.57 \times 10^{-4}}{45.8 \times 10^{-6}} \; 298.15 \times 35.6 = 5.9529 \times 10^4 \text{ atm cc g-mol}^{-1}$$

or 1.4416 kcal g-mole^{-1} and the enthalpy of interaction was

$$\Delta H_i = -4.612 \text{ kcal g-mole}^{-1}$$

both at 298 K. If the calculation is based upon the Stackelberg's method, the result becomes

$$\Delta H_i = -6.023 \text{ kcal g-mole}^{-1}$$

The theory of intermolecular forces can be applied to calculation of the solubility of gases in liquids (Kihara and Jhon, 1970). Ostwald's solubility coefficient C/C_o is expressed by the equation

$$\ln (C/C_o) = -\frac{E}{kT}$$

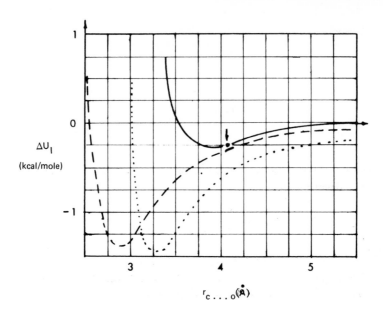

Fig. 2.11 Potential-energy curves for CH_4 - H_2O system.

$$\cdots\cdots\cdots\cdots \quad \begin{matrix} H \\ \diagdown \\ O \cdots\cdots H - C \\ \diagup \\ H \end{matrix} \begin{matrix} H \\ | \\ H H \end{matrix}$$

$$- - - - - - \quad H - C \cdots\cdots H \quad \begin{matrix} H \\ | \\ H H \end{matrix} \quad \begin{matrix} O - H \\ \diagup \end{matrix}$$

$$\underline{\hspace{3cm}} \quad \begin{matrix} H \\ \diagup \\ O \cdots\cdots H - C \\ \diagdown \\ H \end{matrix} \begin{matrix} H \\ | \\ H H \end{matrix}$$

(From Zahradnik et al., 1975.)

where E = activation energy

      k = Boltzmann's constant

      T = absolute temperature, K

The activation energy (E) composed of two parts:

$$E = E_1 + E_2$$

where E_1 is the work necessary to create a cavity for the solute molecule in the solvent having a **surface** tension γ_1 (dyn cm^{-1}). It can be calculated from the surface tension of the solvent and the effective surface area of the solute molecule:

$$E_1 = \gamma_1 \left[S_a + M_a(\sigma_{aa} + b) + \Pi(\sigma_{aa} + b)^2 \right]$$

where S_a = surface area of the solute molecular core

   M_a = mean curvature integrated over the core's surface

$\tfrac{1}{2}(\sigma_{aa} + b)$ = thickness of the molecular core a

The second term E_2 of the activation energy is given by

$$E_2 = f\, n \int_{\sigma_{ab}}^{\infty} \varphi_{ab}(r) \left[S_a + S_b + (2\Pi)^{-1} M_a M_b + 2(M_a + M_b)r + 4\Pi^2 r \right] dr$$

where

$$\varphi_{ab}(r) = 4\,\varepsilon_{ab} \left[(\sigma_{ab}/r)^{12} - (\sigma_{ab}/r)^6 \right]$$

After integration E_2 becomes

$$E_2 = -f\, n\, \varepsilon_{ab} \left[\frac{24}{55}(S_a + S_b) + \frac{M_a M_b}{2\,\Pi}\,\sigma_{ab} + \frac{6}{5}(M_a + M_b)\,\sigma_{ab}^2 + \frac{32\,\Pi\,\sigma_{ab}^3}{9} \right]$$

where f = factor, temperature dependent

   n = number density of solvent molecules, $\overset{\circ}{A}^{-3}$

   σ_{ab} = distance between molecules a and b $\left[= \tfrac{1}{2}(\sigma_a + \sigma_b) \right]$

   ε_{ab} = depth of the energy well $\left[= (\varepsilon_a \varepsilon_b)^{\frac{1}{2}} \right]$

The method presented by Kihara and Jhon (1970) has been applied to calculate the solubility of gases ($c\text{-}C_3H_6$, $CH_3\text{-}CH_3$, Xe, CO_2, Kr, Ar, N_2, and Ne) in cyclohexane between 20 and 40°C and at 1 atm partial pressure. The agreement with the experimental data was satisfactory, illustrating the usefulness of the theory of intermolecular forces to solubility calculations.

   The intermolecular forces acting between the molecules in the solution have been applied to calculate the solubilities in nonpolar-nonpolar, polar-nonpolar, and polar-polar systems (Blanks and Prausnitz, 1964). However, the method is not reliable for systems when hydrogen bonding or other specific interactions are important (e.g., aqueous solutions).

   A method based upon the scaled particle theory has been presented by Pierotti and co-workers (Pierotti, 1963, 1965, 1967, 1976; Pierotti and Liabastre, 1972; Wilhelm and Battino, 1972; Liabastre, 1974; Blank, 1975; Blank and Kinshova, 1975a, b, 1976; Neumann, 1977; De Ligny and van der Veen, 1972, 1975; Monfort et al., 1977; Anik, 1978) for the calculation of the solubilities of nonpolar gases in water using the interaction forces represented by the Lennard-Jones (12,6) potential-energy function. The method developed expresses the Henry's law constant, H:

$$\ln H = \frac{G_{cav}}{R\,T} + \frac{G_{inter}}{R\,T} + \ln\left(\frac{R\,T}{V_{solv}}\right)$$

where the first term, $G_{cav}/(R\,T)$, represents the energy required for the formation of a cavity in the solvent and the second term, $G_{inter}/(R\,T)$, characterizes the energy of interaction of the solute molecules in the cavity of the solvent. The first term can be calculated on the basis of the theory of scaled particles (this expression is slightly different from one that was used by Wilhelm and Battino (1971) for the evaluation of gas solubilities in liquids):

$$\frac{G_{cav}}{R\,T} = \left\{\left(\frac{6\,y}{1-y}\right)\left[2(\sigma_{ab}/\sigma_a)^2 - (\sigma_{ab}/\sigma_b)\right] + \left(\frac{18\,y^2}{(1-y)^2}\right)\left[(\sigma_{ab}/\sigma_b)^2 - (\sigma_{ab}/\sigma_b)\right.\right.$$

$$\left.\left. + \frac{1}{4}\right] - \ln(1-y)\right\} + \Pi\,P\,\sigma_a^3\left[\frac{4}{3}(\sigma_{ab}/\sigma_a)^3 - 2(\sigma_{ab}/\sigma_a)^2 + (\sigma_{ab}/\sigma_a) - \frac{1}{6}\right]$$

where σ_a and σ_b = hard-core diameters of molecules a and b, respectively

σ_{ab} = $\frac{1}{2}(\sigma_a + \sigma_b)$ (see Pesuit, 1978)

y = $\Pi\,\sigma_a^3\,d/6$

d = molecular density of solvent $(= N_A/V_{liq.})$

P = pressure

The second term, $G_{inter}/(R\,T)$, can be estimated from the Lennard-Jones pair potential

$$\frac{G_{inter}}{R\,T} = -5.33\,\frac{\Pi\,d\,C_{disp}}{6\,k\,T\,\sigma_{ab}'^3}$$

where C_{disp} = $4\,\varepsilon_{ab}'(\sigma_{ab}')^6 + \mu_a^2\,\alpha_b$

σ_{ab}' = $\frac{1}{2}(\sigma_a + \sigma_b)\zeta^{-1/6}$

ε_{ab}' = $(\varepsilon_1\,\varepsilon_2)^{\frac{1}{2}}\,\zeta^2$

ζ = $1 + \frac{\alpha_b}{4\,\sigma_b^2}\,\frac{\mu_a^2}{\varepsilon_a\,\sigma_a^3}(\varepsilon_a/\varepsilon_b)^{\frac{1}{2}}$

μ_a = dipole moment of the solvent

α_b = polarizability of the gas (solute)

ε_a = potential-energy well for molecule a

However, in the case of nonspherical and polar compounds, the interaction-free-energy term $G_{inter}/(R\,T)$ had to be modified; that is, in addition to the dispersion term, an inductive interaction energy term is included (Pierotti and Liabastre, 1972):

$$\frac{G_{inter}}{R\,T} = -5.33 \frac{\Pi\,d\,C_{disp}}{6\,k\,T\,\sigma_{ab}^{'3}} - \frac{8\,\Pi\,d\,C_{induc}}{6\,k\,T\,\sigma_{ab}^{'3}}$$

where $\quad C_{induc} = \mu_a^2\,\alpha_b$

μ_a = dipole moment of the solvent

α_b = polarizability of the solute

In the term of the interaction free energy $G_{inter}/(R\,T)$, the polarizability of the solute α_b is a variable, so that the Henry's law constant is a function of polarizability,

$$\ln H = f(\alpha_b)$$

which produces a smooth curve. By extrapolating this curve to $\alpha_b = 0$, we obtain the Henry's law constant or solubility of the hard sphere with a diameter of 2.547 Å in water. That is, if $\alpha_b \to 0$,

$$\lim_{\alpha_b \to 0} (\ln H) = \frac{G_{cav}}{R\,T} + \ln\left(\frac{R\,T}{V_{solv}}\right) = \ln H^o$$

and

$$d \to 2.547\ \text{Å}$$

where H^o is the Henry's law constant for a hard sphere with diameter 2.547 Å.

Finally, the Henry's law constant can be expressed through rearrangement of the expression above:

$$\ln H - \frac{G_{cav}}{R\,T} - \ln\left(\frac{R\,T}{V_{solv}}\right) + \frac{8\,\Pi\,d\,C_{induc}}{6\,k\,T\,\sigma_{ab}^{'3}} = -5.33 \frac{\Pi\,d\,C_{disp}}{6\,k\,T\,\sigma_{ab}^{'3}}$$

where C_{disp} can be expressed with the Lennard-Jones energy parameters ε_a and ε_b for the solvent and solute, respectively:

$$C_{disp} = 4\,\varepsilon_{ab}\,\sigma_{ab}^{'6} = 4\,(\varepsilon_a\,\varepsilon_b)^{\frac{1}{2}}\left(\frac{\sigma_a + \sigma_b}{2}\right)^6$$

where σ_a and σ_b are the distance parameters of the solvent and solute, respectively. Another possible expression for C_{disp} was introduced by Kirkwood-Müller:

$$C_{disp} = \frac{6\, m\, c^2\, \alpha_a\, \alpha_b}{\alpha_a/x_a + \alpha_b/x_b}$$

where m = mass of the electron

          c = velocity of light

α_a and α_b = molecular polarizabilities of the solvent and solute, respectively

x_a and x_b = molecular susceptibilities of the solvent and solute, respectively

        Alternatively, C_{disp} may be estimated by the Slater-Kirkwood expression,

$$C_{disp} = \frac{3\, e\, h}{8\, \Pi} \left(\frac{N\, \alpha_b^2}{m}\right)^{\frac{1}{2}}$$

where e = charge of the electron

          h = Planck's constant

         α_b = polarizability of the solute

          m = mass of the electron

          N = number of the valence electrons

        Inserting the Lennard-Jones function of C_{disp}, the Henry's law constant becomes

$$\ln H - \frac{G_{cav}}{R\,T} - \ln\left(\frac{R\,T}{V_{solv}}\right) + \frac{8\, \Pi\, d\, C_{induc}}{6\, k\, T\, \sigma_{ab}'^3} = -11.163\, \frac{d}{T}\left[(\epsilon_a/k)^{\frac{1}{2}}(\epsilon_b/k)^{\frac{1}{2}}\, \sigma_{ab}'^3\right]$$

By plotting the left-han side of the equation versus $(\epsilon_b/k)^{\frac{1}{2}}\, \sigma_{ab}'^3$, we obtain a straight line with slope equal to (Pierotti, 1976)

$$11.163\, \frac{d}{T}\, (\epsilon_a/k)^{\frac{1}{2}}$$

With the help of the interaction parameters ϵ_a and ϵ_b and the effective diameter σ_{ab}', we can determine the solubility of gases in water. Parameters of water determined by various methods are presented in Table 2.4 (Pierotti and Liabastre, 1972). Values of σ and ϵ/k for halogenated hydrocarbons are listed in Table 2.2 and the Appendix. Because of the relationship between the Henry's law constant (H) and the Gibbs free energy of solution (ΔG_s^o), that is, transfering 1 mole of gaseous solute at unit fugacity to water at unit mole fraction and temperature T:

$$\Delta G_s^o = R\,T\,\ln H = R\,T\,\ln(p_2/x_2)$$

where p_2 = partial pressure of the solute

        x_2 = mole fraction of solute in the solvent

Table 2.4   Parameters σ and ε/k for Water by Different Methods

Method	σ (Å)	ε/k (K)
Pierotti & Liabastre (1972)	2.768	107
Pierotti & Liabastre (1972) using only He and Ne	2.768	77
Scaled-particle theory to compressibility data	2.83	–
Gas-phase transport properties, Stockmayer potential	2.71	775
Second virial coefficient of gas phase	2.65	–
Oxygen-oxygen distance in ice	2.76	–
Fitting clathrate properties using σ_{H_2O} = 2.50	2.50	167
London's equation	–	130
Continumium approximation	–	92

Source: Pierotti & Liabastre, 1972.

it is practical to calculate ΔG_s^o from the solubility determined. ΔG_s^o is a function of temperature:

$$\Delta G_s^o = A + B T + C T^2 + \cdots$$

The standard heat of solution (ΔH_s^o) can also be calculated from the Henry's law constant through the expression

$$\Delta H_s^o = \left[\frac{\partial \ln H}{\partial 1/(R T)} \right]_P$$

(An exact calculation of the heat of solution from solubility data was reported by Williamson (1944)).

Similarly, the standard heat capacity of solution $(\Delta C_{p,s}^o)$ is given by

$$\Delta C_{p,s}^o = \left(\frac{\partial \Delta H_s^o}{\partial T} \right)_P$$

The standard entropy of solution (ΔS_s) is expressed by

$$\Delta S_s = - \left(\frac{\partial \Delta G_s^o}{\partial T} \right)_P$$

Heat of solution, heat capacity of solution, and entropy of solution data are listed in the Appendix for aqueous solutions.

A detailed discussion of the success of scaled-particle theory applied to the calculation of the thermodynamic quantities of aqueous solutions can be found in Pierotti and Liabastre (1972) and Wen and Muccitelli (1979). The solubility of the gases He, Ne, Ar, Kr, Xe, H_2, N_2, O_2, CO, NO, CH_4, CF_4, C_2H_2, C_6H_6, C_2H_4, and other hydrocarbons in water between 0 and 300°C have been investigated (for water at 298 K, σ = 2.704 Å and ϵ/k = 110.23 K) using scaled particle theory.

The scaled-particle theory developed by Pierotti (1963, 1965) has been continued by Lucas and co-workers (Lucas, 1969, 1970, 1972, 1973, 1976; Lucas and Feillolay, 1970; Lucas and Bury, 1976; Neumann, 1977; Wilhelm and Battino, 1972) for several solubility problems. The entropy of solution of polar and nonpolar molecules in water was correlated by the molecular diameter at 4 and 25°C. A definite trend has been shown for the relationship; however, the accuracy of the correlation is not very satisfactory, particularly at 25°C (Lucas, 1970). Correlation of the aromatic hydrocarbons has been different from that for normal hydrocarbons and other substances. The correlated entropy of solution of several substances (excluding aromatic compounds) has been expressed at 4°C by the equation

$$\Delta S_{soln}^{4°}(\text{cal g-mole}^{-1} K^{-1}) = 20.6 - 3.8(a + 2.76) + 0.935(a + 2.76)^2$$

where a is the molecular diameter in angstroms.

The application of the hard-sphere model for the calculation of the thermodynamic properties of aqueous solutions resembles that for aqueous solutions of nonpolar compounds of the same size (Lucas, 1972). The influence of the solute on water structure has been discussed, particularly below and above 4°C. Below 4°C the bonds are stronger between water molecules, and the effect of solute size becomes more significant. However, at higher temperatures (e.g., above 4°C) the bonds decrease and the temperature effect becomes important, and this influence increases with solute size. For further discussion of the structure of water, see Sec. 2.6.

In a further study on the application of scaled-particle theory, Lucas (1973) has shown that this theory makes it possible to ascribe the variation of the solute partial molal properties to the anomalous behavior of compressibility for pure water rather than solute structural effects.

Scaled-particle theory was used to investigate the free energy of transfer for a nonpolar solute from the gaseous state to water using a model of a hypothetical water and a hypothetical liquid. The free-energy-of-transfer values have been calculated from the expression

$$\Delta G_{HS} = R\,T \ln\left(\frac{R\,T}{V}\right) - R\,T \ln(1 - y) + \frac{9\,R\,T}{2}\left[\frac{y^2}{(1 - y)^2} + \frac{2\,y}{3\,(1 - y)}\right]\frac{a_2^2}{a^2}$$

$$+ 3\,R\,T\,\frac{y\,a_2}{(1 - y)\,a}$$

neglecting the effect of dispersion forces. In the expression,

   a = hard-sphere diameter of solvent, Å

   a_2 = hard-sphere diameter of solute, Å

$$V = \text{molar volume of solvent, cc g-mole}^{-1}$$

$$R = \text{gas constant}$$

$$T = \text{absolute temperature, K}$$

$$y = \Pi N a^3/(6 V)$$

$$N = \text{Avogadro's number}$$

The calculated relationship $\Delta G_{HS} = f(a_2)$ is illustrated in Fig. 2.12. The numerical values used in the foregoing expression for water, hypothetical water, and
hypothetical liquid are given in Table 2.5.

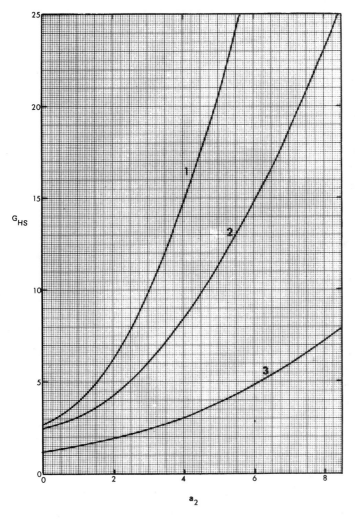

Fig. 2.12 Free energy of transfer of gases into liquids at 25°C. G_{HS} = free energy
of transfer (kcal g-mole^{-1}); a_2 = hard-sphere diameter of solute (Å); 1, in hypothetical liquid; 2, in water; 3, in hypothetical water. (From Lucas, 1976.)

Table 2.5   Molecular Parameters of Water and Hypothetical Water

Solvent	Hard-sphere diameter (Å)	$y = \dfrac{\Pi N a^3}{6 V}$	Molar volume (cc g-mole^{-1})
Water	2.75	0.38	18
Hypothetical water	5.5	0.38	144
Hypothetical liquid	2.75	0.50	13

The free energy of transfer is a function of the Henry's law constant (H):

$$\Delta G = R T \ln H$$

which means that its dependence on the hard-sphere diameter of the solute will express the solubility relationship. The solubility of a nonpolar solute with 4 Å diameter is higher in the hypothetical water than in water and the solubility is lowest in the hypothetical liquid (see Fig. 2.12).

The effects of dispersion forces on the free energy of transfer for nonpolar solutes in water or in nonpolar solvents are of similar magnitude. Because of the larger size of the nonpolar solvent molecules compared with water molecules, the free-energy-of-transfer volumes are negative for the transfer from water to a non-polar solvent (Lucas, 1976).

The solvent size is very important, as the free energy of transfer for the nonpolar solute to another solvent with larger size is again negative, although the negative ΔG value has no specific structural effects. The sign of the entropy of transfer is probably attributable to solvent structural effects, whereas the solvent size effect determines the magnitude of the free energy of transfer of a nonpolar solute molecule from water to another solvent.

The scaled-particle theory used by Pierotti (1963, 1965) for the case of a rigid-sphere solute in liquid water has been criticised by Stillinger (1973). By examining Pierotti's analysis, Stillinger proposed an improvement that incorporates measured surface tension and radial-distribution function for pure water. The improved theory takes account explicitly of the strong, directional hydrogen-bonding forces in water.

The free energy of transfer for a nonpolar solute from the gaseous state to "Stillinger water" was calculated and presented by Lucas (1976) compared to the scaled-particle theory for water, both at 2.75 Å diameter. The plots of the free energy of transfer at 4°C against solute diameter in angstroms showed that the ΔG values are always higher in the case of Stillinger modification in comparison to Pierotti's theory. The free energy of transfer of a nonpolar solvent (a = 5.5 Å) is between the ΔG values of "Stillinger water" (a = 2.7 Å) and the hypothetical Stillinger water (a = 5.5 Å).

Regarding the entropy of transfer for a nonpolar solute from the gaseous state to "Stillinger water" and water at standard temperature and pressure (SPT), the ΔS units are more negative for the latter and they become smallest for nonpolar solvents at 4°C.

The effect of thermodynamic parameters such as enthalpy and entropy have been examined by Lucas and Bury (1976) in connection with the transfer of a nonpolar hydrophobic solute from gas to water at 4°C; the influence of dispersion was neglected. The reduced quantities $\Delta G/a_2^2$, $\Delta H/a_2^2$, and $T\Delta S/a_2^2$ were plotted against the hard-sphere diameter a_2 in angströms. The plots are illustrated in Fig. 2.13.

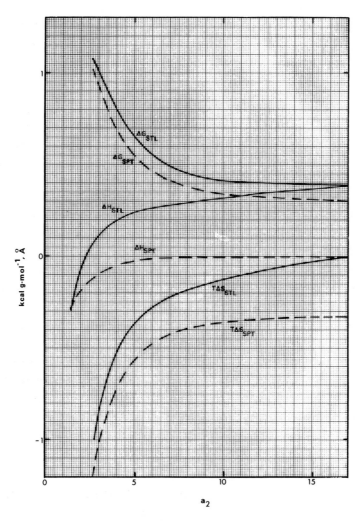

Fig. 2.13 Reduced thermodynamic quantities versus solute hard-sphere diameter at 4°C. a_2 = hard-sphere diameter of solute (Å). (From Lucas & Bury, 1976.)

While the Pierotti's model is good for $a_2 < 1.4 \text{Å}$ and $a_2 > 4 \text{Å}$, it only approximates thermodynamic parameters. The influence of structural features on ΔG values is not as great as on ΔH and $T\Delta S$ values. The solvent structure is dominant for the latter two parameters.

From the analysis Lucas and Bury (1976) concluded that the influence of the solute on water structure may be opposite when the solute is small, when it may disrupt the water structure, or when it is large, all cases in which the high positive enthalpy and small negative entropy are more consistent with a process of breaking H bonds. The early success of scaled-particle theory is due to the small solutes or large solvents used for the test.

The hard-sphere diameter of solute has been used by Lucas (1970) to correlate entropy of dissolution in water at $4°$ and $25°C$. At $25°C$ the correlated expression was

$$\Delta S^{25°}_{soln} = 2.559 + 7.797 \ a$$

where a is the hard-sphere diameter in angstroms. This correlation reveals some similarity with the correlation of solute entropy with molar volume at $25°C$ (see Fig. 2.14). The heat of solution (kcal g-mole^{-1}) of gases was also correlated with the rigid-shere diameter and is presented in Fig. 2.15. The correlation equation is

$$\Delta H^{25°}_{soln} = 4.54472 - 2.17523 \ a$$

The intermolecular interaction energy has been calculated by several investigators (e.g., Simonov et al., 1974; Gorodyskii, 1975) for a large number of molecules using the Onsager or other models.

The molecular interaction between an organic molecule and water was first studied by Nemethy and Scheraga (1962a, b). This quantitative study of the hydrophobic interaction of a hydrocarbon molecule to an aqueous solution was continued by Marcelja et al. (1977), Pratt and Chander (1980), Ben-Naim and Wiff (1980), Ben-Naim and Egel-Thal (1965), Hoffmann and Birnstock (1980), and Ben-Naim and Tenne (1977). They calculated the free energy of transfer and compared it with the experimental values for CH_4, C_2H_6, and C_3H_8. The approximate agreement found for these hydrophobic interactions was quite satisfactory considering the simplicity of the model used. Their method is an improvement over the previous ones in two aspects:

1. The new model does not use a simple relation between the surface area of the solute molecule and the hydrophobic free energy.

2. The contributions to the hydrophobic energy are not additive.

Ben-Naim and Tenne (1977) applied scaled-particle theory to the problem of hydrophobic interaction. The enthalpy, entropy, and free-energy changes during the formation of a close-packed configuration of simple solute molecules from fixed positions at infinite separation. The process involved the combination of experimental

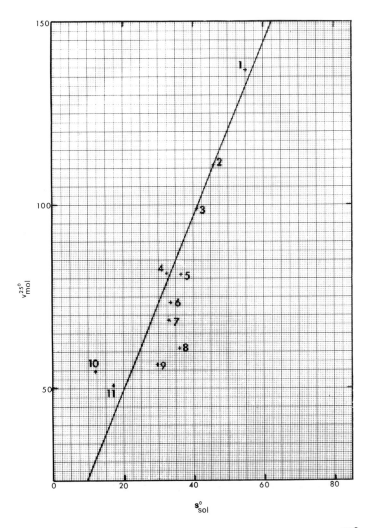

Fig. 2.14 Correlation of solute entropy with molar volume at 25°C. $v_{mol}^{25°}$ = molar volume of solute (cc g-mole^{-1}); S_{sol}^{0} = solute entropy (cal g-mole^{-1} K^{-1}). 1, C_4F_8; 2, $C_2Cl_2F_4$; 3, $CClF_2$-CF_3; 4, CH_3-$CClF_2$; 5, CCl_2F_2; 6, $CHCl_2F$; 7, $CClF_3$; 8, $CHClF_2$; 9, CH_2ClF; 10, CF_4; 11, CHF_3. (From Stepakoff & Modica, 1973.)

results with theoretical calculations based upon scaled-particle theory. The cal-culations were carried out for various solutes in water and in heavy water at diff-erent temperatures. The recommendation based on their work is that the temperature dependence of the hard-core diameter used in scaled-particle theory has a positive temperature dependence.

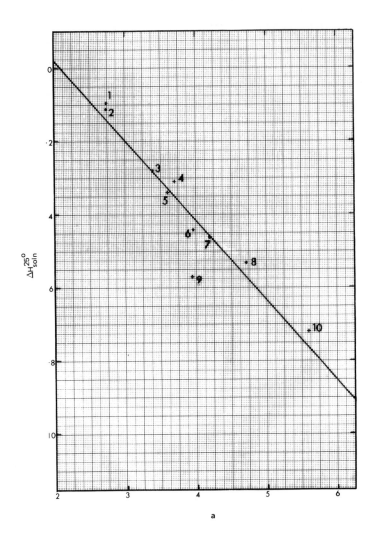

a

Fig. 2.15 Heat of solution versus rigid-sphere diameter at 25°C. $\Delta H_{soln}^{25°}$ = heat of solution (kcal g-mole^{-1}); a = rigid-sphere diameter (Å). 1, H_2; 2, Ne; 3, Ar; 4, CH_4; 5, Kr; 6, Xe; 7, CH_3-CH_3, 8, CH_3-CH_2-CH_2-CH_3; 9, CH_3Cl; 10, $(CH_3)_4C$. (From Lucas, 1970.)

Goldman (1974) applied the statistical mechanical theory based upon Zwanzig's perturbation theory of fluids for the calculation of the solubility of water in benzene, carbon tetrachloride, and cyclohexane. The intermolecular potential energy was calculated from the expression

$$\varphi_{ab}(r) = 4\,\epsilon_{ab}\,f_{ab}\left[(\sigma_{ab}/r)^{12} - (\sigma_{ab}/r)^{6}\right] - \mu_2^{*2}(\alpha_a\,f_{ab})/r^6 \quad \text{for } r > d_{ab}$$

where f_{ab} = dimensionless fitting parameter

α_a = polarizability of the solvent

ε_{ab} = depth of the potential well

r = intermolecular distance between molecular centers

σ_{ab} = distance between two molecules when the potential energy becomes zero

$$\mu_2^* = \mu_g^* \left(\frac{2D + 1}{2D + n_d^2} \cdot \frac{n_d^2 + 2}{3} \right)$$

D = dielectric constant of the solvent

n_d = refractive index of water for sodium light

μ_g^* = dipole moment of water in the gas phase (= 1.85 debyes)

μ_2^* = dipole moment of water in solution

The calculated parameters for the mentioned compounds are listed in Tables 2.6 and 2.7.

Table 2.6 Lennard-Jones (12,6) Parameters

Substance	Polarizability (cc × 10^{25})	σ (Å)	ε/k (K)
C_6H_6	103.2	5.628	335.0
CCl_4	105.0	5.881	327.0
C_6H_{12}	108.7	6.093	324.0
H_2O	–	2.520	775.0
$C_6H_6 - H_2O$	–	4.0740	509.5
$CCl_4 - H_2O$	–	4.2005	503.4
$C_6H_{12} - H_2O$	–	4.3065	501.1

The potential parameters of the pairs have been calculated from the Lorentz-Berthelot equations:

$$\sigma_{ab} = \tfrac{1}{2}(\sigma_a + \sigma_b)$$

and

$$\varepsilon_{ab} = (\varepsilon_a \, \varepsilon_b)^{\frac{1}{2}}$$

More recent methods for the calculation of binary interaction constants have been described by Pesuit (1978)

Source: Goldman, 1974.

The potential parameters for water in the binary systems of organic solvents are given by Goldman and Krishnan (1976). The temperature dependence of the fitting parameter (f_{ab}) is given in Table 2.8.

Table 2.7 Lennard-Jones (6,12), Stockmayer (3,6,12), Kihara, and 6-exp Potentials

Refrigerant formula	Refrigerant number	Lennard-Jones(6,12) potential			Stockmayer(3,6,12) potential				Kihara potential				6-exp potential			
		ε/k	σ	r_m	ε/k	σ	r_m	δ	ε/k	σ	r_m	γ	ε/k	σ	r_m	α
$CFCl_3$	11	479	4.95	5.56	325	5.20	6.95	-0.25	473	4.97	5.58	0.0	474	5.60	5.61	13
CF_2Cl_2	12	223	5.42	6.07	214	5.14	5.75	0.0	225	5.40	6.04	0.0	240	6.00	6.01	15
CF_3Cl	13	186	5.05	5.68	190	5.03	5.65	0.0	188	5.04	5.67	0.0	169	5.83	5.84	13
CF_4	14	144	4.63	5.18	143	4.63	5.18	0.0	141	4.65	5.21	0.0	143	5.21	5.22	14
$CHFCl_2$	21	312	4.99	5.61	221	5.16	6.83	-0.25	235	5.28	5.95	0.0	288	5.70	5.71	14
CHF_2Cl	22	250	4.75	5.34	195	4.85	6.42	-0.25	254	4.75	5.34	0.0	225	5.47	5.48	14
CHF_3	23	272	4.15	4.66	97	4.70	6.06	-1.0	278	4.15	4.60	0.1	237	4.84	4.83	14
$CClF_2-CClF_2$	114	282	5.86	6.58	276	5.90	6.62	0.0	270	5.98	6.64	0.1	247	6.84	6.85	13
$CClF_2-CF_3$	115	192	5.87	6.58	188	5.87	6.58	0.0	186	5.89	6.62	0.0	183	6.64	6.65	14

Source: Nikul'shin & Petriman, 1976.

Table 2.8 Variation of f_{ab} with Temperature

Temperature	Parameter f_{ab}		
(°C)	C_6H_6	CCl_4	C_6H_{12}
10	1.098	1.174	1.018
20	1.073	1.135	0.991
25	1.061	1.116	0.977
30	1.049	1.098	0.964
40	1.026	1.062	0.938

Source: Goldman, 1974.

In earlier works by Liddel and Becker (1956), Howard et al. (1963), and Epley and Drago (1967), it was shown that some chlorinated hydrocarbons form weak hydrogen bonds in solutions. In such cases the geometrical and arithmetical means rules will not produce a satisfactory result, and therefore the presented values must be taken as approximations.

Several investigators (Nelson and de Ligny, 1968b; de Ligny and van der Veen, 1972, 1975; Neff and McQuarrie, 1973; Monfort et al., 1977) used Pierotti's theory for the calculation of the solubility of gases in liquids. They assumed that the interaction entropy is zero and that the interaction enthalpy in the case of a non-polar solute gas embedded in a polar solvent can be described by dispersion, inductive, and repulsive interactions:

$$R\,T\,\ln(p_2/x_2) = R\,T\,\ln\left(\frac{31{,}390\,R\,T}{V_1}\right) + f(T, V_1, \sigma_a, \sigma_b) - 3.556\,R\,\Pi\,d\,\sigma_{ab}^3\,\frac{\varepsilon_{ab}}{k}$$

$$- 1.333\,N\,\Pi\,d\,\frac{\mu_a^2\,\alpha_b}{\sigma_{ab}^3}$$

where x_2 = mole fraction of the solute
       p_2 = partial pressure of the solute

A large number of nonpolar solute/polar solvent systems were studied at 25°C.

A simplified form of the perturbation theory was applied for the calculation of gas solubility in a multicomponent solvent mixture by Tiepel and Gubbins (1972). The theory is based on the principles of statistical mechanics and requires molecular parameters and solvent density to predict the gas solubility.

Jonah and King (1971) assumed the Lennard-Jones type of intermolecular forces for prediction of the temperature dependence of the Henry's law coefficient for the solubilities of gases in nonpolar liquids. The systems investigated included H_2, Ar, CH_4, O_2, N_2, and CO in liquid C_6H_6, CCl_4, and CS_2. No aqueous systems were studied.

By using the Lennard-Jones potential parameters (Leites and Sergeeva, 1973), the solubility of 23 gases in 12 different solvents was calculated at 20°C and 1 atm partial pressure. The linear relationship has been illustrated between $\log \beta$ (β = Bunsen's solubility coefficient) and $(\varepsilon_{0.22}/k)^{\frac{1}{2}}$. (See also Goldman, 1977.)

There are several factors that influence the mutual solubility or miscibility between halogenated hydrocarbons and water (Antropov et al., 1972; Simoniv et al., 1970, 1974; Karyakin and Muradov, 1971; Karyakin et al., 1970):

1. Polarity of the molecules
2. Molecular size
3. pH of the system (result of hydrolysis)
4. Nature of the intermolecular interaction
5. Other factors

Simonov et al. (1974) divided the chlorocarbon – water systems investigated into two groups:

1. Nonpolar weakly hydrolyzed substances with low energy of intermolecular interaction between the molecules
2. Polar strongly hydrolyzed compounds with higher energy of intermolecular interaction

The bond energies formed between halogenated hydrocarbons and water are of two types: weak van der Waals interaction and hydrogen bonding. The bonds between halogenated hydrocarbons and water are weak in the range 0.93 to 1.58 kcal g-mole^{-1} comparing to the bonds between water molecules of 4.5 to 5.0 kcal g-mole^{-1}. The energy of hydrogen bondings aroses from the acceptor behavior of halocarbons toward the protons of the water molecules, that is, the halocarbon molecules can function as a proton acceptor (Karyakin and Muradov, 1971; Simonov et al., 1974). The molecules classified into the two groups listed above were correlated by Simonov et al. (1974) according to their polarity, hydrolyzability, energy of intermolecular interaction, and temperature. The temperature dependence of the energy of intermolecular interaction between six fully chlorinated hydrocarbons and water is illustrated in Fig. 2.16. The straight-line relationship is characteristic between 0 and 150°C. The numerical values are given in Table 2.9 at 40°C, which were determined by infrared spectroscopy.

The increasing or decreasing tendency of the solubility with one or other of the parameters in each group is characteristic for the relationships between these factors listed in Table 2.9. When the acidity increases (i.e., the value of pH decreases), the energy of the molecular interaction will increase due to the formation of hydrogen bond in group 2. Similarly, the miscibility also increases because of the dominant

Table 2.9 Properties of Halogenated Hydrocarbon-Water Systems at 40°C

Group	Halocarbon	pH of the solution (hydrolyzability)	Energy of intermolecular interaction (kcal g-mole^{-1})	Dipole moment (debyes)	Solubility at 40°C (wt %)	
					In H_2O	In Halocarbon
1	CCl_4	6.5	1.16	0	0.0197	0.0250
	C_2Cl_4	5.1	1.0	0	0.0166	0.0166
	C_4Cl_6	5.0	0.93	0	0.0110	0.0059
2	C_3Cl_6	2.5	1.58	0.45	0.0091	0.005
	C_5Cl_6	2.6	1.29	1.05	0.0055	0.0011
	C_5Cl_8	3.0	1.24	1.03	0.0001	0.0006

Source: Simonov et al., 1974.

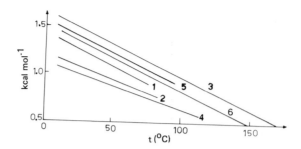

Fig. 2.16 Temperature dependence of the energy of intermolecular interaction.
1, CCl_4; 2, C_2Cl_4; 3, C_3Cl_6, 4, C_4Cl_6; 5, C_5Cl_6; 6, C_5Cl_8.

influenec of pH on the system (Antropov et al., 1972). However, in group 1 the
mutual solubility decreases with the increase of the activity of the system as a
result of the hydrolysis. With hydrolysis the molecular size increases and the nature
of the intermolecular interaction will also change (Hine and Ehrenson, 1958; Hine
and Prosser, 1958; Hine et al., 1958, 1963; Hine, 1975; Hine and Mookerjee, 1975).
The increased acidity is accounted for by the formation of hydronium ions and their
hydration. The correlation between the energy of the intermolecular interaction and
mutual solubility is apparent (see Table 2.9). The low miscibility between halogenated
hydrocarbons and water can be explained through the strength of the bonding. Whereas
the $\varphi(r)$ between water molecules is about 5 kcal g-mole^{-1}, the energy of molecular
interaction between water and halocarbon molecules is only about 1 kcal g-mole^{-1}
(Karyakin et al., 1960) (see Table 2.10). The dissolved water molecules are linked
symmetrically in halocarbons by hydrogen bonds: for example,

$$Cl_3-R-Cl \cdots H-O-H \cdots Cl-R-Cl_3$$

Table 2.10 Intermolecular Bond Energy Between H_2O and CCl_4

Temperature (°C)	Intermolecular bond energy (kcal g-mole^{-1})
-10	1.59
0	1.52
20	1.38
40	1.24
60	1.10

Source: Karyakin et al., 1960.

The mutual solubility between halocarbons and water is a direct function or dependence of the energy of intermolecular interaction between the molecules in question. If φ (r) increases, it will indicate an increasing tendency for the miscibility, too.

The state of water in nonpolar solvents (e.g., carbon tetrachloride, chloroform) has been studied by Karyakin et al. (1970) and Yukhnevich and Karyakin (1964) using infrared spectroscopy. The infrared spectra of these solutions showed four bands, assigned to symmetric and asymmetric vibrations of water molecules: 3550, 3615, 3705, and 3800 cm^{-1}. When the water concentration increased, the intensity of all four bands also increased. The infrared spectra showed that the water molecules dissolved in the nonpolar solvents are mainly in the monomeric state, with hydrogen bonding to the solvent molecules in a microemulsified state. For example, water appears in carbon tetrachloride in three states: monomeric (bound to the solvent), microemulsified, and monomeric (bound to the microemulsion).

The proton-accepting power of halogenated hydrocarbons toward water molecules can be explained by strengthening the energy of the

$$H - O - H \cdots X$$

bond (X = halogen group), which is caused by the electron-donating power to the electron cloud of the molecule from the proton-accepting atom. The hydrogen bond is sensitive to the electronic structure of the halocarbon interacting with water.

The effect of hydrolysis on the solubility of water in halogenated hydrocarbons has been studied by Simonov et al. (1970). Most of these compounds are fairly inert in the absence of moisture. However, if they are in contact with water, hydrolysis can take place, promoted by elevated temperatures and inorganic salts acting as catalysts (Carlisle and Levine, 1932; Carlisle, 1932; Radukov et al., 1971; Hagen and Elphingstone, 1974). The products of the hydrolysis are acids, which cause corrosion on chemical equipment and increase the number of impurities in the chemicals (see the Introduction). The dependence of the solubility of four fully chlorinated hydrocarbons investigated (CCl_4, $CCl_2=CCl_2$, $CCl_2=CCl-CCl=CCl_2$, and $CCl_3-CCl=CCl_2$) show irregularity with the dipole moment. As described before, the solubility of water in halocarbons depends upon several factors, including dipole moment and pH (degree of hydrolysis and the formation of acids). The dependence of water solubility upon the pH of the aqueous extract at 30 and 50°C is shown in Table 2.11. The trend is that the solubility of water in fully chlorinated hydrocarbons increases as the pH of the aqueous extract increases (i.e., when it becomes less acidic).

Mabey and Mill (1978) have examined and compiled the literature data on the hydrolysis of organic compounds in water, including the halogenated hydrocarbons.

Table 2.11   Effect of Degree of Hydrolysis upon Water Solubility

Solvent	pH at 30°C	Solubility of H_2O (wt %)	pH at 50°C	Solubility of H_2O (wt %)
CCl_4	6.60	0.0135	6.4	0.0305
$CCl_2=CCl_2$	6.10	0.0104	5.39	0.0282
$CCl_2=CCl-CCl=CCl_2$	5.30	0.00676	4.90	0.0212
$CCl_3-CCl=CCl_2$	4.10	0.00590	2.00	0.0144

2.2 Polar and Nonpolar Fluids

Whether a rule for the estimation or prediction techniques is acceptable or recommended for solubilities, it is relevant to distinguish between the different sorts of molecules present in the system. In connection with solubility principles, the following types of molecules of solute and solvent have to be considered: polar, nonpolar, associated, nonassociated, those forming hydrogen bonds, and those not forming hydrogen bonds. For example, the regular solution theory developed for non-polar molecules (Hildebrand and Scott, 1962) is not valid for aqueous solutions.

The chemical substances belong either to the polar or to the nonpolar group of fluids. Whether a compound is polar or nonpolar depends upon its dipole moment. The polarity of a solvent molecule represents its ability to interact with the solute molecule. Four factors are responsible for the dipole moment of the molecule:

1. The difference in the dimensions of the atomic orbitals
2. The displacement of the center of gravity of the charge of the electrons toward the more electronegative atom
3. Hybridization, causing the asymmetry of a nonbonding pair of electrons
4. Hybridization is also responsible for the asymmetric atomic orbitals

Consequently, the magnitude of the atomic dipoles depends upon the hybridization parameter. Unfortunately, this cannot be measured experimentally.

There are two types of dipole moments: permanent and induced. The permanent dipole moment originates from the shape or unsymmetric structure of the molecule, whereas if the molecules are subjected to an electric field, the electrons will be displaced and the molecules will show induced dipoles. The relationship between the induced dipole moment μ^i and the strength of the electric field E is given by

$$\mu^i = \alpha E$$

where α is the polarizability of the molecule. The polarizability can be calculated from the dielectric constant or refractive indices (Partington, 1953). The polarizability can also be related to the molar **polarization**:

$$\alpha = \frac{2\ P}{4\ \Pi\ N}$$

where P is the molar polarization in cc $g\text{-mole}^{-1}$. The molar polarization is the sum of the atom, electron, and orientation polarizations:

$$P = P_{atom} + P_{electron} + P_{orientation}$$

For gases and dilute solutions of polar compounds in nonpolar solvents. the following relationship is valid between the molar polarization (P) and the dielectric constant (D):

$$P = \frac{D-1}{D-2}\ \frac{M}{d}\ \ cc\ g\text{-mole}^{-1}$$

This is the Clausius-Mossoti equation. Denbigh (1940) introduced a bond polarizability method for the calculation of molecular polarization from individual bonds (see Table 2.12). The bond polarizabilities are directional and can be either parallel or perpendicular to the bond.

Table 2.12   Bond Polarizabilities According to Denbigh (1940)

Bond	Parallel, $\|\|$ (cc $g\text{-mole}^{-1} \times 10^{-24}$)	Perpendicular, $\perp$ (cc $g\text{-mole}^{-1} \times 10^{-24}$)
C - C	1.88	0.2
C = C	2.86	1.06
C ≡ C	3.54	1.27
C - Cl	3.67	2.08
C - Br	5.40	2.88
C - H	0.79	0.58

Another method is the addition of group contributions for the estimation of mean polarizability presented by Ketelaar (1958) (see Table 2.13). Miller and Savchik (1979) introduced an empirical method to calculate average molecular polarizabilities.

The molar polarization (P) can also be calculated from refractive index values (Teague and Pings, 1968; Reisler et al., 1972; Vuks, 1969):

$$P = \frac{n_{\infty}^{2} - 1}{n_{\infty}^{2} + 2}\ \frac{M}{d}$$

where n_{∞} is the refractive index corresponding to light of infinite wavelength. This is calculated from the dielectric constant:

$$n_{\infty}^{2} = D$$

Table 2.13 Group Contribution to the Polarizability of Molecules

Group	Polarizability (cc g-mole^{-1} $\times$ 10^{-24})	
F -	0.38	
Cl -	2.28	
Br -	3.34	
I -	5.11	
H -	0.42	
= C =	0.93	
- CH$_2$ -	1.77	
Double bond	0.58	extra
Triple bond	0.86	extra
C$_6$H$_5$ -	9.38	

Example:

Compound	Polarizability of molecule (cc g-mole^{-1} $\times$ 10^{-24})	
	Calculated	Landolt-Börnstein (1951)
CCl$_4$	10.15	10.5
CHCl$_3$	8.19	8.23

Source: Ketelaar, 1958.

Molar polarizability was correlated with heat of vaporization (see Myers, 1977).

The permanent dipole moment is zero for typical symmetric molecules, such as Cl_2, CCl_4, and p-dichlorobenzene.

The carbon atom is unique in that it forms chemical compounds with a large number of modifications in which the carbon atoms are linked to each other by covalent bonds. For example, one carbon atom can be combined with four hydrogen atoms to form methane. This molecule contains four single covalent bonds. These bonds are directed toward the corners of a regular tetrahedron. In ethylene there is a double bond between two carbon atoms (i.e., two pairs of electrons), whereas three pairs of electrons are responsible for the triple bond in acetylene. The single and multiple bond lengths are given in Tables 2.14 and 2.15. The van der Waals and covalent bond radii of atoms are compared in Table 2.16.

The electronic structure of carbon is

$$1 s^2 \; 2 s^2, \; 2 p^2$$

which shows that it requires four electrons in the second energy level to become complete (i.e., to be resembled to the electronic configuration of neon).

Table 2.14   Single Bond Lengths of Atoms

	sp^3-C	sp^2-C	sp-C	F	Cl	Br	I
H	1.09	1.08	1.06	0.92	1.27	1.41	1.61
sp^3-C	1.54	1.51	1.46	1.37	1.77	1.94	2.16
sp^2-C	-	1.48	1.43	1.33	1.71	1.87	2.09
sp-C	-	-	1.38	1.29	1.64	1.80	1.99
F	-	-	-	1.42	1.63	1.76	1.80
Cl	-	-	-	-	1.99	2.14	2.32
Br	-	-	-	-	-	2.28	2.48
I	-	-	-	-	-	-	2.67

Source: Hine, 1975.

Table 2.15   Multiple Bond Lengths of Atoms

Bond	Length (Å)
sp^2-C = C-sp^2	1.34
sp^2-C = C-sp	1.31
sp-C = C-sp	1.28
C ≡ C	1.20

Source: Hine, 1975.

Table 2.16   van der Waals and Covalent Bond Radii of Atoms

Atom	van der Waals' radius (Å)	Bond radius (Å)
H-	1.20	0.30
F-	1.47	0.64
Cl-	1.75	0.99
Br-	1.85	1.24
I-	1.98	1.33
C-	1.70	0.77
C=	1.77	0.665
C≡	1.78	0.60

Source: Bondi, 1964; Pauling, 1960.

There are three significant consequences of the covalent bond in a molecule:

1. The covalent bonds are directional.
2. The compound has lower boiling and melting points than electrocovalent substances.
3. These substances are insoluble in polar solvents (e.g., water), but readily soluble in nonpoler liquids (e.g., benzene, diethyl ether).

The distance between two atoms is filled by the two electrons. If the two electrons are in the middle of the two atoms, because they are equally electronegative, the positive and negative charges are balanced and the molecule is nonpolar in contrary to the polar molecule, where one atom is more or less electronegative to the other one. A chemical bond between atoms of different elements normally has a dipole moment acting in the direction of the bond. Meanwhile, a nonpolar molecule is symmetrical and its permanent dipole moment equals to zero.

In the H_2O molecule the two H - O bonds form an angle of approximately 105° and the charge foci do not balance (i.e., the dipole moment is greater than zero). Therefore, the water molecule is polar. Table 2.17 gives the dipole moment of H_2O in various conditions. For further discussion of the water molecule, see Sec. 2.6.

Table 2.17 Dipole Moment of Water

State	Temperature ($^{\circ}C$)	Dipole Moment (debyes)
solid	-70	Varies with time and type of charge
liquid	20	3.11
gas	550-1000	1.80
gas	-2	1.836
gas	0-50	1.853
gas	90	1.86
gas	-	1.884
Benzene	25	1.76
Benzene	25	1.835
Dioxane	25, 50	1.70
Pyridine, acetone	30-50	1.88
Glycerol	30-50	1.93
1,2-Dichloroethane	25	2.43

Source: McClellan, 1974.

The degree of polarity is characterized by the magnitude of the dipole moment. The total dipole moment of a molecule is made up of the vectoral addition of the

dipole moments of the individual groups. The vector additivity of the bond dipoles
has been used to obtain a qualitative picture of the molecules. However, the assumption is that the moment of a bond between a sp^3-hybridized carbon atom and a hydrogen atom is always the same. This should mean that saturated hydrocarbons with a
bond angle of 109° will have zero dipole moment. It is true that most paraffins
have zero dipole moment, but their bond angles are not exactly 109°. The dipole
moments of monohalogenated normal paraffins are plotted in Fig. 2.17. The increasing
effect of molecular size is a characteristic phenomenon. The explanation is based on
the difference in the nuclear charge between the halogen and carbon atoms. The bond
electrons are strongly attracted toward the halogen atom, leaving the positive
charge for the carbon atom, which attracts electrons more strongly than carbon atoms
without a strongly electron-withdrawing substituent. The halogen atom induced a
partial positive charge upon the carbon atom next to it, which in turn induces a
small positive charge on the next carbon atom, and so on. The electron-withdrawing
tendency of the halogen atom passes down on the chain of carbon atoms. The Pauling's
electronegativities of some elements are listed in Table 2.18. The phenomenon mentioned is known as the inductive effect of the halogen atoms. According to the illustrated dependence of the dipole moment of a series of halogenated hydrocarbons
upon the molecular weight (see Fig. 2.17). It is conclusive that the effect of the
halogen atom on the positive charge of the chain of carbon atoms decreases with the
length of the chain. Dewar and Grisdale (1962) estimated an eightfold falloff on
going one atom farther down the chain. For each additional methylene group in the
halogenated normal paraffin molecules, the dipole moment shows an increase. This
tendency has not always been considered during the selection of the dipole moments
of the gaseous molecules (Nelson et al., 1967).

The dipole moments of a large number of halogenated hydrocarbons are listed in
the Appendix. The sources of references were Bingham et al. (1975); Di Giacomo and
Smyth (1955); Franck et al. (1975); Gross (1931); Hill et al. (1969); Ketelaar and
van Meurs (1957a, b); McClellan (1963, 1974); Martin (1979); Miller and Smyth (1956,
1957); Minkin et al. (1970); Nelson et al. (1967); Rathmann et al. (1978); Reid et
al. (1977); Osipov and Minkin (1965); Smith (1955); Smyth (1931, 1955, 1972); Vaughan
(1969); Wesson (1948); Vuks (1969); Weissberger and Rossiter (1972).

The group contribution method has been developed by Smyth (1955) to calculate
the molecular dipole moment of gaseous molecules and molecules in solution; see
Table 2.19 for the group contribution values to halogenated hydrocarbons. The vectorial additivity of two group dipole moments is obtained from the expression

$$\bar{\mu}_{12} = (\bar{\mu}_1 + \bar{\mu}_2 + 2\,\bar{\mu}_1\,\bar{\mu}_2\,\cos\theta)^{\frac{1}{2}}$$

where $\bar{\mu}_1$ and $\bar{\mu}_2$ are the vectorial group moments and θ is the angle between them.
In the case of one short vector, the foregoing expression becomes

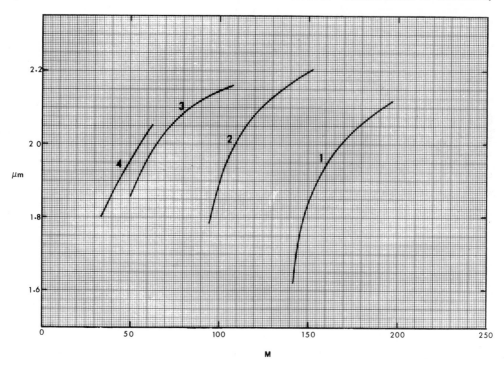

Fig. 2.17 Dipole moments of gases as function of their molecular wights. μ = dipole moment (debyes); M = molecular weight. 1, $C_nH_{2n+1}I$; 2, $C_nH_{2n+1}Br$; 3, $C_nH_{2n+1}Cl$; 4, $C_nH_{2n+1}F$.

$$\bar{\mu} \;=\; 2\, \bar{\mu}_1 \cos(\theta/2)$$

For example, the dipole moment ($\bar{\mu}$) of o-dichlorobenzene can be calculated ($\theta = 60^\circ$):

$$\bar{\mu} \;=\; 2 \;\times 1.55 \cos(60/2) = 2.68 \text{ debyes} \quad\text{or}\quad 2.68 \times 10^{-18} \text{ esu-cm}$$

The value calculated from the experimental dielectric constant for liquid o-chloro-benzene is 2.05 debyes at 20°C. The discrepancy is 30%.

The temperature dependence of dielectric constants of halogenated hydrocarbons is illustrated in Fig. 2.18. Casteel and Sears (1974) fitted the dielectric constant (D) in the following empirical equation

$$D \;=\; A + \frac{B}{T} + \frac{C}{T^2}$$

where T is the absolute temperature and A, B, and C are empirical constants for a substance.

Table 2.18 Pauling's Electronegativity of Elements

Element	Electronegativity
F	4.0
Cl	3.0
Br	2.8
I	2.5
C	2.5
H	2.1
O	3.5
N	3.0
S	2.5
Se	2.4
P	2.1
B	2.0
As	2.0
Si	1.8
Al	1.5
Mg	1.2
Li	1.0
Na	0.9

Source: Pauling, 1960.

Table 2.19 Group Contributions to Permanent Dipole Moments

Group	CH_3-		$C_nH_{2n+1}-$ (n > 1)		C_6H_5-		Angle,
	Gas	Solution	Gas	Solution	Gas	Solution	θ
CH_3-	0	–	0	–	0.36	0.40	180°
Br–	1.80	1.80	2.01	1.90	1.73	1.54	0
Cl–	1.87	1.70	2.05	1.80	1.70	1.58	0
F–	1.81	–	1.92	–	1.59	1.46	0
I–	1.64	1.50	1.87	1.80	1.70	1.30	0

The static dielectric constants of liquified fluoromethanes were measured from their melting points to their normal boiling points (Tremaine and Robinson, 1973). The temperature function was illustrated on a linear scale for CF_4, CHF_3, CH_2F_2, and CH_3F. The temperature and pressure dependence of the dielectric constants

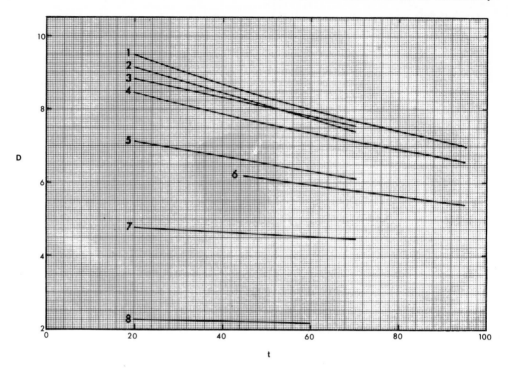

Fig. 2.18 Dielectric constant of halogenated hydrocarbon liquids. D = dielectric constant, dimensionlass; t = temperature (°C). 1, $CH_2Br-CH_2-CH_2Br$; 2, $CH_2Br-(CH_2)_3CH_2Br$; 3, $CH_2Br-(CH_2)_2-CH_2Br$; 4, $CH_2Br-(CH_2)_4-CH_2Br$; 5, $CH_2Br-(CH_2)_7-CH_2Br$; 6. $CH_2Br-(CH_2)_8CH_2Br$ 7, CH_2Br-CH_2Br; 8, CCl_4. (From Ketelaar & van Meurs, 1957a, b.)

of CH_3F and CF_3H was reported by Buckingham et al. (1978).

The order of magnitude of the dipole moment is made up of the proauct of the electronic charge ($\sim 10^{-10}$ esu) and the intermolecular distance ($\sim 10^{-8}$ cm), so that **the result is in the order of** 10^{-18} **esu-cm or 1 debye (1 debye = 3.335 × 10^{-30}C m** = 10^{-18} $g^{\frac{1}{2}}$ $cm^{5/2}$ sec^{-1} = 10^{-18} $dyne^{\frac{1}{2}}$ cm^2; see Hoppe, 1972).

The dipole moment can be calculated from bond moments, by quantum mechanical methods, from dielectric constant, and so on. The various methods were presented by Minkin et al. (1970).

The permanent dipole moments calculated from experimental dielectric constants showed a considerable difference depending on whether it was determined in solution or in the pure gaseous state (Müller, 1933, 1934, 1937). The effect of dipolar interaction on the dielectric constant in the case of solids is very different from that of liquids (Fröhlich, 1946).

The Debye formula (Debye, 1912)

$$\frac{D-1}{D+2}\frac{M}{d} = \frac{4\,\Pi\,N\alpha}{3} + \frac{4\,\Pi\,N\,\mu^2}{9\,k\,T}$$

should be used for gases only. This expression has been developed for random molecules in the absence of externally applied fields. However, if the molecules are widely separated in the solution, the equation gives a rather good approximation, for example, in benzene.

The Debye formula can be rewritten in a form for molar polarization (P):

$$P = A + \frac{B}{T}$$

where

$$P = \frac{D-1}{D+2}\frac{M}{d}$$

$$A = \frac{4\,\Pi\,N\alpha}{3}$$

$$B = \frac{4\,\Pi\,N\,\mu^2}{9\,k}$$

By plotting the molar polarization (P) versus 1/T, a straight line is obtained. The graphical illustration of this relationship is shown in Fig. 2.19 for several halogenated hydrocarbons. The intercept at 1/T = 0 gives the value for A and the slope of the line provides the B value. This method is the most common to calculate dipole moment from the measured dielectric constant values of gases (Di Giacomo and Smyth, 1955). The dielectric constants are usually measured by parallel plates of metal using a high-frequency alternating current in the capacitor (Vaughan et al., 1972; Baker and Smyth, 1939a, b; Smyth, 1931, 1972; Minkin et al., 1970; Smith, 1955, 1975). The gas is filled between the plates. The dielectric constant (D) is the ratio of capacity of the empty C_o and the filled condensor C:

$$D = \frac{C}{C_o}$$

The electric field between the charged plates causes the polarization of the solvent molecules not having own permanent dipoles. The order of the drop in the electric field is proportional to the polarization. From the measured dielectric constant (D) value at 760 mm Hg, the molar polarization (P) can be calculated at $0^{\circ}C$:

$$P = \frac{22{,}414\,(D-1)}{273.15\,(D+2)}$$

or at different temperatures the Mosotti-Clausius equation gives

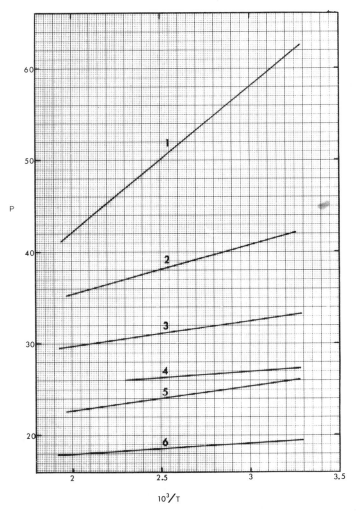

Fig. 2.19 Temperature dependence of molar polarization of vapors. P = molar polarization (cc); T = absolute temperature (K). 1, CHF_3; 2, CF_3I; 3, CF_2Br_2; 4, C_2F_5Cl; 5, CF_3Br; 6, CF_3Cl. (From Di Giacomo & Smyth, 1955.)

$$P = \frac{D-1}{D+2} \frac{M}{d}$$

The polarization and dielectric constant for a large number of liquid substances are listed by Wyman (1936) and Wen and Muccitelli (1979). The dielectric constants of halogenated hydrocarbons are given in the Appendix (see also Eiseman, 1955; Hill et al., 1969; Maryott and Smith, 1951; Reichardt and Dimroth, 1968; Tremaine and

and Robinson, 1973; Weissberger and Rossiter, 1972.) The linear relationship between
polarization and dielectric constants was also studied by Wyman (1936) and Abboud
and Taft (1979).

   There have been several attempts to correlate the dielectric constants of liquids with some physical parameters (Partington, 1954; Paruta, 1963; Paruta et al.,
1962; Laprade et al., 1977; Holmes and Weiser, 1925). In 1931, Danforth (1931) and
later Kirkwood (1936) showed the linear relationship between $\left[(D - 1)/3\right] v_1$ and
the molar density of liquid pentane, where D is the dielectric constant of liquid
pentane and v_1 the molar volume of pentane. Extrapolation to zero density gave
26.5 ml.

   An empirical relationship has been presented by Papazian (1971) between the
dielectric constant and the surface tension of 13 nonpolar liquids, having zero
dipole moments. In a later article Holmes (1973) discussed the theoretical relationships between the parameters surface tension (γ), liquid dielectric constant (D),
and liquid refractive index (n_D) for 18 nonpolar liquids. The six functions of the
surface tension examined correlated well with the dielectric constant. The simplest
relation is just as good as the more complicated ones; therefore, it seems reasonable
to presume that for practical use the first one is recommended: that is,

$$\gamma = 20.9 \, D - 20.5$$

   The dielectric constants were also correlated with the acidity of alcohols
(Dankleff et al., 1968). Greater acidity yields a decrease in the values of dielectric constant. In solutions the dielectric constant and densities depend linearly
on concentration (Osipov et al., 1965).

   McRae (1957), Laszlo et al. (1969), and Raynes (1969) correlated the nuclear
magnetic resonance (NMR) chemical shifts with the refractive index for a large number
of halogenated hydrocarbons. This correlation suggests that there may be a relationship between the chemical shifts and the dielectric constants (McRae, 1957;
Suppan, 1968). The assumption is that the shifts are due to dipole-dipole or ion-
dipole interactions. However, additional molecular interactions may occur between
the solute and solvent, which causes deviation from the linear function assumed,
particularly in the case of hydrogen-bonding anomaly in the system when the transition of an electron takes place. It is worthwhile to bear in mind the following sorts
of interactions between solute and solvent molecules (Suppan, 1968):

1. Dipole-dipole
2. Dispersion forces
3. Dipole/induced dipole
4. Multipole interaction
5. Association, hydrogen bonding
6. Solvent-cage strains

Gordy and Stanfor (1939-1941) showed a successful correlation between solubility and shift in the OD band using Zellhoefer's (1939) data. There were no details regarding the correlation of water solubility with shift in OD band, although the correlation of dielectric constants and solubilizing properties of aqueous solutions was studied by Rebagay and De Luca (1976) and Reichardt and Dimroth (1968), as an addition to the enhanced solubility of various substances as a function of the dielectric constants reported previously. Modern solution theories, their successes and problems were reviewed by Friedman (1973).

The dielectric constants of pure organic liquids were determined by measuring their radio-frequency power using a calibration curve (Malinowski and Garg, 1977). The average difference between the determined values and the reported values was 2.6%.

The theory of electrolytes furnishes the limiting relation between the molar volumes and the dielectric constants of the medium (Redlich, 1963; Redlich and Meyer, 1964; Millero, 1971; Mishchenko, 1972). However, in the expression there are other parameters that are not easily accessible (e.g., the compressibility of the solvent). For further discussion on the properties of aqueous electrolyte solutions, see **Horvath (1982)**.

Paruta and co-workers (Paruta et al., 1962; Paruta, 1963) correlated the dielectric constant with the solubility parameter, obtaining a linear relationship for 25 solvents at 25°C:

$$\delta = 0.22 D + 7.5$$

In this relationship, those solvents that give the best correlations associate primarily through hydrogen bonding, so that the correlation is good enough for nonpolar solvents. Further details will be discussed in connection with regular solution theory.

The correlation of solubility with dielectric constant has been studied by several investigators (e.g., Paruta et al., 1962; Paruta, 1963; Glew, 1952; Laprade et al., 1977). The correlation between the dielectric constants of halogenated hydrocarbons and the solubility of water at 25°C is plotted in Fig. 2.20. Similarly, the relationship between the electrostatic factor, expressed in debyes, and the weight percent solubility of water at 25°C in halogenated hydrocarbons is shown in Fig. 2.21. Both figures show that the correlation can provide only a very approximate value of solubility.

For an approximative calculation for pure polar liquids, Onsager's equation (Onsager, 1936; Hoffman, 1952) gives the relationship among the dipole moment, dielectric constant, and refractive index:

$$\mu^2 = \frac{9 M k T}{4 \Pi N d} \frac{(D - n^2)(2 D + n^2)}{D(n^2 + 2)^2}$$

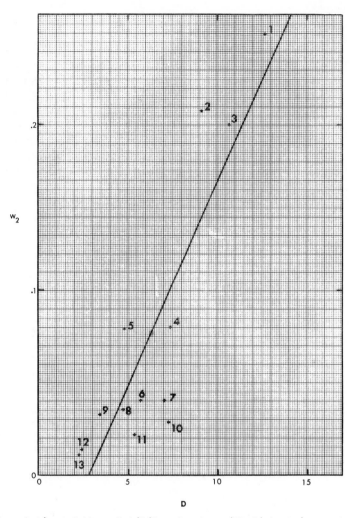

Fig. 2.20 Correlation of the solubility of water with dielectric constants. w_2 = solubility of water in solvents at 25°C (wt %); D = dielectric constant, dimensionless. 1, CH_3Cl; 2, CH_2Cl_2; 3, CH_2Cl-CH_2Cl; 4, $1-C_4H_9Cl$; 5, $CHCl_3$; 6, C_6H_5Cl; 7, $CHBr_2-CHBr_2$; 8, $CCl_2=CH_2$; 9, $CHCl=CCl_2$; 10, CH_2Br_2; 11, CH_2I_2; 12, CCl_4; 13, $CCl_2=CCl_2$.

where μ = dipole moment, $dyne^{\frac{1}{2}} cm^2$ (1 debye = 10^{-18} $dyne^{\frac{1}{2}} cm^2$)

      M = molecular weight

      k = Boltzmann's constant, $dyne\ cm\ K^{-1}$

      T = absolute temperature, K

      N = Avogadro's number, $g-mole^{-1}$

      d = liquid density, $g\ cc^{-1}$

      n = refractive index

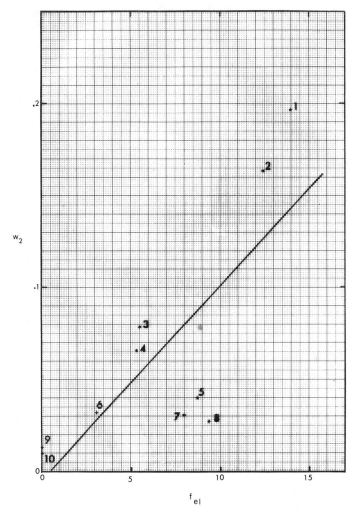

Fig. 2.21 Correlation of the solubility of water with electrostatic factors. f_{el} = electrostatic factor (debyes); w_2 = solubility of water in solvents at 25°C (wt %). 1, CH_2Cl_2; 2, C_2H_5Cl; 3, $CHCl_3$; 4, CH_2Br-CH_2Br; 5, C_6H_5Cl; 6, $CHCl=CCl_2$; 7, C_6H_5F; 8, C_6H_5Br; 9, CCl_4; 10, $CCl_2=CCl_2$.

This expression is an improvement of Debye's equation for the prediction of the static dielectric constant of dense fluids. However, neither the Debye nor the Onsager relationship takes account of the local forces between neighboring dipoles. The experimental values showed that the dielectric constant is sensitive to these forces. Kirkwood (1939a) introduced a modification to the theory by taking the forces into consideration for a spherical model of the substance having dipoles. The field inside the specimen is smaller than that outside by a factor of $3/(D + 2)$. The modified expression is

$$\frac{(D - 1)(2 D + 1)}{3 D} = \frac{4 \Pi N}{V} (\alpha + \frac{g \mu^2}{3 k T})$$

where V = molar volume

     g = Kirkwood dipole correction parameter (Kirkwood, 1946)

     D = dielectric constant

However, this equation is also an approximation only, because it does not become the Onsager relation when $g = 1$. If we assume that the local field is acting upon **the polarizable molecule, the dipole moment of a gas molecule becomes**

$$\frac{(D - n^2)(2 D + n^2)}{D(n^2 + 2)^2} = \frac{4 \Pi N g \mu_g^2}{9 k T V}$$

where μ_g = dipole moment of the gas molecule in debyes
This is the Kirkwood-Fröhlich expression for gaseous molecules (Fröhlich, 1958).

    Although both the Kirkwood (1939a) and Fröhlich (1958) equations are sounder theoretically, the Onsager (1936), Böttcher (1946a, b, and c), and Scholte (1949) expressions have greater practical potential for dipole liquids, needing only limited information as to their structure.

    The Böttcher (1946a, b, and c) equation for a pure dipole liquid is

$$\frac{(D - 1)(2 D + 1)}{12 \Pi D} = \frac{N d}{M} \left[\frac{\alpha}{1 - \alpha f} + \frac{1}{3 k T} \frac{\mu^2}{(1 - f \alpha)^2} \right]$$

where $f = \dfrac{2 D - 2}{r(2 D + 1)}$

    r = radius of the spherical molecule

    α = polarizability of the molecule

In the case of polar molecules in nonpolar solvents, Böttcher (1946a, b, and c) gives a more complex expression. However, in a recent investigation Finsy and van Loon (1976) have determined the permanent dipole moment of 1,1,1-trichloroethane in various nonpolar solvents and concluded that in the case of the calculation of the dipole moment from the dielectric constant for a polar solute in nonpolar solvents, the Onsager-Kirkwood-Fröhlich equation should be applied instead of the Debye's expression. The Onsager-Kirkwood-Fröhlich formulation is as follows:

$$\frac{4 \Pi d g \mu^2}{3 k T} = (\frac{3}{\epsilon_\infty + 2})^2 \frac{2 \epsilon + \epsilon_\infty}{2 \epsilon + \epsilon_1} \left[\frac{2 \epsilon + \epsilon_\infty}{3 \epsilon} \frac{\epsilon - \epsilon_1}{\varphi} - (\epsilon_\infty - \epsilon_1) \right]$$

where ε_1 = dielectric constant of the nonpolar solvent

μ = permanent dipole moment of the polar solute

ε_∞ = high-frequency permittivity

d = density of the pure polar solute

φ = volume fraction in the polar component

ε = dielectric constant of the binary mixture

g = Kirkwood's orientation correlation factor

According to Scholte (1949), the dipole moment of a dipole liquid is described by the expression

$$\frac{D-1}{4\pi} = \frac{N\,d}{M}\left[\frac{3\,D\,\alpha}{(2\,D+1)(1-f_1\alpha_1)} + \frac{D}{D+(1-D)A_\mu}\frac{\mu^2}{3\,k\,T(1-f_\mu\alpha_\mu)^2}\right]$$

where A_μ is a constant that depends on the shape of the molecule and is calculated from the relationship

$$A_\mu = -\frac{1}{p^2-1} + \frac{p}{(p^2-1)^{3/2}}\log_e\left[p+(p^2-1)^{\frac{1}{2}}\right]$$

where $p = a/b$ (a and b are the axes of the spheroid molecule). Furthermore,

$$f_1 = \frac{1}{r^3}\left(\frac{2\,D_o-2}{2\,D_o+1}\right)$$

$$f_\mu = \frac{3}{r^3}\left[\frac{A_\mu(1-A_\mu)(D-1)}{D+(1-D)\,A_\mu}\right]$$

$$\alpha_\mu = \frac{(D_i-1)\,r^3}{3\left[1+(D_i-1)\,A_\mu\right]}$$

$$D_i = \frac{r^3+2\alpha}{r^3-\alpha}$$

Expressions for dilute and concentrated solutions were also given by Scholte (1949).

The calculated dipole moments for liquid 1,2-dibromoethane and 1,3-dibromopropane at various temperatures calculated from the Onsager-Böttcher-Fröhlich equation are given in Fig. 2.22. Leroy et al. (1978) calculated dipole moments of a series of fluoroesters by means of the CNDO method.

The Kirkwood's dielectric theory has been further developed and applied to dilute solutions by Grunwald et al. (1976). Regarding the calculations of the dipole moments of solutions, Partington (1954) gives a full account up to 1952. The experimentally determined values have been recently reported by Stokes and Marsh (1976) and Finsy and van Loon (1976).

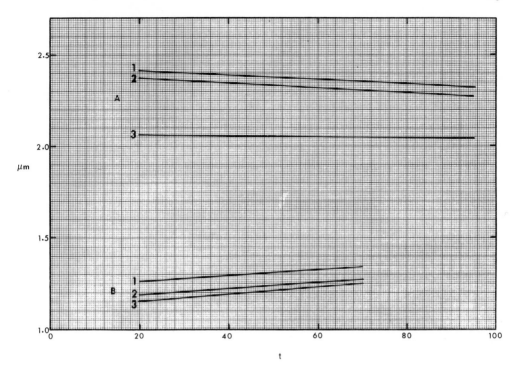

Fig. 2.22 Dipole moment of liquids according to Onsager, Böttcher, and Scholte.
μ = dipole moment (debyes); t = temperature (°C). A, 1,3-dibromopropane; B, 1,2-dibromoethane. 3, Onsager; 2, Böttcher; 1, Scholte. (From Ketelaar & van Meurs, 1957a, b.)

The dielectric constants of polar liquids were plotted against the dipole moments by Gjaldbaek and Anderson (1954) for the study of liquid association.

The relationship between the dielectric constant (D) and the dipole moment (μ_o) of liquids has been studied by Salem (1979). He correlated the square of dipole moment with reduced dielectric constant (D - 1) M/d for 45 organic liquids of various classes. The direct proportionality is expressed by the equation

$$(D - 1) \ M/d \ = \ 0.184 \times 10^{24} \ \mu_o^2$$

The correlated values do not deviate from the theoretical straight line by more than 10 - 15 percent. The largest departures from the formula were observed with liquids that form hydrogen bonds (water, organic acids, alcohols, etc.) and also in solvents whose molecules are capable of ionization by autoprotolysis (e.g., formaldehyde, acetamide, and inorganic acids).

The graphical representation of the Debye equation has been presented by Cole and Cole (1941). By plotting the complex dielectric loss (D'') versus the dielectric constant (D'), a semicircle stretching from $D' = D_s$, $D'' = 0$ in the low-frequency limit to $D' = n^2$, $D'' = 0$ in the high-frequency limit was obtained. This representation of the dielectric constant is known as a "Cole-Cole diagram" (Hasted, 1973). Another graphical illustration is by means of two straight lines:

$$D'' \; \omega \; = \; \frac{-(D' - D_s)}{\tau}$$

$$\frac{D''}{\omega} \; = \; \tau\,(D' - D_\infty)$$

where ω = angular frequency

τ = relaxation time

D_s = static dielectric constant

D_∞ = infinite-frequency dielectric constant

The Cole-Cole empirical equation,

$$D^* - D_\infty \; = \frac{D_s - D_\infty}{1 + (j\,\omega\,\tau)^{1-\alpha}}$$

represents the dispersion and absorption of a large number of liquids and dielectrics. In the equation,

D^* = complex dielectric constant

ω = 2 Π times the frequency

τ = generalized relaxation time

α = width, values between 0 and 1

According to this relationship, the locus of the dielectric constant in the complex plane is a circular arc with end points on the axis of real values and with center below this axis. That is, the center lies below the horizontal $D'' = 0$ axis, on the line drawn from $D'' = 0$, $D' = D$ and making an angle of $\Pi\,\alpha/2$ with the horizontal axis. Fig. 2.23 shows the arcs obtained from the Cole-Cole and rectangular distributions. It is seen that the actual form of a symmetrical distribution function is very difficult to separate from the dielectric data. In the case of liquids, D_s is known from low-frequency measurements and D_∞ can be obtained by extrapolation of the optical index of refraction to zero frequency.

The distribution of the frequency parameter along the arc is dependent upon $f(\omega,\tau)$ and not on α. The usual graphical technique is by two chords of length u and v. The length of chord u is from D_s, 0 to D', D'' and for chord v is from D_∞, 0 to D',D''. If we plot log(u/v) versus log v, the slope of the straight line will give $(1-\alpha)$.

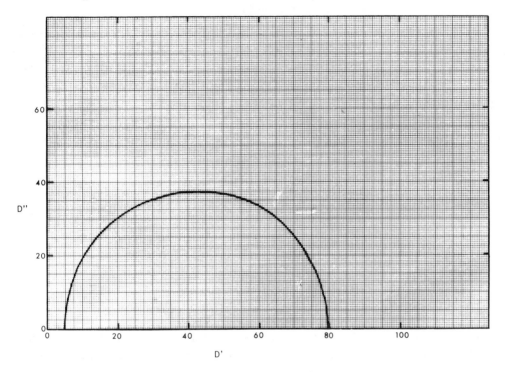

Fig. 2.23 Cole-Cole diagram of water at 25°C. (From Hasted, 1973.)

There are three modifications of the Cole-Cole equation. The first is

$$D^* = \frac{D_s - D_\infty}{1 + j^{1-\alpha}(\omega\tau)^{1-\beta}}$$

The second modification is by Cole-Davidson (Davidson and Cole, 1950):

$$D^* - D_\infty = \frac{D_s - D_\infty}{(1 + j\omega\tau)^\alpha}$$

The third is the Fuoss-Kirkwood analysis (Fuoss and Kirkwood, 1941).

In general, α varies only slightly with temperature, D_s increases gradually, and τ increases exponentially with reciprocal temperature (Davidson and Cole, 1950). Davidson and Cole found that the complex locus of D^* for glycerine is intermediate between the semicircle of Debye's theory and the arc locus of Cole-Cole's work. Plots of the complex dielectric constant D^* for $CBrCl_3$ at low temperatures show a similar phenomenon (von Dahl and Cole, 1966). The Cole-Cole plot for several halogenated hydrocarbons is given by Baba and Kamiyoshi (1977).

The solvents may be classified from several points of view (see, e.g., Ahmed, 1979; Ben-Naim, 1975; Brönsted, 1928; Bell, 1941; Ewell et al., 1944; Joffe and Orchin, 1961; Davis, 1968; Eberson, 1969; Reichardt, 1973; Malesinski, 1967; Riddik and Bunger, 1970; Craver, 1970; Murgulescu and Demetrescu, 1970; Tokura, 1970; Ziolkpwsky, 1970; Covington and Dickinson, 1973; Criss and Salomon, 1973; King, 1973; Barton, 1974; Ulbrich, 1974; Hine, 1975; Dack, 1976; Cosaert, 1971; Gutmann, 1977; Llor and Cortijo, 1977; Furniss et al., 1978; Weisberger, 1963; Karger et al., 1978; Snyder, 1980). However, only the most successful treatments will be considered in this compilation, bearing in mind that the aim here is to explain the mutual solubilities between halogenated hydrocarbons and water. Such field as the solvent effects on chemical phenomena (Amis and Hinton, 1973) is outside the scope of this book.

The solvents may be classified into two main groups: polar and nonpolar. Polar solvents (e.g., water, formic acid) have a large permanent dipole moment. The dipole moments are calculated values, often from the dielectric constant (Smyth, 1931, 1972; Minkin et al., 1970; Smith, 1955, 1975; Schupp and Mecke, 1948; Perram and Anastasiou, 1981). The dielectric constant of a solvent is measured between two electrostatically charged plates of a condenser (Vaughan et al., 1972; Smith, 1955, 1975). Depending on whether the molecules tend strongly or weakly toward the charged plates, the solvent has a high or low dipole moment. The effect is related to the vacuum when the magnitude of the dielectric constant is obtained. The order of the dielectric constant indicates how the solvent molecule orients its dipoles and electric charges. If the temperature is increasing, the dielectric constant decreases, because thermal vibrations interfere with the orientation of the dipoles. This has been demonstrated by Franck et al. (1975) for CH_3F, CF_3H, and other compounds. The temperature dependence of the dielectric constant (D) may be correlated precisely by the following empirical equation:

$$D = A + \frac{B}{T} + \frac{C}{T^2}$$

The dielectric constant at the critical point has been studied by, for example, Mistura (1973). The larger the dielectric constant, the greater is its variation with temperature. Further studies of the effect of temperature on the dielectric constant have been carried out by Miller and Smyth (1957), Ketelaar and van Meurs (1957b), Greenacre and Young (1976), and Younglove (1972). The dielectric constant also decreases with pressure (De Groot, 1949; Danforth, 1931; David et al., 1952). The temperature and pressure dependence of the dielectric constant of water was studied by Srinivasan and Kay (1974).

The dipole moment is proportional to the dielectric constant (see Debye's equation and Abbound and Taft, 1979; Salem, 1979; and Perrman and Anastasiou, 1981). However, in symmetrical substances whose dipole moments are zero, the dielectric constant can be quite high (e.g., CCl_4 has a dielectric constant of 2.24 at 20°C). This phenomenon is explained by displacement of the electrons and positive nuclei within the atoms by the electric field.

Because the force between molecules is reciprocally proportional to the dielectric constant of the fluid, a large dielectric constant indicates less attraction between the two molecules. This is the principle of the dissolving action of a polar solvent. The polar molecules of the solute are attracted by the polar molecules of the solvent. In a polar solvent the ion pairing is minimal. The influence of polarity upon the solubility parameter was studied by Gardon (1966).

The ionizing solvents consist of polar molecules. They are dissolving ionic substances, such as inorganic salts, and the solution contains dissolved cations and anions. The dissolving strength of the ionic solvent toward ionic compounds is measured by the magnitude of its dielectric constant. That is, a high dielectric constant indicates that the solvent is strongly polar and capable of dissolving ionic substances that have attractive forces between the oppositely charged ions. These solutions show electrical conductivity, contrary to the solutions of polar solvents, where the anions and cations have no independent mobility.

Liquids such as H_2O and HF have very hugh dielectric constants because of their strong permanent dipole moment, coupled with hydrogen bonding, so they are strongly ionizing solvents. Because of its hydrogen bonds, ammonia is also a good ionizing solvent, despite its low dielectric-constant value. Liquid water is not only polar but is also an ionizing solvent; consequently, it dissolves ionic and covalent compounds of high polarity such as glycol and methanol. However, water does not dissolve nonpolar molecules that have small dipole moments (e.g., hydrocarbons). A further discussion of the dissolving power of water is given in Sec. 2.5.

A large number of chemicals are nonpolar solvents. These substances have low dielectric constants owing to their small permanent dipole moments and ion-pairing effects. Typical nonpolar compounds are CCl_4, C_6H_6, and p-dichlorobenzene. The cohesive forces are very weak in these substances, and therefore they cannot dissolve polar or ionic compounds with strong forces between the molecules. Consequently, the dissolving power of nonpolar solvents is limited to solutes having low polarity, such as propanol. However, ionic compounds are not soluble in nonpoler solvents.

The line that is usually drawn between polar and nonpolar solvents is at dielectric constants having a value of 30. In this division, CH_3OH belongs to the polar group of compounds, whereas C_2H_5OH is a nonpolar solvent ($\mu = 24$). Furthermore, both polar and nonpolar classes can be subdivided into protic and aprotic subclasses, (i.e., hydrogen-bonded and nonhydrogen-bonded solvents). The molecules of aprotic solvents are more polar than are those of protic solvents. Therefore, the ion-dipole and molecule-dipole solute-solvent interactions are stronger in dipolar aprotic solvents. These molecules are more polarizable than the molecules of protic solvents,

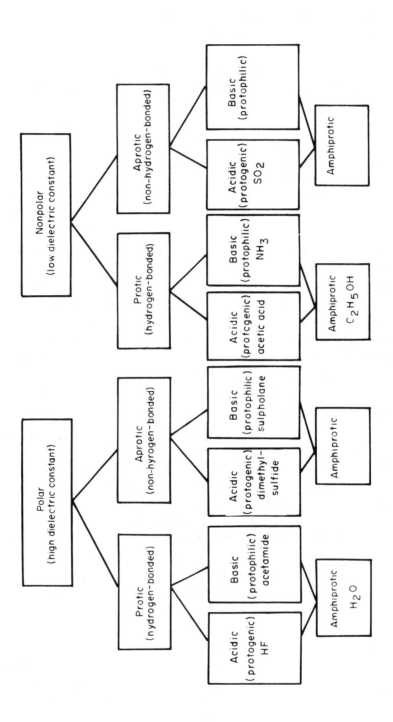

Fig. 2.24 Classification of solvents.

and the London dispersion forces are also more pronounced in dipolar aprotic solvents (Symons, 1977).

The hydrogen-bonded solvents of low dielectric constant may form ion pairs between two ions; however, they are less stable than dipolar aprotic solvents because of competition for hydrogen bonds between the solvent and the ions. These subclasses are further divided into acidic and basic (i.e., protogenic and proto-philic solvents) (Iogansen, 1971). However, when both acidic and basic characteristics are possible, the solvent is considered to be amphiprotic. These amphiprotic solvents make possible self-ionization or autoprotolysis. Fig. 2.24 shows the classification of various solvents described above.

2.3 Association of Molecules

In addition to the treatment of various aspects of polar and nonpolar fluids, a short summary is given here on the association, dimerization, trimerization, and polymerization properties of various substances. Hydrogen bonding is treated in Sec. 2.4. Solvation and regular solutions are discussed in Sec. 2.5.

The association tendency of some molecules means that the monomeric molecule is capable of forming chemical aggregates: that is of self-association. In other words, the molecular association is a union of simple molecules into groups or aggregates of larger molecules. The terms "dimerization," "trimerization," and so on, adopted for molecular association, indicate the number of molecules in an aggregate specimen.

Association between molecules can take place in gaseous and liquid states and in solution states. The molecular association of gases has been explained by both physical and chemical forces between the molecules. The relation of the behavior of the associated species to the ideal gas is a commonly used method. In the case of polymerization, the compressibility factor of the associated gas shows negative deviation, contrary to the dissociation process, when the compressibility factor of the molecule has positive deviation from ideal gas behavior. The association behavior of polar vapors has been discussed by Lambert (1953), for a large number of substances. The association theory has been used by Ginell (1979) for explaining the concept of entropy.

The experimentally measured thermal conductivity of various vapors showed greatly enchanced values. Furthermore, the pressure dependence of the thermal conductivity will indicate an abnormal behavior which is due to the degree of dimerization. The pressure dependence of the vapor heat capcity is also a good indication of dimerization or polymerization (e.g., for alcohols see Kirchnerova, 1974). The virial coefficient values provide a convenient way to calculate the degree of association in the vapor phase; that is, the deviation from the gas laws can be expressed

$$B_{dimer} = -\frac{R\,T}{K_p}$$

where K_p is the dissociation constant of the dimerization reaction $2\,A \rightarrow A_2$. The dimerization constant can be written

$$K_p = \frac{p_A^2}{p_{A_2}} = \frac{y_A^2\, P_{tot}}{y_{A_2}}$$

where y_A and y_{A_2} are the mole fractions of the monomer and dimer, respectively, and P is the total pressure. The effect of dimerization on the value of activity coefficient in the vapor phase was studied by Malijevska (1979).

The observed second virial coefficient B_{obs} in an associated system may be regarded as a superposed parameter on the normal second virial coefficient B:

$$B_{obs} = B + B_{dimer}$$

When the critical properties are known, then by the application of the corresponding-state principle for the second virial coefficient,

$$\frac{B}{V_{cr}} = f(\frac{T}{T_{cr}})$$

one can show the departure of the associated molecules by plotting B/V_{cr} versus T/T_{cr} (Hirschfelder et al., 1942). Lambert (1953) illustrated this departure for polar vapors: methanol, steam, ammonia, ethyl chloride, and acetonitrile.

The second virial coefficient B may be estimated from the generalized Redlich-Kwong equation of state, which yields (Svoboda et al., 1977)

$$B = 0.0867\,(1 - \frac{4.93}{T_r^{1.5}})\,\frac{R\,T_{cr}}{P_{cr}}$$

An empirical correlation of the second virial coefficient for haloalkanes has been reported by Tsonopoulos (1970) and Djordjevic et al. (1980). They also discussed the method for the calculation of the second virial coefficients of mixtures. The second virial coefficients of pure gases and mixtures have been compiled by Dymond and Smith (1979) from the published literature.

The lower carboxylic acids form mostly dimers which are stable even in the gaseous phase, but they do not tend to form larger aggregates. This has been proven by use of equilibrium measurements. The dimeric form of the formic acid has the structure (Karle and Brockway, 1944; Ketelaar, 1958; Herman, 1940; Herman and Hofstadter, 1938, 1939)

These double molecules have a small or no resultant dipole moment and the association thus leads to a reduction of the dielectric constant of the liquid. Because
there is no additional free hydrogen atoms, association goes no further. In the dimer
molecule of formic acid, the distance between O and H atoms is 1.07 $\overset{\circ}{A}$, compared
with 0.97 $\overset{\circ}{A}$ in the monomer molecule. The hydrogen position between the two oxygen
atoms has nonequivalent bond energy levels. The interaction energy between two formic
acid molecules is considerable when the positively charged hydrogen of each hydroxyl
group is opposite an oxygen atom. The thermal examination of this dimer molecule
showed that about 14 kcal g-mole^{-1} is required to split the dimer into two molecules.
Consequently, each hydrogen bond has a bond energy of approximately 6 to 7 kcal g-mole^{-1}.
If one compares the hydrogen bond energy with the covalent bond energies (25 to 100
kcal g-mole^{-1}), it becomes apparent that although the hydrogen bond is very strong
for dipole–dipole interaction, it is a rather weak interaction.

There is some discrepancy between the reported bond energies in the literature.
The systematic errors in the experimental results have been investigated by Shreiber
(1976). The proposed expressions can easily be applied to discover systematic errors.
The experimental densities of the saturated vapors were used to calculate the equilibrium constants of the associated vapors.

The total bond energy due to dimerization in acetic acid is about 15 kcal g-mole^{-1};
that is, the energy required to break one bond is about 7 to 8 kcal g-mole^{-1}. This
is very similar in order to the dimer of formic acid molecules. This is, in fact, much
greater than the thermal energy, which is about 0.6 kcal g-mole^{-1}.

Both the enthalpy (ΔH_{dimer}) and the entropy (ΔS^{O}_{dimer}) of dimerization are
calculated from the equilibrium constant (K_p) for association by plotting $\log_{10} K_p$
versus 1/T, according to the equation

$$\log_{10} K_p \;=\; -\,\frac{\Delta H}{R\,T} \;+\; \frac{\Delta S^{O}}{R}$$

Fig. 2.25 shows the $\log_{10} K_p$ versus 1/T plot for some carboxylic acids. An excellent compilation of the published association constants for single and multicomponent systems was published by Pimentel and McClellan (1960). For comparison, a
few interesting equilibrium constants are presented in Table 2.20.

Tyuzyo (1957) used the energy of vaporization for calculating the degree of
association of saturated fatty acids, aliphatic alcohols, and others.

In an investigation by MacDougall (1936), it was assumed that acetic acid forms
a dimer and a trimer in the vapor phase, whereas Ritter and Simons(1945) found that
there are a dimer and a tetramer present (see also Johnson and Nash, 1950; Taylor,
1951; Weltner, 1955). They argued that if a trimer were to be present, it would
necessitate the breaking of an exothermic bond.

The values obtained for the heat of dimerization of carboxylic acids were large

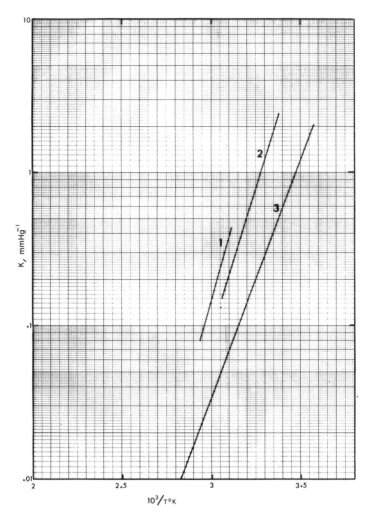

Fig. 2.25 Equilibrium constant (K) of dimerization as function of temperature (T, K).
1 = Propionic acid (MacDougall, 1941); 2 = acetic acid (MacDougall, 1941); 3 = formic
acid (Coolidge, 1928).

compared to other associating molecules (Fenton and Garner, 1930; Ramsperger and
Porter, 1926). Table 2.21 gives the heat and entropies of dimerization of some
substances for comparison. The difference in bond energy is due to the two hydrogen
bonds in the carboxylic acid molecules, whereas there is only one hydrogen bond in
the other compounds.

Some polar substances such as alcohols and acetone, do not give a straight line
when $\log K_p$ is plotted versus 1/T. Also, the curved plots of thermal conductivity

Table 2.20 Equilibrium Constants for Associated Molecules $(atm^{-1})^a$

---

1. $H_2O(g)$ between 40 and 400°C:

$$\log_{10} K = -5.650 + \frac{1250}{T} \quad (atm^{-1})$$

2. NH_3 between −30 and 250°C:

$$\log_{10} K = -5.869 + \frac{967}{T} \quad (atm^{-1})$$

3. CH_3Cl between −34 and 177°C:

$$\log_{10} K = -4.533 + \frac{555}{T} \quad (atm^{-1})$$

4. HF between 0 and 105°C:

$$\log_{10} K = -42.38 + \frac{8660}{T} \quad (mm \ Hg^{-1})$$

5. CH_3OH between 40 and 120°C:

$$K = \frac{100}{R \, T} + \frac{2.148}{R \, T} \exp(\frac{1986}{T}) \quad (atm^{-1})$$

---

[a]T in degrees Kelvin.

Source: Rowlinson, 1949; Jarry & Davis, 1953; Kretschmer & Wieber, 1954.

versus pressure suggest that association is greater than dimerization. The curvature of the pressure variation with density is another phenomenon showing a higher degree of association (e.g., for acetone).

The determination of association is usually carried out through vapor density measurements (Ramsay and Young, 1886). However, instead of vapor density, molecular weight can be determined by depression of the freezing point and elevation of the boiling point. The number of dimer molecules decreases with increasing pressure and decreasing temperature. The hydrogen bonding becomes weaker with rising temperature as a result of the thermal agitation. Therefore, all associated liquids approximate more and more to a normal liquid at high temperatures. The extent of association of solvents was reported by Gjaldbaek and Anderson (1954).

The association in infinitely dilute solution of water in Freon-23 was studied by Khodeeva et al. (1977). The result of the investigation showed that the degree of dissociation of solute dimer (α) in the infinitely dilute critical phase is described by the equation

Table 2.21   Enthalpy and Standard Entropy of Dimer Molecules

Compound	$-\Delta H_{dimer}$ (kcal g-mole^{-1})	$-\Delta S_{dimer}$ (cal g-mole^{-1} K^{-1})
CH_3Cl	3.10	20.7
H_2O	5.76	25.8
NH_3	4.40	26.8
CO_2	2.98	23.4
CH_3CN	5.20	20.6
CH_3CHO	4.60	22.3
CH_3NH_2	3.40	22.0
$C_2H_5NH_2$	3.60	22.0
$(CH_3)_2NH$	3.40	22.0
$(C_2H_5)_2NH$	3.30	20.3
$HCOOCH_3$	4.95	25.1
$HCOOC_2H_5$	4.87	23.5
$HCOOC_3H_7$	5.05	24.5
CH_3COOCH_3	4.5	22.3
$CH_3COOC_2H_5$	4.54	21.3
$C_2H_5COOCH_3$	4.36	21.3

Source: Lambert, 1953.

$$\chi = 1 - \beta N_{2c}^{0.5}$$

where β is a coefficient and N_{2c} is the mole fraction. The association in the critical infinitely dilute phases is much faster than in ordinary infinite dilute solutions. An earlier investigation by Krichevskii et al. (1966) showed that along the critical curve the ratio of the fugacity of the dissolved solute to its mole fraction tends toward infinity as the mole fraction of the solute goes to zero.

The association process is reduced for HF with rising temperature up to 100°C, which can be calculated from the molecular weight as a function of temperature (Slough, 1973; Armitage and Gray, 1963; Armitage et al., 1963; Franck and Mayer, 1959; Long et al., 1943; Maclean et al., 1962; Medvedov, 1963; Spalthoff and Franck, 1957; Vanderzee ans Rodenburg, 1970; Tyuzyo, 1957; Horvath, 1975b). In the vapor phase there is an equilibrium between normal and polymerized HF molecules. In the liquid state, an aggregated molecule has been found with the $(HF)_6$ formula. It is rather difficult to calculate the association of HF vapor, owing to the simultanety of the process (Jarry and Davis, 1953; Spalthoff and Franck, 1957).

In the case of association in pure liquids, the Trouton constant is usually too large (26 instead of 21) and the Eötvös constant too small (1 instead of 2). Furthermore, the viscosity of the associated liquids is much higher compared to similar nonassociated substances under the same conditions. Their liquid viscosities show a nonlinear relationship in a log η versus 1/T plot (Hildebrand, 1977b).

Kirkwood (1946) related the association of molecules in liquids to a correlation parameter (g), which is calculated from the observed dielectric constant, density, and refractive index:

$$g = \frac{9 \, M \, k \, T \, (D - D_\infty)(2 \, D + D_\infty)}{D \, (D_\infty + 2)^2 \, 4 \, \Pi \, N \, d \, \mu_o^2}$$

which is the well-known Kirkwood-Fröhlich equation (Fröhlich, 1958; Kirkwood, 1939a). The extent of association of a polar liquid depends upon g, which is a measure of the short-range effects that hinder orientation of a molecule with respect to its surrounding neighbors. If the g value of a polar substance departs only slightly from unity (i.e., the intermolecular forces are nonspecific or perfectly random), the compound behaves normally and the Kirkwood-Fröhlich equation reduces to the Onsager equation; but if the g value departs significantly from unity (i.e., in liquids, specific intermolecular forces orient neighboring dipoles in a net parallel configuration), the substances are considered to be associated liquids. This is contrary to the net antiparallel configuration of dipoles when the value of g is less than unity.

Obraztsov and Khrustaleva (1973) expressed the degree of association of liquids (g) as a function of their viscosity, based upon the theories of Ewell and Eyring and Wang:

$$g = \frac{2 \, N \, h \, d}{\eta \, M} \exp(1 - \frac{T \, d\eta}{\eta \, dT})$$

where   g = degree of association of liquids, dimensionless
        N = Avogadro's number
        h = Planck's constant
        η = liquid viscosity
        d = liquid density
        M = molecular weight
        T = absolute temperature, K

Table 2.22 gives the g values for a number of liquids at various temperature intervals. The temperature dependence of the degree of association for acetic acid, shown in Table 2.22, corresponds with the values expected. In general, the degree of association decreases with increasing temperature. Obraztsov and Khrustaleva (1973) also showed for water that the degree of association diminishes with an increase in

Table 2.22 Degree of Association of Various Liquids

Liquid	Temperature range (°C)	Degree of association (g)
Nonassociation		
$n-C_5H_{12}$	-100 to 0	0.99
$n-C_6H_{14}$	-60 to 40	1.11
$n-C_7H_{16}$	-20 to 80	1.04
$n-C_8H_{18}$	-40 to 80	0.94
$\left[(CH_3)_2CHCH_2\right]_2O$	0 to 80	1.11
$C_4H_{10}S$	0 to 90	1.03
$C_{16}H_{34}$	30 to 200	1.03
$C_2H_5COC_2H_5$	0 to 100	1.21
$CH_3CH_2CH_2COOCH_3$	0 to 100	1.32
$CH_3COC_3H_7$	0 to 100	1.42
$HCOOC_3H_7$	0 to 80	1.45
CCl_4	10 to 70	1.52
Association		
$C_6H_5CH_3$	-10 to 80	1.73
C_6H_6	10 to 60	1.96
C_6D_6	10 to 60	2.44
C_6H_{12}	15 to 60	2.26
$n-C_3H_7OH$	25	14.7
CH_3COOH	20	4.0
CH_3COOH	30	3.5
CH_3COOH	40	3.0
CH_3COOH	50	3.0
$n-C_4H_9OH$	10	11.8
C_2H_5OH	10	14.9

Source: Obraztsov & Khrustaleva, 1973.

the external pressure. On the assumption that the degree of association is infinite at a supercooled temperature (T_o) and reaches a constant value at the critical temperature (T_{cr}), they introduced the following expression:

$$g = a\left(\frac{T_{cr} - T_o}{T - T_o}\right)^2 + \frac{b\,d}{M}\left[\left(\frac{1}{k\,T}\right)^{\frac{1}{2}} - \frac{d_{cr}}{d}\left(\frac{1}{k\,T_{cr}}\right)^{\frac{1}{2}}\right]$$

where the constants a, b, and T_o can be determined from experimental data. In the case of water at 1 atm pressure, $a = 1.17$, $b = 4.14 \times 10^{-5}$ cm^4 sec^{-1}, and $T_o = 261$ K. The agreement between the two preceding expressions for the degree of association is very good.

Water is regarded as being associated in the liquid state by forming the associated species $(H_2O)_n$. This aggregate consists of a definite number of monomeric H_2O molecules, where n can be as high as 8 in ice. Water and its properties and structure are discussed further in Sec. 2.6.

Regarding the association phenomenon in binary mixtures, the reader is referred to Lambert (1953), Prausnitz (1969), Rowlinson (1959), Guggenheim (1952), Hildebrand and Scott (1950), Prigogine et al. (1957), and Davis (1968). A particularly interesting investigation was presented by Lown and Thirsk (1972) on the effect of the solution vapor pressure on the temperature dependence of the dissociation constant of acetic acid in water. A recent series of investigations on molecular association is that edited by Foster (1975). The thermodynamic aspect of the system has been studied by Sebastiani and Lacquaniti (1967).

To explain the solubility phenomenon between water and nonpolar substances, many investigators carried out extensive studies of whether water is associated or non-associated in the organic phase. Because of the low solubility of water in nonpolar solvents, the measurements required very sensitive techniques for the accurate determination of concentrations. Some of the more common techniques used are:

Solubility

Vapor pressure

Partial molal volume (for a good review on the partial molal volumes of organic
   compounds in water, see Edward et al., 1977)

Cryoscopy

Infrared spectroscopy

Dielectric constant

Radioactive traces

Nuclear magnetic resonance (NMR) method

During the early years of investigations, Peterson and Rodenbush (1928) carried out some cryoscopic work showing the linear function of the freezing-point depression upon the concentration of water in benzene. Further work by Greer (1930) also showed, through the measurement the vapor pressure of the solution and applying Henry's law, that water does not associate in benzene. An infrared study by Greinacher et al. (1955) revealed the absence of any band that might characterize the association of water molecules. See also Risbourg and Liebaert (1967) for further evidences.

A conflict arose after a work by Gordon et al. (1960) was published stating that water molecules are highly associated in benzene and toluene solvents at 60 and 70°C. They measured the solubility of water in these solvents and the density and viscosity of the solutions. They found that the apparent molal volume of water in both solvents decreases with an increase in the water concentration, which they believed was due to the increase in hydrogen bonding between water molecules.

However, contrary to the findings of Gordon et al. (1960), several newer investigations, in agreement with earlier ones, showed that water is not associated to any degree in nonpolar solvents (Högfeldt and Bolander, 1963; Christian, Affsprung, and Johnson, 1963; Christian, Affsprung, Johnson, et al., 1963; Christian et al., 1968; Ackermann, 1964; Yukhnevich and Karyakin, 1964; Johnson et al., 1965, 1966; Masterton and Gendrano, 1966; Müller and Simon, 1967; Masterton and Seiler, 1968; Seiler, 1968; Karyakin et al., 1970). Particularly interesting examinations were those of Masterton and Seiler (1968) and Seiler (1968), who used the same techniques as that used by Gordon et al. (1960); that is, they measured the densities and concentrations of water in the nonpolar solvents. The apparent molal volume of water ϕ_2 was calculated by

$$\phi_2 = \frac{1}{d_o} \left[M_2 - \frac{1000(d - d_o)}{c} \right]$$

where d_o = density of the organic solvent, g cc^{-1}
   d = density of the solution, g cc^{-1}
   M_2 = molecular weight of water, g g-mole^{-1}
   c = concentration of water in the organic solvent, g-mole liter^{-1}

The result indicated that the estimated association of water was less than 10% in carbon tetrachloride, 1,1-dichloroethane, and 1,1,1-trichloroethane and no more than 5% in benzene and 1,2-dichloroethane. A good review of the molecular complexes of water in organic solvents is that of Christian et al. (1968, 1970).

The polymerization of water in solvents has been interpreted as a monomer-trimer or probably monomer-tetramer equilibrium. The water aggregates in nonaqueous solvents with proton-accepting groups were studied by Johnson et al. (1967). Table 2.23 summarizes the literature data on the association of water molecules in organic solvents.

Water obeys Henry's law when it is dissolved in organic solvents in which its solubility is low (i.e., less than 0.05 M); see, for example, Caddock and Davies (1959) for the solubility of water in hydrocarbons. This is due to the low solubilities, which preclude the formation of significant concentrations of water aggregates. They form in significant concentrations in numerous solvents in which high water concentrations exists (Johnson et al., 1967). The experimental evidence indicates that if the mole fraction of water is greater than 0.25, there are high polymers present. Johnson et al. (1966, 1967) have proposed that the important polymeric species are cyclic aggregates, which is also supported by Lin et al. (1965), Jolicoeur and Cabana (1968), and Christian et al. (1970):

Table 2.23 Literature Data for the Association of Water in Organic Solvents

Solvent	Investigator(s)	Method of detection, conclusion, comments, etc.
C_6H_6	Peterson & Rodebush (1928)	Linear dependence of freezing-point depression on water concentration.
	Greer (1930	Solubility of water obeys Henry's law.
	Staveley et al. (1943)	Solubility of water obeys Henry's law.
	Greinacher et al. (1955)	Absence of shift in infrared stretching frequencies.
	Gordon et al. (1960)	Apparent molal volume of water decreased with increasing concentration.
	Högfeldt & Bolander (1963)	Linear relation between activity of water and concentration of water.
	Christian, Affsprung, and Johnson (1963)	Solubility of water obeys Henry's law.
	Christian, Affsprung, Johnson, et al. (1963)	Linear relation between activity of water and water concentration.
	Ackermann (1964)	IR absorption frequency of water does change with concentration of water.
	Johnson et al. (1966)	Linear relation between activity of water and concentration of water.
	Masterton & Gendrano (1966)	Linear relation between activity of water and concentration of water.
	Gregory et al. (1968)	Linear relation between dielectric constant and concentration of water.
	Masterton & Seiler (1968)	No decrease in apparent molal volume with increasing concentration of water.
	Roddy & Coleman (1968)	Solubility of water obeys Henry's law.
	Seiler (1968)	No decrease in apparent molal volume with increasing concentration of water.
	Kirchnerova & Cave (1976)	Solubility of water obeys Henry's law.
	Kirchnerova (1975)	Solubility of water obeys Henry's law.
$C_6H_5CH_3$	Peterson & Rodebush (1928)	Linear dependence of freezing-point depression on water concentration.
	Gordon et al. (1960)	Apparent molal volume of water decreased with increasing concentration.
	Ackerman (1964)	IR absorption frequency of water does change with concentration of water.

Table 2.23 (Continued)

$C_6H_5CH_3$	Johnson et al. (1966)	Linear relation between activity of water and concentration of water.
	Kirchnerova & Cave (1976)	Solubility of water obeys Henry's law.
	Kirchnerova (1975)	Solubility of water obeys Henry's law.
C_6H_{12}	Johnson et al. (1966)	Linear relation between activity of water and concentration of water.
	Roddy & Coleman (1968)	Solubility of water obeys Henry's law.
	Kirchnerova & Cave (1976)	Solubility of water obeys Henry's law.
	Kirchnerova (1975)	Solubility of water obeys Henry's law.
CCl_4	Greinacher et al. (1955)	Absence of shift in infrared stretching frequencies.
	Christian, Affsprung, and Johnson (1963)	IR absorption frequency of water does not change with the concentration of water.
	Johnson et al. (1965)	Solubility of water obeys Henry's law.
	Muller & Simon (1967)	Nuclear magnetic resonance study shows no chemical shift in the solution.
	Masterton & Seiler (1968)	No decrease in apparent molal volume with increasing concentration of water.
	Seiler (1968)	No decrease in apparent molal volume with increasing concentration of water.
	Magnusson (1970)	Absence of shift in infrared stretching frequencies.
	Kirchnerova (1975)	Solubility of water obeys Henry's law.
$CHCl_3$	Staveley et al. (1943)	Solubility of water obeys Henry's law.
	Masterton & Gendrano (1966)	Nonlinear relation between activity of water and concentration of water.
	Kirchnerova & Cave (1976)	Solubility of water disobeys Henry's law.
	Kirchnerova (1975)	Solubility of water disobeys Henry's law.
$C_2H_4Cl_2$ (1,2-)	Lin et al. (1965)	Proton chemical shift of water in magnetic resonance data supports the association.
	Christian et al. (1966)	Solubility of water disobeys Henry's law.
	Johnson et al. (1966)	Nonlinear relation between activity of water and concentration of water.
	Masterton & Gendrano (1966)	Nonlinear relation between activity of water and concentration of water.
	Jolicoeur & Cabana (1968)	Shoulder on the low frequency side of the OH stretching band in infrared spectra.
	Masterton & Seiler (1968)	No decrease in apparent molal volume with increasing concentration of water.

Table 2.23 (Continued)

$C_2H_4Cl_2$ (1,2-)	Seiler (1968)	No decrease in apparent molal volume with increasing concentration of water.
	Kirchnerova & Cave (1976)	Solubility of water disobeys Henry's law.
	Kirchnerova (1975)	Solubility of water disobeys Henry's law.
$C_2H_4Cl_2$ (1,1-)	Masterton & Seiler (1968)	No decrease in apparent molal volume with increasing concentration of water.
	Seiler (1968)	No decrease in apparent molal volume with increasing concentration of water.
CCl_3-CH_3	Masterton & Seiler (1968)	No decrease in apparent molal volume with increasing concentration of water.
	Seiler (1968)	No decrease in apparent molal volume with increasing concentration of water.
C_6H_5Cl	Kirchnerova & Cave (1976)	Solubility of water obeys Henry's law.
	Kirchnerova (1975)	Solubility of water obeys Henry's law.
$o-C_6H_4Cl_2$	Kirchnerova & Cave (1976)	Solubility of water obeys Henry's law.
	Kirchnerova (1975)	Solubility of water obeys Henry's law.
m-Xylene	Ackermann (1964)	IR absorption frequency of water does change with the concentration of water.
p-Xylene	Kirchnerova & Cave (1976)	Solubility of water obeys Henry's law.
	Kirchnerova (1975)	Solubility of water obeys Henry's law.
$n-C_{16}H_{34}$	Kirchnerova & Cave (1976)	Solubility of water obeys Henry's law.
	Kirchnerova (1975)	Solubility of water obeys Henry's law.
$n-C_6H_{14}$	Roddy & Coleman (1968)	Solubility of water obeys Henry's law.
$C_6H_5NO_2$	Staveley et al. (1943)	Solubility of water obeys Henry's law.
$C_6H_5NH_2$	Staveley et al. (1943)	Solubility of water obeys Henry's law.
Dimethyl-aniline	Staveley et al. (1943)	Solubility of water obeys Henry's law.
C_6H_5Br	Staveley et al. (1943)	Solubility of water obeys Henry's law.
Anisole	Staveley et al. (1943)	Solubility of water obeys Henry's law.

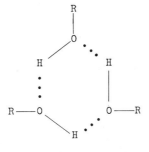

Chain polymers of dimer or trimer have not been found in the solutions. Only monomers and cyclic trimers exist in chlorinated hydrocarbons at 10 and 25°C. The approximate enthalpy of association was -16 kcal g-mole^{-1}, corresponding to about 5 kcal g-mole^{-1} to an individual hydrogen bond in the aggregate. It might seems odd that the association takes place in partially chlorinated hydrocarbons but not in solvents with low dielectric constants. This is due to the lower solubility of the latter solvents. In a plot of the water concentration versus the water activity (see Fig. 2.26), the curves are straight, (see also Table 2.24) when there is no associa- tion, in contrast to curves showing detectable deviation from linearity, which is attributed to the association of water molecules. The activity data for water dissolved in various hydrocarbons and chlorinated hydrocarbons were plotted versus the total concentration of water in a log-log plot (Christian et al., 1970). Although the straight lines imply that water is essentially monomeric in cyclohexane, carbon tetrachloride, biphenylmethane, toluene, benzene, 1,2,3-trichloropropane, 1,2-dichloro- ethane, and 1,1,2,2-tetrachloroethane, there is evidence for positive curvature in the plot for aromatic solvents near saturation. However, the association is merely a few percent of water. The partially chlorinated hydrocarbons are slightly polar, and consequently water is somewhat polymerized in them, particularly in chloroform, 1,2- dichloroethane, 1,1,2,2-tetrachloroethane, and 1,2,3-trichloropropane.

An interesting comparison regarding the activity of carbon tetrachloride in water at various concentrations and 20°C is presented in Table 2.25 and Fig. 2.27 (Platford, 1977). A definite trend is apparent between the activity a_c^w and the activity coefficient γ_c^w: γ_c^w increases as a_c^w decreases. However, the experimental error is about 10 to 15%, which makes this observation questionable. The dimeriza- tion of benzene between 5 and 30°C was studied by Green and Frank (1979) and they concluded that there is no appreciable dimerization in the temperature range examined.

The relationship between the total concentration of water (C_w, mole liter^{-1}) and its activity (a_w, mole fraction) has also been studied by Masterton and Gendrano (1966) and Sahar et al. (1975). They expressed the equilibrium between the monomer and the polymer by

$$n\ H_2O \Longleftrightarrow (H_2O)_n$$

with the equilibrium constant

$$K = \frac{C_n}{C_1^n}$$

where C_1 and C_n represent the concentration (mole liter^{-1}) of monomer and polymer, respectively. The total concentration (C_w) is

$$C_w = C_1 + n\ C_n$$

which can be written

$$C_w = C_1 + n\ K\ C_1^n$$

Table 2.24 Solubilities and Association Constants of Water in Organic Solvents

Solvent	Temperature ($^{\circ}C$)	Solubility of H_2O (mole liter^{-1})	Concentration of monomeric H_2O (mole liter^{-1})	Association constants (mole l^{-1}) K_3	Association constants (mole l^{-1}) K_4
1,1,2,2-Tetrachloroethane	25	0.1010	0.0925	3.6	24.5
1,2-Dichloroethane	25	0.1262	0.1083	4.6	24.5
1,2-Dichloroethane	10	0.0812	0.0650	20.0	172.0
Benzene	25	0.0349	–	–	–
Toluene	25	0.0274	–	–	–
Carbon tetrachloride	25	0.0087	–	–	–
Cyclohexane	25	0.0024	–	–	–

Source: Johnson et al., 1966.

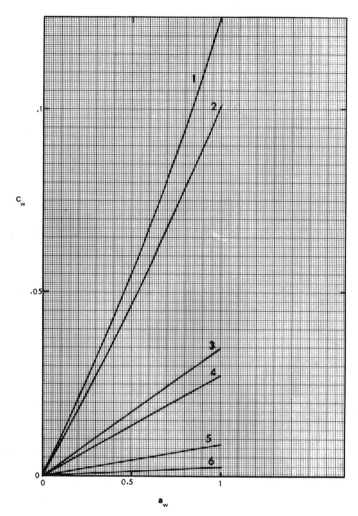

Fig. 2.26 Dependence of formal water concentration on water activity at 25°C. C_w = concentration of water (g-mole liter^{-1}); a_w = activity of water (mole fraction). 1, CH_2Cl-CH_2Cl; 2, $CHCl_2$-$CHCl_2$; 3, C_6H_6; 4, C_6H_5-CH_3; 5, CCl_4; 6, C_6H_{12}. (From Johnson et al., 1966.)

If we assume that water vapor is monomeric and in solutions obeys Henry's law,

$$C_1 = H\,a_w$$

where H is Henry's law constant, the final expression becomes

$$C_w = H\,a_w + n\,K\,H^n\,a_w^n$$

Table 2.25 Activity and Activity Coefficient of CCl_4 in Water at 20°C

Concentration of CCl_4, x_c^w (mole fraction)	Activity of CCl_4, a_c^w	Activity coefficient of CCl_4, $\gamma_c^w = a_c^w/x_c^w$
0.25×10^{-4}	0.332	13,280
0.31×10^{-4}	0.405	13,064
0.48×10^{-4}	0.622	12,958
0.56×10^{-4}	0.630	11,250
0.68×10^{-4}	0.787	11,573
0.90×10^{-4}	1.00	11,111

Source: Platford, 1977.

By plotting C_w versus a_w, it is apparent from this expression that for a monomer in solution, K is zero and the concentration of water C_w is a linear function of its activity a_w. In the case of a polymeric species, the association constant can be calculated from the coefficients of the equation by fitting C_w versus a_w. The least-squares fitting of the experimental data gave the following equations at 25°C:

Monomeric water molecules

Benzene, $C_w = 0.0345\ a_w$

Carbon tetrachloride, $C_w = 0.0087\ a_w$

Cyclohexane, $C_w = 0.0024\ a_w$

Toluene, $C_w = 0.0274\ a_w$

Trimer water molecules

Chloroform, $C_w = 0.0691\ a_w + 0.0045\ a_w^2$

1,2-Dichloroethane, $C_w = 0.1131\ a_w + 0.0138\ a_w^2$

1,1,2,2-Tetrachloroethane, $C_w = 0.074144\ a_w + 0.05248\ a_w^2$

The calculated equilibrium constant (K, liters $mole^{-1}$) for dimer formation (dimerization) at 25°C are obtained using the equation

$$C_w = H\ a_w + n\ K\ H^n\ a_w^n$$

The numerical values are as follows:

Chloroform, $K = 0.47$ liter $mole^{-1}$

1,2-Dichloroethane, $K = 0.54$ liter $mole^{-1}$

1,1,2,2-Tetrachloroethane, $K = 4.77$ liter $mole^{-1}$

The work by Masterton and Gendrano (1966) did not result in concrete suggestions regarding the effect of temperature on the equilibrium constant for association of

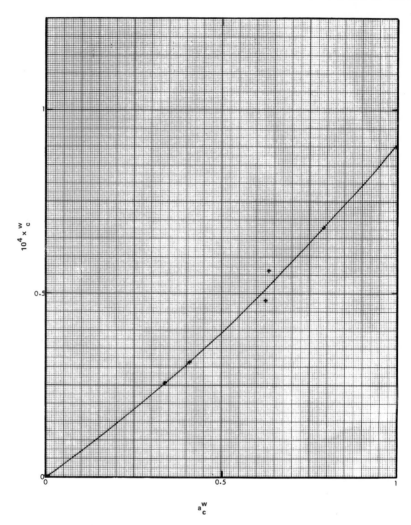

Fig. 2.27 Activity of carbon tetrachloride (a_c^w) versus its concentration (x_c^w = mole fraction) in water at 20°C. (From Platford, 1977.)

water in organic solvents. However, they suggested that a change of 20% in the heat of dilution at 5°C could cause an increase in K to about 0.7 liter mole^{-1}. A larger increase in the temperature dependence of the association has been reported by Christian (1963) and Johnson et al. (1966). The presence of water molecules associated in chloroform solution is only about 6% at 25°C.

Magnusson (1970) examined the process of polymerizing H_2O molecules using infra-red spectrometry at 25°C. He found that 3.6% of the total water molecules in saturated carbon tetrachloride were present as dimer. This has been detected by double-beam cancellation of the monomer absorbance in the 2700 nm stretching region. The calculated dimerization content was 2.2 liters mole^{-1} at 25°C. He observed two OH stretching frequencies at 3693 and 3552 cm^{-1}, confirming that the structure is cyclic with two bent H bonds.

If a solvent has two or more basic sites, it dissolves water probably as monomer and dimer units, whereas in solvents having only one basic site, water dissolves as monomer-trimer or monomer-tetramer units.

The existence of water trimers in organic solvents has also been supported by Glasoe and Schultz (1972). Investigations by infrared spectrometry (Jolicoeur and Cabana, 1968) and by proton magnetic resonance studies (Ödberg et al., 1972; Lin et al., 1965) also provided evidence for the presence of water trimers. The trimers were identified with absorption bands at 3510 and 3350 cm^{-1} in the infrared study of self-association. It has also been proposed that the complex formation between water and organic solvents is due to the formation of a more "icelike" structure (i.e., a more ordered and more highly hydrogen-bonded cage of water around the dissolved solute molecules and/or a complex formation of the type involving the labile Π electrons of the aromatic nucleus) (Bohon and Clausse, 1951). The solvent is capable of forming an associated cage around a solute molecule, thereby holding it in solution. This has been interpreted from the heat and entropy of solution calculated. The high solubility values of water in some organic solvents were attributed by Hildebrand and Scott (1950) to self-association. The apparent and partial molal volumes of water in 1,2-dichloroethane, benzene, 1,1,1-trichloroethane, and carbon tetrachloride have been determined by Seiler (1968) and Masterton and Seiler (1968) (see Table 2.26). By examining the density-concentration data for solutions of water in organic solvents, Masterton and Seiler concluded that there is no evidence for polymerization in the four solutions examined, despite the finds of Gordon et al. (1960). This was supported by the linear dependence of density upon concentration. Furthermore, the fact that the molal volume of water in the four solvents mentioned above is independent of concentration precludes extensive polimerization of water molecules. The fraction of all possible hydrogen bonds that are formed by water molecoles ($f_{h.b.}$) is expressed by

$$\frac{f_{h.b.}}{(1 - f_{h.b.})^2} = \frac{3\,C}{4\,C_s}$$

where C is the concentration of water (mole liter^{-1}) and C_s the concentration of water in the saturated solution (mole liter^{-1}). In this formula, the parameter $f_{h.b.}$ varies from zero in the infinitely diluted solution (no association) to $\frac{1}{8}$ in the saturated solution (maximum association). The experimental results showed that the apparent molal volume of water in the four solvents examined indicates no evidence of decreasing with an increase in water concentration. This is in disagreement with the

Table 2.26 Apparent and Partial Molal Volume of Water in Solvents at 25°C

Solvent	Liquid density ($g\ cc^{-1}$)	Solubility parameter ($cal\ cc^{-1})^{\frac{1}{2}}$	Liquid compressibility (atm^{-1})	Coefficient of thermal expansion (K^{-1})	Partial (apparent) molar volume ($cc\ g\text{-}mole^{-1}$)	Change of partial molal volume of solute ($cc\ g\text{-}mole^{-1}$)	
						Interaction	Cavity formation
$CH_2Cl\text{-}CH_2Cl$	1.246	9.8	0.000083	0.00118	20.1	-14.0	32.5
C_6H_6	0.8736	9.2	0.000095	0.00124	22.1	-12.0	32.0
$CCl_3\text{-}CH_3$	1.329	8.5	0.000113	0.00125	22.3	-16.0	35.1
CCl_4	1.585	8.6	0.000110	0.00123	31.6	-5.0	33.9
H_2O	0.99705	24.0	0.0000465	-	-	-	-

Source: Masterton & Seiler, 1968.

model of Gordon et al. (1960) for the polymerization process:

$$\Delta\phi_2 \quad = \quad -4.82 \ p$$

where $\Delta\phi_2$ is the difference between the apparent molar volume ϕ_2 and its limiting value ϕ_2^0 in the infinitely dilute solution. Their model was based upon specific volume, relative viscosity, and heat of solution measurements of water solutions in benzene and toluene, so that the polymerization of water in benzene and toluene at 25°C reported by Gordon et al. (1960) disagrees with studies based upon cryoscopic and spectroscopic measurements, infrared spectroscopy, and so on. Furthermore, the findings of Masterton and Seiler (1968) for the monomeric water molecules in 1,2-dichloroethane disagree with those of other investigations (e.g., Jolicoeur and Cabana, 1968) showing cyclic trimer water molecules. Further discrepancies between Gordon's results and those of other investigators indicates the differing views on the self-association of water in organic solvents (Masterton and Seiler, 1968).

Solvent-extraction data (Högfeldt and Bolander, 1963) of aqueous solutions of acids by benzene also showed that water is monomeric in the organic phase. The spectroscopic studies of solutions of water in nonpolar solvents showed no evidence of extensive polymerization (Greinacher et al., 1955; Gentric et al., 1970).

Kirchnerova (1975) and Kirchnerova and Cave (1976) investigated the solubility of water in low-dielectric solvents and proposed a new formation constant for 1:1 water-solvent complexes in carbon tetrachloride, benzene, toluene, and p-xylene. The formation constant K_{ws} of the 1:1 complex is expressed when the activity coefficients are unity:

$$K_{ws} \quad = \quad \frac{x_{ws}}{\bar{x}_w \ x_s}$$

where $\bar{x}_w$ is the mole fraction of free water in the saturated solution, x_s the mole fraction of the solvent, and x_{ws} the concentration of 1:1 water-solvent complex expressed as a mole fraction. The estimated values for K_{ws} of the 1:1 complex are presented in Table 2.27, together with other interaction parameters at 25°C. The free energies, enthalpies, and entropies of interaction of methylene groups in water at 25°C were also studied by Blackburn et al. (1980). The investigators proposed a group additivity approach for the calculations.

Kirchnerova and Cave (1976) suggested that their formation constants and 1:1 complexes could be used to predict the solubility of water in nonpolar solvents. In the calculation, a modified version of the Scatchard-Hildebrand equation was used which includes the dispersion, dipole-dipole, and dipole/induced dipole interactions between water and solvent:

$$- R \ T \ (\log_e x_w + \log_e \frac{V_w}{V_s} + 1 - \frac{V_w}{V_s}) \ = \ V_w \ (\delta_w^2 + \delta_s^2 - 2 \ C_{ws})$$

Table 2.27 Interaction Parameters and Formation Constants for Water-Nonpolar Solvent Mixtures at 25°C

Solvent	δ_s^{25} (cal cc^{-1})$^{\frac{1}{2}}$	C_{ws}	C_{ws}/δ_s	$19.86\delta_s =$ $dC_{ws} + PC_{ws}$	$cC_{ws} =$ $C_{ws} - 19.86\delta_s$	Concentration x_w^{25} (mole frac.)	Formation constant K_{ws}
Hexadecane	8.01	160.3	20.0	158.6	1.7	0.00085	-
Cyclohexane	8.19	160.3	19.6	162.2	-1.9	0.00032	-
Carbon tetrachloride	8.58	181.1	21.1	169.9	11.2	0.00085	1.0
p-Xylene	8.77	198.3	22.6	173.6	24.7	0.00269	3.5
Toluene	8.91	202.4	22.7	176.4	26.0	0.00283	3.9
Benzene	9.16	208.7	22.8	181.4	27.3	0.00312	4.2
Chlorobenzene	9.68	211.0	21.8	191.7	19.3	0.00297	-
o-Dichlorobenzene	10.05	211.8	21.1	199.0	12.8	0.00271	-
Chloroform	9.49	223.7	23.6	187.9	35.8	0.00596	-
1,2-Dichloroethane	9.78	235.4	24.1	193.6	41.8	0.01010	-
1,1,2,2-Tetrachloroethane	9.80	232.7	23.7	194.0	38.7	0.01070	-

Source: Kirchnerova & Cave, 1976.

where C_{ws} is the interaction parameter due to the contributions arising from dispersion, polar, and hydrogen-bond interactions. Some values for C_{ws} calculated from the preceding expression are given in Table 2.27. Regarding the adsorption of the 23.43 $(\text{cal cc}^{-1})^{\frac{1}{2}}$ value for the total solubility parameter for water at 25°C, great caution is proposed by Kirchnerova and Cave (1976), particularly for calculation of the solubility of inert solutes in water. This is due to the process, which would not be expected to reduce very much the n-mer distribution in the water. The recommended solubility parameter for water δ_w sould be calculated from the hypothetical model proposed by Scheraga and Owicki (Owicki et al., 1975b), when ΔE_1^v equals to 5.36 kcal g-mole^{-1} at 25°C.

The dissociation of CH_3F molecule in water was studied by Cremaschi et al. (1977) using a molecular model for ion pairs in solution. The existence of three different ion pairs was established and their geometric and electronic structures were discussed. The object was to understand the mechanism of nucleophilic substitution and solvolytic reactions in polar solvents. The MO theory was applied and the calculations were performed using the CNDO/2 method. The existence of ion pairs had been studied previously by Scott (1970), Sneer (1973), Glew and Moelwyn-Hughes (1953), Cremaschi et al. (1972), and Kistenmacher et al. (1974).

Because of the small molecular diameter of CH_3F molecule, a solvent cage was built with a small number of water molecules. The system calculated consisted of CH_3F surrounded by $11 \cdot H_2O$ molecules. Molecular packing of $CH_3F \cdot 11\ H_2O$ for the energy minimum at 1.388 Å is proposed by Cremaschi et al. (1977).

By increasing the C-F distance, several energy minimums have been achieved (1.388, 3.480, and 5.463 Å), and consequently the negative charge on fluorine is increased from -0.185 to -0.531 via -0.255 and -0.516. The plot of the total energy versus C-F distance for $CH_3F \cdot 11\ H_2O$ showed three minimums. However, further work is needed, particularly for the examination of the bond-making and bond-breaking possibilities between nonpolar and water molecules.

The CNDO/2 calculation was carried out for the $CF_4 - H_2O$ system by Schönfeld and Seibt (1976). The stability of the four aggregates was expressed by the bond energy (ΔE, kcal) of $F \cdots HO$:

	$CF_4 \cdot 2H_2O$	$CF_4 \cdot 3H_2O$	$CF_4 \cdot 4H_2O$	$CF_4 \cdot 5H_2O$
ΔE (kcal)	-3.3	-6.5	-10.2	-12.8

Several geometrical proposals have been examined for calculation by CNDO/2:

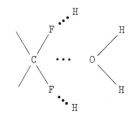

The stability of the $CF_4 \cdot nH_2O$ aggregate was illustrated by the following structures:

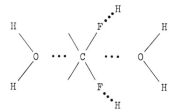

and

The number of H_2O molecules can vary from 1 to 4 (n = 1 to 4). No one model is superior to the others.

Further study of the association of water molecules in three binary mixtures -- benzene - carbon tetrachloride, benzene - cyclohexane, and carbon tetrachloride - cyclohexane -- at 25°C have been carried out by Goldman and Krishnan (1976).

2.4 Hydrogen Bonding

The molecular association of carboxylic acids, alcohlos, H_2O, HF, HCN, and so on, as discussed in the preceding section, is a direct consequence of the formation of strong intermolecular hydrogen bonds. The strength of the hydrogen bonding is about 6 kcal g-mole^{-1}, between the strong interionic and the weak van der Waals forces. The hydrogen bond is not a chemical bond, only a cohesive force. It may be regarded as a special case of donor-acceptor interaction. The energy of a hydrogen bond has to be greater than dipolar or London dispersion force energies.

The hydrogen bond was discovered by Maurice L. Huggins in 1919. An excellent summary of the idea was recently reported by the author (Huggins, 1980). For the classification of the hydrogen bonds, see Emsley (1980).

The hydrogen bonding originates from two sources. First, the donor hydrogen atom has only one orbital electron, and second, the other element or acceptor is a strongly electronegative element, such as fluorine, oxygen, or nitrogen. A list of elements is given in Table 2.18. Hydrogen loses its lone-pair electron and the proton becomes unscreened. Consequently, the polarization and the dipole moment of the bonds (e.g., H-O, H-N, and H-F) are large. Furthermore, the attraction between the unshielded proton and the electronegative atom is very strong. Strong combinations are observed from -OH and =NH species (in H_2O, NH_3, carboxylic acids, and alcohols), whereas the -SH group tends to give weaker hydrogen bonds. Some groups with hydrogen-donating tendency can form hydrogen bonds with -OH and =NH (e.g., acetone and water mixed form hydrogen bonds with one another).

There are several theoretical treatments of the properties of hydrogen-bonded systems, such as electrostatic, valence bond, and molecular orbital (Lin, 1970). The electrostatic treatment of hydrogen bonding was explained either by the dipole inter-actions or by the interaction of point-charge models. The electrostatic description of the hydrogen bond based on point-charge models proved to be more successful. The interaction energy is provided by the interaction of the lone-pair dipole with the proton. Tetrahedral hybrid orbitals can be used to represent the lone pairs. It can be concluded that the NH_3 lone pair will form a stronger hydrogen bond than will the HF lone pair with the H_2O lone-pair intermediate.

The water molecule is represented by tetrahedral structures. Each oxygen atom is surrounded tetrahedrally by four hydrogen atoms. Two hydrogen atoms are covalently linked and they lie close to the oxygen atom at a distance of 1 Å , whereas the other two hydrogen bonds are at a distance of 1.76 Å from the oxygen atom. The interaction energy of water molecules based upon the point-charge model has three orientations:

1. The H-O group of one molecule is directed along the lone-pair orbital direction of the second molecule.

   2. The H-O group of one molecule is directed along the molecular axis of the
      second molecule.

   3. The H-O group of one molecule of the H_2O dimer has a side-by-side
      orientation.

Hydrogen-bond energy calculations based on the point-charge model give the best
answer regarding the order of the energy of the hydrogen bond.

    For the valence-bond treatment of the hydrogen-bonded system, there are three
assumptions:

   1. Pure covalent bond, no charge transfer

   2. Pure ionic bond, no charge transfer

   3. Charge transfer, ions H-O bonding

In case 1, two electrons are evenly distributed between the two atoms. In case 2,
the pair of electrons are on the first and second oxygen atoms, and in case 3, the
second pair of electrons are covalently paired and form a H-O bond. Hydrogen bonding
originates essentially from the electrostatic nature of the compound. When the distance between oxygen atoms is shortened, the energy and O-H distance increase.
When the distance between H and O decreases, the covalent character increases.
The hybrid atomic orbit does not influence the order of energy. It is clear that the
covalent contribution is significant for hydrogen bonds.

    Molecular orbital treatment of hydrogen bonding provides the most satisfactory
approach (Kollman and Allen, 1972). This consists of three parts:

   (1) The 2n Π-electrons contributed by both molecular fragments and the mol-
       ecular orbitals, which are linear contribution of k Π-atomic orbitals

   (2) The four σ electrons from the X-H bond and the lone-pair electrons,
       which are described by the hybridized orbitals

   (3) The nuclei of the atoms and the remaining electrons in inner orbitals.

We have to consider two cases: when the hydrogen bonding is weak and when it is
strong. In the first case, both the deformation of the orbitals and the configuration interactions have to be considered. For strong-hydrogen-bonding systems, the
calculation of the core energy concept is preferable.

    The configutation interaction treatment of the donor-acceptor interaction can
also be used for molecular orbital treatment of hydrogen bonding (Krasnec, 1967).
In this concept, the wave function of the system is approximated by the configuration
interaction. However, as Kollman and Allen (1972) have pointed out, the wave-function
theory does not improve the prediction of the molecular properties. The example is
shown for the calculation of the dipole moment of water using three different bases:
the Slater orbital, the near-Hartree-Fock, and the best extended.

    Substances that form strong hydrogen bondings have the tendency to associate
with other molecules (Zimmermann, 1970; Emsley, 1980). These associated molecules

are held together by hydrogen bonding. As a consequence of the strong hydrogen
bonding, the compounds have unexpectedly high boiling and melting points. The
dielectric constants of these chemicals are high and they are good ionizing solvents.
The hydrogen bonds hold the molecules together, and a considerable energy (in the
form of heat) is required to separate the molecules. However, Kollman and Allen
(1972) have remarked that using the reported calculations on the water dimer in the
literature, the dimerization energy varies between 5 and 12 kcal g-mole^{-1}. Similarly,
the angle of the dimer has been reported to be between 0 and 57° and the O-O
distance between 2.53 and 3.00 Å. Further details regarding the dimer and polymer
molecules of water are given in Sec. 2.6.

The different types of hydrogen bonds were studied by Singh et al. (1966). They
presented an energy scale representing the various type of bonding with numerical
values, expressed in kcal g-mole^{-1}. Some of the energy values are shown in Fig. 2.28.

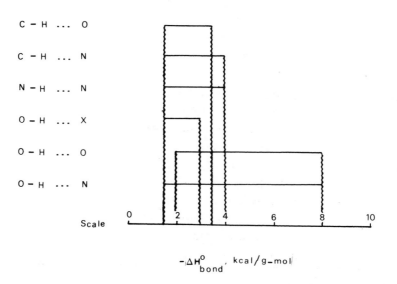

Fig. 2.28   Various bond energies between hydrogen and other atoms.

Table 2.28 Major Works on Hydrogen Bonding

Type of work[a]	Reference
A	Bonner & Choi (1974)
B	Briegleb (1961)
B	Coetzee & Ritchie (1969-1976: Vol. 1)
A	Copley et al. (1938a, b)
B	Covington & Jones (1968)
A	Craver (1970)
R	Davis (1968)
A	Di Paolo & Sandorfy (1974a, b)
R	Eberson (1969)
A	Epley & Drago (1967)
B	Hadzi & Thompson (1959)
B	Hamilton & Ibers (1968)
R	Huggins (1971, 1980)
A	Karyakin & Muradov (1971)
R	Kollman & Allen (1972)
A	Kroon & Kanters (1974)
A	Lieberman (1962)
R	Lin (1970)
R	Luck (1976)
A	Manja et al. (1974)
R	Murthy & Rao (1970)
B	Pimetel & McClellan (1960)
R	Schuster (1973)
B	Schuster et al. (1976)
B	Sokolov & Tschulanovskii (1964)
A	Teresawa et al. (1975)
B	Vinogradov & Linnell (1971)
R	Zimmermann (1970)
B	Zundel (1972)

[a] A = article; B = book; R = review.

There are numerous major or significant works, including books, reports, reviews, and articles on hydrogen bonding, dealing with both theoretical and experimental aspects. A list of references is presented in Table 2.28 in alphabetical order of investigators.

The presence of hydrogen bonding in the chemical substances influence their physical properties (Davies, 1946). Intramolecular hydrogen bonds affect the elec-**tronic structure of the molecule, whereas intermolecular** hydrogen bonding changes the mass, shape, and electronic structure of the molecules in question. In the following discussion only the intermolecular hydrogen bonds will be treated. Regarding intra-molecular hydrogen bonds, the reader is referred to the following comprehensive reviews: Briegleb (1961), Coetzee and Ritchie (1969-1976: Vol. 1), Covington and Jones (1968), Davis (1968), Eberson (1969), Franks (1972-1975: Vol. 1), Hadzi and Thompson (1959), Hamilton and Ibers (1968), Joesten and Schaad (1974), Kollman and Allen (1972), Lin (1970), Luck (1976), Pimentel and McClellan (1960), Schäfer (1976), Schuster et al. (1976), Sokolov and Tschulanovskii (1964), Vinogradov and Linnell (1971), Zimmermann (1970), and Zundel (1972).

The general tendency of melting and boiling points in a homologous series is to increase with rising molecular weight. The hydrogen-bonding effect is clearly indic-ated by the difference between the melting and boiling points of non-hydrogen-bonded substances compared with hydrogen-bonded compounds of equal molecular weight. Table 2.29 compares the melting and boiling points of some of hydrogen-bonded and non-hydrogen-bonded chemicals. The great difference in melting and boiling points is clearly indicated. The kinetic energy required to melt a substance or to bring a liquid up to its normal boiling point is much greater for hydrogen-bonded compounds. This is due to intermolecular hydrogen bonds, which hold the molecules together.

Table 2.29 Comparison of Hydrogen- and Non-Hydrogen-Bond Effects

Substance	Molecular weight	Melting point (^{o}C)	Boiling point (^{o}C)
H_2O	18.015	0.0	100.0
N_2O	44.01	-90.8	-88.5
HF	20.01	-83.1	19.54
NO	30.01	-163.6	-151.8
CH_3COOH	60.05	16.604	117.9
$CH_3CH(CH_3)CH_3$	58.12	-138.3	-0.50
CH_3OH	32.04	-93.9	64.96
CH_3CH_3	30.07	-183.3	-88.63

Cryoscopic measurements have often been used to determine the molecular weight of substances and to establish some information on hydrogen-bond formation or association. The freezing point of a solvent can be lowered by the addition of a nonvolatile solid, assuming that there will be no solute-solvent interaction. The concentration is important in cryoscopic measurements (Anantaraman and Goldman, 1980). Several substances that have been used in the experiment (e.g., alcohols, phenols, etc.) proved the formation of aggregates. The degree of association is particularly high when the solvent is saturated and itself difficultly polarizable. The molecular weight of hydrogen-bonded substances increases as the concentration goes up, whereas constant molecular weight is found in nonassociated compounds. In addition to vapor-pressure measurement, the equilibrium constant of association may be also obtained from cryoscopic measurements. The vapor pressures of hydrogen-bonded substances are lower than those of non-hydrogen-bonded molecules. This is explained by the kinetic energy necessary to overcome the bonding energy of the molecules; the temperature has to be increased to reach the boiling points of normal compounds, such as the halogenated paraffin series, where no hydrogen bonding is present.

The nonideality of a solution is shown by a deviation from Raoult's law. The deviation might be positive or negative, depending on the intermolecular interactions. The negative deviation from Raoult's law and maximum azeotrope is observed in solutions when the components in pure state were non-hydrogen-bonding compounds. If the components of a binary mixture form hydrogen bonds in the pure state, the mixture will probably form a minimum azeotrope or no azeotrope at all. From measurements of vapor-pressure isotherms in solution, activity coefficients were calculated that provide information on hydrogen bonding (Wolf, 1976).

As a consequence of the short interatomic distances in hydrogen-bonded substances, their molar volume is lower compared to normal substances without hydrogen bonding (Batsanov and Pakhomov, 1956); that is hydrogen bonding increases the densities. Bondi (1964) calculated the volume change due to the formation of a hydrogen bond and found it to be -1.08 ml g-mole^{-1} per bond. During the study of hydrogen bonding, it is common to measure the density of the compound. However, the density and molar volume are influenced by structural factors other than the hydrogen bonding (Millero, 1971). In solution the hydrogen bond usually increases the density compared to the arithmetic average obtained from pure components. This effect naturally depends upon the degree of association; therefore, measurement of the solution density gives some indication of the possible association. Further effects have been observed in heat **of mixing, contraction, and thermal expansion in aqueous solutions.**

Hydrogen-bond formation and degree of association are conveniently studied through equations of state (e.g., the virial equation of state). The virial equation indicates the deviation of the gas studied from ideal gas behavior. The most detailed investigation has been carried out on the second virial coefficient, to indicate the presence of hydrogen bonds in materials. In general, the second virial coefficient is a negative value, expressing the molecular attraction between molecules in real gases compared to ideal gases (Getzene, 1976). Comparing the second virial coefficient of a substance obtained from experimental work with that calculated from the equation of state will indicate the order of association in the material (Lambert, 1953):

$$B_{obs} - B_{calc} = -\frac{R\,T}{K_p}\,x^2$$

where B_{obs} = observed second virial coefficient
B_{calc} = calculated second virial coefficient
K_p = mass-action dissociation constant
x = mole fraction of the polar component

The compression of a hydrogen-bonded liquid is more difficult than that of a normal liquid. This is due to the fact that the interatomic distances are much shorter in the former than in the latter molecules (Yokozeki and Bauer, 1975). The adiabatic compressibility of a liquid is a function of its density and acoustic velocity. Both of these variables are affected by the hydrogen bonds of the molecule. Regarding the compressibility of mixtures, see Parshad (1942) and Jacobson (1950).

The viscosities of the hydrogen-bonded compounds are high compared with those of normal substances without hydrogen bonds (Hildebrand, 1977b). This observation is in line with what has been said in general on the effect of hydrogen bonding on physical properties: the hydrogen bond reduces the distance between atoms, the liquid becomes more dense, and its fluidity becomes less or the viscosity increases. This has been observed when associated substances are compared with nonassociated molecules (Kendall, 1944; Ewell and Eyring, 1937; Thomas, 1948a). By comparing the molecular weight dependence of a nonassociated series with that of hydrogen-bonded ones (e.g., n-paraffins compared with n-alcohols), it will be clear that the viscosity of the latter increases more steeply. The number of hydrogen bonds or the degree of association increases the viscosity values significantly (Thomas, 1948a, b, 1960, 1963, 1966; Thomas and Meatyard, 1963). In general, the viscosity measurement is not difficult; however, the interpretation of the viscosity value for hydrogen bonding or association is very obscure because of the complications due to the solute and solvent interaction (Staveley and Taylor, 1956; Vinogradov and Linnell, 1971).

The effect of association upon the surface and interface tension of the substances plays a large roll, particularly in the study of detergents. In solutions that show association between the components, it has been observed that the concentration dependence of the surface tension produces a maximum in contrary to the mimimum observed for nonassociated mixtures. The hydrogen bond influences the stability between two layers. Therefore, both surface and interfacial tensions might be used to study the hydrogen-bond formation in single and multicomponent solutions.

The presence of hydrogen bonds in the molecule lowers the parachor values. Parachor is additive and related to the surface tension and the molar volume. The theoretical explanation of parachor is given by, for example, Burshtein (1972). The lowering effect of hydrogen bonding upon parachor is on the order of 10.

The thermal conductivities of associated substances are high compared to those of nonhydrogen bonded compounds (Lambert, 1953). On the principle of the kinetic theory of gases, there is no expectation of abnormal behavior for real gases. However, in the case of dimerization, trimerization, and polymerization, the observed thermal conductivity of these aggregated molecules in the vapor phase showed greatly enchanced values. In general, hydrogen-bonded compounds conduct heat better than do nonassociated substances (Palmer, 1948). Palmer's explanation of the higher conductivity is based upon two factors: the orientation of the molecules and the type of the mechanism for heat conductivity (Filippov, 1954). The experimental values show that the thermal conductivity of hydrogen-bonded vapors increases relatively rapidly with increased pressure or decreased temperature. As with thermal conductivity, the acoustic and electrical conductivities are higher for hydrogen-bonded substances. The electrical conductivity of refrigerants has been reported by Döring (1977) and Eiseman (1955).

Hydrogen bonding also influences ionization constants, particularly for carboxylic acids. Correlation of ionization constants with dimerization constants has been studied, but clear evidence is still lacking. Some investigators tried to explain the relationship on the basis of intramolecular hydrogen bonding, but further work is required for a satisfactory conclusion.

The refractive index as well as the dielectric constant (permittivity) and dipole moment are influenced by the hydrogen bonds present in the molecules. These quantities are interrelated by several expressions, depending upon the material and condition in question. Hydrogen-bond formation in the aggregate molecule increases its polarity, causing greater dielectric constant and dipole moment values. In general, molecules with intermolecular hydrogen bonding have higher permittivity values than do intramolecularly hydrogen bonded substances. Measurement of the dielectric constant in solutions is the usual method used to gain information regarding the presence of hydrogen bonds. However, it has been observed that an increase in the dielectric constants of simple compounds will follow their dipole moment data (Pimentel and McClellan, 1960). Whereas the dielectric constant of intermolecularly hydrogen bonded compounds is high, for intramolecularly hydrogen bonded molecules it is low. The dipole moment values are affected by the solvent used for the measurement (Müller, 1933, 1934, 1937). This effect has been summarized by Pimentel and McClellan (1960). However, in practice, the dipole moments are determined in solutions of inert solvent, which themselves do not reflect any charges due to hydrogen bonding.

As noted above, in hydrogen-bonded systems additional kinetic energy is required to overcome the hydrogen bonding or to break them. This has been observed in the case of boiling point, vapor pressure, latent heat of vaporization, and so on (Bondi and Simkin, 1957). The effect is indicated by Trouton's constant, which gives higher values for hydrogen-bonded compounds.

Madgin and Biscoe (1927) carried out a detailed investigation on the heat of mixing of a large number of solutions. Based mostly on this work, it has been established that the heat of mixing is a larger negative number for associating systems than for nonassociating systems. Whether the heat of mixing is exothermic or en-dothermic, it depends upon hydrogen-bond formation or breaking during mixing. In the case of hydrogen-bond formation, the heat of mixing is exothermic in most cases, whereas during dissociation processes, the heat of mixing becomes **endothermic**. The heat of mixing has been investigated and reviewed by Wolf (1937), McGlashan (1962), Minto (1975), and Lieberman and Wilhelm (1976). Naturally, hydrogen-bond formation depends upon the properties of the components -- whether they are proton-donor or -acceptor molecules. In other words, the decisive factor is how many hydrogen bonds have been completed or destroyed upon mixing. However, in addition to the heat of mixing in the system, it is important to consider the contribution of the heat of dilution. The solute-water interaction is a rather difficult phenomenon to understand, and the models available are not able to describe adequately all the associated variations. One effect observed by D'Orazio and Wood (1963) were partial molar heat capacities at infinite dilution of gases in water at 25°C. They found that $\Delta\bar{C}^O_{p_2}$ is increased by nonpolar groups and reduced by polar sites, especially those capable of hydrogen bonding. However, the experimental $\Delta\bar{C}^O_{p_2}$ values were not explained by Eley (1939a, b) or Nemethy and Scheraga (1962a, b) using interstitial and cluster models, respectively.

The solubility concept is merely the interaction between like and unlike molecules. Whether the solute and solvent are chemically and physically similar substances will influence the solubility data. If the molecule-molecule forces in the solute and the molecule-solvent forces in the solution are favorable, solubility will take place. The combination of hydrogen bonds between solute and solvent molecules is shown by the good miscibility between water and alcohols, which are also amphoteric in nature. Regarding solute-solvent interaction from the hydrogen-bonding point of view, Pimentel and McClellan (1960) classified the solvents as follows:

A. Proton donor (acid) (e.g., $CHCl_3$, C_2HCl_5): compounds with enough halogen atoms to activate the hydrogen atoms

B. Proton acceptor (base) (e.g., ketones, aldehydes, ethers, tert-amines, esters, olefins, aromatic hydrocarbons)

AB. Both proton donor and acceptor (acid-base) (e.g., water, alcohols, carboxylic acids, primary and secondary amines)

N. Non-hydrogen bonding (e.g., paraffins, CS_2, CCl_4)

An experimental technique to measure the proton- or electron-donor capability of substances has been developed by Craver (1970) through sonic velocity measurement using bonded fibers of paper which form hydrogen bonds with the solvent under examination. The degree of bonding is detected by an increase or decrease in the velocity of sound going through the wetted paper. Water is used as the reference solvent. The relative effects of other solvents are expressed by the relative hydrogen-bonding capabilities of various solvents:

$$\gamma_w = \frac{\text{velocity of sound through water-soaked paper}}{\text{velocity of sound through solvent-soaked paper}} \times 100$$

Values of γ_w for some common solvents are presented in Table 2.30. Compounds with large γ_w values are more capable of forming hydrogen bonds than are solvents with low γ_w values. Craver (1970) has not reported γ_w values for halogenated hydrocarbons, but their values should probably be near to the paraffins, which do not form hydrogen bonding. A compilation of the velocity of sound in halogenated hydrocarbons was reported by Lienert (1975).

Table 2.30 Hydrogen-Bonding Ability of Solvents According to Craver (1970)

Solvent	γ_w	δ^{25} $(cal/cc)^{\frac{1}{2}}$
Water	100	23.53
Aniline	94	11.8
Formamide	94	17.8
Pyridine	80	10.62
Methanol	72	14.50
Glycerin	66	17.69
Dioxane	46	10.13
Ethanol	38	12.78
n-Propanol	35	12.18
n-Butanol	33	11.60
Acetone	29	9.62
Benzene	28	9.16
Toluene	27	8.93
n-Octanol	25	10.30
Diethyl ether	24	7.53
n-Hexane	24	7.27
n-Pentane	24	7.02
n-Octane	24	7.54

Table 2.31 Hydrogen Bonding of Solvents According to Crowley et al. (1966)

Solvent	Hydroge-bonding parameter	Solubility parameter	Dipole moment
Water	39.0	23.5	1.8
Methyl alcohol	18.7	14.5	1.7
Aniline	18.1	11.8	1.5
Pyridine	18.1	10.7	2.2
Acetonitrile	6.3	11.9	3.9
Acrylonitrile	5.7	10.5	3.8
Ethylene dichloride	1.5	9.8	1.1
Methylene chloride	1.5	9.7	1.5
1,1,2-Trichloromethane	1.5	9.6	1.2
Chlorobenzene	1.5	9.5	1.6
Chloroform	1.5	9.3	1.2
1,2-Dichloropropane	1.5	9.0	1.6
n-Butyl bromide	1.5	8.7	2.0
2,2-Dichloropropane	1.5	8.2	2.3
Isobutyl chloride	1.5	8.1	2.1
Carbon disulfide	0.0	10.0	0.0
Benzene	0.0	9.2	0.0
Carbon tetrachloride	0.0	8.6	0.0
Cyclohexane	0.0	8.2	0.0
n-Heptane	0.0	7.4	0.0
n-Hexane	0.0	7.3	0.0
n-Pentane	0.0	7.0	0.0
n-Decane	0.0	6.6	0.0

The hydrogen-bonding ability of solvents have also been examined by Crowley et al. (1966) and Lieberman (1962). Table 2.31 shows the classification for some solvents according to Crowley et al. (1966). It is interesting to note that the hydrogen-bonding parameter is 1.5 for all the halogenated hydrocarbons except CCl_4. Therefore, it would not be too much help to distinguish between these solvents as far as the strength of hydrogen bonding is concerned in connection with their solubility characteristics. Lieberman (1962) introduced three groups of solvents, having high, medium, and low hydrogen bondings, as follows:

High   =   1.4 - 2.0
Medium   =   0.7 - 1.3
Low   =   0.0 - 0.6

Table 2.32 Hydrogen-Bonding Parameters According to Lieberman (1962)

Solvent	Hydrogen-bonding parameter	Solubility parameter
Diethylene glycol	2.0	9.1
Pyridine	1.7	10.7
Methyl alcohol	1.7	14.5
Water	1.7	23.4
Diethyl ether	1.0	7.4
Sec-butyl bromide	1.0	8.1
n-Amyl chloride	1.0	8.3
n-Butyl bromide	1.0	8.7
Acetone	1.0	10.0
Acrylonitrile	1.0	10.5
Aniline	0.7	11.8
Acetonitrile	1.0	11.9
n-Hexane	0.0	7.3
n-Heptane	0.3	7.4
n-Octane	0.3	7.6
Vinyl chloride	0.3	7.8
Isobutyl chloride	0.3	8.1
2,2-Dichloropropane	0.3	8.2
n-Propyl chloride	0.3	8.5
Carbon tetrachloride	0.0	8.6
n-Propyl bromide	0.3	8.9
trans-1,2-Dichloroethylene	0.3	9.0
1,2-Dichloropropane	0.3	9.0
Propylene dichloride	0.3	9.0
cis-1,2-Dichloroethylene	0.3	9.1
Benzene	0.3	9.2
Chloroform	0.3	9.3
Trichloroethylene	0.3	9.3

Some selected solvents with their hydrogen-bonding parameters are presented in Table 2.32 according to Lieberman (1962). This type of classification also lacks selectivity as far as their application for halogenated hydrocarbons is concerned. In addition to this, the hydrogen-bonding parameters of Freon compounds have not been included in the lists presented, which would have made the classification more interesting from the point of view of this book. Consequently, none of the hydrogen-bonding parameters

discussed above can be used to treat the problem of solubility in a quantitative way.
A numerical predictive calculation of solubilities on the proposed parameters is not
yet possible. However, in a large number of cases, the quantitative conditions do
lead to an understanding of the order of solubility in agreement with experimental
data.

The hydrogen-bonding indices of various solvents have been determined by Cosaert
(1971) and plotted versus the solubility parameter at ambient temperatures (see
Fig. 2.28a). Typical groups of solvents are clearly indicated by certain areas in
the graph. Some selected values are listed in Table 2.33.

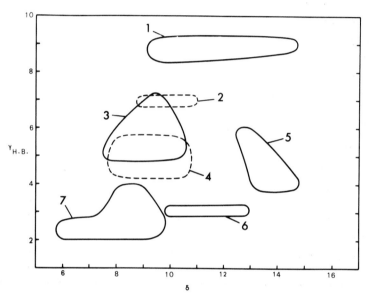

Fig. 2.28a   Hydrogen-bonding index versus solubility parameter. $\gamma_{H.B.}$ = hydrogen-
bonding index; δ = solubility parameter $(cal/cc)^{\frac{1}{2}}$. 1, alcohols; 2, glycol ethers;
3, ketones; 4, organic esters; 5, lactones and cyclic carbonates; 6, nitrohydro-
carbons; 7, hydrocarbons and halogenated hydrocarbons. (From Cosaert, 1971.)

Hayduk and Laudie (1973) successfully introduced a hydrogen-bonding factor for
solubilities of gases in water and associated solvents, based on the ratio of actual
gas solubility to its ideal solubility:

$$\sigma_{H_2O} = \frac{x_2 \text{ (solubility in water, mole fraction)}}{x_2^i \text{ (ideal solubility in water, mole fraction)}}$$

Table 2.33 Hydrogen-Bonding Index According to Cosaert (1971)

Solvent	Solubility parameter	Hydrogen-bonding index
Benzene	9.2	2.2
n-Butyl bromide	8.7	2.7
Carbon tetrachloride	8.6	2.2
Chlorobenzene	9.5	2.7
Chloroform	9.3	2.2
Cyclohexane	8.2	2.2
n-Decane	6.6	2.2
Dichlorodifluoromethane	6.1	2.5
1,2-Dichloropropane	9.0	2.7
2,2-Dichloropropane	8.2	2.7
Ethylene dichloride	9.8	2.7
Freon MF solvent	7.8	2.5
Freon TF solvent	7.2	2.5
n-Heptane	7.4	2.2
n-Hexane	7.3	2.2
Methyl chloroform	8.3	2.2
Methylene chloride	9.7	2.7
n-Pentane	7.0	2.2
Perchloroethylene	9.3	2.2
1,1,2-Trichloroethane	9.6	2.7
Trichloroethylene	9.3	2.5
Trichlorofluoromethane	7.8	2.5
Water	23.5	16.2

They assumed that the large reduction in solubility data from ideal solubility is due to the strong hydrogen bonds in the solvents. The ideal gas solubility is obtained from the vapor pressure of the solute:

$$x_2^i = \frac{P_2 \ \text{(partial pressure of the solute)}}{P_2^o \ \text{(vapor pressure of the solute)}}$$

However, it is very difficult to calculate the hydrogen-bonding effect, which tends to reduce the ideal solubility value of paraffin gases in water by factors of thousands. This is due to the presence of strong hydrogen bonds, which hinder the gas molecules from entering into the solvent. Hayduk and Laudie (1973) considered the hydrogen-bonding factor σ_{H_2O} as a measure of the hydrogen bonding for a particular gas-solvent system. Table 2.34 gives the hydrogen-bonding factors for several

Table 2.34 Hydrogen-Bonding Factors of Halocarbons in Water at 25°C

Solute	x_2^{id}	$x_2^{25°}$	$\sigma_{H_2O} = x_2^{25°}/x_2^{id}$
CCl_3F	0.9399	0.0001621	0.0001725
$CClF_2-CClF_2$	0.4704	0.00001448	0.00003078
CF_3-CClF_2	0.1088	0.000006788	0.00006239
CCl_2F_2	0.1516	0.0000447	0.0002949
$CBrClF_2$	0.3668	0.002118	0.005774
$CClF_3$	0.0280	0.00001656	0.0005914
$CHCl_2F$	0.5444	0.003280	0.006025
$CHClF_2$	0.0945	0.006055	0.006407
CHF_3	0.0212	0.001056	0.049811
CH_3-CH_2Cl	0.6231	0.001573	0.002524
CH_3Br	0.4615	0.002925	0.006338
CH_3Cl	0.1736	0.001907	0.010985
$CH_2=CHCl$	0.2570	0.008273	0.003215

halogenated hydrocarbons in water according to Hayduk and Laudie (1973). They used the hydrogen-bonding factor of water σ_{H_2O} as the basis for calculating the hydrogen-bonding factors in other solvents (e.g., alcohols, acids, acetone). It is expected that a reduction will also take place in other associated solvents. The relationship between σ_{H_2O} and the hydrogen-bonding factors of other associated solvents has been graphically evaluated by Hayduk and Laudie (1973). They also showed the significance of the temperature dependence of the solubility of gases in associated solvents. Whereas the $\log x_2$ versus $\log T$ scale provides a constant slope for nonpolar and slightly polar solvents (i.e., regular solutions) (Hayduk and Castaneda, 1973), in associated solvents the solubility will be represented by curves on the same scales (see Sec. 2.8). However, the important observation is that in both regular and non-regular solutions the gas solubilities tend toward a common solubility at the critical temperature of the solvent. In the case of water, the critical temperature is 647.3 K and the corresponding common solubility expressed as a mole fraction is 1.40×10^{-4}, which has been found through representation of the experimental solubility data. For a further discussion of the temperature dependence of solubility, see Sec. 2.8.

The method of calculating the hydrogen-bonding factors according to Hayduk and Laudie (1973) and Sahgal et al. (1978) is limited to gas-liquid solubility data. The solid-liquid and liquid-liquid solubilities have not been included.

Regarding the strength of the hydrogen bonding in halogenated hydrocarbons and water systems, probably the binary mixture of water in CCl_4 has been most extensively

investigated (Magnusson, 1970; Bonner and Choi, 1974). When water is dissolved in
chlorinated hydrocarbons (e.g., in CCl_4, $CHCl_3$, CH_2Cl_2) the sharp band occuring at
1890 to 1900 nm indicates the almost free $\nu_{OH} + \nu_2$ made in the 1:1 complex
H - O - H $\cdots$ B. The position of this band shows that water is very weakly bonded
in these solvents. When the water-proton(s) bonding becomes stronger, the band is
simultaneously shifted toward a longer wavelength (e.g., 1944 nm in pyridine). The
band also shifts to longer wavelengths as the basicity of the solvent increases.
Carbon tetrachloride certainly does not interact strongly with water, because the
rotational branches of the monomer are prominent even though lines are broadened
beyond resolution. The occurrence of bands in various nm-regions is directly related
to the strength of the hydrogen bonds that are formed. In all halogenated hydrocarbon
solvents, water is largely doubly bonded. This is due to the low solubility of water,
about 0.1 mole fraction or less.

The formation of hydrogen bonds by C-H and hydrogen is not so well accepted
as formation by O-H and N-H. Zellhoefer and co-workers (Zellhoefer, 1939;
Zellhoefer and Copley, 1938; Zellhoefer et al., 1938; Copley et al., 1938a, b)
studied the solubilities of halogenated hydrocarbons in oxygen-containing substances,
and they interpreted the high values obtained as evidence of hydrogen bonding. There
are a large number of workers suggesting the hydrogen bonding of halogenated hydrocarbons in various solvents (see e.g., Pimentel and McClellan, 1960), despite the
statement of Hildebrand (1949b, c). The presence of hydrogen bonds affects the temperature dependence of the solubilities. Usually, the temperature coefficient is
negative; similarly, the heat of mixing becomes negative.

Whether an organic compound is soluble or insoluble in water is determined by
its structure. The binding forces between the solute molecules and those between the
solvent molecules play a considerable role. The associated water molecules are bonded
together by hydrogen bonds between the hydroxyl groups in the liquid state, and one
expects that those substances will be soluble in water that can fit into the water
structure and play the same or similar role as water molecules. Therefore, those
organic substances that possess hydroxyl or/and carboxylic groups tend to be water-
soluble. However, the halogenated hydrocarbons do not possess hydroxyl or carboxylic
groups, and therefore they cannot form associative bonds with associated water molecules.

Mixtures of hydrogen-bonding substances can show either positive or negative
deviations from Raoult's law. Consequently, both maximum and minimum azeotropes are
possible. The effect and prediction of azeotropes in hydrogen-bonding systems has
been well described by Ewell et al. (1944). Further comments on azeotrope formation
are given in Sec. 2.5-4.

The effects of hydrogen bonding on the partition coefficient have been summarized by Pimentel and McClellan (1960), Leo et al. (1971), and Vinogradov and Linnell
(1971).

Table 2.35 Comparison of Physical Properties of Hydrogen- and Non-Hydrogen-Bonded Substances

Substance	Molecular weight	Melting point (°C)	Boiling point (°C)	Density d_4^{20} (g cc^{-1})	n_D^{20}	Surface tension (dyn cm^{-1})	Thermal conductivity (W m^{-1} K^{-1})
H_2O	18.0153	0.0	100.0	1.00	1.3330	72.88	0.603
H_2S	34.08	-85.5	-60.7	0.836	1.374	13.75	0.140
HF	20.01	-83.1	19.54	0.987	1.1574	8.84	0.435
HCl	36.46	-114.8	-84.9	0.905	1.3287	5.1	0.191
CH_3CH_2OH	46.07	-117.3	78.5	0.7893	1.3611	22.39	0.1657
$C_2H_5OC_2H_5$	74.12	-116.2	34.51	0.71378	1.3526	17.10	0.1285
HCOOH	46.03	8.4	100.7	1.220	1.3714	37.67	0.1830
$HCOOC_2H_5$	74.08	-80.5	54.5	0.9168	1.3598	23.84	0.1680
$HCONH_2$	45.04	2.55	111.0	1.1334	1.4472	57.32	0.350
CH_3Br	94.94	-93.6	3.56	1.6755	1.4218	24.20	0.0956

The physical properties of hydrogen- and non-hydrogen-bonded substances are compared in Table 2.35.

The interaction between the solute and solvent molecules are of several types, such as:

Electrostatic (dipole-dipole and ion-dipole)

Dispersion forces

Donor-acceptor

Hydrogen bonding

Structure breaking

Structure making

The nature of solute-solvent interaction is responsible, for example, for the solvation of reactants and the transition state. Furthermore, in nucleophilic reactions there is hydrogen bonding between a protonic solvent and the nucleophile (e.g., alcohols and water). When the nucleophile is uncharged, the hydrogen bonding is weak. The nucleophilic strength of carbon for halide ions in water or alcohols has been studied by Parker (1962), Alexander et al. (1968), and Bunnett (1963). They found that the order of nucleophilic strength is Cl< Br<I in alcohols or water and the reverse in aprotic solvents. In protonic solvents, the hydrogen-bonding solvation is stronger for small nucleophiles. Solvents with dielectric constants greater than 15 (i.e., polar aprotic solvents) are capable of promoting nucleophilic reactions. The hydrocarbons, CCl_4, and so on are nonpolar aprotic solvents.

The solvent concept was well reviewed by Gutmann (1977), with particular emphasis on the clear distinction between nucleophilic and electrophilic solvent properties. Furthermore, the interpretation and application of donicity and acceptor number are discussed. This two-parameter concept is very useful for the explanation of the reactivity of a solute and allow separate characterization of the nucleophilic and electrophilic solvent properties.

2.5 Liquid and Dissolved States

2.5-1 Pure Compounds

Despite the large number of investigations of the liquid state of the matter, this is still considered to be the least known among the three states: solid, liquid, and gas. As a result of intensive studies, there are many books, reviews, reports, and articles on the subject. The reader who is interested in more details is referred to the following works: Adams et al. (1975), Anonymous (1953), Barker (1963a, b), Barton (1974), Boublik et al. (1980), Buckingham et al. (1978), Chen (1971), Cole (1967), Covington and Jones (1968), De Boer and Uhlenbeck (1964), Dreisbach (1966), Egelstaff (1967, 1973, 1979), Eyring and Jhon (1969), Fisher (1964), Flowers and Mendoza (1970),

Frenkel (1955), Frisch and Lebowitz (1964), Frisch and Salsbury (1968), Fürth (1949), Glasstone (1937), Gotoh (1972), Green (1970), Goldberg (1975), Hansen and McDonald (1976), Henderson (1971), Hildebrand (1939, 1953, 1977a, 1978), Hirschfelder et al. (1964), Hughel (1965), Jhon and Eyring (1978), Johnson and Porter (1970), Klotz and Rosenberg (1972), Kohler (1972), Kohler and Wilhelm (1976), Kruus (1977), Lykos (1978), McDonald and Singer (1970, 1973), Maharajh (1973), Marcelja et al. (1977), Marcus (1977), Morrell and Hildebrand (1936), Mountain (1970), Nitta et al. (1977b), Pings (1968), Prigogine et al. (1957), Pryde (1968), Rice and Gray (1965), Rowlinson (1959, 1970), Salsburg (1968), Scatchard (1976), Scheraga (1979), Schuster et al. (1975), Scott (1970), Tabor (1970), Temperley et al. (1968), Temperley and Trevena (1978), Tomkins (1978), Tsykalo et al. (1975), Vavruch (1978), Veselovskii (1975), Watts (1971), Watts and McGee (1977), Woodcock (1971), Woodhead-Galloway (1972), and Yeo (1973). The author(s) or editor(s) are listed in alphabetical order for easier identification.

Although the discussion of the liquid state of pure compounds in the first part of this section is short, the coverage of the dissolved state in the second part is more detailed.

There are three states of matter relevant to solubility and miscibility studies: solid, liquid, and gas (the plasma state is beyond the scope of this book).

In the solid state, the molecules are closely packed in a well-defined position. Consequently, they are not able to move freely as they are in the gaseous state. The crystalline state of a solid (the concept of pitch, glass, and polymers are beyond the interests of this book) is a well-defined form in three dimensions. The molecules maintain a boundary surface and are arranged in a regular geometrical pattern.

The crystal is built up of a large number of units packed side by side in every direction. The unit cells are bounded by plane faces, with sides seldom more than a few angstroms long. Inside the cells there are one, two, or four atoms (of elements) or molecules (of organic substances). The direction of the faces appearing on crystals and the principal planes on which atoms are located within the crystals are controlled by two fundamental laws: the constancy of the angle and the rational indices. The atoms in a crystalline structure are hold together by four type of forces:

1. The homopolar bond binds the atoms though two electrons, which have been derived from one atom or shared by each atoms.

2. In the heteropolar or ionic bond, the atoms gain or lose one or more electrons and become ions with electronic structure similar to the rare gases. The ions will attract each other electrostatically.

3. The metallic bond is represented by positive ions in fixed positions and is held together by valency electrons, which are free to move, providing electrical conductivity in the metals.

4. The van der Waals forces are cohesive attractions and are very weak compared to other forces.

Some typical examples for various type of crystals are tabulated in Table 2.36.

Table 2.36 Various Type of Crystalline Substances

Type of bond	Substance
Homopolar	Diamond
Heteropolar	NaCl
Metallic	PbSe
van der Waals forces	Organic compounds

Relationships among the various phases are clearly illustrated by phase diagrams, plotting the pressure of a pure compound versus temperature. Figs. 2.29 and 2.30 show the phase diagrams for water (Kamb, 1965) and 1,1,2-trichloroethane (Babb and Christian, 1977), respectively.

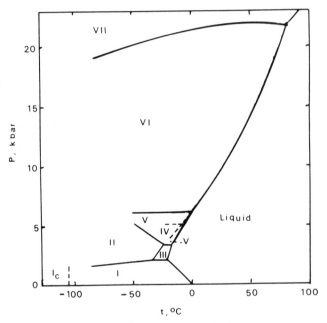

Fig. 2.29 Phase diagram for water. (From Kamb, 1965.)

Contrary to the case in solids, in gases the molecules are randomly distributed and are free to move in all directions, colliding with each other and with the walls. The motion is completely random. The motion of the molecules in the space is proportional to the absolute temperature. The volume of a molecule is very small relative to the space occupied at ordinary temperature and pressure. The perfect or ideal gas

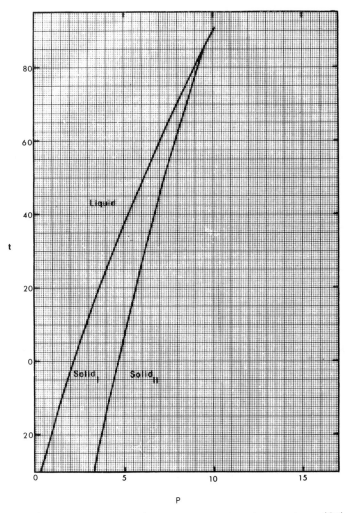

Fig. 2.30 Phase diagram of 1,1,2-trichloroethane. t = temperature (^{o}C); P = pressure (kbars). (From Babb & Cristian, 1977.)

is defined as one that occupies no space or negligible space. The attractive forces between the molecules are vanishingly small, because there are no intermolecular attractions. The gases become more ideal when the pressure is very low and the temperature is high. The kinetic energy and the distribution of gaseous molecules can be calculated by Maxwell's distribution law. An excellent compilation on the molecular theory of gases is that of Hirschfelder et al. (1964).

For 1 g-mole of perfect gas, the following expression is valid:

$$P\,V\ =\ R\,T$$

and for n g–mole, the formula is

$$P V = n R T$$

where P = pressure

V = molar volume

T = absolute temperature

R = gas constant

However, no actual gas is absolutely perfect, but most of the permanent gases (He, H_2, O_2, N_2, etc.) are so nearly perfect under atmospheric pressure that their deviation from perfection can be neglected for practical calculations. There are two causes for the deviation from the ideal–gas law:

1. Molecular attraction when the molecules are far apart.
2. Molecular repulsion when the molecules are very close to one another. The repulsion is due to the molecules being in contact and thus incompressible.

There have been numerous attempts to develop an empirical or semiempirical relationship among the three variables pressure, volume, and temperature for a real fluid; however, until now, all the efforts have produced only a fair approximation of the equation of state. A simple but historically interesting equation introduced by van der Waals (1873) has been used for both liquid and gaseous states:

$$(P + \frac{a}{V^2}) (V - b) = R T$$

where a and b are constants for each pure substance. The constants have been derived from the following considerations. The actual pressure of the gas is less than the pressure of the ideal gas, because the molecules approaching the wall will be attracted back by other molecules close to them. The number of molecules that strike the wall is proportional to the volume of the gas. The total diminition in pressure caused by the attraction between the molecules will therefore be proportional to a/V^2. Because of the volume of the molecules, the actual volume will be less than the total volume. Therefore, the intermolecular volume becomes V − b. At moderate pressure b is four times the actual volume occupied by spherical molecules.

The main discrepancy occurs between the van der Waals equation of state and the experimental pressure–volume–temperature (P-V-T) data close to the critical point. This is caused by the departure from the value calculated by the van der Waals equation for $(R T_{cr})/(V_{cr} P_{cr}) = 2.67$. The value for a real gas is always higher than 2.67, and therefore the calculated values fron the van der Waals equation show large discrepancies.

Many modifications of the van der Waals equation has been attempted or completely new equations of state have been developed. Some of the well-known equations of state are (in chronological order) those developed by:

1. Berthelot (1899)
2. Dieterici (1899)
3. Virial (Onnes, 1901)
4. Whol (1914)
5. Beattie-Bridgeman (1927)
6. Benedict-Webb-Rubin (1940)
7. Redlich-Kwong (1949)
8. Martin-Hou (1955)
9. Hirschfelder et al. (1958)
10. Lee-Kesler (1975)
11. Peng-Robinson (1976)

Whereas the Redlich-Kwong equation of state has only two disposable constants, the Benedict-Webb-Rubin equation of state has eight constants, and the Martin-Hou equation of state has nine constants. A good summary of the various aspects of these equations of state has been given by Reid et al. (1977). As far as the solubility calculations are concerned, the Redlich-Kwong equation has been used successfully by several investigators (De Santis et al., 1974; De Mateo and Kurate, 1975; Evelein et al., 1976; Namiot et al., 1976). A further discussion of the calculations is given in Sec. 2.7.

The Redlich-Kwong equation is the most successful two-constant equation of state (Kemp et al., 1974; Horvath, 1972a, 1974). This equation is widely used in research and development work on new substances or on compounds that have very limited information as to their physical properties. To derive the two constants of the equation, one requires only the critical properties of the compounds. The critical properties of organic substances can be estimated with good accuracy by Lydersen's method from the molecular structure, molecular weight, and normal boiling point (Reid et al., 1977).

The critical point of a pure compound is very important for an understanding of its liquid and vapor coexistence. At the critical point, the demarcation disappears between the liquid and vapor phases. In other words, there is no liquid phase above the critical temperature. Because of the exact definition of the critical state (i.e., the disappearance of the meniscus between the liquid and vapor phases), it has been applied to relate other physical properties to it (Horvath, 1980). The corresponding-state principle is one of the most successful theories based upon the reduced properties, which have been deduced from the critical properties and the actual variables (e.g., temperature, pressure, volume) (Sterbacek et al., 1979). The reduced temperature, pressure, and volume become

$$T_r = \frac{T}{T_{cr}} \qquad\qquad P_r = \frac{P}{P_{cr}} \qquad\qquad V_r = \frac{V}{V_{cr}}$$

It has been found that $T_r - P_r - V_r$ diagrams of most substances give the same lines and that the discrepancy is minimal for normal fluids. That is, the corresponding-state principle gives the same values for the reduced properties of most substances. Those compounds that follow the corresponding-state principle or deviate from it minimally are called normal fluids, in contrast to some others which behave in an anomalous way (e.g., associated compounds). Those substances are usually normal whose molecules are of the same nature (e.g., inert gases or nonpolar organic molecules whose nature as to intermolecular interactions are very similar). The law of corresponding states has been used very successfully to calculate the physical, thermodynamic, and transport properties of substances, both in the liquid and gaseous states(Sterbacek et al., 1979).

The liquid state is intermediate between the solid and gaseous states. Some investigators call it a solid- or gas-like state. The liquids have no definite shape but a well-defined volume. The molecules are quite mobile and are free to move in the liquids. However, the molecules exert forces on one another, which are caused by the attractive and repulsive short-range forces. The attractive forces between the molecules are responsible for the volume or density of liquids. The repulsive forces play a greater roll in determining the distribution function. Liquid theories based upon the hard-sphere model assume no attractive forces between the molecules treated as rigid cores. In a new proposal, a hard-sphere term is combined with an attractive term (Kreglewski et al., 1973). The properties of liquids may be calculated from the interatomic forces and energies, which are the sum of pair interactions in a classical liquid. The energy of a system is the sum of the energies required to take the atoms in pairs. The equations of liquid state have been deduced from pair theory by Egelstaff (1967). His book has been written with these conditions in mind. These equations will enable us to calculate various properties of the liquid. However, in some cases the calculated properties will not be accurate enough, indicating failure of the theory. During the treatment of the models of the pair theory, there are two ways to express the results: to derive expressions for the properties and to calculate the numerical values for the functions being present in the expressions.

Another theory of the liquid state is lattice theory. This is one of the simplest and oldest theories of liquids, which has been modified and improved since its introduction (Guggenheim, 1952). This assumes the solidlike or quasicrystalline state of the liquid. The molecules are in a semistable position and they can only vibrate back and forth in this lattice with some form of regularity. The theory assumes that the spherical molecules are alike in size and shape. In the structure of the liquid, there are the same intermolecular spacing, the same composition, and the same coordination number. The molecules are very close to one another and only the interactions between the nearest neighbors have been considered. Consequenly, the interaction energy and the energy of mixing calculated will be the same.

Lattice theory has been developed from statistical mechanics. With increasing temperature the solid expands and the molecules start to vibrate with an amplitude about the rest position. The order of aplitude depends upon the temperature. The actual forces determining the motion of the molecules depend upon their neighbors. At a well-defined temperature, the violent vibration of the molecules causes a breakdown in the regular lattice structure (i.e., the solid melts). There are two types of forces that influence the position of a molecule. The intermolecular forces tend to keep the molecules in an ordered lattice structure, whereas the kinetic energy of the molecules transferred by thermal agitation tends to destroy the crystalline arrangement of the molecules. Thus the melting point usually increases with rising pressure for organic compounds (see Fig. 2.30 for 1,1,2-trichloroethane; however, water exhibits the opposite behavior). The order in the liquids varies, and if we chose a particular molecule for study, the observations show that the disorder increases among its neighbors according to the distance between them. The nearest neighbors are better organized than those molecules in greater distance. In other words, in liquids there is only a short-range order. With rising temperature, the order is reduced and the liquid becomes a highly disordered solid. The disadvantage of the lattice theory is the overemphasis on regularity in the liquid structure with the regular crystalline state.

A refinement of the lattice theory is the significant structure theory incorporating the physical properties in the model (Eyring and Marchi, 1963). This theory is based on the idea that fluidized vacancies are present in the liquids which are mirrored in the vapor phase. This is caused by rapid shifting of the positions of the molecules, creating empty lattice sites or holes by pushing the molecules out of their competitors. However, this happens only if a molecule possesses enough energy to push the neighboring molecules aside. These holes have an average volume throughout the liquid phase. The holes are also mobile and have gaslike properties. A disadvantage with this theory is that it has not been derived from an exact partition function but by reasoning that the solidlike molecules move into a potential well formed by their neighbors, and the gaslike molecules move into the holes of the liquid, the voids from the porous or pseudocrystalline framework of the liquid. On the basis of the two types of translational motion of molecules, the expressions developed predict the various properties of liquids with a good result. Encouraging results have been reported for the explanation of the law of rectlinear diameters, liquid heat capacity at constant volume, when using the expression

$$C_v = \frac{6 V_s}{V_l} + \frac{3(V_l - V_s)}{V_l}$$

where V_s = molar volume of the solid

V_l = molar volume of the liquid

and other thermodyanmic and transport properties (Eyring and Jhon, 1969). On the
average, there is no other theory for the liquid state that has covered so many different properties of substances with such promising results (Eyring, 1936; Eyring et
al., 1963; Eyring and Hirschfelder, 1937; Eyring and Jhon, 1969; Eyring and Ree, 1961).
The significant structure theory has been used successfully for water (Jhon et al.,
1966), helium (Oh et al., 1977), and binary mixtures.

In 1936, Eyring and co-workers introduced a model for liquid states. This new
theory was a further development of the hole theory (see, e.g., Frenkel, 1955 and
Ishizuka et al., 1980). In the cell model, the molecules in a liquid are confined by
a shell or cell formed by neighboring molecules. The cell formed by neighboring molecules located on regular spaced lattice points. The central molecule is moved according to the force field caused by the neighboring molecules. When the liquid expands,
the volume of each cell increases. The cells have equal volume for all molecules.
Fig. 2.31 shows the cell model of a liquid (Chao and Greenkorn, 1975). The environment of a molecule is not very different from that in a solid, but it is somewhat
freer to wander round. All cells are identical. The configurational integral for the
liquid state is too involved, and therefore it is necessary to introduce some approximations by breaking it up into small units, assuming that each cell contains only
one molecule. In other words, the number of cells is equal to the number of molecules.
Furthermore, if one assume that the free volume in the cell has a complicated shape
and close-packed structure, the equation of state for a liquid becomes

$$\frac{P\,V_1}{N\,k\,T} = \frac{1}{1 - (V_s/V_1)^{1/3}}$$

where   N = Avogadro's number

V$_s$ = molar volume of the solid at 0 K

V$_1$ = molar volume of the liquid

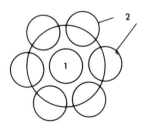

Fig. 2.31 Two-dimensional cell model of a liquid. The central molecule is confined
by a cage of surrounding molecules. 1, Central molecule; 2, neighboring molecules.
(From Chao & Greenkorn, 1975.)

One of the several shortcomings of this expression is that the hard-sphere potential function neglects the effect of the attractive forces. This very simple model fails to give satisfactory results; therefore, numerous improvements have been proposed by Lennard-Jones and Devonshire, Eyring and Hirschfelder, and others. Although much effort has been expended, there is still no good model that can give a satisfactory approximation. The single-occupancy model can be checked by Monte Carlo calculations.

The cell model is based upon ordered structures which do not allow density fluctuations. The central molecule is surrounded by 8 to 12 symmetrically arranged molecules. The number of neighboring molecules define the coordination number. The distance between the molecules is just a little greater than the molecular diameter. The attraction between the molecules varies inversely with the sixth power of the distance of separation. From Fig. 2.31 it is apparent that the enclosed molecule can move only until it touches the neighboring molecules. However, the total cell volume is larger than the free volume available for occupancy by the central molecule. Consequently, the entropy is lower than it would be if the molecule could occupy the entire volume of the cell. The free volume of the cell expresses the freedom of motion of a molecule in a liquid.

Eyring and Hirschfelder (1937) improved this expression by calculating the total potential energy:

$$U = -\frac{a(T)}{V}$$

Through this assumption the final expression becomes

$$\left[P + \frac{a(T)}{V^2}\right](V - V_o^{1/3} V^{2/3}) = RT$$

This semiempirical equation is similar to the van der Waals equation of state.

Lennard-Jones and Devonshire (1937) investigated the free volume of the cell in depth, including the examination of dense gases, liquids, and the critical phenomena. Their theory was based upon the concept that, if the molecules come closer together, the repulsive action dominates, and if they move farther apart, the attraction will be stronger. The potential energy of the central molecule has been calculated for all positions within the cell. Furthermore, it was assumed that the neighboring molecules are in the centers of their cells and the potential field was averaged for symmetry. The simple collision has been replaced by multiple collisions. The free-volume theory of Lennard-Jones and Devonshire (1937) provided a very promising model for the liquid state. Several physical properties have been calculated from the model with results surprisingly close to the experimental data (Gotoh, 1976).

Wentorf et al. (1950) carried out an extensive study on the free-volume approx-

imation by including in the calculations the effects of molecules that are not in
the first coordination shell. However, most of the calculated properties still
showed differences from the experimental values.

The free volume is not a physical or thermodynamic parameter, and therefore
theories based upon this concept will not provide consistent thermodynamic properties for liquids (Hildebrand et al., 1970). Further improvements have been attempted
by various workers; however, as Hirschfelder et al. (1964) pointed out, "little
further progress can be made without removing some of the assumptions inherent in the
fundamental theory."

The cell cluster theory proposed by de Boer (1954) is a further modification of
Lennard-Jones and Devonshire's model. The theory assumes that the cells themselves
are clustered, forming from one, two, three, or more cells. However, proper calculation showed difficulties of carrythrough. In addition to the clusters of cells, Dahler
and Cohen (1960) introduced holes in the structure. This certainly produced only a
minor improvement in the model.

The tunnel theory of Barker (1963b) grew out from the hole theory. The principle
is that one-dimensional chains of molecules move through tunnels composed of neighboring
molecules arranged in parallel lines or cylinders. The tunnels are moved relatively
to each other and the motion of the molecules is random.

The radical distribution function in liquids has been developed by Kirkwood
(1939a, b) and co-workers. The theory is based upon the volume of repulsion forces
rather than attractive forces. X-ray analysis showed promising agreement with results
derived from the two- and three-dimensional interactions. The model assumes that the
liquid molecules are spherical and nonpolar. There is only short-range order in the
structure, when the distance is about the size of the molecular diameter. The number
of molecules in a **spherical** shell of thickness dr is

$$4 \pi r^2 g(r) \, dr$$

where $g(r)$ is the radial distribution function and r the radial distance of molecules from a particular molecule. The principle of Kirkwood's proposal is that the
attractive forces are not significant and the molecules have spherical shape. However,
the radial distribution function has been determined by the repulsive forces. Furthermore, the long-range attractive forces influence the condensed state. Calculation
of the radial distribution function was based on the assumption that the central molecule with only one neighboring molecule was considered at a time and the pairs were
summed up to derive the resultant force on the central molecule. The Kirkwood model
gives good agreement only with the experimental data for fluids of low densities.
Consequently, numerous modifications of the Kirkwood's theory have been suggested.
In 1939, Kirkwood and Boggs attempted to extend the theory to solutions. Despite the
several modifications of the model, little improvement has been achieved (Rowlinson,
1959).

One extension of the Kirkwood theory was introduced by Fisher (1962). This hypersuperposition approximation provides some improvement only as far as the density range was concerned.

Order-disorder in the liquid state has been studied by Bernal (1959, 1964). His proposal involved a geometrical or spatial model in which rigid spheres are irregularly packed as closely as possible. The environment, that is, the immediate neighbours, are steadily changed and the coordination number is constantly varied. The disordered structure of the liquid changes rapidly. The interchange of regular and irregular arrangements of molecules is permanent. Experimental simulation of the liquid model has been examined by Bernal using small steel balls and a dye. From this investigation he was able to determine the number of neighboring balls and the size of cavities between the spheres packed randomly. Bernal's model is a fair one for glass, but there is no complete description of the structure of liquids. With cooling, the molecules withhold their random arrangement, but their relative positions will not change when approaching the freezing point, and the liquid viscosity rapidly increases to infinity. The model does not allow enough expansion of liquids (e.g., CCl_4 can expand 30% and not 15%). A more recent model of the liquid state has been proposed by Pinsker (1972), who presented a graphic geometric image that permits examination of the atomic structure. The structure of the liquid is composed of mirror-image pairs of molecules. The model is constructed graphically from spheres of uniform diameter according to the law of mirror-symmetric tetrahedrons, where 12 spheres surround 1 sphere.

Since the availability of electronic computers for fast calculations of iterative procedures, the simulation of the liquid phase has been attempted by several investigators. The computer simulation technique involves determination of the molecular arrangements in a real liquid by trying all possibilities. The model uses an effective pair potential-energy function and calculates the thermodynamic properties. The results are compared with theoretical data obtained from other models. Some of the more recent ones are reviewed by Barton (1974), Kohler (1972), Kohler and Wilhelm (1976), and Alder and Wainwright (1959).

One of the first methods used was the Monte Carlo technique, which has been applied to the problem of molecular distribution in a liquid to obtain a statistical estimate of the desired quantity by random sampling (Metropolis et al., 1953). In Monte Carlo calculation, classical statistics is assumed, only two-body forces are considered, and the potential field of a molecule is assumed to be spherically symmetric. Because of the great number of configurations needed in the model, the individual configurations appear with probability proportional to the Boltzmann factor, $\exp(-E/kT)$, and are weighted evenly, where E is the potential energy of the system. The Monte Carlo technique gives details on equilibrium properties only, not on transport properties, because the moves of the molecules are not associated with real time (Scheraga, 1979).

The distribution of molecules in liquids has been simulated by the molecular dynamic method (Alder and Wainwright, 1959). This theory permit the calculation of transport properties that are not possible with Monte Carlo calculation. Basically, this method is an approximation of the theory of many-particle systems. The particles are grouped together to form a large cell, which make up to an aggregate by repetition of these cells. The number of particles in the cell is constant, that there is an equilibrium between the leaving and entering particles in the cell. The velocity of the molecules points toward random directions and the successive configurations are obtained by solving simultaneous Newtonian equations of motion after a certain period of time. Molecular dynamics theory is unable to determine the partition function and entropy in the liquid state.

During the last decade, computer simulations of the liquid state became very popular. Several investigators adopted the random assemblies of hard spheres for computer calculations (Adams and Matheson, 1972; Finney, 1970).

The equation of state for a hard-sphere system gives a poor representation of the real liquid state. Therefore, it has been suggested that the hard-sphere potential could be taken as a reference potential and perturbation techniques used (Zwanzig, 1954; Watts, 1971; Watts and McGee, 1977; Boublik et al., 1980). The hard-sphere potential is different from that of true potential because of the attraction and repulsion potentials. However, the spherical reference potential has been replaced by the nonspherical model, owing to the repulsive interaction that dominates in the structure of liquids. The perturbation potential is anisotropic (Goldman, 1979). In second-order perturbation theory, subsequent improvement has been achieved by introducing two new parameters: to control the steepness of the positive potential energy and to alter the depth of the negative well. Although the perturbation theory provides good agreement with the experimental data (Reed and Gubbins, 1973; Gurikov, 1969), it is unsuitable for calculations in the vicinity of the critical point. The perturbation theory is particularly useful to calculate liquid transport properties. This theory expresses the perturbation potential $\phi(r)$ as a power series in $\phi(r)/kT$. The intermolecular potential energy is expressed by

$$\phi(r) = \phi^{ref}(r) + \alpha \, \Psi(r)$$

where $\phi^{ref}(r)$ = hard-sphere potential energy

α = perturbation parameter from 0 to 1

$\Psi(r)$ = perturbation potential

When $\alpha = 0$, there is an unperturbed state, whereas if $\alpha = 1$, there is an actual liquid. In a more realistic system, the Lennard-Jones (6,12) potential is used. In general, perturbation theory gives good quantitative prediction of the transport properties of liquids. The perturbation method as improved by Barker and Henderson (1967) using the Lennard-Jones (6,12) potential function gives a better agreement with the experimental values.

The thermodynamic properties of matter can be calculated from the thermodynamic expressions derived in terms of the partition function. The partition function depends upon the temperature-pressure-volume nature and the number of the components in the system. It also depends upon the quantum states of the system. However, for the evaluation of the partition functions in the equations, it is necessary to introduce some sort of approximations. Because of the large number of molecules in the system, the macroscopic properties of molecules can be successfully approached through statistical methods, dividing into two parts: finding the possibility that the examined system is in a particular quantum state, using the statistical distribution law, and from the distribution law to obtain thermodyanmic expressions representing the properties of matter. The basic partition function ϕ_N for a mole of liquid is

$$\phi_N = \phi_s^{N(V_s/V_1)} \; \phi_g^{N((V - V_s)/V_1)}$$

where N =   Avogadro's number

ϕ_s =   partition function of solidlike degree of freedom

ϕ_g =   partition function of gaslike degree of freedom

This expression enables us to evaluate the thermodynamic properties from the triple-point to the critical point for various substances and mixtures. The partition function ϕ_N can be expressed in term of the Helmholtz free energy (A):

$$A = -k \, T \, \log_e \phi_N$$

where A is a function of temperature and volume:

$$P = -\left(\frac{\partial A}{\partial V}\right)_T$$

Further thermodynamic relationships are

$$S = -\left(\frac{\partial A}{\partial T}\right)_V$$

$$E = -T^2 \left(\frac{\partial (A/T)}{\partial T}\right)_V$$

$$H = E + P\,V$$

$$G = A + P\,V$$

$$C_v = \left(\frac{\partial E}{\partial T}\right)_V$$

$$\alpha = \frac{1}{V}\left(\frac{\partial V}{\partial T}\right)_P$$

$$\beta = -\frac{1}{V}\left(\frac{\partial V}{\partial P}\right)_T$$

$$C_p = C_v + \alpha^2 \frac{T V}{\beta}$$

where   P   =   pressure
        T   =   absolute temperature
        V   =   molar volume
        S   =   entropy
        E   =   total energy
        H   =   enthalpy
        G   =   Gibbs free energy
        C_v =   heat capacity at constant volume
        C_p =   heat capacity at constant pressure
        α =   thermal expansion coefficient
        β =   coefficient of compressibility

The transport properties (viscosity, diffusion, thermal conductivity, and surface tension) and dielectric constants of liquids and mixtures have also been calculated using the expressions deduced from the statistical mechanics (Reed and Gubbins, 1973).

The treatment of the liquid and dissolved states would not be completed without a brief criticism of the mentioned models and theories. Hildebrand and Scott (1962) and Hildebrand (1977a) expressed their opposition aginst the cell, hole, or cage theories. In the argument, they emphasize the calculated values of entropy of vaporization of compounds (Hildebrand, 1939; Nath, 1977). The line spectrum has not been found for undercooled liquid by x-ray scattering at the same temperature as for solids. The artificial model of the liquid state gave a complete disorder picture which indicates that there is no order. The partial molal volumes of solutes are similar in different solvents. The greatest disadvantage of the cell model is the built-in high degree of order. Those models that maintain long-range order fail to give good representations, although, there is some evidence that there is some short-range order in the liquids as indicated on x-ray patterns. There is no doubt that the molecular attractions and repulsions in liquids ought to be between those forces that determine the solidlike and gaslike structures.

2.5-2 Dissolved States

A short review on the various properties effecting the solubility of organic compounds was reported by James (1972). The author discussed some of the more important effects (e.g., like dissolves like, hydrogen bonding, and dipole moments) (see also Williamson, 1966).

The main four solubility theories: Prigogine, Flory-Huggins, Lewis acid, and molecular clustering were reviewed by Tees (1973).

The aqueous solubilities of various substances was investigated by Luneau (1965). Further discussions were reported by Guerrant (1964), Wilhelm et al. (1977), Tsonopoulos (1970), Ohtaki (1975), Palit (1947), and Palit and McBain (1947). A good review on the various aspects of the properties of dilute solutions of nonelectrolytes was reported by Rozen (1969a, b).

Scaled-Particle theory is probably one of the most successful models developed for the liquid state. This theory has been presented by Reiss et al. (1959, 1960), and a good review, with emphasis upon the scaled-particle theory of aqueous and non-aqueous solutions, is that of Pierotti (1976). The principle of this theory is that there are discontinuations in the number of molecules that can occupy a cavity in a liquid as a function of the volume of the cavity. The smallest volume can hold only one molecule, but greater volumes that are random can hold two or three molecules. In other words, there is a discontinuity regarding the number of molecules occupied in the voids of the liquid phase. The developed expression for pressure is based on the nearest-neighbor distribution function of the rigid spheres:

$$P = \rho_o \, k \, T \, \frac{1 + x + x^2}{(1 - x)^3}$$

where $x = \rho_o \, \sigma^3 \, \pi/6$

ρ_o = number density of the molecule

σ = diameter of the rigid sphere

Scaled-particle theory has been used successfully for real fluids, giving good agreement with the experimental data. In the application of scaled-particle theory to solutions, the theory derives the reversible work required to introduce a hard-sphere molecule into a real liquid whose molecules are made up of hard cores. The real intermolecular potentials between the molecules will determine the temperature dependence of the pressure and volume of the system. The applications and theoretical aspects of the scaled-particle theory for solubility determination are discussed in some detail in Sec. 2.1.

Scaled-particle theory was applied to water by Pierotti (1965). The early success was somewhat fortuitous. An improved version introduced by Stillinger (1973) incorporates both measured surface tensions and radial-distribution functions for pure liquids (see also Neumann, 1977).

The Scatchard-Hildebrand regular solution theory is probably the most generally used method for the calculation of mixture properties from pure component properties. This is different from the theory for strictly regular solutions introduced by

Guggenheim (1935, 1952). The regular solution theory is based on van Laar's proposal
that both the excess entropy and the volume of mixing (Battino, 1971 reported a review
on the volume changes on liquid mixing) are equal to zero, and developed by Scatchard
and Hildebrand working independently on solutions. But instead of using the van der
Waals equation of state, they expressed the temperature dependence of pressure at
constant volume as a function of the coefficient of thermal expansion (α) and the
isothermal compressibility (β):

$$T \left(\frac{\partial P}{\partial T} \right)_V = T \left(\frac{\alpha}{\beta} \right)$$

Regarding the correlation of isothermal compressibility with surface tension, the
report by Sahli et al. (1976) is very useful in cases when compressibility data are
not available (see Fig. 2.32). The foregoing relationship is in good agreement with
experimenatl values for nonpolar and slightly polar liquids (Renuncio et al., 1977).
Because the term

$$T \left(\frac{\partial P}{\partial T} \right)_V$$

is far larger than P, we can write

$$\left(\frac{\partial U}{\partial V} \right)_T \approx T \left(\frac{\partial P}{\partial T} \right)_V$$

where U is the internal energy. Hildebrand found that the ratio of the energy of
complete vaporization (isothermal vaporization of the saturated liquid) ΔU and
the molar liquid volume is a function of cohesive-energy density δ(the solubility
parameter):

$$\frac{\Delta U}{V_1} = f(\delta)$$

The molecular factors affecting the solubility parameter were examined by Hansen
(1967) and Peiffer (1980). A good review of the various aspects of solubility parameter is that of Barton (1975). The solubility parameter (δ) can be calculated from
the molar energy of vaporization (ΔU_{vap}):

$$\delta = \left(\frac{\Delta U_{vap}}{V_1} \right)^{\frac{1}{2}}$$

If experimental molar volume (V_1) or density data are not available, they can be
estimated by a method proposed Tyn and Calus (1975). The available liquid density
data of halogenated methanes has been reviewed by Kudchadker et al. (1978). Furthermore, at low vapor pressure, the molar energy of vaporization is derived from the

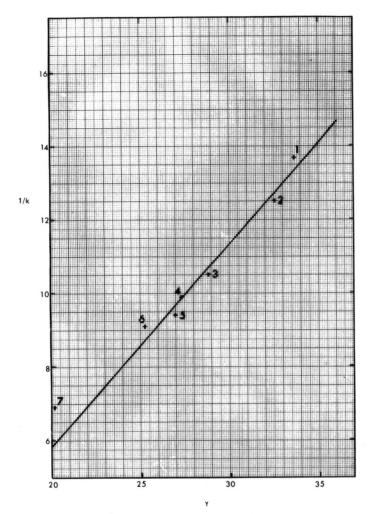

Fig. 2.32 Reciprocal of isothermal compressibility versus surface tension at 25°C. $1/k$ = reciprocal of isothermal compressibility (dyn cm^{-1}); γ = surface tension (dyn cm^{-1}). 1, C_6H_5Cl; 2, CH_2Cl-CH_2Cl; 3, C_6H_6; 4, $CHCl_3$; 5, CCl_4; 6, C_6H_{12}; 7, C_7H_{16}. (From Sahli et al., 1976.)

latent heat of vaporization ΔH_{vap} (see, e.g., Lawson et al., 1978):

$$\Delta U_{vap} = \Delta H_{vap} - R T$$

Hildebrand (1969) correlated ΔU_{vap} with the force constant (ε/k). In this case, where there no experimental ΔH_{vap} data are available, it can be calculated from the

vapor-pressure curve and the Clausius-Clapeyron's equation or from empirical expre-
ssions (e.g., that of Bhattacharyya, 1976 and Lawson, 1979). The latent heat of
vaporization can be correlated with polarizability (Myers, 1977) and cube of the
molecular diameter (Muccitelli, 1978). The temperature dependence of the latent heat
of vaporization is expressed by the equation (Schatzberg, 1963)

$$\Delta H^{T_2}_{vap} = \Delta H^{T_1}_{vap} - \Delta C_p (T_2 - T_1)$$

where ΔC_p is the heat-capacity difference between the liquid and gaseous substance.
The solubility parameter (in addition to the latent heat of vaporization and vapor
pressure) can be calculated from the corresponding-states principle, solubility data,
internal pressure $(\partial E/\partial V)$ (Dack, 1974), equation of state, critical constants,
optical data, dielectric constant, and so on (Hildebrand and Scott, 1950; Hildebrand
et al., 1970). Various methods have also been proposed by Sharma (1976), Lyckman et
al. (1965), Lozadac et al. (1977), Laprade et al. (1977), Sokolova and Pereverzev
(1977), Rawat and Gulati (1977), Ostrenga (1969), Small (1953), Peiffer (1980),
Reineck and Lin (1968), Linford and Thornhill (1977), Lombardo and Missen (1977),
Mitomo and Teshirogi (1979), Snyder (1979), Cave et al. (1980), Fedors (1974), Deak
(1973), Dayantis (1977), Bhattacharyya (1976), Breon et al. (1980), Jayasri and
Yaseen (1980a), Tager and Kolmakova (1980), Weinberg and Yen (1980), Gebelein (1978),
Jayasri and Yaseen (1980b), Kirchnerova (1975), Lawson (1978), Shinoda (1978), Ishida
(1958), Stawiszynski and Zawadzki (1979), Huggins (1972), Karger and Snyder (1976),
Khalil et al. (1976), Kirk-Othmer (1969), Koenhen and Smolders (1975), Konstam and
Feairheller (1970), Tanaka (1971), and Tees (1973). A good review of the various
estimation methods based upon chemical group contribution techniques was reported by
Ahmad and Yaseen (1977, 1974, 1978, 1979, 1980), Burrell (1955), and Jayasri and
Yaseen (1979, 1980b). An excellent discussion of the relationship between heats of
vaporization of binary mixtures as that of Majer et al. (1974).

A particularly simple estimation method when only the chemical structure of the
substance is known was reported by Rawat and Gulati (1977). For normal fluorinated
alkanes, correlation of the solubility parameter resulted in the simple expression:

$$\delta = -1.848 \, r + 6.179$$

where

$$r = \frac{\text{number of } CH_3 \text{ groups in the substance}}{\text{total number of carbon atoms in the substance}}$$

The fluorine-containing group is considered as $-CH_3$. For example, the solubility
parameter of n-fluorobutane is calculated as follows:

$$r = \frac{2}{4} = 0.5$$

$$\delta = -1.848 \times 0.5 + 6.179 = 5.250 \, (cal/cc)^{\frac{1}{2}}$$

Some typical values of solubility parameter of common liquids are (Gardon, 1977a, b; Burrell, 1975; Hansen, 1967; Högfeldt, 1976; Hoy, 1970) listed in Table 2.37.

Table 2.37 Solubility Parameter Values of Common Liquids

Liquid	Solubility parameter $(cal/cc)^{\frac{1}{2}}$
Water	23.2
Aliphatic hydrocarbons	6.7 - 8.2
Aromatic hydrocarbons	8.7 - 9.9
Perfluoro compounds	5.2 - 8.1
Halogen compounds	9.3 - 11.4
Hydroxyl compounds	10.0 - 14.5
Organic acids	10.8 - 12.6
Esters	7.9 - 10.7
Aldehydes, ketones, ethers	7.4 - 10.0
Amines	7.9 - 10.8
$R\text{-}CONH_2$, $R\text{-}NO_2$, $R\text{-}CN$	9.9 - 15.1
Organic sulfur compounds	8.8 - 13.4
Metals	40.0 - 110.0

Source: Gardon, 1977a, b.

Regular solutions were defined by Hildebrand (1929, 1935) and Hildebrand and Wood (1935): "A regular solution is one involving no entropy change when a small amount of one of its components is tranferred to it from an ideal solution of the same composition, the total volume remaining unchanged." Furthermore, Hildebrand et al. (1970) state regarding use of the model: "The regular solution model is applicable to all solutions whose molecules are randomly mixed by thermal agitation, regardless of their disparity in size. This is especially striking in the case of certain extremely unsymmetrical liquid-liquid systems, in which the molal volume of one component is as much as six times that of the other, but whose critical composition can be calculated by our formulation." Application of the regular solution theory for solubility calculations is discussed in a subsequent chapter. Meanwhile, it is worthwhile to mention that this theory does not give a satisfactory treatment of the solubility of halogenated hydrocarbons in water. However, in the case of the solubility of water in halogenated hydrocarbons, the solubility parameter provides a very useful correlation (Schatzberg, 1963; Högfeldt, 1976; Högfeldt and Fredlund, 1970; Jones and Monk, 1963; Krasnoshch'ekova et al., 1977):

$$\log_e x_2 = -\frac{V_2}{RT}(\delta_2 - \delta_1)^2$$

where x_2 =   solubility, mole fraction of water

$\quad\quad$ V_2 =   molar volume of water

δ_1 and δ_2 =   solubility parameter for solvent and water, respectively

The solubility parameter was given as 23.43 $(cal/cc)^{\frac{1}{2}}$ by Kirchnerova (1975) and Kirchnerova and Cave (1976). The authors evaluated the solubility parameter from the solubility of water in selected solvents at 25°C. Högfeldt (1976) calculated an average value of 23.0, and Shields (1976a, b) reported 23.78.

The solubility parameter has been correlated with several other physical properties (e.g., with surface tension (η, ergs cm^{-2}) (Vavruch, 1978; Shinoda, 1978) and molar volume (V_1, cc $g\text{-}mole^{-1}$)). Table 2.38 presents the values for some liquids (Gardon, 1977a, b).

Table 2.38   Correlated Solubility Parameter Values

Liquids and melts	$\delta^2 V_1/\eta$
Monobasic alcohols, organic acids	21.0 – 27.6
Water, glycerol, ethylene glycol, cyclohexanol	18.7 – 19.6
Various organic liquids	11.7 – 15.6
Molten metals	
$\quad$ Na, K, Bi, Pb, Tl, Sn, Ag, Ga, Al, Au, Cu, Fe	12.8 – 15.4
$\quad$ Hg, Cd, Mg, Zn	4.9 – 8.4

Source: Gardon, 1977a, b.

Linford and Thornhil (1977, 1980) have proposed a refinement of the theory of the solubility parameter originally introduced by Hildebrand and Lamoreaux (1974). Hildebrand (1979b) found that the classical regular solution theory requires some correction when the molecules contain methyl group(s). The new parameters increased the solvent power in proportion to the number of methyl groups per molecule.

The solubility parameter has been utilized to obtain several thermodynamic parameters for describing the physical and thermal behavior of liquids and liquid mixtures (Sharma, 1977; Snyder, 1980; Stawiszynski and Zawadzki, 1979; Srebrenik and Cohen, 1976; Wakahayashi et al., 1964).

For the solubility of water in binary nonpolar mixtures, the reader is referred to Högfeldt and Fredlund (1970).

As early as 1926, Tammann (1926) treated the gas solubility process by dividing it into two terms. The first term gives the work as being proportional to

$$4 \pi r^2 \eta$$

where r is the radius of the solute molecule and η the surface tension of the
solvent. The second term expresses the molecular solute-solvent interaction using
Londons formula for attractive forces between nonpolar molecules. This two-step
process of gas solubility has been developed further by Uhlig (1937) and Eley (1939a,
b).

The theory assumes that both the solute molecule and the cavity of the solvent
are spherical, essentially with the same radius r, and a certain reversible work is
required to create a cavity in the solvent large enough to accommodate the solute
molecule. The expression proposed consists of two terms, similar to the suggestion
of Tammann (1926). The first term gives the area of the cavity multiplied by the
surface tension of the solvent (η, ergs cm^{-2}). An excellent compilation by Jasper
(1972) of the surface-tension values available in the literature supersedes an earlier
compilation by Quayle (1953). The second term, which arises because of the interaction
of solvent and solute molecules, accounts for the repulsive or attractive forces at
intermolecular distances that are characteristic of molecular separation in liquids.
The forces depend upon the properties of the solute and solvent molecules, and the
theory of molecular interaction provides the explanation. If we designate ΔE as the
interaction energy the change in energy of the system is transferring a molecule from
the solvent phase to a low pressure gas phase is expressed by

$$\Delta U \quad = \quad 4 \, \Pi \, r^2 \, \eta \quad - \quad \Delta E$$

In this equation it has been assumed that the vapor pressure of the solvent is small
and consequently that the interaction between the solute and solvent molecules in the
gas phase is negligible; that is, for those gases that have limited interaction with
the solvent molecules, ΔE is small. Considering the dissolved water as a gas, ΔE
should remain nearly constant in the homologous series of solvents used, because the
water molecule would always be in an environment of the same groups of halogenated
hydrocarbons.

In the liquid phase, surface tension is a direct measure of intermolecular forces.
Whereas the intermolecular forces are purely dispersion forces in halogenated hydro-
carbons ($\eta_{hal.hydr.} \approx$ 15 to 25 dyn cm^{-1} at 25°C in water), (η_{H_2O} = 71.3 dyn cm^{-1}
at 25°C), in addition to the dispersion forces, there are hydrogen-bonding interac-
tions. The interfacial tension between water and halogenated hydrocarbons is about
50 dyn cm^{-1} at 25°C. The dispersion energy per unit volume of water is rather close
to that of halogenated hydrocarbons. Consequently, the heat of mixing between halogenated hydrocarbons and water is largely due to the changes in the hydrogen-bonding
interaction in water surrounding the solute molecules.

Two equations have been proposed by Gardon (1977a, b) for examination of the
interfacial tension between water and nonpolar liquids:

$$\eta_{12} \quad = \quad \eta_1 \quad + \quad \eta_2 \quad - \quad 2 \, \phi \, (\eta_1 \, \eta_2)^{\frac{1}{2}}$$

where η_{12} = interfacial tension

η_1 and η_2 = surface tension of components 1 and 2, respectively

ϕ = interaction constant

and

$$\eta_{12} = \eta_1 + \eta_2 - 2(\eta_1^d \, \eta_2^d)^{\frac{1}{2}} - 2(\eta_1^p \, \eta_2^p)^{\frac{1}{2}}$$

where the superscripts refer to the dispersion (d) and nondispersion or polar (p) components. The interaction constant (ϕ) has been determined experimentally, being in the range of 0.51 to 0.6, whereas the calculated value is 0.44. The dispersion surface tension of water $\eta_{H_2O}^d$ is 21.3 to 22.6 dyn cm^{-1}.

Grifalco and Good (1957) have examined the possible trend of the interaction constant ϕ in aqueous organic solutions and have found that the mutual solubility increases with ϕ. Mathematically, the single parameter ϕ can be calculated from the fractional polarities p_1 and p_2. This parameter ϕ is common in the mathematical formulas predicting the values of η_{12} and ΔH_{mix}:

$$\Delta H_{mix} = V_1 \, v_2^2 (\delta_1 - \delta_2)^2 + 2 \, (1 - \phi) \, \delta_1 \, \delta_2$$

This is Walker's expression for the heat of mixing.

The interaction constants ϕ and interfacial tension η_{12} of halogenated hydrocarbon – water systems are listed in Table 2.39 (Grifalco and Good, 1957).

The heat of mixing has been determined by Cheesman and Whitaker (1952) for binary liquid mixtures of carbon tetrachloride, chloroform, methylene chloride, and benzene. (A review was reported by McGlashan (1962) and an estimation method by Lai et al. (1978)). When the molecules are sufficiently similar in size and shape and interchangeable on a quasicrystalline lattice, the following expression is applicable:

$$\Delta H_{vap} = x_2(1 - x_2) \, N \, w$$

where N is Avogadro's number and w is the interchange energy.

The concentration distribution between two phases at equilibrium is described by the Maxwell-Boltzmann theory:

$$\frac{C_1}{C_g} = e^{-\Delta U / kT}$$

where C_1 = concentration of the gas in the liquid phase

C_g = concentration of the gas in the gas phase

k = Boltzmann's constant

T = absolute temperature, K

ΔU = free-energy difference in transferring a molecule from the solvent to the gas phase

Table 2.39   Interfacial Constant and Interfacial Tension in Halocarbon-Water Systems at 20°C.

Substance	Interfacial constant, ϕ	Interfacial tension, n_{12} (ergs cm^{-2})
CH_2Cl_2	0.80	28.3
$CHCl_3$	0.76	31.6
CCl_4	0.61	45.0
$CCl_2=CCl_2$	0.59	47.5
Isobutyl chloride	0.88	24.4
t-Butyl chloride	0.91	23.75
Isoamyl chloride	0.98	15.4
$CHBr_3$	0.67	40.9
CBr_4	0.69	38.8
CH_3-CH_2Br	0.78	31.2
CH_2Br-CH_2Br	0.71	36.5
$CHBr_2-CHBr_2$	0.695	38.8
$CH_2Br-CHBr-CH_2Br$	0.69	38.5
CH_2I_2	0.61	48.5
C_6H_5Cl	0.70	37.4
C_6H_5Br	0.69	38.1
C_6H_5I	0.66	41.8

Source: Grifalco & Good, 1957.

The ratio C_1/C_g is the Ostwald coefficient of solubility and the final relationship becomes

$$k\,T\,\log_e s_\gamma = -4.0\ \Pi\,r^2\,\eta\ +\ \Delta E$$

where s_γ is the Ostwald coefficient of solubility. By plotting the surface tension of solvents versus the logarithm of gas solubility, expressed as an Ostwald coefficient, a straight line is obtained for each gas at 20°C. From the slope the molecular radii of the solute molecules are calculated. The molecular radii obtained are in good agreement with the values derived from the van der Waals gas constant (Uhlig, 1937). The intercept of the straight line with the $\log_e s_\gamma$ axis gives the interaction energy ΔE. In general, if the solubility of a gas is small, ΔE is negative; if large, then ΔE is positive. From values of ΔE so determined, the free energies of solution of gases can be calculated. However, the values obtained from the temperature dependence of gas solubility in water showed a notable difference from the data calculated from the foregoing equation. This is due to the anomalous properties of water as a solvent.

Eley (1939a, b, 1953) examined Uhlig's (1937) proposal and stated that "ΔE should really be written ΔG, as strictly it is a Gibbs free energy, but this error does not effect his results." Eley's version of the two-step process is expressed when the internal energy change for solution of a mole of gas,

$$\Delta E = \Delta E_c + \Delta E_A$$

where ΔE_c = positive internal energy change of cavity formation
        ΔE_A = energy liberated when the gas molecule is accommodated in the cavity

Similarly, the entropy changes of solution is given by

$$\Delta S = \Delta S_c + \Delta S_A$$

where ΔS_c = entropy change of 1 mole of cavities of the appropriate size made
                in the liquid
        ΔS_A = entropy change of 1 mole of gas molecules transferred from the gas
                phase into the cavities in the solvent

The temperature coefficient of gas solubility is determined by the magnitude of ΔE_c and ΔE_A:

$$-\Delta E_A > \Delta E_c \quad \text{negative temperature coefficient}$$

$$\Delta E_c > -\Delta E_A \quad \text{positive temperature coefficient}$$

Eley (1939a, b) observed that in the case of gas solubility in water at room temperature, ΔE_c is negligible if the heat of solution is exothermic. This has been also illustrated on a curved plot for $\log_e s_\gamma$ versus $1/T$. At low temperatures the solution is in a quasilattice structure, whereas at higher temperatures it forms a close-packed liquid and for the quasilattice formation a ΔE_c is required. Most gases show negative temperature coefficients in water, and the $\log_e s_\gamma$ versus $1/T$ plots are distinctly curved (Valentiner, 1927). The positive temperature coefficient observed for some gases in water above $100^\circ C$ is due to the large energy required to form cavities at these temperatures, which outweights the solute-solvent attraction.

The thermodynamic values of solution ΔE°, ΔH°, and ΔS° have been calculated from the free-energy change when 1 mole of gas is transferred from an infinite volume of gas at 1 atm pressure to an infinite volume of solution at a concentration of 1 mole liter^{-1} $\Delta\mu^\circ$:

$$-\Delta\mu^\circ = R T \log_e s_\gamma - R T \log_e(0.082\ T)$$

Perona (1979) discussed further the free-energy change during the vaporization of a liquid. The temperature dependence of a gas is expressed by Valentiner's equation:

$$\log_{10} s_\gamma = \frac{A}{T} + B \log_{10}T - C$$

where s_γ is the solubility of a gas expressed by the Ostwald coefficient and A, B, and C are the coefficients of the equation. These coefficients are characteristic for a gas. The entropy of solution is given by

$$-\Delta S^o \quad = \frac{\partial}{\partial T}(\Delta\mu^o) \quad = \quad R\left[-B \quad -B \log_e T + \quad 2.30 \ C \quad + \quad \log_e(0.082 \ T) + 1\right]$$

The heat of solution is

$$\Delta H^o \quad = \quad -T^2 \frac{\partial}{\partial T}\left(\frac{\Delta\mu^o}{T}\right) = \quad R(-2.30 \ A \quad + \quad B \ T \quad - \quad T)$$

The internal energy of solution is expressed by

$$\Delta E^o \quad = \quad \Delta H^o \quad + \quad R \ T$$

ΔS^o and ΔH^o values are both strongly dependent upon temperature (Williamson, 1944). The internal energy change of cavity formation ΔE_c has been expressed as a function of the thermal expansion coefficient $\alpha \ (K^{-1})$ and the compressibility $\beta(atm^{-1})$:

$$\Delta E_c \quad = \quad 0.0241 \ T \ \frac{\alpha}{\beta} \ \bar{V} \quad cal \ g\text{-mole}^{-1}$$

where $\bar{V}$ is the partial molal volume of the gas (cc g-mole^{-1}). Similarly:

$$\Delta S_c \quad = \quad 0.0241 \ \frac{\alpha}{\beta} \ \bar{V} \quad cal \ g\text{-mole}^{-1} \ K^{-1}$$

For organic solvents, ΔE_c is a large positive quantity and not much effected by temperature; for water it is zero at 4°C, but increases strongly with the temperature. This explains why ΔE_c is positive for organic solvents but negative for water, and also why ΔE_c for water becomes much less negative with increasing temperature, corresponding to large positive values of the partial molal heat capacity of the gas in water (Alexander et al., 1971; Arnett et al., 1969).

At 200°C, ΔE_c is about 2.15 kcal g-mole^{-1}, and since the effect of ΔE_A is small, ΔE_c becomes positive. The explanation is that at high temperatures the gas molecules share the quasilattice points of water equally with the water molecules and water now behaves similarly to organic solvents at more normal temperatures, although at 20°C, the difference in the internal energy change $\partial(\Delta E)$ for solution of a gas is roughly 3.5 kcal g-mole^{-1} between water and organic solvents. The corresponding value for $\partial(\Delta S)$ is about 12 cal g-mole^{-1} K^{-1}. Meanwhile, Eley (1939b) stated that ΔS_A for a given gas is much the same whether the solvent is water or an organic solvent. Although at room temperature the gas molecules in water do not share quasilattice points with the water molecules but occupy a special cavity, in organic solvents the gas molecules share the quasilattice points of the solvent molecules at room temperature. The first conclusion involves the smaller free volume and the higher concentration term $\left[(N_S + N_G)/N_G\right]$ already established for the pure solvent. Therefore ΔS_A will not be very different for the same gas in water or organic solvents. Despite the fact that the heat of solution is exothermic and anomalous of small gas molecules in water at constant pressure because of the energy of cavity formation (which is smaller than for organic solvents), still for a large number of gases, ΔE_A is similar for water and organic solvents (Namiot, 1967). ΔE_A is determined primarily by London forces and the nature of the gas molecules, and the effect of solvent is not very relevant. The difference in the entropy of solution of gas molecules in water and in organic solvents is to be attributed to the relatively large increase in the entropy of the organic solvents associated with the formation of a cavity in the solvent for the gas molecule.

The temperature effect upon the heat and entropy of solution is considerably different in water than in organic solvents. This is explained by the temperature effect upon the thermal expansion coefficient α and the compressibility β. Although ΔE_c and ΔS_c are increased strongly in water through the effect of temperature, in organic solvents they do not show a large variation with temperature. The large variation in heat and entropy of solutions in water are quite anomalous, arising from the difference energies required to form a cavity in the solvent in which the gas molecule is accommodated. In organic solvents that are made up of quasilattice points, the energy requirement for cavity formation is considerable less compared to water. At higher temperatures the cavities are less readily formed, so that at 80 to 100°C, water behaves in a manner more like that of nonpolar solvents.

The small free spaces present in water are capable of expansion, causing a small increase in energy and entropy that arises because the structures forming around the cavity compensate for the broken hydrogen bonds in the **expansion process.** Eley (1939a, b) related the energy change ΔE_A to the polarizability of gas molecules. The nearly linear correlation was explained by assuming that the cavities in water are somewhat larger than gas molecules (Eley, 1953). The plot shows a continuously increasing tendency, which might be due to the neglect effect of induction, orientation, and repulsion forces.

Two solubility mechanisms acting against one other were suggested by Bohon and Claussen (1951) to explain the solubilities of nonpolar solutes in water. One is the positive heat of solution acting against the negative heat of solution. Whereas the positive heat of solution is responsible for the heat of cavity formation, the negative heat is due to the formation of a more "icelike" structure around the solute molecule and/or a complex formation involving the labile Π-electrons of the aromatic nucleus. The affinity between the solute and solvent molecules makes it possible to form an icelike structure, or clathrate, which is an associated cage around a solute molecule. In the case of a complex-compound formation between the hydrogen ion and the solute molecule, the effect should be wider than for one single molecule. In both case, the Π-electrons would function as a base and the water as an acid.

Among the aromatic hydrocarbons investigated (e.g., benzene, toluene, ethyl-
benzene, and xylenes), they found that at approximately $18^\circ C$ the heats of solution
are zero. That is, above $18^\circ C$, the heats of solution of liquids (heat of cavitation)
are positive, and below this temperature they are negative (freezing or complex formation). **Because of the high partial molal heat capacity of aromatic hydrocarbons in**
water, their heat of solution will depend strongly on temperature. Bohon and Claussen
(1951) suggested that this might have been caused by stretching of the hydrogen bonds
in the ice shell surrounding the solute molecule. Measurement of partial molal volumes
of solution as a function of temperature and pressure will reveal whether the
suggestion is correct.

Solute-solvent interactions between organic compounds and water have been studied
by Teresawa et al. (1975) through solute-solvent interaction partial molar volume
$\bar{V}_{s-s}^{(HC)}$. Hydrocarbons were considered as standard materials, and all other organic
substances were related to them. It has been assumed that hydrocarbons do not interact strongly with liquid water molecules, and $\bar{V}_{s-s}^{(HC)}$ has been defined by the
expression

$$\bar{V}_{s-s}^{(HC)} = (\bar{V}_{void} - \bar{V}_{void}^{(HC)})_{V_w} = V_w^{(HC)}$$

$$= (\bar{V}^\circ - \bar{V}^{\circ,(HC)})_{V_w} = V_w^{(HC)}$$

where $\bar{V}_{void}$ = void partial molar volume of solute, ml g-mole^{-1}

$\bar{V}_{void}^{(HC)}$ = void partial molar volume of hydrocarbon, ml g-mole^{-1}

$\bar{V}^\circ$ = partial molar volume of solute, ml g-mole^{-1}

$\bar{V}^{\circ,(HC)}$ = partial molar volume of hydrocarbon, ml g-mole^{-1}

In the preceding equation, the partial molar volume and the void partial molar volume
of the solute are compared with those of the corresponding hydrocarbon values. The
calculated $\bar{V}_{s-s}^{(HC)}$ represents the additional void partial molar volume, which is
caused by the hydrophilic molecular properties of a solute, compared to the same for
the hydrocarbon as solute. The value of $\bar{V}_{s-s}^{(HC)}$ is dependent on the differences in
the molecular properties of the solvent.

The absolute values of $\bar{V}_{s-s}^{(HC)}$ show tendencies for slight decreases with lengthening of the alkyl group or addition of a methyl group for the series of substances
(e.g., a homologous series). It has been concluded that the molecular properties of
a solute in a water solvent reflect the value of a partial molar volume (Roux et al.,
1978).

The theory introduced by Uhlig (1937) and Eley (1939a, b) and developed further
by Pierotti (1963, 1965) is now known as the scaled-particle theory for the solubility
of gases in liquids. The energy parameter ε_1/k and the hard-sphere diameter σ_1 of

the solvent molecules are obtainable from vapor viscosities or second virial coefficients and the van der Waals diameter, respectively. Generally, the solvents might
consist of globular, elongated, gaint, apolar, moderately, and strongly polar aprotic
molecules.

The original scaled-particle theory does not take account of the strong directional interactions due to the hydrogen bonding in water. Therefore, Stillinger
(1973) proposed to improve Pierotti's theory by incorporating a radial distribution
function for pure water. A more detailed discussion of scaled-particle theory is
given in Sec. 2.1.

The anomalous entropies of solution of the inert gases in water were explained
by Frank and Evans (1945) by the so-called iceberg formation. In this model the solute
molecules are surrounded by a layer of ordered water molecules. When the temperature
is raised, the iceberg melts, which determines the temperature dependence of entropies
of solution in water. For a further discussion of the icing model, see Sec. 2.6.

The aqueous solutions of fluorinated hydrocarbons have been studied by several
investigators (e.g., Friedman, 1954; Longuet-Higgins, 1951; Hildebrand and Scott,
1962; Muccitelli, 1978; Wen and Muccitelli, 1979). They found that these solutions
do not follow the Barclay-Butler (1938) rule, that is, the linear relationship between
entropy and enthalpy of solutions (see Chap. 5). The low solubilities of fluorinated
gases have been attributed to the unusually large entropy decrease on forming the
solution (Friedman, 1954). Ashton et al. (1968) proposed a revision of the original
thory of Powell and Latimer (1951) for the relationship between entropy of solution
ΔS_{soln} at $25^{\circ}C$ and the volume of the liquid solute at its normal boiling point
$(V_{b.p.})$:

$$-\Delta S_{soln} = a + b V_{b.p.}$$

Ashton et al. (1968) found that $a = 25$ and $b = 0.22$ for the entropy of solution
of nonpolar gases in water at $25^{\circ}C$. A modification of the equation has been suggested
by Miller and Hildebrand (1968), for aqueous solutions in general:

$$-\Delta S_{soln} = a + b V_{b.p.}^{2/3}$$

From the study outlined above, Ashton et al. (1968) concluded that the anomalous large
negative entropy of solution of fluorinated gases does not effect their low solubilities, but that they are affected by the size of the molecules. Kozlova and Korol
(1971) used a statistical relationship for the calculations.

The cohesion forces and phase theory of binary liquid systems were studied by
Staverman (1941a) by considering only fundamental atomic properties such as radius
and electronegativity. This approach is directed particularly toward the understanding
of the phenomenon of mixing and separating of liquids using a simple model of cohesion
forces. Among the various cohesion forces acting between molecules in a liquid-liquid

mixture, Staverman neglected the induction and dispersion forces, which do not
contribute significantly to the heat of mixing. The heat of mixing originates from
orientation energy only, which is characterized by the molecular surface area and the
entropy of association of the positive and negative ends of the molecules. The orientation energy is the most important reason why the heat of mixing is sufficient to
produce immiscibility. The orientation energy E_{orient} is the product of the orientation forces of a molecule x, which varies from point to point on the surface and
touches another molecule at a point x':

$$E_{orient} = x\, x'$$

Whether energy is gained or formed, it depends upon the signs of x and x'. For
example, when a molecule whose positive end associates x_{+2} touches another molecule having negative end x_{-1}, association energy is gained:

$$E_{orient} = x_{+2}\, x_{-1}$$

In addition to the quantities x and x', there is also a need to account for the
entropy of association of the positive S_{+} and negative S_{-} ends of the molecule.
These quantities depend upon the area of the polar surface.

In the case of a polar-nonpolar liquid mixture, the mixing requires much energy
$x_{+1}\, x_{-1}$; therefore, at low temperature there is only limited solubility. However, at
high temperatures entropy becomes a dominating parameter and the liquids are miscible
and form an upper critical mixing temperature. In most cases the solubility curve,
T versus x, is symmetric, which is explained by the energy increase when component 1
is in great dilution in component 2 and the solubility is strongly temperature-
dependent. When component 2 is dissolved in component 1, no dissociation will occur
in complexes of component 1 molecules at low temperatures. Therefore, $E_{orient} = 0$
and the solubility is practically temperature-independent.

Regarding the immiscibility relationship between water and organic compounds,
Staverman gives the following explanation. Since in water $S_{-1} > S_{+1}$ and x_{+1} and x_{-1}
are large, one must mix water with molecules of large x_{-} and small S_{-} to obtain
a closed miscibility gap. Such molecules are ethers, ketones, and pyridine derivatives
having electronegative atoms. The miscibility gap becomes wider when the molecular
surface area increases, as in the case of a homologous series. Consequently, the upper
critical temperature of immiscibility increases and the lower critical temperature
decreases as a result of the increases of energy of solution in water.

The solubility of water in aromatic compounds is higher than in aliphatic hydrocarbons. For example, water is more soluble in benzene than in n-hexane or cyclohexane. The positive heat of mixing of benzene and chloroform indicates that benzene
has a rather large x_{-}, owing to a distribution of electrons with a region of high
density not far from the surface of the molecule. Therefore, aromatic compounds show
a more polar character.

Table 2.40   Thermodynamic Data for Water Solubility at 25°C

Solvent	Solubility of H_2O, x_{H_2O}	ΔG (kcal g-mole^{-1})	ΔU (kcal g-mole^{-1})	ΔS (cal g-mole^{-1}K^{-1})
CCl_4	0.000991	4.09	6.34	7.6
$CHCl_3$	0.006230	3.01	4.15	3.7
CH_2Cl_2	0.00788	2.87	4.59	5.7
CH_3-CHCl_2	0.00531	3.10	5.00	6.3
CH_2Cl-CH_2Cl	0.01025	2.71	4.81	7.0
CH_2Br-CH_2Br	0.00685	2.95	–	–
CH_3-CCl_3	0.00257	3.30	5.30	5.8
$CH_2Cl-CHCl_2$	0.00879	2.80	5.16	8.0
$CHCl_2-CHCl_2$	0.01020	2.71	4.21	5.1
$CH_2Cl-CCl_3$	0.00513	3.12	5.34	7.6
$CHCl_2-CCl_3$	0.00391	3.28	5.14	6.3

Source: Staverman, 1941b.

The miscibility of water and alkyl halides was examined by Staverman (1941b) by determining the mutual solubilities at 0, 25, and 30°C. The free energy, total energy, and entropy of solution of water in great dilution in different alkyl halides have also been calculated at 25°C, using the following expressions (Williamson, 1944; Hollenbeck, 1980):

$$\Delta G = -R\,T\,\log_e x_{H_2O}$$

$$\Delta U = -R\,\frac{\partial(\log_e x_{H_2O})}{\partial(1/T)}$$

and

$$\Delta S = \frac{\Delta U - \Delta G}{T}$$

The values are presented in Table 2.40. Staverman found that the solubilities of water in alkyl halides and of alkyl halides in water are closely correlated. From the experimental results, he concluded that water is about 5 to 10 times more soluble in alkyl halides, expressed as a mole fraction, than alkyl halides in water at the same condition. This is due to the large surface of the organic molecules and the energy required for cavity formation.

When an H atom is replaced by a Cl atom, three effects upon the cohesion energy result:

1. The dispersion energy increases.
2. The symmetry of the charge distribution is changed.
3. The size of the molecule increases.

Whereas items 1 and 3 decrease the solubility with an increasing number of halogen atoms, item 2 increases the solubility, which is confirmed by the solubilities of isomers. Therefore, between two isomers, the largest number of partially halogenated carbon atoms shows the greatest solubility for water: for example,

$$CH_3-CHCl_2 \quad < \quad CH_2Cl-CH_2Cl$$

$$CH_3-CCl_3 \quad < \quad CH_2Cl-CHCl_2$$

$$CH_2Cl-CCl_3 \quad < \quad CHCl_2-CHCl_2$$

The orientation forces of the molecules are influenced by the positively charged spots on the surface. When a carbon atom is partially halogenated, the remaining H atoms are positively charged and can attract water molecules. Such molecules, which do not have positively charged H atoms (e.g., CCl_4, CH_3-CCl_3) show the least solubility for water.

The influence of methy- and methylene-groups on aqueous solubility is twofold: stearic and electronic (Albert, 1973). In general, the methyl-group is water-repelling, it usually lowers solubility, but there are exceptions. To dissolve a substance it is necessary to break the strong hydrogen bonding that exists between water molecules. This is easier for smaller than larger molecules, which can force the water molecules apart.

Another striking case is the "cut-off" effect of $-CH_2$ group in homologous series. The addition of just one more $-CH_2$ group changes the linearity of the solubility dependence with the number of carbon atoms in the molecule.

The methyl group releases electron in any environment, consequently it strengthens a base and weakens an acid. Methyl groups unstabilize a molecule by replacing a hydrogen atom that would otherwise be split out in conjuction with a neighboring atom. The attraction of one molecule for another is greater than its attraction for water molecules. As a result, the crystal lattice energy is increased, and the substance falls out of solution. This however, happens more often among heterocyclic substances.

Liquid-liquid immiscibility originates primarily from cohesion forces (Garett, 1972). The dispersion forces have only a marginal effect for very polar molecules such as water. This is also illustrated by the symbatic nature of the water solubility at the same condition: for example, the solubility of water in chlorinated methanes decreases in the order (molal solubility)

$$CH_3Cl \quad > \quad CH_2Cl_2 \quad > \quad CHCl_3 \quad > \quad CCl_4$$

Similarly, as Svetlanov et al. (1971) have demonstrated, the molal solubility of chloromethanes in water shows the same decreasing trend, with a slight modification:

$$CH_2Cl_2 > CH_3Cl > CHCl_3 > CCl_4$$

It was discussed in great detail in Sec. 2.3 that there is very little or no interaction between water molecules in nonpolar solvents and therefore that water molecules are mostly monomeric in structure. However, experimental evidence shows that there is some sort of attraction between water and organic substances. This has been concluded from the difference in the apparent molal volume of water in various organic solvents. The apparent molal volume has been found to be low in most non-polar solvents, except in carbon tetrachloride. The low values of the apparent molal volumes indicate that there is large water-solvent interaction in these liquids.

The early works of Staverman (1941a, b) and Staveley et al. (1943) reflect the interaction between water and benzene molecules. Later, Högfeldt and Bolander (1963) confirmed the interaction by showing that a 1:1 complex was formed between water and benzene. By examining the solubility of water in aliphatic halides, Staverman suggested that the chlorine atom attached to the hydrocarbon chain activates the C-H bond, and therefore it will interact with water molecules. The interaction and solubility of water will increase as the number of partially halogenated carbon atoms increases. In completely halogenated molecules there is no C-H bond, and therefore the sol-ubility or interaction of these molecules with water would be very small. This is the explanation for the low solubility of water in completely halogenated hydrocarbons.

Solute-solvent interaction between water and chlorinated hydrocarbons was also studied by Seiler (1968). For the calculation of the magnitude of the interaction, he proposed the following scheme:

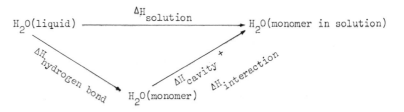

That is, the heat of solution is broken up into three steps. The first step is the formation of water monomers by breaking the hydrogen bonds in liquid water. During the bond breaking, heat is absorbed. The second step involves the cavity formation for the solute molecule in the solvent. This process also involves the heat of absorp-tion. The third step is the heat of interaction between water and the solvent. The equation is

$$\Delta H_{solution} = \Delta H_{hydrogen\ bond} + \Delta H_{cavity} + \Delta H_{interaction}$$

The heat-of-solution values have been calculated from the solubility data using the relation (May et al., 1978a, b):

$$\log_{10} x_2 = \frac{\Delta H_{solution}}{R T} + constant$$

The value of $\Delta H_{solution}$ was calculated from the slope of the $\log_{10} x_2$ versus $1/T$ plot: that is,

$$\Delta H_{solution} = 2.303 \times 1.9872 \times slope$$

The exact calculation of $\Delta H_{solution}$ was discussed by Williamson (1944). The heat of hydrogen bond formation is 6.40 kcal g-mole^{-1} (Pimentel and McClellan, 1960). The heat of cavity formation can be calculated from a method proposed by Pierotti (1963). The calculated values are summarized in Table 2.41 (Seiler, 1968). Solubility and calorimetric data were compared by Cesaro and Russo (1978). Iguchi (1977) discussed the heat and free energy of mixing.

The heat of solution of water in carbon tetrachloride derived by Glasoe and Schultz (1972) deviates from that reported by Seiler (1968) at 25°C. The discrepancy is 17.4%, which is considerable relative to the value Seiler reported. Some other thermodynamic data for the solution of water in carbon tetrachloride are, at 25°C (Glasoe and Schultz, 1972; and Staverman, 1941b, in parentheses):

$$\Delta G_{solution} = -1.27 \quad (4.09) \text{ kcal g-mole}^{-1}$$
$$\Delta S_{solution} = 23.0 \quad (7.6) \text{ cal g-mole}^{-1} K^{-1}$$
$$\Delta H_{solution} = 5.4 \quad (6.34) \text{ kcal g-mole}^{-1}$$

The free energy of solution $\Delta G_{solution}$ is calculated from the expression

$$\Delta G_{solution} = -R T \log_e K_s$$

where K_s is the solubility of water in carbon tetrachloride, expressed in mol liter^{-1}, and the entropy of solution $\Delta S_{solution}$ is obtained from the relation

$$\Delta S_{solution} = -\frac{\Delta G_{solution} - \Delta H_{solution}}{T}$$

The negative sign with the heat of interaction indicates that there is an attraction between the water and nonpolar molecules. The $\Delta H_{interaction}$ values are approximately the same in all cases except for CCl_4. The symbatic character has also been observed for the apparent molar volume of water in these solvents. The interaction energy between water and carbon tetrachloride is lowest, and water has a large expansion of the same among the solvents examined. Positive heat-of-mixing values are always accompanied by expansion.

Table 2.41 Thermodynamic Properties of Interaction Between Water and Nonpolar Solvents at 25°C

Solvent	$\Delta H_{hydrogen\ bond}$ (kcal g-mole^{-1})	ΔH_{cavity} (kcal g-mole^{-1})	$\Delta H_{solution}$ (kcal g-mole^{-1})	$\Delta H_{interaction}$ (kcal g-mole^{-1})
CCl_4	6.40	2.68	6.34 (5.40)[a]	-2.72
CH_3CCl_3	6.40	2.71	5.30	-3.81
CH_3CHCl_2	6.40	2.74	5.00	-4.10
C_6H_6	6.40	2.58	5.10	-4.25
CH_2ClCH_2Cl	6.40	3.15	4.90	-4.46

[a]Reported by Glasoe & Schultz (1972).

Source: Seiler, 1968.

The low solubility of nonpolar solutes in water $(\Delta G_{solution} > 0)$ is explained by the large negative value of $T\Delta S_{solution}$ even though $\Delta H_{solution}$ is negative (i.e., exothermic) (Blandamer and Burgess, 1975). This is due to the structure former behavior of solutes in water. The water molecules are organized around a nonpolar molecule like that found in clathrate hydrates. The water-water interaction is responsible for the positive partial molar heat capacity of the solute, which increases as the hydrophobic content of the solute increases.

In an article, Shinoda (1977) has reported on the water solubility of nonpolar liquids. He concluded that the low solubility values are caused by the iceberg formation of water molecules surrounding solute molecules. An explanation by so-called hydrophobic bonding is elucidated. The hydrophobic interaction was also studied by several workers (Ben-Naim, 1978a, b, 1980; Ben-Naim and Wilf, 1980; Abraham, 1979; Hildebrand, 1979a; Scheraga, 1979; Tanford, 1973, 1979, 1980; Wen and Muccitelli, 1979; Gill and Wadsö, 1976; Schönfeld and Seibt, 1976; Wen and Hung, 1970; Yaacobi and Ben-Naim, 1974; Lyashchenko and Stunzas, 1980; Hofmann and Birnstock, 1980).

The effect of iceberg formation in aqueous solutions of nonpolar compounds on the temperature dependence of solubility was shown by plotting the logarithm of solubility expressed in mole fraction (x_2) versus the reciprocal of the absolute temperature $(1/T)$, according to the expression

$$\Delta H_{solution} = -R \left[\frac{\partial \log_e x_2}{\partial (1/T)} \right]_{satn}$$

where $\Delta H_{solution}$ is the enthalpy of solution and R the gas constant. That is, the slope of the solubility curve multiplied by $-R$ is equal to the enthalpy of solution. For example, the heat of solution (or mixing) of water in perfluorohexane between 2 and $53^{\circ}C$ was expressed by (Shields, 1976a, b):

$$\Delta H_{solution} \ (cal \ mole^{-1} \ solute) = -R \frac{835,818.4}{T(K)} - 5999.96$$

In the case of the water solubility of nonpolar substances, the solubility curve always deviates downward or upward from the straight line in a $\log_e x_2$ versus $1/T$ diagram between, for example, 0 and $300^{\circ}C$. Whereas the solubility of solids decreases more rapidly by lowering the temperature (see Fig. 2.33 for p-dichlorobenzene), the nonpolar liquids show an upward tendency going through a minimum. Above $160^{\circ}C$, the iceberg formation of water disappears and the constant increase in solubility becomes characteristic. The order of magnitude of the enthalpy of solution will vary with the slope of the solubility curve. Above $160^{\circ}C$, the enthalpy of the solution is a large positive value, because of the absence of the iceberg formation. By lowering the temperature, the enthalpy of solution gradually diminishes to negative values when the iceberg formation of water molecules surrounding the solute molecules becomes profound.

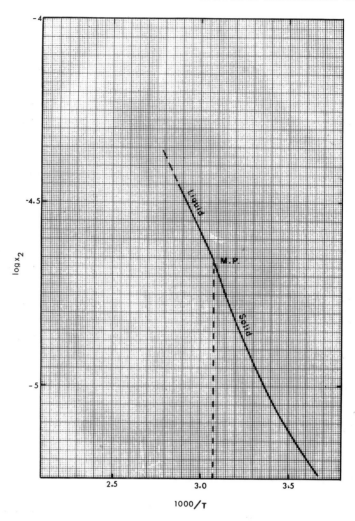

Fig. 2.33 Solubility of p-dichlorobenzene in water. x_2 = solubility in water (mole fraction); T = absolute temperature (K). (From Wauchope & Getzen, 1972.)

This is the result of a large positive value (no iceberg formation and a straight solubility curve) and a large negative enthalpy of iceberg formation (destruction of hydrogen-bonding interactions and the formation of icebergs are the dominant processes). Accurate $\Delta H_{solution}$ values can be obtained from the expression above at temperatures above 160°C when the solubility curve becomes a straight line (i.e., the slope is constant) (see, e.g., May et al., 1978a, b).

The iceberg formation is not the entropy effect. The entropy of solution of a nonpolar substance in water is expressed by

$$\Delta S_{solution} \; = \; -R \, \log_e x_2 \; + \; \Delta S_{hydrogen \; bond} \; + \; n \; \Delta\bar{S}_i$$

where $\Delta S_{solution}$ =   entropy of solution

$-R \, \log_e x_2$ =   entropy of dilution

$\Delta H_{hydrogen \; bond}$ =   partial molal entropy of the solute due to the decrease in hydrogen-bonding interactions

$n \, \Delta\bar{S}_i$ =   partial molal entropy of the solute due to iceberg formation of n molecules of solvent

Above 160°C, the expression becomes

$$\Delta S_{solution} \; = \; \frac{\Delta H_{solution}}{T} \; = \; -R \, \log_e x_2 \; + \; \Delta S_{hydrogen \; bond}$$

Approximate values for $\Delta H_{solution}$ and $\Delta S_{solution}$ are presented in Table 2.42. The ratio

$$\frac{\Delta H_{solution}}{\Delta S_{solution}} \; = \; 360 \rightarrow 380 \text{ K}$$

for these compounds are listed in Table 2.42. It is apparent from this table that the low solubility of nonpolar liquids in water is caused by the large positive enthalpy of mixing, and the enthalpy of iceberg formation is largely canceled by the accompanying entropy decrease at 25°C. The large negative standard entropy of solution also results from iceberg formation.

The heat-of-solution values for benzene in water at 25°C vary considerably from one investigator to another (Lyashchenko and Kalinowska, 1977). Despite the good agreement for the solubility of benzene in water, more and more difficulty arises for the establishment of reliable heat-of-solution data. The discrepancy is expressed by a factor of 5 to 8. It becomes apparent that there is an urgent need for critical examination of heat-of-solution data reported in the literature and to establish theoretically sound relations for checking published values. Some of the collected thermodynamic data for solutions are presented in the Appendix. Further data have been reported by Jones et al. (1957), Herington (1951), Alexander et al. (1971), Rossky and Friedman (1980), and May (1980).

An excellent collection of the literature sources of mixing and excess thermodynamic properties was published by Wisniak and Tamir (1978). The book includes binary, ternary, and multicomponent systems. In addition to the heat, entropy, and Gibbs free energy of solution (mixing), the specific heat, internal energy, and volume of mixing are presented. In the introduction the authors discuss the fundamental relationships and definitions applicable to mixing and excess thermodynamic properties, for example,

Table 2.42 Thermodynamic Data for the Solubility of Liquid Hydrocarbons in Water at 25°C

Solute	$\Delta H_{solution}$ (cal g-mole^{-1})	$\Delta H_{hydrogen\ bond}$ (cal g-mole^{-1})	$n\ \Delta \bar{H}_i$	$\Delta S_{solution}$ (cal g-mole^{-1}K^{-1})	$R\ \log_e x_2$ (cal g-mole^{-1}K^{-1})	$\Delta S_{hydrogen\ bond}$ (cal g-mole^{-1}K^{-1})	$n\ \Delta \bar{S}_i$
C_6H_6	580	8100	-7,520	1.9	15.4	7.2	-20.7
$C_6H_5CH_3$	640	9260	-8,620	2.2	17.9	7.9	-23.6
$C_6H_5C_2H_5$	400	10,500	-10,100	1.4	20.4	8.4	-27.4

Regular solution theory

Barker-Guggenheim lattice theory

Flory-Huggins theory

Wohl theory

Redlich-Kister equation

Wilson equation

NRTL equation

Further selection of data on mixtures was reported by the Thermodynamic Research Center (1975).

The concept of order and disorder in solutions was studied by Patterson (1976) and Patterson and Barbe (1976). Our main interest in this book involves the conclusion they derived for hydrocarbon-water systems. The molecular size and shape are important and sensitive parameters in determining the excess thermodynamic quantities. The small differences of shape has been shown on the heat-of-mixing values. Kozlova and Korol (1971) introduced a linear statistical relation between $\Delta H_{solution}$ and the number of substituents in the benzene ring.

By introducing solute molecules (rare gases or hydrocarbons) into water, an increase of order was observed. At higher temperatures, about above $100^{\circ}C$, the enchainment of structure becomes less. If the thermodynamic properties of transfer -- $\Delta H_{transfer}$, $\Delta S_{transfer}$, and $\Delta G_{transfer}$ -- are controlled by the order-increase contributions, all these values would become negative. However, the experimental results show that $\Delta G_{transfer}$ is large and positive, corresponding to the hydrophobic character of the solute-water interaction. This is explained by the presence of effects other than the order and structure of the system. When molecules of widely different natures are present in a system, such as in a halogenated hydrocarbon-water mixture, $\Delta G_{transfer}$ becomes positive as a result of iceberg formation (Shinoda, 1977; Shinoda and Fujihira, 1968).

2.5-3 Nonideal Solutions: Activity Coefficient

The nonideal behavior of binary liquid systems is usually studied from measurements of their vapor pressure, boiling points, and heat of mixing. The nonideal behavior arises out of intermolecular forces of the two liquids and their properties. The close similarity between the variations of solubility and total vapor pressure with composition of the liquid system suggests that the intermolecular forces, which give rise to the nonideal behavior of the liquid system also give rise to the nonideal behaviour in the solubility.

The pressure-composition relationship in a solution was studied by Raoult (1888) for methyl salicylate in diethyl ether. For this nonelectrolyte solution he found that at constant temperature the partial pressure of solvent (diethyl ether) in the vapor phase (p_1) was proportional to its mole fraction in solution (x_1). This proportionality has been generalized to all ideal systems with various concentrations. Raoult's law expresses the proportionality:

$$p_1 \quad = \quad p_1^o \, x_1$$

where the proportionality constant p_1^o is the vapor pressure of the pure solvent. (The vapor pressures of halogenated hydrocarbons are available in the literature, see West, 1976; Boublik et al., 1973; ASHRAE Tables, 1969; Jordan, 1954; Gallant, 1968-1970; Engineering Science Data Units Item No. 75010, 1975; Landolt-Börnstein, 1960; Lange, 1967; Ambrose, 1977; Ohe, 1976.) The mole fraction is a convenient unit in which to express the concentration. The conversion formulas for the expression or conversion of the various units for concentrations are tabulated in the Appendix. In a binary system it is costumary to index the solvent by 1 and the solute by 2; for example, if the concentrations are expressed as mole fractions, x_1 and x_2 represent the solvent and solute concentrations, respectively.

In a binary system, the total pressure P is the sum of the partial pressures of components (Dalton's law):

$$P \quad = \quad p_1 \, + \, p_2 \quad = \quad p_1^o \, x_1 \, + \, p_2^o \, x_2$$

where

$$x_1 \, + \, x_2 \quad = \quad 1$$

Therefore,

$$P \quad = \quad p_1^o \, x_1 \, + \, p_2^o (1 - x_1)$$

or

$$P \quad = \quad p_2^o \, + \, x_1 (p_1^o \, - \, p_2^o)$$

This expression shows that by plotting the total pressure (P) of a binary solution versus the mole fraction of the solvent (x_1), we obtain a straight-line relationship, similarly to the partial pressure $(p_1$ and $p_2)$ versus mole fraction $(x_1$ and $x_2)$ plots. Although Raoult's law valid only for ideal solutions, it has a very wide application, particularly to determine the deviation from ideality. Furthermore, to use Raoult's law, only the properties of the pure components are required. Therefore, it is costumary to calculate the properties of a binary solution using Raoult's law. The data obtained represent the standard behavior compared to the real behavior of the solution. This is commonly called the deviation from ideality.

In real solutions Raoult's law is not applicable; therefore, it became neccessary to modify it by introducing some sort of parameter that takes care of the deviation:

$$p_1 \quad = \quad \gamma_1 \, x_1 \, p_1^o$$

where γ_1 is the activity coefficient of the solvent. For a real binary solution the expression becomes

$$P = \gamma_1 x_1 p_1^o + \gamma_2 x_2 p_2^o$$

where γ_2 is the activity coefficient of the solute.

Regarding the vapor phase above the solution, Dalton's law gives

$$p_1 = y_1 P$$

where y_1 is the mole fraction of component 1 (solvent) in the vapor phase. Now, if we assume that the vapor phase is ideal (i.e., Dalton's law is applicable), the equilibrium condition between the real solution and its ideal vapor can be expressed by

$$P y_1 = \gamma_1 x_1 p_1^o$$

The assumption of an ideal vapor phase is a good approximation for cases when the pressure above the solution is low (i.e., below 1 atm) and there is no association in the vapor phase. Otherwise, for nonideal vapor and liquid systems at equilibrium, the modified expression becomes

$$P y_1 \Phi_1 = \gamma_1 x_1 p_1^o$$

where Φ_1 is the fugacity coefficient of component 1 (solvent) in the vapor phase and P is the total pressure of the system. At high pressures the deviation from ideality becomes significant and it is necessary to take it into account. The vapor-phase fugacity coefficient Φ_1 is a function of composition (y_1), temperature (T), and pressure (P). There are various thermodynamic expressions for the calculation of Φ_1, for example,

$$\Phi_1 = \frac{f_1^o}{y_1 P} = \exp\left[\frac{1}{R T} \int_0^P \left(\bar{V}_1 - \frac{R T}{P}\right) dP\right]$$

where f_1^o is the fugacity of component 1 (solvent) (Mackay, 1979) reported a good description of the concept of fugacity with practical illustrations.) and $\bar{V}_1$ the partial molar volume of constituent 1. The latter is calculated from the apparent molal volume (Seiler, 1968). A list of $\bar{V}_1$ values is given in Table 2.43. The apparent molal volume of water is independent of concentration. The densities of halogenated hydrocarbon-water systems are compiled in Landolt-Börnstein (1977). A bibliography of liquid density was reported by Hales (1980).

The vapor-phase fugacity coefficient of component i (Φ_i) could be expressed by an alternative expression

$$\log_e \Phi_i = \left(2 \sum_{j=1}^N y_j B_{ij} - B\right) \frac{P}{R T}$$

Table 2.43 Apparent and Partial Molal Volumes

Solvent	Apparent molal volume of H_2O in solvents (ml g-mole^{-1})
1,2-Dichloroethane	20.1
1,1-Dichloroethane	21.7
1,1,1-Trichloroethane	22.1
Carbon tetrachloride	31.6
Benzene	22.1
Methanol	38.25
Ethanol	55.12
Propanol	70.63
n-Butanol	86.48
n-Pentanol	102.88

Solute	Partial molar volume of solute in water at 25°C, $\bar{V}_1$ (ml g-mole^{-1})
Ethyl bromide	66.7
n-Propyl bromide	82.2
Allyl bromide	77.6
Methyl fluoride	35.9
Methyl chloride	46.2
Methyl bromide	53.0
Methyl iodide	63.7

Source: Alexander, 1959; D'Orazio & Wood, 1963; Holland & Moelwyn-Hughes, 1956; Jolicoeur & Lacroix, 1976; Seiler, 1968.

where B is the second virial coefficient (Hayden and O'Connell, 1975). Dymond and Smith (1979) compiled the available second virial coefficients of organic compounds.

Under considerable pressure, the liquid-phase behavior is not represented correctly by the equation

$$p_1 = \gamma_1 x_1 p_1^o$$

Therefore, it was necessary to replace the vapor pressure of component 1 with its fugacity f_1^o:

$$p_1 = \gamma_1 x_1 f_1^o$$

so that in a strongly nonideal system at high pressures, the vapor-liquid equilibrium is expressed by

$$P \, y_1 \, \Phi_1 \;=\; \gamma_1 \, x_1 \, f_1^o$$

where the liquid-phase fugacity is, at temperature T,

$$f_1^o \;=\; \gamma_1^o \, p_1^o \, \exp\!\left(\int_{p_1^o}^{P} \frac{V_1}{R\,T}\,dP\right)$$

where γ_1^o is the fugacity coefficient of pure component 1 and V_1 its molar volume at the same temperature.

The liquid phase fugacity f_1^o can also be calculated from the adjusted vapor pressure p_1^* (Redlich et al., 1952) at temperature T:

$$p_1^* \;=\; \frac{p_1^o \, \gamma_1^o}{\Phi_1} \, \exp\!\left(\int_{p_1^o}^{P} \frac{V_1}{R\,T}\,dP\right)$$

The fugacity of hypothetical liquids at 1 atm pressure has been correlated with reduced temperature (see Table 2.44) (Yen and McKetta, 1962; Prausnitz and Shair, 1961; Shinoda, 1978; Khmara, 1976.)

Alternatively, the reference-state fugacity f_1^o can be obtained from the virial coefficient B_{11} (regarding estimating B_{11}, see Martire, 1966; Svoboda et al., 1977; Djordjevic et al., 1980; and Hirschfelder et al., 1942):

$$f_1^o \;=\; p_1^o \, \exp\!\left[\frac{p_1^o (B_{11} - V_1)}{R\,T}\right]$$

and the vapor-phase fugacity coefficient Φ_1 is

$$\Phi_1 \;=\; \frac{\exp\!\left[2\left(\dfrac{y_1 \, B_{12}}{V_{mix}} + \dfrac{y_2 \, B_{11}}{V_{mix}}\right)\right]}{Z_{mix}}$$

where

$$V_{mix} \;=\; \frac{R\,T + (1 + P\,B_{12}/(R\,T))^{\frac{1}{2}}}{2\,P}$$

$$Z_{mix} \;=\; \frac{P\,V_{mix}}{R\,T}$$

Table 2.44   Correlation of Fugacity of Hypothetical Liquids at 1 atm Pressure

Reduced temperature, T/T_{cr}	Fugacity of hypothetical liquid, f_1^o/P_{cr}
0.7	0.07
0.8	0.2
1.0	0.7
1.2	1.5
1.4	2.4
1.6	3.3
1.8	4.0
2.0	4.8
2.2	5.4
2.4	5.9
2.6	6.1
2.8	6.3
3.0	6.4

Source: Yen & McKetta, 1962; Prausnitz & Shair, 1961.

For the binary mixture,

$$P = \frac{\gamma_1 \, x_1 \, f_1^o}{\Phi_1} + \frac{\gamma_2 \, x_2 \, f_2^o}{\Phi_2}$$

The fugacities of water (f_w^o) and hydrocarbons (f_h^o) were investigated by Lotter et al. (1978) in binary systems. With satisfactory results they calculated the fugacities from the second virial coefficients

$$R \, T \, \log_e f_w^c = R \, T \, \log_e P + P \left[B_{11} + x_2^2 (2 \, B_{12} - B_{11} - B_{22}) \right]$$

and

$$R \, T \, \log_e f_h^o = R \, T \, \log_e P + P \left[B_{22} + x_2^2 (2 \, B_{12} - B_{11} - B_{22}) \right]$$

It is noteworthy that the mixed second virial coefficient B_{12}, greatly differs from the arithmetic means of B_{11} and B_{22}. This fact characterises the appreciable increase of the volume in excess of the additive value when water vapor is mixed with vapors of nonpolar substances. Pesuit (1978) reported a good compilation on the available mixing rules, including the second virial coefficients. For further investigation see Eckert et al. (1976).

Since the early years of study of mixture properties, the aim was to relate the behavior of the pure components to the mixtures, in other words, to predict the non-

ideal properties of the solution from the physical properties of the pure components.
Apart from the difficulties in many cases, some good approximations have been achieved
by considering some phenomenon (e.g., hydrogen bonding, polarity) that enables us to
draw some sort of conclusion regarding the deviation from ideality. In general, the
mixtures of homologous series (e.g., methanol-ethanol, ethane-propane, benzene-
toluene, 1,2-dibromoethane – 1,2-dibromopropane, n-heptane – n-butane, 1,2-dichloro-
ethane – benzene) approach ideal behavior. Similarly, the mixtures of isomer compounds
form ideal or almost ideal systems. A classification of various liquids has been
proposed by Ewell et al. (1944) for the expectation of ideal behaviour. The classification divides the liquids into five groups, based on their hydrogen-bond-forma-
tion tendencies. Whether a hydrogen bond is strong or weak depends on the nature of
the atoms and the coordinating ability of the hydrogen atom between them. On these
principles, the deviation from Raoult's law and the likelihood of azeotropic formation may be predicted. Table 2.45 summarizes the five classes of liquids. The deviation from Raoult's law in mixtures of liquids belonging to various classes is indicated in Table 2.46 with remarks upon the hydrogen bonds involved.

The associated liquids are in classes I and II, indicated by the presence of
OH   and   NH   groups. These liquids have relatively high dielectric constants. The
hydrogen-bond formation is the deciding factor as to whether a liquid or a solid is
soluble in water or tends to dissolve in water, although the length or the number of
CH_2 group in the molecule influences the miscibility. Whereas methanol and ethanol
are completely miscible with water, about 10%   n-butanol and   3%   n-amyl alcohol will
dissolve in water. If a substance forms many hydrogen bonds per molecule (e.g., prot-
ein, carbohydrate, hydrophilic material), then it is soluble in water apart from its
high molecular weight.

The concept of solubility between polar-polar and nonpolar-nonpolar substances
is misleading and should not be used. For example, water (dipole moment = 1.8 debyes)
does not dissolve ethyl iodide (dipole moment = 1.8 debyes); however, ethyl iodide
dissolves naphthalene (dipole moment = 0). Furthermore, water dissolves pyrazine
(dipole moment = 0).

According to Ewell et al. (1944), two major factors determine the solubility
ability of substances: the hydrogen-bond phenomena in mixing and the internal pressure when no hydrogen bonds are formed upon mixing. The internal pressures of some
substances are listed in Table 2.47. The internal pressure of a substance can be
calculated from the expression

$$P_{internal}(atm) \quad = \quad \frac{\Delta H \; - \; R \, T}{V_1}$$

A simple model was proposed by Rosseinsky (1977), relating the surface tension to the
internal pressure (see also Vavruch, 1978). The calculated values agree well with the

Table 2.45 Classification of Liquids by Ewell et al. (1944)

Class I	Class II	Class III	Class IV	Class V
Strong hydrogen bonds	Weak hydrogen bonds	No active hydrogen atom(s)	Active hydrogen atom(s)	No hydrogen-bond forming
Three-dimensional networks	Three-dimensional networks	Donor atoms	No donor atoms	All others
High dielectric constant	High dielectric constant			
Water-soluble	Water-soluble	Water-soluble	Immiscible with water	Immiscible with water
Water, glycol, glycerol, amino alcohols, hydroxylamine, hydroxyacids, polyphenols, amides, etc.	Alcohols, acids, phenols, amines (prim. and sec.), oximes, nitrocompounds (with α-H atom), nitriles (with α-H atoms), ammonia, hydrazine, HF, HCN, etc.	Ethers, ketones, aldehydes, esters, tertiary amines, pyridine compounds, nitrocompounds (no α-H atom), nitriles (no α-H atom), etc.	Chlorinated hydrocarbons, $CHCl_3$, CH_2Cl_2, $C_2H_4Cl_2$, CH_2ClCH_2Cl, $C_3H_5Cl_3$, $CH_2ClCHCl_2$, etc.	Hydrocarbons, CS_2, sulfides, mercaptans, halohydrocarbons not in class IV, iodine, phosphorus, sulfur, etc.

Table 2.46   Deviation from Raoult's Law in Liquid-Liquid Mixtures

Classes	Positive(+) or negative(-) deviation from Raoult's law	Comments upon hydrogen bonding
I-V	Always positive deviation, limited miscibility.	Hydrogen bonds broken only.
II-V	Always positive deviation, limited miscibility.	Hydrogen bonds broken only.
III-IV	Always negative deviation.	Hydrogen bonds formed only.
I-IV	Always positive deviation, limited miscibility.	Hydrogen bonds both broken and formed,
II-IV	Always positive deviation, limited miscibility.	dissociation of classes I and II.
I-I	Mostly positive deviation; if the deviation is	Hydrogen bonds both formed and broken.
I-II	negative, then it gives some maximum azeotropes.	Hydrogen bonds both formed and broken.
I-III	Very complicated groups.	Hydrogen bonds both formed and broken.
II-II	Same as for class I-I.	Hydrogen bonds both formed and broken.
II-III	Same as for class I-I.	Hydrogen bonds both formed and broken.
III-III	Ideal system or positive deviations. Azeotropes are	No hydrogen bonds involved.
III-V	always minimum if forms. Quasi-ideal systems.	No hydrogen bonds involved.
IV-IV	Same as for class III-III.	No hydrogen bonds involved.
IV-V	Same as for class III-III.	No hydrogen bonds involved.
V-V	Same as for class III-III.	No hydrogen bonds involved.

Source: Ewell et al., 1944.

Table 2.47   Internal Pressure of Substances

Substance	Internal pressure (atm)
NaCl	72,900
Hg	38,750
H_2O	16,000
I_2	7,600
S	7,420
P	7,000
Br_2	5,700
CH_3I	4,950
$CHBr_3$	4,220
CS_2	4,140
CH_3COCH_3	3,940
C_6H_5I	3,640
C_6H_6	3,450
C_2H_5Br	3,150
CCl_4	3,070
C_6H_{12}	2,780
$n-C_8H_{18}$	2,420
$n-C_5H_{12}$	2,020
$(CH_3)_2CHCH_2CH_3$	1,842
$(CH_3)_4C$	1,535

Source: Ewell et al., 1944; Dack, 1974.

experimental data. An approximate relation between the internal pressure and solubility parameter was discussed by Renuncio et al. (1977), Getzen (1976), and Dack (1974).

Liquids in class IV are partially miscible in water. The broken hydrogen bonds
are more distinct and the quite weak formation cannot compensate for it.

Class V liquids are very slightly soluble in water. Similarly, water shows a
very limited solubility in these liquids. Hydrogen bonds broken in the mixture are
characteristic for the systems.

When the concentration of one of the components in a binary mixture approaches
100%, its activity coefficient will be close to unity. Approximate values are given
by Hala et al. (1967) for the activity coefficient in the case mentioned (see Table
2.48). Arich et al. (1975) reported the activity coefficients calculated from liquid-
liquid equilibrium data (see also Kikic and Alessi, 1977; Letcher, 1975; Pinder, 1973;
Tassios and van Winkle, 1967; Tikhonova et al., 1976a, b.).

Table 2.48 Activity Coefficients When the Concentration of One Constituent Approaches 100%

	Concentration (mole fraction)						
0.0	0.5	0.8	0.9	0.95	0.99
2	1.19	1.03	1.007	1.002	1.00007
5	1.50	1.07	1.016	1.004	1.00016
10	1.78	1.10	1.023	1.006	1.00023
50	2.66	1.17	1.04	1.01	1.0004

Source: Hala et al., 1967.

The activity coefficient of a solute in a partially soluble system at infinite dilution (γ_2^∞) provides a uniform composition, and it is advantageous to know it (Joffe, 1977). There are usually two ways to derive the activity coefficients at infinite dilution for organic compounds in water. The experimental technique uses a gas-liquid chromatograph fitted with flame ionization detector. Whereas the extrapolation method fitts the activity coefficients calculated from VLE measurements into equations (e.g., van Laar, Redlich-Kister, Wilson, NRTL, and UNIQUAC). However, this extrapolation method gives discrepancies from the experimentally measured data (Mash and Pemberton, 1980). The following table shows the comparison of the values reported for chloroform in water at 25°C:

γ_2^∞	Reference
781	Mash and Pemberton (1980)
846	Donahue and Bartel (1952)
742	Reinders and de Minjer (1947)
934	Chancel and Parmentier (1885)
846	Rex (1906)
570	Pecsar and Martin (1966)
311-578	Hardy (1959)

The temperature dependence of γ_2^∞ for CH_2Cl_2, $CHCl_3$, and CCl_4 in water between 20 and 40°C was studied by Pecsar and Martin (1966). The investigations mentioned demonstrate the urgent need for improvement in the determination of γ_2^∞.

The usual calculation is carried out by using the two-suffix Margules equation

$$\log_{10} \gamma_2^\infty = \frac{\log_{10} \gamma_2}{(1 - x_2)^2}$$

In this expression it has been assumed that there is no dissociation. For further methods of calculations, see Herington (1977) and Helpinstill and Winkle (1968).

In a binary vapor-liquid equilibrium, the activity coefficient varies with the concentration, temperature, and pressure. The activity coefficient characterizes non-electrolyte systems, whether they are ideal or nonideal, and if a system is nonideal, indicates how much it deviates from ideality (Rao et al., 1974; Rao, 1977; Gaube and Koenen, 1979). Therefore, the principal problem is to establish the coefficient. For an ideal solution, the activity coefficient equals unity, but for real systems, it can show values both larger and smaller than unity. A system exhibits a negative deviation when γ_i is less than unity and a positive deviation indicates that the activity coefficient is greater than unity. The order of magnitude of γ_i in some halogenated hydrocarbon-water nonmiscible systems can be on the order of several thousands (Saito and Tanaka, 1965; Pinder, 1973; Tsonopoulos, 1970; Tsonopoulos and Prausnitz, 1971;

Table 2.49 Activity Coefficients of Halogenated Hydrocarbons in Aqueous Solutions

Solute	Temperature (°C)	Activity coefficient γ_2	γ_2^∞
Solids			
Benzene, 1,4-dichloro-	30	53,369.4	53,382.6
	50	55,111.0	55,131.5
Benzene, 1,2,3-trichloro-	25	226,815.9	226,829.7
Benzene, 1,3,5-trichloro-	25	169,033.9	169,044.1
Benzene, 1,4-dibromo-	25	174,174.3	174,180.7
Benzene, 1,4-diiodo-	25	1,641,719.9	1,641,723.4
Toluene, 4-bromo-	25	83,097.1	83,118.9
Liquids			
Benzene, fluoro-	25	3,436.7	3,453.0
	30	3,459.0	3,475.4
Benzene, chloro-	25	12,486.6	12,505.5
	30	12,769.0	12,787.9
	50	6,501.7	6,519.3
Benzene, 1,2-dichloro-	25	56,251.1	56,273.0
	30	47,698.4	47,720.0
	50	38,296.6	38,317.8
Benzene, 1,3-dichloro-	25	66,321.5	66,343.7
	30	58,268.9	58,290.8
	50	43,620.2	43,641.5
Benzene, 1,2,4-trichloro-	25	402,783.9	402,809.8
Benzene, bromo-	25	21,246.8	21,266.7
	30	19,528.1	19,547.9
Benzene, iodo-	25	62,885.1	62,907.1
	30	33,291.1	33,311.9
Toluene, α,α,α-trifluoro-	25	18,010.6	18,030.2
Benzylchloride	30	15,102.4	15,121.7
Benzene, 2-bromo-1-ethyl-	25	263,304.8	263,329.8
Benzene, 2-bromo-1-isopropyl-	25	849,935.6	849,963.0

Source: Tsonopoulos, 1970.

Mackay and Shiu, 1975; Mackay et al., 1980). Some typical values are presented in Table 2.49.

The activity coefficients of hydrocarbons were studied by Leinonen and Mackay (1973) in six binary systems with water. The authors found that the activity coefficients of the components were reduced by the presence of the other hydrocarbons. The implication of these effects were discussed for the multicomponent solubility of hydrocarbons in water.

The influence of electrolytes on the activity coefficient of nonpolar solutes (e.g., benzene) in aqueous salt solutions was studied by McDevit and Long (1952) and Randall and Failey (1927). The authors found that the salting-out and salting-in effects vary greatly among the various electrolytes. The interpretation of activity coefficients in alkane systems was discussed by Petterson et al. (1972).

The temperature dependence of the activity coefficient is given by the relationship (Gaube and Koenen, 1979):

$$\left(\frac{\partial \log_e \gamma_i}{\partial T}\right)_{P,x} = -\frac{\bar{H}_i - H_i^o}{R\,T^2}$$

where $\bar{H}_i$ =   partial molar enthalpy of component   i   in the solution

H_i^o =   molar enthalpy of pure component   i, both at the same   T and P

The sign of the term $\bar{H}_i - H_i^o$ determines whether the system shows negative or positive deviation from ideality. If the sign is positive, the system shows positive deviation, whereas in the case of a negative sign for the term, the system is characterized by negative deviation from Raoult's law. A negative sign indicates that the heat evolved on the mixing of pure components, and γ_i, will increase with temperature [see, e.g., Stepakoff and Modica (1972, 1973) for data on Freon-114 $(CClF_2-CClF_2)$]. On the other hand, if $\bar{H}_i - H_i^o$ is positive, heat is absorbed on mixing and the activity coefficient decreases with rising temperature. It is a general observation that with rising temperature, γ_i values approach unity. The foregoing expression for the temperature dependence of the activity coefficient shows that on a graph of $\log_e \gamma_i$ versus   1/T   is approximately linear. However, for the dependence of γ_i upon temperature in a large temperature interval, integration of the expression is necessary. This is usually carried out through graphical integration.

The temperature dependence of the activity coefficients in binary systems was also studied by Gaube and Koenen (1979). They plotted the excess Gibbs energy of mixing (G^E) versus the excess enthalpy of mixing (H^E) for nonassociated, associated, and solvalized systems

This illustration of the relationship between G^E and H^E is based upon the thermo-dynamic expressions

$$\left[\frac{\partial\left(\frac{G^E}{T}\right)}{\partial\left(\frac{1}{T}\right)} \right]_{n_i,\, P} = H^E$$

and

$$\log_e \gamma_i = \left(\frac{\partial \frac{n_i G^E}{R T}}{\partial n_i} \right)_{n_i,\, T,\, P}$$

Whether the activity coefficients in a binary mixture will approach or depart from unity with increasing temperature, depends upon the thermodynamic quantities G^E and H^E in the above illustration.

Some empirical formulas have been developed by several investigators (e.g., Berg and McKinnis, 1948):

$$\log_{10} \gamma_i = \frac{K(1 - T_r)^{0.43}}{T_r}$$

where K is a constant and T_r the reduced temperature. Sometimes, a very simple assumption might be applied when no experimental data are available:

$$T \log_{10} \gamma_i = \text{constant}$$

In the case of regular solutions, the following equation has been proposed by Hildebrand and Scott (1950):

$$\log_e \left(\frac{a_i}{x_i} \right) = \log_e \gamma_i = \frac{V_i \Phi_j}{R T} (\delta_j - \delta_i)^2$$

where a_i = activity of component i

       V_i = molar volume of component i

       Φ_j = volume fraction of component j

       δ = solubility parameter, $(\text{cal/cc})^{\frac{1}{2}}$

For other empirical methods for the calculation of the activity coefficients, the reader is referred to Hala et al. (1967), Prausnitz (1969), Hine and Mookerjee (1975), Krasovskii and Gritsan (1976), Palmer (1975), Rao (1977), and Redlich (1976).

The pressure dependence of the activity coefficient is given by

$$\left(\frac{\partial \log_e \gamma_i}{\partial P}\right)_{T,x} = \frac{\bar{V}_i - V_i}{R\,T}$$

where $\bar{V}_i$ is the partial molar volume of component i and V_i its molar volume at the same temperature and pressure. Similarly to the expression for temperature dependence, the activity coefficient can be evaluated by integration.

The concentration dependence of the activity coefficient is probably the most interesting phenomenon. The expression relating γ_i to the concentration has been obtained for a binary system from the Gibbs-Duhem equation:

$$x_1 \frac{\partial \log_e \gamma_1}{\partial x_1} + x_2 \frac{\partial \log_e \gamma_2}{\partial x_2} = 0$$

at constant temperature and pressure. This formula originates from the general expression for mixtures:

$$\sum_i x_i \, d \log_e \gamma_i = 0$$

There is no rigorous solution to the Gibbs-Duhem equation; therefore, several methods have been proposed for the best approximation. Rowlinson (1959) gives an account of various propositions. One of the most commonly used methods is the assumption that $\log_e \gamma_i$ can be represented by a power series of the mole fraction (x_i). If we consider the first three terms, the logarithm of the activity coefficient becomes

$$\log_e \gamma_1 = \alpha_2 x_2^2 + \alpha_3 x_2^3 + \alpha_4 x_2^4$$

and the expression for γ_2 becomes

$$\log_e \gamma_2 = (\alpha_2 + \frac{3}{2}\alpha_3 + 2\alpha_4)x_1^2 - (\alpha_3 + \frac{8}{3}\alpha_4)x_1^3 + \alpha_4 x_1^4$$

A very accurate value of $\log_e \gamma_1$ is necessary to obtain precise values of α_3 and α_4 before one can calculate $\log_e \gamma_2$. However, if the deviation from ideality is small, the first term is applied:

$$\log_e \gamma_1 = \alpha_2 x_2^2$$

and

$$\log_e \gamma_2 = \alpha_2 x_1^2$$

These expressions are very similar to those which represent the regular solutions:

$$\log_e \gamma_1 = \frac{\alpha}{R\,T} x_2^2$$

$$\log_e \gamma_2 = \frac{\alpha}{R\,T} x_1^2$$

where α is independent of temperature and can be expressed by

$$\alpha = R\,T\,\alpha_2$$

To express the activity coefficient γ_i as a function of concentration (e.g., as a mole fraction, x_i), numerous empirical equations have been proposed by several investigators. One of the oldest suggestions is that of Margules (1895):

$$\log_e \gamma_1 = x_2^2 \left[A + 2(B - A)\, x_1 \right]$$

and

$$\log_e \gamma_2 = x_1^2 \left[B + 2(A - B)\, x_2 \right]$$

Other expressions were introduced by van Laar (1910):

$$\log_e \gamma_1 = \frac{A\, x_2^2}{(x_1\, A/B + x_2)^2}$$

and

$$\log_e \gamma_2 = \frac{B\, x_1^2}{(x_2\, B/A + x_1)^2}$$

which are still used by several scientists. The expressions of both Margules and van Laar can be derived from Wohl's equation.

A later form of relationship introduced by Redlich and Kister (1948) and Redlich et al. (1952) is known as the Redlich-Kister equation:

$$\log_e \gamma_1 = x_2^2 \left[B + C(3\, x_1 - x_2) + D(x_1 - x_2)(5\, x_1 - x_2) + E(x_1 - x_2)^2 (7\, x_1 - x_2) \right]$$

and

$$\log_e \gamma_2 = x_1^2 \left[B - C(3\, x_2 - x_1) + D(x_2 - x_1)(5\, x_2 - x_1) - E(x_2 - x_1)^2 (7\, x_2 - x_1) \right]$$

For most practical application, it is sufficient to truncate the equation after three terms. By taking the ratio of these equations, we obtain

$$\log_e \left(\frac{\gamma_1}{\gamma_2} \right) = B(x_2 - x_1) + C(6\, x_1\, x_2 - 1) + D(x_2 - x_1)(1 - 8\, x_1\, x_2)$$

If we plot $\log_e(\gamma_1/\gamma_2)$ versus the mole fraction, we obtain the coefficients of the

equation. This is a very simple and practical way to show the consistency of the
data, and produces the equal-area condition (Redlich and Kister, 1948; Redlich et al.,
1952; Herington, 1947, 1950, 1951). To overcome the drawbacks of this method (graphical extrapolation and not utilising total vapor pressure data), Gorbunov (1980)
proposed a thermodynamic test free from these disadvantages.

Gorbunov plotted $p_1 p_2$ (partial pressure of component 1 and 2, respectively)
versus y_1 (vapor phase composition of component 1) and found a stationary point
$(x_1 = x_2 = 0.5)$ corresponding to the maximum of the function $p_1 p_2$ at T = const.
This function always goes through a maximum at the equimolar composition of the
liquid in any binary system consisting of volatile components, according to

$$\frac{d\ (p_1\ p_2)}{d\ y_1} = p^2(x_2 - x_1)$$

The function $p_1 p_2 (x_1)$ has similar properties. The integration of this equation
from the component $A_2 (y_1 = 0)$ to the component $A_1 (y_1 = 1)$ gives

$$\int_0^1 p^2(x_2 - x_1)\ dy_1 = 0 \qquad \text{at} \quad T = \text{const.}$$

where P is the total pressure of the system. This expression contains all the
variables which characterize the vapor-liquid equilibrium. Furthermore, the function
to be integrated takes fully determined values at the boundaries of the composition
range, depending only on the properties of the pure components A_1 and A_2. The area
between the experimental $p^2(x_2 - x_1)$ isotherm, the ordinates $y_1 = 0$ and $y_1 = 1$,
and the corresponding segment of the horizontal axis should vanish if the experimental
data are correct. That is, the areas below and above the horizontal axis respectively
should be equal.

A more widely applied expression for the activity coefficients as a function of
mole fraction was introduced by Wilson (1964). The equation is derived by extending
Flory-Huggins' equation to systems in which the molecular interactions of the
constituents are different. For binary systems the activity coefficients are expressed by

$$\log_e \gamma_1 = -\log_e(x_1 + A_{12}\ x_2) + x_2\left(\frac{A_{12}}{x_1 + A_{12}\ x_2} - \frac{A_{21}}{x_2 + A_{21}\ x_1}\right)$$

and

$$\log_e \gamma_2 = -\log_e(x_2 + A_{21}\ x_1) - x_1\left(\frac{A_{12}}{x_1 + A_{12}\ x_2} - \frac{A_{21}}{x_2 + A_{21}\ x_1}\right)$$

For a multicomponent system, the following general formula is used:

$$\log_e \gamma_k = -\log_e\left(\sum_j x_j A_{kj}\right) + 1 - \sum_i \frac{x_j A_{ik}}{\sum_j x_j A_{ij}}$$

A good article describing the various minimization methods has been published by Desplanches et al. (1975). Further compilations on the Wilson equation and its param-eters have been provided by Ghosh and Chopra (1975), Hirata et al. (1975), Gmehling and Onken (1977), Fredenslund et al. (1977), Landolt-Börnstein (1976b), and Lenoir and Sakata (1978).

The Wilson equation has many very useful features, particularly compared with Margules' and van Laar's equations. However, in several cases it fails to represent the vapor-liquid equilibrium data sufficiently well:

1. When the liquids are immiscible (e.g., two liquid phases)
2. When the activity coefficients plotted versus the mole fractions show a maximum (e.g., $CHCl_3$ - C_2H_5OH system) or a minimum
3. When the Gibbs excess energy of the system is negative (e.g., NH_3 - H_2O)

Consequently, contrary to the wide use of the Wilson equation, it is not suited for halogenated hydrocarbon - water systems. Several modifications and improvements have been suggested since 1964 (e.g., by Chandrasekaran et al., 1976 and Nagata et al., 1975).

One of the proposed equations for liquid-liquid immiscibile systems is the NRTL (non-random, two-liquid) expression introduced by Renon and Prausnitz (1968). This is a three-parameter equation and has been used successfully on binary and ternary liquid-liquid equilibria with partially miscible components (see, e.g., Leach, 1977). The latest published equation is an extension or genralization of Guggenheim's model presented by Abrams and Prausnitz (1975). This empirical expression is called the UNIQUAC (universal quasi chemical) equation. The activity coefficient for a binary system is given by

$$\log_e \gamma_1 = \log_e\left(\frac{\phi_1}{x_1}\right) + \left(\frac{z}{2}\right) q_1 \log_e\left(\frac{\theta_1}{\phi_1}\right) + \phi_2\left(l_1 - \frac{r_1}{r_2} l_2\right) -$$

$$q_1 \log_e(\theta_1 + \theta_2 \tau_{21}) + \theta_2 q_1\left(\frac{\tau_{21}}{\theta_1 + \theta_2 \tau_{21}} - \frac{\tau_{12}}{\theta_2 + \theta_1 \tau_{12}}\right)$$

where $l_1 = \left(\frac{z}{2}\right)(r_1 - q_1) - (r_1 - 1)$

$l_2 = \left(\frac{z}{2}\right)(r_2 - q_1) - (r_2 - 1)$

where r and q are structural parameters calculated from group contributions (Bondi, 1968; Fredenslund et al., 1977; Reid et al., 1977). For component 2, γ_2 can be found by interchanging subscripts 1 and 2. This equation has been applied successfully to a large variety of multicomponent systems. It has only two adjustable parameters per binary system. If one inserts the following equalities in the UNIQUAC equation,

$$x_1 = \phi_1 = \theta_1$$
$$x_2 = \theta_2$$
$$q_1 = q_2 = r_1 = r_2 = 1$$
$$\tau_{12} = A_{21}$$
$$\tau_{21} = A_{12}$$

it will be reduced to the Wilson equation. However, there is no evidence that this simple mathematical manipulation will produce useful results. That is, the Wilson parameters obtained after the reduction will not represent the original vapor-liquid equilibrium data used to fit the UNIQUAC equation. This is not surprising, as both equations are purely empirical.

Flemr (1976) has shown that the local composition equation is not consistent with the Wilson, NRTL, and UNIQUAC models and therefore, that they should be treated as empirical expressions only. The same conclusion was shown independently by McDermott and Ashton (1977). Consequently, Maurer and Prausnitz (1978) introduced a third parameter into the UNIQUAC equation, however, the improvement was only marginal.

A good discussion of the various types of equations used to express vapor-liquid and liquid-liquid equilibria systems is that of Vetere (1977). In this paper, there is a brief criticizm of the expressions available up to date. For further details, see Nicolaides and Eckert (1978), Rogalski and Malanowski (1975), Mattelin and Verhoeye (1975), and Storvick and Sandler (1977).

Because of the usefulness of the activity coefficients to express nonideality of a system and to correlate its behavior at various conditions, several attempts have been carried out to predict their values from the structure or properties of the pure constituents (Kikic and Alessi, 1977; Hine and Mookerjee, 1975). The principle of the group contribution method for estimating the thermodynamic properties of liquid mixtures originates from Langmuir (1925). Activity coefficients were estimated from molecular structures by Wilson and Deal (1962) using the solution-of-group method both for finite concentrations and unfinite dilution. The method was an extension and further development of Pierotti et al.'s (1956, 1959) theory using molecular structure for the determination of activity coefficients, and also includes some earlier work by Deal et al. (1962).

Pierotti et al. (1959) introduced an equation for the correlation of activity coefficients at infinite dilution and 25°C for normal alkylbenzenes in water:

$$\log_{10} \gamma_2^{\infty} \;=\; 3.554 \;+\; 0.622(n_c - 6) \;-\; \frac{0.466}{n_c - 4}$$

where n_c is the total number of carbon atoms in the hydrocarbon molecule.

A correlation was presented by Helpinstill and Van Winkle (1968) for predicting infinite binary activity coefficients (γ_2^{∞}) for saturated, unsaturated, and aromatic hydrocarbons in polar-nonpolar and polar-polar binary systems,

$$R\,T\,\log_e \gamma_2^{\infty} \;=\; V_2 \left[(\lambda_1 - \lambda_2)^2 + (r_1 - r_2)^2 - 2\,\psi_{12} \right] + R\,T\left[\log_e\!\left(\frac{V_2}{V_1}\right) + \left(1 - \frac{V_2}{V_1}\right)\right]$$

where V_1 and V_2 are the molar volumes of component 1 and 2, respectively. The nonpolar (λ) and polar (r) solubility parameters were obtained from the energy of vaporization (ΔU_{vap})

$$\Delta U_{vap} \;=\; \frac{2.303\,R\,T^2\,B}{(t + C)^2} \;-\; R\,T$$

where A, B, and C are the coefficients of the Antoine's vapor pressure equation. For nonpolar species

$$\lambda_1^2 \;=\; \frac{\Delta U_1^{vap}}{V_1} \qquad \text{and} \qquad \lambda_2^2 \;=\; \frac{\Delta U_2^{vap}}{V_2}$$

and for polar solubility parameter (r)

$$r_1 \;=\; \left[\frac{\Delta U_1^{vap}}{V_1} - \lambda_1^2\right]^{\frac{1}{2}} \qquad \text{and} \qquad r_2 \;=\; \left[\frac{\Delta U_2^{vap}}{V_2} - \lambda_2^2\right]^{\frac{1}{2}}$$

The interaction energy (ψ_{12}) was calculated by linear correlation of the rearranged equation.

In an investigation by Wilson and Deal (1962), examples are given for alcohol-water systems (see also Scheller, 1965). In 1968, Deal and Derr (1968) presented an estimation method for activity coefficients, based upon data from structurally related systems. The work has been continued by Derr and Deal (1969) with their analytical-solution-of-group (ASOG) method and some extension of the same by Ronc and Ratcliff (1971). The prediction of vapor-liquid equilibria by ASOG method is well described by Kojima and Tochigi (1979) in a book. There are several examples in the book for illustration of the method.

A group-contribution molecular model of liquids and solutions was proposed by Nitta et al. (1977a, b) for polar and nonpolar liquids and their solutions. The method is based upon the cell model.

The group-contribution method for the calculation of the infinite dilution activity coefficient of one binary constituent from the other was proposed by Wilson (1974). The types of systems investigated included water - alcohol, but no halogenated hydrocarbon - water mixtures were mentioned. To be able to calculate the infinite dilution activity coefficient of one component, one must know the value of the other coefficient. Chao and Greenkorn (1975) and Wilson (1974) discussed the various methods suggested for the prediction of the activity coefficients. The group-contribution method for estimating activity coefficients in nonideal mixtures, including mixtures containing hydrocarbons, alcohols, ketones, amines, nitriles, chlorides, water, and other organic substances in the temperature range 0 to 125°C, has proved to be most successful (Fredenslund et al., 1975, 1976, 1977). The method is called UNIFAC (UNIQUAC functional-group activity coefficients). Group-interaction parameters are tabulated for a large number of groups. The accuracy of the method is checked by comparing the observed and calculated activity coefficients of various systems at infinite dilution (see Table 2.50 for some aqueous partially miscible systems). The most interesting point is the comparison of the activity coefficient of benzene and aniline in water at infinite dilution and 25°C with the values obtained by Tsonopoulos (1970) (see Table 2.50). The discrepancies are an indication for the sensitivity of the activity coefficients calculated from solubility data. The activity coefficients obtained by the UNIFAC method for organic compounds - water systems at various temperatures are listed in Table 2.51. Later developments in the UNIFAC method have been reported by Prausnitz (1977), Skjold-Jørgensen et al. (1979, 1980, and Zakarian et al. (1979).

A prediction method for activity coefficients in binary hydrocarbon systems was reported by Gothard et al. (1976). The interaction energy differences $(\lambda_{12} - \lambda_{11})$ and $(\lambda_{21} - \lambda_{22})$ were correlated with differences in the solubility parameters of the pure components:

$$\lambda_{12} - \lambda_{11} = -400.95 + 0.28129(|\delta_1 - \delta_2|) - 1.41756 \times 10^{-5}(\delta_1 - \delta_2)^2$$

and

$$\lambda_{21} - \lambda_{22} = 186.17 + 0.14849(|\delta_1 - \delta_2|) + 8.329080 \times 10^{-6}(\delta_1 - \delta_2)^2$$

The parameters A_{12} and A_{21} were obtained from the usual expression:

$$A_{12} = \frac{V_2}{V_1} \exp\left(-\frac{\lambda_{12} - \lambda_{11}}{R\,T}\right)$$

Table 2.50 Comparison of Observed and Calculated (UNIFAC) γ_1^∞ Values

| System | Temperature (°C) | Activity coefficient at infinite dilution | | Deviation (%) | γ_1^∞ (calculated) (Tsonopoulos, 1970) |
		γ_1^∞ (observed)	γ_1^∞ (calculated)		
Water–hexadiene	20	226	105	-53.5	–
Hexadiene–water	25	26,900	30,600	12.1	–
Aniline–water	100	80	115	30.4	144.4$^{25°C}$
Benzene–water	25	488	458	-6.6	2,519.6
Water–benzene	25	430	359	-19.8	–

Table 2.51 Activity Coefficients Obtained by the UNIFAC Method

System	Temperature ($^{\circ}$C)	$\log_e \gamma_1$	γ_1	Reference
Water-dichloroethane	19	5.36	212.72	Stephen & Stephen(1963)
Dichloroethane-water	23	6.44	626.41	Stephen & Stephen(1963)
Water-dichloroethane	33	4.79	120.30	Stephen & Stephen(1963)
Dichloroethane-water	33.5	6.39	595.86	Stephen & Stephen(1963)
Water-dichloroethane	53.0	4.11	60.95	Stephen & Stephen(1963)
Dichloroethane-water	56.5	6.19	487.85	Stephen & Stephen(1963)
Water-dichloroethane	69.0	3.56	35.16	Stephen & Stephen(1963)
Dichloroethane-water	72.5	5.98	395.44	Stephen & Stephen(1963)
Tetrachloroethane-water	25	7.56	1919.85	Othmer et al. (1941)
Water-tetrachloroethane	25	4.97	144.03	Othmer et al. (1941)
Water-chlorobenzene	25	6.038	419.05	Stephen & Stephen(1963)
Chlorobenzene-water	25	8.153	3473.78	Stephen & Stephen(1963)
Benzene-water	10	7.886	2659.8	Stephen & Stephen(1963)
	20	7.814	2475.0	Stephen & Stephen(1963)
	30	7.651	2102.7	Stephen & Stephen(1963)
	50	7.458	1733.7	Stephen & Stephen(1963)
	70	7.199	1338.1	Stephen & Stephen(1963)
	25	7.826	2504.9	Black et al. (1948)
Water-benzene	10	6.463	641.0	Stephen & Stephen(1963)
	20	6.135	461.7	Stephen & Stephen(1963)
	30	5.422	226.3	Stephen & Stephen(1963)
	50	4.505	90.5	Stephen & Stephen(1963)
	70	4.144	63.1	Stephen & Stephen(1963)
	10	6.642	766.6	Black et al. (1948)
	20	6.272	529.5	Black et al. (1948)
	26	6.056	426.7	Black et al. (1948)

and

$$A_{21} = \frac{V_1}{V_2} \exp\left(-\frac{\lambda_{21} - \lambda_{22}}{R\,T}\right)$$

These parameters give the activity coefficients according to the Wilson equations:

$$\log_e \gamma_1 = -\log_e(x_1 + A_{12} x_2) + x_2 \left(\frac{A_{12}}{x_1 + A_{12} x_2} - \frac{A_{21}}{x_2 + A_{21} x_1} \right)$$

and

$$\log_e \gamma_2 = -\log_e(x_2 + A_{21} x_1) - x_1 \left(\frac{A_{12}}{x_1 + A_{12} x_2} - \frac{A_{21}}{x_2 + A_{21} x_1} \right)$$

The reliability of the method proposed has been examined by regression analysis for the interaction energy differences:

$$\lambda_{21} - \lambda_{22} = 423.96 + 0.83263(\lambda_{12} - \lambda_{11}) + 2.366 \times 10^{-4}(\lambda_{12} - \lambda_{11})^2$$

The root-mean-square deviation was 96 J g-mole^{-1}. For further details on the various prediction and correlation methods for Wilson's parameters, see Dewan et al. (1978), Gothard et al. (1975), Hiranuma (1976), Hiranuma and Honma (1975), Miyahara et al. (1970), Nagahama et al. (1971), and Surovy and Dojcansky (1976).

The Wilson A_{12} and A_{21} parameters can be calculated from the activity coefficients at infinite dilution (γ_1^∞ and γ_2^∞), according to Kurtyka and Kurtyka (1979):

$$\log_e \gamma_1^\infty = -\log_e A_{12} + 1 - A_{21}$$

and

$$\log_e \gamma_2^\infty = -\log_e A_{12} + 1 - A_{12}$$

The activity coefficients of various groups of aromatic hydrocarbons in water were also calculated by Tsonopoulos (1970), using the group contribution method for $\log_{10} \gamma_2^\infty$, at 25°C. The additivity of the correction factors are not valid for substituted acids and compounds forming strong intermolecular hydrogen bonds. The variation of the activity coefficient between 0 and 50°C was small and so was neglected. The correction factor Δ was defined by the formula

$$\Delta = \log_{10} \gamma_2^\infty(\text{hydrocarbon derivative}) - \log_{10} \gamma_2^\infty(\text{hydrocarbon with same number of carbon atoms and structure})$$

where the correlation equation for monocyclic hydrocarbons is

$$\log_{10} \gamma_2^\infty = 3.3918 + 0.58259(n_c - 6)$$

where n_c is the number of carbon atoms in the molecule. For halogenated aromatic compounds, $\log_{10} \gamma_2^\infty$ has been calculated from the foregoing equation and the $\log_{10} \gamma_2^\infty$ values presented in Table 2.52. The logarithms of the activity coefficients per halogen atom at infinite dilution have been calculated and are compiled in Table 2.53 (Tsonopoulos, 1970; Tsonopoulos and Prausnitz, 1971). The size of the halogen group is

Table 2.52 Activity Coefficients of Halocarbons in Water at $25^{\circ}C$

Solute	x_2	$\log_{10}\gamma_2^{\infty}$	γ_2^{∞}
C_6H_6	4.02×10^{-4}	3.3918	2,464.9
C_6H_5F	2.94×10^{-4}	3.5318	3,402.5
C_6H_5Cl	8.09×10^{-5}	4.0918	12,353.8
C_6H_5Br	4.88×10^{-5}	4.3118	20,502.2
C_6H_5I	1.62×10^{-5}	4.7918	61,915.6
$C_6H_4F_2$	2.13×10^{-4}	3.6718	6,496.8
$C_6H_4Cl_2$	1.62×10^{-5}	4.7918	61,915.6
$C_6H_4Br_2$	5.86×10^{-6}	5.2318	170,529.7
$C_6H_4I_2$	6.43×10^{-7}	6.1918	1,555,249.2
$C_6H_3F_3$	1.54×10^{-4}	3.8118	6,483.4
$C_6H_3Cl_3$	3.22×10^{-6}	5.4918	310,313.0
$C_6H_3Br_3$	7.05×10^{-7}	6.1518	1,418,404.2
$C_6H_3I_3$	2.56×10^{-8}	7.5918	39,066,094.8
$C_6H_2F_4$	1.12×10^{-4}	3.9518	8,949.5
$C_6H_2Cl_4$	6.43×10^{-7}	6.1918	1,555,249.2
$C_6H_2Br_4$	8.48×10^{-8}	7.0718	11,797,772.0
$C_6H_2I_4$	1.02×10^{-9}	8.9918	981,295,935.5
C_6HF_5	8.09×10^{-5}	4.0918	12,353.8
C_6HCl_5	1.28×10^{-7}	6.8918	7,794,710.7
C_6HBr_5	1.02×10^{-8}	7.9918	98,129,593.6
C_6HI_5	4.06×10^{-11}	10.3918	24,649,039,460.0
C_6F_6	5.86×10^{-5}	4.2318	17,053.0
C_6Cl_6	2.56×10^{-8}	7.5918	39,066,094.8
C_6Br_6	1.23×10^{-9}	8.9118	816,206,407.9
C_6I_6	1.62×10^{-12}	11.7918	619,155,877,600.0

Source: Tsonopoulos, 1970.

Table 2.53 Group Contribution to the Logarithm of the Activity Coefficients

Group	Logarithm of activity coefficient at infinite dilution $\Delta \log_{10} \gamma_W^\infty$
F-	0.14
Cl-	0.70
Br-	0.92
I-	1.40

Source: Tsonopoulos, 1970; Tsonopoulos and Prausnitz, 1971.

a function of the group contribution. The effect of halogen atoms is additive and practically independent of molecular structure.

For example, the activity coefficient for 1,4-dibromobenzene can be calculated at 25°C as follows:

$$\log_{10} \gamma_2^\infty = 3.3918 + 0.58259(6 - 6) + 2 \times 0.92 = 5.2318$$

and

$$\gamma_2^\infty = 170,529.7$$

For polymethylbenzenes with saturated side chains, the correlation equation becomes

$$\log_{10} \gamma_2^\infty = 3.8227 + 0.33044(n_c - 6)$$

where $8 < n_c < 12$. When the side chains are branched, the expression is

$$\log_{10} \gamma_2^\infty = 3.2589 + 0.53267(n_c - 6)$$

where $n_c > 9$. The corrections to $\log_{10} \gamma_2^\infty$ per unsaturated bond are

double bond (=) in side chain: -0.30

triple bond (≡) in side chain: -0.46

For condensed polycyclic hydrocarbons, the following equation is proposed:

$$\log_{10} \gamma_2^\infty = 3.3950 + 0.35794(n_c - 6)$$

Derivatives of hydrocarbons were also correlated using a correction factor Δ:

$$\Delta = \log_{10} \gamma_2^\infty(\text{hydrocarbon derivative}) - \log_{10} \gamma_2^\infty(\text{hydrocarbon with the same } n_c \text{ and structure})$$

Further details on the correlation and tabulated correction factors for several groups are available from Tsonopoulos (1970) and Tsonopoulos and Prausnitz (1971).

The method described above has limitations; for example, it does not provide guidance for the calculation of different isomers or di-, tri-, and tetrahalogenated benzenes.

The total surface area proved to be a very useful parameter for the correlation of molecular properties. It is the sum of individual atoms or group surface area contributions. Mackay et al. (1980) calculated the molecular surface area for 37 chlorinated biphenyls according to the method of Valvani et al. (1973). The interatomic distances between various atoms or groups in the molecule were:

Aromatic	C-C	1.40 Å
Aromatic	C-H	1.08
Aromatic-aliphatic	C-C	1.54
Aliphatic	C-C	1.53
Aliphatic	C-Cl	1.70

The van der Waals radii used were:

Aromatic	C	1.70 Å
Aromatic	H	1.20
Methyl or methylene group		2.0
Chlorine	Cl	1.80
Aromatic	C-C-C angles and H-C-C angles	= 120°

The activity coefficients were calculated from the solubility data and plotted against the total surface area on a semilogarithmic plot. The straight line was represented by the equation:

$$\log_e \gamma_i = 0.0815 \text{ TSA} - 2.78$$

where TSA is the total surface area.

In the case of a solid solubility in a liquid, the activity of the solute ($a_2 = \gamma_2 x_2$) is calculated from the following formula at low pressure (Prausnitz, 1969):

$$\log_e a_2 = \frac{\Delta H_{melt}^{tp}}{R \, T_{tp}}\left(\frac{T - T_{tp}}{T}\right) + \frac{\Delta H_{tr}^{tr}}{R \, T_{tr}}\left(\frac{T - T_{tr}}{T}\right) - \frac{\Delta C_{Pmelt}^{tp}}{R}\left[\frac{T - T_{tp}}{T} + \log_e\left(\frac{T_{tp}}{T}\right)\right]$$

where T_{tp} and T_{tr} are the absolute temperature (degrees Kelvin) at the triple and transition points, respectively, and ΔH_{melt} and ΔH_{tr} are the heat of melting and transition, respectively. ΔC_{Pmelt} is the heat capacity at constant pressure and at the melting point.

The activity coefficient and limiting activity coefficient of a solid in a liquid are calculated from the foregoing relationships. The numerical values for halogenated hydrocarbon - water systems are tabulated in Table 2.52. In some cases the magnitude of γ_2 is in the order of 10^6 (Mackay and Shiu, 1977; Dexter and Pavlou, 1978; Mackay et al., 1980). As was mentioned above, the activity coefficient is too sensitive to solubilities. The heat capcity ($\Delta C_{P_{melt}}$) is sometimes negligible, but it can contribute (i.e., it can increase γ_2 values by 10%).

In the case of liquid-liquid solubility, particularly for the solubility of water in liquid halogenated hydrocarbons (mole fraction of the halogenated hydrocarbon >> 0.99), the activity coefficient of water (γ_2^w) can be approximated from the solubility data, using the expression

$$\gamma_2^w = \frac{1}{x_2^w}$$

where x_2^w is the mole fraction of water in the organic liquid. Typical example is the solubility of water in CCl_2F_2 at various temperatures (Zhukoborskii, 1973) (see Table 2.54). The activity coefficient of water in various solvents has been reported by Prosyanov et al. (1974) at the boiling point of the solvents (see Table 2.55). While several investigators have stated that the simple equation above should not be used for the calculation of the activity coefficient of liquid halogenated hydrocarbon in water, that is, from the solubility of halogenated hydrocarbons in water,

$$\gamma_1^{hh} \neq \frac{1}{x_1^{hh}}$$

where hh indicates the halogenated hydrocarbon, Mackay et al. (1975) and Mackay and Shiu (1975a, b) have shown that a good approximation can be obtained contrary to previous reports. Furthermore, a simple but very useful correlation was reported by Mackay and Shiu (1975b, 1977) and Dexter and Pavlou (1978) for the activity coefficients of hydrocarbons in aqueous solution at 25°C and 1 atm pressure (see Fig. 2.34). The numerical values are given in Table 2.56.

Example: Solubility of benzene in water at 25°C
_____

The solubility of benzene in water is 1780 g m^{-3} at 25°C (Claxton, 1961). This corresponds to a mole fraction of benzene in the aqueous phase:

$$x_a = \frac{1780/78.11}{1780/78.11 + 10^6/18.0153} = 4.10 \times 10^{-4}$$

The activity coefficient of benzene in the aqueous phase (γ_a) is obtained from the fugacity (f):

$$f = x_a \gamma_a f_R$$

where f_R is the reference fugacity. In the system of aqueous equilibrium of benzene,

Table   2.54   Activity Coefficient in the H_2O - CCl_2F_2 System

Temperature	Solubility of H_2O in CCl_2F_2		Activity coefficient of H_2O,	$\Delta H_{solution}$
(^oC)	ppm by weight	mole fraction	$\gamma_2^w = 1/x_2^w$	(kcal g-mole^{-1})
-30	3.0	0.0000202	49,600	
				8.57
-20	6.1	0.0000411	24,400	
				8.73
-10	12.0	0.0000796	12,600	
				9.85
0	24.0	0.0001593	6,280	
				9.22
10	43.2	0.0002903	3,450	
				8.10
20	70.1	0.0004710	2,120	
				9.20
30	118.0	0.0007930	1,260	
				8.50
40	185.0	0.001240	807	

Source: Zhukoborskii, 1973.

Table   2.55   Activity Coefficients and Heat of Solution of Water in Solvents

Solvent	Boiling point (^oC)	Activity coefficient of H_2O at the boiling point	Heat of solution of H_2O (kcal g-mole^{-1})
CH_2Cl_2	40.0	67	5.90
CCl_4	76.54	296	5.88
$CHCl_2$-$CHCl_2$	146.2	19	6.30
$CHCl=CCl_2$	87.0	63	6.50
$CCl_2=CCl_2$	121.0	273	7.10
C_6H_5Cl	132.0	17	6.30

Source: Prosyanov et al., 1974.

the following relation is valid:

$$f \;=\; x_b\, \gamma_b\, f_R \;=\; x_a\, \gamma_a\, f_R$$

where the subscripts a and b refer to the aqueous and benzene phases, respectively.
(The fugacity ratio for solid hydrocarbons have been correlated with temperature by
Mackay and Shiu (1977)). From this expression it follows that

$$\gamma_a \;=\; \frac{x_b\, \gamma_b}{x_a}$$

However, according to the Raoult's law convention, if the mutual solubility is very

Table 2.56 Solubility and Activity Coefficients of Halocarbons in Water at $25^\circ C$
            and 1 atm Pressure

Solute	Molecular weight	Solubility in water, x_2 (mole fraction)	Vapor pressure, p (atm)	Activity coefficient, γ_2
CH_4	16.04	2.71×10^{-5}	269	137.5
C_2H_6	30.07	3.62×10^{-5}	39.4	702
C_3H_8	44.11	2.55×10^{-5}	9.29	4,230
$n-C_4H_{10}$	58.13	1.90×10^{-5}	2.40	21,900
$n-C_5H_{12}$	72.15	9.61×10^{-6}	0.675	104,000
$n-C_6H_{14}$	86.18	1.99×10^{-6}	0.205	504,000
$n-C_7H_{16}$	100.21	5.27×10^{-7}	0.0603	1,900,000
$n-C_8H_{18}$	114.23	1.04×10^{-7}	0.0186	9,620,000
$n-C_9H_{20}$	128.26	3.09×10^{-8}	5.64×10^{-3}	32.4×10^6
$n-C_{10}H_{22}$	142.29	6.58×10^{-9}	1.73×10^{-3}	15.8×10^7
$n-C_{12}H_{26}$	170.34	3.91×10^{-10}	1.55×10^{-4}	25.6×10^8
$n-C_{14}H_{30}$	198.40	2.00×10^{-10}	1.26×10^{-4}	50.1×10^8
C_2H_4	28.05	8.41×10^{-5}	59.91	199
C_3H_6	42.08	8.56×10^{-5}	11.29	1,150
$1-C_4H_8$	56.12	7.13×10^{-5}	2.933	5,070
$1-C_5H_{10}$	70.14	3.80×10^{-5}	0.839	26,000
$1-C_6H_{12}$	84.16	1.07×10^{-5}	0.245	93,500
$2-C_7H_{14}$	98.19	2.75×10^{-6}	0.0637	364,000
$1-C_8H_{16}$	112.22	4.33×10^{-7}	0.0229	2,310,000
Benzene	78.12	4.10×10^{-4}	0.125	2,400
Toluene	92.15	1.01×10^{-4}	0.0374	9,900
Ethyl benzene	106.17	2.58×10^{-5}	0.0125	38,800
o-Xylene	106.17	2.97×10^{-5}	8.71×10^{-3}	33,700
i-Propylbenzene	120.20	7.49×10^{-6}	6.03×10^{-3}	134,000
Naphthalene	128.19	4.83×10^{-6}	1.14×10^{-4}	76,900
Biphenyl	154.21	8.74×10^{-7}	7.45×10^{-5}	475,000
Acenaphthene	154.21	4.53×10^{-7}	3.97×10^{-5}	319,000
Fluorene	166.23	2.06×10^{-7}	1.64×10^{-5}	795,000
Antracene	178.24	7.58×10^{-9}	5.04×10^{-5}	1,814,000
Phenantrene	178.24	1.19×10^{-7}	4.53×10^{-6}	1,820,000

Source: Mackay & Shiu, 1975a, b.

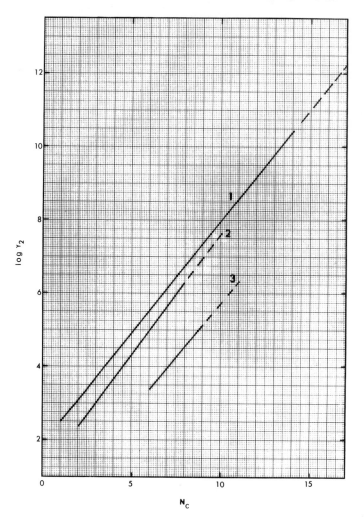

Fig. 2.34 Correlation of activity coefficients with number of carbon atoms. γ_2 = activity coefficient of solute; N_c = number of carbon atoms in the compounds. 1, n-Paraffins: $\log_{10} \gamma_2 = 1.84574 + 0.61236 \, N_c$; 2, n-Olefins: $\log_{10} \gamma_2 = 1.04271 + 0.659317 \, N_c$; 3, Aromatics: $\log_{10} \gamma_2 = -8.82158 \times 10^{-2} + 0.580539 \, N_c$. (from Mackay and Shiu, 1975a, b, 1977.)

small, the more concentrated phase will so closely resemble the pure substance that its vapor pressure will not be appreciably modified by the solvent (Christie and Crisp, 1967),

$$\gamma_b = 1 \quad \text{when} \quad x_b = 1$$

which suggests that in very dilute aqueous solutions both values are approximately
unity; therefore:

$$\gamma_a \;=\; \frac{1}{x_a}$$

Applying this expression for the aqueous phase of benzene, we obtain

$$\gamma_a \;=\; \frac{1}{4.10 \times 10^{-4}} \;=\; 2439$$

The Henry's constant (H, atm g-mole^{-1} m^{-3}) is calculated from the vapor pressure
of benzene ($p^{25^\circ C} = 0.125$ atm), and the concentration of benzene in the aqueous
phase:

$$C \;=\; \frac{1780}{78.11} \;=\; 22.788 \text{ g-mole m}^{-3}$$

so that the result becomes

$$H \;=\; \frac{P}{C} \;=\; \frac{0.125}{22.788} \;=\; 5.49 \times 10^{-3} \text{ atm g-mole}^{-1} \text{ m}^{-3}$$

The value of $\gamma_a f_R$ is now calculated by

$$\gamma_a f_R \;=\; \frac{p}{x_a} \;=\; \frac{0.125}{4.10 \times 10^{-4}} \;=\; 305 \text{ atm}$$

The order of $\gamma_a f_R$ gives an indication for solubility. Large value means that the
solute is less soluble in the solvent and partitions preferentially into the vapor
phase.

   A good review of the methods for calculating Henry's law constant from solubility
was reported by Friend and Alder (1957). The authors also presented the calculation
of liquid-vapor equilibrium constants from solubility data. There are several worked
examples in the article. Similarly, Prausnitz (1958) also reported a method for estimating of gas-liquid phase equilibrium at high pressures from the modified regular-
solution theory. Prausnitz discusses the resulting equations for estimating the solubilities of gases and the temperature coefficients of solubility.

   The activity coefficients of aromatic compounds are considerably lower than the
corresponding n-alkanes or n-olefins with the same number of carbon atoms in the molecule. This indicates that the aromatics have less tendency to evaporate.

   The activity coefficient of $CClF_2-CClF_2$ in the aqueous phase as a function of
temperature has been reported by Stepakoff and Modica (1973) (see Table 2.57). The
activities and activity coefficients in the tetrahydrofuran-water system were reported
by Pinder (1973). The activity coefficient varied between 53 and 0.81 in the temperature interval of 63 and 93°C.

   The limiting activity coefficient of water γ_2^∞ in the organic liquids can be

Table 2.57 Activity Coefficient of $CClF_2-CClF_2$ in Water

Temperature (°C)	Vapor pressure (psia)	Gas solubility (ppm atm^{-1})	Liquid solubility (ppm)	Activity coefficient
-3.88	10.95	347	258	36,700
0.0	12.81	295	256	36,900
25	31.32	104	221	42,900
51	67.35	46	210	45,200

Source: Stepakoff & Modica, 1973.

derived from the two-suffix Margules equation:

$$\log_{10} \gamma_2^{\infty} = \frac{\log_{10} \gamma_2^{w}}{(1 - x_2^{w})^2}$$

The activity coefficient of the solute at infinite dilution provides a rough measure of $G^E(x_2 = 0.5)$, and hence an indication of the likelihood that phase separation will occur in the system concerned (Copp and Everett, 1953). To enable us to calculate the activity coefficients γ_1 and γ_2 of a two-phase liquid-liquid mixture from mutual solubility data, we require some sort of relationship between the solubility and these activity coefficient values. Such well-known equations as the van Laar, Margules, and Scatchard-Hamer have been proposed by Severance et al. (1963), Brian (1965), and Carlson and Colburn (1942). More recent suggestions refer to the extended Wilson equation (Palmer and Smith, 1972; Chanrasekaran et al., 1976; Nagata et al., 1975) and the NRTL equation (Renon and Prausnitz, 1968), each with three adjustable parameters. These three parameter equations raise some difficulties, such as the requirement for a highly accurate value for the third parameter, which depends upon the nature of the system. Therefore, additional data are needed comparing to the two-parameter equations. Several investigators selected the van Laar equation as the best choice, being accurate enough for relative simplicity (Brian, 1965). The mutual solubility is given by van Laar's equation:

$$\log_e \gamma_1 = \frac{A}{(1 + A x_1/(B x_2))^2}$$

and

$$\log_e \gamma_2 = \frac{B}{(1 + B x_2/(A x_1))^2}$$

where

$$A = \log_e \gamma_1^{\infty}$$
$$B = \log_e \gamma_2^{\infty}$$

The mutual solubility becomes

$$A = \frac{\log_e(x_1^{hh}/x_1^{w})}{(1 + A\, x_1^{w}/(B\, x_2^{w}))^{-2} - (1 + A\, x_1^{hh}/(B\, x_2^{hh}))^{-2}}$$

and

$$B = \frac{\log_e(x_2^{hh}/x_2^{w})}{(1 + B\, x_2^{w}/(A\, x_1^{w}))^{-2} - (1 + B\, x_2^{hh}/(A\, x_1^{hh}))^{-2}}$$

Corresponding expressions based upon the Scatchard-Hamer equation were reported by Severance et al. (1963). These expressions can be used in combination with

$$\log_e \gamma_2^{\infty} = \frac{\log_e \gamma_2}{(1 - x_2)^2}$$

to derive the limiting activity coefficients for systems of halogenated hydrocarbons with water when the mole fraction of water is very close to unity. The accuracy of the activity coefficients depends upon the error involved in the determination of the mutual solubility data. Because the activity coefficient is inversely proportional to the mole fraction, in a liquid-liquid system it goes over a maximum between 15 and 20°C. (Regarding the temperature dependence of the solubility, see Sec. 2.8.) In addition to the accuracy of the mutual solubility data, the accuracy of the activity coefficients depends upon the suitability of the van Laar equation to represent liquid-liquid mutual solubilities.

The success of the van Laar equation for mutual solubility has been tested by Tsonopoulos (1970) by examining the system phenol - water with the aid of the vapor-liquid equilibrium data and the second virial coefficient (to calculate the vapor-phase deviation from ideality). For the plot of γ_2^{∞} versus $1/T$, a break was observed not only for the van Laar equation but also for the three-parameter Wilson equation. At low temperatures an error of 8% was obtained for $\log_{10} \gamma_2^{\infty}$'s. In general, Tsonopoulos (1970) found that the van Laar equation gives satisfactory results for methylated phenols and water mixtures. Satisfaction was also found for water in iso-butanol, aniline, n-butanol, and propylene oxide systems (see Brian, 1965). A computer program for calculation of van Laar's constant from mutual solubility data is presented in Table 2.58 (Hala et al., 1967). The graphical determination was reported by Missen (1978).

The activity coefficients of a system is very useful for the calculation of the excess properties, such as the excess free energy $\bar{G}_i^{E}$, excess volume $\bar{V}_i^{E}$, excess entropy $\bar{S}_i^{E}$, excess enthalpy $\bar{H}_i^{E}$, and excess heat capacity $\bar{C}_{pi}^{E}$. By differentiation of $\bar{G}_i^{E}$,

$$\bar{G}_i^{E} = R\,T \log_e \gamma_i$$

Table 2.58   VLE Data from Mutual Solubility Using Van Laar's Equation

```
10     PRINT "VLE DATA FROM MUTUAL SOLUBILITY USING VAN LAAR'S EQUATION."
20     PRINT "**********************************************************"
30     PRINT
40     PRINT "REF: HATA ET AL. 1967, P. 106."
50     PRINT
60     DIM A$(72),X(13)
70     RESTORE 90
80     MAT READ X
90     DATA 0,0.05,0.1,0.2,0.3,0.4,0.5,0.6,0.7,0.8,0.9,0.95,1
100    READ N
110    REM N = NUMBER OF SYSTEMS.
120    P=760
130    FOR I=1 TO N
140    READ A$?T9,X1,X2
150    REM A$ = NAME OF THE SYSTEM.
160    REM T9 = TEMPERATURE, *C.
170    REM X1 = MOLE FRACTION OF COMPONENT 1 IN COMPONENT 2. (PHASE 2).
180    REM X2 = MOLE FRACTION OF COMPONENT 2 IN COMPONENT 1. (PHASE 1).
190    READ A1,B1,C1
200    READ A2,B2,C2
210    REM A1,B1,C1,A2,B2,C2 = ANTOINE COEFFS., MMHG, *C., LOG TO BASE 10.
220    T1=(X1/(1-X1)+X2/(1-X2))
230    T2=((LOG(X2/X1))/LOG(10))/((LOG((1-X1)/(1-X2)))/LOG(10))
240    T3=X1/(1-X1)+X2/(1-X2)
250    T4=(2*X1*X2)/((1-X1)*(1-X2))*T2
260    R=(T1*T2)/(T3-T4)
270    REM R = A/B WHERE A & B = VAN LAAR'S CONSTANTS.
280    T5=(LOG(X2/X1))/LOG(10)
290    T6=1/((1+R*X1/(1-X1))+2)
300    T7=1/((1+R*X2/(1-X2))+2)
310    A=T5/(T6-T7)
320    B=A/R
330    PRINT A$;" &";T9;"*C."
340    PRINT ":::::::::::::::::::::::::::::::::::::::::::::::::::"
350    PRINT
360    PRINT "VAN LAAR'S CONSTANTS:"
370    PRINT "                      ";"A = ";A
380    PRINT "                      ";"B = ";B
```

Table 2.58 (Continued)

```
390   PRINT
400   PRINT "TEMPERATURE, PRESSURE, ACTIVITY COEFFICIENTS, LIQ-VAP. COMPOSITION,"
410   PRINT "    *C.       MMHG.       G1.      G2.      X1.      Y1."
420   PRINT
430   FOR J=1 TO 13
440   IF X(J)=0 THEN 580
445   IF X(J)=1 THEN 642
450   X2=1-X(J)
460   T=T-0.001
464   GOTO 470
466   T=T+0.001
470   P1=EXP(LOG(10)*(A1+B1/(C1+T)))
480   P2=EXP(LOG(10)*(A2+B2/(C2+T)))
490   F=G1*X(J)*P1+G2*X2*P2-760
500   IF F>0.1 THEN 460
504   IF F<-0.1 THEN 466
510   G1=A/(1+A*X(J)/(B*(1-X(J))))↑2
520   G1=EXP(LOG(10)*G1)
530   G2=B/(1+B*(1-X(J))/(A*X(J)))↑2
540   G2=EXP(LOG(10)*G2)
550   Y1=G1*X(J)*P1/P
560   Y2=G2*(1-X(J))*P2/P
570   GOTO 650
580   P2=760
590   G1=1
600   G2=1
610   X2=1-X(J)
620   Y1=0
630   Y2=1
640   T=B2/((LOG(P2))/LOG(10)-A2)-C2
641   GOTO 650
642   P1= 760
643   G1=1
644   G2=1
645   X2=1-X(J)
646   Y1=1
647   Y2=0
648   T=B1/((LOG(P1))/LOG(10)-A1)-C1
650   IMAGE 3X,3D.D,5X,4D.D,4X,4D.3D,3X,4D.3D,3X,D.4D,3X,2D.4D
```

Table 2.58 (Continued)

```
660   PRINT USING 650;T,P,G1,G2,X(J),Y1
670   NEXT J
680   PRINT
690   NEXT I
700   PRINT
710   DATA 1
720   DATA "WATER - N-BUTANOL AT 760 MMHG.",110,0.9788,0.6759
730   DATA 8.10765,-1750.29,235,7.54472,-1405.87,183.908
740   END
```

VLE DATA FROM MUTUAL SOLUBILITY USING VAN LAAR'S EQUATION.

****

REF: HALA ET AL. 1967, P. 106.

WATER - N-BUTANOL AT 760 MMHG. & 110 *C.

:::

VAN LAAR'S CONSTANTS:
                         A = 0.334771
                         B = 1.62583

TEMPERATURE, PRESSURE, ACTIVITY COEFFICIENTS, LIQ-VAP. COMPOSITION,

*C.	MMHG.	G1.	G2.	X1.	Y1.
117.5	760.0	1.000	1.000	0.0000	0.0000
116.4	760.0	2.126	1.000	0.0500	0.1872
110.4	760.0	2.089	1.002	0.1000	0.3013
104.9	760.0	2.008	1.009	0.2000	0.4798
100.8	760.0	1.917	1.025	0.3000	0.5955
97.8	760.0	1.815	1.056	0.4000	0.6727
95.5	760.0	1.699	1.115	0.5000	0.7244
93.9	760.0	1.568	1.232	0.6000	0.7576
93.1	760.0	1.421	1.483	0.7000	0.7765
93.0	760.0	1.261	2.146	0.8000	0.7853
93.9	760.0	1.099	4.852	0.9000	0.7966
95.6	760.0	1.032	10.746	0.9500	0.8408
99.9	760.0	1.000	1.000	1.0000	1.0000

we obtain the excess molar functions:

$$\left(\frac{\partial \bar{G}_i^E}{\partial p}\right)_{T,x} = \bar{V}_i^E = R\,T\left(\frac{\partial \log_e \gamma_i}{\partial p}\right)_{T,x}$$

$$-\left(\frac{\partial \bar{G}_i^E}{\partial T}\right)_{p,x} = \bar{S}_i^E = -R\left(\frac{\partial T \log_e \gamma_i}{\partial T}\right)_{p,x}$$

$$-T^2\left(\frac{\partial \bar{G}_i^E/T}{\partial T}\right)_{p,x} = \bar{H}_i^E = -R\,T^2\left(\frac{\partial \log_e \gamma_i}{\partial T}\right)_{p,x}$$

$$-T\left(\frac{\partial^2 \bar{G}_i^E}{\partial T^2}\right)_{p,x} = \bar{C}_{p_i}^E = -2\,R\,T\left(\frac{\partial \log_e \gamma_i}{\partial T}\right)_{p,x} - R\,T^2\left(\frac{\partial^2 \log_e \gamma_i}{\partial T^2}\right)_{p,x}$$

When some or all of the excess properties are available from experimental measurements, they can be used to extrapolate equilibrium data; and vice verse, equilibrium data can be useful to derive volume and heat effects. Furthermore, the excess functions of **volume, enthalpy, and heat capacity are equal to the corresponding** functions of mixing, because in these cases the functions of mixing for any ideal solution are zero:

$$V_i^{mix} = \bar{V}_i^E$$

$$H_i^{mix} = \bar{H}_i^E$$

$$C_{p_i}^{mix} = \bar{C}_{p_i}^E$$

The volume changes of mixing for binary mixtures of liquids was reviewed by Battino (1971). The volume change associated with hydrophobic interaction was studied by Morild (1980).

2.5-4   The Phenomenon of Azeotropy

Azeotropes are a common phenomenon in organic systems (Kurtyka, 1975). For example, as documented by Horsley (1973), of 765 binary systems, 637 (i.e., 86%) were azeotropic. This is an indication of the common occurence of azeotropy in aqueous systems. Horsley's (1973) compilation is a valuable source of azeotropic data. Azeotropy and other theoretical problems of vapor-liquid equilibrium in homogeneous liquid systems have been well described by Malesinski (1965). Azeotropic data for hydrocarbons and sulfur compounds have been compiled by Claxton (1958).

The condition for an azeotrope in a thermodynamic system is that the composition

of each phase, the temperature, and the pressure remain constant and uniform throughout the system. The formation of an azeotrope depends upon two factors:

1. The difference in the normal boiling point between the two substances
2. The magnitude of the deviation from Raoult's law

If the boiling-point difference between the two constituents is small, the deviation
from Raoult's law must be small before an azeotrope will form. Or vice verse, if the
deviation from Raoult's law is small, the boiling-point difference also has to be small,
giving the condition for formation of an azeotrope (Ewell et al., 1944).

In most binary liquid systems, the azeotropic boiling point is lower than that
of either component, since positive deviation from Raoult's law occurs more frequently,
particularly for halogenated hydrocarbon – water systems (Eduljee and Tiwari, 1976).
If a system shows a positive deviation from Raoult's law, the point of maximum vapor
pressure will give a minimum-boiling-point composition (Flynn, 1975). A liquid of this
composition will boil at constant temperature without altering its composition (i.e.,
this is a constant-boiling mixture or azeotrope). If the azeotrope is distilled, its
composition will remain unchanged because the vapor and liquid compositions are identical:

$$x_i = y_i$$

The occurrence of an azeotrope is of great importance in distillation, because
it is not possible to separate a mixture into its pure constituents under the given
conditions (Gordon and Bright, 1939). Basically, there are two different types of
azeotropes:

1. Homogeneous (homoazeotrope)
2. Heterogeneous (heteroazeotrope)

Similarly, nonazeotropes, called zeotropic systems, are divided into homozeotropes
and heterozeotropes. The halogenated hydrocarbon – water systems belong to the heteroazeotropes; consequently, more emphasis will be directed toward them (see Table 2.59).
These systems constitute a limited miscibility or partial mutual solubility in the
liquid phases. Limited solubility always gives heterogeneous minimum azeotropes,
whereas maximum azeotropes are homogeneous.

When information is not available on azeotropy, it is necessary to calculate it
with good accuracy. At a given temperature, the total pressure of a heterogeneous
system is equal to the sum of the vapor pressure of the pure compounds. By plotting,
for example in a binary system, the vapor pressure of the two constituents versus the
temperature, it is easy to read off the temperature at which the sum of the vapor
pressure of the two components becomes 760 mm Hg. This is the azeotropic boiling
point of the binary heterogeneous system. Similarly, the azeotropic boiling point can
be calculated at any other pressures. The azeotropic concentration is also calculated

Table 2.59 Azeotropic Data for Halogenated Hydrocarbon/Water Systems

Halogenated hydrocarbon	Molecular weight	Boiling point ($^{\circ}$C)	Azeotropic data	
			B.p. ($^{\circ}$C)	Wt % H_2O
$CHCl_3$	119.38	61.0	56.1	2.8
CH_2Cl_2	84.93	43.5	38.1	1.5
CCl_4	153.82	76.54	66.0	4.1
$CCl_2F-CClF_2$	187.38	47.5	44.5	1.0
$CCl_2=CCl_2$	165.83	121.0	88.5	17.2
$CHCl=CCl_2$	131.39	86.2	73.4	7.02
$CHCl_2-CCl_3$	202.30	162.0	95.9	-
c-$CHCl=CHCl$	96.94	60.2	55.3	3.35
t-$CHCl=CHCl$	96.94	48.35	45.3	1.9
$CHCl_2-CH_2Cl$	133.41	113.8	86.0	16.4
CH_2Cl-CH_2Cl	98.96	83.5	72.28	9.2
150 mm Hg	98.96	-	33.5	4.9
75 mm Hg	98.96	-	19.0	4.9
CH_3-CH_2Br	108.97	38.4	37.0	1.3 vol %
CH_3-CH_2I	155.97	70.0	60.0	3.5 vol %
$CH_2=CH-CH_2Cl$	76.53	44.9	43.0	2.2
$CH_3-CH=CHCl$	76.53	37.4	33.0	0.9
$CH_2=CH-CH_2I$	167.98	102.0	80.7	10.0
$CH_2Cl-CHCl-CH_3$	112.99	97.0	78.0	12.0
$CH_2Cl-CH_2-CH_3$	78.54	46.6	44.0	2.2
$CH_3-CHCl-CH_3$	78.54	36.5	33.6	1.2
$CHCl=C(CH_3)-CH_3$	90.55	68.1	61.9	7.5
$CH_3-(CH_2)_2-CH_2Cl$	92.57	77.9	68.1	6.6
$CH_2Cl-CH(CH_3)-CH_3$	92.57	68.8	61.6	3.3
$CH_2I-CH(CH_3)-CH_3$	184.02	122.5	95.96	21.0 vol %
$CH_2Cl-(CH_2)_3-CH_3$	106.60	108.35	82.0	-
Chlorobenzene	112.56	131.8	90.2	28.4
$CH_2Cl-(CH_2)_4-CH_3$	120.62	134.5	91.8	29.7
p-Chlorotoluene	126.59	163.5	95.0	-
1-Chloro-2-ethylhexane	148.68	173.0	97.3	55.0
Chlorodecane	176.73	210.6	99.7	84.0

Source: Horsley, 1973; Prahl & Mathes, 1934.

from the vapor pressure of the pure components at the azeotropic boiling point of the system:

$$x_1^{Az} = \frac{p_1^o}{p_1^o + p_2^o}$$

The correlation of azeotropic data has been of interest to a number of investigators. The graphical method has had the most success. The early work of Mair et al. (1941) has been developed further by Skolnik (1948) for a homologous series of hydrocarbon - hydrocarbon and alcohol - hydrocarbon solutions. A more recent correlation was reported by Seymour et al. (1977) for alcohol - alkane systems.

The early work of Meissner and Greenfield (1948) on acid - hydrocarbon and acid - halogenated hydrocarbon systems has been extended by Johnson and Modonis (1959) to other combinations of organic liquids, where the deviations from Raoult's law can be predicted from the hydrogen-bonding characteristics of the components as proposed by Ewell et al. (1944). The correlation proposed covered a large number of substances; however, in this book, the emphasis is on two classes of systems:

Classes I to IV

Classes I to V

Water belongs to class I, the halogenated hydrocarbons belong to class IV, and class V covers the halogenated aromatics (see Table 2.60).

In the graphical presentation of the curves for the various combinations of classes, the difference between the normal boiling points of the pure compounds (D) was plotted versus Z, where

$$Z = T_{low\ boiler} - T_{azeotrope}$$

for a minimum-boiling azeotrope, and

$$Z = T_{azeotrope} - T_{high\ boiler}$$

for a maximum-boiling azeotrope. The linear equations are:

$$D = 55 - 3.0\ Z \quad \text{for} \quad \text{classes I - IV} \quad Z = 18.33$$

and

$$D = 75 - 4.0\ Z \quad \text{for} \quad \text{classes I - V} \quad Z = 18.75$$

The curves are plotted in Fig. 2.35.

For binary halogenated hydrocarbon systems Li (1977) introduced a simple method for the prediction of the binary azeotropes using a rectilinear equation. The method has not been tested for aqueous systems.

A good review of the various proposals up to 1976 has been given by Eduljee and Tiwari (1976). See also Iguchi (1977), Tamir (1981a, b), Kurtyka and Kurtyka (1980),

Table 2.60   Classification of Compounds after Johnson & Modonis (1959)

Class IV (halogenated hydrocarbons)	Class V (halogenated aromatics)
CH_3I	C_6H_5Br
CH_2Cl_2	$1,4-C_6H_4Br_2$
$CHBrCl_2$	$1,3-C_6H_4Cl_2$
CH_2Br-CH_2Br	$1,4-C_6H_4ICH_3$
$CHCl_3$	$1,2-C_6H_4BrCH_3$
$CHBr_3$	1-Bromonaphthalene
CCl_4	1-Chloronaphthalene
CH_2Br-CH_3	
$CHCl_2-CH_3$	
CH_2Cl-CH_2Cl	
$CHCl=CCl_2$	
$CHCl_2-CHCl_2$	
$CHCl_2-CCl_3$	
$CH_2Cl-CH_2-CH_3$	
$CH_3-CHCl-CH_3$	
$CH_3-CHBr-CH_3$	
$CH_3-CHI-CH_3$	
$CH_2Cl-CHCl-CH_3$	
$CH_2Cl-CHCl-CH_2Cl$	
$CH_2Cl-CH_2CH_2Cl$	
$CH_2Cl-CH_2-CH_2-CH_3$	
$CH_3-CHBr-CH_2-CH_3$	
$CH_2Br-CH(CH_3)-CH_3$	
$CH_3-CHI-CH_2-CH_3$	
$CH_2I-CH_2-CH(CH_3)-CH_3$	

Othmer (1963), and Pruett (1971). A flow chart of the computer program for the prediction of azeotropic behavior has been presented by Nakanishi (1968). A method for the
quick estimation of azeotropic data away from the critical state was reported by
Mukhopadhyay (1975). The calculation requires the boiling points and the latent heats
of vaporization of the pure compounds and an expression for the activity coefficient
as a function of mole fraction (e.g., van Laar, Redlich-Kister, or Wilson equations).
Some other work not mentioned by Eduljee and Tiwari (1976, 1977) is that of Barboi et
al. (1971), calculating the azeotropic limit through choosing a suitable azeotropic
additive, and Gibbard and Emptage (1975), examining the variation of the azeotropic

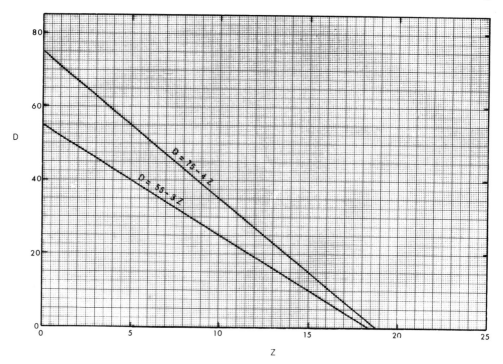

Fig. 2.35 Correlation of azeotrope boiling points of water – halogenated hydrocarbon systems. D = boiling-point difference (K); Z = boiling azeotrope (K). For a
minimum-boiling azeotrope: $Z = T_{low\ boiler} - T_{azeotrope}$; for a maximum-boiling
azeotrope: $Z = T_{azeotrope} - T_{high\ boiler}$. (From Johnson & Modonis, 1959.)

composition and temperature with pressure. The pressure effect upon azeotropic systems
was investigated by Nutting and Horsley (1973) using a graphical method.

    Because our main interest in this book relates to halogenated hydrocarbon –
water systems, we shall concentrate on the correlation of azeotropic data for these
solutions. From the normal boiling points of the constituents and the azeotropic
concentration, expressed as a mole fraction, Eduljee and Tiwari (1976, 1977) derived
the following expression:

$$x_2 = \frac{(S_1 S_2)^{\frac{1}{2}} - S_1}{S_2 - S_1} + \frac{(S_1 S_2)}{2 W} (T_1 - T_2)$$

where x_2 = azeotropic composition, mole fraction
        S = molar entropy of vaporization, cal g-mole^{-1} K^{-1}
        W = interaction parameter, cal g-mole^{-1}
        T = normal boiling point, K

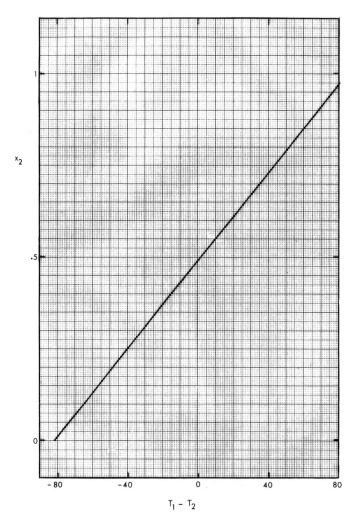

Fig. 2.36 Correlation of azeotropic composition. x_2 = azeotropic composition (mole fraction); $(T_1 - T_2)$ = normal boiling-point difference (K); $x_2 = 0.49 + 0.006(T_1 - T_2)$. (From Eduljee & Tiwari, 1976.)

By plotting $(T_1 - T_2)$ versus x_2, we obtain a straight line with slope $(S_1 S_2)^{\frac{1}{2}}/(2 \, W)$ and intercept $((S_1 S_2)^{\frac{1}{2}} - S_1)/(S_2 - S_1)$. The plot for halogenated hydrocarbon-water systems is illustrated in Fig. 2.36. The correlated binary azeotropic data of the various classes of mixtures have been compiled by Eduljee and Tiwari (1977).

The correlation of azeotropic temperature for halogenated hydrocarbon - water systems has been carried out using the following equation:

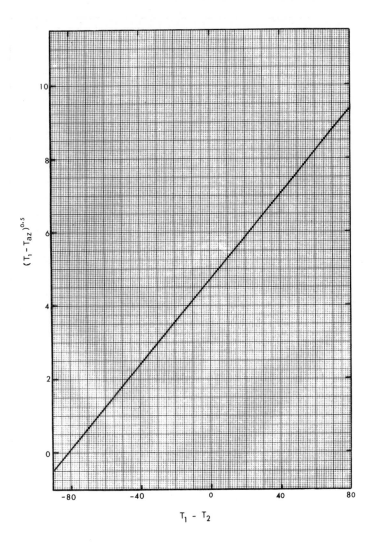

Fig. 2.37 Correlation of azeotropic temperature. $(T_1 - T_2)$ = normal boiling-point difference (K); T_{az} = azeotropic temperature (K). $(T_1 - T_{az})^{\frac{1}{2}}$ = 4.74 + 0.058$(T_1 - T_2)$. (From Eduljee & Tiwari, 1976.)

$$(T_1 - T_{az})^{\frac{1}{2}} = (\frac{W}{S_1})^{\frac{1}{2}} \frac{(S_1 S_2)^{\frac{1}{2}} - S_1}{(S_2 - S_1)} + (\frac{W}{S_1})^{\frac{1}{2}} \left[(S_1 S_2)^{\frac{1}{2}} \frac{(T_1 - T_2)}{2 W} \right]$$

where T_{az} is the azeotropic temperature in degrees Kelvin. The linear relationship between $(T_1 - T_{az})^{\frac{1}{2}}$ and $(T_1 - T_2)$ is shown in Fig. 2.37 for the system mentioned. According to Eduljee and Tiwari (1976), the azeotropic range for water with chlorinated hydrocarbons having active hydrogen or with halogenated hydrocarbons can be determined with satisfaction. The lower and upper limits of the azeotropic temperature of the mentioned systems are as follows:

System	Temperature ($^\circ$C)	
	Lower limit	Upper limit
Water – halogenated hydrocarbon	–74	93
Water – chlorinated hydrocarbon with active hydrogen	–68	84

A computer program calculating the azeotropic condition for halogenated hydrocarbon – water systems is given in Table 2.61. In Table 2.62, the experimentally determined azeotropic concentrations are compared with the calculated values for heterogeneous systems.

Regarding the accuracy and reliability of azeotropic data determined experimentally, Prosyanov et al. (1973b) give three different values for the azeotropic composition of a 1,2-dichloroethane – water system at 1 atm pressure. All three values are different from those given by Horsley (1973) (see Table 2.63).

The 1,2-dichloroethane – water binary system is heteroazeotropic, as mentioned above. The composition of the system was studied by Udovenko and Aleksandrova (1960) at 30, 45, and 60°C (see Table 2.64). With increasing temperature, both the vapor pressure and the water content will increase.

The pressure dependence of composition (x_2 in mol %) and boiling point (t_{az}, K) of binary heteroazeotropes was studied by Draiko (1969). The composition and boiling points of binary azeotropes: hydrocarbons – water were determined at various pressures. The equations correlated were for temperature

$$\log_{10} P_{az}(\text{mm Hg}) = A + \frac{B}{230} + t_{az}$$

and for concentration

$$\log_{10} x_2 = C + \frac{D}{230} + t_{az}$$

where A, B, C, and D are coefficients. The author listed the numerical values of these coefficients for 7 binary systems of hydrocarbon – water (isopentane, n-pentane, isoprene, benzol, toluene, ethylbenzene, and styrene). The effect of pressure was studied from 180 to 860 mm Hg.

Further correlation between the pressure and temperature was reported by Tamir (1981b). The azeotropc composition was also correlated with the temperature. The correlating equations were first introduced by Malesinski (1965).

The reader who is interested in more details on the phenomenon of azeotropy is referred to the following works: Johnson and Modonis (1959), Malesinski (1965), Othmer

Table 2.61 Azeotropic Data for Water-Halogenated Hydrocarbon Systems

---

```
100    PRINT "AZEOTROPIC DATA FOR WATER-HALOGENATED HYDROCARBON SYSTEMS."
110    PRINT "*************************************************************"
120    PRINT
130    PRINT "REFERENCE TO THE METHOD: HORSLEY (1973)."
140    PRINT "REFERNCE TO ANTOINE CONSTANTS:"
150    PRINT "                              WICHTERIE & LINEK (1971)"
160    PRINT
170    DIM A$(72),B$(72)
180    PRINT "HALOGENATED HYDROCARBON,    AZEOTROPIC DATA,"
190    PRINT "        NAME.              B.P.,*C.  MOL.% WATER.  WT.% WATER."
200    PRINT
210    READ N
220    REM N = NUMBER OF BINARY SYSTEMS.
230    FOR J = 1 TO N
240    READ A$,M2,A2,B2,C2
250    REM A$ = NAME OF THE HALOGENATED HYDROCARBON.
260    REM M2 = MOL.WT. OF THE HALOGENATED HYDROCARBON.
270    REM A2,B2,C2 = ANTOINE CONSTANTS OF HALOCARBON, MMHG, *C., TO LOG 10.
280    A1 = 8.07131
290    B1 = 1730.63
300    C1 = 233.426
310    REM A1,B1,C1 = ANTOINE CONSTANTS OF WATER, MMHG, *C., TO LOG 10.
320    READ P
330    REM P = TOTAL PRESSURE OF THE SYSTEM, MAXIMUM 1 ATM.
9000   REM ROOTNL--LOCATES THE ROOTS OF A NON-LINEAR FUNCTION.
9006   X = 100
9016   GOSUB 9050
9018   LET X1 = X
9020   LET X = Y
9022   FOR I = 1 TO 25
9024   LET X2 = X1
9026   LET X1 = X
9028   LET Y1 = Y
9030   GOSUB 9050
9033   IF Y<0.05 AND Y>= 0 THEN 9054
9034   IF Y = Y1 THEN 9054
9036   LET X = (X2*Y-X*Y1)/(Y-Y1)
9038   NEXT I
```

Table 2.61 (Continued)

```
9048  STOP
9050  Y=P-EXP(LOG(10)*(A1-B1/(C1+X)))-EXP(LOG(10)*(A2-B2/(C2+X)))
9052  RETURN
9054  PRINT
9058  GOTO 9062
9062  P3=EXP(LOG(10)*(A1-B1/(C1+X)))
9064  REM P3= PRESSURE OF WATER AT AZEOTROPIC TEMPERATURE, MMHG.
9066  P4=EXP(LOG(10)*(A2-B2/(C2+X)))
9068  REM P4= PRESSURE OF HALOCARBON AT AZEOTROPIC TEMPERATURE, MMHG.
9070  M = 100*P3/(P3+P4)
9072  REM M = AZEOTROPIC CONCENTRATION OF WATER, MOL.%.
9074  W2=M*18.0153/(0.01*M*18.0153+(1-0.01*M)*M2)
9076  REM W2 = AZEOTROPIC CONCENTRATION OF WATER, WT.%.
9082  IMAGE 25A,2X,3D.2D,6X,2D.2D,6X,3D.2D
9085  PRINT USING 9082;A$,X,M,W2
9110  NEXT J
```

(1963), Nakanishi (1968), Hayworth (1969), Schuberth and Kränke (1970), Gibbard and Emptage (1975), Pruett (1971), Nutting and Horsley (1973), Petukhov (1975), Eduljee and Tiwari (1976, 1977), Kudrjawzewa et al. (1977), and Kirk-Othmer (1978).

2.5-5 Partial Miscibility of Liquids

The partial miscibility of liquids was already studied by Alexejew (1886) almost a century ago. Despite the great number of works dealing with water - hydrocarbon immiscibility, the partial miscibility between water and halogenated hydrocarbons has not been dealt with in great detail. However, both groups of organic substances -- hydrocarbons and halogenated hydrocarbons -- belong to the nonpolar or slightly polar compounds and show symbatic tendencies. Ewell et al. (1944) assigned them to classes IV and V.

     Some of the more relevant works that deal in depth with water - hydrocarbon systems are summarized below. The phase equilibrium in the 1-butene - water system and correlation of the hydrocarbon - water solubility data were studied by Leland et al. (1954). Hydrocarbon - water mutual solubilities at high temperatures under vapor-liquid equilibrium conditions have been investigated by Guerrant (1964). The phase equilibria of partially miscible mixtures of hydrocarbons and water have been examined by Burd (1968), Sørensen and Arlt (1979), Sørensen et al. (1979), Skripka (1976, 1979a), Skripka and Boksha (1976), Karapet'yants (1976), Rebert and Hayworth (1967), Robert

Table 2.62   Comparison of the Calculated and Experimental Azeotropic Data

	Azeotropic data				
		Calculated		Experimental	Difference
Halocarbon	B.p. ($^{\circ}$C)	Mol % H_2O	Wt % H_2O	wt % H_2O	(%)
CH_3Cl	-23.98	0.08	0.03	-	-
CH_2Cl_2	38.82	6.82	1.53	1.5	-2.00
$CHCl_3$	55.94	16.21	2.84	2.8	-1.43
CCl_4	67.00	26.90	4.13	-	-
$CCl_2F-CClF_2$	44.83	9.35	0.98	1.0	-2.00
$CCl_2=CCl_2$	100.00	65.42	17.05	17.2	0.87
$CHCl=CCl_2$	73.13	35.08	6.90	7.2	4.17
$CHCl_2-CCl_3$	95.78	85.81	35.01	-	-
$CHCl=CHCl$ (cis)	54.25	14.95	3.16	3.35	5.67
$CHCl=CHCl$ (trans)	45.07	9.47	1.91	1.90	-0.53
$CHCl_2-CHCl_2$	100.00	60.35	17.05	16.4	-3.96
CH_2Cl-CH_2Cl (760 mm Hg)	71.01	32.04	7.90	9.2	14.13
CH_2Cl-CH_2Cl (150 mm Hg)	32.27	24.07	5.46	4.9	-11.43
CH_2Cl-CH_2Cl (75 mm Hg)	18.19	20.81	4.56	4.9	6.94
CH_3-CH_2Br	36.61	6.05	1.05	1.3 vol %	-
CH_3-CH_2I	63.80	23.33	3.39	3.5 vol %	-
$CH_2=CH-CH_2Cl$	35.23	5.61	1.38	2.20	37.27
$CH_3-CH=CHCl$	42.28	8.19	2.06	0.9	-128.89
$CH_2Cl-CHCl-CH_3$	100.00	47.61	12.66	12.0	-5.50
$CH_2Cl-CH_2-CH_3$	43.88	8.90	2.19	2.2	0.45
$CH_3-CHCl-CH_3$	34.75	5.46	1.31	1.2	-9.17
$CH_3-CH_2-CH_2-CH_2Cl$	67.77	27.84	6.98	6.6	-5.76
$CH_2Cl-CH(CH_3)-CH_3$	61.61	21.12	4.95	3.3	-50.00
$CH_2I-CH(CH_3)-CH_3$	100.00	65.75	15.82	21.0 vol %	-
				Average:	8.40

and Kay (1959), and Reed and McKetta (1959).

A widely used method for rough estimates of possible miscibility between two liquids is the comparison of the solubility parameter values for the two constituents. However, this comparative method is applicable only to nonpolar systems, which behave as regular solutions.

Murray and Mason (1952) related the miscibility in liquid-liquid systems to the

Table 2.63 Discrepancy in Azeotropic Data for Water - 1,2-Dichloroethane System

| Pressure (mm Hg) | Azeotropic data | | Reference |
	Boiling point (°C)	Concentration (wt % H_2O)	
760	72.28	9.2	Horsley (1973)
760	72.0	8.7	Prosyanov et al. (1973b)
760	72.0	10.8	Udovenko & Aleksandrova (1960)
760	72.0	7.7	Baranaev et al. (1954)

Table 2.64 Composition of Binary Azeotropic Mixtures of 1,2-Dichloroethane - Water

Temperature (°C)	Pressure (mm Hg)	Concentration of water (wt %)
30	130.8	5.0
45	261.0	6.6
60	492.0	7.9
72	760.0	10.8

Source: Udovenko & Aleksandrova, 1960.

intermolecular potentials. If there is a significant difference between the intermolecular potentials of the two constituents, limited miscibility is expected (Howarth, 1974). By increasing the thermal motion, the mutual solubility will increase and the total intermolecular potential energy of each phase will decrease, causing expansion and absorption of heat. The change is due to the replacement of 1-1 and 2-2 neighbors by 1-2 neighbors. On increasing the temperature, the liquid-liquid solubility curve approaches maximum and both the expansion and absorption of heat increase. The two liquid phases separate when the temperature of the homogeneous mixture is lowered to a point below the critical region. The explanation given by Zimm (1950) is that upon cooling, aggregation of similar molecules takes place and the aggregates increase in size until the gravitational field separates them into two phases (see also Garett, 1972 and Semenchenko and Efuni, 1974).

In miscibility studies, particularly of polar - nonpolar systems, several factors are significant at the point where complete or partial miscibility takes place after two liquids are mixed. In the theoretical treatment of the phase equilibrium (when the chemical potentials or fugacities of each substance in the phases are equal), miscibility is determined by the intermolecular forces, that is, the attractive and repulsive forces between like and unlike molecules. Intermolecular forces are discussed further in Sec. 2.1.

In real solutions the deviation from ideal solution behavior is due to the chemical nature, shape, size, and mass of the molecules. The quantitative measure of the deviation is expressed by the activity coefficients γ_i. This is a calculated value. In ideal solutions $\gamma_i = 1$, whereas in real solutions the activity coefficient can show values greater or smaller than unity. In the first case, $\gamma_i > 1$, there is a positive deviation from ideal solution behavior, and in the second case, $\gamma_i < 1$, there is a negative deviation. Another way to express the nonideality of a binary system is to examine the sign of the excess Gibbs energy of mixing:

$$\Delta G^M = R T(x_1 \log_e x_1 + x_2 \log_e x_2 + x_1 \log_e \gamma_1 + x_2 \log_e \gamma_2)$$

If ΔG^M is positive, the system shows a positive deviation from Raoult's law, whereas a negative sign indicates a negative deviation.

The condition for a stable phase is that the excess Gibbs free energy must be positive (see, e.g., Copp and Everett, 1953):

$$\frac{\partial^2 G^M}{\partial x_2^2} > 0$$

If the Gibbs free energy as a function of concentration shoes a maximum (i.e., is concave downward), the free energy of the system can decrease by separation into two phases. Further analysis of miscibility is found in standard textbooks (e.g., Hala et al., 1967; King, 1969; Hildebrand et al., 1970; Prigogine and Defay, 1967; Prigogine et al., 1957).

In highly diluted solutions the activity coefficient can be as great as 10^5 when the solute molecules differ considerably from their environment (i.e., from their own pure liquid). In such cases, when the system exhibits a large deviation from ideality, the components do not completely dissolve into each other; in other words, they are only partially miscible. All halogenated hydrocarbon - water systems fall into this category. Typical cases are the systems of chloroform - water (Reinders and De Minjer, 1947), 1,2-dichloroethane - water (Baranaev et al., 1954; Davies et al., 1949; Udovenko and Fatkulina, 1952a, b), 1,1,2,2-tetrachloroethane - water (Hollo and Lengyel, 1960), and dichloromethane - water (Sadovnikova et al., 1972), which are illustrated in Figs. 3.38 to 2.41. The system of toluene - water was reported by Selinger (1979). See the Appendix for further illustrations. In table 2.65 a bibliography is compiled for published vapor-liquid equilibrium systems involving water and halogenated hydrocarbons.

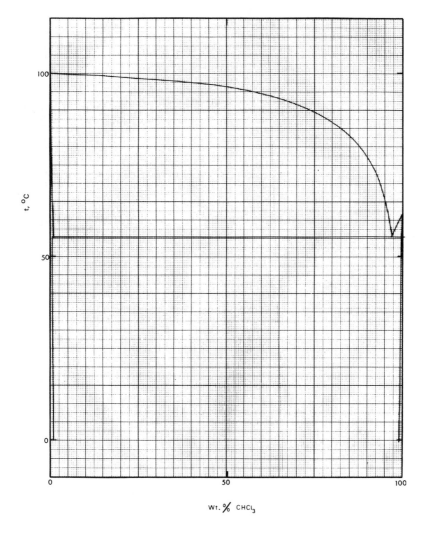

Fig. 2.38 Vapor - liquid equilibrium in water - chloroform system at 1 atm pressure.
(From Reinders & De Minjer, 1947.)

Some halogenated hydrocarbons with a great number of carbon atoms show such a
small mutual solubility with water that they may be considered as substancially
insoluble systems (Horvath, 1974).

Hydrocarbons and halogenated hydrocarbons can exist in water in colloidal,
micellar, or particulate form in appreciable quantities. The degree of accommodation
(difference between the determined and equilibrium concentration of solute in water)
depends on several parameters, e.g., settling time, the way by which the solids are

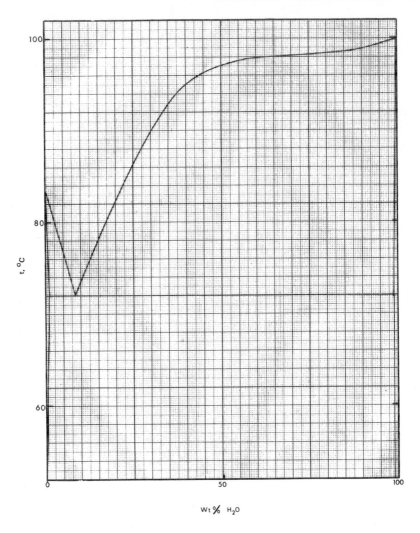

Fig. 2.39 Vapor – liquid equilibrium in water – 1,2-dichloroethane system at 1 atm pressure. (From Baranaev et al., 1954; Davies et al., 1949; Udovenko & Fatkulina, 1952a, b).

introduced into the water, filtration pore size, etc.

The accommodation of C_{20}-C_{33} n-alkanes in distilled and natural water systems was reported by Peake and Hodgson (1966) and Boehm and Quinn (1975). The difference between solubility and accommodation in distilled water of C_4 to C_8 and C_{20} to C_{33} at room temperature was illustrated by plotting of the logarithm of aqueous solubility (mg liter^{-1}) versus the molar volume of alkanes at 20°C (ml g-mole^{-1}). The plot shows that the extrapolation from low (C_4 to C_8) to high number of carbon atoms is not possibl

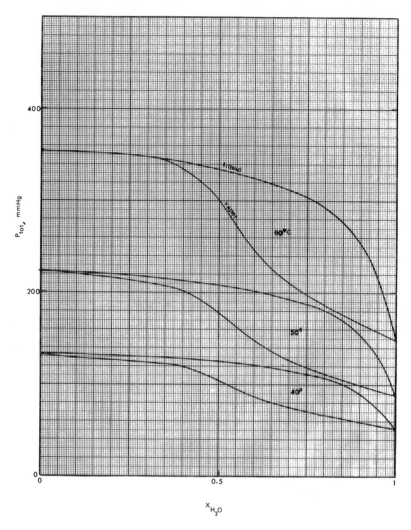

Fig. 2.40 Vapor – liquid equilibrium in water – 1,2-dichloroethane system at 40, 50, and 60°C. (From Udovenko & Fatkulina, 1952a, b.)

Hydrocarbons in the C_{20}-C_{33} range are accommodated in distilled water in amounts much greater than would be expected by simple solubility considerations.

Peake and Hodgson (1967) also reported the accommodation of C_{12}-C_{26} n-alkanes in distilled water, in addition to the previous report.

For example, values of a few parts per billion for the accommodation of high-molecular-weight paraffins in water have been determined by a gas chromatograph equipped with a flame-ionization detector (Sutton and Calder, 1974; Boehm and Quinn, 1975). If the two liquids are completely immiscible, the vapor pressures of the com-

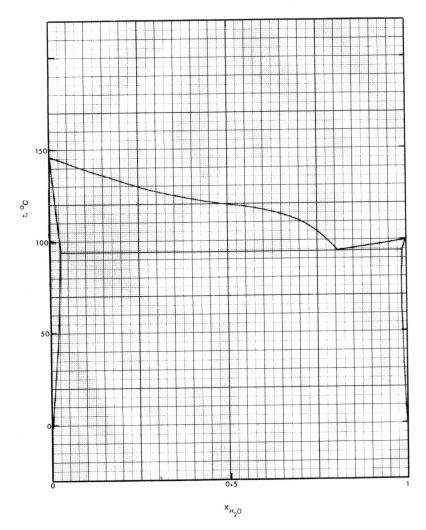

Fig. 2.41 Vapor - liquid equilibrium in water - 1,1,2,2-tetrachloroethane at 1 atm pressure. (From Hollo & Lengyel, 1960.)

ponents will not be influenced by each other, so that the total pressure of the system will be the sum of the pure constituents' vapor pressures. The boiling-point temperature will remain the same until two separate liquid phases exist. The vapor composition also remains constant.

In a system consisting of two liquid phases and one vapor phase, the condition for phase equilibrium is that the fugacity of component i is equal in all three phases; that is,

$$(f_i)_{L_1} \quad = \quad (f_i)_{L_2} \quad = \quad (f_i)_V$$

Table 2.65 Bibliography for Vapor-Liquid Equilibrium for Halogenated Hydrocarbon-
        Water Systems

System	Reference
CCl_4-H_2O	Stolyarenko et al. (1976), Prosyanov et al. (1973a, 1974)
$CHCl_3-H_2O$	De Minjer (1939), Reinders & De Minjer (1947)
$CH_2Cl_2-H_2O$	Prosyanov et al. (1973a, 1974), Sadovnikova et al. (1972), Khanina et al. (1978)
$CCl_2=CCl_2-H_2O$	Prosyanov et al. (1973a, 1974)
$CHCl=CCl_2-H_2O$	Prosyanov et al. (1973a)
$CHBr_2-CHBr_2 - H_2O$	Gooch et al. (1972)
$CHCl_2-CHCl_2 - H_2O$	Hollo & Lengyel (1960), Prosyanov et al. (1973a, 1974)
$CH_2=CHCl - H_2O$	Ethyl Corporation (1968), Prosyanov et al. (1973a, 1974)
$CH_2Cl-CH_2Cl - H_2O$	Baranaev et al. (1954), Bakin (1971), Davies et al. (1949), Prosyanov et al. (1973a, b), Udovenko & Fatkulina (1952b)
$C_6H_5Cl - H_2O$	Prosyanov et al. (1974)
$C_6H_5F - H_2O$	Götze et al. (1975), Jockers (1976), Jockers & Schneider (1978)
$1,4-C_6H_4F_2 - H_2O$	Jockers (1976), Jockers & Schneider (1978)

where f_i = fugacity of component i

L_1 and L_2 = liquid phases 1 and 2, respectively

V = vapor phase

The liquid-phase fugacities can be expressed by

$$(f_i)_{L_1} = \gamma_1 x_1$$

and

$$(f_i)_{L_2} = \gamma_2 x_2$$

in the case of a binary system, where γ_1 and γ_2 are the activity coefficients and x_1 and x_2 are the mole fractions of components 1 and 2, respectivelly. The activity coefficients are very useful for quantitative treatments of real systems showing immiscible liquid phases. The two-phase liquid-liquid immiscibility is determined by the thermodynamic stability criteria (Prigogine and Defay, 1967). The necessary condition for immiscible liquid phases is that the equation

$$\frac{\partial \mu_1}{\partial x_1} = 0$$

should have two roots in the composition range $0 < x_1 < 1$. In this expression, μ_1 is the chemical potential of component 1, derived from

$$\mu_1 = \mu_1^o + R T \log_e \gamma_1 x_1$$

where μ_1^o is the chemical potential of the pure liquid (reference condition). All parameters in this equation are taken at temperature T.

To determine wether the critical solution temperature refer to the upper or lower consolute point, the following expressions (Prigogine and Defay, 1967) can be used:

For the upper critical point:

$$\frac{\partial^2 G^E}{\partial x_2^2} = -\frac{R T}{x_1 x_2} < 0$$

$$\frac{\partial^2 H^E}{\partial x_2^2} < 0$$

$$\frac{\partial^2 S^E}{\partial x_2^2} < \frac{R}{x_1 x_2}$$

For the lower critical point:

$$\frac{\partial^2 G^E}{\partial x_2^2} = -\frac{R T}{x_1 x_2} < 0$$

$$\frac{\partial^2 H^E}{\partial x_2^2} > 0$$

$$\frac{\partial^2 S^E}{\partial x_2^2} > \frac{R}{x_1 x_2}$$

where G^E is the molar excess Gibbs free energy and relates to the enthalpy and entropy through the expression

$$G^E = H^E - T S^E$$

The simplification of these expressions for phase separation and their restriction has been discussed further by Haas (1953), Connolly (1966), Guggenheim (1953), and Herington (1953).

Brandani (1974, 1975) has proposed use of the van Laar equation to predict the existence of an azeotrope or a region of partial miscibility. However, his method is not recommended for highly asymmetric systems such as the solutions of polar or associated compounds in nonpolar solvents. By combining the expressions

$$\frac{\partial \mu_1}{\partial x_1} = 0$$

and

$$\mu_1 = \mu_1^o + R T \log_e \gamma_1 x_1$$

with van Laar's equation,

$$\log_e \gamma_1 = \frac{A x_2^2}{(x_2 - x_1 A/B)^2}$$

we obtain

$$\frac{1}{x_1} - \frac{2 x_2 A^2/B}{(x_2 + x_1 A/B)^3} = 0$$

This equation has two rootes between $0 < x_1 < 1$ if the following two conditions are satisfied:

$$\frac{A}{B} \geqslant 1$$

and

$$A > \frac{(A/B) + (A/B)* ^3}{2(A/B)* (A/B)* + 1 (A/B)}$$

Consequently, two liquid phases appear. In the expression, $(A/B)*$ indicates the integer part of A/B only. Whether two liquid phases will or will not form depends upon the latter two relationships.

A more rigorous expression (inequality) has been reported by, for example, Heidemann (1975) for the prediction of the miscibility gap:

$$A > \frac{2 \left[1 - A/B + (A/B)^2\right]^{3/2} - \left[2 - 3\,A/B - 3(A/B)^2 + 2(A/B)^3\right]}{2(A/B)}$$

The critical condition can be calculated from this equation by replacing the inequality with the equality sign. Furthermore, the equation

$$x_1 = \frac{\left[1 - A/B + (A/B)^2\right]^{\frac{1}{2}} - 1}{A/B - 1}$$

gives the consolute composition. These two expressions are valid for any positive A/B value (Treybal, 1963).

If vapor-liquid equilibrium data are available for a binary system, the condition for partial miscibility can also be expressed using the Redlich-Kister empirical equation (Jaques and Lee, 1966):

$$48\,D\,x^4 - (12\,C + 96\,D)\,x^3 + (2\,B + 18\,C + 58\,D)\,x^2 - (2\,B + 6\,C + 10\,D)\,x + 1 = 0$$

If this expression has two roots in the composition range $0 < x < 1$, one or more of the following conditions are satisfied:

$$B - D > 2$$
$$B + 3^{\frac{1}{2}} |C| + D > 3$$
$$2\,B + 21^{\frac{1}{2}} |C| + 5\,D > 9.6$$

This treatment by Jaques and Lee (1966) of the condition for partial miscibility is similar to those by Brandani (1974, 1975) and Heidemann (1975); that is, experimental vapor-liquid equilibrium data are necessary in order to calculate the empirical constants. The vapor-liquid equilibrium data were represented by the Redlich-Kister empirical equation.

For the determination of the stability of a mixture (i.e., whether it splits into two phases), Ragaine et al. (1974) presented an iterative procedure to solve the systems of equations expressing the conditions required.

Carlson and Colburn (1942) showed how to calculate vapor-liquid equilibrium data from mutual solubility, using van Laar, Margules, and Scatchard-Hamer equations. The coefficients A and B of the Margules and Scatchard-Hamer equation have been determined somewhat more conveniently using matrix algebra (Severance et al., 1963). These expressions are easily handled on electronic computers. An extension of the Wilson equation to partially miscible systems has been reported by Nagata et al. (1975).

The liquid-liquid immiscibility of systems can also be explained by the large activity coefficients, as described in Sec. 2.5-3. Such systems are, for example, mixtures of water with hydrocarbons, which can form hydrogen bonds (King, 1969).

Generally speaking, the miscibility or immiscibility of water with other liquids is also very dependent on temperature and pressure (Moriyoshi and Aoki, 1978). At ambient temperature and atmospheric pressure, miscibility is very limited between water and halogenated hydrocarbons. In other words, the mutual solubility is usually less than 2 wt %. The van Laar constants A and B, may be calculated from mutual solubility data.

In a liquid-liquid system, the variation of solubility with temperature depends upon the sign of the heat of mixing or solution. A negative heat of solution (exothermic) indicates a lower critical solution temperature (LCST), whereas the upper critical solution temperature (UCST) is characterized by a positive heat of solution (endothermic). This is the maximum temperature when the two phases become identical or at which it is possible to obtain two liquid phases at a given pressure. For example, the heat of solution of benzene in water changes sign at 289 K (Gill et al., 1975) and is at its minimum solubility value at this temperature. Because of the change of sign of the heat of solution, both upper and lower CSTs will occur with the same two phases. Molecular interactions explain the phenomenon. That is, at the LCST, the attractive forces between molecules are predominant, whereas at high temperatures, the intermolecular interaction is repulsive, the heat of mixing is positive, and the solution is nearly random at the UCST. The two liquid and vapor phases for a fluorobenzene – water system have been reported by Götze et al. (1975) (see Fig. 2.73 in Sec. 2.8). The critical locus of this system is very similar to that for the benzene – water system (Connelly, 1966). If the three-phase line does not intersect the critical locus in a system, the UCST exists. Meanwhile, the LCST is most frequent in polar and associated systems, which are characterized by specific attractive forces. Study of these forces aids in understanding phase equilibria. A comprehensive discussion of intermolecular interactions is given in Sec. 2.1.

The early studies of mutual solubilities by Alexejew (1886) have been continued by numerous investigators. The main interest for us is Alexejew's rule of the rectilinear diameter:

$$\frac{x' + x''}{2} = a + bT$$

That is, the arithmetical mean of the compositions in both liquid phases is a linear function of temperature in a binary system. The interaction of this straight line with the equilibrium curve gives the consolute point.

Zakharov (1967) continued the work on the relations of the binodal curve and conjugate composition polytherms for 20 aqueous systems with limited miscibility. He derived the consolute temperature and concentration for 20 organic compound – water systems by joining together the conjugate concentrations on the binodal curve with straight lines in the temperature - weight fraction graphs. The systems examined revealed the following phenomena:

1. Systems with upper consolute point (e.g., aniline - water)
2. Systems with both upper and lower consolute points outside the diagram (e.g., methylpyridine - water)
3. Systems with both upper and lower consolute points inside the diagram (e.g., butyl alcohol - water)

With the aid of the consolute point, it is possible to complete the binodal curves if it is not known completely. The following equation was proposed by Zakharov (1967) for the description of the binodal curve:

$$t = t_o + \frac{a}{w(1 - w)} + b\ (w_o - w)$$

where
- t = temperature, $^{\circ}C$
- t_o = consolute temperature of the system, $^{\circ}C$
- w = composition of the system, weight fraction
- w_o = composition of the consolute point, weight fraction
- a and b = coefficients, typical values for a system

The consolute point data are given in Table 2.66 for a few typical systems. The equation satisfactorily describes the binodal curves for the 20 binary systems investigated.

The various aspects of phase separation have been well treated by Copp and Everett (1953) using the relationships between excess thermodynamic functions. The condition of upper and lower critical solution phenomena has been examined and a very simple proposal has been suggested for the determination of the critical temperature; that is, the excess free energy at 0.5 mole fraction reaches a value of about T cal g-mole^{-1}. This is a well-known fact for regular solutions, but it also valid for all systems with G^E curves of closely similar shapes. Regular solutions have only an upper consolute point, whereas the lower consolute point appears in systems that show a large negative excess entropy accompanied by a relatively small negative excess enthalpy. Because the excess entropy is negative, G^E will increase and S^E will fall with increasing temperature. This is due to the positive C_p^E for systems with lower consolute points. With rising temperature the slope of G^E decreases, the curvature of G^E increases with larger c_p^E, and thus G^E will not reach the critical point. The necessary condition for a closed solubility loop is that c_p^E be positive.

Semenchenko and Efuni (1974) proposed an empirical equation for the representation of the general shape of the closed solubility curves for binary systems of limited miscibility. The model equation of an ellipse contains two parameters a and b, which are individual to every system:

Table 2.66 Consolute Point Data for Binodal Curves of Organic-Water Systems

System	Temperature (°C)	Concentration, w_o (weight fraction)		Coefficients of the equation	
		Aqueous phase	Organic phase	a	b
Aniline - water	208	-1.8	-	4	1
Phenol - water	90	0.05	-	46.4	-128
Furfural - water	138	-4.85	-	1.58	-4
Isobutyl alcohol - water	158	-0.05	-	31	-73
Butyl alcohol - water	144	-	0.09	52.8	-160
Butyl alcohol - water	-50	-	-0.06	22.8	90

Source: Zakharov, 1967.

$$\tau^2 + \kappa^2 = 1$$

where τ = $(T - A)/a$

       κ = $(x - B)/b$

       A = $\frac{1}{2}(T_u - T_l)$

       B = $\frac{1}{2}(x_1 + x_2)$

where   a   and   b   are the major and minor semiaxes and T_u and T_l the upper and lower consolute temperatures, respectively. The authors reported the reduced τ–κ diagram for   16 binary systems of limiting miscibility.

Copp and Everett (1953) summarized the requirement for the lower consolute temperature for a system as follows:

1. Large negative excess entropy
2. Small negative heat of mixing
3. If possible, the excess heat capacity should be below the critical value

The approximate condition adopted for phase separation will occur at temperature T_{cr}, when the excess Gibbs energy at $x_2 = 0.5$ becomes equal to $\frac{1}{2}$ R T (Copp, 1953). Curves representing the variation of G^E $(x_2 = 0.5)$ with temperature for systems of alcohol with water were illustrated. It has been observed that G^E $(x_2 = 0.5)$ increases by a constant amount for each additional $-CH_2-$ group. For example, Everett (1953) showed that the ideal free energy of mixing for an equimolar mixture is -0.693 R T, which means that G^E $(x_2 = 0.5)$ cannot exceed 0.693 R T   without formation of two phases. Critical values of G^E $(x_2 = 0.5)$ have been given by Guggenheim (1952) (see Table   2.67).

Table   2.67   Critical Values of the Gibbs Free Energy of Mixing

Type of solution	Critical value, $\dfrac{G^E\ (x_2 = 0.5)}{R\ T}$
Hildebrand's regular solution	0.5
Guggenheim's strictly regular solutions:	
First approximation	0.522
Triplets approximation	0.525
Quadrupoles approximation	0.528
Associated solution (Barker & Fock, 1953)	0.60

Source: Guggenheim, 1952.

By dissolving a long-chain hydrocarbon in water, negative heat and entropy changes will take place. The entropy effect is predominant; therefore, the addition of a $-CH_2-$ group to the molecule will add slightly to the excess free energy of the system. With an increase in the molecular weight of the solute molecule (e.g., by adding a $-CH_2-$ group to it), the partial molal heat capcity of the aqueous solution will enchain. The consequence is that the lower critical temperature of the system decreases and tends to close the solubility loop.

Assuming that the rectilinear diameter law is valid for the pressure effect upon mutual solubility,

$$\frac{x' + x''}{2} = a + b\,p$$

where   p = pressure in the system

         x' = mole fraction of solute in the first liquid phase

        x" = mole fraction of solute in the other liquid phase

  a and b = coefficients, typical values for a binary system

extrapolation of the solute concentration to the maximum pressure read from the solubility isotherm will give the critical solution point. The pressure effect upon mutual solubility between 2-butanol and water has been studied by Moriyoshi et al. (1977) and Brocks (1952). Closed solubility loops were shown to contract with increasing pressure and to completely disappear at the hypercritical solution point. Both the critical pressure and critical concentration curves were plotted as a function of the critical solution temperature. In addition to the graphs, these functions were also expressed in terms of volume, enthalpy, entropy, and concentration.

The effect of pressure on the mutual solubility in binary systems was studied by Timmermans and Lewin (1953) and Connolly (1966). The investigation showed that in systems where V^E is negative, G^E increases as the total pressure decreases (see also Copp, 1953).

Regarding the possibile phase separation in a binary system, G^E $(x_2 = 0.5)$, and consequently $R\,T\,\log_e \gamma_2^\infty$, give a good indication. The experimentally found value for γ_2^∞ was 13, which showed a miscibility gap at 25°C (Copp and Everett, 1953). In terms of G^E, this corresponds to a value of 1600 cal g-mole^{-1}. Organic substances that are incompletely miscible with water at 25°C will exhibit $R\,T\,\log_e \gamma_2^\infty$ values greater than 1600 cal g-mole^{-1}. Five homologous series of organic compounds have been examined by Copp and Everett (1953) in water by plotting the number of carbon atoms in the molecule versus $R\,T\,\log_e \gamma_2^\infty$. In the system examined, all G^E (x_2) curves showed closely related shapes. The $-CH_2-$ group increases reflected an increment of about 800 cal g-mole^{-1} to the partial molar excess free energy in the series of saturated hydrocarbons (2 to 4 carbon atoms), acetates (2 to 4 carbon atoms), n-aliphatic alcohols (1 to 8 carbon atoms), aliphatic acids (1 to 3 carbon atoms),

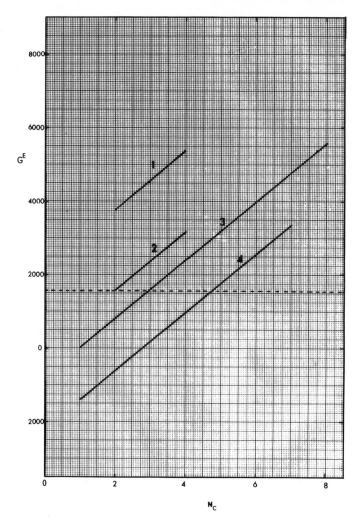

Fig. 2.42 Values of $R\,T\,\log_e\,\gamma_2^{\infty}$ in aqueous solutions versus number of carbon atoms at 25°C. 1 = Saturated hydrocarbons; 2 = acetates; 3 = n-aliphatic alcohols and acids; 4 = aliphatic amines. G^E(cal g-mole^{-1}) = $R\,T\,\log_e\,\gamma_2^{\infty}$; N_c = number of carbon atoms. (From Copp & Everett, 1953.)

pyridines (5 to 7 carbon atoms), and aliphatic amines (1 to 7 carbon atoms) in aqueous solutions (see Fig. 2.42).

The relative effect upon solubility of the various groups in the foregoing substances is shown by the following values:

$$-C \overset{\displaystyle O}{\underset{\displaystyle O}{\diagdown}} \qquad\qquad 0.0 \quad \text{cal g-mole}^{-1}$$

$$-COOH \qquad\qquad -600 \quad \text{cal g-mole}^{-1}$$

$$-OH \qquad\qquad -600 \quad \text{cal g-mole}^{-1}$$

$$=N- \qquad\qquad -2000 \quad \text{cal g-mole}^{-1}$$

$$-NH_2, \quad >NH, \quad \equiv N \qquad -2100 \quad \text{cal g-mole}^{-1}$$

Similar types of regularities have been observed in the free energies of mixing.

The relationship between activity coefficient and solubility in aqueous solutions of organic substances has been studied by Andon et al. (1953). The investigation concluded that an increase in the activity coefficient tends to reduce the solubility. This finding was further commented by Copp and Everett (1953), particularly for partially miscible systems. The excess free energy versus concentration plot indicated an intercept between the $x_2 = 1$ ordinate and the tangent to the G^E curve at $x_2 = 1$. At this point

$$\mu_2^E \;=\; R\,T\,\log_e \gamma_2^\infty$$

which is illustrated schematically in Fig. 2.43, where γ_2^∞ is the activity coefficient of the solute at infinite dilution and μ_2^E the excess chemical potential when $x_2 = 0$. The infinite activity coefficients is given by

$$\gamma_2^\infty \;=\; \frac{H}{p_2^o}$$

where H is the Henry's law constant and p_2^o the vapor pressure of the pure solute. The value of G^E depends upon the concentration, usually expressed as a mole fraction. At $x_2 = 0.5$, the value of the corresponding G^E will depend upon the steepness of the slope, which is measured by $R\,T\,\log_e \gamma_2^\infty$ (see Fig. 2.43). In other words, $R\,T\,\log_e \gamma_2^\infty$ is proportional to G^E:

$$R\,T\,\log_e \gamma_2^\infty \;\sim\; G^E \,(x_2 = 0.5)$$

In connection with partially miscible liquids, the halogenated hydroacrbon - water systems show very little mutual solubility. In both liquid phases the concentration of the solute does not exceed 2 wt %, similarly to hydrocarbon - water systems. It was an early attempt to develop some sort of method to calculate the concentrations in both liquid phases.

A more successful method has been proposed by Polak and Lu (1973) for hydrocarbon - water systems. The principle of the work was based on the Redlich-Kister equation and its expansion,

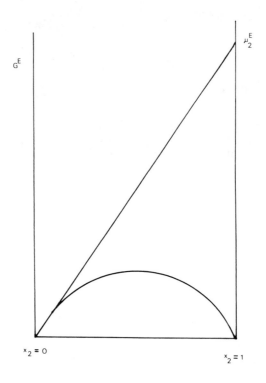

Fig. 2.43 Excess free energy as function of concentration. $\mu_2^E = R T \log_e \gamma_2^\infty$.

$$G^E = R T x_{HC}(1 - x_{HC}) \left[A + B(1 - 2 x_{HC}) \right]$$

where G^E =   molar excess Gibbs energy

    x_{HC} =   mole fraction of hydrocarbon

A and B =   parameters

These parameters have been correlated with the molar volume of the liquid hydrocarbons and the group contributions obtained from mutual solubility data determined experimentally.

Because of the very low mutual solubility between the hydrocarbons and water, the following expression is often used:

$$x_{HC} \sim \frac{1}{\gamma_{HC}^\infty}$$

and at the limiting values at $x_{HC} \to 0$ it becomes

$$\log_e \gamma_{HC}^{\infty} = A + B$$

or

$$x_{HC} = \frac{1}{e^{(A + B)}}$$

and for water

$$\overline{x}_w = \frac{1}{e^{(A - B)}}$$

A similar method could also be used starting from the van Laar equation.

The relationship between the limiting activity coefficients and parameters A
and B has been also applied by Pierotti et al. (1959), Deal et al. (1962), and
Deal and Derr (1968) through correlation with group contributions for homologous
series of solvents. The group-contribution method of Irmann (1965), together with
other methods, will be treated in Sec. 2.7.

In general, the reader who requires further details on the theoretical aspects
of phase-separation behavior is referred to the following works: Staverman (1941a,
b), Barker and Fock (1953), Landolt-Börnstein (1962), Gmelins (1964), Guerrant (1964),
Zakharov (1967), Nitta et al. (1974), Moriyoshi et al. (1977), Schneider (1978),
Kehlen and Hienzsch (1976), Skripka (1976, 1979a), Skripka and Boksha (1976),
Sørensen and Arlt (1979), and Sørensen et al. (1979).

2.5-6 Hydrates of Halogenated Hydrocarbons

It has been observed that some monomer gases and volatile liquids are able to form
solid hydrates, often above the freezing point of water. The structure of these
hydrates resembles the structure of ice. They are stable and have uniform and re-
producible composition (Von Stackelberg, 1949, 1954a, b, 1956; Von Stackelberg and
Meuthen, 1958; Von Stackelberg and Müller, 1954; Villard, 1890). Good reviews of
gas hydrates are those of Byk and Fomina (1968), Jeffrey and McMullan (1967), and
van der Waals and Platteeux (1959).

The main interest in this book is halogenated hydrocatbons, particularly their
methane, ethane, and propane series. Study of these hydrates is connected with several
important processes and with fundamental research concerns, such as the investigation
of hydrate formation in biological systems and a better understanding of the structure
of water molecules. Some of the relevant industrial processes worth mentioning are
the study of occasional clogs in gas pipelines in cold weather, hydrate formation in
refrigerating cycles, and the cyclic formation and decomposition of hydrates in the
process of saline water purification. The hydrate process for demineralizing seawater
has been discussed by several investigators (Barduhn et al., 1960, 1962; Knox et al.,

Table 2.68 Structure of Solid Hydrates

Description	Structure	
	I	II
Cubic structure	–	Face-centered
Lattice constant (cubic cell) ($\mathring{A}$)	12	17
Density d_{struct} (g cc^{-1})	0.797	0.787
$d = d_{struct} \left(1 + \dfrac{\text{mol. wt.}}{18n}\right)\left(\dfrac{\text{lattice const. calc.}}{\text{lattice const. actu.}}\right)^3$		
Number of H_2O molecules in the unit cell	46	136
Number of small voids in the unit cell	2	16
Diameter of voids ($\mathring{A}$)	5.2	4.8
Number of larger voids in the unit cell	6	8
Diameter of voids ($\mathring{A}$)	5.9	6.9
Coordination number of small voids	20	16
Coordination number of larger voids	24	28
Empirical formula		
Only larger voids are filled	$7.67H_2O$	$17H_2O$
All voids are filled	$5.75H_2O$	$5.67H_2O$
Examples:	$CHClF_2$, CHF_3, $CH_2=CHF$, CH_3-CHF_2	CCl_2F_2, $CBrF_3$, $CHCl_3$, C_2H_5Cl, CH_3I, $CFCl_3$, CH_3-CF_2Cl

1961; Briggs and Barduhn, 1963). Another application, studied by Cheng and Pinder (1976), involved the artificial freezing of ground by liquid hydrate formers.

By examination of the structure of solid hydrates by the x-ray diffraction method, two distinct types of crystal structure have been reported: structures I and II (Von Stackelberg, 1949, 1954a, b, 1956; Von Stackelberg and Meinhold, 1954; Von Stackelberg and Meuthen, 1958; Claussen, 1951; Claussen and Polglase, 1952). The two structures are compared in Table 2.68.

A compilation of the various compositions of the hydrates of halogenated hydro-

Table 2.69 Composition and Properties of Solid Hydrates

Hydrate	Moles H_2O (n)	Critical decomposition Temperature ($^\circ$C)	Pressure (bars)	Reference
CCl_3F (R 11)	16.6	8.5	0.872	Wittstruck et al. (1961)
CCl_2F_2 (R 12)	15.6	12.1	4.4518	"
$CBrF_3$ (R 13B1)	15.6	11.0	–	"
$CHClF_2$ (R 22)	12.6	16.3	1.1239	"
CH_2ClF (R 31)	7.98	17.89	2.8623	Barduhn et al.(1962)
Cyclopropane (RC 290)	17.0	17.0	5.9592	Chen & Pinder (1976)
CH_3-CHF_2 (R 152a)	–	15.3	4.5087	"
CH_3-CF_2Cl (R 142b)	17.18	13.09	2.3237	"
CF_2ClBr (R 12B1)	16.57	9.9	1.6971	"
CHF_2Br (R 22B1)	–	9.87	2.6823	"
CF_2Br_2 (R 12B2)	–	4.9	0.5146	"
$CH_2=CHCl$ (R 1140)	–	1.15	1.8211	"
$CHCl_2F$ (R 21)	16.8	8.69	1.0132	"
CH_3Br (R 40B1)	7.89	14.73	1.5345	"
CF_4 (R 14)	6	20.4	–	von Stackelberg(1949)
CH_3F (R 41)	6	18.0	–	"
CHF_3 (R 23)	6	21.8	–	"
CH_2F_2 (R 32)	6	10.5	–	"
C_2H_5F (R 161)	6	22.8	–	"
CH_3Cl (R 40)	7.2	20.4	4.865	"
C_2H_5Cl (R 160)	15	4.8	0.787	"
C_2H_5Br	15	1.4	0.221	"
CH_2Cl_2 (R 30)	15	1.7	0.213	"
CH_3I	15	4.3	0.233	"
CH_3-CHCl_2	15	1.5	0.093	"
$CHCl_3$ (R 20)	15	1.6	–	"
CF_3I	17	–	–	Stupin & Tevikov (1976)

carbons and the critical decomposition temperature and pressure of these hydrates is given in Table 2.69. At the critical decomposition point, the four phases -- gas, liquid solvent, liquid water, and solid hydrate -- are in equilibrium. In addition to the upper invariant point (critical decomposition point), there is a lower one where the four phases gas, liquid water, solid hydrate, and ice are also in equilib-

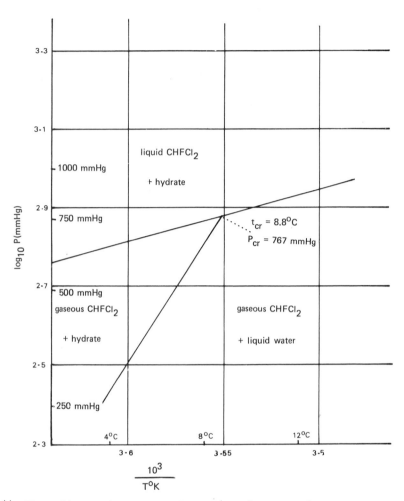

Fig. 2.44 Phase diagram for water – CHFCl$_2$ (R 21) system. (From Barduhn et al., 1960.)

rium. The temperature of the lower invariant point in the phase diagram is always just below 0°C. A bibliography of published phase diagram for halogenated hydrocarbon – water systems is given in Table 2.70. A typical phase diagram for H$_2$O – CHFCl$_2$ is shown in Fig. 2.44.

Regarding the construction of phase diagrams and the determination of hydrate compositions, the reader is referred to Barduhn et al. (1962), John (1974), Ng and Robinson (1976), Wittstruck et al. (1961), and Robinson and Ng (1975).

The various aspects of properties, behavior, and so on, have been studied by numerous workers (e.g., entropy of hydration (Butler and Reid, 1936), examination of

Table 2.70 Bibliography of Published Phase Diagrams for Halocarbon-Water Systems

Solvent	References
CF_3I	Stupin & Tevikov (1976)
CH_3Br (R 40Bl)	Barduhn et al. (1960, 1962), Klausutis (1961), Briggs et al. (1962)
$CHCl_2F$ (R 21)	Barduhn et al. (1960, 1962), Banks et al. (1954), Briggs et al. (1962)
CH_2ClF (R 31)	Barduhn et al. (1960, 1962, 1976), Klausutis (1961), Briggs et al. (1962)
CCl_2F_2 (R 12)	Chinworth & Katz (1947), Wittstruck et al. (1961), Hashizume (1964), Mel'tser & Smirnov (1968)
CCl_3F (R 11)	Chinworth & Katz (1947), Cheng & Pinder (1976)
$CHClF_2$ (R 22)	Chinworth & Katz (1947), Hashizume (1964)
CH_3Cl (R 40)	Chinworth & Katz (1947), Glew & Moelwyn-Hughes (1953)
CH_3-CClF_2 (R 142b)	Briggs & Barduhn (1963), Carey (1965), Barduhn et al. (1976), Briggs et al. (1962)
$CClF_2Br$ (R 12Bl)	Briggs & Barduhn (1963), Glew (1960)
$CHCl_3$ (R 20)	Tammann & Krige (1925)

clathrate hydrates by ultrasonic velocity (Baumgartner and Atkinson, 1971), cluster analysis (Hansch et al., 1973)).

Additional information concerning gas hydrates can be found in the following publications: Hagan (1962), Mandelcorn (1959, 1964), Cramer (1955), Swern (1957), van der Waals and Platteeux (1959), Huang et al. (1965), Palmer (1950), Kirchnerova (1974), Kass et al. (1965), American Petroleum Institute (1970), Jeffrey and McMullan (1967), Vlahakis et al. (1969), Winsor (1954), Davies (1977), Glew and Moelwyn-Hughes (1953), and Briggs et al. (1962).

The isotopic effect in the enthalpies of formation of the aqueous clathrates of CCl_2F_2 and $CHClF_2$ has been studied by Stupin et al. (1978).

2.6 Structure and Properties of Water and Halogenated Hydrocarbons

2.6-1 Water

Water is the most important substance on the earth. It is essential to all biological
processes for the support of life. The presence of water is one of the necessary
condition of life. The human body contains about 65 wt % water. In other words,
"no water -- no life." Fortunately, our reserves of water are unlimited. The total
estimated water on the earth is about 22×10^{23} g. Because of its importance and
wide distribution, its behavior, properties, and so on, in various conditions and
processes are of interest (Selinger, 1979).

The foundation for the modern theory on the structure of water was laid at the
end of the 19th century, when it became apparent that water consists of aggregates
or complexes of monomeric molecules. Although our present knowledge of the gaseous
and solid water phases finds good agreement among scientists, the structure of liquid
water causes controversy, and there is still a long way to go until a satisfactory
theory for the explanation and understanding of the many anomalous properties and
behaviors of liquid water becomes available.

In the gaseous phase, water consists of monomer molecules with a few dimers
(Ashwell et al., 1974; Dyke et al., 1977). Higher aggregates, such as trimers and
tetramers, are very difficult to detect. The monomer water molecule is symmetrically
bent and the H-O-H angle is 104.52°. The H-O bond length is 0.957 Å.

In the solid phase there are nine forms of ice; however, most of these are stable
only at high pressures (see Figs. 2.29 and 2.45). In common ice the water molecule
is hydrogen-bonded to the four nearest neighbors, forming a tetrahedral geometry
around each oxygen with a distance of 2.76 Å. The H-O-H bond angle in ice is the
tetrahedral angle 109.4°, which is different from the angle in the free H_2O molecule, which is 104.52° (see Fig. 2.46).

Our information is most complete for the hexagonal structure of Ice Ih. Another
form of ice is the cubic structure, which is metastable. A phase diagram of water,
showing the pressure effect upon temperature, has been presented by, for example,
Kamb (1965), Ross et al. (1977). The various modifications of ice have been determined by x-ray, electron, or neutron diffraction or x-ray powder study, or by infrared
and Raman spectroscopy (Fox and Martin, 1940). Ice II has a rhombohedral structure,
whereas Ices III, V, and VI have tetrahedrally coordinated structures. In Ice
VII the oxygen atoms are arranged in a body-centered-cubic form. In contrast to the
hydrogen-bonded organic substances, in ice the stronger hydrogen bonds lead to a
decrease in the density of the structure (Fletcher, 1970).

Depending upon the molecular dimensions, the various gas molecules in ice can
easily fit into the vacancies in the crystal lattice (Namiot and Gorodetskaya, 1970).

The structure and properties of liquid water is most relevant to an understanding
of the various factors affecting water's solubility and miscibility with other mol-

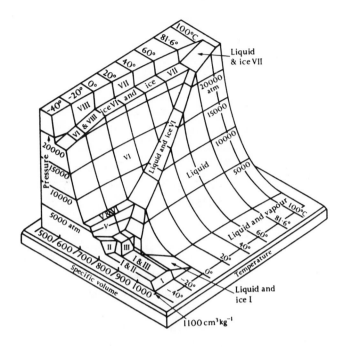

Fig. 2.45 Three-dimensional phase diagram of solid, liquid, and vapor water.

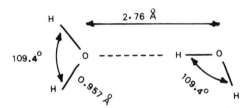

Fig. 2.46 Bond angles and atomic distances in ice.

ecules. The structure of liquid water is very complex, and a comprehensive discussion of the subject is beyond the scope of this book. However, the reader who desires a more detailed understanding of the molecular structure of water is referred to the following reviews and books: Adams et al. (1975), Andersen (1975), Anonymous (1974), Arakawa et al. (1977), Barton (1974, 1975), Ben-Naim (1974a, b), Benson (1978), Blandamer (1970, 1977), Bonner (1971), Camp and Meserve (1974), Ciaccio (1971-1973), Croxton (1975), Dack (1975b), Dorsey (1968), Dyke et al. (1977), Eisenberg and Kauzmann (1969), Erdey-Gruz (1974), Eucken (1947, 1948-1949), Ewing (1975), Eyring and Jhon (1969), Eyring et al. (1963), Fabuss and Korosi (1968), Franks (1967, 1972-

1979: Vols. 1-6), Freier (1978), Getzen and Ward (1971), Getzen (1970), Gibbs (1977), Gibbs et al. (1973), Gingold (1973), Gmelins (1963, 1964), Hamilton and Ibers (1968), Hasted (1973), Holtzer and Emerson (1969), Honeyborne et al. (1971), Horne (1969, 1970, 1972a, b), Hvidt (1978), Jellinek (1972), Jhon and Eyring (1978), Jhon et al. (1966), Kalman (1974), Kanno et al. (1975), Kavanau (1964), Kay (1973), Kirk-Othmer (1970, Vol. 21), Kohler (1972), Konda and Yamamoto (1977), Luck (1970, 1974, 1976), Lyashchenko and Stunzas (1980), Nemethy and Scheraga (1962a, b), Olofsson and Olofsson (1977), Pierotti and Liabastre (1972), Prigogine and Rice (1975), Samoilov (1965), Sarkisov et al. (1974), Sarma and Ahluwalia (1973), Stillinger (1973, 1975), Swain and Bader (1960), Symons (1972), Tanford (1973, 1980), Wada (1979), Wall and Hornig (1965), and Weres and Rice (1972). In addition, there are two comprehensive bibliographies giving references to various aspects of water: Dooms (1975) and Hawkins (1975, 1976).

In the liquid state, water molecules are in continual motion and in an instantaneous structure, the molecules are neither random, as in the gas phase, nor regular, as in the ice phase. The instantaneous structure of liquid water lies between these two forms. The H-O-H bond angle is $104.52°$ in the liquid phase, like it is in the vapor phase. If one assumes a single water molecule having a rigid geometry, then according to Pauling, the molecule has the following geometry (see Fig. 2.47), where the van der Waals radii of oxygen and hydrogen are 1.4 Å and 1.2 Å, respectively. The equilibrium O-H bond length is 0.957 Å. The tetrahedral geometry of the water molecule originates from the two O-H bond directions and the lone-pair electrons, one above and one below the molecular plane, created by the hybridization of the 2s and 2p orbitals of the oxygen atoms (see Fig. 2.48). This distorted electronic cloud of the water molecule is important to an understanding of its abnormal properties. The negative poles opposite the hydrogen atoms are responsible for the hydrogen bonding with other molecules and for the large dielectric constant value, 1.84 debyes. The negative electrification attract the positive charges of nearby protons, which results in association.

There are several theories, used with more or less success, as to the structure of liquid water. Each of the proposed theories has some good features, but none of them can provide a full explanation for the anomalous behavior of water. The investigation by Bernal and Fowler (1933) was the first to indicate a more promising approach to an understanding of the real structure of liquid water. Their proposed model is based on the various physical properties of water that are related to the anomalous behavior. They established that the water molecules are not arranged together tightly, because in this case the density of liquid water would be 1.8 g cc^{-1} instead of 1.0 g cc^{-1}. This is also supported by the radius of the monomeric water molecule, equal to 1.41 Å. Consequently, there are definite cavities between the molecules. These molecules are held together with hydrogen bonding, which also prevents

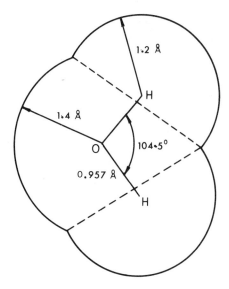

Fig. 2.47   Geometry of water molecule according to Pauling.

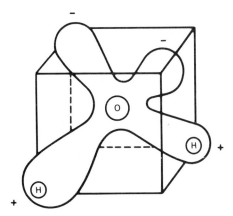

Fig. 2.48   The tetrahedral geometry of the water molecule.

too-tight packing in the liquid phase. The water molecules are arranged in groups of tetrahedral symmetry as in ice, and the coordination number is four. As an explanation for the greater volume of ice compared to liquid water at 4°C, they assumed that the tridymite-like structure of ice changed to a quartzlike form (radius = 4.5 Å), which is more compact than the tridymite modification (radius = 4.2 Å). However, x-ray

diffraction measurements of water indicated that there are three modifications of
water: I, II, and III. Close to the melting point, the tridymite structure dominates. At 4°C the quartzlike geometry is responsible for the high density; that
is, the tridymite-quartz transformation influences the unusual density of water
between 0 and 4°C. At higher temperatures the water molecules fit a tighter structure, particularly near the critical temperature.

The theory of bent hydrogen bonds was introduced by Pople (1951). The explanation for the density difference between ice and liquid water and the high heat capacity of water is related to the distortion of the structure caused by the bending of
hydrogen bonds. Whereas in ice the hydrogen bonds are restrained from bend because
of the crystal lattice, in liquid water the bonds can bend. As a result of bending,
the molecular structure becomes distorted, and the number of neighboring molecules
increases. Pople's was the first theory to suggest the flexibility of hydrogen bonds.
However, the disadvantage of this model lies in its disregard for the translational
movement of molecules.

Another theory of water structure has been proposed by Samoilov (1946, 1965).
In this model the cavity spaces in the crystal lattice of ice are filled by single
H_2O molecules not having hydrogen bonds. The equilibrium position of the water molecules will be shifted when the temperature increases, and the average number of
coordination will increase, together with the density. The occupation of the cavities
increases with increased temperature and pressure. The following anomalous properties
of liquid water have been explained by Samoilov (1946, 1965) through his model: density maximum at 4°C; large heat capacity, large heat of evaporation, thermal expansion coefficient, and viscosity. In principle, this theory described the abnormal
properties of water to the fact that its cavities are filled by single water molecules, not because of association. Samoilov's model has been very popular among
Russian scientists, who have suggested numerous additions to it. Following Samoilov's
theory, Forslind (1952) proposed that there are several imperfections in the cavities.
Some molecules leave their positions due to the effect of heat, and consequently
imperfections appear in the structure.

Pauling's dodecahedral cage model (Pauling, 1960) is based on the assumption
that liquid water is a hydrate of itself. Examination of the structure by x-ray crys-
tallography showed that there is no significant difference between the molecular distances and bond angles in the crystalline hydrates and in ice. According to Pauling,
in the water structure the pentagon dodecahedrons are built from 21 water molecules,
and each dodecahedron is joined to its neighbors by three hydrogen bonds or by the
common pentagonal plane, or by chains of hydrogen bonds forming bridges between the
molecules. Furthermore, there are some free monomeric water molecules in the structure.
This hydrate model is more stable than the structures previously proposed. With
Pauling's model the temperature and pressure effect upon physical properties can be

described within a given interval. However, the model assumes a too-long-range order in the liquid state, contrary to the x-ray diffraction measurements.

The molecular association of water molecules has been considered by several investigators (e.g., Eucken, 1946, 1947, 1948-1949). The association of water molecules helps greatly to explain several aspects of the anomalous properties of water. However, the various associates, such as $(H_2O)_2$, $(H_2O)_3$, ..., $(H_2O)_8$, will not explain the structure of liquid water. Furthermore, the theory of molecular association for liquid water is in contradiction with the spatial structure of water, and there is no association of definite composition in the liquid state. There are no enormous ordered units in the structure.

A more successful model has been introduced by Frank and Wen (1957), Frank (1958, 1963, 1965), and developed further by Nemethy and Scheraga (1962a, b). In this model tha water molecules are attached to each other by a hydrogen bond; however, while the attachment is rather slow between two monomer molecules, the third and subsequent molecules can more easily join the cluster that is available (Owicki et al., 1975). The explanation is that the lone-pair electrons in the L shell of the oxygen atom are more locatized in the water structure and the tetrahedral sp^3 hybridization is more far reaching. Consequently, molecules that are already hydrogen-bonded have a greater tendency for further hydrogen bonding with other monomeric water molecules. On the other hand, if the cluster loses one molecule (i.e., the number of hydrogen bonds started to break), this will create a tendency for decomposition of the cluster. The dynamic nature of the cluster lies in its building and decomposing ability, creating the flickering clusters of water structure (see Fig. 2.49). Between the clusters, there are one or two layers of monomeric water molecules.

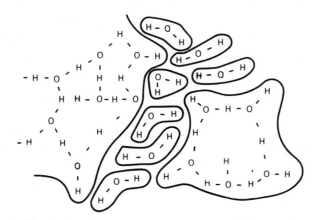

Fig. 2.49 Flickering cluster model of water molecules.

The clusters and monomeric water molecules are held together by van der Waals forces. The lifetime of the clusters is about 10^{-10} sec. The number of water molecules in a cluster varies from 65 to 12 in the temperature range 0 to 100°C (Nemethy and Scheraga, 1962a, b). At ambient temperature an average cluster contains about 40 water molecules. During cluster formation the potential energy decreases by 1 kcal g-mole^{-1} on the average. The anomalous maximum in liquid density at 4°C is due to thermal expansion in addition to the destruction of the less-dense clusters. Furthermore, cluster breakup or cluster buildup is well illustrated through the liquid viscosity of water. When the clusters break up, the viscosity decreases, and when the pressure increases, the water becomes a normal liquid. However, the temperature has a great effect upon the destruction of clusters than does the pressure. While liquid water exhibits many of the structural features of Ice I, there are still some significant deviations from icelike behavior. For example, the ability to supercool below its freezing point indicates that the clusters have a different structure from that of Ice I. From statistical thermodynamical calculations, Nemethy and Scheraga (1962a, b) determined that 70% of the liquid water consists of clusters. The flickering cluster theory has been studied by a large number of investigators, and several improvements and modifications have been proposed (Buijs and Choppin, 1963-1964; Miller, 1963; Hornig, 1964; Vand and Senior, 1965; Davis and Litovitz, 1965; Luck, 1965, 1967; Ageno, 1967; Bonner and Woolsey, 1968; Hagler et al., 1972; Mruzik, 1977; Kurant et al., 1972; Binder, 1975; Namiot, 1967; Lentz et al., 1974; Owicki et al., 1975; Scheraga, 1979).

The significant-structure theory has been applied to water and heavy water by Jhon et al. (1966). The assumption in the model is that the cagelike clusters of Ice I are dispersed in an Ice-III-like structure, and there is an equilibrium between them. During the melting the Ice-I-like structure transforms into the denser Ice-III-like structure, a contraction appears in the volume upon melting, and at the same time, fluidized vacancies are introduced in the structure. The minimum volume is at 4°C, when most of Ice-I-like structure is destroyed. Above 4°C the water behaves as a normal liquid. The calculated results for entropy of vaporization, vapor pressure, **molar volume, heat capacity, and critical data are in good agreement with experimental** values.

It was mentioned at the beginning of the chapter that there is no single theory that can explain satisfactorily all the anomalous behavior of liquid water, and therefore there is a tendency to continue to develop new models or to improve existing ones. The obvious aim is to find a model that will explain the properties of water as determined experimentally.

During the 1960s, there was a great controversy regarding the existence of polywater. This is an indication of the extensive search for an explanation of the anomalous properties of water (Allen, 1973; Derjaguin and Churaev, 1973; Dyke et al., 1977).

Because of the great complexity of the problem, many models are able to cope with only some of the aspects of the behavior. In other words, the proposed models by various workers will simulate only a few properties of liquid water. They help to build up a better understanding of the anomalous properties of liquid water. More recent propositions include those of Rahman and Stillinger (1971), Stillinger and Rahman (1972), Ben-Naim (1972), Hagler et al. (1972), Stillinger (1973), and Hvidt (1978). A good survey of the more recent models has been presented by Nemethy (1974).

One thing has been settled by agreement among scientists: that new models or future proposals have to satisfy the existence of both broken and unbroken hydrogen bonds in cold water. There are denser and bulkier patches in cold water than in most normal liquids. Liquid water has macroscopic anomalies because of its molecular structure. These paculiarities of water are:

1. The maximum density is at about 4°C.
2. The pressure dependence of the coefficient of thermal expansion shows a positive slope.
3. The heat capacity at constant pressure is very high between 0 and 100°C, and nearly constant.
4. The pressure dependence of viscosity shows a minimum below 30°C.
5. Supercooled water has anomalous heat capacities (Angell and Tucker, 1973).

Experimental measurements of liquid water have resulted in evidence that its density increases from 0°C up to 4°C and then decreases progressively (Wada, 1979; Macdonald et al., 1974). In other words, the volume of water shrinks with temperature until 4°C. A similar tendency was observed for heavy and tritiated waters.

Another peculiarity of the properties of water is the positive slope of the curve when the coefficient of thermal expansion $-(1/V)(\partial V/\partial T)_p$ is plotted versus the pressure (p). For normal liquids the slope is negative (Hvidt, 1978).

The heat capacity of liquid water at constant pressure is almost constant between 0 and 100°C. In addition, the heat-capacity values are very high compared to similar liquids (Benson, 1978).

The fourth macroscopic anomaly of liquid water relates to the minimum viscosity values when plotted versus pressure below 30°C. At higher temperatures water behaves like a normal fluid.

Some interesting aspects of peculiarities in the behavior of water were observed at a temperature in the neighborhood of 35°C. A sharp step at 36°C was identified for the position of the absorption peak near 1.15 μm in the infrared spectrum of water (Luck, 1965). An inflection at 32°C was found for the curve of peak intensity of the infrared absorption at 2100 cm^{-1} (Salama and Goring, 1966). A discontiunity was discovered at 35°C in the activation energy of the spin-lattice relaxation process (Voronovich et al., 1969).

Other anomalies reported for the temperature dependence of the physical properties of water at about 35°C include those relating to liquid density (Lavergne
and Drost-Hansen, 1956; Drost-Hanse, 1975), surface tension (Franks and Ives, 1960),
lack of a minimum in the viscosity - pressure isotherms (Bett and Cappi, 1965), a
minimum in the heat capacity (Ginning and Furukawa, 1953), and a break in the thermal
conductivity curve (Frontas'ev, 1956).

These peculiar physical properties of water at about 35°C are due to its
unusual structure. However, no successful explanation has yet been found to explain
this anomalous behavior of water through models or mechanisms. The principal difficulty is that of establishing a quantitative relationship between the physical properties and the structure of liquid water.

To study the structure of water molecules, the solubility of neon in water was
examined by Borina (1977) and Namiot (1967) in the temperature range 15 to 41°C.
The anomalous character of liquid water as a solvent for nonpolar gases was demonstrated not by comparison of the solubilities of these gases in water with their
solubilities in organic liquids, but by comparison of their solubilities in water
with the solubilities of the gases in crystal hydrates. A distinct likeness between
crystal hydrates and solutions of gases in water is now established by several investigators. According to Namiot, the solubility of nonpolar small molecules in water
should differ little from their solubility in ice. This explanation is somewhat different from that of Frank and Evans proposed for describing the solubility of nonpolar substances in water. Because of the small atomic radius of neon, the solute
molecules were able to move freely in the icelike skeleton of water without causing
any distortion in the structure. It was expected that because of the free mobility
of solute molecules in the cavities of solvent, no anomalious behavior should be
observed. However, the experimental results indicated that there is a structural
rearrangement in the water at about 35°C with a relaxation time of about 15 hours.
That is, about 15 hr is required for the achievement of equilibrium in water. The
instability or, in other words, the transition from one state to the other in water
at around this temperature has been proved by experimental solubility data. Although
the establishment of a reliable explanation of the relationship of these changes to
the structure of water awaits further study, and despite the earlier statement by
Frontas'ev (1956), attributing the anomalous characteristics of water, such as the
density, compressibility, heat capacity, and other properties, to the peculiarities
of the water structure and the dissociation energy is in agreement with Eigen's theory.
The anomalous dependencies of gas solubility in water at various temperatures and
pressures were studied by Drost-Hansen (1975).

The physical properties of liquid water are summarized in the Appendix. A detailed discussion of the properties of water and steam was presented at the 9th
International Conference on the Properties of Water and Steam (Straub and Scheffler,

1980), and the reader interested in a particular property of water is referred to
the publication cited, which contains excellent reviews and critical examinations
of various aspects of the problem. See also previous proceedings of the International
Conference, and Vukalovitch (1967) and Dorsey (1968). The effect of pressure and
temperature on the conductivity and ionic dissociation of water up to 100 kbars and
1000°C were reported by Holzapfel (1969). The effect of inorganic salts on the properties of water (aqueous electrolyte solutions) is outside the scope of this book.
Compilations are available elsewhere (e.g., Horvath, 1982).

2.6-2 Halogenated Hydrocarbons

For an understanding of the interaction or mutual solubilities or miscibilities
between halogenated hydrocarbons and water, the physical and chemical properties of
these substances have to be examined. It is appropriate that after our discussion of
the structure and properties of water, a brief summary be given here of the structure,
behavior, and properties of halogenated hydrocarbons. Detailed treatments of the
properties and behavior of halogenated hydrocarbons are those of Slesser and Schram
(1951), Brockway (1937), Burgin et al. (1941), Schiemann (1951), Haszeldine and
Sharge (1951), Huckel (1958), Simons (1950-1965), Lovelace et al. (1958), Patrick
(1969, 1971), Sharpe (1972), Larsen (1969), Stacey et al. (1963), Tomanovskaya and
Kolotova (1970), Badyl'kes (1974), Jolles (1966), Scott (1958), Grosse and Cady (1947),
Fowler et al. (1947), Sanders (1979), Phillips Petroleum (1966), McBee et al. (1947),
Musgrave and Smith (1949), Swinton (1976), Hudlicky (1976), Banks (1979), Green (1969),
and Yokozeki and Bauer (1975).

A critical selection of the ideal gas thermodynamic properties of halogenated
methanes has been reported by Kudchadker and Kudchadker (1978).

The halogen atoms are in group VII of the periodic table and their electronic
configurations have one fewer electron than do the noble gases. They can form a single
covalent bond by sharing an electron or by gaining a single electron to form a uni-
negative ion. The various properties of halogenated atoms, molecules, and ions are
given in Table 2.71 (De La Mare, 1976).

Among the halogen elements, fluorine shows a significant irregularity from the
rest of the elements in the group (Prydz and Straty, 1970). The irregularity occurs
in the increment in nuclear charge and in the shielding of the effects of this charge
on both the bonding and nonbonding shells of electrons. An electrostatic repulsion
offsets an electron that tries to accommodate itself too close to the atomic nucleus.
Consequently, both the F-F bond and the electron affinity become less in the fluorine
molecule compared to the chlorine molecule.

Whereas fluorine, chlorine, and bromine are nonmetallic, iodine is considered to
be an element with metallic character. As far as the stereochemistry of the compounds
of halogens is concerned, fluorine shows a significant difference from the other

Table 2.71 Properties of Halogen Atoms, Molecules, and Ions

Element	Electronic structure	Electron affinity at 25°C (mol kJ^{-1})	Electroneg- ativity	Ionization enthalpy (kJ mol^{-1})	Covalent radius(pm)	van der Waals radius (pm)	Bond enthalpy (kJ mol^{-1}) X$_2$	C-X
Fluorine	$1s^2 2s^2 2p^5$	339	4.1	1682	71	131	159	485
Chlorine	$Ne3s^2 3p^5$	355	2.8	1255	99	181	243	331
Bromine	$Ar3d^{10} 4s^2 4p^5$	331	2.7	1242	114	196	192	276
Iodine	$Kr4d^{10} 5s^2 5p^5$	301	2.2	1008	133	222	151	238
Astatine	$Xe4f^{14} 5d^{10} 6s^2 6p^5$	297	2.0	920	140	230	127	–

halogens. In fluorine compounds the highest oxidation state is virtually zero, with a coordination number of two in covalently bonded systems, although chlorine, bromine, and iodine can have higher oxidation states and a higher coordination number when covalently bounded to the central atom. The covalent bonding in halogen compounds is described by two electrons. In compounds in which the bonding shell contains more than eight electrons, there are two explanations: consideration of the d-orbital which is involved in bonding hybridization, and the assumption that in the stabilization of a bond, fewer than two electrons per bond contribute to the bonding.

Among the halogen molecules, fluorine is the most reactive because of its low bond strength. It readily dissociates into atoms, and the fragments then take part in chain reactions, burning up organic compounds with HF and fluorocarbons as products. These reactions are exothermic. The reactivity of Cl_2, Br_2, and I_2 gradually diminishes relatively to the F_2 molecule. This is due to the weak bonds between these halogens and other elements and the stric congestion in the products of the reactions.

The solubility of iodine in various solvents has been investigated extensively by Hildebrand and co-workers (Hildebrand and Scott, 1950; Hildebrand et al., 1970). Solutions of iodine gave various colors which were examined spectroscopically and change-transfer complexes between the solvents and halogen were established. The resonance hybrids, have an excited state, provide intensive absorption bonds and often lie in the visible region of the spectrum. Color changes have been particularly intensive in iodine solutions. Chlorine and bromine also form charge-transfer complexes with nucleophilic solvents, usually 1:1 complexes, which are stable at low temperatures, but color changes are not likely in these solutions. The complex formation takes place either with σ-donors, such as halogen compounds, amines, and sulfides, or with π-electron donors, such as aromatic compounds (e.g., Br_2 with benzene). This molecular complex is rather weak energetically.

The structural changes in the substances affect the reaction rates; for example, the ionization of CH_3Cl in water is much faster than that for CH_3Br. In contrast, the ionization of t-butyl bromide in water at $25°C$, is slow, but the ionization of t-butyl chloride is rapid.

The electron-withdrawing effect of halogens decreases with increasing size. The electron-releasing influence is greatest for fluorine. In electrophilic aromatic substitution, this effect influences the substitution in ortho or para directions. Chlorine, bromine, and iodine have electron-withdrawing effects, but fluorine can act as an electron release in some situations.

The lone pairs of electrons of halogens can be used for intermolecular coordination with a reaction center. This is sometimes called the "anchimetric effect" or "neighboring-group participation." In this capacity fluorine is the least and iodine the most effective of the halogens.

The high stability of fluorocarbons is explained by the short interatomic distace between fluorine and carbon atoms, and the strength of the bond joining them together.

The bond length and bond energies of several halogenated hydrocarbons are presented in Table 2.72. The bond-length and bond-energy values are linked to the size of the atom in the molecule. This is also concluded from Table 2.72. If the molecule is completely fluorinated, the bond energy has shortened and the energy increased in relation to other compounds.

The chemical and physical properties of halogenated hydrocarbons are responsible for the biological and toxic properties of these substances. Particularly, the stability of the C-F bond is the most important factor. The lack of biological activity is due to C-F stability. A good review of the toxicity and biological action of fluorocarbons is that of Clayton (1967) and Sanders (1979).

The intermolecular forces of liquid fluorinated hydrocarbons are considerably smaller compared with other liquids at ambient temperatures, and they show an enormous deviation from Raoult's law in solutions (Hildebrand, 1953). They are nonpolar liquids with low solubility parameters; consequently they are immiscible with other nonpolar liquids at ordinary temperature (Scott, 1948; Mukerjee and Yang, 1976).

Generally speaking, the reactivity of bromine compounds lies between chlorine and iodine compounds. Chlorine, bromine, and iodine compounds are more reactive than are fluorine substances. Furthermore, aliphatic chlorine, bromine, and iodine compounds are more reactive than their corresponding aromatic derivatives.

The boiling points and densities of halogenated aliphatic hydrocarbons vary in the same order (see Table 2.73). A bibliography of liquid density has been reported by Hales (1980). The boiling points of isomers vary in the order

$$primary > secondary > tertiary$$

Lawson (1979) and Lawson et al. (1978) reviewed the methods for estimation of boiling points, vapor pressures, latent heats of vaporization, and solubility of oxygen in fluorochemicals. For example, the normal boiling points of halogenated butanes were found to be as follows:

	F	Cl	Br	I
n-Butyl-	32.5	78.44	101.6	130.53
sec-Butyl-	-	68.25	91.25	120.0
tert-Butyl-	-	52.0	73.25	100.0

The reactivity of the aliphatic halogen compounds shows the following pattern:

$$I > Br > Cl > F$$

In the case of the Friedel-Grafts reaction, which is an exception, the order of reactivity is reversed.

Table 2.72   Bond Lengths and Bond Energies in Halogenated Methanes

Compounds	X = F			X = Cl			X = Br			X = I
	C-F (Å)	C-H (Å)	C-F (kcal)	C-Cl (Å)	C-H (Å)	C-Cl (kcal)	C-Br (Å)	C-H (Å)	C-Br (kcal)	C-I (Å)
CX_4	1.317	–	116	1.765	–	78.3	1.942	–	66.6	–
CX_3H	1.332	1.098	115	1.767	1.073	78.3	1.930	1.068	66.1	–
CX_2H_2	1.358	1.092	110	1.772	1.068	77.9	1.935	1.091	65.5	–
CXH_3	1.384	1.109	107	1.780	1.11	78.0	1.939	1.110	66.6	–
CF_3Cl	1.328	–	115	1.745	–	76.5	–	–	–	–
CF_2Cl_2	1.355	–	111	1.74	–	77.1	–	–	–	–
$CFCl_3$	1.40	–	109	1.76	–	77.7	–	–	–	–
CF_4	1.335	–	116	–	–	–	–	–	–	–
CF_3Br	1.330	–	–	–	–	–	1.908	–	–	–
$CFBr_3$	1.44	–	–	–	–	–	1.91	–	–	–
CF_3I	1.332	–	–	–	–	–	–	–	–	2.134
$CHClF_2$	1.35	1.09	–	1.74	–	–	–	–	–	–
CH_2ClF	1.378	1.078	–	1.759	–	–	–	–	–	–
$CHCl_2F$	1.41	1.06	–	1.73	–	–	–	–	–	–

Source: Stacey et al., 1963.

Table 2.73 Boiling Points of Halogenated Aliphatic Hydrocarbons (°C)

Compound	X			
	F	Cl	Br	I
CH_3-X	-78.4	-24.2	3.56	42.4
C_2H_5-X	-37.7	12.27	38.40	72.3
n-C_3H_7-X	2.50	46.60	71.0	102.45
n-C_4H_9-X	32.5	78.44	101.6	130.53
n-C_5H_{11}-X	62.8	107.8	129.6	157.0
n-C_6H_{13}-X	91.5	134.5	155.3	181.33

The physical properties of aromatic halogen compounds (e.g., normal boiling point, melting point, density, refractive index) are similar to those of the aliphatic substances (see Table 2.74).

Table 2.74 Liquid Density of Halogenated Aliphatic Hydrocarbons, $d_4^{20°C}$

Compound	X			
	F	Cl	Br	I
CH_3-X	0.5786	0.9159	1.6755	2.2790
C_2H_5-X	0.7182	0.8978	1.4604	1.9358
n-C_3H_7-X	0.7956	0.8909	1.3537	1.7489
n-C_4H_9-X	0.7789	0.8862	1.2758	1.6154
n-C_5H_{11}-X	0.7907	0.8818	1.2182	1.5161
n-C_6H_{13}-X	0.7995	0.8785	1.1744	1.4397

The introduction of halogen atoms into organic molecules changes their fire-retardant characteristics. Most of the halogenated hydrocarbons are nonflammable.

The fluorocarbons are by far the most inert compounds among the halogenated hydrocarbons. The reactivity of fluorocarbons increases from CF_4 to C_2F_6, and so on, that is, as the number of carbon atoms increases. However, this is only relative, because even very long chain fluorocarbons are very resistant to chemical attack. Monocyclic fluorocarbons are more stable against chemical attack than are acyclic fluorocarbons. Hexafluoroethane is thermally stable up to 840°C. Longer-chain fluorocarbons are stable at 400 to 500°C in the presence of CaF_2.

The surface tensions of fluorocarbons are lowest among the halocarbons. The fluorocarbons probably have the lowest values recorded for organic compounds. The cyclic compounds have slightly higher values than those of acyclic substances with the same number of carbon atoms. The low surface-tension values indicate the low intermolecular forces present in the molecules.

The solubilities of fluorocarbons show quite a remarkable behavior (Muccitelli, 1978; Scott, 1948). Fluorocarbons are only alightly soluble in water, and water solubility in fluorocarbons is also very low (Sanders, 1979). The presence of hydrogen atom in the molecule increases the solubility values. The solubility and miscibility between halogenated hydrocarbons and those of hydrocarbons were studied extensively by Hildebrand and Scott (1950), Gilmour (1965), and Mukerjee and Yang (1976). The calculated solubility parameters for these compounds do not differ a lot from one another, and consequently they are miscible with each other in all proportions. The regular solution theory has also been applied to predict their solubilities and miscibilities in nonpolar systems. The acyclic fluorocarbons are completely miscible with aliphatic hydrocarbons; however, partial miscibility occurs with liquids having solubility parameter values higher than 8.5 (e.g., benzene, acetone). The aromatic compounds of halocarbons have higher solubility parameter values and so are miscible with a wide range of solvents. Most of the systems of acyclic fluorocarbons with aliphatic hydrocarbons are miscible in all proportions at 25°C, but the system shows large positive deviation from Raoult's law, being far from ideal.

A comprehensive study on all aspects relevant to aqueous solubility of gaseous fluorocarbons was compiled by Muccitelli (1978). In addition to the experimental measurement of the aqueous solubilities of CF_4, C_2F_6, C_3F_8, and $c-C_4F_8$, he summarized the state-of-the-art of gas solubility, gas-water interaction, hydrophobic effect, and so on.

The molecular structure of a substance is not the determining factor for an anesthetic agent. However, physical properties, such as vapor pressure, solubility in water and oil, and various partition coefficients, are responsible for the activity of general inhalation anesthetics. The action of an anesthetic is primarily a physical rather than a chemical process. For example, narcosis is produced by the solubility of the agent in the lipid portion of the cell membrane. Consequently, narcosis is a solubility phenomenon. By Raoult's law, the mole fraction of the anesthetic agent dissolved in the membrane is proportional to the ratio of the partial and vapor pressures of the agent:

$$x_2 \sim \frac{p_2}{p_2^o}$$

where x_2 is the solubility expressed in mole fraction, p_2 the partial pressure of the solute, and p_2^o the vapor pressure of the pure solute. If the partial pressure of the substance is lower than its vapor pressure, no anasthetic property is expected.

Another factor often used to express solubility is Hildebrand's solubility parameter (see Sec. 2.5-2 for more details). The relation between the solubility parameter (δ_2) and relative saturation (p_2/p_2^o) is expressed by

$$\log_e \frac{p_2}{p_2^o} = \frac{V_{liq}(\delta_{agent} - \delta_{membrane})^2}{R\,T}$$

If the difference between the solubility parameters of the agent and the membrane decreases, the partial pressure (p_2) will also be lowered, and the result will be a reduction in the potential.

Various aspects of fluorine compounds in anesthesiology have been reviewed by Larsen (1969).

The moisture content of a refrigerant is a very important factor with regard to its application and usefulness (Rhodes, 1947) (see the Introduction). Most of the completely halogenated hydrocarbons, particularly perfluorocompounds, dissolve at about 100 ppm by weight water at ambient temperature, but with increasing numbers of hydrogen atoms in the molecule, the water solubility increases, up to 1000 ppm by weight (e.g., $CHClF_2$).

According to Sanders (1979), the solubility of a fluorocarbon in water decreases when the chlorine is replaced by fluorine in the molecule. The solubility of a fluorocarbon in water increases when chlorine is replaced by hydrogen. The solubility of water in fluorocarbons is not affected nearly as much by relpacement of chlorine by fluorine or hydrogen as that of fluorocarbons in water.

The high content of water in the refrigerant generates several problems (e.g., icing of the expansion orifice). A more recent requirement for refrigerants is that they have a limited moisture content -- about 25 ppm by weight or less. Some manufacturers of refrigeration equipment claim 10 ppm by weight moisture limits. With a very low water content, the stability of refrigerants increases and the length of their application is prolonged.

Most refrigerants are colorless and their acidity is rarely greater than 1 ppm of acid contamination. An additional source of contamination for a refrigerant is the content of permanent gases (e.g., air). The usual figures are between 0.5 and 1.0 vol %. The purity of refrigerant is very high. Commercially marketed products exceed 99.9 wt % purity.

The outstanding properties of mixed refrigerants, for example, R 502 ($CHClF_2/$ $CClF_2-CF_3$; 48.8/51.2 wt %) have been recognized. The solubility of water in liquid R 502 has been reported (du Pont, 1969c), however, the theoretical aspects of mixtures of halogenated hydrocarbons are beyond the scope of this book.

Among the many applications of halogenated hydrocarbons, the requirement for high purity is not great. This is due to their use as, for example, propellants, when they are mixed with other products, such as laquers, paints, deodorants, and perfumes.

However, there are some specific uses for which higher purity than 99.9 wt % is demanded (e.g., for dielectrics or as polymerization intermediates).

The completely halogenated hydrocarbons of paraffin series are nonflammable. However, the presence of double bonds or increasing numbers of hydrogen atoms in the molecule will increase the flammability. Flammability limits have been determined for several halogenated hydrocarbons (e.g., CH_3-CClF_2, $CClF=CF_2$, $CF_2=CF_2$). The autoignition temperature of $CF_2=CF_2$, for example, is as low as $180°C$.

It was mentioned previously that commercial halocarbon compounds are quite stable but that the olefin series of halogenated hydrocarbons shows some instability. Consequently, during storage there is some risk of polymerization by an active impurity (e.g., oxygen). The reaction is exothermic and the hazard that arises in the risk of explosion. $CF_2=CF_2$ and $CH_2=CHF$, particularly, need to be handled with cautions. The addition of inhibitors provides a certain margin of safety against explosion.

As far as toxicity is concerned, halogenated hydrocarbons extend over a wide range from safety to toxicity: some substances are safe, whereas others are very toxic. The reader interested in the toxicity aspects of halocarbons is referred to Stacey et al. (1963), who provides good coverage of the toxicity of these compounds. The correlation of the toxicity to algae of halogenated hydrocarbons with their physical-chemical properties was reported by Hutchinson et al. (1980). They found a wide range of toxicities among 38 compounds when the data were expressed as the percentage of saturation causing a 50 percent reduction in photosynthesis. In other words, the aqueous solubilities were used to correlate toxicity.

2.7 Prediction, Estimation, and Correlation of Solubility Data

Despite the great demand for solubility data by scientists and engineers, experimental values published in the open literature are very limited. Regarding the availability of solubility data for halogenated hydrocarbons in water, Beilstein (1958-1964) cites 1369 compounds up to six carbon atoms, of which only 61 have information as to their solubility in water, mostly for a single temperature only (Horvath, 1976b). Therefore, it is apparent that users of the solubility data have two choices: to determine or to estimate the required solubilities in water. In the chemical industry, there is generally not enough time to make experimental measurements; therefore, estimation techniques are adopted most frequently, particularly by chemical and design engineers who have no access to the laboratory facilities. Good discussion of the need for prediction and extrapolation techniques are given by Wei (1977), Merriman (1977), and Shinoda (1978). The need for data is also discussed generally by O'Reilly and Edmonds (1977), James (1972), and Stockmayer (1978).

The varous techniques applied for prediction, estimation, and correlation of solubility data vary with several factors, including:

1. The systems under consideration
2. The time available for the calculations
3. The parameters (e.g., physical properties) of the solvent and solute needed
    for the calculations
4. The investigator's experience with the various methods available
5. The accuracy of the result required

In this discussion, emphasis is directed to halogenated hydrocarbon - water systems, and only a brief summary is given of those techniques that have not been applied for these systems, although a study would be worthwhile.

The time factor is important, particularly in industry when the design engineer needs data in the midst of his calculations. There is no time to wait several days to derive the information required or to order a laboratory determination. In the case of a new compound not available from a commercial supplier, the purification and preparation might require considerable effort and cost.

In some cases the parameters are not readily available for the calculations and an extra effort is required to obtain them from the literature. Such parameters are the physical and thermodynamic properties of the solute and solvent. The properties needed most often are melting point, boiling point, and density (or volume). One has to be sure that states are established for both the solute and the solvent: in other words, whether the system in question is a solid-liquid, a liquid-liquid, or a gas-liquid mixture. The state of a substance will naturally depend upon the conditions, that is, the temperature and pressure. Consequently, the estimation method will be selected after the establishment of the sort of solubility we are dealing with. The main object is to choose a method that will not require values for the physical properties which are not readily available from standard handbooks.

The experience of the investigator plays a large role in the selection of the best method. In the following descriptions, there will be detailed discussions of the various techniques and their applications, often with results given for comparison.

The accuracy and reliability of the estimated, predicted, or correlated solubility data will certainly depend upon the method used. The primary aim is to obtain the best data for a particular job, but as mentioned above, time and other factors will always create some difficulties, so that at the end a compromise will be needed. It is essenatial that the goal be clear, that is, that what the solubility data are to be used for has been delineated.

In addition to the miscellaneous techniques applicable for estimation, prediction, and correlation procedures, seven main groups of methods are available:

1. Correlation of water solubility of substances with chemical structure --
    group-contribution methods
    a. Irmann (1965)
    b. Korenman et al. (1971)

c. Polak and Lu (1973)

d. Davis and co-workers (1970-1974)

2. Correlation of solubility with the constant-increment principle

a. Mercel (1937), Boyer (1960)

b. Robb (1966), Breusch and Kirkali (1968), Bell (1973)

c. Saracco and Spaccamela Marchetti (1958)

d. Schatzberg (1963)

e. Horvath (1974, 1976b)

3. Calculation of solubility from connectivity index

a. Randic (1975)

b. Kier et al. (1975-1977)

4. Calculation of solubility by means of factor analysis

a. de Ligny et al. (1976)

5. Calculation of solubility from the molecular surface area

a. Hanssen et al. (1968), Rheineck and Lin (1968), McGowan (1952, 1954)

b. McAuliffe (1963-1969), Glew (1952)

c. Hansch et al. (1968, 1975), Hermann (1971, 1972, 1977)

d. Harris (1971), Harris et al. (1973)

e. Reynolds et al. (1974)

f. Cohen and Regnier (1970-1976)

g. Teresawa et al. (1975)

h. Amidon et al. (1974-1976), Anik (1978), Yalkowsky et al. (1972-1980)

6. Calculation of solubility from equations of state

a. De Santis et al. (1974)

b. De Mateo and Kurata (1975)

c. Evelein et al. (1976)

d. Namiot et al. (1976)

e. Heidemann (1974)

7. Miscellaneous estimation, prediction, and correlation methods for solubility data

a. Bushinskii et al. (1974)

b. Leland et al. (1955)

c. Lindenberg (1956)

d. Miller (1966)

e. Parnov (1969)

f. Boggs and Buck (1958)

g. Hildebrand and Scott (1950)

h. Jones and Monk (1963)

i. Koskas and Durandet (1967)

j. Prausnitz and Shair (1961)

k. Wing (1956), Wing and Johnston (1957)

l. Hayduk and Laudie (1973)

m. Leites and Sergeeva (1973)

n. Nauruzov (1975)

o. Kuznetsova and Rashidov (1977)

p. Cysewski and Prausnitz (1976)

q. Gotoh (1976)

r. Othmer (1965)

In a critical review, Merriman (1977) has discussed the various aspects of the solubility of gases in CCl_2F_2, including correlation and estimation methods.

2.7-1 Correlation of Solubility with the Group-Contribution methods

The idea of additivity rules for the establishment of physical, thermodynamic, and transport properties of substances has received a considerable attention during the last 20 to 30 years. The methods are based on group-contribution principles. The molecular structure of a chemical substance is composed of a basic group and substitution of other groups and atoms or radicals. Organic compounds have been studied with particular success, and the critical properties (Lydersen, 1955), heat of formation, entropy, and heat capacity (Benson et al., 1969), have been derived with good accuracy. The principal advantage of most of group-contribution methods is that they do not require more than structural information on the substance for the establishment of the needed property.

The various additivity rules were used for many, many years for single compounds. The first-group contribution method for the practical calculation of water solubility values was not introduced until 1965 (Irmann, 1965). Irmann's correlation procedure involves the application of the following relation between the solubility of organic substances in water (s, g/g water) and the structure of the solute at $25°C$:

$$-\log_{10} s = a + \sum_i b_i n_i + \sum_j c_j n_j$$

where   s   =   solubility in water at $25°C$, expressed in g/g water

a   =   group contribution to binding type (see Table   2.75a)

b_i =   group contribution to atom type (see Table   2.75b)

c_j =   group contribution to structure element (see Table   2.75c)

n   =   number of contributing groups

The various group-contribution values have been derived from experimental data and compiled by Irmann (1965) (see Table   2.75). This correlation technique has been tested on a large number of organic compounds, including numerous halogenated hydro-

Table 2.75 Group Contributions by Irmann (1965)

Description		Group contribution
1. Type of compound		
C_6H_6	Aromatic compound	0.50
X, H, =C	Unsaturated halogenated aliphatic with H but no F atoms	0.50
F, H, (Cl), –C	Saturated halogenated aliphatic with F and H atoms	0.50
X, H, –C	Saturated halogenated aliphatic without F atom	0.90
X, –C, F, (X), –C	Saturated perhalogens with F atom and saturated aliphatic compounds without H atom	1.25
H, C	Pure aliphatic hydrocarbon	1.50
2. Type and number of atoms		
C	–	0.25
H	–	0.125
F	In aromatic compounds	0.19
	In saturated hydrocarbons	0.28
Cl	In aromatic and unsaturated compounds	0.675
	In saturated compounds	0.375
Br	In aromatic and unsaturated compounds	0.795
	In saturated compounds	0.495
I	In aromatic and unsaturated compounds	1.125
	In saturated compounds	0.825
3. Structural elements or groups		
–C=C–	Double bond in aliphatic compounds	–0.35
–C≡C–	Triple bond in aliphatic compounds	–1.05
X X >ĊH–ĊH$_2$	Saturated halocarbons with H atom(s)	–0.30
C C | | –C–C–, –C–R | | C C	Aliphatic branches or monosubstitution	–0.10

carbons. The accuracy achieved for 204 substances studied was expressed as percentage of solubility (s, g/g water):

       68% of the substances have been correlated within ±5%

       89% of the substances have been correlated within ±15%

It is strongly recommended that first-time users solve a few examples from the tables presented before attempting this techniques for compounds not listed. This method of calculation is valid for solid-, liquid-, and gaseous-state substances at 25°C. In the calculation there are no differences between cis- and trans- or ortho-, metha-, and para- modifications of compounds. In table 2.76 the experimental and calculated solubility data are compared. Further examinations and comparisons of the calculated solubility data have been published by Rübelt (1969) and Weil et al. (1974).

The logarithm of the solubility (S, g-mole/1000 g water) of liquid aliphatic compounds in water was expressed by Korenman et al. (1971) as the sum of the contributions (s$_i$) by individual constituent groups of the molecules:

$$\log_{10} S = \sum s_i$$

Using the values of s$_i$ obtained from experimental data, the authors calculated the solubilities of more than 70 compounds of different classes (Hydrocarbons, acids, alcohols, nitro compounds, ketones, ethers, halogenated hydrocarbons, etc.). The correlation coefficient for straight-chain compounds is 0.992 and for isomers is 0.983. Values of s$_i$ are tabulated for standard compounds in Table 2.77. In addition to the limitations mentioned for the method of Irmann (1965), that is, for cis- and trans- substances, Korenman et al.'s (1971) method does not make any allowance for the branching on a specific position on the carbon chain of the molecule, and cyclic compounds have not been included in the correlation. The experimental and calculated solubilities of hydrocarbons and halogenated hydrocarbons are compared in Table 2.78.

Polak and Lu (1973) developed a method to correlate the Redlich-Kister coefficients (A and B) with the molar volume of liquid organic compounds and group contributions:

$$A = C_{A_1} V_i + \sum_{i=2}^{7} C_{A_i} n_i$$

$$B = C_{B_1} V_i + \sum_{i=2}^{7} C_{B_i} n_i$$

where V_i = liquid molar volume, cc g-mole^{-1}

n_i = number of groups i present in the molecule

C_{A_i} = coefficient, tabulated from experimental data (see Table 2.79)

C_{B_i} = coefficient, tabulated from experimental data (see Table 2.79)

The mutual solubilities of organic substances and water can be calculated from the expressions

Table 2.76 Solubility of Halogenated Hydrocarbons in Water at 25°C

| Solute | Solubility in water g/1000 g H_2O | | Deviation |
	Experimental	Calculated	(%)
$CHBr_3$	3.1	3.47	11.94
CH_2Br_2	11.8	8.13	-31.10
CH_3-CH_2Br	9.0	6.03	-33.00
CH_2Br-CH_2Br	4.2	5.13	22.14
$CH_2Br-CHBr-CH_3$	1.43	1.62	13.29
$CH_3-CH_2-CH_2Br$	2.4	1.91	-20.42
$CH_3-CHBr-CH_3$	3.0	2.40	-20.00
$CH_3-CH_2-CH_2-CH_2Br$	0.60	0.60	0.00
CCl_4	0.78	1.00	28.21
$CHCl_3$	7.7	7.94	3.12
CH_2Cl_2	16.2	14.1	-12.96
CH_2BrCl	9.0	10.7	18.89
$CCl_2=CHCl$	1.1	1.59	44.55
$CHCl_2-CCl_3$	0.5	0.79	58.00
$cis-CHCl=CHCl$	7.7	5.62	-27.01
$trans-CHCl=CHCl$	6.3	5.62	-10.79
$CH_2Cl-CCl_3$	1.1	1.41	28.18
$CHCl_2-CHCl_2$	2.9	2.82	-2.76
CH_3-CCl_3	1.3	1.26	-3.08
$CH_2Cl-CHCl_2$	4.6	5.01	8.91
$CHCl_2-CH_3$	5.0	4.47	-10.60
CH_2Cl-CH_2Cl	8.6	8.91	3.60
CH_2Cl-CH_2Br	6.7	6.76	0.90
$CH_2Cl-CHCl-CH_3$	2.80	2.82	0.71
$CH_2Cl-CH_2-CH_2Cl$	2.73	2.82	3.30
$CH_2Cl-CH_2-CH_3$	2.50	2.51	0.40
$CH_3-CHCl-CH_3$	3.0	3.16	5.33
$CHCl_2-CH_2-CH_2-CH_3$	0.5	0.447	-10.60
$CH_2Cl-CH_2-CH_2-CH_3$	0.66	0.794	20.30
$CH_3-CHCl-CH_2-CH_3$	1.00	1.00	0.00
$CH_2Cl-CH_2-CH_2-CH_2-CH_3$	0.20	0.251	25.50
$CH_3-CHCl-CH_2-CH_2-CH_3$	0.25	0.316	26.40
$CH_3-CH_2-CHCl-CH_2-CH_3$	0.25	0.316	26.40

Table 2.76 (Continued)

CCl_2F-CCl_2F	0.13	0.155	19.23
CH_2I_2	1.2	1.78	48.33
CH_3I	13.7	8.91	-34.96
CH_3-CH_2I	4.0	2.82	-29.50
$CH_3-CH_2-CH_2I$	1.0	0.891	-10.90
$CH_3-CHI-CH_3$	1.35	1.12	-17.04
$C_6H_3Cl_3$	0.025	0.0398	59.20
$o-C_6H_4Cl_2$	0.145	0.141	-2.76
$m-C_6H_4Cl_2$	0.123	0.141	14.63
C_6H_5Br	0.41	0.380	-7.32
C_6H_5Cl	0.50	0.501	0.20
C_6H_5F	1.55	1.549	-0.06
C_6H_5I	0.18	0.178	-1.11

Average: 4.47

$$x_{HC} = \frac{1}{e^{(A + B)}}$$

$$x_W = \frac{1}{e^{(A - B)}}$$

where x_{HC} = solubility of the organic compound in water, mole fraction

x_W = solubility of water in the organic solvent, mole fraction

For the development of the correlation by Polak and Lu (1973), mutual solubilities were used. Unfortunately, at the time no coefficients for halogenated hydrocarbons were available. Furthermore, neither the cis- and trans- nor the ortho-, para-, and metha- positions were treated in the method proposed. The comparison of the calculated and experimental solubility data is presented in Table 2.80.

The Redlich-Kister equation was also used by Bittrich et al. (1979) to calculate the partial immiscibility between hydrocarbons and water. The two parameters (A and B) of the equation were obtained from mutual solubility data.

The three group-contribution methods of Irmann (1965), Korenman et al. (1971), and Polak and Lu (1973) have been used for estimating solubilities of halogenated hydrocarbons in water. The values obtained are compared with the selected solubility

Table 2.77   Individual Contributions for the Calculation of Solubility at $20^{\circ}C$

Group	Standard compound	s_i (g-mole/1000 g H_2O)
$=CH_2$	Series of ethers	-0.56
$-CH_3$	Hexane	-0.86
$\equiv CH$	Diisopropyl ketone	-0.22
$=C=$	Extrapolation	0.10
$-COO-$	Ethyl acetate	2.23
$-O-$	Ethyl propyl ether	2.74
$=CO$	Butyl methyl ketone	2.62
$-OH$	Heptanol	2.40
$-COOH$	**Valeric acid**	2.06
$-NO_2$	Nitroethane	1.20
$-Cl$	Chloropropane	0.54
$-Br$	Bromopropane	0.28
$-I$	Iodopropane	0.22

Source: Korenman et al., 1971.

data in Tables 2.81 to 2.83. Whereas the experimental values in Tables 2.76, 2.78, and 2.80 have been adopted from the work of Irmann (1965), Korenman et al. (1971), and Polak and Lu (1973), respectively, the comparison in Tables 2.81 to 2.83 are related to selected solubility data recommended at present. It must be remembered that each of the methods discussed above used different sources of experimental data to obtain correlations, and consequently the correlations were based upon the data chosen and not upon the best values available at present. Therefore, Tables 2.81 to 2.83 give information on the discrepancies that might be involved by using the group-contribution data presented by the investigators. By recorrelating the group contributions using the latest selected data, an improvement would certainly be achieved for each method.

Of the three methods, the most accurate results will be obtained by using the correlation of Polak and Lu (1973). However, the group contributions are available at 0 and $25^{\circ}C$ only for paraffinic and aromatic hydrocarbons. The correlations for deriving the coefficients for halogenated hydrocarbons and in general for organic compounds at other temperatures still have not been published. However, if the mutual

Table 2.78   Solubility of Organic Compounds in Water at 20°C

Solute	Molecular weight	Solubility in water (g/1000 g H_2O)		Deviation (%)
		Experimental	Calculated	
$n\text{-}C_5H_{12}$	72.15	0.0387	0.0287	-25.84
$n\text{-}C_6H_{14}$	86.18	0.00945	0.00945	0.00
$n\text{-}C_7H_{16}$	100.21	0.00296	0.00303	2.36
$n\text{-}C_8H_{18}$	114.23	0.000657	0.000950	44.60
CH_3Br	94.94	13.1	24.97	90.61
CH_3I	141.94	14.20	11.81	-16.83
C_2H_5Cl	64.52	9.11	8.51	-6.59
C_2H_5Br	108.97	9.71	7.89	-18.74
C_2H_5I	155.97	3.92	3.57	-8.93
$n\text{-}C_3H_7Cl$	78.54	2.85	2.85	0.00
$n\text{-}C_3H_7Br$	123.00	2.45	2.45	0.00
$n\text{-}C_3H_7I$	169.99	1.07	1.07	0.00
$n\text{-}C_4H_9Cl$	92.57	0.671	0.93	38.60
$n\text{-}C_4H_9Br$	137.03	0.509	0.75	47.35
$n\text{-}C_4H_9I$	184.03	0.211	0.32	51.66
$CH_3\text{-}CHCl\text{-}CH_3$	78.54	3.43	3.13	-8.75
$CH_3\text{-}CH_2\text{-}CH(CH_3)Cl$	92.57	0.926	1.02	10.15
$CH_3\text{-}CHBr\text{-}CH_3$	123.00	2.88	2.69	-6.60
$CH_3\text{-}CH(CH_3)\text{-}CH_2Br$	137.03	0.509	0.83	63.06
$CH_3\text{-}CH(CH_3)\text{-}CH_2\text{-}CH_2Br$	151.06	0.195	0.25	28.21
			Average:	14.22

Source: Korenman et al., 1971.

solubilities between the organic substances and water are available at the required temperatures, the calculation of the coefficient has been carried out as described by Polak and Lu (1973). The great advantage of this method is that it provides group contributions not only to the solubility of hydrocarbons in water, but vice verse:

Table 2.79 Coefficients C_{Ai} and C_{Bi} for Calculating Mutual Solubility

Volume or group	25°C		0°C	
	C_{Ai}	C_{Bi}	C_{Ai}	C_{Bi}
Volume	0.02306	0.02058	-0.02040	0.01779
CH_3-	2.97571	-0.95356	4.63484	-1.21691
$-CH_2-$	0.32934	0.43558	1.02098	0.48634
$-CH-$	-2.44926	1.68675	-2.70958	2.09649
$-C-$	-5.31870	2.87305	-6.55307	3.56926
$=CH-$ (in benzene)	0.79215	-0.14996	1.49667	-0.18018
$=C-$ (in benzene)	-1.89802	1.08183	-2.08991	1.40298

Source: Polak & Lu, 1973.

Table 2.80   Solubility of Organic Compounds in Water at 25°C

Solute	Solubility in water (ppm by weight)		Deviation (%)
	S_{expt}	S_{calc}	
n-Pentane	47.6	44.6	-6.30
n-Hexane	12.4	12.6	1.61
n-Heptane	3.37	3.41	1.19
n-Octane	0.85	0.92	8.24
2-Methylbutane	49.6	55.2	11.29
2-Methylpentane	15.7	15.6	-0.64
3-Methylpentane	17.9	17.2	-3.91
3-Methylhexane	4.95	4.62	-6.67
2,2-Dimethylbutane	23.8	23.0	-3.36
2,3-Dimethylbutane	22.5	22.0	-2.22
2,4-Dimethylpentane	5.50	5.26	-4.36
2,2,4-Trimethylpentane	2.05	2.11	2.93
2,3,4-Trimethylpentane	2.30	2.38	3.48
2,2,5-Trimethylhexane	0.54	0.54	0.00
Benzene	1755	1857	5.81
Toluene	573	583	1.75
Ethylbenzene	177	154	-12.99
o-Xylene	213	204	-4.23
m-Xylene	162	185	14.20
p-Xylene	185	181	-2.16
		Average:	0.18

Source: Polak & Lu, 1973.

that is, for the solubility of water in hydrocarbons.

The methods of both Irmann (1965) and Korenman et al. (1971) provide groups for the solubility of halogenated hydrocarbons in water, but not vice versa. The application of Korenman et al.'s (1971) method is easier and in most cases the results are marginally better than those obtained using Irmann's method. The average deviation for **the 16 compounds examined (see Table 2.83) is less for Irmann's method,** but the individual deviations are slightly better for Korenman's correlation. Consequently, an equal weight is given to both methods for the calculation of the solubility of halogenated hydrocarbons in water.

Table 2.81 Comparison of the Solubility of Hydrocarbons in Water by Various Methods at 25°C (g/1000 g H_2O)

Solute	Experimental (Polak & Lu, 1973)	Irmann (1965)	Deviation (%)	Korenman et al. (1971)	Deviation (%)	Polak & Lu, (1973)	Deviation (%)
n-Pentane	0.0476	0.0562	18.07	0.0287	-39.71	0.0446	-6.30
n-Hexane	0.0124	0.0178	43.55	0.00945	-23.79	0.0126	1.61
n-Heptane	0.00337	0.00562	66.77	0.00303	-10.09	0.00341	1.19
n-Octane	0.000850	0.00178	109.41	0.000950	11.76	0.00092	8.24
2-Methylbutane	0.0496	0.0708	42.74	0.0315	-36.49	0.0552	11.29
2-Methylpentane	0.0157	0.0224	42.68	0.0104	-33.76	0.0156	-0.64
3-Methylpentane	0.0179	0.0224	25.14	0.0104	-41.90	0.0172	-3.91
3-Methylhexane	0.00495	0.00708	43.03	0.00332	-32.93	0.00462	-6.67
2,2-Dimethylbutane	0.0238	0.0282	18.49	0.0109	-54.20	0.0230	-3.36
2,3-Dimethylbutane	0.0225	0.0282	25.33	0.0114	-49.33	0.0220	-2.22
2,4-Dimethylpentane	0.00550	0.00891	62.00	0.00364	-33.82	0.00526	-4.36
2,2,4-Trimethylpentane	0.00205	0.00355	73.17	0.00120	-41.46	0.00211	2.93
2,3,4-Trimethylpentane	0.00230	0.00355	54.35	0.00125	-45.65	0.00238	3.48
2,2,5-Trimethylhexane	0.000540	0.00112	107.41	0.000370	-31.48	0.000540	0.00
Average			52.30		33.06		0.09

Table 2.82 Comparison of the Solubility of Aromatic Hydrocarbons in Water by Various Methods at 25°C (g/1000 g H_2O)

Solute	Experimental (Polak & Lu, 1973)	Irmann (1965)	Deviation (%)	Polak & Lu (1973)	Deviation (%)
Benzene	1.755	1.778	1.31	1.857	5.81
Toluene	0.573	0.562	-1.92	0.583	1.75
Ethylbenzene	0.177	0.178	0.57	0.154	-12.99
o-Xylene	0.213	0.178	-16.43	0.204	-4.23
m-Xylene	0.162	0.178	9.88	0.185	14.20
p-Xylene	0.185	0.178	-3.78	0.181	-2.16
Average			1.73		0.40

Table 2.83 Comparison of the Solubility of Halogenated Hydrocarbons in Water by Various Methods at 25°C (g/1000 g H_2O)

Solute	Experimental (selected)	Irmann (1965)	Deviation (%)	Korenman et al. (1971)	Deviation (%)
CH_3Br	15.46	19.05	23.22	24.97	61.51
CH_3I	14.09	8.91	-36.76	11.81	-16.18
C_2H_5Cl	5.64	7.94	40.78	8.51	50.89
C_2H_5Br	9.02	6.03	-33.15	7.89	-12.53
C_2H_5I	4.06	2.82	-30.54	3.57	-12.07
$n-C_3H_7Cl$	2.95	2.51	-14.92	2.85	-3.39
$n-C_3H_7Br$	2.43	1.19	-21.40	2.45	0.82
$n-C_3H_7I$	1.05	0.89	-15.24	1.07	1.90
$n-C_4H_9Cl$	0.615	0.794	29.11	0.93	51.22
$n-C_4H_9Br$	0.549	0.603	9.84	0.75	36.61
$n-C_4H_9I$	0.182	0.28 (18°C)	53.85	0.32	75.82
$CH_3-CH(Cl)-CH_3$	2.95	3.16	7.12	3.13	6.10
$CH_3-CH(CH_3)-CH_2Cl$	0.920	1.00 (13°C)	8.70	1.02	10.87
$CH_3-CHBr-CH_3$	3.096	2.40	-22.48	2.69	-13.11
$CH_3-CH(CH_3)-CH_2Br$	0.502	0.759 (18°C)	51.20	0.83	65.34
$CH_3-CH(CH_3)-CH_2-CH_2Br$	0.205	0.240 (17°C)	17.07	0.25	21.95
Average			4.15		20.36

The correlation of solubility and partition coefficients with chemical structure was also studied by Davis and co-workers (Davis, 1970, 1973a-e; Davis et al., 1972, 1974; Davis and Mukhayer, 1972), with particular emphasis on the methylene and methyl groups in solutions (Davis et al., 1972; Davis, 1973a). The basic assumption is that the free energy of the solution process is additive and consists of independent contributions from the constituent functional groups. Davis and co-workers examined the activity coefficient, excess free energy, and partition coefficient by summing contributions for the different groups comprising the molecule. The application of the group-contribution method to the calculation of the various properties of solutions had been reported previously in several works (Butler, 1937, 1962; Pierotti et al., 1956, 1959; Deal and Derr, 1968; Higuchi and Davis, 1970; Irmann, 1965; Hansch, 1971; Copp and Everett, 1953; Nelson and de Ligny, 1968a; Black et al., 1963; Martire, 1966; Monfort et al., 1977; Chao et al., 1967; Wilson and Deal, 1962; Tsonopoulos and Prausnitz, 1971; Kakovsky, 1957; Saracco and Spaccamela Marchetti, 1958; Spaccamela Marchetti and Saracco, 1958; Ratouis and Dode, 1965; Robb, 1966; Blackburn et al., 1980; Breusch and Kirkali, 1968; Yalkowsky, Flynn, and Amidon, 1972; Langmuir, 1925; Laprade et al., 1977).

The methylene-group contribution in a homologous series of substances has been calculated by Davis et al. (1972). They found that the excess free energy for CH_2 in water is 849 ± 40 cal g-mole^{-1}; however, there was some evidence that the increment depends to some extent on the nature of the polar grouping. The reported value by Butler and Harrower (1937) was 1600 cal g-mole; that is, greater by a factor of 2. The methylene-group contributions to the thermodynamic properties of transfer of a hydrocarbon from the pure state to the water phase were calculated by Davis et al. (1974) from the linear relationship between the thermodynamic values and the carbon numbers. The following average values were proposed at $25^\circ C$:

$$\Delta G_{CH_2} = 800 \text{ cal g-mole}^{-1}$$
$$\Delta H_{CH_2} = 230 \text{ cal g-mole}^{-1}$$
$$\Delta S_{CH_2} = 1.76 \text{ cal g-mole}^{-1} \text{ K}^{-1}$$

These values differ considerably from those obtained by Kakovsky (1957):

$$\Delta G_{CH_2} = 860 \text{ cal g-mole}^{-1}$$
$$\Delta H_{CH_2} = -360 \text{ cal g-mole}^{-1}$$
$$\Delta S_{CH_2} = -4.1 \text{ cal g-mole}^{-1} \text{ K}^{-1}$$

and in a plot of ΔG versus n (number of carbon atoms), the CH_2 increment was found by Nelson and de Ligny (1968a, b) to be about 850 cal g-mole^{-1}. More recently Amidon and Anik (1980) reported 788-908 cal g-mole^{-1} for aliphatic hydrocarbons, or

29 cal Å^{-2} on a surface-area basis. This is very similar to that found previously for aromatic hydrocarbons (28 cal Å^{-2}). The corresponding heat of solution was equivalent to 502 cal per CH_2 group (CH_2 = 31.8 Å^2). Consequently, all three parameters -- free energy (ΔG), enthalpy (ΔH), and entropy (ΔS) of solution -- are linearly related to the chain length. However, the accuracy of the latter two quantities are not conclusive (Nelson and de Ligny, 1968b; Davis et al., 1974; Rytting et al., 1978; Amidon and Anik, 1980).

The limiting activity coefficients at infinite dilution (γ_i^∞) in water were calculated either from the Raoult's law and extrapolation, or from the excess partial **molar free energy of solution at infinite dilution:**

$$\Delta G_i^E = R T \log_e \gamma_i^\infty$$

Although ΔG_i^E is the sum of individual contributions, it originates from the characteristic groupings of the solute or solvent structure. A good summary of the method developed by the Shell Development Group has been given by Davis et al. (1974). The mean value for $\Delta \log_{10} \gamma^\infty$ was reported to be 0.62 ± 0.03 for the methylene-group contributions in water at 25°C (Davis et al., 1972). The graphical comparison of $\log_{10} \gamma^\infty$ versus the number of carbon atoms shows linear plots, which suggests that there is a regularity for each additional CH_2 group (Deal and Derr, 1968). Regularity was also found by Kakovsky (1957) for the solubility of a homologous series in water at 25°C. Each lengthening of the chain by one CH_2 link causes a 4.34-fold decrease, regardless of the character of the polar group of the homologous series.

Choosing the pure solute as a standard state is not always convenient; therefore, Davis and co-workers (Davis et al., 1972: Davis, 1973b; Rytting et al., 1978) proposed an infinitely dilute solute in a reference hydrocarbon solvent. Furthermore, instead of the limiting law of Raoult, Henry's limiting law was preferred, when the activity coefficient approaches 1 ($\gamma_i \to 1$) as the concentration decreases to zero ($x_i \to 0$), and the Henry's law becomes

$$H_i = \gamma_i^\infty P_i^o$$

where γ_i^∞ =   activity coefficient at infinite dilution

   P_i^o =   vapor pressure of the pure solute

The Henry's law constants were calculated· by Davis et al. (1972) for a number of series of organic substances in water, and they plotted the values obtained versus the number of carbon atoms in the molecules. The graphs show parallel straight-line correlation, illustrating the additive group-contribution effect. The substances investigated included alkanes, alkanals, alkyl amines, alkanols, alkyl ethers, and alkyl formate esters. The gradient of the line gives the methylene-group contribution and that of the functional group. The extrapolation of the straight lines to zero carbon number gives the Henry's law coefficient for water, which is lower than the experimental value. This value is even smaller in the case when water is the solvent (see also Rytting et al., 1978).

Further to the study of methylene-group contributions, Davis (1973a) investigated the methyl-group contribution and compared them with the methylene-group values. The difference between the contributions of these two groups was illustrated for molar volume, parachor, surface area, cross-sectional area, and so on. In most cases the values were greater for the CH_3 group than for the CH_2 group. The free energy of transfer of the CH_3 group from water to organic solvent was 2.00 kcal g-mole^{-1} compared to a value of 850 cal g-mole^{-1} for the CH_2 group (Davis, 1973a).

The various halogen groups in solubility processes and their effects on activity and partition coefficients were covered in the work of Davis (1973c), as were the thermodynamics of hydroxyl and carboxyl groups in solutions of drug molecules (Davis, 1973d). Davis pointed out that by substituting halogen atoms for hydrogen atoms in the aromatic molecules, the activity coefficients would be higher and, as a result, the water solubility would decrease.

Guseva and Parnov (1964) studied the variation of the solubility of monocyclic arenes in water with the number of carbon atoms in the molecule. At 25°C, the solubility diminishes fairly smoothly with increase in the molecular weight, both for straight chain homologues and branched-chain homologues of benzene. The higher solubility of isomers with the longer side chain compared with that of the corresponding isomers with two or more methyl substituents may be to some extent explained by postulating that a linear configuration of the molecule facilitates their entry into vacancies in the structure of water.

2.7-2 Correlation of Solubility with the Constant-Increment Principle

In order to develop a simple model for the prediction of solubilities of substances, numerous studies have been directed toward the constant-increment principle in homologous series of compounds. Basically, the idea is the same as in the method used by Davis and co-workers, but the emphasis is more specific for homologous series of substances. The trends observed in various investigations proved to be an extremely useful method for prediction of the properties of members of homologous series. Most frequently used in the semilogarithmic expression

$$\log_e x_2 \; = \; a \; + \; b \, C_n$$

where x_2 =   solubility of solute in the solvent, mole fraction
   C_n =   number of carbon atoms in the molecule
 a and b   =   coefficients for a typical series of substances

which was derived from the Clausius-Clapeyron equation,

$$\log_e x_2 = \frac{\Delta H_{fus}}{R}\left(\frac{1}{T_o} - \frac{1}{T}\right)$$

where ΔH_{fus} =   heat of fusion, cal g-mole^{-1}

  T_o =   freezing point, K

The heat of fusion of a homologous series is a linear function of the number of carbon atoms in the molecule (C_n):

$$\Delta H_{fus} = a_1 + b_1 C_n$$

and at the freezing point (T_o, K):

$$\frac{\Delta H_{fus}}{T_o} = a_2 + b_2 C_n$$

(For estimating ΔH_{fus}, see Rao, 1975). By substituting the latter two equations into the Clausius-Clapeyron equation, we obtain

$$\log_e x_2 = \frac{1}{R}(a_2 + b_2 C_n) - \frac{1}{R\,T}(a_1 + b_1 C_n)$$

or

$$\log_e x_2 = \left(-\frac{a_1}{R\,T} + \frac{a_2}{R}\right) + \left(-\frac{b_1}{R\,T} + \frac{b_2}{R}\right) C_n$$

At constant temperature the terms within parantheses become constant, giving

$$\log_e x_2 = a + b\,C_n$$

This simple relation between the solubility (x_2) and the number of carbon atoms (C_n) in the chain of solutes has been used by several investigators for fatty acids, alcohols, alkanes, and so on (see, e.g., Bell, 1973; Boucher and Skau, 1954; Copp and Everett, 1953; Kabadi and Danner, 1979; May, 1980; Peake and Hodgson, 1967; Ralston and Hoerr, 1942; Skau and Boucher, 1954; Tokunaga et al., 1980; and Ufnalski, 1978). Among a large number of organic compounds, Saracco and Spaccamela Marchetti (1958) and Spaccamela Marchetti and Saracco (1958) included several monohalogenated hydrocarbons in the correlation procedure.

Kabadi and Danner (1979) correlated the aqueous solubilities of a large number of homologous series at 25°C, using this equation. The results were more accurate than those given by the UNIFAC (Fredenslund et al., 1977) and Leinones's methods (Leinonen et al., 1971). The correlation did not include halogenated compounds. Price (1973) also used the above expression for the correlation of aqueous solubilities of three

series; alkylated benzenes, methyl alkanes, and alkylated cyclopentanes. Further
investigations have been carried out by Klevens (1950), Pierotti et al. (1956, 1959),
Robb (1966), Breusch and Kirkali (1968), Mercel (1937), Boyer (1960), May (1980),
McAuliffe (1980), and Franks (1966). However, owing to the lack of solubility data
at the same temperature, the solubility values investigated have been used at various
temperatures, assuming that $\pm 3^{\circ}C$ would not effect the solubility significantly.

According to McAuliffe (1963), a plot of $\log_{10} x_2$ versus molar volume gives
a straight line inside a limited number of substances. For example, in the case of
hydrocarbon solubility in water, the linear relationship is valid for the C_5 to C_{10}
members. Later Franks (1966) reported that the correlation could be simplified by
plotting $\log_{10} x_2$ versus the carbon number (C_n) for liquid-liquid systems. However,
the logarithm of solubility of a large number of organic compounds in water was correlated successfully by Natarajan and Venkatachalam (1972) and Burgess and Haines
(1978) using the molar volume of the solutes at $25^{\circ}C$ (see the Appendix, Figs. A8
to A13.). Similarly, the aqueous solubilities of C_4 to C_8 and C_{20} to C_{33} hydrocarbons (mg liter^{-1}) was correlated by the molar volumes of the alkanes at $20^{\circ}C$
(cc g-mole^{-1}) (Peake and Hodgson, 1966). Baker (1959) has correlated the aqueous
solubilities of single-ring aromatics with molar volumes, however with less success.

Fig. 2.50 illustrates the solubility (S) of alkyl acetates and iodides in
water expressed in g-mole liter^{-1} as a function of the number of carbon atoms (N_c)
on a semilogarithm graph. The linear relationship is very convincing, but as Mercel
(1937) pointed out, the linearity could not be established for carboxylic acids. May
(1980) showed that the correlation of the aqueous solubilities of polycyclic aromatic
hydrocarbons with carbon number, molar volume, and molecular length is not accurate.
This is strongly supported the solubilities of isomers.

The solubility of water in several normal alkanes from C_7 to C_{16} was studied
by Schatzberg (1963), and three different correlations were presented. In one of the
correlations proposed, the solubility, expressed as a mole fraction or in parts per
million (ppm) by weight was plotted against the molecular weight on a semilogarithmic
scale. According to the graphs, the water solubilities are directly proportional to
molecular weight when expressed as a mole fraction and inversely proportional when
expressed on a weight basis. In both cases the least-squares correlation produced a
good fit at 25 and $40^{\circ}C$. Similar correlations could probably be carried out at other
temperatures.

In a log-log type correlation, Horvath (1973) examined the linear relationship
between solubility expressed as a mole fraction (x_2) and as a molecular weight (M),
according to the expression

$$x_2 = a M^b$$

for the aqueous solubilities of homologous series of organic substances. In the equa-

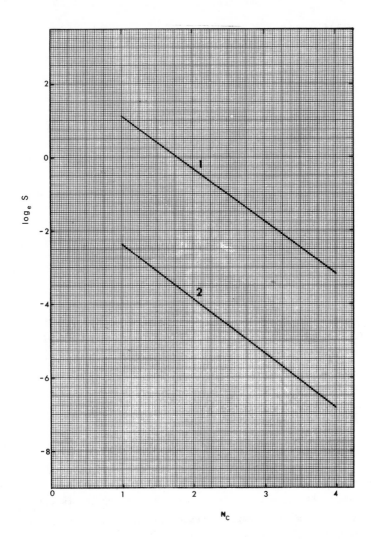

Fig. 2.50 Solubility in water at 20°C. S = solubility in water (g-mole liter^{-1}); N_c = number of carbon atoms in the compound. 1, Alkyl acetates; 2, alkyl iodides. (From Mercel, 1937.)

tion above, a and b are the correlation coefficients. In the double-logarithmic plot, the following homologous sereies were presented (see Fig. 2.51):

1. Normal alkanes
2. Iodinated normal alkanes
3. Normal alkyl acetates
4. Chlorinated normal alkanes
5. Normal aliphatic alcohols
6. Fatty acids

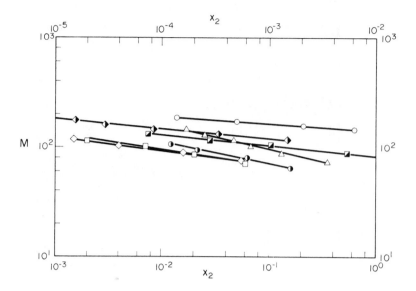

Fig. 2.51 Solubility x_2 (lower and upper horizontal axes) of members of the homologous series plotted against molecular weight M (vertical axis). △ , Normal alkanes
in methanol (lower scale); □ , normal alkanes in water (upper scale); ○ , iodinated
normal alkanes in water (upper scale); ◇ , normal alkyl acetates in water (lower scale);
◕ , chlorinated normal alkanes in water (upper scale); ◪ , normal aliphatic alcohols
in water (upper scale); ◆ , fatty acids in water (upper scale). (From Horvath, 1973.)

Later, the log-log relationship between the water solubility of halogenated
hydrocarbons (x_2) and their molecular weight (M) was studied further by Horvath
(1975a) and a great number of typical series proposed for this simple correlation
method at 25°C. The correlation of

$$C_n H_{2n+1} X$$

where n = 1 to 10 and X = F, Cl, Br, and I, is illustrated in Fig. 2.52. Another
typical series proposed for the correlation was

$$C_n H_{2n} X_2$$
$$C_n H_{2n-1} X_3$$
$$C_n H_{2n-1} X \quad \text{etc.}$$

It is important to point out that this linear relationship is valid only for liquid
solutes at 25°C.

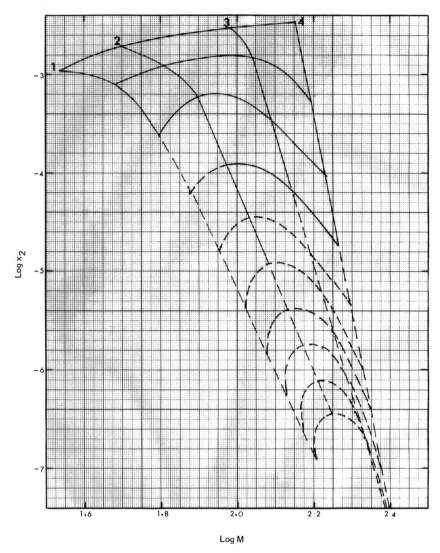

Fig. 2.52 Solubility of normal monohalogenated hydrocarbons in water at 25°C.
x_2 = solubility of monohalogenated hydrocarbon (mole fraction); M = molecular weight
of monohalogenated hydrocarbon. 1, $C_nH_{2n+1}F$: n = 1-10; 2, $C_nH_{2n+1}Cl$: n = 1-10; 3,
$C_nH_{2n+1}Br$: n = 1-10; 4, $C_nH_{2n+1}I$: n = 1-8. (From Horvath, 1975a.)

The solubility of a large number of insecticides in water at 25°C has been
determined by Weil et al. (1974). These insecticides are chlorinated benzenes, cyclo-
hexanes, and biphenyls. By plotting the number of chlorine atoms in the molecule versus
the logarithm of solubility in water at 25°C (expressing the solubility in µg liter^{-1}
a linear correlation was found for chlorinated byphenyl molecules. The straight line

was crossed through the solubility of diphenyl (8×10^3 µg liter^{-1}) and decachloro-phenyl (16×10^{-3} µg liter^{-1}) at 25°C on a semilogarithmic graph.

2.7- 3 Calculation of Solubility from Connectivity Index

Most molecular properties depend on the shape and size of the molecule, so in the past a great effort was given to establishing the relationships that exist between the topological and physical properties of a molecule. Following earlier achievements on the subject, Randic (1975) showed the relationships between the Branching index (later called the connectivity index) of the molecules and normal boiling points, enthalpy of formation, surface area, and so on, for alkanes. Some of the branching connectivity indices are listed in Table 2.84.

The relationship between molecular connectivity and physicochemical properties and biological activities is shown in Fig. 2.53. The additive and constitutive cha-racter of the molecules (i.e., the molecular size and shape) are affected by the connectivity index. This has been correlated successfully for many physicochemical properties and used in physicochemical-activity studies (Kier and Hall, 1976b). It has been found that if a series of compounds correlated with a connectivity index versus biological response, a similar correlation of additive and constitutive molecular property showed for the physical properties.

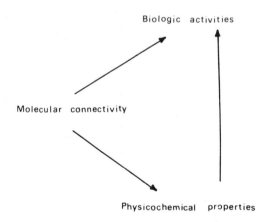

Fig. 2.53 Relationships among molecular connectivity, physicochemical properties, and biological activities.

In subsequent articles by Kier and co-workers (Kier, Murray, et al., 1975; Kier, Hall, et al., 1975; Hall et al., 1975; Murray et al., 1975, 1976; Kier et al., 1976; Di Paolo and Kier, 1977; Kier and Hall, 1976a, b, 1977a, 1978, 1979; Hall and Kier, 1977a, b, 1978a, b; Kier et al., 1977; Cammarata, 1979; Di Paolo et al., 1977,

Table 2.84 Branching or Connectivity Indices of Compounds

Compound	Connectivity index
Methane	0.0000
Ethane	1.0000
Propane	1.4142
n-Butane	1.9142
n-Pentane	2.4142
n-Hexane	2.9142
n-Heptane	3.4142
n-Octane	3.9142
Methanol	1.000
Ethanol	1.414
n-Propanol	1.914
n-Butanol	2.414
n-Pentanol	2.914
n-Hexanol	3.414
n-Heptanol	3.914
n-Octanol	4.414
Acetone	1.732
Ethyl ether	2.414
Benzene	3.000
Toluene	2.893
Aniline	2.893
Nitrobenzene	3.804
Pyridine	2.500
Cyclohexane	3.000
Methylcyclohexane	3.394
Cycloheptane	3.500
Cyclooctane	4.000
Ethylene chloride	1.914
1,2-Dibromoethylene	1.914
1,2-Dichloroethylene	1.914
Chloroform	1.732
Ethylchloride	1.414
Ethylbromide	1.414
Ethyliodide	1.414
Isopropyl chloride	1.732
Isopropyl bromide	1.732

Table 2.84 (Continued)

Isopropyl iodide	1.732
n-Oropylchloride	1.914
n-Propylbromide	1.914
n-Propyliodide	1.914
2-Methyl-1-chloropropane	2.270
2-Methyl-1-bromopropane	2.270
2-Methyl-1-iodopropane	2.270
tert-Butylchloride	2.000
tert-Butylbromide	2.000
tert-Butyliodide	2.000
n-Butylchloride	2.414
n-Butylbromide	2.414
n-Butyliodide	2.414
2-Chloropentane	2.770
2-Bromopentane	2.770
2-Iodopentane	2.770
n-Propylfluoride	1.914
n-Butylfluoride	2.414
n-Pentylfluoride	2.914

Source: Randic, 1975; Kier, Hall, Murray, and Randic, 1975.

1979; Gupta and Singh, 1979), the molecular conncetivity index was used to correlate cavity surface area, molecular polarizability, molar refractivity, parachor, **anesthetic potency, boiling point, partition coefficient, biological activity, density,** minimum inhibitory, minimum killing concentration, electron density, odor of molecule, solubility in water, and other properties. The various methods have been compiled and described in a book by Kier and Hall (1976b).

The method of calculating connectivity index is relatively simple and the only knowledge of the molecular skeleton is required. The hydrogen atoms are suppressed. The attached carbon atoms are assigned according to the number of connections, and the connectivity index is obtained from the reciprocal square root of the product of the numbers associated with the two atoms of the bond. For example, the connectivity index (χ) of 2,4-dimethylpentane is calculated by

$$C_1 \xrightarrow{\dfrac{1}{(1 \times 3)^{\frac{1}{2}}}} C_3 \xrightarrow{\dfrac{1}{(3 \times 2)^{\frac{1}{2}}}} C_2 \xrightarrow{\dfrac{1}{(2 \times 3)^{\frac{1}{2}}}} C_3 \xrightarrow{\dfrac{1}{(1 \times 3)^{\frac{1}{2}}}} C_1$$

$$\Big|\ \dfrac{1}{(1 \times 3)^{\frac{1}{2}}} \qquad \qquad \Big|\ \dfrac{1}{(1 \times 3)^{\frac{1}{2}}}$$

$$C_1 \qquad \qquad \qquad C_1$$

$$\chi = 4 \times \frac{1}{(1 \times 3)^{\frac{1}{2}}} + 2 \times \frac{1}{(2 \times 3)^{\frac{1}{2}}} = 3.125$$

Another example is that for ethyl chloride:

$$C_1 \underset{(1 \times 2)^{\frac{1}{2}}}{\overline{}} C_2 \underset{(1 \times 2)^{\frac{1}{2}}}{\overline{}} Cl$$

$$\chi = \frac{1}{(1 \times 2)^{\frac{1}{2}}} + \frac{1}{(1 \times 2)^{\frac{1}{2}}} = 1.414$$

Halogen atoms are taken as carbon atoms. Consequently, there is no difference between the halogen atoms connected to the molecule, so the connectivity index values are the same for these molecules.

The solubility of alcohols (from C_4 upward) in water was correlated with the connectivity index on a semilogarithmic scale (Hall et al., 1975):

$$\log_e S(\text{molal}) = 6.702 - 2.666 \, \chi$$

Similarly, for hydrocarbons in water, the expression becomes

$$\log_e S(\text{molal}) = -1.505 - 2.533 \, \chi$$

The connectivity index provides a simple computational method for the calculation of important characteristics of a molecule (Kier and Hall, 1976b). In this book the main interest lies in its application to solubility data. The application of connectivity-index principles has not been developed for the solubility of halogenated hydrocarbons in water, or vice versa; however, the forces of interaction involved would be different from hydrocarbons and alcohols (Kier, 1976). It is more likely that a second connectivity term will be necessary to add to the expanded connectivity series, or that the valence deltas for the halogens should be utilized.

The multiple regression equation for the combined hydrocarbon and alcohol series has been shown by Hall et al. (1975):

$$\log_e S(\text{molal}) = -1.516 - 2.528 \, \chi - 3.961 \, C_{OH} - 10.13 \, Q$$

where C_{OH} is the contribution of the C-OH bond; that, is C_{OH} is 1 for an alcohol and zero for a hydrocarbon. Similarly, Q is 1 for alcohols and zero for a hydrocarbon. As mentioned above, it is likely that a similar type of expression will represent the solubility of halogenated hydrocarbons in water at 25°C.

2.7-4   Calculation of Solubility by Means of Factor Analysis

The solubility of gases in liquids is considered to be known functions of hard sphere diameters σ_1 and σ_2 of solvent and solute respectively and the force constants ϵ_1/k and ϵ_2/k (Pierotti, 1963, 1965, 1967, 1976; de Ligny and van der Veen, 1972, 1975). Furthermore, it has been established that the solution process consists of two steps:

    1.   Creation of a cavity in the solvent

    2.   Introducing the solute molecule into the cavity

That is,

$$\log_e \frac{p_2}{x_2} \;=\; f_c(\sigma_1) \;-\; f_i(\sigma_1, \epsilon_1/k, \epsilon_2/k)$$

where p_2 is the partial pressure of the solute and x_2 the concentration (mole fraction) of solute in solvent.

    However, the failure of the above equation inspired de Ligny et al. (1976) for the introduction of a new type of equation for the prediction of solubility,

$$y_{g,s} \;=\; \sum_{j=1}^{n} G_j \, S_j$$

where g and s denotes gas and solvent, respectively, G and S are adjustable parameters, depending only on the identity of the gas and solvent, respectively. The number of n is large enough for reducing the residuals $(r_{g,s,n})$,

$$y_{g,s} \;-\; \sum_{j=1}^{n} G_j \, S_j \;=\; r_{g,s,n}$$

The values of G and S can be found by factor analysis. For simplicity the above equation has been reduced to

$$y \;=\; G_1 \, S_1 \;+\; G_2 \, S_2$$

where y is the experimental datum.

    The authors correlated 20 gases in 39 solvents. The size of both the solute and solvent molecules had a wide range. The equation allows the calculation of all missing data on the investigated gases and solvents.

2.7-5 Calculation of Solubility from Molecular Surface Area

The thermodynamic properties of a solute in solution have been related to the volume
of the solute molecule by a number of workers (Hanssens et al., 1968, Rheineck and
Lin, 1968; McGowan, 1952; Deno and Berkheimer, 1960; McAuliffe, 1966; Ashton et al.,
1968). The relationship between the solubility and molar volume, particularly, had
many practical applications (McAuliffe, 1963, 1966, 1969, 1980; McAulif, 1970; Glew,
1952; Sutton and Calder, 1975). However, in the present correlation relating to the
solubility of hydrocarbons in water as a function of their molar volumes, it appears
that there is not a similar dependence on volume for the straight-chain compounds as
there is for branched chains and cyclic substances (Harris et al., 1973). It appears
that the compounds investigated were selected from symmetrical molecules or that the
investigation was limited to homologous series of substances.

The correlation of solubility with the van der Waals diameter has been demonstrated by Franks et al. (1963) for benzene derivatives at $25^{\circ}C$. The $\log_{10} x_2$
versus van der Waals diameter resulted in a straight-line correlation.

To improve the foregoing proposed correlations, Hermann (1971, 1972, 1975, 1977)
examined the significant structure theory of liquids (Eyring and Jhon, 1969) and
correlated the hydrocarbon solubility in water with solvent cavity surface area. The
size of the solvent cavity, being large enough to accomodate a solute molecule, was
calculated and expressed as a function of hydrocarbon solubility in water. In other
words, the number of water molecules that can be packed around the hydrocarbon molecule is a measure of solubility. The cavities computed by Hermann (1972) were nonspherical and depended on detailed hydrocarbon shape.

The cavity surface area was defined as a two-dimensional space available to a
water molecule in contact with the solute molecule. Fig. 2.54 illustrates the cavity
surface area $A_{RW}(T)$. For n conformations for hydrocarbon R in water, the average
surface area is expressed by the equation

$$A_{RW}(T) = \sum_{i=1}^{n} \frac{A_{RW_i}(T) \exp\left[-\dfrac{\Delta G_{oi} + b_o A_{RW_i}(T)}{k\,T}\right]}{\displaystyle\sum_{j=1}^{n} \exp\left[-\dfrac{\Delta G_{oi} + b_o A_{RW_i}(T)}{k\,T}\right]}$$

where ΔG_{oi} is the difference between the molecular free energy for conformation i
and an arbitrary standard conformation. Hermann (1972) reported a good correlation
between the logarithmic values for solubility and cavity surface area A_{HW} for
alkanes, cycloalkanes, and alkylbenzenes at $25^{\circ}C$ without taking account of variations in the cavity radius of curvature. The correlated equations are for alkanes and
cycloalkanes at $25^{\circ}C$:

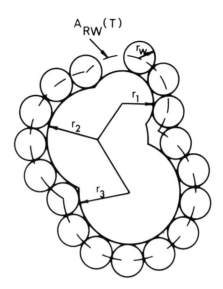

Fig. 2.54 Cavity surface with area $A_{RW}(T)$. (From Hermann, 1972.)

$$-\log_{10} S \;=\; 0.25847 \;+\; 0.01171\; A_{HW}$$

and for alkylbenzenes at $25^{\circ}C$:

$$-\log_{10} S \;=\; -2.7689 \;+\; 0.0184\; A_{HW}$$

where S is the molal solubility in water. Hermann (1972) concluded from his results
that cavity surface area is a better parameter than experimental molar volume in
correlating solubilities. Furthermore, the method of calculating cavity surface area
is simple comparing to the π values (proportional to the free energy of transfer of
a function from one phase to another) used by Hansch et al. (1968, 1975) and Hansch
and Anderson (1967). Different values are required for computing π values for normal
alkane CH_2 groups and cyclohexane CH_2 groups, and a correction due to branching
is required. Therefore, the simplicity and availability of the cavity surface area
method make it preferable. A computer program has been written for the calculation of
$A_{RW}(T)$, which requires the following input data:

    Bond lengths and angles
    Atomic radii
    Radius of a spherical solvent molecule

The atomic radii were taken from Bondi (1964). In principle, Hermann (1972) and Hansch et al. (1968) showed that the solubility of organic substances in water is an additive-constitutive property.

Following the cavity-surface-area approach of Hermann (1971, 1972, 1975) to the free energy of transfer for the organic substances in aqueous solutions, Harris (1971) and Harris et al. (1973) demonstrated the interfacial solute-solvent interactions in the aqueous phase by correlating the free energy of transfer of organic molecules with the relative surface areas at 30°C. The relative surface areas were determined from the hydrocarbon group; tert-butyl has been taken as unity. The group areas were obtained by using atomic models and contacting spheres corresponding to hydrogen atoms in diameter. The result was derived from the number of contacting spheres with the atomic model in relation to the surface area of the tert-butyl group. According to the observation, a direct relationship was found between the free energy of transfer $(\Delta\Delta G_{30})$ and the relative surface areas (A_{RS}) of all groups investigated at 30°C:

$$\Delta\Delta G_{30} \text{ (cal g-mole}^{-1}) \quad = \quad 962.55 \; - \; 3917.59 \; A_{RS}$$

Fig. 2.55 shows the straight-line relationship. The numerical values are presented in Table 2.85.

Table 2.85 Free Energy of Transfer as Function of Relative Surface Areas at 30°C

Group	$\Delta\Delta G_{30}$ (cal g-mole^{-1})	A_{RS}
Cyclohexyl	−4405	1.33
Neopentyl	−4036	1.29
Cyclopentyl	−3699	1.20
sec-Butyl	−3376	1.15
tert-Butyl	−3136	1.00
Isopropyl	−2507	0.90
n-Pentyl	−4580	1.42

Source: Harris et al., 1973.

The theoretical explanation of the solubility process can be described in three steps:

1. Removal of the solute molecule from its environment

2. Cavity formation in the solvent to which the solute molecule is being transferred

3. Introduction of this molecule into the cavity

Consequently, the thermodynamic properties of the solute in the solution will depend
upon the size of the solute molecule and the magnitude of the molecular interaction
energies. The foregoing expression and Fig. 2.55 illustrate the relationship using
relative surface areas instead of absolute values. According to this relations, those
groups that have relative surface areas will the same free energy of transfer.

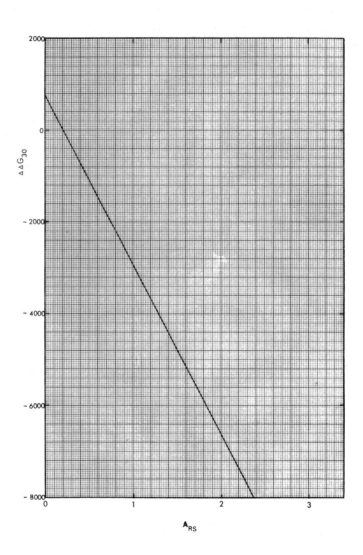

Fig. 2.55 Free energy of transfer at 30°C versus relative surface areas of groups.
$\Delta\Delta G_{30}$ = free energy of transfer (cal g-mole^{-1}); A_{RS} = relative surface area. (From
Harris, 1971; Harris et al., 1973.)

A comparison of the absolute enthalpies of transfer for groups shows a definite trend. The interaction between water and butoxy groups is stronger than that between water and bromo, chloro, or fluoro groups. This is also reftected by the free energies of transfer between these groups (see Table 2.86). Therefore, it is convincing that the strength of the specific solute-solvent interaction is well represented by the free-energy values. The interfacial interactions between solute and water molecules are of great importance for an understanding of the thermodynamic properties of aqueous solutions.

Table 2.86 Free Energy of Transfer and Enthalpies of Groups

Group	$\Delta\Delta G_{30}$ (cal g-mole^{-1})	ΔH (cal g-mole^{-1})
Butoxy	−2053	−92
Bromo	−807	−1789
Chloro	−439	−1203
Fluoro	−38	−449

Source: Harris, 1971.

In examining the free energy of transfer of the halogen groups, Hansch and Anderson (1967) and Harris (1971) came to the same conclusion: that the numerical values decrease in the order

$$F^- < Cl^- < Br^- < I^-$$

That is, iodine is most negative. This is also followed by the size of the group; that is, the free energy of transfer tends to become more negative as the size of the group increases. Although the enthalpy and entropy show a gradual decrease in the order

$$F^- > Cl^- > Br^- > I^-$$

the fluoro group shows a different behavior when attached to an aliphatic substance than when attached to an aromatic substance. In the case of an aromatic compound, the fluoro group withdraws electrons from the ring, allowing it to interact more readily with the water molecules. This is also reflected by the positive enthalpy and entropy of transfer values of aromatic fluoro groups compared with the negative values of aliphatic substituents. Generally speaking, the presence of a halogen group will decrease the water solubility of a molecule more for aliphatic compounds than for aromatic compounds.

The methods of empirical correlation between hydrophobic free energy and aqueous cavity area proposed by Hermann (1972) and by Harris et al. (1973) have both been critically examined by Reynolds et al. (1974). Whereas Hermann used the absolute cavity surface areas, Harris et al. correlated the free energy of transfer with the relative surface areas. Assuming that the surface area of CH_4 is 152.4 $Å^2$, a conversion factor of 192 $Å^2$ per relative area unit is obtained, which corresponds to 21 cal g-mole^{-1} $Å^{-2}$ at 25°C between hydrophobic free energy and the cavity surface area (Reynolds et al., 1974) for aliphatic hydrocarbons; for aromatic hydrocarbons the corresponding value is 20 cal g-mole^{-1} $Å^{-2}$. These figures can also be applied to cyclic and branched hydrocarbons and to series of n-alkanes.

A theoretical model for hydrophobicity has not yet been developed, and the reported relationship between free energy and cavity surface areas is purely empirical in nature (Rossky and Friedman, 1980). In correlating unitary free energy of transfer of hydrocarbons from the pure liquid to aqueous solution at 25°C with the cavity surface area, Hermann (1972) reported a slope of 33 cal g-mole^{-1} $Å^{-2}$, whereas Reynolds et al. (1974) obtained a value of 25 cal g-mole^{-1} $Å^{-2}$. In the case of Reynolds et al., the straight line does not lead to the zero area for a solute of zero hydrophobic free energy (see Fig. 2.56). The explanation might involve the experimental error involved or the theoretical reason for nonlinearity (Reynolds et al., 1974), (e.g., the shape of the cavity). Furthermore, the regularity in enthalpy and entropy values is not a condition for hydrophobic effect relatively to the contact areas. The hydrophobicity is connected with high heat capacity, which is explained by the presence of water molecules surrounding the solute, and it is possible that there are two or more states of nearly the same free energy but with different enthalpies and entropies (Anik, 1978).

The solubility of hydrocarbons in water was expressed by Cohen and Regnier (1976) as a function of the molecular geometry of the hydrocarbon in question:

$$-\log_{10} S = 0.00328\ S_C + 0.02044\ S_H - 2.24335$$

where S = solubility in water at 25°C, molarity

        S_C = surface area of carbon atom(s), $Å^2$

        S_H = surface area of hydrogen atom(s), $Å^2$

The geometries of any molecules can be calculated using a computer program called GEMO (Cohon, 1971). Alternatively, the molar area of a substance might be calculated at the liquid-vapor interface from the surface tension and energy of evaporation (Jain et al., 1976).

According to Cohen and Regnier, the proceding equation may be used to calculate the solubility of all hydrocarbons in water at 25°C. Unfortunately, there is no indication that further development, which would include a modified expression capable of calculating the solubility of halogenated hydrocarbons in water, is on the way.

The solute-solvent interaction, particularly in the case of hydrocarbon and

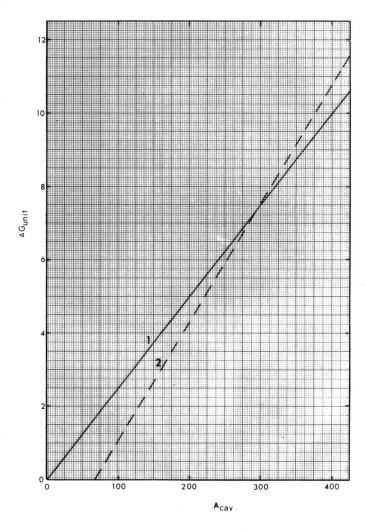

A_{cav}

Fig. 2.56 Free energy of transfer of hydrocarbons to aqueous solution at 25°C. ΔG_{unit} = unitary free energy of transfer (kcal g-mole^{-1}); A_{cav} = cavity surface area ($\mathring{A}^2$). 1, Hermann (1972; 2, Reynods et al. (1974). (From Reynolds et al., 1974.)

water interactions has been studied by Teresawa et al. (1975). Because of the weak interaction between hydrocarbons and solvent water molecules, the solute-solvent interaction partial molar volume $\bar{V}_{s-s}^{(HC)}$ is defined by the equation

$$\bar{V}_{s-s}^{(HC)} = (\bar{V}_{void} - \bar{V}_{void}^{(HC)})_{V_w = V_w^{(HC)}}$$

where $\bar{V}_{void}$ =   void partial volume

   $\bar{V}_{void}^{(HC)}$ =   void partial molar volume of the hydrocarbon

   V_w =   van der Waals volume

The physical meaning of $\bar{V}_{s-s}^{(HC)}$ refers to the additional void partial molar volume, which is caused by the hydrophilic molecular property of a solute, compared to a hydrocarbon with exactly the same value used as the solute. Furthermore, the numerical value of $\bar{V}_{s-s}^{(HC)}$ depends on the differences between the molecular properties of the solvent. The absolute values of $\bar{V}_{s-s}^{(HC)}$ show a trend toward slight decreases with upon lengthening the alkyl group or adding a methylene group for the alcohols and glycols. The $\bar{V}_{s-s}^{(HC)}$'s for alkyl bromides are given in Table 2.87 (Teresawa et al.,

Table 2.87   Solute-Solvent Interaction Partial Molar Volumes at 25°C

Solute	$\bar{V}_{s-s}^{(HC)}$ (cc g-mole^{-1})
Ethyl bromide	-2.0
n-Propyl bromide	-2.1
Allyl bromide	-1.3

Source: Teresawa et al., 1975.

1975). The solute-solvent interaction partial molar volume $\bar{V}_{s-s}^{(HC)}$ is calculated from the expression

$$\bar{V}_{s-s}^{(HC)} = \bar{V}^o - V_w - \bar{V}_{void}^{(HC)}$$

where $\bar{V}_{s-s}^{(HC)}$ =   void partial molar volume, ml g-mole^{-1}

   $\bar{V}^o$ =   partial molar volume, ml g-mole^{-1}

   V_w =   van der Waals volume, ml g-mole^{-1} (see Bondi, 1964)

The numerical values for these parameters for bromo compounds at 25°C are given in Table 2.88 (Teresawa et al., 1975). A closer examination of $\bar{V}^o$ values for several solutes in various solvents reveals that these values reflect the molecular properties of a solute in aqueous solution.

   The partial molal volume of benzene derivatives in water was reported by Perron and Desnoyers (1979) at 25°C. A group contribution method for both partial molal volume and partial molal heat capacity has been proposed.

   The original proposal of Langmuir (1925) for correlating solubility with molecular surface area has been studied by Uhling (1937), Eley (1939a, b), Hermann (1971,

Table 2.88 Various Volume Parameters of Bromo Compounds at 25°C

Compound	$\bar{V}^{o}$ (ml g-mole^{-1})	V_{w} (ml g-mole^{-1})	$\bar{V}^{(HC)}_{void}$ (ml g-mole^{-1})
Ethyl bromide	66.7	38.3	30.4
n-Propyl bromide	82.2	48.5	35.8
Allyl bromide	77.6	45.0	33.9

Source: Teresawa et al., 1975.

1972), and the method has continued to be improved by Amidon and co-workers (Amidon et al., 1974, 1975; Amidon and Anik, 1976); Yalkowsky, Flynn, and Amidon, 1972; Yalkowsky, Flynn, and Slunick, 1972; Yalkowsky et al., 1975, 1976; Yalkowsky and Zografi, 1972; Yalkowsky and Valvani, 1980; Valvani et al., 1976; Cohen and Connors, 1970; Cohen and Regnier, 1976; Anik, 1978; Samaha, 1979; Cammarata, 1979; Mackay et al., 1980). The principle of the model is based on the proposal that the number of water molecules that can be packed around a solute molecule is determinated by the molecular surface area of the solute. Therefore, the molecular surface area of a solute is a very important parameter for solubility calculations in water. In general, the total molecular surface area can be divided into polar and nonpolar components. Multiplying the respective surface areas by the appropriate free energy per unit area term will give the solubility value.

Amidon and co-workers used a slightly modified version of Hermann's (1972) computer program, which considers a molecule as a collection of spheres with radius located in the center of each sphere. In addition to the radius of the solute, the radius of water (1.5 Å) was used. The planar section of the total surface for ethyl chloride is shown schematically in Fig. 2.57. The molecular dimensions were calculated from the van der Waals radii (C = 1.6 Å, H = 1.2 Å, O = 1.4 Å, F = 1.35 Å, Cl = 1.8 Å, Br = 1.95 Å, I = 2.15 Å), bond angles, and distances. The total area of a molecule has been taken as the sum of the group contributions (e.g., hydrocarbon- and chlorine-group portions for chlorinated hydrocarbons).

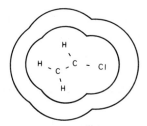

Fig. 2.57 Schematic picture of the total surface of ethyl chloride.

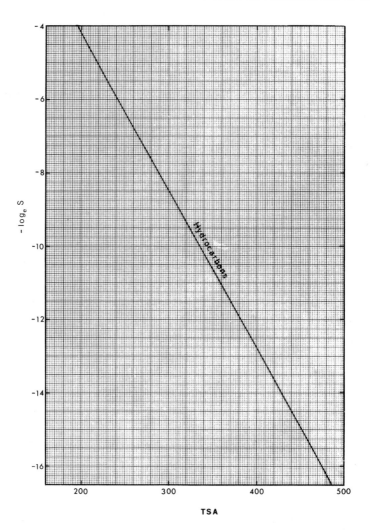

Fig. 2.58 Correlation of aqueous solubility with total surface area at 25°C. S = solubility of hydrocarbons in water (g-mole liter^{-1}); TSA = total surface area of solute ($\AA^2$). (From Amidon et al., 1974.)

Regression analyse have been performed by Amidon and co-workers on a large number of substances, particularly for hydrocarbons and alcohols. The presented expression is

$$\log_e S \ \ = \ -0.043 \ \text{HYSA} \ + \ 8.0031 \ \text{IOH} \ - \ 0.0586 \ \text{OHSA} \ + \ 4.420$$

where S   = molar solubility in water (see Fig. 2.58)
       TSA = total surface area, $\AA^2$
       OHSA = surface area of the OH group, $\AA^2$
       IOH = equals unity for alcohols and zero for hydrocarbons

The foregoing equation is illustrated in Fig. 2.58 for hydrocarbons. Furthermore, the surface area of hydrocarbons is expressed by

$$HYSA = TSA - OHSA$$

According to Amidon and co-workers, the correlated data provide an excellent estimate of the solubilities of alcohols and hydrocarbons in water at $25^{\circ}C$.

The great advantage of the method discussed is that no additional correction is required for branching, cyclization, positional isomers, and so on.

The molecular-surface approach for calculating aqueous solubilities at $25^{\circ}C$ has been extended from hydrocarbons and alcohols (Amidon et al., 1974; Valvani et al., 1976; Amidon and Anik, 1976) to various types of substances: amines (Yalkowsky et al., 1975, 1976), ethers, ketones, aldehydes, esters, carboxylic acids, and olefins (Amidon et al., 1975; Samaha, 1979), and polychlorinated biphenyls (Mackay et al., 1980). However, none of the articles listed dealt with aliphatic halogenated hydrocarbons. In a personal communication, Amidon (1976) proposed the following expression for chlorinated hydrocarbons:

$$\log_e S = -0.04922 \text{ HYSA} - 0.0436 \text{ CLSA} + 1.086 \text{ DM} + 6.339$$

where CLSA = total chlorine surface area, $\overset{o}{A}^2$
      DM = dipole moment of the solute, debyes
      HYSA = TSA - CLSA

Values for chlorinated hydrocarbons supplied by Amidon (1976) are reproduces in Table 2.89.

The average error for 25 solubilities investigated was 16.43%. The maximum error was 249.7%. The conclusion regarding the method is that it is simple if one has access a computer program for calcualting molecular surface areas, but that otherwise it is a rather time-consuming task to derive them. The error involved in the calculations are a little too high. Further improvement should be welcome for halogenated hydrocarbons.

In a recent article, Yalkowsky et al. (1979) reported the solubility of 26 mono- and multi-halogenated benzenes in water at $25^{\circ}C$. The molar solubility (S_w) was correlated with the melting points (MP) and the total surface areas (TSA) according to the expression

$$\log_{10} S_w = -0.0095(\text{MP}) - 0.0282(\text{TSA}) + 1.42$$

or alternatively if the octanol-water partition coefficient (PC) is used

$$\log_{10} S_w = -0.01(\text{MP}) - 0.88 \log_{10}(\text{PC}) - 0.012$$

In both cases the correlation coefficient is greater than 0.9877 and the standard

Table 2.89 Surface-Area Parameters for the Calculation of Solubility in Water at 25°C

Solute	Molecular weight	Total surface area ($\mathring{A}^2$)	Chlorine surface area ($\mathring{A}^2$)	Dipole moment (debyes)	Solubility in water at 25°C (molar) Experimental	Calculated	Error (%)
CCl_4	153.82	260.69	260.69	0	0.005	0.00656	31.2
$CHCl_3$	119.38	238.68	215.18	1.07	0.0645	0.04796	-25.7
CH_2Cl_2	84.93	213.49	157.64	1.54	0.191	0.19977	4.6
$CHCl_2-CCl_3$	202.30	302.47	284.84	0.92	0.00247	0.00212	-14.1
CCl_3-CCl_3	236.74	318.07	318.07	0	0.000211	0.00054	154.7
CCl_3-CH_2Cl	167.85	284.48	242.63	1.2	0.00655	0.00676	3.3
$CHCl_2-CHCl_2$	167.85	286.73	251.52	1.29	0.0173	0.00702	-59.4
CH_3-CCl_3	133.41	263.93	189.91	2.03	0.00974	0.03406	249.7
$CHCl_2-CH_2Cl$	133.41	266.48	202.05	1.25	0.0345	0.01379	-60.0
CH_3-CHCl_2	98.96	242.97	139.06	2.07	0.0505	0.07499	48.5
CH_2Cl-CH_2Cl	98.96	246.27	152.68	1.34	0.0869	0.03114	-64.2

$CCl_3-CCl_2-CCl_3$	319.66	330.03	330.03	0.46	0.000487	0.00053	7.9
$CH_2Cl-CHCl-CH_2Cl$	147.43	287.51	189.44	1.63	0.0129	0.00689	-46.6
$CH_3-CHCl-CH_2Cl$	112.99	268.64	135.44	1.60	0.0248	0.01247	-49.7
$CH_2Cl-CH_2-CH_2Cl$	112.99	272.14	143.30	2.09	0.0228	0.01867	-18.1
$CH_2Cl-CH_2-CH_3$	78.54	249.88	75.39	2.05	0.03183	0.03651	14.7
$CH_3-CHCl-CH_3$	78.54	246.38	67.53	2.17	0.03819	0.04728	23.8
$CH_3-CH_2-CH_2-CHCl_2$	127.03	301.57	130.68	2.21	0.00393	0.00466	18.5
$CH_3-CHCl-CHCl-CH_3$	127.03	291.11	118.32	2.22	0.00275	0.00735	167.1
$CH_3-CH_2-CH_2-CH_2Cl$	92.57	282.62	67.34	2.06	0.00712	0.00741	4.0
$CH_3-CH_2-CHCl-CH_3$	92.57	273.22	59.16	2.14	0.01080	0.01165	7.9
$CH_2Cl-CH(CH_3)-CH_3$	92.57	267.98	61.03	2.06	0.00994	0.01397	40.6
$CH_2Cl-CH_2-CH_2-CH_2-CH_3$	106.60	314.40	76.29	2.14	0.00187	0.00169	-9.6
$CH_3-CHCl-CH_2-CH_2-CH_3$	106.60	305.03	59.14	1.92	0.00234	0.00192	-18.1
$CH_3-CH_2-CHCl-CH_2-CH_3$	106.60	300.08	50.76	1.92	0.00234	0.00233	-0.3

Average: 16.43

error is less than 0.271. Furthermore, Yalkowsky and co-workers proposed the following relationship between molar solubility (S_w) and the activity coefficient of
solute in water (γ_w):

$$\log_{10} S_w = -0.01 \ (MP) - 25 \ - \log_{10} \gamma_w$$

The above equations were also used by Yalkowsky and Valvani (1979) for correlating the aqueous molar solubility of 31 polycyclic aromatic hydrocarbons and indan.
The authors also presented a relationship between total surface area (TSA) and
octanol-water partition coefficient (PC):

$$\log_{10} (PC) = 0.0303(TSA) - 1.389$$

The method of calculating molecular surface areas has been computerized by
Hermann (1972) and slightly modified by Amidon (1974). Further discussion on the
method was reported by Valvani et al. (1976). As mentioned above, the method of
calculation is not simple without the computer program, and it is too cumbersome for
a scientist or engineer to evaluate a single value that he needs for his work. However,
the method is suitable for the investigation of a large number of substances with
various characteristics (Anik, 1978).

For a quick and practical application, the surface area of a typical series of
substances can be successfully correlated with molecular weight. A few examples are
presented in Fig. 2.59 and in Tables 2.90 to 2.93. It seems from the correlated
substances that the aliphatic series produce parallel lines and that the cyclic compounds show a different slope from the others. This simple relationship could be used
to approximate surface areas from least-squares fitting of the members of a typical
series. The first-degree polynomonals of four series of substances are given in Table
2.94.

The relationship between the aqueous solubility $(S, \ \mu\text{mole liter}^{-1})$ and the
experimental n-octanol-water partition coefficients (K) was studied by several workers. Chiou et al. (1977) reported the following empirical equation

$$\log_{10} K = 5.00 - 0.670 \log_{10} S$$

for a wide variety of chemicals including aliphatic and aromatic hydrocarbons, aromatic acids, organochlorine and organophosphate pesticides, arid polychlorinated biphenyls.
The magnitude of the solubility varied from 10^{-3} to 10^{4} ppm, whereas the magnitude
in the partition coeffcient was between 10 and 10^{7}. The correlation of the bio-
concentration factor with solubility was also included in the investigation.

The aqueous solubility of many organic compounds can be predicted from their
melting points, entropies of fusion, and octanol-water partition coefficients
(Yalkowsky, 1977). The melting point of a large number of disubstituted benzenes were
reported by Martin et al. (1979). They showed that the entropy of fusion ΔS_f is in-

Table   2.90   Surface Areas of Paraffin Series as Function of Molecular Weight

Solute molecule	Molecular weight	Total surface area ($Å^2$) Calculated	Correlated	Deviation (%)
Methane	16.04	152.36	156.514	-2.65
Ethane	30.07	191.52	189.029	1.32
Propane	44.11	223.35	221.566	0.80
n-Butane	58.13	255.2	254.058	0.45
n-Pentane	72.15	287.0	286.55	0.16
n-Hexane	86.18	319.0	319.064	-0.02
n-Heptane	100.21	351.0	351.579	-0.16
n-Octane	114.23	383.0	384.071	-0.28
n-Nonane	128.26	418.8[a]	–	–
n-Decane	142.29	446.7[a]	–	–

[a]Extrapolated values from the correlated equation.

Table   2.91   Surface Areas of n-Alcohol Series as Function of Molecular Weight

Solute molecule	Molecular weight	Total surface area ($Å^2$) Calculated	Correlated	Deviation (%)
n-Butanol	74.12	272.1	271.836	0.097
n-Pentanol	88.15	303.9	303.754	0.048
n-Hexanol	102.18	335.7	335.672	0.008
n-Heptanol	116.21	367.5	367.59	-0.025
n-Octanol	130.23	399.4	399.486	-0.021
n-Nonanol	144.26	431.2	431.404	-0.047
n-Decanol	158.29	463.0	463.322	-0.069
n-Dodecanol	186.32	527.0	527.09	-0.017
n-Tetradecanol	214.40	591.0	590.972	0.005
n-Pentadecanol	228.43	623.0	622.89	0.018
n-Hexadecanol	242.45	655.0	654.785	0.033

Table 2.92 Surface Areas of 2-Alcohol Series as Function of Molecular Weight

Solute molecule	Molecular weight	Total surface area ($\mathring{A}^2$) Calculated	Correlated	Deviation (%)
2-Butanol	74.12	264.1	264.086	0.005
2-Pentanol	88.15	295.9	295.879	0.007
2-Hexanol	102.18	327.7	327.672	0.009
2-Heptanol	116.21	359.5[a]	–	–
2-Octanol	130.23	391.0	391.235	-0.060
2-Nonanol	144.26	423.2	423.028	0.041
2-Decanol	158.29	454.8[a]	–	–

[a]Interpolated or extrapolated values from the correlated equation.

Table 2.93 Surface Areas of Monocyclic Hydrocarbon Series as Function of Molecular Weight

Solute molecule	Molecular weight	Total surface area ($\mathring{A}^2$) Calculated	Correlated	Deviation (%)
Cyclopropane	42.08	211.2[a]	–	–
Cyclobutane	56.12	233.7[a]	–	–
Cyclopentane	70.14	255.40	256.103	-0.274
Cyclohexane	84.16	279.10	278.53	0.205
Cycloheptane	98.19	301.94	300.972	0.322
Cyclooctane	112.22	322.58	323.415	0.258
Cyclononane	126.25	345.9[a]	–	–
Cyclodecane	140.28	368.3[a]	–	–

[a]Extrapolated values from the correlated equation.

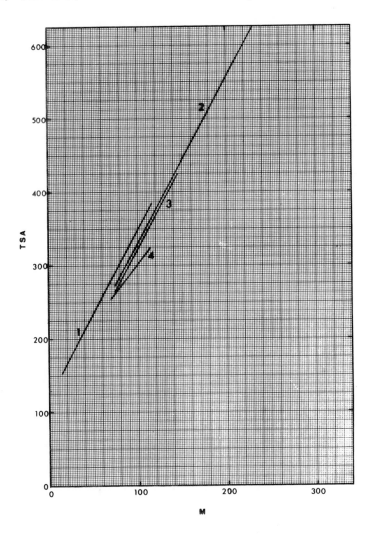

Fig. 2.59 Molecular surface areas as function of molecular weight. TSA = total surface area (Å^2); M = molecular weight. 1, Paraffins; 2, n-alcohols; 3, 2-alcohols; 4, monocyclic hydrocarbons.

dependent of the number of hydrogen bonds, dipole moment, and other measures of attractive electronic factors. It is approximately constant, having a value of 12.7 ± 2 cal g-mole^{-1} K^{-1}. However, Yalkowsky (1979) showed that ΔS_f equals to 13.5 cal g-mole^{-1} K^{-1} for rigid molecules and 13.5 + 2.5(n - 5) cal g-mole^{-1} K^{-1} for flexible molecules, where n is the total number of chain atoms (exclusive of protons).

Table 2.94 Correlation of Molecular Surface Areas of Typical Series with Molecular Weight

Series	Polynominal	Standard error of estimate
Methane – n–Decane	x ($Å^2$) = 119.341 + 2.31752 (mol. wt.)	2.22244
n–Butanol – n–Hexadecanol	x ($Å^2$) = 103.214 + 2.27499 (mol. wt.)	0.188449
2–Butanol – 2–Decanol	x ($Å^2$) = 96.1266 + 2.26605 (mol. wt.)	0.169594
Cyclopropane – Cyclododecane	x ($Å^2$) = 143.906 + 1.59962 (mol. wt.)	1.10748

2.7-6 Calculation of Solubility from Equation of State

Since the introduction of the equation of state for the representation of pressure-volume-temperature (P-V-T) relationships in single-compound systems, several attempts have been made to extend their application to binary and multicomponent systems. The various proposals have been more or less successful, but the reliability of the equation of the state or its accuracy for mixtures has not been satisfactory until recently, particularly when dealing with liquid-liquid immiscibility. This is due to the lack of a good understanding of the liquid state of the compounds. Consequently, it became popular to develop empirical equations or semiempirical expressions to fit P-V-T data. In mixtures, the interaction parameters have been introduced with many modified versions. The success of an equation of state for P-V-T-x data depends upon the reliability of the mixing rule applied (Reid et al., 1977).

Pesuit (1978) reported a good compilation on the binary interaction constants for mixtures with a wide range in component properties. The calculation of the mixed virial coefficients (B_{12}) and also the mixed coefficients (a_{12}) of the Redlich-Kwong equation of state for the calculation of the solubility of a nonpolar gas in water was reported by Skipka (1979b).

Whereas the coefficients of an equation of state are related to molecular interaction between the same sorts of molecules, the mixing rules propose to deal with the intermolecular forces and interactions between unlike molecules, such as those of polar and nonpolar substances.

The phase behavior of mixtures of water and nonpolar substances was studied by Namiot et al. (1976) using the Redlich-Kwong equation of state. The cross-interaction coefficient a_{12} has a significant influence upon the results of calculations on phase equilibria in the three-phase region and at the critical curve. Whereas the dipole-dipole interaction is strong in the system of polar molecules, in water and nonpolar systems of molecules there is a weaker inductive interaction. Consequently, the observed value of a_{12} is appreciably lower than the geometrical mean of a_1 and a_2. The fugacity of component 1 (f_1) has been derived from the expression

$$\log_e f_1 = \log_e(p\, x_1) + \log_e \frac{V}{V-b} + \frac{b_1}{V-b} + \log_e \frac{R\,T}{P\,V}$$

$$- \frac{2\left[x_1\, a_1 + (1-x_1)\, a_{12}\right]}{b\, R\, T^{3/2}} \log_e \frac{V+b}{V} + \frac{a\, b_1}{b^2\, R\, T^{3/2}}(\log_e \frac{V+b}{V} - \frac{b}{V+b})$$

where a_1 and b_1 =   Redlich-Kwong coefficients of component 1

         x_1 and x_2 =   mole fraction of components 1 and 2, respectively

         P, V, T   =   pressure, volume, and temperature, respectively

            a = $a_1\, x_1^2 + 2\, a_{12}\, x_1\, x_2 + a_2\, x_2^2$

            b = $b_1\, x_1 + b_2\, x_2$

A similar expression is applicable for the fugacity of component 2 (f_2).

Several binary systems of water with nonpolar substances have been investigated by Namiot et al. (1976), Wenzel and Rupp (1979), Skripka (1979b), and Baumgaertner et al. (1980). The aqueous binary systems contained helium, neon, argon, nitrogen, carbon dioxide, methane, ethane, propane, n-butane, n-pentane, n-hexane, cyclohexane, benzene, and so on. Despite the selection of coefficient a_{12} so that one of the calculated parameters agrees with the corresponding experimental value, the adequacy of the quantitative description of the phase behavior of the binary aqueous systems by the Redlich-Kwong equation of stste was strongly criticized by Namiot et al. (1976). They found that the representation of the experimental data is rather poor. However, it has been shown that the Redlich-Kwong equation gives a better representation than does the van der Waals equation for water-nonpolar compound systems. Therefore, there are several proposals for modifications of the Redlich-Kwong equation of state for the immiscible systems under consideration.

Because of the inadequacy of the original Redlich-Kwong equation of state for polar-nonpolar compound systems, de Santis et al. (1974) expressed the coefficient a in the equation

$$P = \frac{R T}{V - b} - \frac{a(T)}{T^{\frac{1}{2}}(V + b) V}$$

as a function of temperature

$$a(T) = a^{(0)} + a^{(1)}(T)$$

for the representation of aqueous gas mixtures in the region 25 to 700°C and pressures up to 1500 bars. However, even this modification of the equation of state could provide only an approximate representation of the volumetric properties of gaseous water. In the expression above, $a^{(0)}$ is responsible for the intermolecular attraction due to dispersion (London) forces, and the temperature-dependent coefficient $a^{(1)}$ reflects intermolecular attraction due to hydrogen bonds, permanent dipoles, and quadrupoles. For simple nonpolar gases, $a^{(1)} = 0$.

When the gaseous mixtures contained water, carbon dioxide, argon, nitrogen, methane, ethane, or n-butane, the following mixing rules were applied:

$$b = y_1 b_1 + y_2 b_2$$

$$a = y_1^2 a_1 + y_2^2 a_2 + 2 y_1 y_2 a_{12}$$

where a_{12} represents the intermolecular forces between dissimilar molecules, and calculated according to type of mixture:

Component 1 (a_1)	Component 2 (a_2)	a_{12}
Nonpolar	Nonpolar	$(a_1 a_2)^{\frac{1}{2}}$
Nonpolar	H_2O, CO_2	$(a_1 a_2^{(0)})^{\frac{1}{2}}$
Water	CO_2	$(a_1^{(0)} a_2^{(0)})^{\frac{1}{2}} \dfrac{R^2 T^{5/2} K}{2}$

where K is an equilibrium constant for the reaction

$$H_2O \quad + \quad CO_2 \quad \rightleftharpoons \quad complex$$

The vapor-phase composition y_1 was calculated from the relationship

$$y_1 \quad = \quad \frac{(1 - x_2) f_{pure\ 1}}{\phi_1 P}$$

where $f_{pure\ 1}$ = fugacity of pure component 1

x_2 = liquid-phase mole fraction, or solubility

ϕ_1 = fugacity coefficient of component 1

P = total pressure of the system

The method reported here is not higly accurate, but it can provide a good estimate for aqueous gas mixtures.

The correlation of the phase behavior in the systems $H_2S - H_2O$ and $CO_2 - H_2O$ was studied by Evelein et al. (1976), using Soave's modification of the Redlich-Kwong equation of state. Furthermore, the coefficients a and b were further adjusted for water; for component i they are

$$a_i(T) \quad = \quad 0.42742 \frac{R^2 T_{cr_i}^2}{P_{cr_i}} \left[1 + m_i (1 - T_{r_i}^{0.5}) \right]^2$$

where m_i is a function of the acentric factor ω_i:

$$m_i \quad = \quad 0.480 \ + \ 1.574\ \omega_i \ - \ 0.176\ \omega_i^2$$

and

$$b_i \quad = \quad 0.08664 \frac{R\ T_{cr_i}}{P_{cr_i}}$$

The following mixing rules were used for the aqueous mixtures:

$$a = \sum_j \sum_i (a_i a_j)^{0.5} (1 - k_{ij}) x_i x_j$$

and

$$b = \sum_j \sum_i \frac{b_i + b_j}{2} (1 - c_{ij}) x_i x_j$$

where k_{ij} and c_{ij} are empirical interaction parameters and are independent of temperature. These values were calculated by Evelein et al. (1976):

System	k_{ij}	c_{ij}
$H_2S - H_2O$	0.163	0.08
$CO_2 - H_2O$	0.280	0.195

With the aid of the foregoing mixing rules, the fugacity of the ith component was obtained by

$$\log_e\left(\frac{f_i}{x_i P}\right) = -\log_e(Z - B) + \left[\frac{\sum_j x_j (1 - c_{ij})(B_i + B_j)}{B} - 1\right]\left[Z - 1 + \frac{A}{B} \log_e\left(\frac{Z + B}{B}\right)\right]$$
$$- \frac{2 A_i^{0.5} \sum_j x_j (1 - k_{ij}) A_j^{0.5}}{B} \log_e\left(\frac{Z + B}{2}\right)$$

where

$$A = \frac{a P}{R^2 T^2}$$

$$B = \frac{b P}{R T}$$

$$Z = \text{compressibility factor}$$

The calculated three-phase pressures agrees very well with the experimental values for the two binary aqueous systems mentioned. The three-phase lines were predicted in close agreement with the experimental data from 70 atm at 311 K to 200 atm and 444 K for the $H_2S - H_2O$ system and from 500 atm at 298 K to 1000 atm at 523 K. However, at higher pressures, the inadequacy of the result becomes apparent.

De Mateo and Kurata (1975) used the Redlich-Kwong equation of state to correlate and predict the solubility of solid hydrocarbons in liquid methane with sufficient accuracy. The dimensionless Redlich-Kwong coefficients Ω_a and Ω_b were found by solving the simultaneous system of equations by trial and error in a saturated condition:

$$P = \frac{R\,T}{V - b} - \frac{a}{T^{\frac{1}{2}}(V + b)\,V}$$

$$\log_e\left(\frac{f}{P}\right) = (Z - 1) - \log_e\left(Z - \frac{b\,P}{R\,T}\right) - \frac{a}{b\,T^{3/2}}\log_e\left(1 + \frac{b\,P}{Z\,R\,T}\right)$$

where $\quad Z = \dfrac{P\,V}{R\,T}$

and the coefficientas a and b were calculated from the critical properties:

$$a = \Omega_a\frac{R^2\,T_{cr}^{2.5}}{P_{cr}}$$

$$b = \Omega_b\frac{R\,T_{cr}}{P_{cr}}$$

The Ω_a and Ω_b coefficients obtained were fitted and expressed by fifth-degree polynominal equations. The fugacity coefficient was calculated from the fugacity of the saturated vapor and the virial equation of state. The properties of mixtures were derived from the mixing rules:

$$a = x_1^2\,a_1 + x_2^2\,a_2 + 2\,x_1\,x_2\,a_{12}$$

and

$$b = x_1\,b_1 + x_2\,b_2$$

where

$$a_{12} = \left(\frac{\Omega_1 + \Omega_2}{2}\right)\frac{R^2\,T_{cr_{12}}^{2.5}}{P_{cr_{12}}}$$

$$T_{cr_{12}} = (1 - k_{12})(T_{cr_1}\,T_{cr_2})^{\frac{1}{2}}$$

$$P_{cr_{12}} = \frac{1}{2}(Z_{cr_1} + Z_{cr_2})\frac{R\,T_{cr_{12}}}{V_{cr_{12}}}$$

$$V_{cr_{12}} = \frac{1}{8}(V_{cr_1}^{1/3} + V_{cr_2}^{1/3})^3$$

$$k_{12} = \text{interaction parameter, computed from experimental data of mixtures}$$

Finally, the fugacity coefficient of component 1 (f_1) was obtained from the expression

$$\log_e(\frac{f_1}{x_1 \, P}) \;=\; (Z - 1)\frac{b_1}{b} \;-\; \log_e(Z - \frac{b \, P}{R \, T})$$

$$-\; \frac{a}{R \, T^{3/2} \, b} \left[\frac{2 \, (x_1 \, a_1 + x_2 \, a_{12})}{a} - \frac{b_1}{b} \right] \log_e(1 + \frac{b \, P}{Z \, R \, T})$$

De Mateo and Kurata (1975), using an iterative procedure, calculated the solubility of solid hydrocarbons, such as n-butane, n-pentane, n-hexane, n-heptane, benzene, and toulene in liquid methane from the following equations:

1. The proceding equation
2. The Redlich-Kwong equation of state
3. The expression for fugacity given above following the Redlich-Kwong equation of state

$$\frac{f_1^s}{f_1^l} \;=\; \frac{f_1}{P \, x_1}(\frac{P}{f_1^l}) \, x_1$$

and

$$\log_e(\frac{f_1^s}{f_1^l}) \;=\; A \;+\; \frac{B}{T} \;+\; C \log_e T \;+\; D \, T \;+\; E \, T^2 \;+\; \frac{v_1^s - v_1^l}{R \, T_{triple}} \, P$$

The total pressure of the system was the vapor pressure of methane, except for the solubility of benzene in methane.

According to De Mateo and Kurata, this method of calculating solubility from the equation of state is sufficient for most technical purposes. However, no attempt to calculate phase behavior and mutual solubilities in halogenated hydrocarbon – water systems has been reported. It is most likely that some sort of modification would be needed, because of the iceberg formation around the molecules, which is peculiar to water.

Several investigators found that the original Redlich-Kwong equation of state cannot, without some sort of modification, provide an accurate description of phase behavior in binary or multicomponent systems with phase separation. Emphasis is directed toward the modified expressions for the Redlich-Kwong coefficients and the mixing rules. Good discussions on the various modifications and mixing rules were reported by Reid et al. (1977), Djordjevic et al. (1977), and Oellrich et al. (1977).

The three-phase equilibria of water in normal paraffins has been studied by Heidemann (1974) using Wilson's version of the Redlich-Kwong equation of state. He found that in the system methane – n-butane – water, both the liquid hydrocarbon and the vapor phase were predicted with satisfactory precision.

The a and b coefficients of the Redlich-Kwong equation are expressed by

$$b = \sum_i x_i b_i$$

and

$$a = \sum_i \sum_j x_i x_j a_{ij}$$

where

$$b_i = 0.0865 \frac{R T_{cr}}{P_{cr}}$$

and

$$a_{ij} = 2.467 b_i \left(\frac{T_{cr}^j k_{ij}}{T} \right)^{0.12} \left[1 + (1.45 + 1.62\ \omega_j)\left(\frac{T_{cr}^j k_{ij}}{T} - 1 \right) \right]$$

$$+ 2.467 b_j \left(\frac{T_{cr}^i k_{ij}}{T} \right)^{0.12} \left[1 + (1.45 + 1.62\ \omega_i)\left(\frac{T_{cr}^i k_{ij}}{T} - 1 \right) \right]$$

where k_{ij} is the interaction parameter, found by fitting binary data. For the fore-going system, k_{ij} = 0.50. The Redlich-Kwong equation was used for both the liquid and vapor phases. The fugacity of component i is obtained by

$$\log_e f_i = \log_e \left(\frac{x_i R T}{V - b} \right) + (Z - 1)\frac{b_i}{b} + (b_i \sum_j \sum_k x_j x_k a_{jk} \frac{1}{b^2}$$

$$-2 \sum_j x_j a_{ij} \frac{1}{b}) \log_e \left(\frac{V + b}{V} \right)$$

The condition for an equilibrium between two phases I and II is that

$$f_i^I = f_i^{II}$$

The interaction parameter k_{ij} has been derived from the mutual solubility data between n-butane and water. The optimized value of 0.5 can be used with resonably accuracy over the temperature range 300 to 400 K.

The great advantage of this approach is that a single equation of state is used to describe all three phases, that is, liquid-liquid-vapor equilibrium.

Masuoka et al. (1979) used the modified BWR equation of state of Lee and Kesler (1975) for calculating of solid-liquid equilibria, that is, the solubility of solid hydrocarbons and inorganics in liquid hydrocarbons. However, the procedure is not recommended for systems in which immiscible liquids form or solid solubility varies greatly with small temperature changes.

In a recent article Baumgaertner et al. (1980) proposed a modified van der Waals type equation for the calculation of phase equilibrium for aqueous systems with low mutual solubility. According to the investigators, the results obtained were satisfactory for different types of aqueous systems over a wide range of temperature and pressure.

2.7-7 Miscellaneous Estimation, Prediction, and Correlation Methods for Solubility
           Data

Aqueous solubility is a very important parameter for correlating other properties or behaviors in chemistry, biology, and pharmacology (see, e.g., Metcalf, 1962; Metcalf et al., 1975).

In a good review article, Wilhelm et al. (1977) pointed out the great practical advantages of the empirical and semiempirical correlations when a limited amount of information is available. They were particularly concerned with the application of semiempirical correlation of solubility, expressed as a mole fraction (x_2) with various properties, such as energy of vaporization at the normal boiling point of the solute gas:

$$R \, T \log_e x_2 \;=\; f(\Delta U_2^{vap})$$

polarizability α_2:

$$\log_e x_2 \;=\; f(\alpha_2)$$

hard-sphere diameter σ_2:

$$\log_e x_2 \;=\; f(\sigma_2)$$

normal boiling point $T_{b.p._2}$:

$$\log_e x_2 \;=\; f(T_{b.p._2})$$

and critical temperature T_{cr_2}:

$$\log_e x_2 \;=\; f(T_{cr_2})$$

The graphs presented reflect a considerable scatter of data for the solubility of gases in water at 1 atm partial pressure and 25°C. A wide scattering of data was also observed when the solubility of water in several halogenated hydrocarbons was plotted versus the heat of vaporization (see Fig. 2.60), critical temperature (see Fig. 2.61), and surface tension (see Fig. 2.62). These plots show that there is no simple or linear relationship between solubility and the physical properties mentioned.

Despite the considerable scatter shown by Wilhelm et al. (1977) in the solubility values when $\log_e x_2$ was plotted versus the normal boiling point of gases at 1 atm pressure and 25°C, Almgren et al. (1979) reported a linear correlation between the

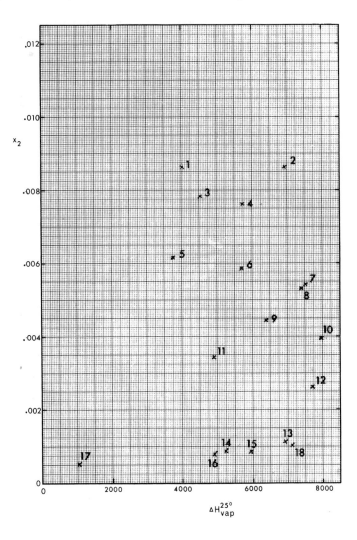

Fig. 2.60 Heat of vaporization versus solubility of water in halocarbons at 25°C.
x_2 = solubility of water (mole fraction); $\Delta H_{vap}^{25°}$ = latent heat of vaporization at 25°C
(cal g-mole^{-1}). 1, CCl_2F_2; 2, CH_2Cl_2; 3, CH_3Cl; 4, $CHCl_2F$; 5, $CHClF_2$; 6, C_2H_5Cl; 7,
$CHCl_3$; 8, $CHCl_2CH_3$; 9, $CHCl_2CF_3$; 10, $CHCl_2CCl_3$; 11, $CHCl=CH_2$; 12, CH_3CCl_3; 13,
CCl_2FCClF_2; 14, $CClF_2CClF_2$; 15, CCl_3F; 16, $CBrClF_2$; 17, $CClF_3$; 18, CCl_4.

logarithm of the aqueous solubility of aromatic hydrocarbons (S, mole dm^{-3}) and the
normal boiling point ($t_{b.p.}$, °C) of the arenes

$$\log_{10} S(\text{mole dm}^{-3}) = 0.0138\, t_{b.p.}(°C) + 0.76$$

In the investigation Almgren et al. (1979) correlated 32 aromatic hydrocarbons with

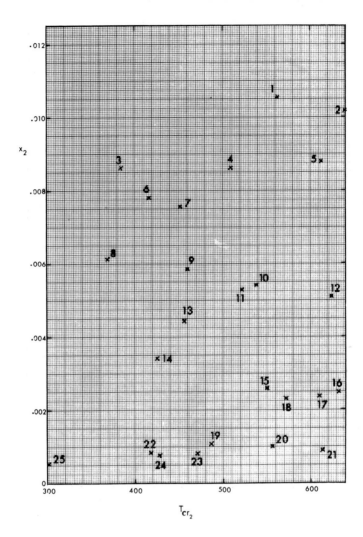

Fig. 2.61 Critical temperature versus solubility of water in halocarbons at 25°C.
x_2 = solubility of water (mole fraction); T_{cr2} = critical temperature (K). 1, CH_2ClCH_2Cl
2, $CHCl_2CHCl_2$; 3, CCl_2F_2; 4, CH_2Cl_2; 5, $CHCl_2CH_2Cl$; 6, CH_3Cl; 7, $CHCl_2F$; 8, $CHClF_2$;
9, C_2H_5Cl; 10, $CHCl_3$; 11, $CHCl_2CH_3$; 12, CH_2ClCCl_3; 13, $CHCl_2CF_3$; 14, $CHCl=CH_2$; 15,
CH_3CCl_3; 16, C_6H_5Cl; 17, C_6H_5Br; 18, $CHCl=CCl_2$; 19, CCl_2FCClF_2; 20, CCl_4; 21, $CCl_2=CCl_2$;
22, $CClF_2CClF_2$; 23, CCl_3F; 24, $CBrClF_2$; 25, $CClF_3$.

a reasonable success.

    Attempt has been made to correlate the aqueous solubilities of halogenated
benzenes at 25°C, reported by Yalkowsky et al. (1979). The equation obtained by a
linear regression

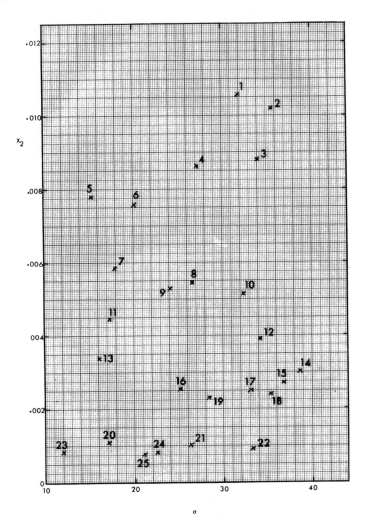

Fig. 2.62 Surface tension versus solubility of water in halocarbons at 25°C. x_2 = solubility of water (mole fraction); σ = surface tension (dyn cm^{-1}). 1, CH_2ClCH_2Cl; 2, $CHCl_2CHCl_2$; 3, $CH_2ClCHCl_2$; 4, CH_2Cl_2; 5, CH_3Cl; 6, $CHCl_2F$; 7, C_2H_5Cl; 8, $CHCl_3$; 9, $CHCl_2CH_3$; 10, CH_2ClCCl_3; 11, $CHCl_2CF_3$; 12, $CHCl_2CCl_3$; 13, $CHCl=CH_2$; 14, C_6H_5I; 15, o-$C_6H_4Cl_2$; 16, CH_3CCl_3; 17, C_6H_5Cl; 18, C_6H_5Br; 19, $CHCl=CCl_2$; 20, CCl_2FCClF_2; 21, CCl_4; 22, $CCl_2=CCl_2$; 23, $CClF_2CClF_2$; 24, CCl_3F; 25, $CBrClF_2$.

$$\log_{10} S(\text{mole dm}^{-3}) = -0.400347 - 0.015573\ t_{b.p.}(°C)$$

was very much different from the above expression. The comparison between the experimental and correlated values are shown in Table 2.94a. The percentage differences

Table 2.94a   Correlation of Molar Solubility of Aromatic Hydrocarbons with Their
             Normal Boiling Points

Solute	Yalkowsky et al. (1979)		Correlated[a]	Deviation
	$\log_{10}S$	S	S	(%)
Benzene, pentachloro-	-5.65	2.24×10^{-6}	1.93×10^{-5}	762
1,2,3,4-tetrachloro-	-4.70	2.00×10^{-5}	4.41×10^{-5}	121
1,2,3,5-tetrachloro-	-4.79	1.62×10^{-5}	5.87×10^{-5}	262
1,2,4,5-tetrachloro-	-5.56	2.75×10^{-6}	6.20×10^{-5}	2155
1,2,4-tribromo-	-4.50	3.16×10^{-5}	2.08×10^{-5}	-34
1,2,3-trichloro-	-3.76	1.74×10^{-4}	1.57×10^{-4}	-10
1,2,4-trichloro-	-3.72	1.91×10^{-4}	1.88×10^{-4}	-2
1,2-dibromo-	-3.50	3.16×10^{-4}	1.25×10^{-4}	-60
1,3-dibromo-	-3.38	4.17×10^{-4}	1.60×10^{-4}	-62
1,4-dibromo-	-4.07	8.51×10^{-5}	1.57×10^{-4}	84
1,2-dichloro-	-3.20	6.31×10^{-4}	6.15×10^{-4}	-3
1,3-dichloro-	-3.09	8.13×10^{-4}	8.05×10^{-4}	-1
1,4-dichloro-	-3.21	6.17×10^{-4}	7.76×10^{-4}	26
1,3-difluoro-	-2.00	1.00×10^{-2}	2.03×10^{-2}	103
1,4-difluoro-	-1.97	1.07×10^{-2}	1.64×10^{-2}	53
1,2-diiodo-	-4.24	5.37×10^{-5}	1.40×10^{-5}	-74
1,3-diiodo-	-4.57	2.69×10^{-5}	1.45×10^{-5}	-46
bromo-	-2.64	2.29×10^{-3}	1.48×10^{-3}	-35
chloro-	-2.35	4.47×10^{-3}	3.50×10^{-3}	-22
fluoro-	-1.79	1.62×10^{-2}	1.88×10^{-2}	16
iodo-	-2.95	1.12×10^{-3}	4.65×10^{-4}	-58
...	-1.64	2.29×10^{-2}	2.25×10^{-2}	-2
1-bromo-2-chloro-	-3.19	6.46×10^{-4}	2.65×10^{-4}	-59
1-bromo-3-chloro-	-3.21	6.17×10^{-4}	3.53×10^{-4}	-43
1-chloro-2-iodo-	-3.54	2.86×10^{-4}	8.87×10^{-5}	-69
1-chloro-3-iodo-	-3.55	2.82×10^{-4}	1.04×10^{-4}	-63
1-chloro-4-iodo-	-4.03	9.33×10^{-5}	1.16×10^{-4}	24

Average: 110

[a] $\log_{10}S = -0.400347 - 1.55726 \times 10^{-2} t_{b.p.}$ (°C)

Source: Almgren et al., 1979.

were too high for practical application.

The solubility of water in several normal alkanes from C_7 to C_{16} has been studied by Schatzberg (1963). Three linear correlations were presented. First, the solubility expressed as a mole fraction was plotted versus the square of the solubility parameter difference on a semilogarithmic scale, after Hildebrand and Scott (1950). The second correlation was based on Maxwell-Boltzmann distribution theory; that is, the Ostwald solubility coefficient for water, plotted versus the surface tension of solvents, should produce a straight line on a semilogarithmic scale. The interaction energy (E), calculated from the Maxwell-Boltzmann equation,

$$\log_{10} \gamma = - \frac{4 \, \Pi \, r^2 \, \sigma}{2.303 \, k \, T} + \frac{E}{2.303 \, k \, T}$$

was -782 and -485 cal g-mole^{-1} at 25 and 40°C, respectively. According to Le Chatelier's rule, the negative interaction energies account in part for the increase in solubility of water with increasing temperature. In the expression above, γ is the Ostwald solubility coefficient, r the radius of the gas molecule, σ the surface tension of the solvent, k the Boltzmann constant, and T the absolute temperature in degrees Kelvin. The water radii, also calculated from the equation, were 1.37 and 1.36 Å at 25° and 40°C, respectively. In the third method of correlation, the solubility of water, expressed as a mole fraction or as parts per million (ppm) by weight was plotted on semilogarithmic paper against the alkane molecular weight. This was discussed in Sec. 2.7-2.

Ostwald's solubility coefficient has also been related to the surface tension and other molecular parameters for inert gases in polar and nonpolar solvents (Kuznetsova, 1975). The expressions derived from the theoretical considerations result in solubility data in satisfactory agreement with experimental values. The same conclusion has been reached by Saylor and Battino (1958) regarding the solubility of the rare gases in simple benzene derivatives. That is, there is a linear relationship between the logarithm of Ostwald's coefficient and the surface tension of the solvent at 25°C.

The solubility of completely halogenated hydrocarbons in water is much less than that of partially halogenated hydrocarbons. By increasing the number of hydrogen atoms in the molecule through replacing the halogen atoms, the solubility will increase, sometimes several orders of magnitude. To understand the trend in solubility processes, the existence of a C-H←F bond in fluorinated hydrocarbons has been studied (Zellhoefer et al., 1938; Marvel et al., 1940) in relation to the oxygen dipole of the water molecule (Pierotti, 1965). On the basis of the theoretical models proposed by Hirschfelder et al. (1964) and Pierotti (1965), the solubility of halogenated hydrocarbons in water at 25°C and 1 atm pressure was correlated with molar polarizability on a semilogarithmic scale (Stepakoff and Modica, 1973). A definite

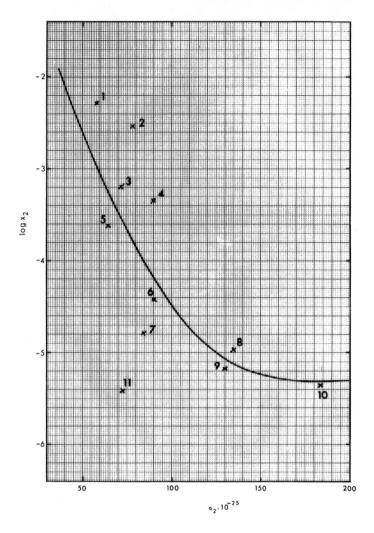

Fig. 2.63 Correlation of water solubility with molar polarizability at 25°C. x_2 = solubility in water (mole fraction); α_2 = molar polarizability (cc). 1, CH_2ClF; 2, $CHCl_2F$; 3, $CHClF_2$; 4, CH_3CClF_2; 5, CHF_3; 6, CCl_2F_2; 7, $CClF_3$; 8, $C_2Cl_2F_4$; 9, $CClF_2CF_3$; 10, C_4F_8; 11, CF_4. (From Stepakoff and Modica, 1973.)

trend was shown, but the method for correlating or predicting aqueous solubilities of Freons or chlorinated hydrocarbons will not produce accurate-enough solubility data (see Figs. 2.63 and 2.64 and Tables 2.95 and 2.96).

Bushinskii et al. (1974) introduced a correlation method between the solubility of organic solids (S, g cc^{-1}) in liquids and the free energy of intermolecular interaction of the solvent (E, cal g-mole^{-1}):

Table 2.95 Correlation of Solubility of Halocarbons in Water with Molar Polarization at 25°C

Freon No.	Formula	Solubility in H_2O at 25°C, x_2	$\log_{10} x_2$	Molar polarization (cc $\times 10^{25}$)
12	CCl_2F_2	3.9216×10^{-5}	-4.41	90
13	$CClF_3$	1.6556×10^{-5}	-4.78	84
14	CF_4	3.8911×10^{-6}	-5.41	72
22	$CHClF_2$	6.2893×10^{-4}	-3.20	72
23	CHF_3	2.4631×10^{-4}	-3.61	65
115	$CClF_2CF_3$	6.8027×10^{-6}	-5.17	130
318	C_4F_8	4.5045×10^{-6}	-5.35	184
114	$C_2Cl_2F_4$	1.0941×10^{-5}	-4.96	135
31	CH_2ClF	5.2910×10^{-3}	-2.28	58
21	$CHCl_2F$	2.9326×10^{-3}	-2.53	78
142b	CH_3CClF_2	4.4643×10^{-4}	-3.35	90

Source: Stepakoff & Modica, 1973.

Table 2.96 Solubility of Halocarbons in Water Versus Polarizability at 25°C

Solute	Molar polarization (cc $\times 10^{25}$) [a]	Solubility in water at 25°C, x_2	$\log_{10} x_2$
CH_3Cl	45.6	0.001907	-2.7196
CH_3Br	55.5	0.002925	-2.5339
CH_2Cl_2	64.8	0.002793	-2.5539
$CHCl_3$	82.3	0.001203	-2.9197
CCl_4	105.0	0.00009298	-4.0316
C_2H_5Cl	64.0	0.001573	-2.8033
C_6H_5Cl	122.5	0.00008819	-4.0546
$p-C_6H_4Cl_2$	144.7	0.00001099	-4.9590
$m-C_6H_4Cl_2$	142.3	0.00001521	-4.8179
$o-C_6H_4Cl_2$	141.7	0.00001135	-4.9450

[a] Landolt-Börnstein, 1951.

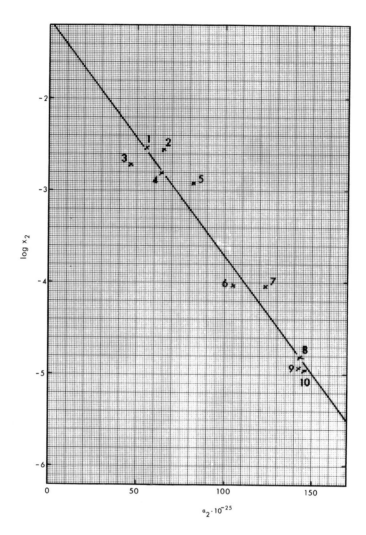

Fig. 2.64 Correlation of water solubility with molar polarizability at 25°C. x_2 = solubility in water (mole fraction); α_2 = molar polarizability (cc). 1, CH_3Br; 2, CH_2Cl_2; 3, CH_3Cl; 4, C_2H_5Cl; 5, $CHCl_3$; 6, CCl_4; 7, C_6H_5Cl; 8, $m\text{-}C_6H_4Cl_2$; 9, $o\text{-}C_6H_4Cl_2$; 10, $p\text{-}C_6H_4Cl_2$.

$$-\log_{10} S \;=\; A \;+\; B(E\, v_{mol})^{\frac{1}{2}}$$

where v_{mol} =   molar volume of the solvent at constant temperature

A and B   =   constants

The free energy of intermolecular interaction is calculated from the expression

$$E = R T \log_{10}\left(\frac{R T}{P v_{mol}}\right) - R T$$

This correlation has been used with success for the solubility of solid compounds in polar solvents, such as C_1 to C_6 alcohols, but its reliability has not been tested for aqueous solutions. The numerical data have been calculated for p-dichlorobenzene using the foregoing expressions (see Table 2.97) and are plotted in Fig. 2.65.

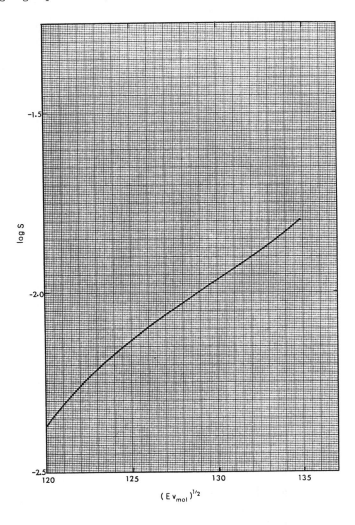

Fig. 2.65 Correlation of solubility with free energy of intermolecular interaction. S = solubility of p-dichlorobenzene in water (g/100 cc H_2O); E = free energy of inter-molecular interaction (cal g-mole^{-1}); v_{mol} = molar volume (cc g-mole^{-1}). (From Bushinskii et al., 1974.)

Table 2.97 Correlation of Solubility with Free Energy of Intermolecular Interaction for Solid p-Dichlorobenzene in H_2O

Temperature (°C)	Solubility of p-dichlorobenzene in H_2O, S (g/100 cc H_2O)	Molar volume of H_2O, v_{mol} (cc g-mole^{-1})	E (cal g-mole^{-1})	$(E \, v_{mol})^{\frac{1}{2}}$	$\log_{10} S$
0	0.00443	18.0177	802.7759	120.2671	-2.3536
10	0.00632	18.0202	840.9180	123.0996	-2.1993
20	0.00804	18.0472	879.0188	125.9517	-2.0947
25	0.00894	18.0682	898.0640	127.3829	-2.0487
30	0.00994	18.0935	917.1096	128.8166	-2.0026
40	0.01236	18.1561	955.1999	131.6917	-1.9080
50	0.01566	18.2329	993.2924	134.5756	-1.8052

Source: Bushinskii et al., 1974.

The plot is not a straight line between 0 and 50°C. A simplified version of the correlation was applied by Burgess and Haines (1978) for solubilities of derivatives of 1,10-phenanthroline.

Another relationship has been developed by Leland et al. (1955) between the equilibrium solubilities in very dilute solutions and the thermodynamic and volumetric properties of the pure substances:

$$x_2 = (\frac{V_1}{V_2}) \exp \left[-\left[1 - \frac{V_2}{V_1} + \frac{V_2(C_{11} - 2 C_{12} + C_{22})}{R T} \right] \right]$$

where C_{11}, C_{22}, and C_{12} = intermolecular potential energy between solvents, solute, and unlike molecules, respectively

V_1 and V_2 = molar volume of solvent and solute, respectively, at the condition of the system $(ft^3 \ lb^{-1})$

x_2 = solubility of water in the organic phase, mole fraction

The intermolecular potential energy for the solvent (C_{11}) was calculated from

$$C_{11} = \left[\frac{(U_o)_1 - (U_p)_1}{V_1} \right]_T \quad Btu \ ft^{-3}$$

where $(U_o)_1$ = molar internal energy of pure component 1 at the temperature of the system and at zero pressure

$(U_p)_1$ = molar internal energy of pure component 1 at the temperature and pressure of the system

The value of C_{12} has been derived from

$$C_{12} = (C_{11} C_{22})^{\frac{1}{2}} \quad Btu \ ft^{-3}$$

The molar internal energies $(U_o)_1$, $(U_p)_1$, $(U_o)_2$, and $(U_p)_2$ are expressed as a function of enthalpies (H_o): for example,

$$(U_o)_1 = (H_o)_1 - R T \quad Btu \ lb^{-1}$$

$$(U_p)_1 = (H_p)_1 - P V_1 \quad Btu \ lb^{-1}$$

Leland et al. (1955) calculated the solubilities of water in liquid methane, ethane, propane, butane, pentane, and 1-butene at various temperatures and pressures. The differences between the calculated and experimental figures were almost as great as some of the larger deviations reported in the literature. However, the calculation results in greater error as the state of the hydrocarbon solvent approaches ideal condition.

Table 2.98 Correlation of Water Solubilities by Molar Volume

Solute	Temperature ($^{\circ}$C)	$\log_{10}\left(\dfrac{1}{\phi}\right)$	Solubility in H_2O (wt %)		Error (%)
			Calculated	Selected	
Benzene	25	2.6820	0.18225	0.178	2.42
Toluene	25	3.1705	0.05840	0.0515	13.40
Ethylbenzene	25	3.6245	0.02054	0.0152	35.13
o-Xylene	25	3.6281	0.02035	0.0175	16.29
m-Xylene	25	3.6354	0.01997	0.0196	1.89
p-Xylene	25	3.6486	0.01930	0.0190	1.58
Styrene	25	3.4145	0.03480	0.0310	12.26
Biphenyl	25	4.5328	0.00292	0.00075	289.33
Naphthalene	25	3.9084	0.01192	0.3440	-96.53
Fluorobenzene	30	2.8357	0.14858	0.1538	-3.38
Chlorobenzene	30	3.0566	0.09659	0.0550	75.36
Bromobenzene	30	3.1473	0.10597	0.0446	137.60
Iodobenzene	30	3.3265	0.08594	0.03399	152.84
1-Bromopropane	30	2.7601	0.23337	0.2464	-5.28
1-Bromobutane	30	3.2221	0.07596	0.05322	42.73
1-Trichloroethane	20	2.9466	0.15267	0.15539	-1.74
Pentachloroethane	20	3.5486	0.04756	0.050228	-5.32
1-Chloropropene	20	2.4685	0.31847	0.3600	-11.53
n-Pentane	16	3.3829	0.02610	0.0360	-27.50
n-Hexane	16	3.8141	0.01018	0.0095	7.16
n-Heptane	16	4.2553	0.00382	0.0050	-23.60
n-Octane	16	4.7109	0.00137	0.0020	-31.50
Dichloromethane	30	1.9985	1.31390	1.2605	4.24
Chloroform	25	2.4376	0.54086	0.7920	-31.71
Sym-Dichloroethane	20	2.3894	0.51112	0.8524	-40.04
1,1-Dichloroethane	20	2.5341	0.34405	0.4843	-28.96
1-Chloropropane	20	2.6420	0.20379	0.2713	-24.88
1,3-Dichloropropane	25	2.8599	0.16331	0.2723	-40.03
1,2,2-Trichloroethane	20	2.7705	0.24449	0.4420	-44.66
1-Chlorobutane	20	3.1018	0.07022	0.06424	9.31
Ethyl chloride	0	2.1288	0.68472	0.5842	17.20
Carbon tetrachloride	30	2.9152	0.19212	0.07913	142.79
Dibromomethane	30	2.1497	1.74028	1.1574	50.36
Bromoform	30	2.6506	0.64052	0.3180	101.42
Ethyl iodide	30	2.4626	0.66026	0.4133	59.75

Table 2.98 (Continued)

Propyl iodide	30	2.9342	0.20178	0.1029	96.09
Diiodomethane	30	2.4563	1.14811	0.1340	756.80
Carbon disulfide	20	1.8656	1.71727	0.2940	484.11

Average: 55.09

Source: Lindenberg, 1956.

Lindenberg (1956) reported a simple expression giving the relation between the molecular volume of the solute (V_2) and the solubility in water of various hydrocarbons (ϕ) expressed as a volume fraction. In the empirical expression

$$\log_{10}(\frac{1}{\phi}) = 0.03 \, V_2$$

V_2 is the molar volume of the solute, expressed in cc g-mole^{-1}. The 36 solutes investigated included several halogenated hydrocarbons. The results are shown in Table 2.98.

A similar type of investigation was reported by Natarajan and Venkatachalam (1972) for the solubility of olefins in water. A semilogarithmic plot of solubility versus the molar volume of the solutes produced a straight line.

More recently Epshtein and Nizhnii (1979a, b, c) correlated the aqueous solubility of 200 organic compounds, expressed in $\log(1/x_2)$, with the van der Waals volume of the solute molecules. The statistical analysis of various classes of substances showed that the standard deviation in the calculation of $\log(1/x_2)$ from corresponding experimental values was 0.431. The method may be used in calculating the solubility (expressed as a mole fraction) of diversified structures.

The aqueous solubility of nonelectrolytes was correlated with molecular diameter (Å) (Jockers, 1976). However, no regularity could be found.

The solubility of gases at 25°C expressed as a logarithm of the mole fraction gave a straight line when plotted against the force constants of the gases (Miller, 1966). The correlation was tested for several gases (He, Ne, H_2, N_2, O_2, Ar, CF_4, CH_4, Kr, CO_2, SF_6, and Xe) in various solvents (C_6H_6, perfluoro-n-heptane, CS_2, CCl_4, C_7H_{16}, H_2O, and i-C_8H_{18}). Fig. 2.66 shows the relation for the solubility of several gases, including halogenated hydrocarbons in water at 25°C.

Parnov (1969) related the solubility of organic compounds in water to the dielectric constant of water and the temperature:

$$3 = \frac{273}{T} + (\frac{E_t}{E_o})^3 + (\frac{C_t}{C_o})^{1/3}$$

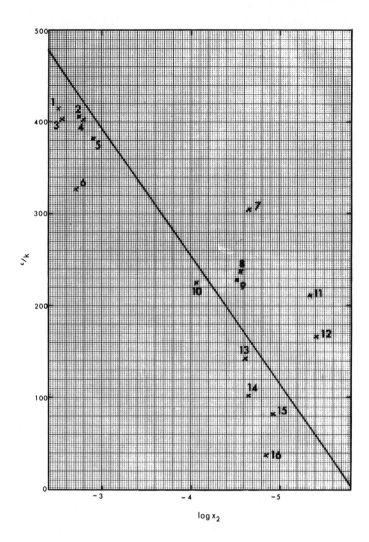

Fig. 2.66 Correlation of solubility with force constant at 25°C. ε/k = force constant (K); x_2 = solubility in water (mole fraction). 1, CH_3Br; 2, CH_3I; 3, CH_2Cl_2; 4, C_2H_5Cl; 5, $CHCl_3$; 6, CH_3Cl; 7, $n-C_4H_{10}$; 8, $n-C_3H_8$; 9, C_2H_6; 10, C_2H_4; 11, SF_6; 12, CF_4; 13, CH_4; 14, O_2; 15, N_2; 16, H_2. The correlated equation was: $\log_{10} x_2$ = $-5.81876 + 0.00717336 \, \varepsilon/k$. (From Miller, 1966.)

where E_t and E_o =   dielectric constants of water at t and 0°C, respectively
C_t and C_o =   solubilities of organic substances (wt %) at t and 0°C,
                      respectively

This expression is satisfactory not only for hydrocarbon-water systems but for other systems as well. Using this equation, the extrapolated solubility data of 1,1,2,2-

tetrabromoethane ($CHBr_2$-$CHBr_2$) in water are compared with experimental values in Table 2.99. The errors presented in the table indicate that the right-hand side of the equation is not equal to 3, but is a function of temperature.

Table 2.99 Correlation of Solubilities in Water with Dielectric Constant

Temperature	Solubility of $CHBr_2$-$CHBr_2$ in H_2O (wt %)		Error
($^{\circ}$C)	Experimental	Calculated	(%)
0	0.05230	0.05230	0.0
10	0.05660	0.08384	48.13
20	0.06294	0.12009	90.80
25	0.06754	0.1394	106.40
30	0.07234	0.1594	120.35
40	0.08750	0.2002	128.80
50	0.1059	0.2416	128.14
60	0.1261	0.2826	124.11
70	0.1560	0.3228	106.92
80	0.1938	0.3616	86.58
90	0.2456	0.3988	62.38
100	0.3064	0.4344	41.78
		Average:	87.03

Source: Parnov, 1969.

The solubility of CH_3Cl, CH_2FCl, CHF_2Cl, and CF_2Cl_2 in water at various temperatures has been investigated by Boggs and Buck (1958). They found that there is a rough correlation between the solubility of the fluorochloromethanes in water and their dipole moments (see Fig. 2.67). Furthermore, a plot of the heat of solution (ΔH_{sol}) of these substances versus temperature resulted in straight lines within experimental error (see Fig. 2.68). The heat of solution of halogenated hydrocarbons in water is presented in the Appendix.

The concept of solubility parameters, developed by Hildebrand and Scatchard (Hildebrand and Scott, 1950, 1962; Archer and Hildebrand, 1963; Hildebrand et al., 1970) has been used with great success to explain the interaction energies between molecules and to correlate and predict solubilities in nonpolar or nearly nonpolar solutes in solvents (see Sec. 2.5-2 for further details on regular solution theory). And despite the fact that regular solution theory is not valid for aqueous solutions, several investigators have reported work on the application of solubility parameters to correlate and predict solubilities in aqueous solutions.

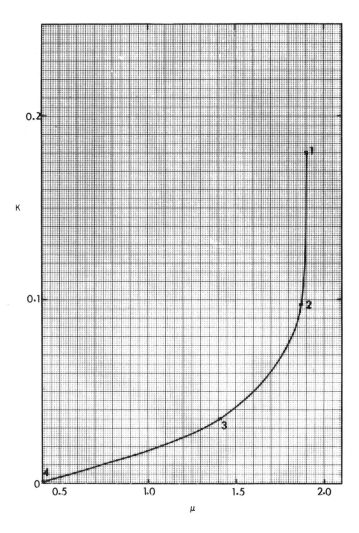

K

μ

Fig. 2.67 Correlation of solubility at 25°C with dipole moment. K = solubility in water (g-mole liter^{-1} atm^{-1}); μ = dipole moment (debye). 1, CH_2FCl; 2, CH_3Cl; 3, CHF_2Cl; 4, CF_3Cl. The correlated equation was: K (g-mole liter^{-1} atm^{-1}) = 0.012218 $μ^{3.6447}$. (From Boggs & Buck, 1958.)

Using Hildebrand's concept of solubility parameters, Jones and Monk (1963) have studied the solubility of water in some organic liquids. The expression derived,

$$\log_{10} S_2 = \frac{V_2}{V_1} - 1 - \frac{V_2(\delta_1 - \delta_2)^2}{RT}$$

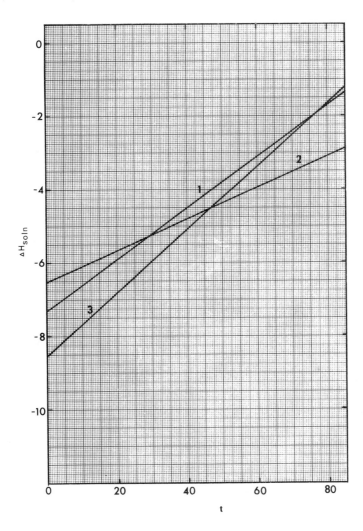

Fig. 2.68 Heat of solution in water as function of temperature. ΔH_{soln} = heat of solution (kcal g-mole^{-1}); t = temperature (°C). 1, CH_3Cl; 2, CH_2FCl; 3, CHF_2Cl. (From Boggs & Buck, 1958.)

where V_1 and V_2 = molar volumes of solvent and solute, respectively

δ_1 and δ_2 = solubility parameters of solvent and solute, respectively $(cal/cc)^{\frac{1}{2}}$

S_2 = solubility of water, cc H_2O/cc solvent

T = absolute temperature (K)

provides an estimation method for the solubility of water in organic solvents. The

Table   2.100   Calculation of the Solubility of Water in Aromatic Hydrocarbons at 25°C.

| Solvent | Solubility of water (cc H_2O/cc solvent) | | Error |
	Experimental	Calculated	(%)
Benzene	0.00057	0.0005968	4.70
Toluene	0.00040	0.0004010	0.25
Ethylbenzene	0.00037	0.0003676	-0.65
n-Butylbenzene	0.00035	0.0003400	-2.86
Chlorobenzene	0.00034	0.0003419	0.56
Bromobenzene	0.00041	0.0004655	13.54
Iodobenzene	0.00049	0.0004869	-0.63
o-Dichlorobenzene	0.00032	0.0003238	1.19
		Average:	2.01

Source: Jones & Monk, 1963.

estimation produced a better result for chloro-, bromo-, iodo-, and o-dichlorobenzene than for toluene, ethylbenzene, and n-butylbenzene. The great advantage of
this procedure is that the required properties (density and vapor pressure) of the
solvents are readily available in the majority of cases. The predicted solubility of
water in halogenated hydrocarbons is compared with experimental values in Table
2.100.

Similar expressions were derived by Koskas and Durandet (1967) for the mutual
solubilities between water and hydrocarbons:

1.   Solubility of hydrocarbons (component 1) in water (component 2):

$$\log_{10}\left(\frac{1}{x_1}\right) = A + 0.4343\frac{V_1}{RT}(\delta_w - \delta_H)^2$$

2.   Solubility of water (component 2) in hydrocarbons (component 1):

$$\log_{10}\left(\frac{1}{x_2}\right) = A + 0.4343\frac{V_2}{RT}(\delta_w - \delta_H)^2$$

where

$$A = \log_{10}\left(\frac{V_2}{V_1}\right) + 0.4343\left(1 - \frac{V_2}{V_1}\right)$$

(see also Lawson, 1979). Both expressions predict the solubilities with moderate
accuracy; however, no halogenated hydrocarbon - water systems were included in the

Table 2.101 Calculation of the Solubility of Water in Hydrocarbons at 25°C

| Solvent | Solubility of water (mole fraction) | | Error |
	Experimental	Calculated	(%)
n-Pentane	0.000425	0.0004676	10.0
n-Hexane	0.00069	0.0006716	2.67
n-Heptane	0.00087	0.0008714	1.63
n-Octane	0.00113	0.001076	4.8
i-Propane	0.000481	0.000472	2.0
1-Butene	0.00142	0.001726	21.5
2-Butene	0.00153	0.001704	11.3
1,3-Butadiene	0.00225	0.001915	15.0
1,5-Hexadiene	0.0055	0.004839	12.0
Cyclohexane	0.00059	0.00059	0.0
Benzene	0.0026	0.003178	22.0
Toluene	0.00362	0.002961	18.2
		Average:	10.09

Source: Koskas & Durandet, 1967.

comparison (see Tables 2.101 and 2.102).

Fleury and Hayduk (1975) used the principle of regular solutions for the calculation of gas solubilities in a range of solvents. The expression used depends upon the molecular sizes of the solute and solvents. For solutions when the molecules are approximately equal in size, the following relationship was used:

$$-\log_e x_2 \;=\; -\log_e x_2^{id} \;+\; \frac{V_2(\delta_1 - \delta_2)^2}{R\,T}$$

By introducing the volume fraction (ϕ) in the expression above, Szczepaniec-Cieciak et al. (1977) calculated the solubility of organic solids in liquids.

The expression for regular solutions was also used by Högfeldt and Fredlund (1970) for the calculation of water solubility in mixtures of organic solvents at 25°C. For example, for a ternary system,

$$\log_e x_{H_2O} \;=\; -\frac{V_{H_2O}}{R\,T}(\delta_{H_2O} - \delta_m)^2$$

where $\delta_m \;=\; \phi_2\,\delta_2 + \phi_3\,\delta_3$

$\phi \;=\;$ volume fraction

Table 2.102 Calculation of the Solubility of Hydrocarbons in Water at 25°C

| Solute | Solubility in water (mole fraction) | | Error |
	Experimental	Calculated	(%)
1-Pentene	0.000039	0.00004099	5.1
3-Methyl-1-butene	0.0000347	0.00003424	1.3
1-Hexene	0.000011	0.00001292	17.0
1,3-Butadiene	0.00026	0.0002258	13.0
Cyclopentane	0.000040	0.0000431	7.8
Cyclohexane	0.0000118	0.0000126	6.7
Benzene	0.00041	0.000332	19.0
Toluene	0.000105	0.0000969	7.7
o-Xylene	0.0000297	0.0000357	20.0
n-Pentane	0.0000096	0.00000756	21.0
n-Hexane	0.0000020	0.00000202	1.0
n-Heptane	0.000000518	0.000000533	3.0
n-Octane	0.000000104	0.0000001396	34.0
		Average:	12.05

Source: Koskas & Durandet, 1967.

The regular solution theory of Hildebrand for the solubility of gases in liquids has been reviewed by Reed (1970) and Shinoda (1978). Blandamer et al. (1974) correlated the solubilities of gases in binary aqueous mixtures with some success.

Prausnitz and Shair (1961) correlated the gase solubilities in water at 100°C by plotting $\log_{10}(\gamma_2/V_2^L)$ versus δ_2, where

γ_2 = activity coefficient of the solute gas

V_2^L = molar volume of the solute, cc g-mole^{-1}

δ_2 = solubility parameter of the solute $(cal/cc)^{\frac{1}{2}}$

Because of the structural features of water, the correlation becomes more and more unreliable at lower temperatures, giving rise to hydrates. Fig. 2.69 shows these relationship for several gases at 25°C and 1 atm partial pressure. The numerical values are tabulated in Table 2.103.

The solubility of water in aromatic halides was studied by Wing (1956) and Wing and Johnston (1957), using the semiempirical formula of Hildebrand:

$$\log_{10} \phi_2 + \phi_1(1 - \frac{V_2}{V_1}) + \frac{V_2 \phi_1^2 (\delta_2 - \delta_1)^2}{R T} = 0$$

Table 2.103 Correlation of Gas Solubility in Water at 25°C and 1 atm Partial Pressure with Their Solubility Parameters[a]

Gas	Solubility parameter, δ_2 $(\text{cal cc}^{-1})^{\frac{1}{2}}$	$(\delta_w - \delta_2)^2$ (cal cc^{-1})	$\dfrac{\log_e \gamma_2}{v_2^L} = \dfrac{(\delta_w - \delta_2)^2 \phi_1^2}{R\,T}$
N_2	2.58	433.47	0.7316
O_2	4.00	376.36	0.6352
CO	3.13	410.87	0.6935
CO_2	6.00	302.76	0.5110
Cl_2	8.70	216.09	0.3647
CH_4	5.68	313.99	0.5300
C_2H_4	6.60	282.24	0.4764
C_2H_6	6.60	282.24	0.4764

[a] $\phi_1^2 = 1.0$; $\delta_w = 23.4 \ (\text{cal cc}^{-1})^{\frac{1}{2}}$ (Hayduk & Laudie, 1973).

Source: Prausnitz & Shair, 1961.

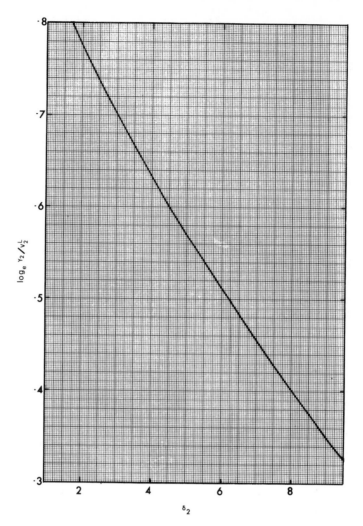

Fig. 2.69 Correlation of gas solubility in water with their solubility parameter at 25°C and 1 atm partial pressure. γ_2 = activity coefficient of solute gas in water; v_2^L = molar volume of solute (cc g-mole^{-1}); δ_2 = solubility parameter of solute (cal/cc)$^{\frac{1}{2}}$. (From Prausnitz & Shair, 1961.)

where ϕ_1 and ϕ_2 = volume fraction of solvent and solute, respectively

V_1 and V_2 = molar volume of solvent and solute, respectively

δ_1 and δ_2 = solubility parameter of solvent and solute, respectively

The expression above has been improved by splitting the solubility parameters into two parts; one part represents the dispersion effect and the other parameter is due to dipole-dipole interactions. The modified expression becomes

$$\log_e \phi_1 \ + \ \phi_2(1 - \frac{V_1}{V_2}) \ + \ \frac{V_1 \, \phi_2^2}{R \, T} \left[(\delta_1 - \delta_2)^2 \ + \ (W_1 - W_2)^2 \right] \ = \ 0$$

where W is the polarity parameter. Wing concluded from the experimental and correlated values that there is a general decrease in solubility of water as the polarity of the solvent increases. The predicted solubilities of water in several halogenated hydrocarbons are presented and compared with experimental values in Table
2.104.

The relation between the various H-bonding factors has been used by Hayduk and
Laudie (1973) to estimate of gas solubilities in water at 1 atm partial pressure.
The H-bonding factor was defined as the ratio between the ideal and real-gas solubilities in water, both expressed as a mole fraction. For example, the H-bonding
factor for the solubility of vinyl chloride in water at 25°C and 1 atm partial pressure is calculated as follows:

$$\sigma_{solvent} \ = \ \frac{x_2}{x_2^{id}} \ = \ \frac{7.98 \times 10^{-4}}{0.257} \ = \ 0.0031$$

The reduction in the water solubility of gases compared with that in nonpolar solvents is due to the powerful forces between water molecules. Hayduk and Laudie (1973)
have established that there is a definite correlation between the H-bonding factors
($\sigma_{solvent}$) in water. The H-bonding factors are discussed further in Sec. 2.4.

The solubility of a gas in a solvent can be predicted within 25% provided that
the solubility is known in at least one of the following solvents: chlorobenzene,
butanol, ethanol, methanol, acetone, acetic acid, and ethylene glycol. Diagrams for
H- bonding factors have been published by Hayduk and Laudie (1973). A further discussion and recommendations have been provided by Iguchi (1974).

The solubility of gases in liquids (including water) has been calculated at 25°C
and 1 atm pressure by Leites and Sergeeva (1973) using several simplifying assumptions. The Bunsen's solubility coefficient (β) was expressed as a function of
$\epsilon_{0.22}/k$, where $\epsilon_{0.22}$ is the Lennard-Jones potential parameter and k the Boltzmann's
constant:

$$\log_e \beta \ = \ a \ + \ b(\frac{\epsilon_{0.22}}{k})^{\frac{1}{2}}$$

The constants a and b vary with the type of solvent. The formula was tested for
23 gases in 12 solvents, including water. However, no halogenated hydrocarbon gas
was included in the study.

A genarel relationship between the solubility of water in n-alkanes at various
temperatures has been proposed by Nauruzov (1975):

Table 2.104 Solubility of Water in Various Solvents at 25°C

Solvent	Molar volume (cc g-mole^{-1})	Solubility parameter (cal cc^{-1})$^{\frac{1}{2}}$	Solubility of water at 25°C. (cc water/100 cc solvent)				
			Experimental	Calculated (Hildebrand)	Error (%)	Calculated (Wing)	Error (%)
Benzene	89.41	9.1	0.0554	0.152	174.4	0.0878	58.5
Toluene	106.39	8.9	0.0334	0.081	141.8	0.0320	-4.2
Fluorobenzene	93.85	9.3	0.0316	0.104	229.1	0.0451	42.7
Chlorobenzene	101.68	9.6	0.0360	0.125	247.2	0.0545	51.4
Bromobenzene	105.03	9.8	0.0424	0.157	270.3	0.0616	45.3
Iodobenzene	111.85	9.8	0.0504	0.143	183.7	0.0764	51.6
o-Dichlorobenzene	112.65	9.8	0.0309	0.154	398.4	0.0517	67.3
Water	18.02	23.5	-	-	-	-	-
Average					235.0		44.7

Source: Wing, 1956; Wing & Johnston, 1957.

$$S_M^t = 71.03(V_M^t - 21.72 - 15.58\ n)^{5.123} \times 10^{-10}$$

where S_M^t = solubility of water, g-mole liter^{-1} at t$^{\circ}$C

  V_M^t = molar volume of solvent, cc g-mole^{-1} at t$^{\circ}$C

  n = number of carbon atoms in the homologous series

The validity of the expression for other homologous series has not been tested.

A scheme for the calculation of the solubility of water in nonpolar solvents was proposed by Kuznetsova and Rashidov (1977) from the partition coefficient of water (K_w) in the organic (C_{org}) and vapor phases (C_{vap}):

$$K_w = \frac{C_{org}}{C_{vap}}$$

Assuming that the solubility of the nonpolar solvent in water is negligible (in fact, it is not), the vapor pressure of the water in the system is in good approximation equal to the vapor pressure of pure water. The vapor pressure of water at 25°C is 23.76 mm Hg. At this condition 100 cc of water vapor will correspond to

$$C_{vap} = \frac{(100 \times 23.76)/760}{82 \times 298}$$

that is, the number of moles of water (C_{vap}). The number of moles of water in 100 cc of nonpolar solvent (C_{org}) at equilibrium with water is

$$C_{org} = \frac{(100 \times 23.76)/760}{82 \times 298}\ K_w$$

Covering the number of moles of water to grams per 100 cc of the solvent (S), the expression becomes

$$S = K_w \frac{(100 \times 23.76 \times 18)/760}{82 \times 298} = 0.002303\ \beta$$

A comparison of the calculated and experimental values has been tabulated for 10 water-nonpolar solvent systems at 25°C. The solubilities calculated and determined experimentally agree satisfactorily, which shows the feasibility of the method.

A close relationship between aqueous solubility of a substance and its lipophilicity was recognized by Hansch et al. (1968) and consequently they correlated the solubilities with partition coefficients for a variety of organic liquids.

In a recent publication, based on a thesis (Dexter, 1976), Dexter and Pavlou (1978) correlated the aqueous solubilities, expressed in g liter^{-1} on a semilogarith-

mic scale with parachor. However, the idea originates from McGowan (1956). The homologous series included saturated n-alkanes (C_{5-8}), monounsaturated n-alkenes (C_{5-7}),
monounsaturated cykloalkenes (C_{5-7}), saturated cykloalkanes (C_{5-9}), chlorinated
aromatics (chlorobenzene, 1,2-, and 1,3-dichlorobenzenes, and p,p'-DDT), and polychlorinated biphenyls (n = 2 to 7). A good review of the various aspects of the
parachors of organic compounds with tabulated data has been provided by Quayle (1953).

Further correlations were proposed by π values (Leo et al., 1971) and F values
(Nys and Rekker, 1973).

Cysewski and Prausnitz (1976) used a semiempirical correlation for estimating
gas solubilities in water. The method is based upon Alder's perturbed hard-sphere
equation of state. The perturbation was calculated by molecular dynamics using the
square-well potential. The correlation provides resonable estimates of solubilities
over a wide range of temperatures for a variety of gases in water. The expression
for Henry's law constant is, however, too elaborate and requires a computer program
for the calculation.

Similarly, a more complex expression was reported by Gotoh (1976) for the calculation of solubilities of nonreacting gases in liquids for the free-volume theory.
The Henry's law constant was obtained from the critical properties of the solute and
solvent and their acentric factor.

Othmer (1965), Othmer and Chen (1968), and Othmer and Roszkowski (1949) correlated the Henry's law constant (H) on a logarithmic scale as a straight-line function of the vapor pressure of water (P) at the same temperature:

$$\log_{10} H \ = \ m \, \log_{10} P \ + \ C$$

The coefficients m and C, can be evaluated as a function of the concentration of
the solute (x_2). The expression was applied for the calculation of the solubility of
gases in water.

2.8 Temperature and Pressure Dependence of Solubility Data

The solubility data are sometimes illustrated through solubility curves -- plots of
temperature against composition, representing the saturated solution (i.e., the limits
of complete solubility). In several published works the curves are presented without
reference to the pressure of the system (see, e.g., McGovern, 1943). In such cases the
solubility curves normally represent the solubility values at atmospheric pressure or
at the saturated vapor pressure of the solutions considered (see, e.g., van Ness,
1964).

An excellent collection of solubility data is illustrated through solubility
curves in Landolt-Börnstein (1962, 1976a), giving details on the pressure in addition
to the temperature dependence of the solubility for various binary systems. The figures
represent the best collection of solubility curves available at present (for more on

Table 2.105   Compound Index of Solubility Curves: Solubility of Gases in Liquids

Solute (gas)	Fig. Number
Air	11, 20, 72
Ar	2, 3, 6, 47
As	33, 97
B_2H_6	188
Br_2	157
CO	39, 95, 98, 128, 186, 191-193
CO_2	34-35, 99-109, 129-131, 179-187, 191
Cl_2	15-18, 67
Cl_2O	24
CH_4	40-41, 110-111, 140, 186, 189, 195-196
C_2H_2	114-115, 132, 186, 190
C_2H_4	38, 42, 112-113, 186, 195-196
C_2H_4O	43-44
C_2H_6	45, 186
C_3H_3N	46
C_3H_4	186
C_3H_6	116, 186, 195-196
C_3H_8	45, 186
C_4H_2	186
C_4H_4	186
C_4H_8	195-196
C_4H_{10}	45
C_6H_{14}	197
H_2	9-10, 49-58, 117, 119, 135-136, 147-154, 186
HBr	25
HCl	23, 73-74, 163-164, 166-167
HCN	36-37
HF	21, 162
HI	22
H_2O	138, 158, 160-161
H_2S	27-28, 75, 77, 139
H_2Se	26
He	1, 4, 6, 134, 194
Kr	6-7
N_2	11, 19, 55, 68-71, 122-124, 133, 158-159, 186
NH_3	30, 85, 125, 137, 175-176

Table 2.105 (Continued)

NO	32, 95, 178
N_2O	31, 86-94, 126-127, 177
Ne	5-6
O_2	11-13, 59-66, 120-121, 155, 186
O_3	14, 156
PH_3	96
Rn	6, 48, 118, 141-146
SO_2	29, 76, 78-84, 165, 168-174
UF_6	198
Xe	6, 8

Source: Landolt-Börnstein, 1962.

work in progress, see Chapter 3, Table 3.2, and Kertes, 1974-1981). A compound index for these systems is given in Tables 2.105 and 2.106. Because of the limited availability of solubility data for halogenated hydrocarbons and water before 1962, illustrations of these systems are restricted to the solubility of water vapor in six refrigerants between -50 and 50°C (see Fig. 161 in Landolt-Börnstein, 1962). Since the publication of 1962 volume, more solubility data have become available in the literature and elsewhere; these are analyzed, discussed, and selected in the 1976 edition of Landolt-Börnstein.

The theory of the temperature and pressure dependence of solubility has been treated in several good works (e.g., Hildebrand, 1924; Hildebrand and Scott, 1950, 1962; Hildebrand et al., 1970; Hibbard and Schalle, 1952; Himmelblau, 1960; King, 1969; Parusnitz, 1969; Craubner, 1976; Battino and Clever, 1966; Gerrard, 1976; Landolt-Börnstein, 1976a, b; Merriman, 1977; Shinoda, 1978; Getzen, 1976; Gokcen, 1973; Gokcen and Chang, 1973, 1975; Tsiklis, 1968).

A particularly good compilation is that of Landolt-Börnstein (1976a), which treats the solubilities of gases in liquids. The work is based upon published solubility data available in the literature. The majority of solubility data are presented in graphical form, where the solubility is expressed either in g-mole kg^{-1} bar^{-1} or $N\ m^3\ t^{-1}\ at^{-1}$ and plotted versus the temperature (°C) on semilogarithmic scales. Table 2.106 summarizes the halogenated hydrocarbon – water systems presented in the book (Landolt-Börnstein, 1976a).

The van't Hoff equation

$$\frac{d \log_e K_p}{dT} = \frac{\Delta H}{R\ T^2}$$

gives the relationship between the temperature (T, K) and the equilibrium constant

Table 2.106 Solubility Data for Halogenated Hydrocarbons in Water

System with H_2O	Fig. Number	Page Number	Solubility of gas/vapor in water	Temperature range ($^{\circ}C$)
CH_2FCl	2	54	gas	10 - 83
$CHFCl_2$	2	54	gas	0 - 52
CHF_2Cl	2	54	gas	7 - 82
CF_3Cl	-	56	gas	10 & 60
CHF_2Br	-	56	gas	-
CH_3Cl	2	54	gas	7 - 68
CH_2Cl_2	2	54	vapor	0 - 33
$CHCl_3$	2	54	vapor	0 - 62
CCl_4	1	32-33	vapor	0 - 68
	2	54	vapor	0 - 68
	27	158	vapor	10 - 53
C_2F_4	-	57	gas	-
C_2H_5Cl	2	54	vapor	17 - 44
$1,1-C_2H_4Cl_2$	-	57	vapor	0 - 30
$1,2-C_2H_4Cl_2$	2	54	vapor	0 - 72
C_2H_3Cl	2	54	gas	0 - 72
	4c	8	gas	0 - 374
	2.11	67	gas	0 - 75
$1,1,1-C_2H_3Cl_3$	2	54	vapor	0 - 68
C_2HCl_3	2	54	vapor	0 - 72
C_2Cl_4	2	54	vapor	10 - 72
CH_3Br	2	54	vapor	8 - 32
CH_2Br_2	2	54	vapor	0 - 35
C_2H_5Br	-	57	vapor	0 - 30
CH_3I	-	57	vapor	0 - 30
C_2H_5I	-	57	vapor	0 - 30
C_3F_6	-	57	vapor	-
C_3H_5Cl	2	54	vapor	22 - 72
CF_4	1	33	gas	0 - 52

Source: Landolt-Börnstein, 1976a.

(K_p) at constant pressure. ΔH is the increase in enthalpy. Because

$$K_p = K_n P^{\Delta n}$$

the logarithms gives

$$\log_e K_p = \log_e K_n + \Delta n \log_e P$$

After differentiation with respect to T, we obtain

$$\frac{d \log_e K_p}{dT} = \frac{d \log_e K_n}{dT} + \frac{\Delta n \, d \log_e P}{dT} = \frac{\Delta H}{R\,T}$$

where Δn is the number of moles increased. At constant pressure the total pressure (P) is independent of the temperature, and

$$\frac{d \log_e P}{dT} = 0$$

Consequently, the final expression becomes

$$\left(\frac{d \log_e K_n}{dT} \right)_P = \frac{\Delta H}{R\,T}$$

where the equilibrium constant (K_n) is given in terms of the mole fraction.

Clarke and Glew (1966) made a comparison of equilibrium equations for the temperature dependence of $\log_e K_p$. Their recommendations are based upon the following reasons:

1. Thermodynamically meaningful
2. They can include any number of independent variables to fit data
3. The equations can be readily solved
4. The constants are related to the standard thermodynamic function change
5. They provide predictive relations for both values and the standard errors
    of standard thermodynamic function changes at any desired temperature

In a saturated solution, an equilibrium exists between the pure substance and its solution, and therefore the equilibrium constant (K_n) is proportional to the activity (a_i), and the equation above can be written

$$\left(\frac{d \log_e a_i}{dT} \right)_P = \frac{\Delta H}{R\,T^2}$$

This expression may be integrated, first, if we assume that ΔH does not vary with temperature, and second, when ΔH varies with temperature, according to the equation

$$\Delta H = \Delta H_o + a T + b T^2 + c T^3$$

In the first case, upon integration we obtain

$$\log_e a_i = - \frac{\Delta H}{R T} + I$$

and in the second case

$$\log_e a_i = \frac{1}{R}(- \frac{\Delta H_o}{T} + a \log_e T + b T + \frac{c T^2}{2}) + I'$$

where I and I' are the integration constants $(I \neq I')$.

If the solution is ideal or the activity can be calculad from the mole fraction, the activity a_i may be replaced by the mole fraction x_i.

In the following discussion, three cases are considered in more detail: the solubility of solids, liquids, and gases in liquids as a function of temperature and pressure.

2.8-1 Solubility of Solids in Liquids

The solubility of a solid in a liquid as a function of temperature is described by the van't Hoff equation:

$$\left(\frac{d \log_e a_i}{dT} \right)_P = \frac{\Delta H_m}{R T}$$

where ΔH_m is the heat of melting. The heat of fusion (ΔH_f) for 63 organic compounds has been correlated by Rao (1975). The correlation based on the use of the temperature of fusion (T_f, K) and the normal boiling point $(T_{b.p.}, K)$ or critical temperature (T_{cr}, K):

$$\Delta H_f = A(\frac{T_f^2}{T_{b.p.}})$$

and

$$\Delta H_f = A(\frac{T_f^2}{T_{cr}})$$

The second expression is recommended for olefin hydrocarbons only. The value of A varies for different homologous series,

	A
Paraffin hydrocarbons up to C_{20}	0.385
Paraffin hydrocarbons above C_{20}	0.246
Olefins (second expression)	0.80

	A
Acids and aromatics	0.21
Halobenzenes	0.115
Halogenated hydrocarbons	0.127

The average absolute deviation was 5.1%.

If we assume that the heat of melting is independent of temperature, which is a fair approximation if the actual temperature is near the melting point of the solid, after integration the equation above becomes (e.g., Biggar and Riggs, 1974; May et al., 1978b; MacKay et al., 1980)

$$\log_e a_i = \frac{\Delta H_m}{R} \left(\frac{1}{T_m} - \frac{1}{T} \right) = \frac{\Delta H_m}{R} \left(\frac{T - T_m}{T\, T_m} \right)$$

where T_m is the melting temperature of the solid. In this expression we also assumed that there is no difference between the heat capacity of the solid and liquid, which is not the case. The heat of melting is a function of temperature and heat capcity:

$$\Delta H_m = \Delta H_m^{m \cdot p \cdot} - (c_p^l - c_p^s)(T_m - T)$$

where $\Delta H_m^{m \cdot p \cdot}$ is the heat of melting at the melting point and c_p^l and c_p^s are the molar heat capacities of liquid and solid, respectively. The temperature dependence of heats of melting was also studied by Shenkin et al. (1979). The final expression for the solubility of a solid in a liquid at temperature T (K) becomes (e.g., Lewis and Randall, 1961; Hildebrand and Scott, 1950; Prausnitz, 1965; Riggs and Diefendorf, 1979):

$$\log_e a_i = \frac{\Delta H_m^{m \cdot p \cdot}}{R} \left(\frac{T - T_m}{T\, T_m} \right) + \frac{c_p^l - c_p^s}{R} \left(\frac{T_m - T}{T} - \log_e \left(\frac{T_m}{T} \right) \right)$$

In an ideal solution, a_i may be substituted by x^i or $x_i\, \gamma_i$ in equilibrium with the saturated solution.

Hildebrand (1952) has introduced a graphical representation of the solubilities of nonelectrolyte solids in solutions from which chemical interactions are absent, by plotting the logarithm of mole fraction against the logarithm of absolute temperature, obtaining practically straight lines except near the melting point of the solute. However, this linear relationship is valid only for regular solutions. Aqueous solutions of the halogenated hydrocarbons do not give straight solubility curves.

In regular solutions, Hildebrand and Scott (1950) expressed the solubility of solids as follows:

$$\log_e \left(\frac{x_2^i}{x_2} \right) = \frac{V_2\, \phi_1^2 (\delta_1 - \delta_2)^2}{R\, T}$$

where δ_1 and δ_2 = solubility parameter of solvent and solute, respectively

x_2^i = solubility in ideal solutions, mole fraction

x_2 = solubility in regular solutions, mole fraction

V_2 = molar volume of the solute, cc g-mole^{-1}

T = absolute temperature, K

ϕ_1 and ϕ_2 = volume fraction of solvent and solute, respectively

The ideal solubility is obtained from the expression

$$\log_{10} x_2^i = \log_{10} \phi_2 + V_2 \phi_1^2 \left(\frac{(\delta_2 - \delta_1)^2}{2.3025\ R\ T}\right) + 0.434\ \phi_1 \left(1 - \frac{V_2}{V_1}\right)$$

However, for a more accurate calculation, the solid solubility can be obtained in a regular solution from the following equation (Hildebrand and Scott, 1950: Boulegue, 1978):

$$\log_{10}\left(\frac{1}{x_2}\right) = \frac{\Delta H_m^{m \cdot p \cdot}(T_m - T)}{2.3025\ R\ T_m\ T} - \frac{c_p^l - c_p^s}{2.3025\ R}\left(\frac{T_m - T}{T}\right) + \frac{c_p^l - c_p^s}{R} \log_e\left(\frac{T_m}{T}\right)$$

$$+ \frac{V_2 \phi_1^2 (\delta_1 - \delta_2)^2}{2.3025\ R\ T}$$

A very simple, but surprisingly accurate representation of the temperature dependence of solid solubility in water is described by the expression (e.g., Biggar and Riggs, 1974; Washburn, 1929; May et al., 1978b; May, 1980)(see Fig. 2.33)

$$\log_e x_i = F\left(\frac{1}{T}\right)$$

which was derived from the equation

$$\log_e a_i = \frac{\Delta H_m}{R}\left(\frac{1}{T_m} - \frac{1}{T}\right)$$

This relationship shows a good approximation for solubility in a moderate temperature interval. Fig. 2.70 shows typical plots of solubility for several solid organic compounds in water at various temperatures. However, for a wider temperature interval, the temperature dependence is more complex and a more elaborate expression is required for the representation of the experimental data (Getzen, 1970; Getzen and Ward, 1971; Wauchope, 1970; Wauchope and Getzen, 1972):

$$R \log_e x_2^s + \frac{\Delta H_m}{T} - 0.000408(T - 291.15)^2 = -c + b\ T$$

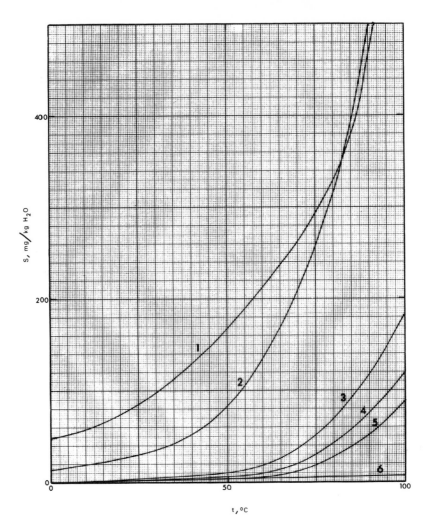

Fig. 2.70 Solubility of organic solids in water as function of temperature. 1 = p-Dichlorobenzene; 2 = naphthalene; 3 = acenaphthalene; 4 = fluorene; 5 = phenanthrene; 6 = antracene. (From Wauchope, 1970.)

By plotting the left-hand side of this equation versus the absolute temperature, a straight line can be obtained. In the expression

x_2^s =   solubility of the solid in the liquid, mole fraction

ΔH_m =   heat of melting

a and b   =   coefficients of the equation

The well-known **Valentiner** equation (Valentiner, 1927)

$$\log_e (x_2)_{sat} \;=\; a \;+\; b \log_e T \;+\; \frac{c}{T}$$

has been derived from the van't Hoff isochore (Moelwyn-Hughes, 1964) and applied to the fitting of solid solubility data (see also Glew and Robertson, 1956).

The fugacity ratio as a function of temperature has been studied by Mackay and Shiu, (1977) and Mackay et al. (1980).

The solubility of organic solid substances has been correlated by Bushinskii et al. (1974) using the free energy of intermolecular interaction of the solvents on a semilogarithmic plot:

$$-\log_{10} S \;=\; A \;+\; \left\{ B \; v_{mol} \left[R\,T \log_{10} \left(\frac{R\,T}{P\,v_{mol}} \right) \;-\; R\,T \right] \right\}^{\frac{1}{2}}$$

where S =   solubility of the solid, g/cc solvent

   v_{mol} =   molar volume of the solvent, cc g-mole^{-1}

   P =   equilibrium vapor pressure, atm

 A and B =   coefficients, typical for a substance

This expression has been tested for the representation of the solubility of solid p-dichlorobenzene in water at various temperatures (se Fig. 2.65).

The solubility of solid organic substances in water can also be expressed in terms of thermodynamic parameters:

$$\log_e x_2 \;=\; \frac{\Delta H_m \;-\; \Delta G_{ex}}{R\,T}$$

where ΔG_{ex} is a function of temperature (Wauchope, 1970):

$$\Delta G_{ex} \;=\; -a \;+\; c\,T \;-\; b\,T^2 \;-\; (0.000408\,T)(T \;-\; 291.15)^2$$

If only one solubility data is known, the usual approximation is to draw a straight line through the known point and the zero solubility at the melting point temperature in a $\log_e x_i$ versus $1/T$ diagram.

The pressure effect on the solubility of solids in liquids has been discussed by Hildebrand and Scott (1950), Bradley et al. (1973), Suzuki et al. (1974), and Zipp (1973). Suzuki et al. (1974) found, that the solubility of naphthalene decreases in water with increasing pressure, while Bradley et al. (1973) showed that the solubility of benzene and toluene in water increases with pressure up to 1.2 kbars. However, above this pressure the solubility of toluene reaches a maximum similarly to n-propyl benzene (Zipp, 1973).

At equilibrium, the fugacity of component i is the same in the solid and solution:

$$f_i^s = f_i^l$$

The pressure dependence of the solubility in the system is expressed by

$$\frac{d \log_e x_i}{dP} = \frac{v^s - v^l}{R T \left(\dfrac{\partial \log_e f_i^s}{\partial \log_e x_i}\right)_P}$$

The magnitude of $\partial \log_e f_i^s / \partial \log_e x_i$ determines whether the solubility follows Raoult's law ($\gamma = 1$) or deviates from it in either the positive or negative direction. In the case of positive deviation, the solubility decreases, whereas at negative deviation the solubility increases with rising pressure. This is clearly shown by the expression above.

In addition to the temperature and pressure dependence of the solid solubility in liquids, the solubility also depends upon the particle size (Hildebrand and Scott, 1950). The fugacity of the liquid f^l has been related to the fugacity of the solid f^s by the following equation (Guggenheim, 1952):

$$R T \log_{10}\left(\frac{f^l}{f^s}\right) = \frac{2 V \gamma}{r}$$

where γ =   surface tension of the liquid

   V =   molar volume of the liquid

   r =   radius of the particle (dispersed)

The effect of particle size of the solid upon its solubility has been discussed further by Dundon and Mack (1923), Hulett (1901), Bowman et al. (1960), Pedersen and Brown (1976a, b), Fürer and Geiger (1977), Kaneniwa and Watari (1978), Kaneniwa et al. (1978), and Morokhov et al. (1979).

2.8-2 Solubility of Liquids in Liquids

Two liquids, each with different chemical composition, can form a single- or two-phase solution after mixing. Wether the solubility or miscibility between the two liquids is partial or complete depends on the likeness or unlikeness of the molecules of the two chemical compounds. For practical reasons, the greatest amount present in the mixture is called the solvent and the remaining components (smaller amounts) are the solutes. A theoretical treatment of miscibility and immiscibility is given in Sec. 2.5-5. The halogenated hydrocarbon and water systems are only partially miscible without exeptions. A typical system is chloroform – water, illustrated in Fig. 2.38. If the liquids are completely immiscible, the vapor pressure of the components will be influenced by each other, so that the total pressure of the system will be the sum of the pure components' vapor pressures. The boiling temperature will remain the same until the two separate liquid phases exist. The vapor composition also remains constant.

The liquid-liquid immiscibility of systems can also be explained qualitatively by the large activity coefficients, as described in Sec. 2.5-3 (see, e.g., King, 1969). Such systems form hydrogen bonds, and a typical example is the mixture of water with hydrocarbons (King, 1969).

Generally speaking, the miscibility or immiscibility of water with other liquids is also very dependent on temperature. At constant pressure, the slope of the liquid-liquid equilibrium curve (temperature versus concentration) is given by the sign of the expression (Prigogine and Defay, 1967)

$$x_1' (H_1' - H_1'') + (1 - x_1')(H_2' - H_2'')$$

or since

$$\mu_i' = \mu_i''$$

and

$$H_i' - T S_i' = H_i'' - T S''_i$$

alternatively, by the sign of the expression

$$x_1' (S_1' - S_1'') + (1 - x_1')(S_2' - S_2'')$$

where x_1 =   mole fraction of component 1,

H_i =   partial molar heat content of component i

S_i =   partial molar entropy of component i

μ_i =   chemical potential of component i

The prime and double prime refer to the first and second liquid phases, respectively. The first expression gives the molar heat of transfer and the second gives the molar entropy of transfer upon the slope of the equilibrium curve. At ambient temperature and atmospheric pressure, the miscibility is very limited between water and halogenated hydrocarbons; in other words, the mutual solubility is less than 2 wt %. Miscibility and immiscibility are discussed further in Sec. 2.5-5. In a liquid-liquid system, the variation of solubility with temperature depends on the sign of the heat of mixing or solution. The negative heat of solution (exothermic) indicates a lower critical solution temperature (LCST), whereas the upper critical solution temperature (UCST) is characterized by a positive heat of solution (endothermic). Tabulated heat-of-solution data for halogenated hydrocarbons in water are given in the Appendix. This is the maximum temperature at which it is possible to obtain two liquid phases at the given pressure, or the temperature at which the two liquid phases become identical. For example, the heat of solution of benzene in water changes sign at 289 K (Gill et al.,

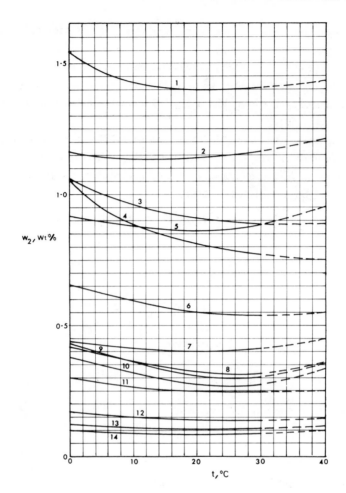

Fig. 2.71 Solubility of halogenated hydrocarbons in water at saturated pressure. t = temperature (°C); w_2 = solubility in water (wt %). 1, CH_3I; 2, CH_2Br_2; 3, C_2H_5Br; 4, $CHCl_3$; 5, CH_2Cl-CH_2Cl; 6, $CHCl_2-CH_3$; 7, C_2H_5I; 8, $i-C_3H_7Br$; 9, $i-C_3H_7Cl$; 10, $n-C_3H_7Cl$; 11, $n-C_3H_7Br$; 12, $i-C_3H_7I$; 13, $n-C_3H_7I$; 14, CCl_4. (From Rex, 1906.)

1975; May, 1980) and the minimum value of solubility occurs at this temperature (Nango et al., 1980). The minimum solubility occurs for most halogenated hydrocarbon-water systems between 0 and 40°C (see Figs. 2.71 and 2.72)(Rex, 1906; McGovern, 1943). The minima in the solubility of alkane derivatives in water have been studied by Nishino and Nakamura (1978). Because of the change of sign of the heat of solution, both upper and lower CSTs will occur with the same two phases. The molecular interactions give the explanation for the phenomena. That is, at the LCST, the attractive

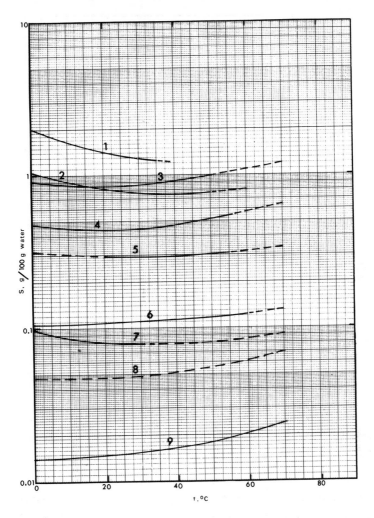

Fig. 2.72 Solubility of chlorinated solvents in water at saturated pressure. S = solubility (g solvent/100 g water); t = temperature (°C). 1, CH_2Cl_2; 2, $CHCl_3$; 3, $CHCl_2-CH_3$; 4, $CHCl_2-CH_2Cl$; 5, $CHCl_2-CHCl_2$; 6, $CCl_2=CHCl$; 7, CCl_4; 8, CCl_3-CHCl_2; 9, $CCl_2=CCl_2$. (From McGovern, 1943.)

forces between molecules are predominant, whereas at high temperatures the intermolecular interaction is repulsive, the heat of mixing is positive, and the solution is nearly random. The condition for a closed temperature versus concentration curve is that the heat of mixing is negative at low temperatures and positive at higher temperatures. Consequently, a lower and an upper CST will occur. This is the case when interaction between unlike molecules is repulsive for a majority of relative orientations and attractive for a few. The positive heat of mixing is at higher temperatures,

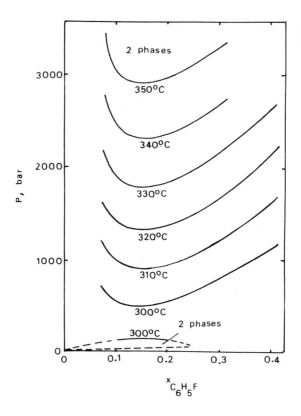

Fig. 2.73 Solubility of fluorobenzene in water at high pressures. (From Götze et at., 1975.)

when the orientations become random and the interactions are repulsive, whereas the attractive orientations are favored at lower temperatures, giving rise to a negative heat of mixing. A general description of the various models has been given by Barker and Fock (1953). They also discussed the classical example of the coexistence curve for nicotine – water system. Further examples were reported by Semenchenko and Efuni, (1974).

The one vapor and two liquid phases for a fluorobenzene – water system is shown in Fig. 2.73 (Götze et al., 1975; Jockers, 1976). The numerical values are given in Table 2.107. The critical locus of this system is very similar to the banzene – water system. If the three-phase line does not intersect the critical locus in a system first, the UCST exists. The LCST is most frequent in polar and associated systems characterized by specific attractive forces. The study of these forces aids the understanding of phase equilibria. A comprehensive discussion on the intermolecular interactions is presented in Sec. 2.1.

Table 2.107 Solubility of C_6H_5F in Water

Temp. (°C)	$x_{C_6H_5F}$ = 0.10	Total pressure (bars)				
	0.10	0.13	0.17	0.20	0.30	0.40
300	580	515	515	590	810	1120
310	1000	930	930	960	1180	1620
320	1440	1380	1380	1400	1650	2160
330	1900	1810	1810	1880	2170	2620
340	2470	2320	2320	2390	2700	3250
350	3080	2900	2900	2980	3300	3700
300 (saturated)	120	140	140	130	–	–

Source: Götze et al., 1975.

The pressure has only a small effect on the thermodynamic properties of condensed phases. For example, an increase of pressure of 250 atm alters the critical temperature by only 1.6°C (Prigogine and Defay, 1967).

Similar to the case of solid-liquid solubility, Getzen and co-workers (Getzen, 1970; Getzen and Ward, 1971; Wauchope, 1970; Wauchope and Getzen, 1972) showed the temperature dependence of solubility of nonpolar liquids in water:

$$R \log_{10} x_2^s + \frac{\Delta H_m^o}{T_m} - (0.000408)(T - 291.15)^2 = -C$$

where x_2^s = mole fraction of solute in liquid

T = absolute temperature, K

T_m = melting point of the solute, K

ΔH_m^o = enthalpy of melting per mole of pure solute at the melting point

C = empirical constant

This equation represented well the liquid-liquid solubility for nonpolar molecule – water systems between 0 and 43°C. It is sufficient to obtain only one solubility data point from experimental measurements for the establishment of constant C. This constant is empirical and shows a reasonably ordered behavior when plotted versus the formula weight for a homologous series (Getzen and Ward, 1971). This correlation makes it possible to establish C values when no solubility data are available. Once the C constant is known, the solubility as a function of temperature can be calculated, using the melting point and the heat of melting of the solute.

The expression above can be rewritten

$$T = 291.15 + \left[\frac{R \log_{10} x_2^s + \Delta H_m^o/T_m + C}{0.000408} \right]^{\frac{1}{2}}$$

The right-hand side of this equation is a linear function of temperature and the straight line goes through the origin at 0 K. A graphical illustration and data for the solubility of liquid p-dichlorobenzene in water is given in Fig. 2.74 and Table 2.108. The straight-line representation is valid in only a limited temperature interval, when the solubility curve does not show a minimum, contrary to the case of 1,1,2,2-tetrabromoethane – water system (see Fig. 2.74 and Table 2.109).

Getzen (1970) pointed out that the treatment of the solubilities presented is limited to systems that follows Henry's law. This is generally the case for nonpolar compounds at and above room temperature. The changes in heat capacity differences (ΔC_p) with temperature and the variation of liquid volume have not been considered. However, from the results obtained, Getzen concluded that the aqueous solutions of nonpolar molecules show the highly structured character of water and the decrease in

Table 2.108 Solubility of p-Dichlorobenzene in Water at Saturated Pressure[a]

Temperature		Solubility in H_2O, x_2^s	$R \log_e x_2^s$	$R \log_e x_2^s + \dfrac{\Delta H_m^o}{T_m} + C$
K	°C	(mole fraction)		
332.35	59.2	2.5735×10^{-5}	-21.0011	1997.0128
333.85	60.7	2.6408×10^{-5}	-20.9498	2122.7536
338.25	65.1	2.8640×10^{-5}	-20.7886	2517.9601
338.35	65.2	2.8695×10^{-5}	-20.7847	2527.3051
346.55	73.4	3.4106×10^{-5}	-20.4414	3368.7436
373.15	100.0	7.2227×10^{-5}	-18.9503	7023.5277

0.000408   + 291.15

[a] ΔH_m^o = 4.34 kcal g-mole^{-1}; ΔS_m^o = 13.3071 cal g-mole^{-1} K^{-1}; T_m = 326.14 K; C = 8.39.

Source: Wauchope, 1970.

Table 2.109 Solubility of 1,1,2,2-Tetrabromoethane in Water at Saturated Pressure[a]

| Temperature | | Solubility in H_2O, x_2^s | $R \log_e x_2^s$ | $R \log_e x_2^s + \dfrac{\Delta H_m^o}{T_m} + C$ |
K	°C	(mole fraction)		$0.000408 \qquad + 291.15$
274.15	1	2.7257×10^{-5}	-20.8869	580.0695
283.15	10	2.6027×10^{-5}	-20.9787	355.1544
298.15	25	2.5947×10^{-5}	-20.9848	340.1597
313.15	40	2.8371×10^{-5}	-20.8073	775.1811
333.15	60	3.6898×10^{-5}	-20.2851	2,055.1867
353.15	80	5.6553×10^{-5}	-19.4364	4,135.1343
373.15	100	10.2152×10^{-5}	-18.2614	7,015.1784
473.15	200	23.0744×10^{-3}	-7.4902	33,415.1787

[a] $\dfrac{\Delta H_m^o}{T_m} = \Delta S_m^o = 10.2088$ cal g-mole^{-1} K^{-1} ; $C = \dfrac{10.796}{21.0048}$.

Source: Gooch, 1971.

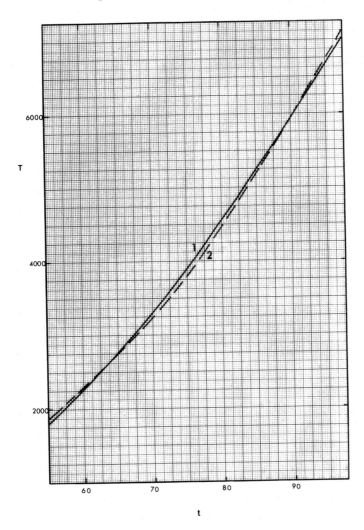

Fig. 2.74 Solubility of p-dichlorobenzene and 1,1,2,2-tetrabromoethane in water.
1, p-Dichlorobenzene; 2, 1,1,2,2-tetrabromoethane. t = temperature (°C);

$$T = \frac{R \log_e x_2^s + \Delta H_m^o/T_m + C}{0.000408} + 291.15 \quad \text{(From Wauchope, 1970; Gooch, 1971.).}$$

this structure with increasing temperature. The unique character of water as a solvent
is strongly dependent upon temperature.

In their investigations, Getzen (1970, 1976), Getzen and Ward (1971), and
Wauchope and Getzen (1972) neglected the heat-capacity corrections. However, these
corrections are needed for the endothermic interaction of the π-cloud with water, or
because the heats of fusion themselves are small relative to ΔH_{soln}^o for aromatic

compounds. It has been stated (Wauchope, 1970) that the temperature dependence of
liquid solubility in water is due to the great variety of vapor pressures of substances equilibrated with water. Consequently, the heat capacities of these solutions
will be considered. These are also indicated by the variation of heats of solution
with temperature (Timimi, 1974). The heat-capacity difference ($\Delta C_{p,\ soln}$) varies
greatly with temperature (Robertson, 1967; Wauchope, 1970) (see Table 2.110),
which is also apparent from the excess standard enthalpies of solution data. Wauchope's
(1970) calculation of the derived $\Delta C_{p,soln}$ values for liquid hydrocarbons in water
showed approximately similar values at the same temperature. Meanwhile, Robertson
(1967) pointed out that the available data are subjected to considerable error. The
temperature dependence of solubility is a result of the structural modifications in
water, and therefore more energy is required for cavity formation and for the decrease
in rigidity of the lattice (see Sec. 2.1).

Table 2.110 Heat-Capacity Difference of Solution of Liquid p-Dichlorobenzene in H_2O

Temperature (°C)	Heat-capacity difference of solution, $\Delta C_{p,soln}$[a] (cal g-mole^{-1} K^{-1})
0	51
50	99
100	161

[a] $\Delta C_{p,soln}$ = 6 × 0.000408 T^2 − 4 × 0.000408 T × 291.15 = 0.00244 T^2 − 0.475157 T.
(Source: Wauchope, 1970.)

The temperature dependence of the solubility of water in perfluorohexane between
2 and 53°C was expressed by the empirical equation (Shields, 1976a, b):

$$S_T \ (\mu g/g \ n\text{-}C_6F_{14}) = 0.053253(\exp(\frac{417909.2}{T^2} - \frac{5999.96}{T} + 7.28368))$$

where T is the absolute temperature (K).
The temperature dependence of liquid-liquid solubility has been investigated by
Gill et al. (1976). Their method involves the calculation of the minimum solubility
temperature from calorimetric measurement of the enthalpy of solution at infinite
dilution. The minimum solubility temperatures have also been studied by Nishino and
Nakamura (1978). The values obtained at different temperatures have been fitted into
a polynominal:

$$\Delta H_{soln}^{\infty} = a + b(T - 273.15) + c(T - 273.15)^2$$

where ΔH_{soln}^{∞} =   enthalpy of solution at infinite dilution

T   =   absolute temperature, K

a, b, c   =   coefficients of the equation

The change in the heat capacity was obtained by differentiation of the expression above:

$$\Delta C_{p,soln}^{\infty} = \frac{\left(d\ \Delta H_{soln}^{\infty}\right)}{dT} = b + 2c(T - 273.15)$$

where $\Delta C_{p,soln}^{\infty}$ is the change in heat capacity of a pure liquid, going to an infinite dilute aqueous solution. The calculation and interpretation of $\Delta \ddot{C}_{p,soln}^{\infty}$ have been discussed further by Konicek and Wadsö (1971), Kusano et al. (1973), Gill et al. (1975), Bergström and Olofsson (1975), and Timimi (1974).

The temperature dependence of ΔH_{soln}^{∞} is very large and varies from negative to positive values within a small temperature range; consequently, very large errors may be caused by an erroneous interpretation. Furthermore, there are large discrepancies between most published values by different investigators. As a conclusion, the success of the solubility calculations from heat of solution data depends very much upon the reliability and accuracy of the ΔH_{soln}^{∞} values. This has been emphasized by Gill et al. (1976).

The minimum solubility temperature (T_{min}) may be calculated from ΔH_{soln}^{∞} and $\Delta C_{p,soln}^{\infty}$ values (see the equation below). The temperature dependence of the heat of solution can also te expressed by

$$\Delta H_{soln}^{\infty} = \Delta H_{soln}^{\infty}(298.15) + \Delta C_{p,soln}^{\infty}(T - 298.15)$$

where ΔH_{soln}^{∞} =   enthalpy of solution at infinite dilution

$\Delta H_{soln}^{\infty}(298.15)$ =   enthalpy of solution at infinite dilution and 298.15 K

ΔH_{soln}^{∞} is zero at the minimum solubility temperature (T_{min}); therefore,

$$\Delta H_{soln}^{\infty}(298.15) + \Delta C_{p,soln}^{\infty}(T - 298.15) = 0$$

which becomes after rearrangement:

$$T_{min} = 298.15 - \frac{\Delta H_{soln}^{\infty}(298.15)}{\Delta C_{p,soln}^{\infty}}$$

Gill et al. (1976) have calculated the minimum solubility temperature for several hydrocarbons in water and concluded that the accuracy was about ± 3 K.

From the calorimetric values and the calculated minimum solubility temperature, the single solubility value can be extrapolated to other temperatures within a moderate temperature interval:

$$\log_{10} \left(\frac{x_2}{x_{2,min}} \right) = \frac{\Delta C_{p,soln}^{\infty}}{R} \left[\log_{10} \left(\frac{T}{T_{min}} \right) - 1 \right]$$

where x_2 =   solubility at temperature T, mole fraction

$x_{2,min}$ =   solubility at the minimum solubility temperature (T_{min}), mole fraction

Gill et al. (1976) have pointed out that the calculated $\Delta C_{p,soln}^{\infty}$ values reported by several investigators were unreliable. Furthermore, their own calorimetric technique (flow microcalorimetry) was not suitable for compounds more volatile than pentane (boiling point 36ºC). The calculation procedure can be simplified by assuming that $\Delta C_{p,soln}^{\infty}$ is constant over a 20 K range of temperature, that is, as far as the ΔH_{soln}^{∞} is linear with temperature (Gill et al., 1975). This assumption has been reported by several investigators (see, e.g., Bohon and Claussen, 1951, in Fig. 2.75). However, for a wide temperature interval, a systematic deviation will appear between the experimental and calculated solubility data, which is due to the dependence of the heat-capacity change with temperature (Robertson, 1967).

In the solubility calculation using the foregoing equation, it has been assumed that the liquid organic phase is pure; that is, the activity coefficient of the solute is unity, and $\Delta C_{p,soln}^{\infty}$ is constant within the calculated temperature interval. For slightly soluble solvents, the method described above gives an easy and practical way to extrapolate a single measured value to another temperature. At the present time there is no equation available that can represent the liquid-liquid solubility curve from the triple point up to the critical temperature.

The only published information available on the pressure effect on liquid-liquid binary systems of a halogenated hydrocarbon with water is that for a fluorobenzene - water system up to 3500 bars of pressure (Götze et al., 1975)(see Table 2.111 and Fig. 2.73; see also Fig. A7 for a $H_2O - C_6H_4F_2$ system). The critical curves of the fluorobenzene - water system is compared with other aqueous solutions of nonpolar compounds in Fig. 2.76 (Jockers, 1976). Because of the expected resemblance between the aqueous systems of benzene, toluene, and fluorobenzene, a detailed examination is needed to explain the various phenomena that occur at liquid-liquid solubility under pressure. The pressure dependence of the solubility of benzene and toluene in water has been investigated by Bradley et al. (1973). The mutual solubility of 2-butanol - water and i-butanol - water systems has been extensively studied under high pressures (Moriyoshi et al., 1975, 1977), which is also useful for the understanding of heterogeneous liquid-liquid equilibria under pressure. The clear and simple illustrations using isobars provide an easy understanding of the pressure effect (see Fig. 2.77).

It is important to compare the pressure effect on the two systems benzene - water and toluene - water, because they show a prominent difference when their solubilities

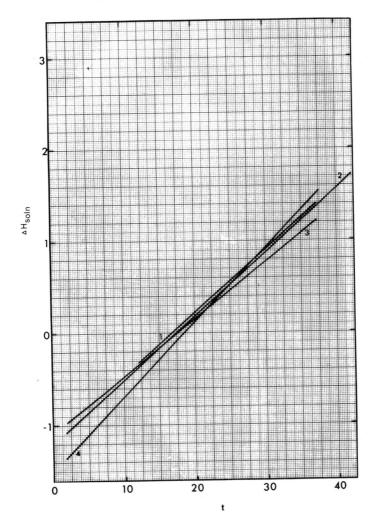

Fig. 2.75 Temperature dependence of the heat of solution in water. ΔH_{soln} = heat of solution (kcal g-mole^{-1}); t = temperature ($^{\circ}$C). 1, Benzene; 2, toluene; 3, ethyl-benzene; 4, m-xylene. (From Bohon & Claussen, 1951.)

are expressed as a logarithm of the mole fraction plotted versus the pressure. Fig. 2.78 illustrates the two sets of solubility curves. The pressure effects were determined up to several hundreds of bars. Whereas the solubility curves are linear for the solubility of benzene in water (i.e., the change in the partial molar volume ($-\Delta \bar{V}_2$) is roughly constant with varying pressure (except at low pressures)), the effect of pressure on the structure of water decreases at high pressures for the toluene – water system (i.e., $\Delta \bar{V}_2$ value becomes zero and then changes sign at sufficiently high

Table 2.111 Critical Curve of Water – Fluorobenzene System

Temperature (°C)	Pressure (bars)	
295	250	250
300	505	140
320	1320	150
340	2290	165
360	3410	180
374		221

Source: Götze et al., 1975.

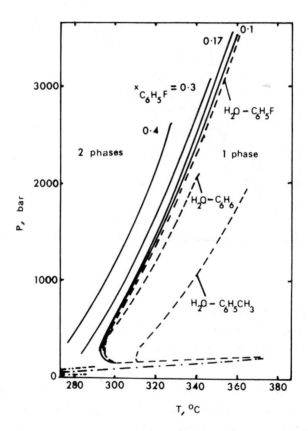

Fig. 2.76 Critical curves and isopleths of aqueous nonmiscible systems. Isopleths (constant concentrations) for C_6H_5F – H_2O system (mole fraction): $x_{C_6H_5F} = 0.4$; $x_{C_6H_5F} = 0.3$; $x_{C_6H_5F} = 0.17$; $x_{C_6H_5F} = 0.1$.

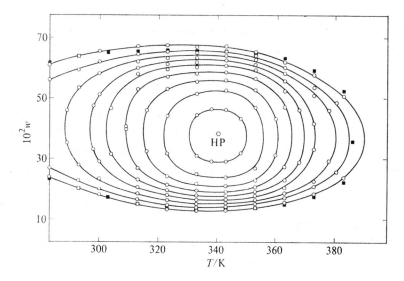

Fig. 2.77 Isobaric solubility curves in 2-butanol - water system. w = mass fraction of 2-butanol; T = absolute temperature (K). Isobars from outside to center refer to those at 1, 100, 200, 300, 400, 500, 700, and 800 atm. o, this work; ·, literature at 1 atm. (From Moriyoshi et al., 1975.)

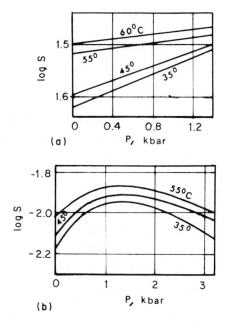

Fig. 2.78 Effect of pressure on the solubility of benzene (a) and toluene (b) in water. S = solubility in water (g-mole liter^{-1}); P = pressure (kbars); t = temperature (°C). (From Bradley et al., 1973.)

pressures). The difference in behavior between these two systems is also expressed through the difference in the partial molal heat content of the solutes:

$$\Delta \bar{H}_2 \quad = \quad \bar{H}_2^{aq} \quad - \quad \bar{H}_2^{solute}$$

plotted versus the pressure. The $\Delta \bar{H}_2$ for benzene is linear, whereas for toluene it shows a minimum. The explanation is given by the changes taking place during the contact between solute and solvent molecules, such as heat of solution, breakdown of the water structure, and cavity formation. Molecular interactions are discussed fully in Sec. 2.1.

In a binary system of water (1) and a halogenated hydrocarbon (2), forming two phases, A (rich in water) and B (rich in halogenated hydrocarbon), the isothermic variation with pressure of the mole fraction x_2 of halogenated hydrocarbon in the aqueous layer has been given by Bradley et al. (1973):

$$\left(\frac{\partial x_2}{\partial p} \right)^A_T \quad = \quad \frac{\Delta \bar{V}_1 \, x_1^B \; + \; \Delta \bar{V}_2 \, x_2^B}{x_2 \left(\dfrac{\partial^2 g}{\partial x_2^2} \right)^A_T}$$

or

$$\left(\frac{\partial \log_e x_2}{\partial p} \right)^A_T \quad \simeq \quad - \frac{\Delta \bar{V}_2}{R \, T}$$

where g = Gibbs free energy per mole

x_2 = $x_2^{aq} - x_2^{solute}$, mole fraction difference

$\Delta \bar{V}$ = change in the partial molal volume

From this equation it is clear that for constant $\Delta \bar{V}_2$ values, the isothermal solubility curve will be linear (e.g., for benzene), whereas if the sign of $\Delta \bar{V}_2$ changes with varying pressure, the solubility curve will show a minimum (e.g., for toluene) (Bradley et al., 1973).

A more complicated solubility diagram for liquid-liquid solubility is given in Fig. 2.77 for a 2-butanol - water system (Moriyoshi et al., 1975). The solubility isotherms for this system follow satisfactorily the rectilinear diameter law with respect to pressure, represented by the equation

$$\frac{x_2^{aq} \; + \; x_2^{org}}{2} \quad = \quad A \; + \; B \, p$$

where x_2^{aq} = mole fraction of 2-butanol in water-rich phase

x_2^{org} = mole fraction of 2-butanol in organic-rich phase

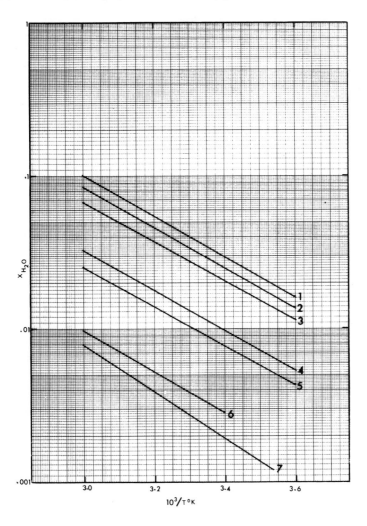

Fig. 2.79 Solubility of water (mole fraction) in organic liquids. 1, C_6H_6; 2, C_6H_5Cl; 3, $o-C_6H_4Cl_2$; 4, $1,2,4-C_6H_3Cl_3$; 5, CCl_4; 6, C_6H_{12}; 7, C_6H_5Br.

p   =   pressure

A and B =   coefficients of the equation

In dilute solutions the van't Hoff equation

$$\left(\frac{d \log_{10} x_2}{dT}\right)_P = \frac{\Delta H_{soln}}{R T}$$

where ΔH_{soln} is the heat of solution of a dilute solute, provides a simple and quick way to express the temperature dependence of the solubility of water in halogenated hydrocarbons. That is, a plot of $\log_{10} x_2$ versus $1/T$ gives a straight line for a moderate temperature interval (Hibbard and Schalla, 1952). This is illustrated in Fig. 2.79 for several water - organic compound systems, including halogenated hydrocarbons. From the slope of the line, the ΔH_{soln} can be determined. A list of estimated ΔH_{soln}'s for several halogenated compounds is given by Robertson (1967), Wen and Muccitelli (1979), and Muccitelli (1978). A complete list of ΔH_{soln}'s is provided in the Appendix. The heat-of-solution values, about 20 to 25 kJ g-mole^{-1}, indicate the assumption that hydrogen bonds are involved in the systems.

Price (1973) plotted the logarithm of the aqueous solubilities (0.1 to 500 ppm) of n-paraffins (pentane through nonane) in water as a function of temperature (25 to 175°C) at system's pressure. It is evident from the figures that the data points give a line with a much smaller slope than the slope of the line defined by the higher temperature data points. These two different slopes strongly suggest two different solution mechanisms. The mechanism operative at lower temperatures is probably some form of an aggregation phenomena (see also Sec. 2.5-2). However, it appears that the plots are actually hyperbolic-like curves and the transition from one solubility mechanism to the other is gradual.

For the solubility of water in chlorinated solvents, expressed in g H_2O/100 g solvent plotted versus the temperature on a semilogarithmic scale, almost parallel curves can be obtained (see Fig. 2.80). As shown in Fig. 2.79, the reciprocal temperature scale against the logarithm of solubility is preferable, because the straight lines provide easier extrapolation if required. This has also been shown by Hoot et al. (1957), Hoot (1957), and Zakuskina et al. (1975) for the solubility of water in hydrocarbons (see Fig. 2.81). The focal point is significant when the solubility lines are extrapolated.

Several investigators (e.g., Wing and Johnston, 1957; Jones and Monk, 1963) have used the semiempirical formula of Hildebrand et al. (1970):

$$R T \log_{10}\left(\frac{a_2}{x_2}\right) = V_2 \, \phi_1^2 (\delta_2 - \delta_1)^2$$

to fit the solubility data. This expression has been slightly modified:

$$\log_{10} x_2 = \log_{10} \phi_2 + 0.434 \, \phi_1 \left(1 - \frac{V_2}{V_1}\right) + V_2 \, \phi_1^2 \frac{(\delta_2 - \delta_1)^2}{2.3025 \, R T}$$

where ϕ_1 and ϕ_2 = volume fraction of solvent and solute, respectively

δ_1 and δ_2 = solubility parameter of solvent and solute, respectively

V_1 and V_2 = molar volume of solvent and solute, respectively

x_2 = solubility of solute in the solvent, mole fraction

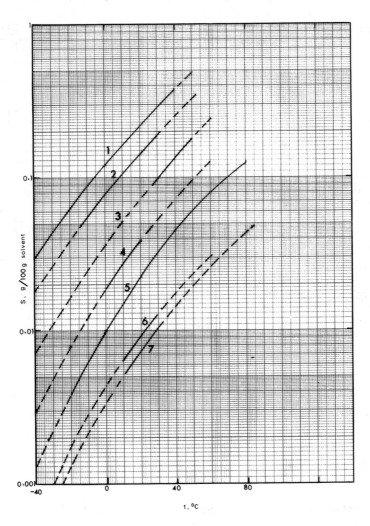

Fig. 2.80 Solubility of water in chlorinated solvents at saturated pressure. S = solubility (g H_2O/100 g chlorinated solvent); t = temperature (°C). 1, CH_3Cl; 2, CH_2Cl_2; 3, $CHCl_3$; 4, CH_2Cl-CH_2Cl; 5, $CCl_2=CHCl$; 6, CCl_4; 7, $CCl_2=CCl_2$. (From McGovern, 1943.)

to calculate the solubility parameter of water. The δ_{H_2O} found at 25°C varied with the nature of the solvent (Wing and Johnston, 1957). This is due to the dipole-dipole effect between the solute and solvent molecules. There is a general decrease in the solubility of water as the polarity of the solvents increases. This could be caused by the difference in polarity interaction between the polar solvent and water. That is, the solubility is reduced by the strong polar solvent, showing a larger difference

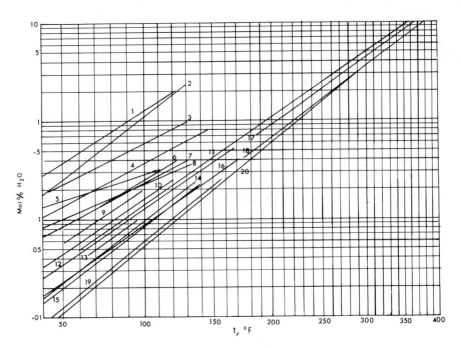

Fig. 2.81 Solubility of water in hydrocarbons at various temperatures (t, °F).
1, Heptene-1; 2, hexadiene-1,5; 3, styrene; 4, benzene; 5, butadiene-1,3; 6, i-butene;
7, butene-1; 8, butene-2; 9, n-octane; 10, n-heptane; 11, lube oil (mol. wt. = 400);
12, n-hexane; 13, cyclohexane; 14, i-pentane; 15, n-pentane; 16, propane; 17, kerosine
naphtha (mol. wt. = 150); 18, kerosine naphtha (mol. wt. = 170); 19, i-butane; 20,
n-butane. (From Hoot, 1957; Hoot et al., 1957.)

in the interaction (orientation) of the solute and solvent dipoles.

The solubility parameter for water at 25°C for several aromatic hydrocarbon –
water systems was found to be 24.8 $(cal/cc)^{\frac{1}{2}}$ by Wing and Johnston (1957) and 24.9
$(cal/cc)^{\frac{1}{2}}$ by Jones and Monk (1963). A tabulation of the solubility parameters of
halogenated hydrocarbons is given in the Appendix.

Despite the fact that the solubility parameter is a function of temperature, no
attempt was made to express the solubility in terms of the solubility parameter as a
function of temperature. In other words, by introducing various temperatures and solubility parameters in the foregoing equation, the solubility values obtained will
differ significantly from the experimental solubility data.

2.8-3 Solubility of Gases in Liquids

In a binary system consisting of components 1 and 2, where either component is in
the liquid or the gaseous state, Raoult's law gives the empirical relation between the

mole fraction (x_i) and the vapor pressure (p_i^o), assuming an ideal solution:

$$p_1 = x_1 \, p_1^o$$

$$p_2 = x_2 \, p_2^o$$

and

$$p_1 + p_2 = P_{tot}$$

where p_1 and p_2 = partial pressures of components 1 and 2, respectively

P_{tot} = total pressure of the system

However, the simple relation among the partial pressure (p_i), mole fraction (x_i), and vapor pressure (p_i^o) is valid only in ideal solutions, where the molecular sizes and the attractive and/or repulsive forces of the two components are about the same, the heat of mixing is zero, and there is no change in the total volume. Furthermore, the molecules are nonpolar, and there is no concentration difference at the surface of the solution (Christian and Tucker, 1977) or molecular association between unlike molecules. It follows from the simple equation given above that the solubility of a particular gas in ideal solution of any solvent is always the same. These conditions are never fullfilled for a binary mixture of an organic substance with water; consequently, Raoult's law cannot be applied. But Raoult's law holds fairly well with nonpolar gases in nonpolar liquids (e.g., benzene in toluene or alcohols and paraffin hydrocarbon gases in paraffin oils).

Comparing the experimental with the ideal (using Raoult's law) solubility of nonpolar gases (such as most of the halogenated hydrocarbons) in water (polar solvent), it is apparent that the actual solubility values are about 1% of their solubilities in ideal solutions. The ideal solubility of several halogenated hydrocarbons in water at 25°C and 1 atm partial pressure is compared in Table 2.112 with the actual solubilities. The temperature dependence of solubility between 0 and 374°C at 1 atm partial pressure for halogenated gases is illustrated in Fig. 2.82.

Because attractive and/or repulsive forces between the molecules in an ideal solution are approximately the same, the temperature will not affect the ideality, which is also supported by experimental evidence. The ideal behavoir of a solution is independent of temperature. In other words, the increasing temperature has a greate and greater influence on nonideal solutions which deviate more and more from ideal behavior (see also Sec. 2.5-3).

Similarly, the pressure does not effect the ideality of a solution; consequently, it remains ideal throughout a range of pressure. The force required to liberate the molecules from an ideal solution will not be affected by the applied pressure to the same extent as it would be in a pure liquid.

In reality, there are no ideal solutions, but practically, many are so nearly

Table 2.112 Comparison of Ideal and Actual Solubilities of Halogenated Hydrocarbon Gases in Water at 25°C and 1 atm Partial Pressure

Solute gas	Boiling point (°C)	Solubility in water (mole fraction)		Deviation (%)
		Ideal	Actual	
CH_3Br	3.52	0.4614	0.002925	15,774
CH_3Cl	-28.78	0.1736	0.001907	9,103
$CHClF_2$	-40.92	0.0945	0.0006055	15,607
CHF_3	-82.01	0.0211	0.001056	1,998
$CHCl=CH_2$	-13.35	0.2573	0.0008273	31,101

ideal that for engineering purposes the assumption is considered (e.g., mixtures of members of the homologous series of organic compounds).

For nonideal liquid solutions, Raoult's law gives incorrect results and apparently does not represent the solubility data determined experimentally. Over a modest concentration range or in very dilute solutions, Henry's law is often applied:

$$p_2 = H x_2$$

where H = Henry's law constant, usually expressed in atm/mole fraction

p_2 = partial pressure of the solute, atm

x_2 = concentration of the solute in the solvent, mole fraction

This equation states that a gas solubility depends upon its partial pressure above the liquid phase. In other words, the increasing pressure of the gas will increase the mass dissolved (Jhon et al., 1972). Because the halogenated hydrocarbons are less soluble gases in water, it can be expected that they will follow Henry's law up to an equilibrium partial pressure of 1 atm if the mole fraction does not exceed 0.02 (King, 1969) or if the H values are very high. The Henry's law constant (H) is only temperature-dependent; by replacing the partial pressure of the gas (p_2) with its fugacity (f_2), the expression becomes more general:

$$f_2 = H x_2$$

and H will be independent of both temperature and pressure. This equation may be used at elevated pressures for slightly soluble gases. The fugacity values may be calculated from P-V-T data for a real gas: for example, from the expression

$$R T \log_{10}\left(\frac{f_2}{P}\right) = \int_0^P \left(V - \frac{R T}{P}\right)dT$$

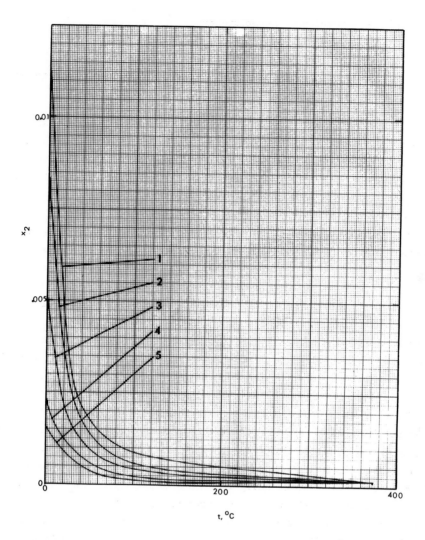

Fig. 2.82 Solubility of gaseous halogenated hydrocarbons in water at 1 atm partial pressure. x_2 = solubility in water (mole fraction); t = temperature (°C). 1, CH_3I; 2, CH_3Br; 3, CH_3Cl; 4, CH_3F; 5, $CHCl=CH_2$.

However, neither the equation above is satisfactory at high pressures.

At elevated pressures the fugacity of the liquid becomes pressure-dependent and the gas cannot be assumed to be ideal. Krichevsky and Kasarnovsky (1935) have introduced an expression for nonpolar gases disssolved in water:

$$\log_{10}\left(\frac{f^G}{x_2}\right) = \log_{10} H' + \frac{\bar{v}_2^L P}{R T}$$

where f^G = fugacity of the solute gas

x_2 = mole fraction of the solute in the solvent

H' = modified Henry's law constant

$\bar{v}_2^L$ = **partial molar volume of the solute dissolved in the solvent**

P = total pressure of the system

By plotting $\log_{10}(f^G/x_2)$ versus P for the solubility of H_2 and N_2 in water at constant temperature, a straight line was obtained up to 1000 atm pressure. Similarly, Weiss (1974) reported straight lines for the solubility of CO_2 in water up to 500 atm. The validity of the foregoing expression has been further supported by Krichevskii and Efremova (1948) at high pressures (e.g., for the solubility of H_2 and N_2 in benzene). However, the correlation is limited to slightly soluble solutes. In the case of high solubility values (e.g., for the solubility of ammonia in water), the discrepancy between the correlated and measured data increases. Further study on the solubility of gases at high pressures was reported by Müller (1975) and Tsiklis (1968).

Kritchevsky and Iliinskaya (1945) obtained different experimental and calculated values for the partial molar volume of H_2 , N_2 , CO_2 , CO, O_2 , air, and CH_4 in water, methanol, and benzene using the foregoing equation in the calculations. The partial molar volume $(\bar{v}_2^L)$ of slightly soluble solutes can also be estimated from their partial molar volumes in similar solvents, where data are available. The partial molar volumes of a large number of solutes in water are listed in the Appendix.

The foregoing equation has been further modified by Lachowicz et al. (1955) for the solubility of nonpolar gases in nonpolar solvents, giving more accurate data for the calculation of the solubility of H_2 and D_2 in n-heptane and n-octane up to 700 atm pressure. An equation developed by Michels et al. (1936) is very similar to the one above, using the experimentally determined solubility of CH_4 in water and electrolyte solutions up to 300 atm pressure. Orentlicher and Prausnitz (1964) also slightly modified the equation by also taking into account deviations from Henry's law due to the effects of composition and total pressure. The straight-line relation between $\log_{10}(f^G/x_2)$ and total pressure (P) was successfully confirmed for the solubility of H_2 in a large number of cryogenic solvents at high pressures. However, for most practical applications, the original equation by Krichevsky and Kasarnovsky (1935) gives satisfactory correlation.

The solubility of gases in water from 0 to the critical point of water was studied by Himmelblau (1960). The solubility expressed as logarithm of the Henry's law constant (H) was plotted versus the reciprocal of the absolute temperature. The equation was fitted by a least squares technique,

$$\log_{10}\left(\frac{H}{H_{max}}\right) = 1.142 - 2.846\left(\frac{1}{T}\right)^* + 2.486\left(\frac{1}{T}\right)^{*2} - 0.976\left(\frac{1}{T}\right)^{*3} + 0.2\left(\frac{1}{T}\right)$$

where

$$\left(\frac{1}{T}\right)^* = \frac{(1/T) - (1/T_{cr})}{(1/T_{max}) - (1/T_{cr})}$$

T_{cr} is the critical temperature of water (647 K), $1/T_{max}$ is the value of $1/T$ at H_{max} (H_{max} = maximum value of H versus $1/T$ curve). In the investigation the author examined and fitted the solubilities of O_2, N_2, H_2, He, Xe, and CH_4 in water.

The solubility of water in compressed gaseous hydrocarbons can be correlated using the expression developed by Rigby and Prausnitz (1968) for the solubility of water in compressed N_2, Ar, and CH_4 up to 100 atm pressure:

$$\frac{P\,y_2}{p_w^o} = A\,P + 1$$

where   P   =   total pressure of the system

y_2 =   mole fraction of water in the gaseous phase

p_w^o =   vapor pressure of liquid water

A   =   constant, depends on the gas studied and temperature of the system

This empirical expression is without theoretical significance. Gardiner (1970) has derived a very similar relationship, showing some thermodynamic justification for

$$\frac{P\,x_2}{p_w^o} = 1 + \frac{P(v_1^o - 2\,B_{12} - B_{22})}{R\,T}$$

where v_1^o =   molar volume of pure water

B_{12} and B_{22} =   second virial coefficients of mixture and water, respectively

The second virial coefficients of pure gases and mixtures were compiled by Dymond and Smith (1979). A fairly complete bibliography of all papers published between 1920 and 1967 has been surveyed by Mason and Spurling (1969).

For the solubility of water in compressed hydrogen between 50 and 300°C, Maslennikova et al. (1976) used Krichevskii-Khazanova's equation, which provided good correlation possibilities in several respects. The straight-line relationship between the solubility expressed as a mole fraction of water was correlated with other variables, such as total pressure, and density. Graphical illustrations were presented for x_2 P versus P, $\log_{10} x_2$ versus $1/T$, $\log_{10} m$ versus d (density), and so on (here m is the amount of water in the compressed hydrogen (g cc^{-1}) and d is the density of hydrogen in the solution, g-mole cc^{-1})).

In gas-liquid equilibrium conditions, the chemical potentials (μ_i) of the ith component are equal in both phases; that is,

$$\mu_2 \text{ (gas)} = \mu_2 \text{ (liquid)}$$

The chemical potential of a gas in equilibrium with a liquid can be expressed by the relationship

$$\mu_2 \text{ (gas)} = \mu_2^o \text{ (gas)} + R T \log_e P_2$$

where μ_2^o (gas) = standard-state chemical potential of the gas

P_2 = equilibrium vapor pressure of the solute above the solution

R = gas constant

Similarly, the chemical potential of a liquid is described by the equation

$$\mu_2 \text{ (liquid)} = \mu_2^o \text{ (liquid)} + R T \log_e(\gamma_2 x_2)$$

where μ_2^o (liquid) = standard-state chemical potential of liquid

x_2 = mole fraction of solute in the solution

γ_2 = activity coefficient of the solute in the solution

Combining the three equations above, we obtain

$$\mu_2^o \text{ (gas)} + R T \log_e P_2 = \mu_2^o \text{ (liquid)} + R T \log_e(\gamma_2 x_2)$$

If Henry's law is obeyed by the solution, then x_2/P_2 is only temperature-dependent; like the two standard chemical potentials, it is independent of the concentration. Consequently, the activity coefficients are also constant and equal to the value for the infinite dilute solution γ_2^∞.

Henry's law expresses the relationship between the solubility of a gas in liquid and the partial pressure of the gas above the solution:

$$H = \frac{p_2}{x_2}$$

where H = Henry's law constant (independent of concentration, but dependent upon the temperature)

p_2 = partial pressure of solute above the solution, atm

x_2 = solubility of solute in the solvent, mole fraction

Regarding the determination of H with good accuracy, see, for example, Fredenslund and Grausø (1975).

The temperature dependence of Henry's law constant was correlated by Benson and Krause (1976) using the empirical expression:

$$\log_e(\frac{1}{H}) = A(\frac{T_1}{T} - 1) + B(\frac{T_1}{T} - 1)^2$$

where T_1 is the absolute temperature (K) at which the Henry's law coefficient hypo-

thetically would be unity, A is a dimensionless constant (= 36.855), and B is a
linear function of the square root of the force constant (ϵ/k) of the gas.

The expression is particularly accurate for the representation of the solubility
of nonpolar gases in water within 0 and 50°C.

Gokcen (1973) and Gokcen and Chang (1973, 1975) carried out a critical evaluation of the existing methods of representing the temperature dependence of sparingly
soluble nonelectrolytes in water. The authors also discussed the effect of highly
polar and associated solvents. In a narrow temperature interval of 50°C, the solubility was accurately represented by the Valentiner's equation

$$\log_{10} x_2 = \frac{A}{T} + B \log_{10} T + C$$

where A, B, and C are typical coefficients for a binary system.

The free energy of solution (ΔG^o_{soln}) is given by the relation

$$\Delta G^o_{soln} = R T \log_e H$$

For a gas at 1 atm partial pressure this equation becomes

$$\Delta G^o_{soln} = -R T \log_e x_2$$

The free energy of solution (ΔG^o_{soln}) is defined as follows:

$$\Delta G^o_{soln} = T \Delta S^o_{subl} - \Delta H^o_{vap} + \Delta G^o_{ex}$$

where $\quad \Delta H^o_{vap} \quad =$ standard heat of vaporization of the pure solute

$\quad \Delta S^o_{subl} \quad =$ entropy of sublimation of the pure solute

$\quad \Delta G^o_{ex} \quad =$ correction term, depends upon the temperature

Getzen (1970) expressed the correction term (ΔG^o_{ex}) as a function of temperature:

$$\Delta G^o_{ex} = -a + c T - (0.000408\ T)(T - 291.15)^2$$

where a and c are constants, specific for a solute. From these equations we can
express the solubility (x_2) as

$$\log_e x_2 = \frac{1}{R T}(\Delta H^o_{vap} - T \Delta S^o_{subl} - \Delta G^o_{ex})$$

at 1 atm partial pressure. This equation can also be expressed

$$\log_e x_2 = \frac{1}{R T}\left[\Delta H^o_{vap} - T \Delta S^o_{subl} + a - c T + (0.000408\ T)(T - 291.15)^2\right]$$

This relationship gives the gas solubility, expressed as a mole fraction, as a function of temperature. In addition to ΔH^O_{vap} and ΔS^O_{subl}, a minimum of two solubility values are required to fit the equation for the calculation of gas solubility data at various temperatures.

Like the free energy of solution ΔG^O_{soln}, the other thermodynamic relationships are expressed as a function of temperature (Wauchope, 1970):

$$\Delta H^O_{soln} = \Delta H^O_m - \Delta H^O_{subl} + b\, T^2 + 2(0.000408\, T^2)(T - 291.15)$$

and

$$\Delta S^O_{soln} = -\Delta S^O_m - \Delta S^O_{vap} + 2(0.000408\, T^2)(T - 291.15)$$

where
ΔH^O_m = heat of fusion

ΔS^O_m = entropy of fusion

ΔH^O_{subl} = heat of sublimation

ΔS^O_{vap} = entropy of vaporisation

Enthalpy and entropy of solution of nonpolar solutes in water diverge strikingly from the normal behavior of regular solutions (Klots and Benson, 1963). This abnormality is due to the iceberg formation around the solute molecules in water. More information on the microscopic icebergs around solute molecules and hydrogen-bond breaking is given in Sec. 2.5-2. Enthalpy-of-solution data are listed in the Appendix.

In regular solutions (see Hildebrand and Scott, 1962), the enthalpy and entropy of solutions are expressed by

$$\Delta H^{soln}_2 = V_2\, \phi_1^2\, B'$$

and

$$\Delta S^{soln}_2 = -R \log_e x_2$$

where
B' = empirical constant

V_2 = molar volume of the solute

ϕ_1 = volume fraction of the solvent

x_2 = mole fraction of the solute in the solution

If in a solution of halogenated hydrocarbons in water, n moles of water form icebergs and n' units of hydrogen bonds are destroyed per mole of solutes, the enthalpy and entropy of solutions become

$$\Delta \bar{H}^{soln}_2 = V_2\, \phi_1^2\, B' - n\, \Delta \bar{H}^f_i + n'\, \Delta \bar{H}_h$$

and

$$\Delta \bar{S}^{soln}_2 = -R \log_e x_2 - n\, \Delta \bar{S}^f_i + n'\, \Delta \bar{S}_h$$

where $-n \, \Delta\bar{H}_i^f$ and $-n \, \Delta\bar{S}_i^f$ =   partial molal enthalpy and entropy changes of

solute due to iceberg formation of n moles of

the surrounding water

$n' \, \Delta\bar{H}_h$ and $n' \, \Delta\bar{S}_h$ =   partial molal enthalpy and entropy changes of

solute due to hydrogen-bond breaking

Shinoda and Fujihira (1968) showed that iceberg formation is predominant in aqueous solutions. If the temperature decreases, iceberg formation proceeds (i.e., n increases). They compared the solubility of hypothetical liquid alkanes in water assuming no iceberg formation with the experimental data. Although the temperature dependence of the hypothetical solubility

$$- \; \frac{\partial \log_e x_2}{\partial (1/T)}$$

gave straight lines, the experimental data decreased upon temperature lowering and the solubility increased relative to the iceberg formation of water. At high temperature, both the hypothetical and experimental solubility curves become asymptotic. Iceberg formation reduces the standard entropy of solute, whereas hydrogen-bond breaking increases the enthalpy and entropy of solute. These two abnormalities cancel each other out at high temperatures.

The enthalpy of mixing is a large positive value, as expected from the intermolecular forces. The small negative enthalpy of solution $\Delta\bar{H}_2^{soln}$ results from a large positive enthalpy of mixing and a large negative enthalpy of iceberg formation. The small solubility of halogenated hydrocarbons in water is mainly a result of the large positive enthalpy of mixing, but the enthalpy decrease due to the iceberg formation of water largely cancels it out. The solubility of nonpolar solutes in water is promoted by the iceberg formation of water molecules (Klots and Benson, 1963).

The temperature dependence of solubility of gases in regular solutions is represented by the relationship

$$\log_{10}(\frac{x_2}{x_1}) \;\; = \;\; \frac{\log_{10}(x_o/x_1) \, \log_{10}(T_2/T_1)}{\log_{10}(T_{cr}/T_1)}$$

where T_{cr} =   critical temperature, K

$$\log_{10}(10^4 \, x_o) \;\; = \;\; 2.265 \; - \; 0.134 \, \delta_1$$

δ_1 =   solubility parameter, $(cal/cc)^{\frac{1}{2}}$

In other words, there is a linear relation between the log solubility expressed as a mole fraction versus the log absolute temperature (Fleury and Hayduk, 1975; Hayduk and Castaneda, 1973; Hayduk and Cheng, 1970; Hayduk and Buckley, 1971; Sahgal et al., 1978). However, it has been pointed out that the application of regular-solution theory to polar or associated solvents, or to highly soluble gases, is simply inappropriate.

An investigation by Miller and Hildebrand (1968; see also Hayduk and Laudie, 1973) has shown that by plotting $\log_{10} x_2$ versus $\log_{10} T$, where x_2 is the solubility of a gas in water at 1 atm partial pressure, slightly concave isotherms are obtained, in contrast to regular solutions, for which the lines on a $\log_{10} x_2$ -- $\log_{10} T$ diagram were straight, that is, constant slopes (e.g., solubility of gases in n-heptane)(Hayduk and Laudie, 1973). The $\log_{10} x_2$ versus $\log_{10} T$ plots originate from Hildebrand and Scott (1950; see also Hildebrand, 1965, and Miller and Hildebrand, 1968), who applied the linear correlation for gases slightly soluble in nonpolar solvents (i.e., regular solutions). A more recent investigation is that reported by Keevil et al. (1978).

By plotting $\log_{10} x_2$ versus $\log_{10} T$, Hayduk and co-workers (Hayduk and Buckley, 1971; Hayduk and Castaneda, 1973; Hayduk and Cheng, 1970; Hayduk and Laudie, 1973, 1974a, b) found that the critical temperature of the solvent, all solubility curves merge together (see also Beutier and Renon, 1978). For the solubility of 20 gases in water at 1 atm partial pressure, the common solubility at the critical temperature of water (i.e., 374°C) is

$$x_2 = 1.40 \times 10^{-4} \text{ mole fraction}$$

This is a very significant discovery. The digram is shown in Fig. 2.83; the corresponding diagram for the solubility of gases in n-heptane (regular solution) is shown in Fig. 2.84. The difficulty arising during the extrapolation of water solubility of gases is apparent from Fig. 2.83. However, with the help of the critical solubility value, we are able to reduce the number of data points required for the completion of the solubility isotherm from 0 to 374°C. It must be emphasized that this diagram is valid only at 1 atm partial pressure.

The extrapolation of bilogarithmic curves of gas solubilities versus absolute temperature to the solvent critical temperature was re-examined by Beutier and Renon (1978). They showed that there is an exact value of the reference solubility derived from thermodynamic considerations. The limiting law justifies the reference solubility of Hayduk and co-workers. However, there is a considerable disagreement in the numerical values reported by Hayduk and Laudie (1974a) and Beutier and Renon (1978) for the reference solubility of gases in water at 1 atm partial pressure. Consequently, further investigations are needed in order to establish the exact value of the reference solubility. It is likely that accurate solubility data close to the critical point of water will revive that the reference solubility is not a constant value for all gases.

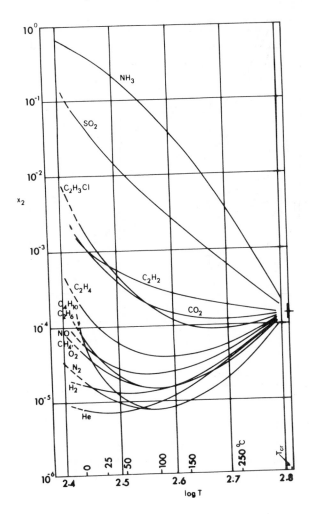

Fig. 2.83 Solubility of gases in water at 1 atm partial pressure. x_2 = solubility in water (mole fraction); T = absolute temperature (K). (From Hayduk and Laudie, 1973.)

A good illustration for the total pressure effect upon solubility of nonpolar gases was reported by Hayduk and Laudie (1974b) (see Fig. 2.85). The logarithm of solubility, expressed as a mole fraction, was plotted versus the logarithm of the total pressure (atm). The straight-line isotherms are typical for this type of diagram up to at least 20 atm total pressure, above the three-phase region.

The solubility of halogenated hydrocarbon gases in water at various temperatures and moderate pressures (up to 5 atm) has been studied by Horvath (1976b). The simple

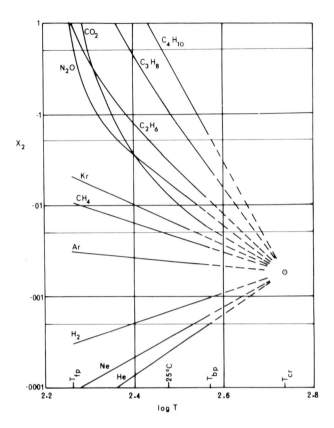

Fig. 2.84 Solubility of gases in n-heptane at 1 atm partial pressure. x_2 = solubility in n-heptane (mole fraction); T = absolute temperature (K). (From Hayduk and Castaneda, 1973.)

graphical representation presented provides a quick and sufficiently accurate method for checking the consistency of the available solubility data or to extrapolate or interpolate them for other conditions. The method includes plotting the available solubility data into three graphs. In the first figure (see Fig. 2.86) the total pressure is plotted versus the solubility, expressed as a mole fraction, for various isotherms. These isotherms are straight lines and intersect with the three-phase boundary line at the saturation pressures. On the other hand, the extrapolation of the isotherms coincide at the origin. However, the experimental values do not meet at the origin, because of the vapor pressure of water at x_2 = 0; they are only hypothetical values in the graphs, very close to the origin. The second graph (see Fig. 2.87) the solubility (x_2) is plotted versus the temperature at constant total pressure. The third figure (see Fig. 2.88) shows $\log_e x_2$ plotted versus $\log_e P$.

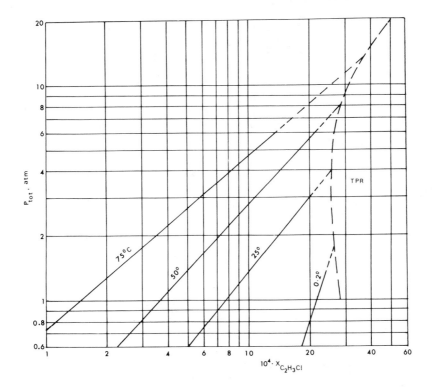

Fig. 2.85 Solubility of vinyl chloride in water at 0.2, 25, 50, and 75°C. P_{tot} = total pressure of the system (atm); $x_{C_2H_3Cl}$ = solubility of vinyl chloride (mole fraction); TPR = three-phase region. (From Hayduk and Laudie, 1974b.)

All the isotherms in this plot are also straight lines. The intersection is consistent at the boiling point in all three figures.

Watanabe and Ôkamoto (1978) reported the solubility of tetrafluoroethylene in water between 0 and 40°C and up to 30 kg cm^{-2} guage pressure. The solubility of tetrafluoroethylene increases with pressure, whereas the solubility decreases with temperature.

The temperature and pressure effects have also been studied by Schröder (1973)

The solubility of numerous gases in water at various temperatures has been illustrated in Landolt-Börnstein (1976a)(see Fig. 2.89). The solubility curves are plotted on a semilogarothmic scale. The temperature ranges from 0 to 150°C on the linear scale, whereas the solubility was expressed in two units on the logarithmic scale: g-mole kg^{-1} bar^{-1} and N m^3 ton^{-1} atm^{-1}.

Similarly, the solubility of steam in various hydrocarbons was plotted on a semilogarithmic scale (see Fig. 2.90)(Landolt-Börnstein, 1976a).

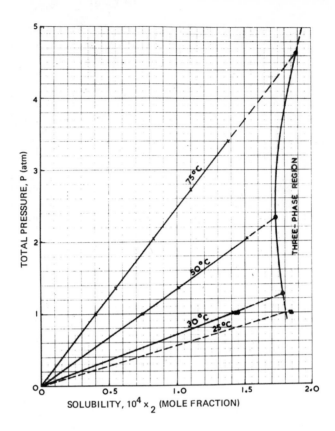

Fig. 2.86 Solubility of CFCl$_3$ (R 11) in water: pressure versus solubility. (From Horvath, 1976b.)

Both Figs. 2.89 and 2.90 are excellent graphical presentations of solubility data. The trends are especially clear on the graphs.

Whereas the representation, that is, correlation, interpolation, and extrapolation, of liquid-liquid solubility with temperature is relatively simple when the logarithm of solubility versus the reciprocal of the absolute temperature are plotted (see, e.g., Figs. 2.79 and 2.81), the temperature dependence of gas solubilities is more difficult to correlate. A typical example is shown in Fig. 2.91 for the solubility of vinyl chloride in water from 0 to 374°C. No simple expression could correlate these solubility data at such a wide temperature interval.

However, Berens (1974) showed that Henry's law is followed by the solubility data at 30 and 50°C. The plot of the pressure ratio $(p_{V_c}/p_{V_c}^o)$ versus solubility (S, mg vinyl chloride g^{-1} solvent) resulted in a straight line in the narrow temperature interval.

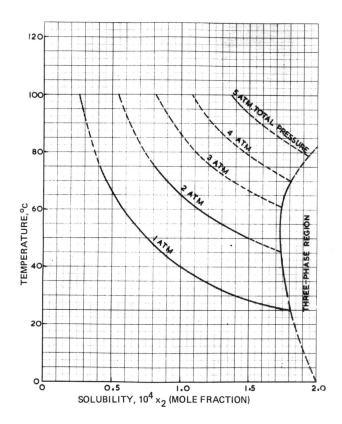

Fig. 2.87 Solubility of $CFCl_3$ (R 11) in water: temperature versus solubility. (From Horvath, 1976b.)

The temperature and pressure dependence of the gas solubilities in liquids has also been studied by Gerrard (1972, 1973, 1976), who introduced the "reference-line diagram" for the illustration of solubility data. The diagram is constructed from the vapor-pressure values of the solute at the specified temperatures. The vertical axes represent the vapor-pressure data using a linear scale; the horizontal scale is divided into 10 equal units and represents the mole ratio of the solute in the solvent (number of moles of solute dissolved in 1 mole of solvent). The diagram consists of a reference line (or diagonal), which is drawn from the origin to the actual vapor-pressure value at the specified temperature on the vertical axel at the right-hand side. The reference lines for vinyl chloride at various temperatures are illustrated in Fig. 2.92. The solvent is water. The diagonal lines start from $N_{VC} = 0$ at the left bottom corner and end at p_{VC}^{vap} of the specified temperatures when $N_{VC} = 1$. The solubility values obtained from Fig. 2.92 represent the ideal-gas solubilities (see also Table 2.113). Whereas the difference between the partial and total pre-

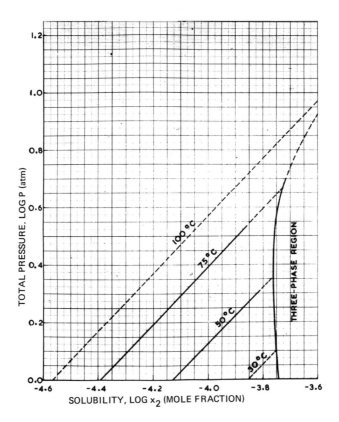

Fig. 2.88 Solubility of $CFCl_3$ (R 11) in water: log P versus log x_2. (From Horvath, 1976b.)

ssure data at the same temperature is not great, the ideal gas solubility is significantly higher. This is apparent from the discussion in Sec. 2.5-3.

Gerrard's reference-line diagram is different from that which has been introduced by Johnson et al. (1954) for the correlation of solubility data. In the proposed
method Johnson et al. plotted the solubilities of compounds against the solubility of
a reference compound whose chemical structure was similar to the former. The method
of correlation uses a logarithmic scale against the temperature. The correlation is
useful for interpolating and extrapolating whenever two solubilities at different
temperatures are known and when no further data are available.

For a narrow temperature interval Douabul and Riley (1979) used a four-term
version of Weiss' equation (Weiss, 1970, 1971) from which the salinity terms had been
omitted

$$\log_e S \text{ (mole liter}^{-1}) \ = \ A \ + \ 100 \ B/T \ + \ C \ \log_e(T/100)$$

Table 2.113 Solubility of Vinyl Chloride in Water at Various Pressures

Temperature	Vapor pressure of vinyl chloride		Solubility of vinyl chloride in water (mole fraction)		
(°C)	mm Hg	atm	Ideal, x_{id}	At 1 atm partial pressure	At 1 atm total pressure
0.2	1,307.9	1.7209	0.5811	0.00158	0.00217
25	2,970.9	3.9091	0.2558	0.000798	0.000779
50	5,920.4	7.7899	0.1284	0.000410	0.000368
75	10,576.7	13.9168	0.07186	0.000225	0.000146
100	17,304.4	22.7690	0.04392	0.000140	0.0000450

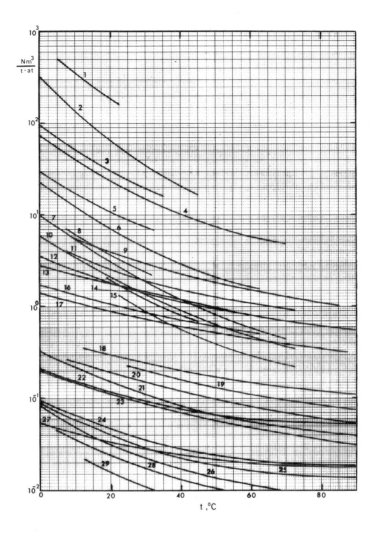

Fig. 2.89 Solubility of gases and vapors in water at 1 at partial pressure.
1, Ethylene oxide, C_2H_4O; 2, naphthalene, $C_{10}H_8$*; 3, dibromomethane, CH_2Br_2*; 4,
1,2-dichloroethane, $C_2Cl_2H_4$*; 5, dichloromethane, CH_2Cl_2*; 6, chloroform, $CHCl_3$*;
7, 1,1,2-trichloroethane, $C_2H_3Cl_3$*; 8, bromomethane, CH_3Br; 9, chlorofluoromethane,
CH_2ClF; 10, trichloroethylene, C_2HCl_3; 11, chloromethane, CH_3Cl; 12, methyl acetylene,
C_3H_4; 13, ethyl acetylene, C_4H_6; 14, ethyl chloride, C_2H_5Cl; 15, allyl chloride,
C_3H_5Cl*; 16, acetylene, C_2H_2; 17, vinyl acetylene, C_4H_4; 18, cyclopropane, C_3H_6;
19, 1,3-butadiene, C_4H_6; 20, cyclohexane, C_6H_{12}*; 21, propylene, C_3H_6; 22, ethylene,
C_2H_4; 23, n-heptane, C_7H_{16}*; 24, ethane, C_2H_6; 25, propane, C_3H_8; 26, n-butane, C_4H_{10};
27, methane, CH_4; 28, isobutane, C_4H_{10}; 29, neopentane, C_5H_{12}*. 1 N m^3/(ton at) =
0.045497 g-mole kg^{-1} bar^{-1}; 1 at = 0.980665 bar = 735.56 mm Hg. (From Landolt-
Börnstein, 1976a.)

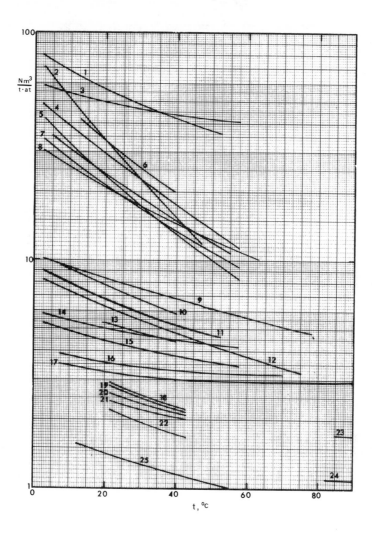

Fig. 2.90 Solubility of steam in hydrocarbons between 0 and 90°C. 1, 1-Pentene; 2, 1,3-butadiene; 3, 1,5-hexadiene; 4, n-octane; 5, 2-butene; 6, benzene; 7, 1-butene; 8, 2-methylpropene; 9, propane; 10; n-octane; 11, n-heptane; 12, n-hexane; 13, cyclohexane; 14, 2-methylbutane; 15, n-pentane; 16, 2-methylpropane; 17, n-butane; 18, n-decane; 19, n-undecane; 20, n-dodecane; 21, n-tridecane; 22, n-hexadecane; 23, kerosine naphtha (mol. wt. = 150); 24, kerosine (mol. wt. = 170). 1 N m^3/(ton at) = 0.045497 g-mole kg^{-1} bar^{-1}; 1 at = 0.980665 bar = 735.56 mm Hg. (From Landolt-Börnstein, 1976a.)

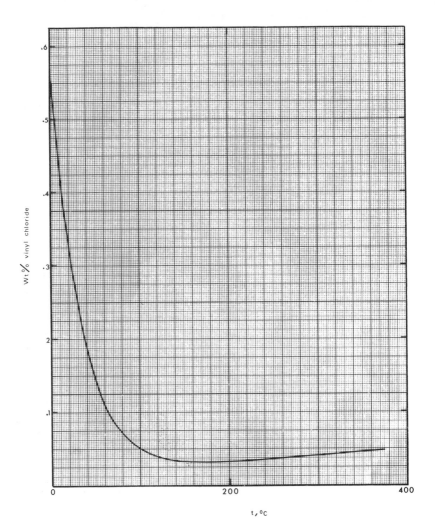

Fig. 2.91 Solubility of vinyl chloride in water at 1 atm partial pressure. (From
Hayduk and Laudie, 1974a, b; 1973.)

where S is the gas solubility, T is the absolute temperature (K), and A, B, and C
are coefficients of the equation. The same expression was used by Wen and Muccitelli
(1979) for the representation of the solubility of CF_4, C_2F_6, C_3F_8, and $c-C_4F_8$ in
water between 5 and 30°C and 1 atm system pressure. However, the solubility S was
replaced by the Henry's law constant p_2/x_2. By this substitution the equation can
represent both the temperature and partial pressure dependence of the gas solubility.
The coefficients of the equation for the mentioned gases are given by the authors.

In fact the equation was originally proposed by Valentiner (1927).

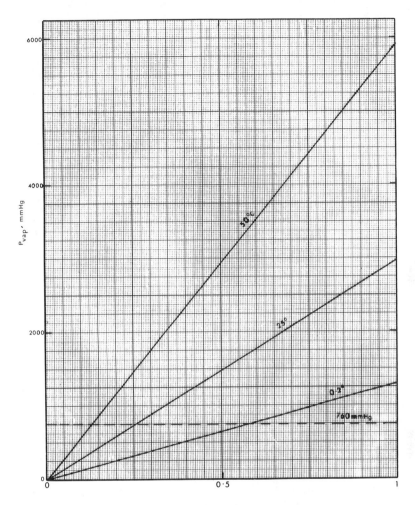

Mole ratio of vinyl chloride

Fig. 2.92 Reference-line diagram for the solubility of vinyl chloride in water. (From Gerrard, 1976.)

Kim and Brückl (1978) used the scaled particle theory for the selection of input parameters to express the temperature dependence of aqueous solubilities of inert gases. The theory is found to be reasonably applicable for the solution of inert gases in the temperature range considered.

When the gas solubility at one temperature is known, then the Jonah-King equation can be applied to predict gas solubilities over a range of temperatures (King et al., 1979). However, the equations derived are recommended for nonpolar solutes and solvents only.

In addition to the temperature and pressure effect upon the solubility of gases in solvents, the effect of the density of the solvent upon solubility has been investigated by Rusanov (1972). The derived general formula

$$\Delta N_2 = k T \rho_2^{(\alpha)} \left(\frac{d \log_e f_1^{(\alpha)}}{dP} + \frac{d \log_e \rho_1^{(\alpha)}}{dP} \right) - \frac{\rho_2^{(\alpha)}}{\rho_1^{(\alpha)}}$$

where ΔN_2 =   number of excess solvent molecules around the solute molecule

          k =   Boltzmann's constant

          T =   absolute temperature, K

          $\rho_1^{(\alpha)}$ and $\rho_2^{(\alpha)}$ =   number of solvent and solute molecules per unit volume, respectively

          $f_1^{(\alpha)}$ =   concentration activity coefficient of solvent

          P =   pressure

is suitable for the calculation of the increased concentration of solvent molecules in the vicinity of the solute molecules.

At low solubility the expression becomes somewhat simplier:

$$\Delta N_2 = X_2^{(\alpha)} k T \rho_2^{(\alpha)} \frac{d \log_e \rho_1^{(\alpha)}}{d \log_e \rho_2^{(\alpha)}} - \frac{\rho_2^{(\alpha)}}{\rho_1^{(\beta)}}$$

where $X_2^{(\alpha)}$ =   isothermal compressibility of the pure solvent

          $\rho_1^{(\beta)}$ =   number of solvent molecules per unit volume of solute

An interesting paper dealing with the effect of anesthetics on the solubility of a nonpolar compound in water was presented before the New York Academy of Medicine in 1976 (Buffington and Turndorf, 1976). Although in the experiment the solubility of benzene in water was decreased by the addition of nitrous oxide, cyclopropane, and ethyl chloride, it is expected that a general decrease of solubility of nonpolar substances will follow this phenomenon.

The effect of high magnetic field on the vapor-liquid equilibrium data for a nitrogen-oxygen system has been reported (Biddulph and Meachin, 1978); however, no information was found for the effect on solubility data.

A comprehensive work on the influence of isotopy on solubility was compiled by Rabinovich (1970).

Chapter 3
Sources of Information on Solubility

Solubility data are very widely published in the literature. Because of the broad spectrum of their applications, numerous publications deal with various aspects of solubility. Most handbooks in physical science, technology, and medicine list solubility data. Some of the most commonly used handbooks in technical libraries are described below.

Handbook of Chemistry and Physics (Hodgman et al., 1956; Weast, 1976). This publication primarily serves the needs of chemists and physicists, but the data presented are very useful to other scientists and engineers as well. The compilation of data is carried out by collaborators and contributors, mostly professional chemists and physicists. Each year a new edition is published by the Chemical Rubber Company. In earlier editions Section C contained numerical values on organic compounds and their solubilities in water; however, in recent editions there is only qualitative information on solubility (e.g., soluble, very soluble, insoluble, etc.) in water. The data listed are from Beilsteins Handbuch. The only quantitative solubility data are given for refrigerants R 11, R 12, R 13, R 13B1, R 14, R 21, R 22, R 23, R 112, R 113, R 114, R 115, and R 502 at 25°C and 1 atm. There is no source of references to these data; consequently, they have not been used in the correlations.

Handbook of Chemistry (Lange, 1973). This book is very similar in content and format to the preceding publication, but its publication is not as frequent. The sources of the data presented are not given. Mostly, single solubility values are available in this work at ambient temperature. Some tabulated solubilities of the most common gases in water are also included, but not the halogenated hydrocarbons. In the preface the editor states that "every effort has been made to select the most reliable information and to record it with accuracy," but because of the obscure nature of the literature references to the solubility data, they have not been included in the correlation procedures.

Chemical Engineers' Handbook (Perry and Chilton, 1973). Almost a hundred engineers, chemists, and other specialists have contributed to the 5th edition of this publication. The section editors and contributors are recognized experts in their field and hold positions in academic and industrial life. In Section 3, "Physical and Chemical Data," solubilities are given at one single temperature only, without references. In Section 14, "Gas Absorption," there are several tables and graphs on the

solubility of gases in water; however, not one halogenated hydrocarbon has been included. Furthermore, in Section 12, "Refrigeration," there is no information at all
on the solubilities of refrigerants.

Taschenbuch für Chemiker und Physiker (Lax, 1967). The 3rd edition of this handbook, published in German, appears in three parts. Part 1 contains solubility data,
usually at one single temperature only. The tabulated gas solubilities do not include
halogenated hydrocarbons. Sources of data are not given for the values reported. However, the reader can easily recognize the similarity between this book and Landolt-
Börnstein (1962).

American Institute of Physics Handbook (Gray, 1963). This publication of the
American Institute of Physics has been compiled by section editors working at scientific organizations and universities. Sources of data are the open literature,
private communications, and theoretical and experimental work at various scientific
establishments. Unfortunately, the solubility values presented are very limited, and
no refrigerants or other halogenated hydrocarbons are included among the substances
whose solubility values in water are given.

Handbook of Tables for Applied Engineering Science (Bolz and Tuve, 1970). Like
the Handbook of Chemistry and Physics, this handbook is published by the Chemical
Rubber Company. It consists of 10 sections. The data included are from unpublished
material and primary literature (e.g., "Courtesy of the Compressed Gas Association";
Courtesy of the American Society of Heating, Refrigerating and Air-Conditioning Engineers"). However, some tables have been "compiled from several sources." Single
solubility values are given for refrigerants R 12, R 13, R 22, R 115, and CH_3Cl at
77°F and 1 atm pressure. There is no information as to how the presented solubility
values have been derived from the published data available. The publication is not
intended to be a critical review.

Physical Properties of Chemical Compounds (Dreisbach, 1955-1961). Three numbers
of the Advances in Chemistry Series have been published by the American Chemical Society in cooperation with the Dow Chemical Company. The solubility data contributed
originate from the Dow Company's own work, without an indication that the values have
been critically evaluated or selected. Only single solubility data are given at 25°C,
sometimes both the solubility of the halogenated hydrocarbon in water and the solubility of water in the halogenated hydrocarbon. Unfortunately, the numerical values
have been given for only a limited number of substances.

Matheson Gas Data Book (Matheson Company, 1966). The 5th edition of this book
presents the physical properties of 130 different gases, including their solubilities
in water at a single temperature and pressure only. It is stated in the foreword that
the information presented "is intended to convey the best available data on gases,"
however, that "no guarantee or warranty is implied or intended." Sources of solubility
data are not cited. There is no indication whether the values reported are expermental
or calculated.

Tables of Physical and Chemical Constants (Kaye and Laby, 1973). This compilation of physical and chemical data of compounds is the 14th edition under the direction of an Editorial Committee. "The material is of interest in more than one branch
of science and not principally to specialists." References to sources are included
for detailed data, but the solubility values reported are not referred to. The publishers state that "neither the publishers nor the editors or contributors can accept
any liability or responsibility for the accuracy of the information contained in the
tables." As far as the solubility and miscibility between halogenated hydrocarbons
and water are concerned, there are only a few single values at ambient temperature
without references.

In general, all nine handbooks mentioned above are very poor sources of information on solubility and miscibility between halogenated hydrocarbons and water, consequently, most of the solubility data reported in these books have been used only
as cross-references or alternative sources and not as reliable and recommended solubility values suitable for the final selection and compilation of data recommended
for correlation and presentation in Chapters 7 and 8.

In addition to the handbooks, there are significant collections and compilations
of solubility data in multivolume handbooks, such as those described next.

International Critical Tables of Numerical Data, Physics, Chemistry, and Technology (Washburn, 1928, 1929). This multivolume handbook, consisting of seven volumes
and an index, was compiled by contributors for the U.S. National Academy of Sciences.
The data presented are no more recent than 1929. The coverage and tabulation are very
comprehensive up to 1929, and all data are referred to the source of reference. Almost
all published solubility data for halogenated hydrocarbon and water systems are included in Volume III, which was published in 1928. It is very convenient, using the
index to allocate the required solubility values. This comprehensive work has been
used to cross-check all references and solubility data published before 1928. Because
of the late production and application of mixed halogenated hydrocarbons, particularly
refrigerants (chlorofluorohydrocarbons), there are no solubility data in these volumes
on the substances mentioned. Only the following compounds are listed under the sol-
ubility section: CH_3Cl, $CHCl_3$, CH_2Br_2, CH_2Cl_2, CH_3I, $CHCl_2-CH_3$, CH_2Cl-CH_2Cl, C_2H_5Cl,
C_2H_5Br, C_2H_5I, C_3H_7Br, C_3H_7Cl, C_3H_7I, and so on. There is no comment on the selection
or recommendation of the reported values.

Landolt-Börnstein Zahlenwerte und Funktionen aus Physik, Chemie, Astronomie, und
Technik (Landolt-Börnstein, 1962). The latest, or 6th edition of this comprehensive
work aimed to cover numerical property values in several areas of physical science.
Sources of data are the published literature. The list of references is very complete
and each value is provided with a reference(s). All relevant solubility data for
halogenated hydrocarbon and water systems appear in Volume 2, Part 2b (1962). There is
no description of the method of selection or evaluation of the tabulated numerical

values. The comprehensiveness and quality of the data depend upon the contributors
and section editors. This multivolume handbook is very useful when cross-checking
published sources and numerical values obtained from other works. There are no data
in this volume more recent than 1961.

Landolt-Börnstein Zahlenwerte und Funktionen aus Physik, Chemie, Astronomie, und
Technik (Landolt-Börnstein, 1976a). The book covers both tabulated and graphical
representations of solubility data. The literature was searched and collected through
1975. Although the compilation is prepared to a high standard, several solubility
data published in the literature have been overlooked on this publication. Most solubilities are expressed in g-mole kg^{-1} bar^{-1} and/or N m^3 ton^{-1} atm^{-1}, although some
other units of solubility are also used.

The introduction consists of a description of Henry's law, followed by the conversion formulas for various solubility coefficients and conversion factors. The
graphical representation of the principles and characteristics of gas solubilities
in liquids is treated as proposed by Hayduk and co-workers. Although the introduction
deals with the relationships between the solubility and the heat and entropy of solutions, no numerical data are available in the book. A short discussion is also included for the regular-solution, hard-sphere, and perturbation theories.

The available mixing rules are summarized for solubilities in liquid mixtures,
particularly when the regular solution theory is applicable: that is, the application
of solubility parameters and Henry's law coefficients.

The graphical presentation of the solubility data is very good. The figures are
very well printed, so that the distinction between the sections and solubility curves
are clear. But it must be pointed out that in a few figures, too many solubility
curves are compressed unneccessarily.

The book does not include liquid-liquid solubility data.

The figures and tables for the solubility of gases in liquids are organized in
the following order:

Solubility in water and heavy water
Solubility in organic liquids
Solubility in mineral and olive oils
Solubility in HF, NH_3, N_2H_4, CH_3NH-NH_2, $(CH_3)_2N-NH_2$, N_2O_4, and CS_2
Solubility in aqueous solutions

At the end of the book there are alphabetical and formula indexes for both the
gases and liquids mentioned in the tables and figures. The various halogenated hydrocarbon gases treated in this volume are listed in Table 2.106. No critical examination or comments are given for these figures or tables. In most of the figures, the
solubility curves were plotted on a semilogarithmic scale, log solubility versus temperature.

All text, comments, and notes are in German.

Landolt-Börnstein Zahlenwerte und Funktionen aus Naturwissenschaften und Technik (Landolt-Börnstein, 1976b). This volume of the multivolume handbook series deals with the heat of mixing and solution in aqueous and nonaqueous systems. There is a good description in the introduction of the integral and differential enthalpy of solutions and the integral and differential enthalpy of dilutions. In addition to the thermodynamic relationships outlined, the molar enthalpy of mixing is discussed. The data are both tabulated and plotted for aqueous and nonaqueous systems. The tables and graphs are organized in the following order:

Inorganic substances - water

Organic substances - water

Polymer compounds - water

Ternary and multicomponent systems - water

Inorganic substances - inorganic substances

Inorganic substances - organic substances

Organic substances - organic substances

Polymer compounds - inorganic substances

Polymer substances - organic substances

Ternary and multicomponent systems not containing water

At the end of this volume is a formula index for the compounds included in the compilation. The thermodynamic data are given in two different units: cal g-mole^{-1} and J g-mole^{-1}.

Both the clarity and typography of the tables and figures are very good. The introduction is written both German and English; however, the footnotes and comments to the tables and figures are in German only.

Landolt-Börnstein Zahlenwerte und Funktionen aus Physik, Chemie, Astronomie, und Technik (Landolt-Börnstein, 1980). This is the latest volume on the solubility of gases in liquids at high pressures. After the description of the fundamental principles of the solubility in the Introduction, the presentation of the literature data are followed in tables and diagrams.

In the Introduction the thermodynamic principles of the equailibrium is treated between a liquid and a vapor phase. The relationships of Henry, Gibbs-Duhem, van Laar, Redlich-Kister, Wilson, Redlich-Kwong, Benedict-Webb-Bubin, Peng-Robinson, Redlich-Kwong-Soave, and others are described. Furthermore, there is a short consideration given for some relevant topics (e.g., distribution coefficients, VLE diagrams, ideal and nonideal systems, solubility gap, critical points and critical curves, supercritical states, thermodynamic conditions of equilibrium, activity and activity coefficient, fugacity, and the convergence-pressure-concept.)

The solubility data reported through tables and diagrams are divided into three chapters:

1. Solubility of elements (gases) in liquids
2. Solubility of gases in inorganic liquids
3. Solubility of gases in organic liquids

The solubility of gases in water is not included in the compilation.

The illustrations are described with comments and details of references are cited for further study if needed. However, there are no evaluation or selection procedures described in the book, except in the Introduction, where the compilation method is mentioned.

Gas Solubilities: Widespread Applications (Gerrard, 1980). In this book the author presented a large number of published solubility data for illustration of the widespread applications of the solubility of gases in liquids at various conditions. The examples were taken from a wide variety of operations in medical research, anaestesiology, pharmacology, oceonography, aerosol technology, biology, chemical engineering and environmental pollution.

Complete chapters were adopted for describing the solubility of one or the other industrially important gases. The solubility of SO_2, Cl_2, H_2S, hydrides, F_2, hydrogen halides, NO, and PH_3 in various liquids occupies a complete chapter. In addition to the numerous examples taken from the literature, the author discusses the effect of temperature on gas solubility, Hildebrand's solubility parameter theory, and the theory of making and filling holes.

The main objective of the book is to show how the Reference Line (R-line) procedure can be applied to calculate the solubility of gases in liquids. This procedure is not restricted to polar or nonpolar compounds and chemical or physical reactions, however, the irreversible reactions are excluded. This book is a continuation of the first book of the author "Solubility of Gases and Liquids. A Graphical Approach," published in 1976.

There is very little in the book regarding the solubility of gases in water. However, the author states that "a detailed analysis of gas-water systems would require a volume itself."

Beilsteins Handbuch der Organischen Chemie (1958-1964). The 4th edition of this work is a very comprehensive encyclopedia of organic compounds compiled by the Beilstein Institut and presented in German. There are three volumes of this series in addition to earlier editions, which include all the halogenated hydrocarbons relevant to this book: Vol. I, Part 1 (1958); Vol. V, Part 1 (1963) and Part 2 (1964). However, as far as the solubilities of substances are concerned, there is only partial information -- mostly single solubility values -- in these volumes. There are 1369 halogenated hydrocarbons listed in the three volumes up to six carbon atoms, but only 61 compounds have some information on their solubilities in water, mostly at a single temperature (Horvath, 1976b). All solubility data cited are furnished with references.

Kirk-Othmer Encyclopedia of Chemical Technology (Kirk-Othmer, 1964-1970). This is the 2nd edition of the standard work on the chemical industry, its methods, processes, equipment, and materials. The information is descriptive, with tabulated physical properties of the substances included. Volumes 3, 5, 9, 11, and 18 contain some solubility data for halogenated hydrocarbons without full bibliography and list of references.

Several volumes of the 3rd edition are already published in 1981.

Ullmanns Encyklopädie der technischen Chemie (Ullmann, 1951-1972). This compilation is very similar to the Kirk-Othmer Encyclopedia, but in German. References to the reported solubility values are not included. As far as the solubility data are concerned, like the Kirk-Othmer Encyclopedia, there is no descriprion of the method used for selection or evaluation. Furthermore, the solubility values are not comprehensive or complete.

However, solubility data, particularly the latest published results have to be retrieved from various abstracts, such as:

Chemical Abstracts
Physical Abstracts
Theoretical Chemical Engineering Abstracts
Nuclear Science Abstracts
Analytical Abstract
Pollution Abstract
Water Quality Abstracts
Citation Index
Chemisches Zentralblatt
Referativny Zhurnal
Dissertation Abstracts

Larger manufacturers issue industrial bulletins, brochures, pamphlets, and so on, on their products, which very often contain useful information on solubility data not published elsewhere. It is difficult to obtain these publications, because, strictly speaking, they are available only to customers of the products. Libraries do not hold copies of these bulletins, and the only way to obtain them is by direct contact with the manufacturers. Several very useful bulletins have been issued by Farbwerke Hoechst AG, du Pont de Nemours and Co., Inc., Dow Chemical Co., Allied Chemical Corp., Ethyl Corporation, Pechiney-Saint-Gobain, and others.

Patents, particularly U.S. patents, often contain the results of solubility measurements in connection with solvents, refrigerants, propellants, and so on, their applications or the properties that make them superior to other substances. Several compounds, mainly halogenated hydrocarbons, are used in the pharmacological industry as anesthetic media, and their solubility in water, blood, and tissues is very important. Consequently, numerous pharmaceutical publications (loose-leaf notes, etc.) contain solubility data in connection with the application and properties of these substances.

Solubility data were also reported on filmstrip (Howard and Loscalzo, 1978).

However, the most convenient sources of information on solubility data are special compilations of solubilities in tabulated form. All of these hanbooks are more than a decade old and require updating. As most of the halogenated hydrocarbons have been developed and marketed during the last 10 to 20 years, the compilations available at present do not include the majority of these hydrocarbons. The following books on solubility data are available:

Solubilities of Organic Compounds (Seidell, 1941)

Solubilities of Inorganic and Organic Compounds (Seidell and Linke, 1952)

Solubilities of Inorganic and Metal-Organic Compounds (Seidell, 1953)

Solubilities of Inorganic and Metal-Organic Compounds (Linke, 1958-1965)

Solubilities of Inorganic and Organic Compounds (Silcock, 1979)

Solubility in Inorganic Two-Component Systems (Broul et al., 1980)

Reference Book on Solubility (Kafarov, 1967-1968)

Handbook of Solubility (Kogan et al., 1961-1970)

Solubilities of Inorganic and Organic Compounds (Stephen and Stephen, 1963)

Organic Solvents. Physical Properties and Methods of Purification (Riddich and
    Bunger, 1970)

Furthermore, several handbooks on solvents in general list solubility data, such as:

Solvents Guide (Marsden, 1963)

The Handbook of Solvents (Scheflan and Jacobs, 1953)

Industrial Solvents (Mellan, 1950)

Industrial Solvents Handbook (Mellan, 1977)

Organic Solvents. Physical Properties and Methods of Purification (Weissberger,
    1955)

NPIRI Raw Materials Data Book (Fetsko, 1974)

Handbook of Solvents (Mellan, 1957)

The Technology of Solvents and Plastics (Doolittle, 1954)

The Technology of Solvents (Jordan, 1937)

Solvents (Durrans, 1950)

Physicochemical Properties of Solutions (Lilich and Mogilёv, 1964)

Source Book of Industrial Solvents (Mellan, 1957)

Physical Chemistry of Solutions (Stakhanova et al., 1970)

Die wichtigasten Lösungs- und Weichmachungsmittel (Fritz, 1957)

Lösungsmittel und Weichmachungsmittel (Gnamm and Sommer, 1958)

Industrial Hygiene and Technology (Patty, 1962)

Handbook of Analysis of Organic Solvents (Sedivec and Flek, 1976)

Gas Encyclopaedia ("L'Air Liquide," 1976)

Hoechst Solvents (Farbwerke Hoechst, 1969)

Books and brochures on anesthetic agents also contain information on solubility data, mainly solubilities in water, blood, and tissues, for example:

Uptake and Distribution of Anesthetic Agents (Papper and Kitz, 1963)

Halothane (Sadove and Wellace, 1962)

Halothane (Stephen and Little, 1961)

Toxicological Research (Karpov, 1964)

Physical and Chemical Data on Anesthetics (Secher, 1971)

Apparently, a comprehensive search for solubility and miscibility data between halogenated hydrocarbons and water requires time and a large technical library providing sources of literature with many aspects of science, technology, and medicine. Consequently, it is extremely difficult and time-consuming to allocate all available solubility data for a requested compound.

To search for solubility data, the recommended procedure is as follows. The first step is to consult the easily available handbooks, such as those described above. These handbooks provide a considerable number of solubility data, but mostly at a single temperature (20 or 25°C) only. Second, the search should be continued in special compilations on solubility if available and in multivolume handbooks. The most frequently used publications are listed above. Finally, if the search in handbooks, special handbooks on solubility, and multivolume handbooks proved to be unsuccessful, then it is necessary to search the various indices of abstracts outlined above. It is always a possibility that the solubility has been measured quite recently, and therefore it was not available when the books mentioned were printed.

The time required to allocate the required information can take anything from a few minutes to several days, depending on the nature of the solubility problem and the availability of library facilities.

The cost of the retrieval of solubility data will depend primarily on the time spent on the search. The cost of the publication, often a photocopy, is irrelevant to the cost of the researcher's time. Consequently, a special handbook on solubility can provide a quick access to the data needed and reduces the cost of the retrieval process. It is important that a handbook have a good index, preferably a formula index in addition to the name index, which will prevent any confusion regarding the nomenclature of the organic compounds which a not very experienced investigator might otherwise experience.

It has happened too often that after an extensive search for solubility data in various abstracts, the result was negative, because several key words have been overlooked. The key word "Soly" does not cover the whole field of solubility. There are other words that have to be examined, for example:

Absorption

Bunsen coeff.

Diffusion

Henry's coeff.

Mass transfer

Ostwald coeff.

Surface tension

System(s), etc.

For example, the solubility of vinyl chloride in water as function of temperature between 25 and 75°C was first published in a paper entitled "Prediction of Diffusion Coefficients for Nonelectrolytes in Dilute Aqueous Solutions" (Hayduk and Laudie, 1974a).

Regarding the availability of solubility data for halogenated hydrocarbons in water, or vice versa, a typical illustration is compiled from Beilstein (1958-1964). These three volumes of the multivolume handbook cite 1369 halogenated hydrocarbons having one to six carbon atoms, of which only 61 compounds have some information on the solubility in water, mostly at a single temperature only. Furthermore, not one of the halogenated hydrocarbons has experimentally determined solubility data from the triple point to the critical point (Horvath, 1974). Therefore, the solubility data belong to one of the most expensive properties of the halogenated compounds. Regarding the cost of experimental measurements, see Chapter 1.

This discussion of sources for solubility data would not be complete without mentioning some of the periodical that are likely to contain solubility data. Primarily, journals that are still being published have the greatest potential. Some of these periodicals are:

Journal of Chemical and Engineering Data

Journal of Chemical Thermodynamics

American Institute of Chemical Engineers Journal

Journal of Solution Chemistry

Journal of Physical and Chemical Reference Data

Chemie-Ingenieur-Technik

Chemical Reviews

Zhurnal Fizicheskoi Khimii (Russian Journal of Physical Chemistry)

Canadian Journal of Chemical Engineering

Canadian Journal of Chemistry

Chemical Engineering Science

Chemical Engineering

Industrial Engineering Chemistry

Industrial Engineering Chemistry, Fundamentals

Industrial Engineering Chemistry, Product, Research and Development

Fluid Phase Equilibria

Journal of Pharmaceutical Science

Anaesthesia

Bulletin of the Chemical Society of Japan

Zeitschrift für Physikalische Chemie, Neue Folge

Zhurnal Prikladnoi Khimii (Russian Journal of Applied Chemistry)

Journal of Applied Chemistry and Biotechnology (Journal of Chemical Tech. Biotechnol.)

Journal of American Chemical Society

From the books and periodicals mentioned, it is apparent that knowledge of the German and Russian languages is desirable to understand the solubility data presented. Very difficult problems can occur by bad presentation of the tables or figures, especially when the condition of the solubility is not given in the tables or figure heading but is included in the text. It is not unusual, even in the English literature, that the reader has to read through a complete article to determine whether, for example, the solubility of a gas in water has been reported at 1 atm partial or total pressure, with the solubility expressed in grams of solvent or grams of solution. Expressions and units of solubility are discussed further in Chapter 9.

Finally, the following publications provide useful discussions of the various estimation methods for solubilities in general: Merriman (1977), Hildebrand and Scott (1962), Hildebrand et al. (1970), Gerrard (1976, 1980), Battino and Clever (1966), Wilhelm et al. (1977), Clever and Han (1980), Kertes et al. (1975), Dack (1975b, 1976), Landolt-Börnstein (1976a), Prausnitz (1969), Davis et al. (1974), Markham and Kobe (1941), Tsonopoulos (1970), Guerrant (1964), Paruta (1963), King (1969), Wei (1977), Shinoda (1978), Körösy (1937), Lachowicz and Weale (1958), Karapet'yants (1973), James (1972), Wilhelm and Battino (1973), Linford and Thornhill (1980), Muccitelli (1978), and Samaha (1979).

Major works on vapor-liquid equilibria are summarized in Table 3.1.

As mentioned in the Introduction the need for compilation of solubility data has been recognized by several large organizations. Consequently, IUPAC launched a very ambitious project for the publication of a comprehensive numerical work of Solubility Data Series (CODATA, 1976a, b, c, 1979). The critically-evaluated solubility data will cover all physical and biological systems of approximately 80 to 100 volumes. The series are organized into five broad classifications according to physical states of the solute and solvents. The five subject areas are:

Table 3.1 Major Works on Vapor-Liquid Equilibria: Theory, Data

References	Title	Comments
American Institute of Chemical Engineers (1979)	Nonideal Vapor-Liquid and Liquid-Liquid Equilibria in Theory and in the Real World	Theory
American Petroleum Institute (1970)	Technical Data Book -- Petroleum Refining, 2nd ed.	Graphs
Chu et al. (1950)	Distillation Equilibrium Data	Data
Chu et al. (1956)	Vapor-Liquid Equilibrium Data	Data
Döring et al. (1980)	Vapor-Liquid Equilibria for Mixtures of Low Boiling Fluids	Data
Dummett (1969)	International Symposium on Distillation	Theory
Francis (1963)	Liquid-Liquid Equilibriums	Theory and bibliography
Francis (1972)	Handbook for Components in Solvent Extraction	Theory and bibliography
Fredenslund et al. (1977)	Vapor-Liquid Equilibria Using UNIFAC: A Group-Contribution Method	Theory
Gmehling & Onken (1977)	Vapor-Liquid Equilibrium Data Collection: Aqueous-Organic Systems	Theory and Data
Gothard (1975)	Echilibre lichid-vapori	Theory
Hála et al. (1967)	Vapor-Liquid Equilibrium, 2nd ed.	Bibliography and theory
Hala et al. (1968)	Vapor-Liquid Equilibrium Data at Normal Pressures	Data
Hicks (1978)	Bibliography of Thermodynamic Quantities for Binary Fluid Mixtures	Bibliography
Hirata et al. (1975)	Computer Aided Data Book of Vapor-Liquid Equilibria	Theory and data
Hiza et al. (1975)	Equilibrium Properties of Fluid Mixtures	Bibliography
Holland (1963)	Multicomponent Distillation	Theory
Katz et al. (1959)	Handbook of Natural Gas Engineering	Theory
King (1969)	Phase Equilibrium in Mixtures	Theory
Kogan & Friedman (1961)	Handbuch der Dampf-Flussigkeits-gleichgewichte	Data

Table 3.1 (Continued)

References	Title	Comments
Kogan et al. (1961-1970)	Vapor-Liquid Equilibria	Data
Köpsel (1974)	Angewählte rechnerische Methoden der Verfahrenstechnik, Berechnung von fluiden Mischphasen und Mischphasen-gleichgewichten	Theory
Landolt-Börnstein (1960)	Eigenschaften der Materie in Ihren Aggregatzuständen	Data
Landolt-Börnstein (1962, 1964)	Eigenschaften der Materie in Ihren Aggregatzuständen. Lösungsgleich-gewichte I and II	Data
Landolt-Börnstein (1975)	Thermodynamic Equilibria of Boiling Mixtures	Data and theory.
Landolt-Börnstein (1980a, b)	Gleichgewicht der Absorption von Gasen in Flüssigkeiten	Data and theory
Maczynski (1976, 1978)	Verified Vapor-Liquid Equilibrium Data	Data
Malanowski (1974)	Rownowaga ciecz-para	Theory
National Physical Laboratory (1979)	Chemical Thermodynamic Data on Fluids and Fluid Mixtures	Theory
Null (1970)	Phase Equilibrium in Process Design	Theory
Oellrich et al. (1973)	Vapor-Liquid Equilibria	Bibliography
Perry & Chilton (1973)	Chemical Engineers' Handbook, 5th ed.	Theory
Prausnitz (1969)	Molecular Thermodynamics of Fluid-Phase Equilibria	Theory
Prausnitz & Chueh (1968)	Computer Calculations for High-Pressure Vapor-Liquid Equilibria	Theory
Prausnitz et al. (1967)	Computer Calculations for Multi-component Vapor-Liquid Equilibria	Theory
Prausnitz et al. (1980)	Computer Calculations for Multi-component Vapor-Liquid and Liquid-Liquid Equilibria	Theory
Prigogine et al. (1957)	The Molecular Theory of Solutions	Theory
Reid et al. (1977)	The Properties of Gases and Liquids	Theory
Reisman (1970)	Phase Equilibria	Theory

Table 3.1 (Continued)

References	Title	Comments
Renon (1966)	Thermodynamic Properties of Nonideal Liquid Mixtures	Theory
Renon et al. (1971)	Calcul sur ordinateur des équilibres liquide-vapeur et liquide-liquide	Theory
Rogalski & Malanowski (1975)	Comparison of Different Methods for Correlation of Vapor-Liquid Equilibria	Theory
Rowlinson (1959)	Liquid and Liquid Mixtures, 2nd ed.	Bibliography and theory
Sørensen & Arlt (1979)	Liquid-Liquid Equilibrium Data Collection	Theory and data
Sørensen et al. (1979)	Liquid-Liquid Equilibrium Data: Their Retrieval, Correlation, and Prediction.	Theory
Storvick & Sandler (1977)	Phase Equilibria and Fluid Properties in the Chemical Industry: Estimation and Correlation	Theory
Timmermans (1960)	The Physicochemical Constants of Binary Systems in Concentrated Solutions, 4 vols	Data
Van Horn (1966)	A Study of Low Temperature Vapor-Liquid Equilibria of Light Hydrocarbons in Hydrocarbon Solvents	Theory
Van Ness (1964)	Classical Thermodynamics of Nonelectrolyte Solutions	Theory
Washburn (1928)	International Critical Tables, Vol. 3	Data
Wichterle et al. (1973-1976)	Vapor-Liquid Equilibrium Data Bibliography	Bibliography
Wisniak & Tamir (1980)	Liquid-Liquid Equilibrium	Bibliography

1. Solubility of gases in liquids
2. Solubility of gases in solids
3. Solubility of liquids and liquids
4. Solubility of solids in liquids
5. Solubility of solids in solids

The editor-in-chief is Professor A.S. Kertes who is assisted by 150 internationally-recognized experts, making up the Board of editors, compilers, and evaluators.

Each volume consists of an introduction followed by the compilation, critical evaluation, and recommended values.

The quality and excellence of the series are well reflected by the reviewers:

J. Am. Chem. Soc. 1980, 102, 3665

Bull. Chem. Thermodyn. 1979, 2, 496

Fluid Phase Equilibria 1980, 4(3/4), 313-317

New Technical Books 1980, 65(3), 81-82

J. Chem. Soc., Faraday Trans. I 1980, 76, 1630

J. Chem. Thermodyn. 1980, 12, 607-608

A list of volumes published and under preparation is given in Table 3.2.

Table 3.2 IUPAC Solubility Data Series (Pergamon Press)

Volume	Author(s)	Subject
1	Clever, H.L.	Helium and Neon
2	Clever, H.L.	Krypton, Xenon, and Radon
3	Salomon, M.	Silver Azide, Cyanide, Cyanamides, Cyanates, Selenocyanate, and Thiocyanate
4	Clever, H.L.	Argon
5	Battino, R.	Oxygen and Ozone
6	Lorimer, J.W.,	Alkaline-earth Sulfates
7	Woolley, E.M.	Silver Halides
8	Farrell, P.	Mono- and Disaccharides
9	Cohen-Adad, R.	Alkali-metal Chlorides
10	Bauman, J.E.	Alkali- and Alkaline-earth Metal Oxides and Hydroxides
11	Scrosati, B. & Vincent, C.A.	Alkali-metal, Alkaline-earth Metal, and Ammonium Halides
12	Galus, Z. & Guminsky, C.	Metals in Mercury
13	Young, C.L.	Oxides of Nitrogen, Sulfur, and Chlorine
14	Battino, R.	Nitrogen
15	Clever, H.L. & Gerrard, W.	Hydrogen Halides
16	Horvath, A.L. & Getzen, F.W.	Halogenated Benzenes
17	Wilhelm, E. & Young, C.L.	Hydrogen, Deuterium, Fluorine, and Chlorine
18	Popovych, O.	Tetraphenylborates

Chapter 4
Methods of Collection and Compilation
Of Solubility Data

The technical literature containing relevant information on the solubility and miscibility between halogenated hydrocarbons and water can be divided into two groups:
primary and secondary sources. The primary sources of data or unorganized new material
are periodicals (journals), patents, bulletins of governmental and scientific establishments, dissertations, and industrial and manufacturers' technical pamphlets and
brochures. The secondary sources are compiled or organized materials of previously
published sources such as periodical reviews (e.g., Chemical Abstrats, Applied Science
and Technology Index, Engineering Index, Industrial Art Index, Citation Index,
Chemisches Zentralblatt, Referativny Zhurnal), tabular compilations (e.g., Handbook
of Chemistry and Physics; Taschenbuch für Chemiker und Physiker; Landolt-Börnstein
Zahlenwerte und Funktionen aus Physik, Chemie, Astronomie, und Technik; International
Critical Tables), encyclopedias (e.g., Kirk-Othmer Encyclopedia of Chemical Technology,
Ullmanns Encyklopädie der technischen Chemie, Kratkaya Khimicheskaya entsiklopediya),
treatises (e.g., Beilsteins Handbuch der Organischen Chemie, Organic Chlorine Compounds), monographs, dictionaries, formularies and recipes, textbooks, and handbooks.

The method of collection of solubility and miscibility data involved a systematic search throughout the secondary sources, starting with the abstracts listed
in Chapter 3. The comprehensiveness of the collected data has been cross-checked by
the tabular compilations, encyclopedias, treatises, and so on. In numerous secondary
sources, there were no references to the origin of the given solubility data (e.g.,
Handbook of Chemistry and Physics, Matheson Gas Data Book, Kirk-Othmer Encyclopedia
of Chemical Technology). Because of the obscure origin of these values, with some
exceptions they have been used only for the record and for comparison purposes, and
were not included in the correlations procedure. The principal object of the compilation was to collect all published solubility data regardless of the origin, accuracy, reliability, and so on, of the values. Whether the collected data are acceptable for use in the correlation procedure is dicussed in detail in Chapter 5.

It was mentioned in Chapter 3 that solubility data are also available in industrial bulletins, pamphlets, leaflets, and brochures issued by manufacturers of chemicals. The significance of some of these sources of data are that they are not available elsewhere. On the other hand, their reliability is doubtful, because they generally lack descriptive information on experimental procedures. In addition to the

uncertainty of the data, these publications usually state that "no responsibility is assumed for the accuracy of the data presented." The solubility values in various sources are discussed further in Chapter 5.

To obtain industrial literature with data on solubility and miscibility between halogenated organic compounds and water, the following manufacturers were contacted:

Allied Chemical Corporation

American Cyanamide Corporation

American Ink Maker

British Petroleum Chemicals (U.K.) Limited

Commercial Solvents Corporation

Diamond Shamrock Chemical Corporation

Dow Chemical Company

E.I. du Pont de Nemours & Company, Inc.

Eastman Organic Chemicals, Eastman Kodak Co.

Ethyl Corporation

Farbwerke Hoechst AG

Henkel GmbH.

Hooker Chemicals & Plastics Corporation

IHARA Chemical Industries Co. Ltd.

Imperial Chemical Industries Limited

Mobil Chemical Co.

Pechiney-Saint-Gobain

Pennwalt Corporation

Phillips Petroleum Co.

Pittsburg Plate Glass Industries, Inc.

Procter & Gamble, Industrial Chemical Division

Reilly Tar & Chemical Corporation

Shell Chemical Company

Stauffer Chemical Company

Union Carbide Corporation

Some of the listed manufacturers above provided useful information and forwarded technical literature on their products; others, however, refused to disclose their unpublished solubility data or have not responded to correspondence. The manufacturers from which solubility and miscibility data were obtained are acknowledged in the Preface.

The collection and evaluation of the solubility data published are not an easy task. For example, an excellent survey of the solubility of Ar, Cl_2, HCl, NO, N_2, N_2O_3, NO_2, and O_2 in water was reported by Wilson (1959).

The collected data have been tabulated according to the formula index of the

substances. A typical example of tabulation is illustrated in Table 4.1 for the
solubility of chloroform in water. The solubility is given in the same units as found
in the literature. However, in this book only selected and recommended solubility data
are plotted or tabulated and are presented in Chapters 7 and 8.

It is convenient to express solubility data as a mole fraction for the theoretical investigations, including the application of the similarity-symmetry-analogy
principle (Horvath, 1975a) and correlation procedures, whereas, chiefly in chemical
engineering practice the percent is a practical unit and so is most often used in
solubility calculations.

Information on solubility can be found in various forms: single value, tabulation
of numerical data, graphical form, nomograms, and equations when the solubility is
expressed as a function of some other physical quantity (e.g., temperature, pressure
(partial or total), density, number of carbon atoms). In some cases the solubility is
described by "slightly soluble" (Carothers et al., 1931, 1932). For example, the solubilities of chloroprene and 4-chlorobutadiene-1,2 are given approximately only as
"slightly soluble" in water. There are several reports on the approximate solubility
values (e.g., El Khishen, 1948). Furthermore, in some older articles "insolubility"
was reported for several halogenated hydrocarbons in water (e.g., Juvala, 1930;
Müller and Hüther, 1931; Swarts, 1933; Prelog, 1936).

The most frequently used form is the tabulation of values; however, several
sources of solubility and miscibility data are presented in graphical form only (e.g.,
McGovern (1943) for chlorinated solvents). Most of these figures appeared in reduced
form in journals and it was desirable to enlarge the graphs so that the accuracy of
reading could be improved. Chapter 6 contains further detail on various aspects of the
accuracy of the solubility data required.

Another form of graphical presentation is the nomograph (Mapstone, 1952; Chernyshev, 1969). Chernyshev illustrated the solubility of water in refrigerants using
nomographs only. This is a very compact presentation of solubility data, but the
accuracy of reading is very subjective; in other words, two independent investigators
might read off different values from the same nomograph. The subjective nature of the
data is also characteristic for all type of graphical representations.

Some articles give the solubility data in the form of an equation, particularly
in expressing the solubility as a function of temperature (e.g., Simonov et al., 1970):

$$\log_{10} c_{H_2O}^t = A + Bt$$

Apparently, this equation represents fitted or smoothed data that have been adopted by
the investigator to express the final results. Simple equations are able to fit the
data quite accurately in a moderate temperature interval, whereas the correlation of
data for a large temperature interval (e.g., from the triple point of the water to its

Table 4.1   Solubility of Chloroform in Water

Temperature (°C)	Solubility	Reference	Page number
15	0.859 g/100 g H_2O	Jones et al. (1957)	188
20	0.815 wt %	Riddick & Bunger (1970)	351
37	3.80 Ostwald coeff.	Papper & Kitz (1963)	10
25	5.6 N cm^3/g H_2O	Lax (1967)	1206
20	0.82 wt %	Lax (1967)	1233
20	0.80 wt %	Marsden (1963)	136
0	1.062 g/100 g H_2O	Kirk-Othmer (1964b)	120
20	0.895 "	"	"
30	0.776 "	"	"
20	0.815 wt %	Lange (1967)	483
15	0.852 g/100 g H_2O	Gross & Saylor (1931)	1750
30	0.771 "	"	"
0	0.00148 mole fraction	Staverman (1941b)	838
20	0.00123 "	"	"
30	0.00116 "	"	"
0	1.062 g/100 g H_2O	Rex (1906)	365
10	0.895 "	"	"
20	0.822 "	"	"
30	0.776 "	"	"
15	0.482 g/100 g H_2O	Svetlanov et al. (1971)	877
30	0.320 "	"	"
45	0.205 "	"	"
60	0.130 "	"	"
25	1.0 g/125 g H_2O	Wright & Schaffer (1932)	408
30	0.00646 g-mole/100 g H_2O	van Arkel & Vles (1936)	409
15	1.0 g/100 g H_2O	Hogman et al. (1956)	855
20	0.666 parts/100 parts H_2O	Secher (1971)	23
20	0.8 wt %	Hoechst Solv. (1975)	171
0	9.87 g/1000 cc H_2O	Chancel & Parmantier (1885a, b)	774
3.2	8.90 "	"	"
17.4	7.12 "	"	"
29.4	7.05 "	"	"
41.6	7.12 "	"	"
54.9	7.75 "	"	"
22	0.42 cc/100 cc H_2O	Herz (1898)	2671

Table 4.1 (Continued)

16–18	0.55	g/100 g sat. sol.	Winterstein & Hirschber (1927)	172
0	0.98	wt %	Reinders & De Minjer (1947)	577
25	0.90	"	"	"
42	0.71	"	"	"
60	0.75	"	"	"
0	1.05	g/100 g H$_2$O	McGovern (1943)	1234
10	0.91	"	"	"
20	0.83	"	"	"
25	0.79	"	"	"
30	0.78	"	"	"
40	0.75	"	"	"
50	0.77	"	"	"
60	0.80	"	"	"
20	0.82	wt %	Mellan (1957)	124
20	0.82	wt %	Mellan & Chilton (1950)	319
20	0.82	g/100 g H$_2$O	Perry (1973)	3–28
0	0.98	wt %	Washburn (1928)	387
1.45	Solid hydrate		"	"
10	0.86	wt %	"	"
20	0.80	"	"	"
30	0.76	"	"	"
40	0.735	"	"	"
50	0.745	"	"	"
55	0.77	"	"	"
20	0.80	wt %	Doolittle (1935)	411
20	0.80	wt %	Durrans (1950)	169
15	1.0	g/100 ml H$_2$O	Scheflan & Jacobs (1953)	209
20	0.82	"	"	"
15	0.852	g/100 g H$_2$O	Huntress (1948)	545
20	0.80	"	"	"
25	0.79	g/100 g H$_2$O	Asinger (1968)	279
20	0.80	wt %	Evans (1936)	208
15	0.84	wt %	Stephen & Stephen (1963)	55
20	0.80	"	"	"
30	0.76	"	"	"
0	0.983	"	"	370
3.2	0.887	"	"	"
17.4	0.710	"	"	"

Table 4.1 (Continued)

29.4	0.703	wt %	Stephen & Stephen (1963)	370
41.6	0.710	"	"	"
54.9	0.773	"	"	"
0	1.052	"	"	"
10	0.888	"	"	"
20	0.815	"	"	"
30	0.770	"	"	"
25	0.792	wt %	Miller (1969)	1301
20	0.80	wt %	Gori (1925)	283
0	0.98	wt %	Landolt-Börnstein (1962)	3-400
10	0.87	"	"	"
20	0.80	"	"	"
30	0.74	"	"	"
40	0.72	"	"	"
50	0.74	"	"	"
56	0.78	"	"	"
37	4.0	Ostwald coeff.	Allott et al. (1973)	294
20	0.79	wt %	Antonov (1907)	372
15	1.0	wt %	Antropov et al. (1972)	311
25	0.9	cc/100 cc H_2O	Booth & Everson (1948)	1491
25	0.75	wt %	Conti et al. (1960)	301
56.1	0.80	"	"	"
25	0.80	wt %	Gladis (1960)	43
40	4.6	Ostwald coeff.	Larson et al. (1962b)	686
37	3.8	"	"	"
30	4.17	"	"	"
20	7.7	"	"	"
10	5.9	"	"	"
20	0.80	g/100 g H_2O	Smith (1932)	1401
37	4.0	Ostwald coeff.	Steward et al. (1973)	282
30	0.82	g/liter solution	Van Arkel (1946)	81
20	7.5	g/liter H_2O	Salkowski (1920)	191
15	1.0	g/100 cc H_2O	Patty (1962)	831
15	1.0	wt %	Gunther et al. (1968)	1
20	0.75	g/100 cc H_2O	Macintosh et al. (1963)	135
37	405	cc (STP)/100 cc H_2O	"	156
20	0.815	wt %	Interdyne (1976)	1
20	820	ppm by wt	Pearson & McConnell (1975)	305

Table 4.1 (Continued)

25	0.82	wt %	Mitchell & Smith (1977)	415
20	0.8	wt %	Nathan (1978)	93
20	0.0675	g-mole/liter H_2O	Pavlovskaya et al. (1977)	2230
20	8200	ppm by wt	Selenka & Baner (1978)	242
38	3.8	Ostwald coeff.	Grant (1978)	164
20	0.82	wt %	du Pont (1966e)	3
25	7840	ppm by wt	Dilling (1977)	405
1.5	10,300	"	"	"
25	7950	ppm by wt	Chiou et al. (1977)	475
25	1270	mg/100 cc H_2O	Aref'eva et al. (1979)	2
0	10510	ppm by wt	Andelman (1978)	183
10	8870	"	"	"
20	8100	"	"	"
25	7800	"	"	"
30	7600	"	"	"
20	0.80	wt %	Matthews (1975, 1979)	565
25	7950	g/m^3 H_2O	Afghan & Mackay (1980)	581
25	7950	g/m^3 H_2O	Hutchinson et al. (1980)	577

critical point) will require a more complex expression which will fit the experimental data with acceptable accuracy (this is discussed in Sec. 2.8). At present there is no equation or expression that can represent the solubility of a gas in water as a function of both temperature and pressure between the triple point and the critical point of water, giving a reasonable accuracy for the system. Some of the difficulties of establishing reliable relationships among solubility, temperature, and pressure are discussed in Sec. 2.5-2 (Horvath et al., 1973).

Principally, there are two ways to search the literature for solubility data of a substance in another substance:

1. Looking for the solubility under the compound's name
2. Searching for the compound under the solubility heading

The second method is very commonly used when one has access to books or compilations of solubility data. However, in the case of Chemical Abstracts, for example, the result would not be successful. Therefore, the practical method is the first one, that is, looking for references for solubility under the compound's name. Naturally, this involves an enormous amount of work for the compiler who is attempting to collect solubility values for halogenated hydrocarbon - water systems.

With a few illustrations it becomes apparent that the size of the work is immense. Take the simplest series, methane with substitutions of hydrogen for fluorine, chlorine, bromine, and/or iodine atom(s). The total number of combinations in the methane series is already 70. The total number of combinations in the methane and ethylene series of halogenated hydrocarbons up to six carbon atoms in the molecule are tabulated in Table 4.2. The symbol X represents either hydrogen and/or halogen atom(s) in various combinations.

The vast number of combination possibilities of halogenated hydrocarbons will surely surprise many readers. It is understandable that it is a practically impossible task to search through the literature for the mutual solubilities between water and halogenated hydrocarbons, taking one compound after the other. An indiscriminant search might not produce more information then could be obtained by a specialist with a carefully planned program with selected search procedures.

The solubility data presented in Chapters 7 and 8 are the result of about 10 years of search, collection, and compilation of literature data, and correspondence with many scientists and manufacturers of halogenated hydrocarbons.

Table 4.2 Number of Halogenated Hydrocarbons in Various Series[a]

Paraffin series

Methane	CX_4	70
Ethane	CX_3-CX_3	595
n-Propane	$CX_3-CX_2-CX_3$	8,970
n-Butane	$CX_3-CX_2-CX_2-CX_3$	133,875
n-Pentane	$CX_3-CX_2-CX_2-CX_2-CX_3$	2,008,125
n-Hexane	$CX_3-CX_2-CX_2-CX_2-CX_2-CX_3$	30,121,875

Total: 32,273,510

Olefin series (including cis- and trans- modifications)

Ethene	$CX_2=CX_2$	165
Propene	$CX_2=CX-CX_3$	4,375
n-Butene	$CX_2=CX-CX_2-CX_3$	65,625
n-Pentene	$CX_2=CX-CX_2-CX_2-CX_3$	984,375
n-Hexene	$CX_2=CX-CX_2-CX_2-CX_2-CX_3$	14,765,625

15,820,165

---

[a] $X = H, F, Cl, Br, I.$

Chapter 5

Discrepancies and Critical
Evaluation of Data

It happens frequently that after a comprehensive search of the technical literature, the investigator finds several sources for the same query with discrepancies of varying degrees between the retrieved data (Stockmayer, 1978; Horvath et al., 1973; Metanomski, 1980; Fair, 1980; CODATA, 1972, 1975, 1977, 1979, 1981; American Chemical Society Committee on Environmental Improvement, 1980).

The difficulties and problems in retrival of numerical data have been recognized since the beginning of this century. This has been discussed during several international meetings and conferences. A major step was achieved in 1966 when CODATA was established for collecting, collating, evaluating, and dissemination of data to the international arena (see CODATA Bulletins and Newsletters).

In September 1979, the American Chemical Society organized a symposium on techniques and Problems in Retrieval of Numerical data." The collection of symposium papers was published in the Journal of Chemical Information and Computer Sciences 1980, Vol. 20, No. 3. The different aspects of the current state of the art in numerical data retrieval were presented by Metanomski (1980), Hawkins (1980), Carter (1980), and Kirschenbaum (1980).

What are the ways to deal with discrepancies, or how can one critically evaluate the available numerical data? Before starting the selection procedure, it is imperative to clarify the objective; in other words, what are the solubility data to be used for? This is the essence of the evaluation technique. Obviously, for a rough or approximate calculation, no more than two or three significant figures are required, whereas for a highly precise instrument, more accurate data, four or more significant figures, will be required.

Another example is the effect of the accuracy of the data on the design and cost of a project (Fair, 1980). Chemical industries are steadily becoming more concerned with the disposal of industrial wastes into rivers and seas (Gurnham, 1965; Sitting, 1969; Andelman, 1978; Petrakis and Weiss, 1980; Lu and Metcalf, 1975; Banks, 1979; Matthews, 1975; Robertson et al., 1980). When the organic substances occur in water as suspensions or sediments, the separation is relatively easy. Furthermore, most halogenated organic solvents form a separate phase with water, because they are nonpolar compounds and very weakly soluble in water. However, because of the toxic character of many halogenated organic compounds for marine life and so on (see the Introduction), local authorities have enacted strict regulations and controls. Wastewaters containing various halogenated organic compounds, often in chemical processes as intermediates or by-products, cannot be disposed of without control. It is therefore important to know the maximum allowable concentrations of these substances in water at various temperatures.

Rivers and seas receive large quantities of wastes high in organic matter, causing significant depletion of oxygen (Sutton and Calder, 1974). Continuous analysis and controls are applied to the determination of water quality, with particular regard to the concentration of organic matter and dissolved oxygen. More than 700 specific organic chemicals have been identified in various drinking water supplies in the United States (Environmental Protection Agency, 1978). The maximum trihalomethane levels ranged as high as 0.695 and 0.784 mg liter^{-1} in terminal samples. This is much higher than the proposed Maximum Contaminant Level of 0.1 mg liter^{-1}. The odor and taste of water are detectable at various concentrations, depending upon the dissolved substance. However, the concentration that is inactive according to hygienic-toxicological tests and most often used in water quality control is the maximum permissible or allowable concentration. This concentration is a characteristic factor of water quality. Some of the typical recommendations for halogenated hydrocarbons are presented in Table 5.1 (Horvath, 1975a; Bukovskii et al., 1977).

After the decision regarding the accuracy requirement of the solubility data (accuracy is discussed further in Chapter 6), one can proceed with the evaluation of the available values.

The published numerical data can be listed in two different groups:

1. Experimental data
2. Estimated, predicted, correlated, or smoothed, interpolated, and extrapolated data

The procedure for critical evaluation of the experimental data is to examine the general aspects of the investigation (absolute reliability): in other words, to find answers to the following questions regarding the experiment:

when -- how -- why -- who -- what way?

The evaluator looks into the circumstantial evidence of the reliability of the experiment by examining the following aspects in great detail:

1. Date of experiment
2. Purity of sample(s) used
3. Reliability of the apparatus (setup)
4. Reliability of the applied technique
5. Competence of the investigator(s)

In addition to these aspects of the experiments, there are other methods used to examine the relative reliability of the reported data or final results -- by the

Table 5.1 Maximum Permissible Concentrations (M.P.C.) in Reservoir Waters

Formula	Halogenated hydrocarbons	M.P.C. (mg liter^{-1})	References	Chemical Abstracts
CH_2Cl_2	Methane, dichloro-	7.5	Tugarinova et al. (1965)	66:22046j
CCl_4	Methane, tetrachloro-	0.3	Kutepov (1968)	68:62489d
CCl_2F_2	Methane, dichlorodifluoro-	10.0	Bukovskii et al. (1977)	-
$CHClF_2$	Methane, chlorodifluoro-	10.0	Bukovskii et al. (1977)	-
$CCl_2=CHCl$	Ethene, trichloro-	0.5	Miklashevskii et al. (1962); Bukovskii et al. (1977)	60:15589c -
CH_2Cl-CH_2Cl	Ethane, 1,2-dichloro-	2.0	Bukovskii et al. (1977)	-
$CHCl_2-CHCl_2$	Ethane, 1,1,2,2-tetrachloro	0.2	Tugarinova et al. (1962)	60:15588h
CCl_3-CCl_3	Ethane, hexachloro-	0.01	Tugariniva et al. (1962)	60:15588h
$C_3H_3Cl_5$	Propane, pentachloro-	0.03	Bukovskii et al. (1977)	-
$CHCl=CH-CH_2Cl$	Propene, 1,3-dichloro-	0.4	Bukivskii et al. (1977)	-
$CH_2=CCl-CH_2Cl$	Propene, 2,3-dichloro-	0.4	Bukovskii et al. (1977)	-
$CH_2Cl-CHCl-CH_3$	Propane, 1,2-dichloro-	0.4	Bukovskii et al. (1977)	-
$C_3H_4Cl_2$	Propene, dichloro-	0.4	Bukovskii et al. (1977)	-
$CH_2Cl-CH=CH_2$	Propene, 1-chloro-	0.4	Bukovskii et al. (1977)	-
$C_3H_4Cl_4$	Propane, tetrachloro-	0.01	Bukovskii et al. (1977)	-
$C_3H_5Cl_3$	Propane, trichloro-	0.07	Bukovskii et al. (1977)	-
$CF_3-CH_2-CH_2Cl$	Propane, 1,1,1-trifluoro-3-chloro-	0.1	Selyuzhitskii (1963, 1967) Bukovskii et al. (1977)	60:13017f 69:38565a
$CCl_2=CCl-CCl=CCl_2$	1,3-Butadiene, hexachloro-	0.01	Bukovskii et al. (1977)	-
$CH_2=CH-CCl=CH_2$	1,3-Butadiene, 3-chloro-	0.1	Bukovskii et al. (1977)	-
$CH_2=CCl-CCl=CH_2$	1,3-Butadiene, 2,3-dichloro	0.03	Bukovskii et al. (1977)	-

Table 5.1 (Continued)

$CH_3-CCl=CH-CH_2Cl$	2-Butene, 2,4-dichloro-	0.05	Bukovskii et al. (1977)	-
$CHCl=C(CH_3)-CH_2Cl$	1-Propene, 2-methyl-3-chloro-	0.4	Bukovskii et al. (1977)	-
$CH_2=C(CH_3)-CHCl_2$	1-Propene, 2-methyl-3,3-dichloro-	0.4	Bukovskii et al. (1977)	-
$(CH_3)_2CCl-CH_2Cl$	Propane, 1-methyl, 2,3-dichloro-	0.4	Bukovskii et al. (1977)	-
$CH_2Cl-CHCl-CHCl-CH_3$	Butane, 1,2,3-trichloro-	0.02	Bukovskii et al. (1977)	-
$C_4H_4Cl_6$	Butane, hexachloro-	0.01	Bukovskii et al. (1977)	-
$C_4H_5Cl_5$	Butane, pentachloro-	0.02	Bukovskii et al. (1977)	-
$C_4H_6Cl_2$	Butene, dichloro-	0.05	Khachat-Ryan (1962)	61:442b
C_5Cl_6	2-Penten-4-yne, hexachloro-	0.001	Bukovskii et al. (1977)	-
$C_5H_8Cl_4$	Pentane, tetrachloro-	0.005	Bukovskii et al. (1977)	-
$C_5H_7Cl_3$	Pentene, trichloro-	0.04	Bukovskii et al. (1977)	-
$C_5H_5Br_2Cl_3$	Pentene, dibromotrichloro-	0.04	Bukovskii et al. (1977)	-
C_6H_5Cl	Benzene, chloro-	0.02	Varshavskaya (1968)	70:22828y
$C_6H_{11}Cl$	Cyclohexane, chloro-	0.05	Orlovskii (1963)	61:14343a
			Lisovskaya et al. (1964)	62:11521d
			Bukovskii et al. (1977)	-
$C_6H_6Cl_6$	Cyclohexane, hexachloro-	0.02	Bukovskii et al. (1977)	-
$C_6H_{10}Cl_2$	Cyclohexane, dichloro-	0.02	Kostovetskii et al. (1962)	60:15589d
			Lisovskaya et al. (1964)	62:11521d
$C_6H_4Cl_2$	Benzene, dichloro-	0.03	Gurfein & Pavlova (1960)	56:7061d
$C_6H_4Cl_2$	Benzene, o- and p-dichloro-	0.002	Varshavskaya (1967, 1968)	70:90586t
				70:22828y
$C_6H_3Cl_3$	Benzene, trichloro-	0.03	Gurfein & Pavlova (1960)	56:7061d
			Meleshchenko (1960a, b)	55:4831f
$C_6H_2Cl_4$	Benzene, 1,2,4,5-tetrachloro-	0.02	Fomenko (1965)	62:8831b
C_6Cl_6	Benzene, hexachloro-	0.05	Gurfein & Pavlova (1960)	56:7061d
			McKee & Wolf (1963)	

application of:

6. Thermodynamic and other consistency tests
7. Relating the solubility of the examined compound to other substances
8. Property trends in homologous or similar series

Obviously, there are connections between the different categories (e.g., date of experiment and purity of sample). Before the introduction of chromatography in chemical analysis in the mid 1930s, the purity of samples could not be checked very accurately, particularly the impurities of organic compounds. For example, the purification of bromine was not very successful until the discovery that permanganate or chromic acid oxidizes the impurities, so that at present 99.99% pure bromine is available in commercial quantity. Consequently, during bromination reactions, fewer impurities will be carried over into the products.

Data published before 1900 are quite common in the technical literature, and often no attempt has been made to repeat the experiments since the date of investigation. Consequently, in this case there is no choice, and the available values have been used despite their date of publication (e.g., solubility of $n\text{-}C_3H_7F$ in water)(Meslans, 1894). In particular, organic compounds used in pharmacology have solubility data determined almost a century ago.

It is not unusual that there is no description regarding the purity of the sample in question. There are often no information as to whether analytical or technical-grade reagent was used or what the impurities were in the solvent. The compiler has to use discretion in evaluating the weight of the data, based on his past experience. The solubility between halogenated hydrocarbons and water is very small and the presence of impurities can cause a change in solubility results. A comprehensive description of the experiment includes the minimum specific purity of the chemicals, expressed in weight or mole percentage and giving the source of a substance or the name and address of the manufacturer. There are several solvents that are instable in pure form, and therefore the manufacturer adds stabilizer to the compounds, which can be as high as 5%. In this case, it is imperative to state in the description of the experiment what the level of stabilizer was. In particular, commercial solvents (technical grade), marketed in bulk quantities, require specifications as to their purity.

There is also a close connection between the reliability of the apparatus and the technique applied, because the choice of setup often depends on the method used in the experiment. Some of the typical methods and apparatus are outlined in Table 5.2 for the solubility of gases in liquids. The versetality of the apparatus and methods are characteristic for solubility measurements. Some of the methods are rapid and sufficiently accurate for most solubility determinations, but the success of the method will also depend on several other factors, such as the temperature range of

Table 5.2 Methods and Apparatus for the Determination of the Solubility of Gases in Liquids

Method	References
Rapid	Dymond & Hildebrand (1967), Lopest (1957), Slobodin et al. (1964)
Manometric	Burrows & Preece (1953), Koonce & Kobayashi (1964), Krichevskii & Sorina (1958), Zampachova (1962)
Equilibration	Bodor et al. (1957), Brown & Wasik (1974), Kobatake & Hildebrand (1961), Kogan et al. (1963), Swain & Thornton (1962), Tsiklis & Svetlova (1958), Wasik & Brown (1973)
Microgasometric	Douglas (1964), Scholander (1947), Wheatland & Smith (1955)
Extraction	Baldwin & Daniel (1952), Van Slyke & Neill (1924)
Modified Ostwald	Cook & Hanson (1957), Markham & Kobe (1941)
Saturation	Klots & Benson (1963), Lannung (1930), Morrison & Billett (1948), Novak & Conway (1973)
Dissolution	Ben-Naim & Baer (1963), Moudgil et al. (1974)
Corrosion-resistant	Cox & Head (1962)
Low-temperature	Karasz & Halsey (1958)
Radioactive tracer	Caddock & Davies (1959)

the experiment, the corrosiveness of the system, and the pressure range. Each apparatus (setup) has advantages and disadvantages, and the opinion of an expert is required to select one or the other for solubility measurements. A detailed discussion of the various methods and apparatus for the determination of solubility and miscibility between halogenated hydrocarbons and water is presented in Chapter 1.

Another example of the applied technique is the selection of the method of analysis. Which is the most reliable method of analysis: GLC, polarography, titration, radioactive isotopic labeling, ultraviolet spectrophotometry, macroreticular resin bed (Chey and Calder, 1972), or other? The importance of the method of analysis is illustrated by Sellers (1971) for the determination of the water content of organic samples using GLC and CaH_2 methods. The water content in ethylene dichloride sample was 0.124 ± 0.004 vol % by the GLC method, whereas it was 0.129 ± 0.005 vol % using the CaH_2 method under the same conditions. The difference is 3.88%, which might be significant in some cases. Wheatland and Smith (1955) gives an example of the comparison of gasometric and titrimetric methods for the determination of the solubility of oxygen in water. Farkas (1965) compared the methods for the determination of hydrocarbon-in-water solubilities. Pierotti and Liabastre (1972) compared the various measuring methods for the solubility of benzene in water. Wing (1956) and Wing and Johnston (1957) discussed in great detail the derived solubility data by various methods, emphasizing particularly the analytical methods. A comparison of the micro-particle dispersion procedure with the turbidity method was reported by Fürer and Geiger (1977) for the determination of water solubilities of commercial pesticides. A few serious discrepancies were found in comparison with the literature data. In particular, some misinterpretation was found for crystalline solutes in water.

In a recent article Schwarz and Miller (1980) discussed the various methods applied for the determination of the aqueous solubilities of organic liquids. Whilst the earlier method was based on the measurement of the decrease in volume of the pure solute phase after equilibration, the more recent determinations of the concentration of the solution are by interferometry, UV spectrophotometry, spectrofluorometry, liquid chromatography, gas chromatography head space and spectrofluorometric analysis, and elution chromatography.

The authors determined the aqueous solubilities of several halogenated hydrocarbons at various temperatures using both UV spectrophotometry and elution chromatography in comparison with literature data.

The average experimental error of both methods was 4%. With a few exeptions they found good agreement with the published solubilities. However, in the case of m-dichlorobenzene, the discrepancy was 15% at 20°C between the chromatographic and UV determined values. It is worth noticing that Schwarz and Miller were not using the more recent data reported for comparison.

The various analytical methods are discussed further in Chapter 1.

The competence of the investigator is very important. There are well-known experts in many different fields of science and technology. Their investigations and final results are obviously more reliable than those of others who have not gained as much experience in the subject. Investigations and publications of the National Bureau of Standards, Lawrence Radiation Laboratory, Thermochemical Laboratory (Lund),

Max-Plank Institute, National Physical Laboratory, among others, are reliable sources
of information. There are several distinct specialists who have achieved high standards or have published books and articles in the field of solubility; they include
(in alphabetical order):

    Battino, R.
    Gerrard, W.
    Gjaldbaek, J. C.
    Glew, D. N.
    Hayduk, W.
    Hildebrand, J. H.
    Moelwyn-Hughes, E. A.
    Prausnitz, J. M.
    Scott, R. L.
    Wilhelm, E.
    Yalkowsky, S. H.

It is not difficult to determine whether an investigator is an expert on a subject.
By looking up the author index in the latest collective volume of Chemical Abstracts,
the published works of an author can be found. From the published works, it is not a
very time consuming job to determine the experience of the scientist. Most reasearch
workers working in the field of solubility are familiar with the name of individual
scientists such as those listed above. The selection committees of symposiums or
conferences use a similar method to invite papers from eminent scientists.

    The evaluation of the validity of experimental or reported data has been accom-
plished mainly by testing for self-consistency, consistency with known thermodynamic
relations, and agreement with other measurements of the same or a different kind. The
specific evaluation methods used depend, of course, on the different types of data
considered (e.g., the creditability was reduced for those measurements in which the
initial or final state of the sample was in question).

    In 1936, on the basis of a large number of experimental data, Evans and Polanyi
(1936) found that there is a linear relationship between entropy (ΔS_s) and heat (ΔH_s)
of solution for a number of solutes (solids and liquids) in various solvents (polar
and nonpolar). Among the 16 solutes were benzene, ethylene dibromide, nitrobenzene,
naphthalene, and benzoic acid. Some of the solvents investigated were carbon tetrachloride, dichloroethane, chloroform, pyridine, acetone, ethyl ether, acetic acid,
aniline, toluene, and bromoform. However, neither gaseous solutes nor water as solvent
or solute were included in the study. The linear relationship was expressed by the
equation

$$\frac{\Delta S_s}{R} = a\ \Delta H_s + b$$

where R is the gas constant and a and b are constants. The heat and entropy changes
were measured with the solute in its normal solid state as zero. In these cases the
solute molecules were larger than the solvent molecules and Bell (1937) suggested
that the relation might hold when the solute molecules are smaller.

The enthalpy-entropy relationship has several useful implications in organic
chemistry (Leffler, 1955).

In aqueous solutions it was shown by Butler and Reid (1936), Evans (1937), and
Butler (1937) that the entropies of solution from the vapor phases of a number of
alcohols and amines are a linear function of the heats of solution. Furthermore, there
is a relation between the entropies and heats of solution of gaseous solutes in water.
The gas solubility data (H_2, N_2, O_2, CO, CH_4, C_2H_2, C_2H_6, etc.) of Horiuti (1931)
were examined by Bell (1937), and he found that in each of the five solvents (CCl_4,
C_6H_6, C_6H_5Cl, acetone, methyl acetate) there was an accurate linear relation between
the heats and entropies of solution. However, some criticism was expressed by Evans
and Polanyi (1936) on the application of the large solute molecules treated.

Barclay and Butler (1938) examined in great detail the various aspects of the
linear relationship between the entropy and heat of solution, criticizing previous
works on the subject, particularly the calculation method for the entropy of solution
from the expression

$$\Delta S_s = - \frac{d(R\, T\, \log_{10} x_2)}{dT}$$

This equation is valid if the activity coefficient does not change greatly with the
concentration and temperature. In concentrated solutions the activity coefficient
varies greatly with the temperature and concentration, and consequently the foregoing
equation is acceptable only when the solubility is small over the whole temperature
range considered.

In the case of the comparable size of solute and solvent molecules, the free
energy of solution can be conveniently calculated from the expression

$$\Delta G_s = R\, T\, \log(p_{part}/x_2)$$

where p_{part} is the partial pressure of the solute at T temperature (K). The corresponding entropy change is

$$\Delta S_s = - \frac{d(\Delta G_s)}{dT}$$

and the heat of solution (ΔH_s) is calculated directly from ΔG_s and ΔS_s at the
required temperature (T, K):

$$\Delta H_s = \Delta G_s - T\, \Delta S_s$$

The values derived for several solutes (SO_2, NH_3, CCl_4, C_6H_6, $CHCl_3$, C_6H_5Cl, etc.)
in acetone and ethyle alcohol solvents were used to plot $T \Delta S_s$ versus ΔH_s on linear
scales. The divergance from a unique relation between $T \Delta S_s$ and ΔH_s is in no case
very great. As a good approximation, it cannot be doubted that up to heats of solution
of about 12 kcal g-mole^{-1}, such a single straight-line relation exists.

Molecules dissolved in associated liquids, such as water or alcohols, cause
abnormally small entropy values. This is connected with the entropy of cavity formation in the associated liquid. When a nonpolar molecule is brought into an associated
liquid, a cavity must be formed, which may be expected to have a low surface entropy,
and thus when the molecule is removed from the liquid, in addition to the normal
entropy of vaporization, it will be necessary to restore the loss of entropy in the
form of the cavity. Hence, the entropy of vaporization of solutes from the solution
will be exceptionally great.

The experimental values of ΔS_s and ΔH_s for numerous groups of substances in
dilute solution, at the same temperature and between the same standard states, are
more or less accurately represented by an equation of the type

$$\Delta S_s = A + B \Delta H_s$$

(Frank, 1945a, b; Frank and Evans, 1945). The generality of the relationship as a
rule for unassociated liquids is supported by the points for inorganic liquids. To
fit so miscellaneous a group of liquids with a single representation will undoubtedly
raise several points:

1. Corrections for gas imperfection
2. Separate straight lines for different groups of substances
3. The relationship between the group lines and the structural features of the
   molecules

A detailed discussion of the various aspects was presented by Frank (1945b) and Frank
and Evans (1945). Water forms frozen patches or microscopic icebergs around nonpolar
solute molecules, the extent of the iceberg increasing with the size of the solute
molecule. Such icebergs are apparently also formed around the nonpolar parts of the
molecules of polar substances such as alcohols and amines dissolved in water. Consequently, the increasing insolubility of large nonpolar molecules is an entropy
effect. Further discussion of the subject appears in Chapter 2.

Frank and Evans (1945) repeated Bell's (1937) calculations for the temperatures
25 and 40°C. The data plotted, except for SO_2, gave full support to the Barclay-Butler
rule. However, a relationship still exists between ΔS_s and ΔH_s for SO_2 in various
solvents using the Horiuti (1931) data. The relative sizes of solvent and solute
molecules determine the type of regularity, so that a good straight line should be
obtained for a series of solutes all of which have small molecules, in a solvent made

up of larger molecules, or for a single solute of large molecular size in a series of smaller-molecule solvents.

The applicability of the Barclay-Butler rule for aqueous solutions of nonelectrolytes was examined in detail by Frank and Evans (1945). They found one rather good straight line for nonpolar solutes in water, and another that represented the alcohols and amines, with several solutes of intermediate polarity falling between these two lines. From the ΔS_s versus ΔH_s plot, the following four features are significant (see Fig. 5.1):

1. ΔS_s values for all solutions are too high, with about 10 entropy units
2. The slope of the line is about half that of the "standard" line
3. The line is far away from the ΔS_s of pure water
4. These solutions respond differently to temperature changes

All those irregular behaviors are related in some unique way to the properties of water. The dissolution of the rare gases and nonpolar molecules in water at room temperature modifies the water structure in the direction of greater crystallinity (i.e., a microscopic iceberg is built around it). The extent of the iceberg depends on the size of the foreign atom. During the dissolution of a rare gas or nonpolar molecule in water, heat and entropy are lost, more than in nonpolar solvents. The loss of the heat and entropy add to the heat and entropy of solution, causing larger values than expected.

The existence of crystalline hydrates of rare gases and hydrocarbon molecules in cold water supports the iceberg formation theory. On the other hand, iceberg formation helps to explain the insolubility of nonpolar compounds in water when the dissolution involves greater entropy loss.

Using data whose accuracy is known with a high degree of confidence, Franks (1972-1975: Vols. 2,3) constructed an improved Barclay-Butler plot for the heat and entropy of solution of alcohol - water systems. The lines obtained were parallel to those of hydrocarbons, but displaced from them. The structural contribution to ΔH_s and ΔS_s in water was satisfactorily accounted for by this formalism.

Flid and Golynetz (1959) investigated the solubility of acetylene in aqueous solutions of electrolytes in relation to the temperature and concentration of the salt. They found that the nature of the variations of ΔH_s and ΔS_s with temperature is identical, indicating an interrelation between ΔH_s and ΔS_s, expressed by the linear relation

$$\frac{\Delta S_s}{R} = \frac{a\,\Delta H_s}{R\,T} + b$$

similar to that of Evans and Polanyi (1936) (see above). The solubility of acetylene was studied in a larger number of aqueous electrolyte solutions of H_2SO_4, Li_2SO_4, $LiCl$, $NaNO_3$, $NaCl$, K_2SO_4, KNO_3, KCl, KBr, NH_4Cl, $ZnCl_2$, $CdCl_2$, $CdBr_2$, CdI_2, $MgSO_4$,

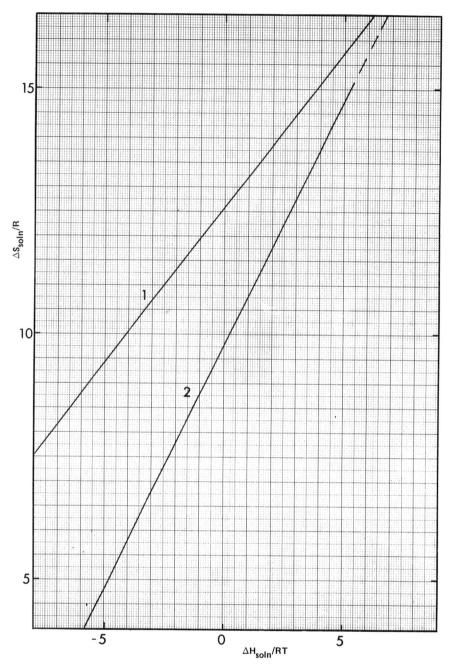

Fig. 5.1 Barclay-Butler plot for nonpolar solutes in aqueous solutions at 25°C.
ΔS_{soln} = entropy of solution in water (cal g-mole^{-1} K^{-1}); ΔH_{soln} = heat of solution
in water (kcal g-mole^{-1}); R = gas constant (cal g-mole^{-1} K^{-1}); T = absolute temperature (K). 1 = Frank & Evans (1945); 2 = Frank & Franks (1968).

$ZnSO_4$, $CdSO_4$, $MnSO_4$, $FeSO_4$, $NiSO_4$, $Al_2(SO_4)_3$, $Cr_2(SO_4)_3$, $Fe_2(SO_4)_3$, and others. The differences between the ΔH_s and ΔS_s values calculated with the foregoing equation and those from experimental data do not exceed, in most cases, 0.3 to 0.4 entropy unit. The value of a is approximately unity and that of b varies between 7 and 9.

Glew and Robertson (1956) used series of alcohols and alkyl benzenes to demonstrate the linear relationship between ΔH_s and ΔS_s at 25°C.

The calculated thermodynamic variables ΔH_s and ΔS_s from the solubility data at 25°C were plotted on linear scales by Moelwyn-Hughes (1971). The plotted values followed the expression

$$T \; \Delta S_s \; = \; A \; + \; B \; \Delta H_s$$

where A and B are constants.

The Barclay-Butler plot was also used by Treger et al. (1964) for the solubility of allyl chloride in water and aqueous hydrochloric acid solutions (Fig. 5.2). The free energy of solution (ΔG_s) was calculated from

$$\Delta G_s \; = \; - \; R \; T \; \log_{10} K \quad (mol/mol \; H_2O)$$

The heat of solution of the gas, including the heat of condensation, was obtained from

$$\Delta H_s \; = \; \Delta H^{298} + a \; (T - 298)$$

where ΔH^{298} and a have been derived from

$$\frac{d \; \log_{10}(\gamma_2/\gamma_1)}{d \; T} = \frac{\Delta H^{298} + a(T - 298)}{R \; T^2}$$

by integrating over the available temperature intervals. The entropy of solution (ΔS_s) was calculated from the well-know expression

$$\Delta S_s \; = \; \frac{\Delta H_s - \Delta G_s}{T}$$

The Barclay-Butler plot was also used by Svetlanov et al. (1971) to show the approximately linear relationship between the enthalpy (ΔH_s) and entropy of solution (ΔS_s) for the solubility of chloromethanes (CH_3Cl, CH_2Cl_2, $CHCl_3$, and CCl_4) and 1,2-dichloroethane in water and aqueous hydrochloric acid solutions (10 and 20 wt.%):

$$\frac{\Delta S_s}{R} \; = \; a \; \frac{\Delta H_s}{R \; T} + b$$

where a and b are constants. From the calculations they found that a = 1 and b = -7.

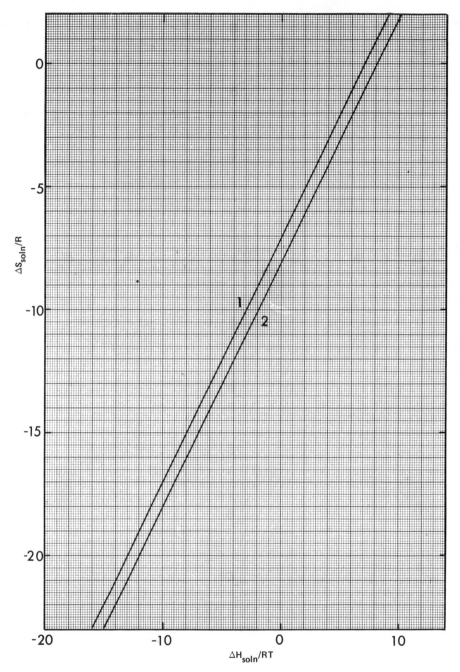

Fig. 5.2 Entropy and heat of solution relation for chlorocarbons in water. ΔS_{soln} = entropy of solution in water (cal g-mole^{-1} K^{-1}); ΔH_{soln} = heat of solution in water (kcal g-mole^{-1}); R = gas constant (cal g-mole^{-1} K^{-1}); T = absolute temperature (K).

1 = Svetlanov et al. (1971): $\Delta S_{soln}/R = (\Delta H_{soln}/R\,T) - 7$;

2 = Treger et al. (1964): $\Delta S_{soln}/R = (\Delta H_{soln}/R\,T) - 8$.

Comparison of the values of ΔH_s and ΔS_s calculated from the experimental solubility data for the solubility of chloroethanes in water and aqueous hydrochloric acid solutions show that regardless of the nature of the solution and temperature, the relation between the heat and entropy of solutions is in all cases approximately linear, as first established by Evans and Polanyi (1936).

The approximate nature of the Barclay-Butler rule for fluorinated hydrocarbon - water systems has been illustrated by Wen and Muccitelli (1979). However, the plot shows the good correlation for the members of the homologous series, e.g., rare-gase series, alkane series, and perfluorocarbon series.

The entropy and enthalpy of solutions have been estimated and correlated by several investigators. Ashton et al. (1968) correlated the entropy of solution (ΔS_s) of nonpolar compounds -- CF_4, SF_6, NF_3, and Ar -- in water at 25°C, using the liquid molar volume of the solvent ($V_{b.p.}$) at its normal boiling point:

$$- \Delta S_s = c + 0.22 \ V_{b.p.}$$

where c is a constant. See also Miller and Hildebrand (1968), who used a similar linear correlation between the entropy of solution of numerous gases in water at 25°C and 1 atm pressure, with the molar volume ($V_{b.p.}^{2/3}$) of the solvent at its normal boiling point.

Glew and Moelwyn-Hughes (1953) calculated the heat of solution (ΔH_s) of gaseous CH_3F, CH_3Cl, CH_3Br, and CH_3I in water from the van't Hoff isochore. However, the approximate expression

$$\Delta H_s = 2.0303 \ R \ c - b \ R \ T$$

is by no means exact. The enthalpy of solution of numerous compounds in water is tabulated in the Appendix.

The temperature dependence of the heat of solution of gaseous CF_4, C_2F_6, C_3F_8, and $c-C_4F_8$ in water was studied by Wen and Muccitelli (1979). The linear relationship between ΔH_s and T (°C) was illustrated between 5 and 30°C.

The heat and entropy of solution of a large number of selected gases and liquids in water at 25°C and 1 atm partial and saturated pressures, respectively, have been correlated according to the Barclay-Butler rule and presented in Figs. 5.3 and 5.4. The data have been taken from Miller (1966) (see Table 5.3) and from Liabastre (1974) (see Table 5.4).

The enthalpy - entropy relationship in general has been reviewed by Exner (1973).

In addition to the discussion above of the thermodynamic relationship between ΔH_s and ΔS_s, which is a very useful way to check the thermodynamic consistency of the solubility data according to the Barclay-Butler rule, there are several other relationships between the solubility and other parameters (e.g., solubility parameter, force constants, intermolecular forces, intermolecular potential energy, dielectric constant; see Chapter 2).

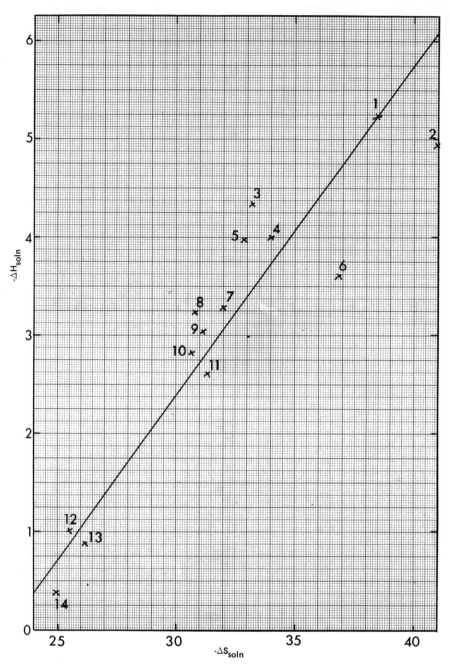

Fig. 5.3 Barclay-Butler plot for the solubility of gases in water at 25°C. ΔH_{soln} = heat of solution in water (kcal g-mole^{-1}); ΔS_{soln} = entropy of solution in water (cal g-mole^{-1} K^{-1}). The correlated equation was: $-\Delta S_{soln}$ = 22.8313 + 2.98885($-\Delta H_{soln}$). 1 = C_3H_8; 2 = SF_6; 3 = Xe; 4 = C_2H_6; 5 = C_2H_4; 6 = CF_4; 7 = CH_4; 8 = Kr; 9 = Ar; 10 = O_2; 11 = N_2; 12 = H_2; 13 = Ne; 14 = He. (From Miller, 1966.)

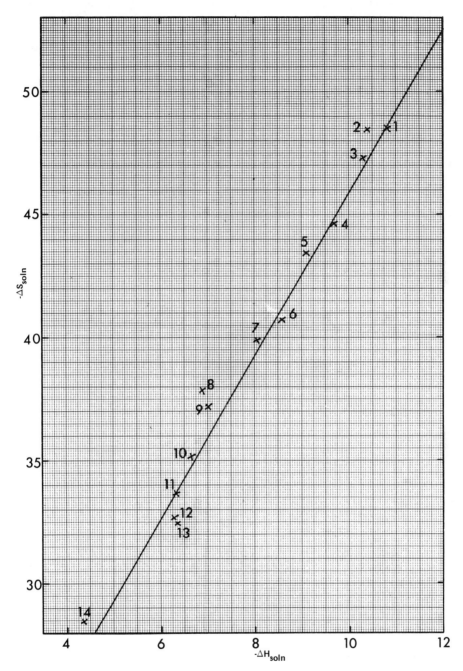

Fig. 5.4 Barclay-Butler plot for the solubility in water at 25°C. ΔS_{soln} = entropy of solution in water (cal g-mole^{-1} K^{-1}); ΔH_{soln} = heat of solution in water (kcal g-mole^{-1}). The correlated equation was: $-\Delta S_{soln}$ = 12.7563 + 3.31066($-\Delta H_{soln}$).
1 = $C_5H_{11}Br$; 2 = $C_5H_{11}I$; 3 = $C_5H_{11}Cl$; 4 = C_4H_9Br and C_4H_9I; 5 = C_4H_9Cl; 6 = C_3H_7Br and C_3H_7I; 7 = C_3H_7Cl; 8 = CCl_4; 9 = C_2H_5Br and C_2H_5I; 10 = C_2H_5Cl; 11 = CH_3Br;
13 = CH_3I; 14 = CH_3F. (From Liabastre, 1974.)

The solubility trend of various homologous series of organic substances in solvents at 25°C have been studied by Horvath (1973) using the expression

$$x_2 = a M^b$$

where x_2 = solubility, mole fraction

M   = molecular weight of the solute

a and b = coefficients of the equation

This correlation provides a simple method for comparing the solubility data of the members of a typical homologous series (see Fig. 2.51).

This very generalized procedure for the solubility of any homologous series in various solvents has been also investigated in more detail for the solubility of typical series of halogenated hydrocarbons in water at 25°C (Horvath, 1975a), such as

$$CX_4, \ C_2X_6, \ C_3X_8, \ C_4X_{10}, \ \cdots$$
$$CH_2X_2, \ CH_2XCH_2X, \ CH_2XCH_2CH_2X, \ \cdots$$
$$CHX_3, \ CHX_2CH_2X, \ CHX_2CH_2CH_2X, \ \cdots$$
$$CH_2{=}CHX, \ CH_3CH{=}CHX, \ CH_3CH_2CH{=}CHX, \ \cdots$$
$$\text{etc.}$$

where X represents a halogen atom. The log-log relationship between the solubility in water, expressed as a mole fraction (x_2), and the molecular weight (M) of the solute is illustrated in Fig. 2.52 for the typical series of halogenated hydrocarbons at 25°C. However, the straight-line relation is valid only for those substances that are

Table 5.3 Solubility of Gases in Water at 25°C and 1 atm Partial Pressure

Gas	Solubility in H_2O mole fraction (x_2)	$-\Delta H_s$ (kcal g-mole^{-1})	$-\Delta S_s$ (cal g-mole^{-1} K^{-1})	ΔG_s (kcal g-mole^{-1})
He	6.83×10^{-6}	0.37	24.9	7.048
Ne	8.23×10^{-6}	0.88	26.2	6.938
Ar	2.54×10^{-5}	3.04	31.2	6.267
Kr	4.32×10^{-5}	3.22	30.8	5.957
Xe	7.71×10^{-5}	4.32	33.3	5.608
H_2	1.42×10^{-5}	1.00	25.5	6.607
N_2	1.19×10^{-5}	2.61	31.3	6.723
O_2	2.31×10^{-5}	2.83	30.7	6.327
CF_4	3.79×10^{-6}	3.60	36.9	7.394
SF_6	4.42×10^{-6}	4.92	41.0	7.305
CH_4	2.48×10^{-5}	3.27	32.0	6.092
C_2H_6	3.10×10^{-5}	3.99	34.0	6.154
C_2H_4	8.74×10^{-5}	3.96	31.9	5.540
C_3H_8	2.73×10^{-5}	5.21	38.4	6.231
$n\text{-}C_4H_{10}$	2.17×10^{-5}	6.04	41.6	6.366

Source: Miller, 1966.

Table 5.4 Free Energy and Heat of Solution of Halocarbons in Water at 25°C.

Solute	ΔG_s (cal g-mole^{-1})	$-\Delta H_s$ (cal g-mole^{-1})	$-\Delta S_s$ (cal g-mole^{-1} K^{-1})
CH_3Cl	3,723	6,300	33.62
CCl_4	4,376	6,900	37.82
CH_3F	4,073	4,404	28.43
CH_3Cl	3,707	5,670	31.45
CH_3Br	3,453	6,275	32.63
CH_3I	3.365	6,325	32.50
CH_3Cl	3,619	5,700	31.26
C_2H_5Cl	3,782	6,700	35.16
C_3H_7Cl	3,783	8,100	39.86
C_4H_9Cl	3,827	9,100	43.36
$C_5H_{11}Cl$	3,790	10,300	47.26
CH_3Br	3,277	6,400	32.46
C_2H_5Br	4,079	7,000	37.16
C_3H_7Br	3,522	8,600	40.66
C_4H_9Br	3,615	9,700	44.66
$C_5H_{11}Br$	3,648	10,800	48.46
CH_3I	3,277	6,400	32.46
C_2H_5I	4,079	7,000	37.16
C_3H_7I	3,522	8,600	40.66
C_4H_9I	3,615	9,700	44.66
$C_5H_{11}I$	3,648	10,800	48.46

Source: Liabastre, 1974.

liquids at 25°C and atmospheric pressure, whereas the gases do not follow the rule. The average deviation for the correlated 11 compounds was 0.1%.

A very similar relation was developed by Boyer (1960) for the solubilities of gases in a group of polar solvents, such as alcohols. He found that the solubility of each gas in the alcohols could be expressed by means of the equation

$$\log_{10} x_2 = a + b \log_{10} C_n$$

where x_2 =   solubility in an alcohol, mole fraction
        C_n =   number of carbon atoms in the alcohol chain
    a and b   =   coefficients, characteristic of each gas

A correlation method between the various substances also provides a consistency check. All those values that show odd behavior in comparison to similar compounds will be suspect as to their consistence with other substances.

When the solubility data for a single solute in a solvent are available at various temperatures and pressures, the graphical illustration of the data will often reveal any incocnsistency. This has been investigated by Horvath (1975b) and described in Sec. 2.8. The three forms of graphical representations of the solubility values of sparingly soluble substances in water that are presented show the way for checking the available published data (see Figs. 2.86, 2.87 and 2.88). Any significant deviation from the linear relationship should show inconsistency and doubt about the reliability of the data.

In Chapters 7 and 8 are a large number of compiled and critically evaluated solubility data, which can be used in combination with the estimation and prediction methods described for the prediction of solubility values of halogenated hydrocarbons not previously investigated. These figures and tables are based on a comprehensive collection and selection of the published literature data up to a recent date. Further details on the solubility values are given as footnote so to tables and figures.

In the following description, a short summary is given on techniques and methods of selecting solubility data from several sources and deriving the recommended values.

1. <u>Solubility of water in bromobenzene.</u> There are five sets of experimental solubility data between the narrow temperature interval 15 to 35°C. These solubility values are illustrated in Fig. 5.5. The single data point by Hutchison and Lyon (1943) has been considered too high and therefore has been omitted from further consideration. The rest of the data have been correlated using the well-established expression (see Figs. 2.79 and 5.6 and Table 5.5):

$$\log_{10} x_2 = 2.48981 - \frac{1531.7}{T}$$

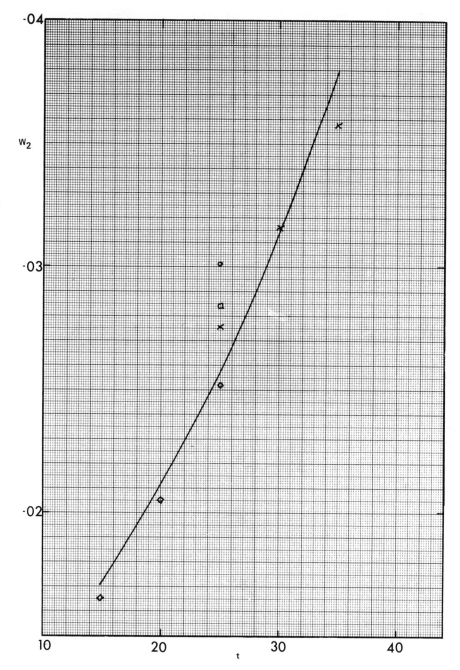

Fig. 5.5 Solubility of water in bromobenzene at atmospheric pressure. W_2 = solubility of water in bromobenzene (wt %); t = temperature (°C). o, Hutchison & Lyon (1943); □ , Wing & Johnston (1957); ×, Jones & Monk (1963); ◊ , Bell (1932).

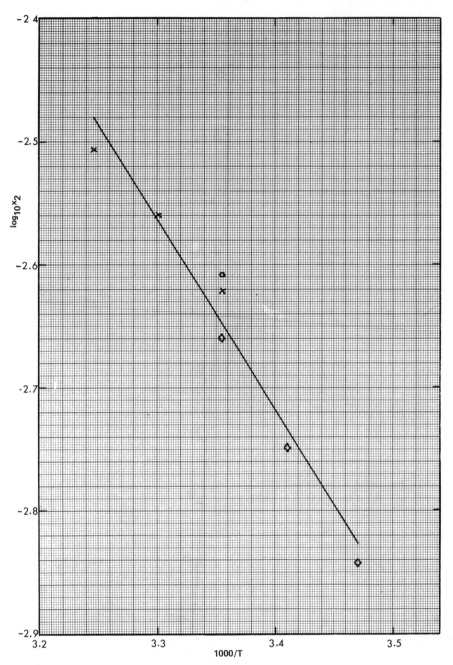

Fig. 5.6 Correlation of water solubility in bromobenzene at saturated pressure.
x_2 = solubility of water in bromobenzene (mole fraction); T = absolute temperature
(K). o, Hutchison & Lyon (1943); □ , Wing & Johnston (1975); × , Jones & Monk (1963);
◊, Bell (1932).

Table 5.5   Solubility of Water in Bromobenzene at Saturated Pressure

Temperature		1000	Solubility of water in C_6H_5Br			Reference
°C	K	T	wt %	Mole fraction, x_2	$\log_{10} x_2$	
15	288.15	3.4704	0.0165	0.001436	-2.8428	Bell (1932)
20	293.15	3.4112	0.0205	0.001784	-2.7486	Bell (1932)
25	298.15	3.3540	0.0252	0.002192	-2.6591	Bell (1932)
25	298.15	3.3540	0.0275	0.002392	-2.6213	Jones & Monk (1963)
30	303.15	3.2987	0.0316	0.002748	-2.5611	Jones & Monk (1963)
35	308.15	3.2452	0.0358	0.003112	-2.5070	Jones & Monk (1963)
25	298.15	3.3540	0.0284	0.002470	-2.6073	Wing (1956)
25	298.15	3.3540	0.0301	0.002617	-2.5821	Hutchison & Lyon (1943)

where x_2 =   solubility of water in bromobenzene, mole fraction

         T   =   absolute temperature, K

In the correlation, equal weight has been given to the three sets of data by Wing and Johnston (1957), Jones and Monk (1963), and Bell (1932). The data presented by Wing and Johnston (1957) originate from a Ph. D. dissertation by Wing (1956). Table 5.6 gives the recommended solubility data between 15 and 35°C. A moderate extrapolation of Fig. 5.6 should not involve great error; this is a well-established fact.

Table 5.6 Solubility of Water in Bromobenzene at Saturated Pressure (Recommended Data)

Temperature		Solubility of H_2O in C_6H_5Br	
°C	K	Wt %	Mole fraction
15	288.15	0.01716	0.001494
20	293.15	0.02115	0.001841
25	298.15	0.02589	0.002252
30	303.15	0.03148	0.002737
35	308.15	0.03804	0.003306

2. **Solubility of dibromomethane in water.** Five experimental works were reported between 1906 and 1963 on the solubility of CH_2Br_2 in H_2O in the temperature interval 0 to 30°C (Rex, 1906; Gross and Saylor, 1931; Van Arkel and Vles. 1936; Booth and Everson, 1948; O'Connell, 1963) (see Table 5.7 and Fig. 5.7).

There is a very good agreement between the solubility data of Gross and Saylor (1931) and those of Van Arkel and Vles (1936) at 30°C, which receives the highest weight in the correlation procedure. The solubility values of O'Connell (1963) and Booth and Everson (1948) have been not considered, because of their gross deviation from the rest of the data. Consequently, three sets of data, by Rex (1906), Gross and Saylor (1931), and Van Arkel and Vles (1936), have been correlated using a normal third-degree polynominal:

$$W_2 \text{ (wt \%)} = a + b t + c t^2 + d t^3$$

where W_2 =   solubility of CH_2Br_2 in H_2O, wt %

         t   =   temperature, °C

a, b, c, d =   coefficients of the polynominal

The experimental and correlated solubility data are presented in Fig. 5.7. The recommended values are tabulated in Table 5.8 between 0 and 30°C.

Table 5.7   Solubility of CH_2Br_2 in Water at Saturated Pressure

Temperature		Solubility of CH_2Br_2 in H_2O		Reference
°C	K		Wt %	
0	273.15	1.173 g/100 g H_2O	1.1594	Rex (1906)
10	283.15	1.146 g/100 g H_2O	1.1330	Rex (1906)
20	293.15	1.148 g/100 g H_2O	1.1350	Rex (1906)
30	303.15	1.176 g/100 g H_2O	1.1574	Rex (1906)
15	288.15	11.70 g/1000 g H_2O	1.1565	Gross & Saylor (1931)
30	303.15	11.93 g/1000 g H_2O	1.1789	Gross & Saylor (1931)
30	303.15	0.00686 g-mole/100 g	1.1786	Van Arkel & Vles (1936)
25	298.15	0.70 ml/100 ml H_2O	1.7174	Booth & Everson (1948)
25	298.15	1.10 g/100 g H_2O	1.0880	O'Connell (1963)

Table 5.8 Solubility of Dibromomethane in Water at Saturated Pressure (Recommended)

Temperature		Solubility of CH_2Br_2 in H_2O	
°C	K	Wt %[a]	Mole fraction
0	273.15	1.1584	0.001213
5	278.15	1.1484	0.001202
10	283.15	1.1421	0.001196
15	288.15	1.1404	0.001194
20	293.15	1.1441	0.001198
25	298.15	1.1541	0.001208
30	303.15	1.1713	0.001227

$^a W_2$ (wt %) = $1.15839 - 2.31622 \times 10^{-3} t + 5.6995 \times 10^{-5} t^2 + 1.15173 \times 10^{-6} t^3,$

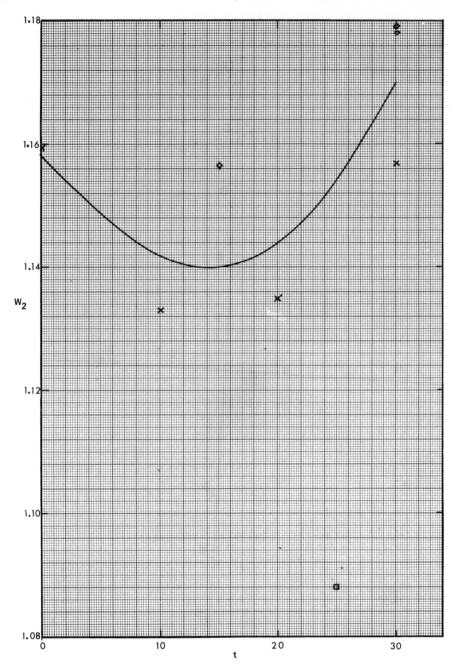

Fig. 5.7 Solubility of dibromomethane in water at saturated pressure. W_2 = solubility of dibromomethane in water (wt %); t = temperature ($^{\circ}$C). ×, Rex (1906);
◊, Gross & Saylor (1931); o, Van Arkel & Vles (1936); □, O'Connell (1963).

3. <u>Solubility of tribromomethane in water</u>. The temperature interval from 0 to
30°C is covered by five experimentally determined sets of solubility data -- by
Squire and Caines (1905), Gross and Saylor (1931), Van Arkel and Vles (1936), Booth
and Everson (1948), and O'Connell (1963) (see Table 5.9). The single data point at
0°C ("cold") from a secondary source (Lange, 1967) was disregarded because of the
uncertainity of its origin. The solubility value of Booth and Everson (1948) is only
approximate; therefore, it was omitted from further consideration. Similarly, the
reported data by Squire and Caines (1905) are too low, so have also been disregarded.
The remaining four data points, by Gross and Saylor (1931), Van Arkel and Vles (1936),
and O'Connell (1963), are plotted in Fig. 5.8.

The correlation of these four values has been carried out using a normal second-
degree polynominal. The recommmended solubility data are tabulated in Table 5.10.

4. <u>Solubility of chlorobenzene in water</u>. Experimentally measured solubility data
were reported in 11 references for chlorobenzene in water between 5 and 280°C.
However, the values by Vorozcov and Kobelev (1938) between 240 and 280°C were
illustrated in graphs without numerical data. Furthermore, the pressure was not stated;
therefore, the data have not been included in the correlation procedure. The solubility
data reported by Othmer et al. (1941) and Booth and Everson (1948) have been left out
because the former is too high and the latter is too low relatively to the other sources.
Furthermore, the solubility values of Van Arkel and Vles (1936) and Dreisbach (1955-
1961: Ser. 15) were reported without any detail or information regarding their determination, so they have also been disregarded in the correlation.

The solubility data determined by Gross and Saylor (1931), Andrews and Keefer
(1950), Kisarov (1962), Chey and Calder (1972), and Vesala (1974) (see Table 5.11
and Fig. 5.9) have been correlated using a normal third-degree polynomonal:

$$W_2 \text{ (wt \%)} = 4.71538 \times 10^{-2} - 5.09717 \times 10^{-4} t + 2.25008 \times 10^{-5} t^2$$
$$+ 8.42385 \times 10^{-8} t^3$$

where t is the temperature in °C. This polynominal has been forced through the
solubility value of 0.0498 wt % C_6H_5Cl at 25°C, being the most reliable data
point in the correlation. The result is illustrated in Fig. 5.9 between 20 and
90°C. The recommended numerical values are given in Table 5.12.

The solubility data reported by Nelson and Smit (1978) have been excluded from
the evaluation because of the uncertainty of the values presented in relation to
other measurements. Their solubility data are the lowest reported to date.

The solubility data reported by Yalkowsky et al. (1979) for halogenated benzenes
in water, have been included in Table A11 (see Appendix) without critical examination.
Similarly, several solubility data reported during 1979 and 1980 have been listed
in Table A11 only. A more up-to-date evaluation of halogenated benzenes will be
published in Vol. 16 of the IUPAC Solubility Data Series (see Table 3.2).

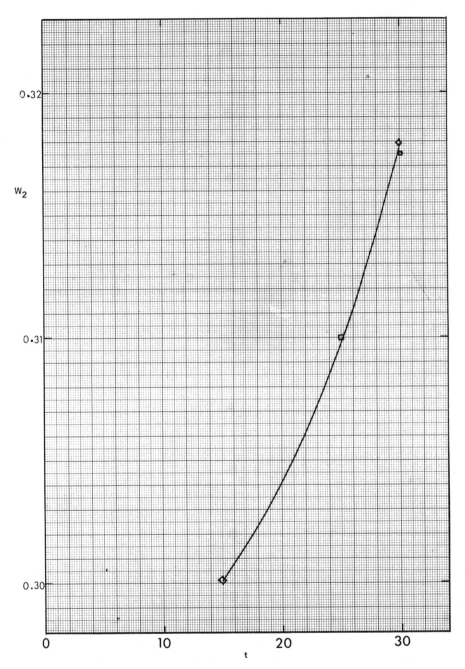

Fig. 5.8 Solubility of tribromomethane in water at saturated pressure. W_2 = solubility of tribromomethane in water (wt %); t = temperature ($^{\circ}$C). $\lozenge$, Gross & Saylor (1931); o, Van Arkel & Vles (1936); $\square$, O'Connell (1963).

Table 5.9 Solubility of Tribromomethane in Water at Saturated Pressure

Temperature		Solubility of $CHBr_3$ in H_2O		
°C	K		Wt %	Reference
20	293.15	1.0 g/800 g H_2O	0.1248	Squire & Caines (1905)
15	288.15	3.01 g/1000 g H_2O	0.3001	Gross & Saylor (1931)
30	303.15	3.19 g/1000 g H_2O	0.3180	Gross & Saylor (1931)
30	303.15	0.00126 g-mole/100 g H_2O	0.3175	Van Arkel & Vles (1936)
25	298.15	<0.02 ml/100 ml H_2O	<0.0578	Booth & Everson (1948)
25	298.15	0.311 g/100 g H_2O	0.3100	O'Connell (1963)
∿0	∿273.15	0.1 g/100 g H_2O	0.1000	Lange (1967)

Table 5.10 Solubility of Tribromomethane in Water at Saturated Pressure (Recommended Data)

Temperature		Solubility of $CHBr_3$ in H_2O	
°C	K	Wt %[a]	Mole fraction
15	288.15	0.3001	0.0002145
20	293.15	0.3041	0.0002174
25	298.15	0.3100	0.0002216
30	303.15	0.3178	0.0002272

[a] W_2 (wt %) = $0.29925 - 5.03323 \times 10^{-4} t + 3.73332 \times 10^{-5} t^2$ (t = °C)

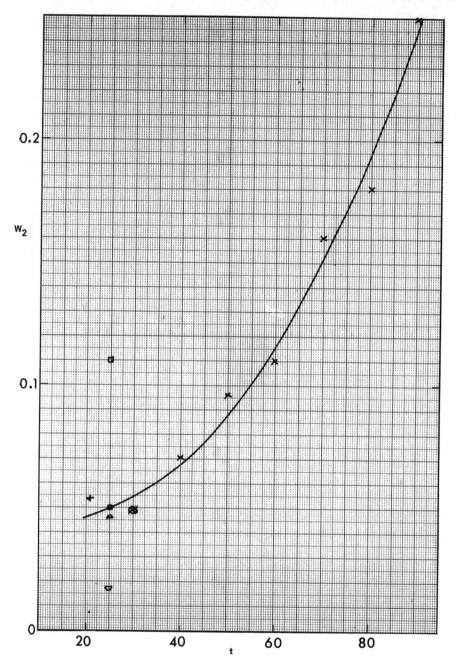

Fig. 5.9 Solubility of chlorobenzene in water at saturated pressure. W_2 = solubility of chlorobenzene in water (wt %); t = temperature (°C). o, Gross & Saylor (1931); □, Othmer et al. (1941); ▽, Booth & Everson (1948); ◊ , Andrews & Keefer (1950); ×, Kisarov (1962); +, Chey & Calder (1972); ⌒, Vesala (1974).

Table 5.11 Solubility of Chlorobenzene in Water at Saturated Pressure from Various Sources

Temperature °C	K	Solubility of C_6H_5Cl in H_2O (wt %)	Reference
30	303.15	0.0488	Gross & Saylor (1931)
25.5	298.65	0.11	Othmer et al. (1941)
25	298.15	0.0181	Booth & Everson (1948)
25	298.15	0.050	Andrews & Keefer (1950)
25	298.15	0.050	Dreisbach (1955-1961: Ser. 15)
30	303.15	0.049	Kisarov (1962)
40	313.15	0.0705	Kisarov (1962)
50	323.15	0.0960	Kisarov (1962)
60	333.15	0.1100	Kisarov (1962)
70	343.15	0.1605	Kisarov (1962)
80	353.15	0.1805	Kisarov (1962)
90	363.15	0.2500	Kisarov (1962)
21	294.15	0.0534	Chey & Calder (1972)
24.85	298.0	0.04626	Vesala (1974)
5	278.15	0.0040	Nelson & Smit (1978)
25	298.15	0.01068	Nelson & Smit (1978)
35	308.15	0.02673	Nelson & Smit (1978)
45	318.15	0.04004	Nelson & Smit (1978)

Table 5.12 Solubility of Chlorobenzene in Water at Saturated Pressure (Recommended
          Data)

Temperature		Solubility of C_6H_5Cl in water	
°C	K	Wt %[a]	Mole fraction
20	293.15	0.04663	0.00007467
25	298.15	0.04979	0.00007972
30	303.15	0.05439	0.00008709
35	308.15	0.06049	0.00009686
40	313.15	0.06816	0.0001092
45	318.15	0.07746	0.0001241
50	323.15	0.08845	0.0001417
55	328.15	0.1012	0.0001621
60	333.15	0.1158	0.0001855
65	338.15	0.1322	0.0002119
70	343.15	0.1506	0.0002414
75	348.15	0.1710	0.0002741
80	353.15	0.1935	0.0003102
85	358.15	0.2181	0.0003498
90	363.15	0.2450	0.0003929

[a] W_2 (wt %) = $4.71538 \times 10^{-2} - 5.09717 \times 10^{-4}t + 2.25008 \times 10^{-5}t^2 +$
$8.42385 \times 10^{-8}t^3$, where t is temperature (°C).

Chapter 6
Accuracy and Reliability
Of Solubility Data

Users of solubility data are always interested in the reliability and accuracy of the data available in various literature sources. In other words, how good are the data? To give a satisfactory answer to this question, it is neccessary to consider some aspects of obtaining the values.

Almost all solubility data presented in Chapters 7 and 8 are based upon experimental results reported in the published literature. As a result of the collection of literature articles dealing with solubility determinations, several sources have been found for the solubility values. If the purity of the sample and the experimental technique applied were the same in the various works, the results should not be different. However, for several reasons, we find smaller or larger discrepancies among the reported values for the same system, particularly when several years intervened between experiments. Regarding the importance of the purity of samples, see the discussion in Chapter 1. The various techniques used for solubility determinations are outlined in Chapter 1. A comparison of the results obtained by different analytical methods is discussed in Chapter 5. Here the emphasis is directed toward the reliability of the solubility data presented in Chapters 7 and 8.

The main feature of the solubility data in Chapters 7 and 8 is consistency. No single value has been adopted for the compilation without looking at its consistency with other substances by means of some sort of empirical or semiempirical relationship. Unfortunately, at present there are no rigorous relationships that can be used to check relative solubilities of various halogenated compounds in water, or vice versa. Therefore, more and more empirical expressions were and are under development for improving the existing ones.

The greatest risk is taken when there is only one experimental determination available for a substance. How good then, is the result? There are two ways to check the order of magnitude:

1. By estimating the solubility using empirical methods (e.g., group contribution, molecular surface area, connectivity index, scaled particle theory)
2. Relating the solubility value to the solubility of another member in a homologous series or other compounds which show a certain type of trend with the chemical in the same solvent

All these aspects have been kept in mind during compilation of Chapters 7 and 8, so
it is unlikely that any sort of serious error occurs in the data presented.

Some of the solubility data have been correlated using a normal polynominal for
a limited temperature interval. These equations should be used only in the temperature
range indicated. The equations play two major rolls: to provide data users with
interpolated data with good accuracy, and for use in computer programmes, if needed.

No attempt has been made to express the accuracy of data in percentages, but
the deviation of various experimental values from the fitted data are illustrated in
several figures. The solubility data given in Chapters 7 and 8 are the recommended
values resulting from the compilation. More emphasis is directed toward consistency
than toward the absolute accuracy of the experimental values.

Several solubility curves are presented in Chapter 7. The object of these is to
demonstrate the trends, relative solubilities, and other aspects typical for some
types of compounds. The solubility graphs indicate the risks involved when one attempts
to extrapolate the reported values. In some cases, extrapolation is quite safe for
the desired temperature, whereas in others, the risk is very high and extrapolation
is not recommended. The shapes of solubility curves for homologous series are very
typical. If one member of the series shows some sort of ambiguity, there must be an
error in the data produced.

The units of solubilities used in Chapters 7 and 8 are mole fraction, g liter^{-1},
and weight percentage. Conversion to other units can be accomplished using the conversion formula given in Chapter 9. The mole fraction is a scientific unit and widely
used by scientists for research and development work. The weight percentage is a
practical unit for design and process engineers, who require solubility values so
expressed.

Special attention is required for the expression of the pressure effect upon gas
solubilities in water at various temperatures. In most cases there are two possibilities: to express the solubility values at a certain total or partial pressure. Numerous solubility data for gaseous solutes are presented in the literature without reference to the pressure, whether total or partial. This difficulty is completely elim-
inated in the present compilation. Each table or graph heading clearly indicates
whether the pressure is partial or total for the system in question.

It is quite common among scientists and engineers to relate data to some sort of
reliability code system, which gives an indication of how good the data are. The various coding systems are very subjective indeed, but they still provide the data users
with a guide. Some workers are very reluctant to give high reliability codes for experimental data, and estimated or predicted data are placed very low in the coding scale.
This is not a very good practice, because some estimation methods based upon well-
established experimental data could provide solubility data with good accuracy despite
the fact that they were not experimentally determined. In fact, some experimental

results show quite a large discrepancy, depending upon several factors. It is the
view of the author that data published as the result of an estimation method should
not be considered without examination of the way they have been derived. There are
numerous physical properties that were determined experimentally in the past which
are less reliable than the latest data obtained by an estimation or prediction method.
A great advantage of some of the estimation methods in consistency, which is sometimes
more important than the absolute accuracy of single data points.

The IUPAC Solubility Data Project classified the precision of solubility data
into six groups:

Class	Precision interval (%)
A	0.1
B	0.1 - 0.5
C	0.5 - 1.5
D	1.5 - 4.0
E	4.0 - 10
F	10 - 50

This type of classification is useful for cases in which there are several or many
experimental data for selection and critical evaluation, but a difficulty appears
when there are very limited data and the discrepancy is considerable between the
reported values. In such cases it becomes necessary to relate the data to other substances. This type of consideration is not included in the classification procedure
mentioned above. According to the latest reports, the IUPAC has abandoned this classification system.

This chapter would not be complete without a brief discussion of the difference
between precision and accuracy of numerical and graphical data representation. A
simple but very effective illustration is provided by the determination of the concentration of a dissolved compound in water by the GLC method. The recorder plots
the concentration given by the height of the pick versus the time. The shape of the
pick illustrates the precision of the measurements (see Figs. 6.1 and 6.2). The
accuracy is expressed by the extent of bias caused by a systematic error (see Figs.
6.3 and 6.4). Further illustrations on the extent of measurement quality are given in
Figs. 6.5 and 6.6 (Mashiko et al., 1978).

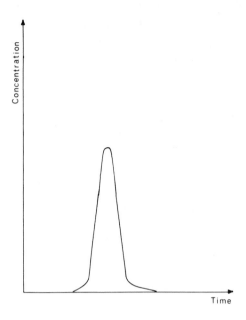

Fig. 6.1 Measurement in high precision.   Fig. 6.2 Measurement in low precision.

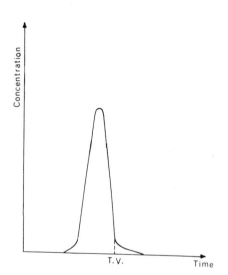
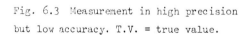
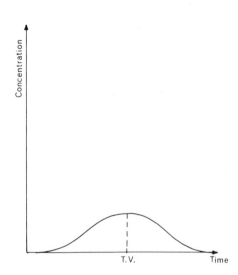

Fig. 6.3 Measurement in high precision   Fig. 6.4 Measurement in high accuracy
but low accuracy. T.V. = true value.   but low precision. T.V. = true value.

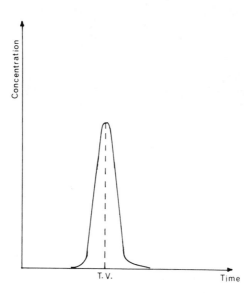

Fig. 6.5 Measurement in high precision and high accuracy. T.V. = true value.

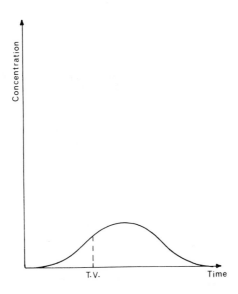

Fig. 6.6 Measurement in low precision and low accuracy. T.V. = true value.

Chapter 7

Graphs

The most illustrative way to present physical data when there are two or more variables is by graphs, drawings, or diagrams. Just as with other sorts of presentations (e.g., tabulation, functions) graphs have several advantages as well as disadvantages. The graphs play a particularly important roll when one is looking for trends between various quantities which are functions of the same variable.

In solubility studies temperature is the most important variable, and great effort has been spent on the study of the temperature dependence of solid, liquid, and gas solubilities. The main difficulty is to be representative for a very wide temperature interval, that is, from the triple point to the critical point in the case of liquid-liquid solubility. This problem is well illustrated by various diagrams. The selection of the scales, whether linear, logarithmic, or other, that are most suitable for the representation is important for the clarity of the trends one is looking for. Linearity, parallelisms, and focusing or convergence play a very significant roll during the investigation of treands and regularities in solubility data. The shape and form of a curve suggest the extrapolation and interpolation possibilities. The main aims during the selection of scales and relationships are that the plotted curves should resemble a straight line, which will provide an excellent opportunity for extrapolation or interpolation with good accuracy. However, extrapolation is always a risky process, and a comparison of similar curves will provide a sounder picture of the relative trends of the functions examined.

The graphical representation of solubility provides a concise form of data on a single sheet. Whereas in tables the temperature steps have to be decided, which makes interpolation and probably plotting or correlation necessary, in disgrams the temperature can be read off directly and the interpolation does not causes difficulty.

However, the figures read off a curve are subjective, depending upon the scales and the accuracy or number of significant figures required. This disadvantage can be overcome by large-scale diagrams; however, these might not be economically feasible or practical for wide distribution. Therefore, one must compromise in the choice between tabular or graphical presentation of solubility data. A large number of data are not accurate enough for more than three significant figures, which could be given in diagrams with satisfactory accuracy.

In the diagrams that follow, the solubility data can be read off with three sig-

nificant figures, which is sufficiently accurate in relation to the experimental
techniques used for their determination.

Figs. 7.1 to 7.12 illustrate the temperature and pressure effects upon the
solubilities of gaseous halogenated hydrocarbons in water. Each graph shows three
isotherms between 25 and 75°C for refrigerants $CBrF_3$, $CClF_3$, CCl_2F_2, CCl_3F, $CClF_2H$,
CCl_2FH, CF_3H, CF_4, C_2ClF_5, $C_2Cl_2F_4$, $C_2Cl_3F_3$, and C_4F_8.

The solubility of water in halogenated hydrocarbons is given in Figs. 7.13 to
7.15 at saturated pressure. In these three disgrams the solubility was plotted on
a logarithmic scale versus the reciprocal of the absolute temperature. The curves
obtained are represented by a straight line for each refrigerant investigated: R 21,
R 22, R 11, R 12, R 113, R 123, R 214, R 215, R 216, and R 216B2. The source of the
data is du Pont (1966c).

The temperature dependence of the solubility of water in halogenated hydrocarbons
can be most usefully represented on this type of graph, giving an opportunity for
cautious extrapolation. The solubility data for water tabulated in Chapter 8 can be
plotted in the same way.

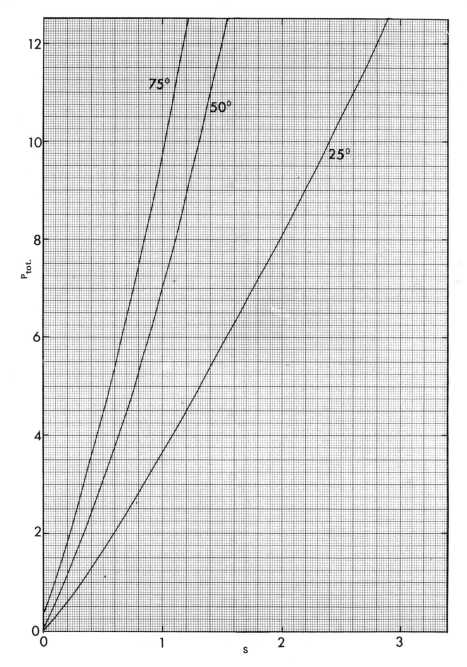

Fig. 7.1 Solubility of CBrF$_3$ in water at 25, 50, and 75°C. S = solubility of
CBrF$_3$ in water (g liter^{-1}); P$_{tot}$ = total pressure (atm). (From du Pont, 1966c.)

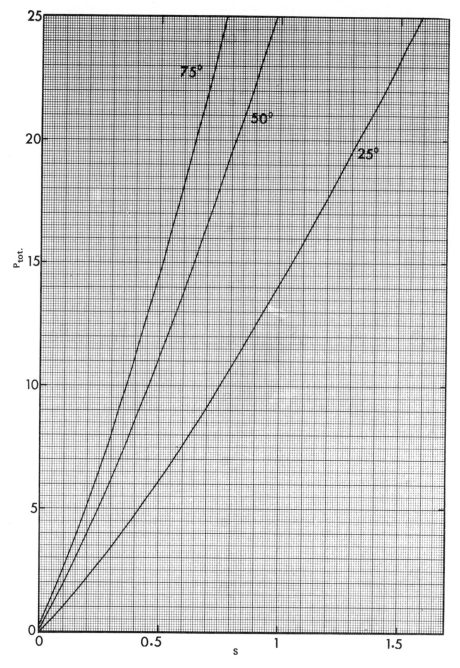

Fig. 7.2 Solubility of $CClF_3$ in water at 25, 50, and 75°C. S = solubility of $CClF_3$ in water (g liter^{-1}); P_{tot} = total pressure (atm). (From du Pont, 1966c.)

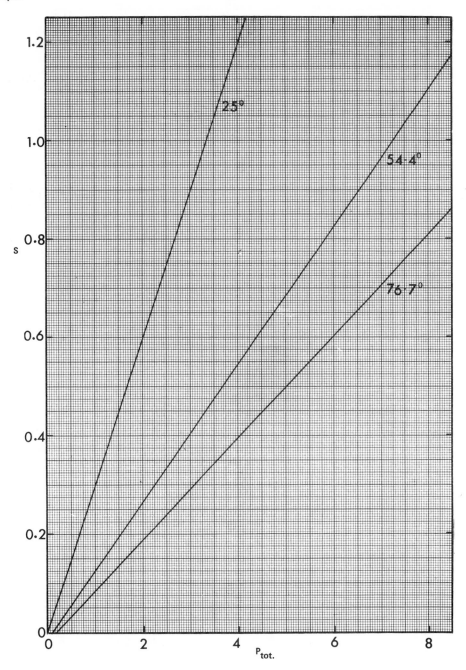

Fig. 7.3 Solubility of CCl_2F_2 in water at 25, 54.4, and 76.7°C. S = solubility of CCl_2F_2 in water (g liter^{-1}); P_{tot} = total pressure (atm). (From du Pont, 1966c.)

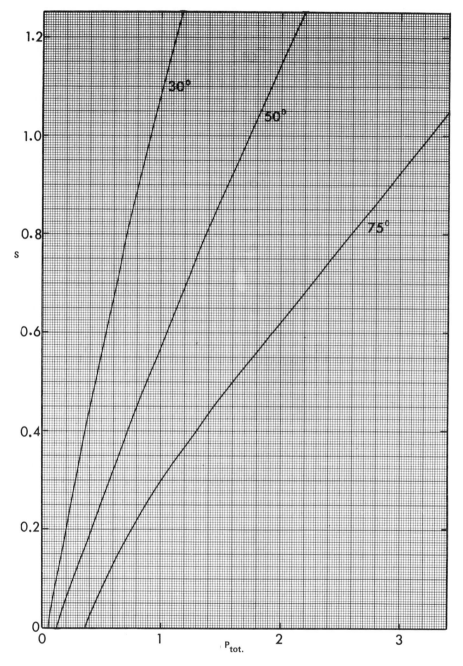

Fig. 7.4 Solubility of CCl_3F in water at 30, 50, and 75°C. S = solubility of CCl_3F in water (g liter^{-1}); P_{tot} = total pressure (atm). (From du Pont, 1966c.)

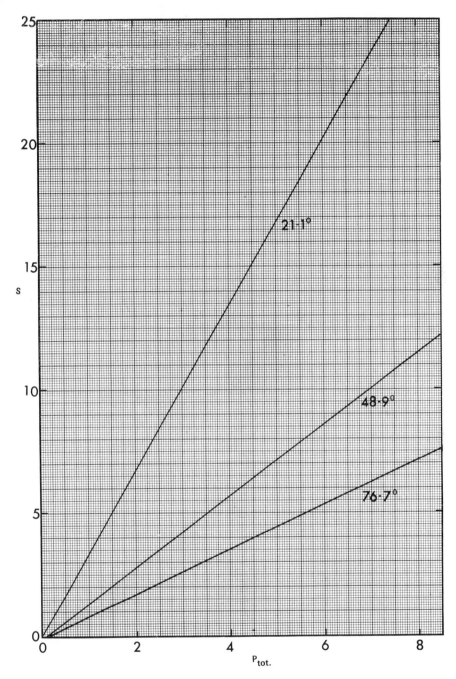

Fig. 7.5 Solubility of $CClF_2H$ in water at 21.1, 48.9, and 76.7°C. S = solubility of $CClF_2H$ in water (g liter^{-1}); P_{tot} = total pressure (atm). (From du Pont, 1966c.)

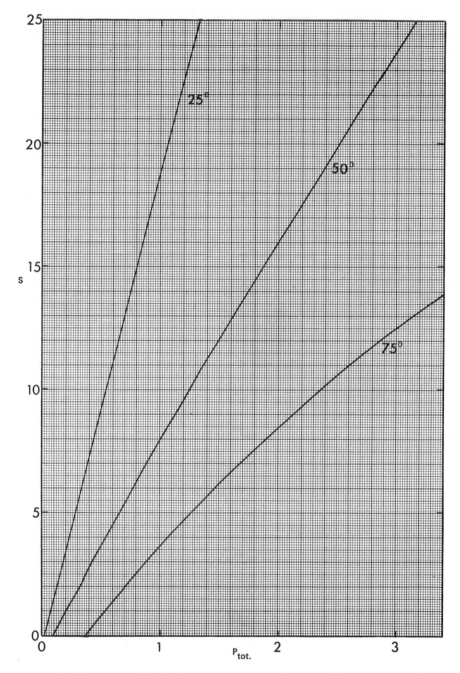

Fig. 7.6 Solubility of CCl_2FH in water at 25, 50, and 75°C. S = solubility of CCl_2FH in water (g liter^{-1}); P_{tot} = total pressure (atm). (From du Pont, 1966c.)

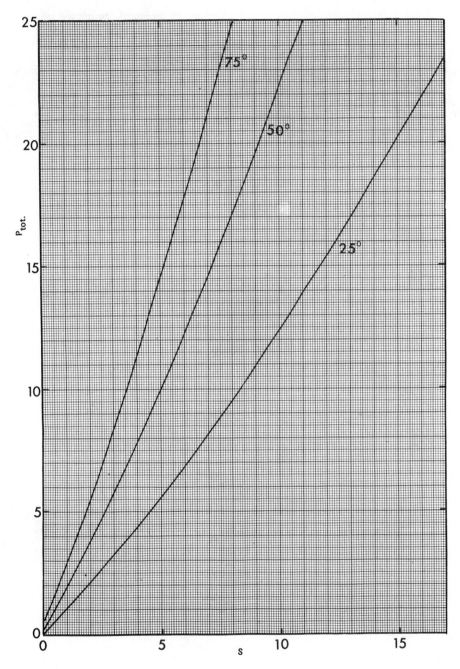

Fig. 7.7 Solubility of CF_3H in water at 25, 50, and 75°C. S = solubility of CF_3H in water (g liter^{-1}); P_{tot} = total pressure (atm). (From du Pont, 1966c.)

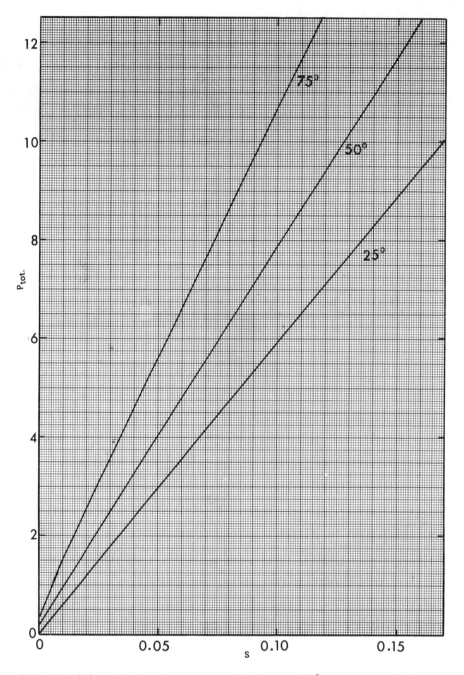

Fig. 7.8 Solubility of CF_4 in water at 25, 50, and 75°C. S = solubility of CF_4 in water (g liter^{-1}); P_{tot} = total pressure (atm). (From du Pont, 1966c.)

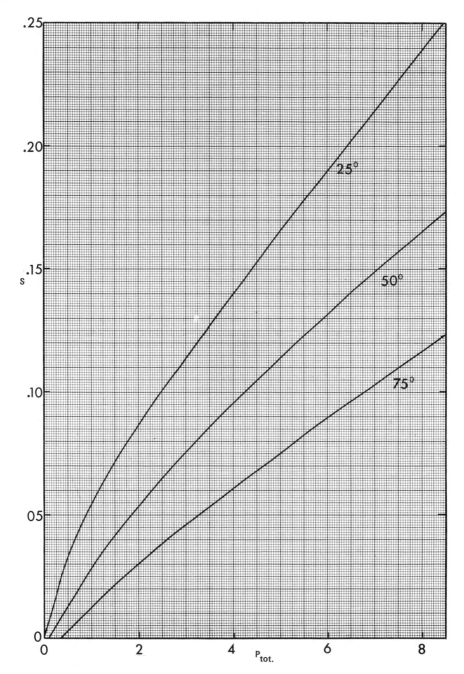

Fig. 7.9 Solubility of $CClF_2-CF_3$ in water at 25, 50, and 75°C. S = solubility of $CClF_2-CF_3$ in water (g liter^{-1}); P_{tot} = total pressure (atm). (From du Pont, 1966c.)

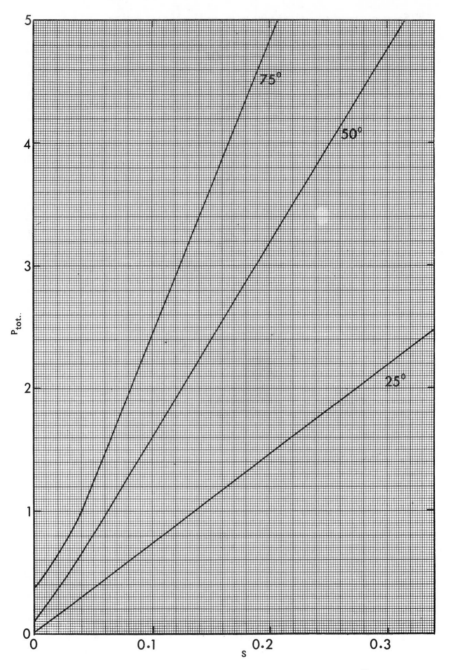

Fig. 7.10 Solubility of $CClF_2-CClF_2$ in water at 25, 50, and 75°C. S = solubility of $CClF_2-CClF_2$ in water (g liter^{-1}); P_{tot} = total pressure (atm). (From du Pont, 1966c.)

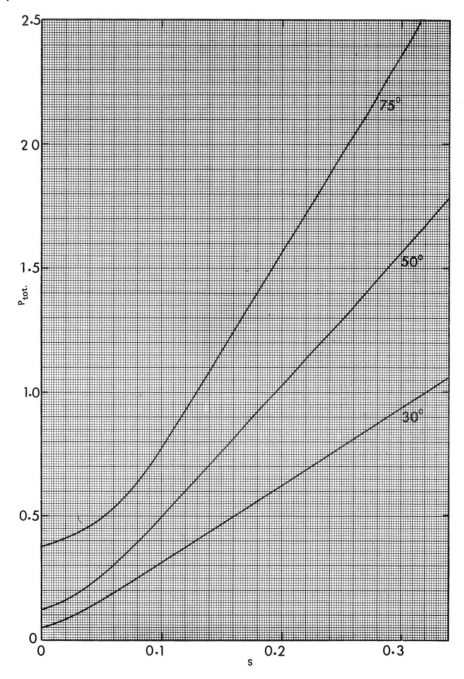

Fig. 7.11 Solubility of $CCl_2F-CClF_2$ in water at 30, 50, and 75°C. S = solubility of $CCl_2F-CClF_2$ in water (g liter^{-1}); P_{tot} = total pressure (atm). (From du Pont, 1966c.)

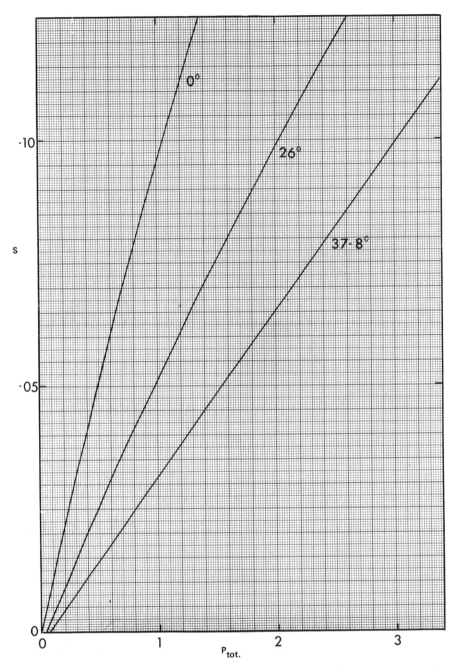

Fig. 7.12 Solubility of C_4F_8 in water at 0, 26, and 37.8°C. S = solubility of C_4F_8 in water (g liter^{-1}); P_{tot} = total pressure (atm). (From du Pont, 1966c.)

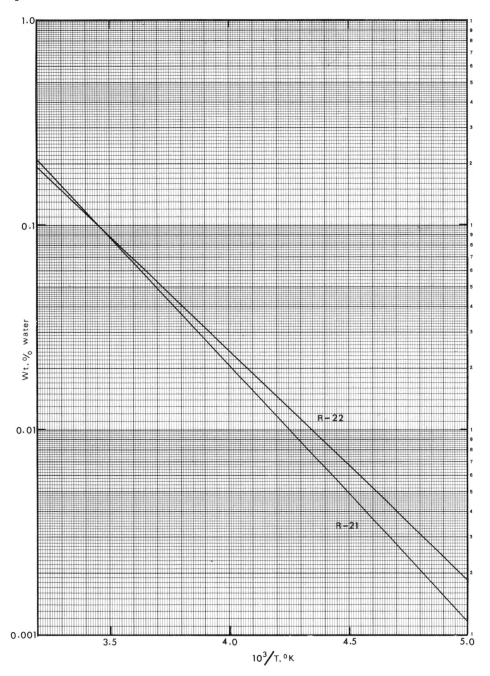

Fig. 7.13 Solubility of water in R 21 and R 22 at saturated pressure.
$CHClF_2$ (R 22); $CHCl_2F$ (R 21).

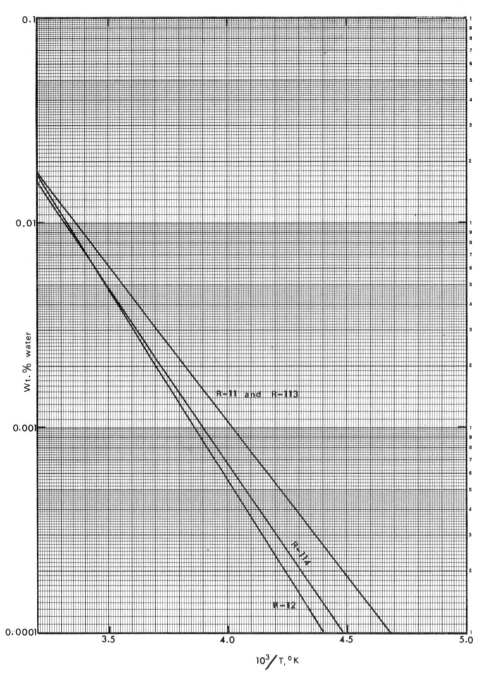

Fig. 7.14 Solubility of water in R 11, R 12, R 113, and R 114 at saturated pressure. CCl_3F (R 11); CCl_2F_2 (R 12); CCl_2F-$CClF_2$ (R 113); $CClF_2$-$CClF_2$ (R 114).

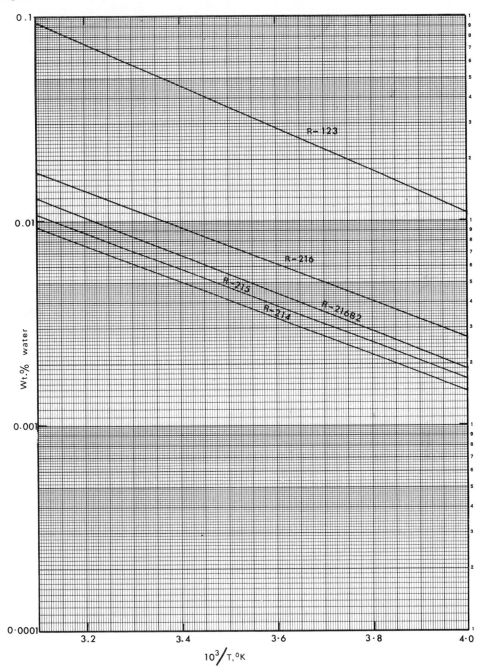

Fig. 7.15 Solubility of water in refrigerants at saturated pressure.
$CHCl_2$-CF_3 (R 123); $CClF_2$-$CClF$-CF_3 (R 216); $CBrF_2$-$CBrF$-CF_3 (R 216B2); CCl_3-CF_2-CF_3 (R 215); CCl_3-CF_2-$CClF_2$ (R 214).

Chapter 8
Tabulated Data

In the following tables, the selected and recommended solubility data are compiled and presented. Each table is self-explanatory and can be understood without difficulty.

In the first part the solubility of halogenated hydrocarbons in water is tabulated, followed by the solubility of water in halogenated hydrocarbons. However, before the tabulation of the data there is a formula index for both parts for easy retrieval of the solubility data.

The formula indexes consist of three columns: the first character is the carbon atom(s) followed by the other atoms in alphabetical order. The second column gives the chemical structure of the molecule, followed by its chemical name in the third column.

The fundamental physical properties of halogenated hydrocarbons are also given in each table. This information is important to users for understanding the solid, liquid, and gas temperature range of the substances. The melting and boiling points of each substance are given with the very few exceptions of when the data were not available in the literature. The molecular weight and density provide the data needed for converting the solubility values into other units.

The solubilities are expressed in weight percentage and mole fraction as a function of temperature at saturated or other pressure as specified (partial or total).

When the solubility is known for a wide temperature interval, the weight percentage (W_2) is fitted into a polynominal equation as a function of temperature (t, $^{\circ}C$) expressed in degrees Celsius:

$$W_2 \text{ (wt \%)} = a + b t \quad c t^2 + d t^3 + \dots$$

The degree of the polynominal equation has been chosen to best fit the available data. The standard error of estimate for W_2 is also given for the fitting.

No reliable expression is available at present that can present the solubility data as a function of both the temperature and pressure, and therefore no attempt of fitting has been made. However, if the solubility data are needed for use in computer programs, the practical way is to adopt Henry's law for a limited temperature interval. This law is always valid if the temperature interval chosen is small enough. By dividing the full temperature range into small intervals, Henry's law will be sufficiently accurate for most work.

Formula Index: Solubility of Halogenated Hydrocarbons in H_2O

$CBrClF_2$	$CBrClF_2$	Methane, bromochlorodifluoro-
$CBrClH_2$	$CBrClH_2$	Methane, bromochloro-
$CBrF_3$	$CBrF_3$	Methane, bromotrifluoro-
$CBrH_3$	$CBrH_3$	Methane, bromo-
CBr_2H_2	CBr_2H_2	Methane, dibromo-
CBr_3F	CBr_3F	Methane, tribromofluoro-
CBr_3H	CBr_3H	Methane, tribromo-
CBr_4	CBr_4	Methane, tetrabromo-
$CClFH_2$	$CClFH_2$	Methane, chlorofluoro-
$CClF_2H$	$CClF_2H$	Methane, chlorodifluoro-
$CClF_3$	$CClF_3$	Methane, chlorotrifluoro-
$CClH_3$	$CClH_3$	Methane, chloro-
CCl_2FH	CCl_2FH	Methane, dichlorofluoro-
CCl_2F_2	CCl_2F_2	Methane, dichlorodifluoro-
CCl_2H_2	CCl_2H_2	Methane, dichloro-
CCl_3F	CCl_3F	Methane, trichlorofluoro-
CCl_3H	CCl_3H	Methane, trichloro-
CCl_4	CCl_4	Methane, tetrachloro-
CFH_3	CFH_3	Methane, fluoro-
CF_3H	CF_3H	Methane, trifluoro-
CF_4	CF_4	Methane, tetrafluoro-
CHI_3	CHI_3	Methane, triiodo-
CH_2I_2	CH_2I_2	Methane, diiodo-
CH_3I	CH_3I	Methane, iodo-
CI_4	CI_4	Methane, tetraiodo-
C_2BrClF_3H	$CBrClH-CF_3$	Ethane, 1-bromo-1-chloro-2,2,2-trifluoro-
C_2BrClH_4	$CBrH_2-CClH_2$	Ethane, 1-bromo-2-chloro-
C_2BrF_4H	$CBrFH-CF_3$	Ethane, 1-bromo-1,2,2,2-tetrafluoro-
C_2BrH_5	$CBrH_2-CH_3$	Ethane, bromo-
$C_2Br_2H_2$	$CBrH=CBrH$	Ethene, cis-1,2-dibromo-
$C_2Br_2H_4$	$CBrH_2-CBrH_2$	Ethane, 1,2-dibromo-
$C_2Br_4H_2$	CBr_2H-CBr_2H	Ethane, 1,1,2,2-tetrabromo-
C_2ClFH_4	$CClH_2-CFH_2$	Ethane, 1-chloro-2-fluoro-
$C_2ClF_2H_3$	$CClF_2-CH_3$	Ethane, 1-chloro-1,1-difluoro-
C_2ClF_5	$CClF_2-CF_3$	Ethane, chloropentafluoro-
C_2ClH_3	$CClH=CH_2$	Ethene, chloro-
C_2ClH_5	$CClH_2-CH_3$	Ethane, chloro-
$C_2Cl_2F_4$	$CClF_2-CClF_2$	Ethane, 1,2-dichloro-1,1,2,2-tetrafluoro-
$C_2Cl_2H_2$	$CCl_2=CH_2$	Ethene, 1,1-dichloro-

$C_2Cl_2H_2$	CClH=CClH	Ethene, cis-1,2-dichloro-
$C_2Cl_2H_2$	CClH=CClH	Ethene, trans-1,2-dichloro-
$C_2Cl_2H_4$	CCl_2H-CH_3	Ethane, 1,1-dichloro-
$C_2Cl_2H_4$	$CClH_2-CClH_2$	Ethane, 1,2-dichloro-
$C_2Cl_3F_3$	$CCl_2F-CClF_2$	Ethane, 1,1,2-trichloro-1,2,2-trifluoro-
C_2Cl_3H	$CClH=CCl_2$	Ethene, trichloro-
$C_2Cl_3H_3$	CCl_3-CH_3	Ethane, 1,1,1-trichloro-
$C_2Cl_3H_3$	$CCl_2H-CClH_2$	Ethane, 1,1,2-trichloro-
C_2Cl_4	$CCl_2=CCl_2$	Ethene, tetrachloro-
$C_2Cl_4F_2$	CCl_2F-CCl_2F	Ethane, 1,2-difluoro-1,1,2,2-tetrachloro-
$C_2Cl_4H_2$	CCl_3-CClH_2	Ethane, 1,1,1,2-tetrachloro-
$C_2Cl_4H_2$	CCl_2H-CCl_2H	Ethane, 1,1,2,2-tetrachloro-
C_2Cl_5H	CCl_3-CCl_2H	Ethane, pentachloro-
C_2Cl_6	CCl_3-CCl_3	Ethane, hexachloro-
C_2FH_3	$CFH=CH_2$	Ethene, fluoro-
C_2FH_5	CFH_2-CH_3	Ethane, fluoro-
$C_2F_2H_2$	$CF_2=CH_2$	Ethene, 1,1-difluoro-
$C_2F_2H_4$	CF_2H-CH_3	Ethane, 1,1-difluoro-
C_2F_4	$CF_2=CF_2$	Ethene, tetrafluoro-
$C_2F_4H_2$	CF_3-CFH_2	Ethane, 1,1,1,2-tetrafluoro-
$C_2H_2I_2$	CHI=CHI	Ethene, cis-1,2-diiodo-
$C_2H_2I_2$	CHI=CHI	Ethene, trans-1,2-diiodo-
C_2H_5I	CH_3-CH_2I	Ethane, iodo-
C_3BrClH_6	$CBrH_2-CH_2-CClH_2$	Propane, 1-bromo-3-chloro-
C_3BrH_7	$CH_2Br-CH_2-CH_3$	Propane, 1-bromo-
C_3BrH_7	$CH_3-CBrH-CH_3$	Propane, 2-bromo-
$C_3Br_2ClH_5$	$CBrH_2-CBrH-CClH_2$	Propane, 1,2-dibromo-3-chloro-
$C_3Br_2H_6$	$CBrH_2-CBrH-CH_3$	Propane, 1,2-dibromo-
$C_3Br_2H_6$	$CBrH_2-CH_2-CBrH_2$	Propane, 1,3-dibromo-
$C_3ClF_3H_4$	$CClH_2-CH_2-CF_3$	Propane, 1-chloro-3,3,3-trifluoro-
C_3ClH_5	$CClH_2-CH=CH_2$	1-Propene, 3-chloro-
C_3ClH_7	$CClH_2-CH_2-CH_3$	Propane, 1-chloro-
C_3ClH_7	$CH_3-CClH-CH_3$	Propane, 2-chloro-
$C_3Cl_2H_4$	$CClH=CH-CClH_2$	1-Propene, cis-1,3-dichloro-
$C_3Cl_2H_6$	$CClH_2-CClH-CH_3$	Propane, 1,2-dichloro-
$C_3Cl_2H_6$	$CClH_2-CH_2-CClH_2$	Propane, 1,3-dichloro-
$C_3Cl_3H_5$	$CClH_2-CClH-CClH_2$	Propane, 1,2,3-trichloro-
C_3Cl_6	$CCl_3-CCl=CCl_2$	1-Propene, hexachloro-
C_3FH_5	$CFH_2-CH=CH_2$	1-Propene, 3-fluoro-
C_3FH_7	$CFH_2-CH_2-CH_3$	Propane, 1-fluoro-
C_3FH_7	$CH_3-CFH-CH_3$	Propane, 2-fluoro-

C_3F_6	$CF_2=CF-CF_3$	Propene, hexafluoro-
C_3F_8	$CF_3-CF_2-CF_3$	Propane, octafluoro-
C_3H_7I	$CH_3-CH_2-CH_2I$	Propane, 1-iodo-
C_3H_7I	$CH_3-CHI-CH_3$	Propane, 2-iodo-
C_4BrH_9	$CBrH_2-CH_2-CH_2-CH_3$	n-Butane, 1-bromo-
C_4BrH_9	$CBrH_2-CH(CH_3)-CH_3$	Propane, 1-bromo-2-methyl-
C_4ClH_9	$CClH_2-CH_2-CH_2-CH_3$	n-Butane, 1-chloro-
C_4ClH_9	$CH_3-CClH-CH_2-CH_3$	n-Butane, 2-chloro-
C_4ClH_9	$CClH_2-CH(CH_3)-CH_3$	Propane, 1-chloro-2-methyl-
$C_4Cl_2H_8$	$CCl_2H-CH_2-CH_2-CH_3$	n-Butane, 1,1-dichloro-
$C_4Cl_2H_8$	$CH_3-CClH-CClH-CH_3$	n-Butane, 2,3-dichloro-
C_4Cl_5H	$CCl_2=C=CCl-CCl_2H$	1,2-Butadiene, 1,1,3,4,4-pentachloro-
C_4Cl_6	$CCl_2=CCl-CCl=CCl_2$	1,3-Butadiene, hexachloro-
C_4F_8	$\overline{CF_2-CF_2-CF_2-CF_2}$	Cyclobutane, octafluoro-
C_4H_9I	$CH_2I-CH_2-CH_2-CH_3$	n-Butane, 1-iodo-
C_5BrH_{11}	$CBrH_2-CH_2-CH(CH_3)-CH_3$	n-Butane, 1-bromo-3-methyl-
C_5ClH_{11}	$CClH_2-CH_2-CH_2-CH_2-CH_3$	n-Pentane, 1-chloro-
C_5ClH_{11}	$CH_3-CClH-CH_2-CH_2-CH_3$	n-Pentane, 2-chloro-
C_5ClH_{11}	$CH_3-CH_2-CClH-CH_2-CH_3$	n-Pentane, 3-chloro-
C_5ClH_{11}	$CH_3-CCl(CH_3)-CH_2-CH_3$	n-Butane, 2-chloro-2-methyl-
$C_5Cl_2H_{10}$	$CH_3-CCl(CH_3)-CClH-CH_3$	n-Butane, 2,3-dichloro-2-methyl-
C_5Cl_6	$CCl \equiv C-CCl=CCl-CCl_3$	2-Butene-4-yne, hexachloro-
C_5Cl_6	$CCl_2-CCl=CCl-CCl=CCl$	1,3-Cyclopentadiene, hexachloro-
C_5Cl_8	$CCl_2=CCl-CCl_2-CCl=CCl_2$	1,4-Pentadiene, octachloro-
C_6BrClH_4	C_6BrClH_4	Benzene, 1-bromo-2-chloro-
C_6BrClH_4	C_6BrClH_4	Benzene, 1-bromo-3-chloro-
C_6BrClH_4	C_6BrClH_4	Benzene, 1-bromo-4-chloro-
C_6BrH_4I	C_6BrH_4I	Benzene, 1-bromo-4-iodo-
C_6BrH_5	C_6BrH_5	Benzene, bromo-
$C_6Br_2H_4$	$C_6Br_2H_4$	Benzene, 1,2-dibromo-
$C_6Br_2H_4$	$C_6Br_2H_4$	Benzene, 1,3-dibromo-
$C_6Br_2H_4$	$C_6Br_2H_4$	Benzene, 1,4-dibromo-
$C_6Br_3H_3$	$C_6Br_3H_3$	Benzene, 1,2,3-tribromo-
$C_6Br_3H_3$	$C_6Br_3H_3$	Benzene, 1,2,4-tribromo-
$C_6Br_3H_3$	$C_6Br_3H_3$	Benzene, 1,3,5-tribromo-
$C_6Br_4H_2$	$C_6Br_4H_2$	Benzene, 1,2,4,5-tetrabromo-
C_6ClH_4I	C_6ClH_4I	Benzene, 1-chloro-2-iodo-
C_6ClH_4I	C_6ClH_4I	Benzene, 1-chloro-3-iodo-
C_6ClH_4I	C_6ClH_4I	Benzene, 1-chloro-4-iodo-
C_6ClH_5	C_6ClH_5	Benzene, chloro-
C_6ClH_{13}	$CClH_2-(CH_2)_4-CH_3$	n-Hexane, 1-chloro-

$C_6Cl_2H_4$	$C_6Cl_2H_4$	Benzene, 1,2-dichloro-
$C_6Cl_2H_4$	$C_6Cl_2H_4$	Benzene, 1,3-dichloro-
$C_6Cl_2H_4$	$C_6Cl_2H_4$	Benzene, 1,4-dichloro-
$C_6Cl_3H_3$	$C_6Cl_3H_3$	Benzene, 1,2,3-trichloro-
$C_6Cl_3H_3$	$C_6Cl_3H_3$	Benzene, 1,2,4-trichloro-
$C_6Cl_3H_3$	$C_6Cl_3H_3$	Benzene, 1,3,5-trichloro-
$C_6Cl_4H_2$	$C_6Cl_4H_2$	Benzene, 1,2,3,4-tetrachloro-
$C_6Cl_4H_2$	$C_6Cl_4H_2$	Benzene, 1,2,3,5-tetrachloro-
$C_6Cl_4H_2$	$C_6Cl_4H_2$	Benzene, 1,2,4,5-tetrachloro-
C_6Cl_5H	C_6Cl_5H	Benzene, pentachloro-
C_6Cl_6	C_6Cl_6	Benzene, hexachloro-
$C_6Cl_6H_6$	$C_6Cl_6H_6$	Cyclohexane, α-1,2,3,4,5,6-hexachloro-
$C_6Cl_6H_6$	$C_6Cl_6H_6$	Cyclohexane, β-1,2,3,4,5,6-hexachloro-
$C_6Cl_6H_6$	$C_6Cl_6H_6$	Cyclohexane, δ-1,2,3,4,5,6-hexachloro-
$C_6Cl_6H_6$	$C_6Cl_6H_6$	Cyclohexane, γ-1,2,3,4,5,6-hexachloro-
C_6FH_5	C_6FH_5	Benzene, fluoro-
$C_6F_2H_4$	$C_6F_2H_4$	Benzene, 1,2-difluoro-
$C_6F_2H_4$	$C_6F_2H_4$	Benzene, 1,3-difluoro-
$C_6H_4I_2$	$C_6H_4I_2$	Benzene, 1,2-diiodo-
$C_6H_4I_2$	$C_6H_4I_2$	Benzene, 1,3-diiodo-
$C_6H_4I_2$	$C_6H_4I_2$	Benzene, 1,4-diiodo-
C_6H_5I	C_6H_5I	Benzene, iodo-

CBrClF$_2$[a]

Methane, bromochlorodifluoro-   (R 12B1)

Temperature, t	Solubility of CBrClF$_2$ in H$_2$O at saturated pressure	
(°C)	Wt %, W$_2$[b]	Mole fraction, x$_2$
-20	0.1347	0.0001469
-10	0.0942	0.0001027
0	0.0747	0.0000814
10	0.3495	0.0003819

[a]M.wt. = 165.3648; m.p. = -159.5°C; b.p. = -3.3°C; d^{15} = 1.850 g cc^{-1}.

[b]W$_2$ = 0.0747 + 0.00821t + 1.4715 x 10^{-3}t^2 + 4.555 x 10^{-5}t^3; standard error of estimate for W$_2$ = 0.0012.

CBrClH$_2$[a]

Methane, bromochloro-   (R 30B1)

Temperature, t	Solubility of CBrClH$_2$ in H$_2$O at saturated pressure	
(°C)	Wt %, W$_2$	Mole fraction, x$_2$
25	1.4778	0.002084

[a]M.wt. = 129.39; m.p. = -86.5°C; b.p. = 68.11°C; d$_4^{20}$ = 1.9344 g cc^{-1}.

CBrF$_3$[a]

Methane, bromotrifluoro-   (R 13B1)

Total pressure	Solubility of CBrF$_3$ in H$_2$O (g/100 cc H$_2$O)		
(atm)	25°C	50°C	75°C
1.00	0.032	0.017	0.008
1.70	0.050	0.028	·0.019
3.40	0.094	0.054	0.038
6.80	0.172	0.098	0.074
10.20	0.242	0.134	0.103
13.61	0.307	0.165	0.130

[a]M.wt. = 148.92; m.p. = -168.0°C; b.p. = -59.0°C; d$_4^{25}$ = 1.538 g cc^{-1}.

CBrH$_3$[a]

Methane, bromo-   (R 40B1)

Temperature, t	Solubility of CBrH$_3$ in H$_2$O at 1 atm partial pressure	
(°C)	Wt %, W$_2$[b]	Mole fraction, x$_2$
5	3.5000	0.006835
10	2.7997	0.005436
20	1.8401	0.003545
25	1.5223	0.002925
30	1.2804	0.002455
40	0.9517	0.001820
50	0.7430	0.001418
60	0.5919	0.001128
70	0.4757	0.0009061
80	0.4030	0.0007673

[a]M.wt. = 94.94; m.p. = -93.6°C; b.p. = 3.56°C; d_4^{20} = 1.6755 g cc^{-1}.

[b]W$_2$ = 4.39429 - 0.20089t + 4.6927 x $10^{-3}t^2$ - 5.84027 x $10^{-5}t^3$ + 3.5081 x $10^{-7}t^4$ - 7.3870 x $10^{-10}t^5$; standard error of estimate for W$_2$ = 0.02239.

CBr$_2$H$_2$[a]

Methane, dibromo-   (R 30B2)

Temperature, t	Solubility of CBr$_2$H$_2$ in H$_2$O at saturated pressure	
(°C)	Wt %, W$_2$[b]	Mole fraction, x$_2$
0	1.1594	0.001214
10	1.1330	0.001186
20	1.1350	0.001188
25	1.1442	0.001198
30	1.1574	0.001212

[a]M.wt. = 173.85; m.p. = -52.55°C; b.p. = 97.0°C; d_4^{20} = 2.4970 g cc^{-1}.

[b]W$_2$ = 1.1594 - 4.3267 x $10^{-3}t$ + 1.8201 x $10^{-4}t^2$ - 1.3335 x $10^{-6}t^3$; standard error of estimate for W$_2$ = 0.0010.

CBr_3F^a

Methane, tribromofluoro-   (R 11B3)

Temperature, t	Solubility of CBr_3F in H_2O at saturated pressure	
(°C)	Wt %, W_2	Mole fraction, x_2
25	0.040	0.00002662

[a]M.wt. = 270.74; m.p. = -74.5°C; b.p. = 106.0°C; d_4^{20} = 2.7648 g cc^{-1}.

CBr_3H^a

Methane, tribromo-   (R 20B3)

Temperature, t	Solubility of CBr_3H in H_2O at saturated pressure	
(°C)	Wt %, W_2 [b]	Mole fraction, x_2
15	0.3001	0.0002145
20	0.3051	0.0002181
25	0.3100	0.0002216
30	0.3180	0.0002273

[a]M.wt. = 252.75; m.p. = 8.30°C; b.p. = 149.5°C; d_4^{20} = 2.8899 g cc^{-1}.

[b]W_2 = 0.2525 + 6.0835 x 10^{-3}t - 2.5801 x $10^{-4}t^2$ + 4.2667 x $10^{-6}t^3$; standard error of estimate for W_2 = 0.0011.

CBr_4^a

Methane, tetrabromo-   (R 10B4)

Temperature, t	Solubility of CBr_4 in H_2O at saturated pressure	
(°C)	Wt %, W_2	Mole fraction, x_2
30	0.0240	0.00001303

[a]M.wt. = 331.65; m.p. = 90.5°C; b.p. = 189.5°C; d_4^{20} = 2.9609 g cc^{-1}.

CClFH$_2$[a]

Methane, chlorofluoro-   (R 31)

Temperature, t	Solubility of CClFH$_2$ in H$_2$O at 1 atm partial pressure	
($^\circ$C)	Wt %, W$_2$[b]	Mole fraction, x$_2$
10	1.6840	0.004486
20	1.2294	0.003264
25	1.0522	0.002790
30	0.9042	0.002395
40	0.6830	0.001806
50	0.5403	0.001427
60	0.4506	0.001189
70	0.3886	0.001025
80	0.3285	0.000866

[a]M.wt. = 68.48; m.p. = -133.0°C; b.p. = -9.1°C; d^{20} = 1.271 g cc^{-1}.

[b]W$_2$ = 2.2937 - 6.9564 x 10^{-2}t + 9.0228 x 10^{-4}t^2 - 4.2473 x 10^{-6}t^3; standard error of estimate for W$_2$ = 0.

CClF$_2$H[a]

Methane, chlorodifluoro-   (R 22)

Temperature, t	Solubility of CClF$_2$H in H$_2$O at 1 atm partial pressure	
($^\circ$C)	Wt %, W$_2$[b]	Mole fraction, x$_2$
10	0.5296	0.001108
20	0.3510	0.0007333
25	0.2899	0.0006055
30	0.2428	0.0005068
40	0.1783	0.0003721
50	0.1402	0.0002926
60	0.1180	0.0002461
70	0.1052	0.0002193
80	0.09728	0.0002028

[a]M.wt. = 86.47; m.p. = -146.5°C; b.p. = -40.8°C; d^{-69} = 1.4906 g cc^{-1}.

[b]W$_2$ = 0.8171 - 3.5818 x 10^{-2}t + 8.0081 x 10^{-4}t^2 - 1.0075 x 10^{-5}t^3 + 6.9823 x 10^{-8}t^4 - 2.0786 x 10^{-10}t^5; standard error of estimate for W$_2$ = 0.

$CClF_2H$ (Continued)

Total pressure	Solubility of $CClF_2H$ in H_2O (g/100 cc H_2O)						
(atm)	5°C	21.1°C	25°C	48.9°C	50°C	75°C	76.7°C
1.00	0.792	0.334	0.277	0.143	0.136	0.087	0.089
1.70	1.378	0.569	0.471	0.241	0.231	0.148	0.151
3.40	2.800	1.14	0.942	0.479	0.461	0.296	0.300
5.10	4.222	1.71	1.413	0.725	0.692	0.444	0.450
6.80	-	2.28	1.884	0.971	0.923	0.592	0.599
9.53	-	-	2.641	-	1.293	0.829	-
10.39	-	-	2.80	-	1.410	0.904	-
19.33	-	-	-	-	2.40	1.682	-
33.34	-	-	-	-	-	2.10	-

$CClF_3$ [a]

Methane, chlorotrifluoro-   (R 13)

Total pressure	Solubility of $CClF_3$ in H_2O (g/100 cc H_2O)		
(atm)	25°C	50°C	75°C
1.00	0.009	0.005	0.004
3.40	0.030	0.017	0.013
6.80	0.055	0.033	0.026
10.20	0.077	0.047	0.037
13.61	0.097	0.059	0.047
17.01	0.116	0.071	0.056
20.41	0.135	0.083	0.066
23.81	0.154	0.095	0.075

[a]M.wt. = 104.46; m.p. = -181.0°C; b.p. = -81.1°C; d^{-30} = 1.298 g cc^{-1}.

$CC1H_3{}^a$

Methane, chloro- (R 40)

Temperature, t	Solubility of $CClH_3$ in H_2O at 1 atm partial pressure	
(°C)	Wt %, $W_2{}^b$	Mole fraction, x_2
0	1.4019	0.005048
10	0.9113	0.003271
20	0.6274	0.002248
25	0.5325	0.001907
30	0.4579	0.001639
40	0.3490	0.001248
50	0.2742	0.000980
60	0.2228	0.000796
70	0.1891	0.000675
80	0.1609	0.000575

[a]M.wt. = 50.49; m.p. = -97.73°C; b.p. = -24.2°C; d_4^{20} = 0.9159 g cc^{-1}.

[b]W_2 = 1.4019 - 6.3652 x 10^{-2}t + 1.7177 x $10^{-3}t^2$ - 2.8262 x $10^{-5}t^3$ + 2.5369 x $10^{-7}t^4$ - 9.3470 x $10^{-10}t^5$; standard error of estimate for W_2 = 0.0038139.

CCl_2FH^a

Methane, dichlorofluoro- (R 21)

Total pressure	Solubility of CCl_2FH in H_2O (g/100 cc H_2O)		
(atm)	25°C	50°C	75°C
0.068	0.0737	0.0279	0.0149
1.00	1.88	0.798	0.418
1.36	2.56	1.09	0.519
2.04	-	1.63	0.854
2.72	-	2.17	1.14
3.40	-	2.72	1.42

[a]M.wt. = 102.92; m.p. = -135.0°C; b.p. = 9.0°C; d^9 = 1.3724 g cc^{-1}.

CCl_2F_2 [a]
Methane, dichlorodifluoro-   (R 12)

Total pressure	Solubility of CCl_2F_2 in H_2O (g/100 cc H_2O)		
(atm)	25°C	54.4°C	76.7°C
1.00	0.030	0.012	0.008
1.70	0.050	0.022	0.014
3.40	0.100	0.046	0.034
5.10	0.151	0.069	0.052
6.80	-	0.093	0.068

[a]M.wt. = 120.91; m.p. = -158.0°C; b.p. = -29.8°C; d^{57} = 1.1834 g cc^{-1}.

CCl_2H_2 [a]
Methane, dichloro-   (R 30)

Temperature, t	Solubility of CCl_2H_2 in H_2O at saturated pressure	
(°C)	Wt %, W_2 [b]	Mole fraction, x_2
0	1.9610	0.004225
10	1.5938	0.003424
20	1.3702	0.002938
25	1.3030	0.002793
30	1.2605	0.002701
40	1.2350	0.002645
50	1.2640	0.002708

[a]M.wt. = 84.93; m.p. = -95.1°C; b.p. = 40.0°C; d_4^{20} = 1.3266 g cc^{-1}.

[b]W_2 = 1.961 - 4.4883 x 10^{-2}t + 8.6617 x 10^{-4}t^2 - 4.9463 x 10^{-6}t^3; standard error of estimate for W_2 = 0.00082.

CCl_3F^a

Methane, trichlorofluoro-   (R 11)

Total pressure	Solubility of CCl_3F in H_2O (g/100 cc H_2O)		
(atm)	30°C	50°C	75°C
1.00	0.108	0.057	0.031
1.36	-	0.078	0.042
2.04	-	0.117	0.063
2.72	-	-	0.084
3.40	-	-	0.105

[a]M.wt. = 137.37; m.p. = -111.0°C; b.p. = 23.82°C; d_4^{25} = 1.467 g cc^{-1}.

CCl_3H^a

Methane, trichloro-   (R 20)

Temperature, t	Solubility of CCl_3H in H_2O at saturated pressure	
(°C)	Wt %, W_2[b]	Mole fraction, x_2
0	0.9950	0.001514
10	0.8983	0.001366
20	0.8216	0.001249
25	0.7920	0.001203
30	0.7689	0.001168
40	0.7440	0.001130
50	0.7511	0.001141
60	0.7940	0.001206

[a]M.wt. = 119.38; m.p. = -63.5°C; b.p. = 61.7°C; d_4^{20} = 1.4832 g cc^{-1}.

[b]W_2 = 0.995 - 1.0531 x 10^{-2}t + 7.9819 x 10^{-5}t^2 + 6.6431 x 10^{-7}t^3; standard error of estimate for W_2 = 0.0010.

CCl$_4$[a]

Methane, tetrachloro-   (R 10)

Temperature, t	Solubility of CCl$_4$ in H$_2$O at saturated pressure	
(°C)	Wt %, W$_2$[b]	Mole fraction, x$_2$
0	.0.09784	0.0001147
10	0.06826	0.0001011
20	0.08048	0.00009432
25	0.07934	0.00009298
30	0.07913	0.00009275
40	0.08086	0.00009477
50	0.08430	0.00009880
60	0.08807	0.0001032
70	0.09081	0.0001064

[a]M.wt. = 153.82; m.p. = -22.99°C; b.p. = 76.54°C; d$_4^{20}$ = 1.5940 g cc^{-1}.

[b]W$_2$ = 9.7842 x 10^{-2} - 1.4942 x 10^{-3}t + 3.5854 x 10^{-5}t^2 - 2.2775 x 10^{-7}t^3; standard error of estimate for W$_2$ = 0.000588.

CFH$_3$[a]

Methane, fluoro-   (R 41)

Temperature, t	Solubility of CFH$_3$ in H$_2$O at 1 atm partial pressure	
(°C)	Wt %, W$_2$[b]	Mole fraction, x$_2$
0	0.4227	0.002242
10	0.3002	0.001592
20	0.2266	0.001201
25	0.2001	0.001061
30	0.1782	0.0009440
40	0.1435	0.0007600
50	0.1184	0.0006273
60	0.1021	0.0005406
70	0.09201	0.0004873
80	0.08005	0.0004239

[a]M.wt. = 34.03; m.p. = -141.8°C; b.p. = -78.4°C; d$_4^{20}$ = 0.5786 g cc^{-1}.

[b]W$_2$ = 0.4227 - 1.5872 x 10^{-2}t + 4.3600 x 10^{-4}t^2 - 8.2401 x 10^{-6}t^3 + 8.7318 x 10^{-8}t^4 - 3.7260 x 10^{-10}t^5; standard error of estimate for W$_2$ = 0.001259.

CF_3H^a

Methane, trifluoro-   (R 23)

Total pressure	Solubility of CF_3H in H_2O (g/100 cc H_2O)		
(atm)	25°C	50°C	75°C
1.00	0.09	0.05	0.02
3.40	0.32	0.18	0.12
6.80	0.59	0.35	0.24
10.20	0.85	0.50	0.36
13.61	1.08	0.65	0.47
17.01	1.30	0.80	0.58
20.41	1.50	0.93	0.68

Temperature, t	Solubility of CF_3H in H_2O at 1 atm partial pressure	
(°C)	Wt %, W_2	Mole fraction, x_2
25	0.4087	0.001056

[a]M.wt. = 70.01; m.p. = -160.0°C; b.p. = -82.2°C; d^{-100} = 1.52 g cc^{-1}.

CF_4^a

Methane, tetrafluoro-   (R 14)

Temperature, t	Solubility of CF_4 in H_2O at 1 atm partial pressure	
(°C)	Wt %, W_2^b	Mole fraction, x_2
0	0.003892	0.00000797
10	0.002700	0.00000553
20	0.002100	0.00000430
25	0.001877	0.00000384
30	0.001700	0.00000348
40	0.001500	0.00000307
50	0.001300	0.00000266

[a]M.wt. = 88.01; m.p. = -150.0°C; b.p. = -129.0°C; d^0 = 3.034 g cc^{-1}.

[b]$W_2 = 3.8919 \times 10^{-3} - 1.8349 \times 10^{-4}t + 9.1815 \times 10^{-6}t^2 - 3.3593 \times 10^{-7}t^3 + 6.5654 \times 10^{-9}t^4 - 4.9325 \times 10^{-11}t^5$; standard error of estimate for W_2 = 0.00136.

CF_4 (Continued)

Total pressure	Solubility of CF_4 in H_2O (g/100 cc H_2O)		
(atm)	25°C	50°C	75°C
1.00	0.0016	0.0010	0.0005
1.70	0.0028	0.0019	0.0013
3.40	0.0056	0.0041	0.0028
5.10	0.0085	0.0064	0.0045
6.80	0.0114	0.0086	0.0062
8.50	0.0144	0.0108	0.0078
10.20	0.0172	0.0130	0.0095

CHI_3[a]
Methane, triiodo-

Temperature, t	Solubility of CHI_3 in H_2O at saturated pressure	
(°C)	Wt %, W_2	Mole fraction, x_2
25	0.012	0.000005477

[a]M.wt. = 393.73; m.p. = 123.0°C; b.p. = 218.0°C; d_4^{20} = 4.008 g cc^{-1}.

CH_2I_2[a]
Methane, diiodo-

Temperature, t	Solubility of CH_2I_2 in H_2O at saturated pressure	
(°C)	Wt %, W_2[b]	Mole fraction, x_2
0	0.1570	0.0001058
10	0.1475	0.00009935
20	0.1380	0.00009294
25	0.1350	0.00009090
30	0.1340	0.00009024

[a]M.wt. = 267.84; m.p. = 6.1°C; b.p. = 182.0°C; d_4^{20} = 3.3254 g cc^{-1}.

[b]W_2 = 0.157 - 7.6667 x 10^{-4}t - 2.7499 x $10^{-5}t^2$ + 9.1666 x $10^{-7}t^3$; standard error of estimate for W_2 = 0.00085.

CH_3I[a]

Methane, iodo-

Temperature, t	Solubility of CH_3I in H_2O at saturated pressure	
(°C)	Wt %, W_2[b]	Mole fraction, x_2
0	1.5386	0.001979
10	1.4345	0.001844
20	1.3848	0.001780
25	1.3894	0.001785
30	1.4185	0.001823
40	1.5646	0.002013

[a]M.wt. = 141.94; m.p. = -66.45°C; b.p. = 42.4°C; d_4^{20} = 2.279 g cc^{-1}.

[b]W_2 = 1.53863 - 0.012169t + 1.2714 x $10^{-4}t^2$ + 4.8334 x $10^{-6}t^3$; standard error of estimate for W_2 = 0.00065.

CI_4[a]

Methane, tetraiodo-

Temperature, t	Solubility of CI_4 in H_2O at saturated pressure	
(°C)	Wt %, W_2	Mole fraction, x_2
20	None	-

[a]M.wt. = 519.63; m.p. = 171.0°C; b.p. = subl. 130°C; d^{20} = 4.23 g cc^{-1}.

$CBrClH-CF_3$[a]

Ethane, 1-bromo-1-chloro-2,2,2-trifluoro-   (R 123B1)

Temperature, t	Solubility of C_2BrClF_3H in H_2O at saturated pressure	
(°C)	Wt %, W_2	Mole fraction, x_2
25	0.372	0.0003406

[a]M.wt. = 197.39; m.p. = -118.27°C; b.p. = 50.30°C; d_4^{20} = 1.776 g cc^{-1}.

$CBrH_2-CClH_2$ [a]

Ethane, 1-bromo-2-chloro-

Temperature, t	Solubility of C_2BrClH_4 in H_2O at saturated pressure	
(°C)	Wt %, W_2	Mole fraction, x_2
30	0.683	0.0008631

[a]M.wt. = 143.42; m.p. = -16.7°C; b.p. = 107.0°C; d_4^{20} = 1.7392 g cc^{-1}.

$CBrFH-CF_3$ [a]

Ethane, 1-bromo-1,2,2,2-tetrafluoro-

Temperature, t	Solubility of C_2BrF_4H in H_2O at 1 atm partial pressure	
(°C)	Wt %, W_2	Mole fraction, x_2
37	0.040	0.00004019

[a]M.wt. = 181.0; b.p. = 8.66°C; d_4^0 = 1.862 g cc^{-1}.

$CBrH_2-CH_3$ [a]

Ethane, bromo-

Temperature, t	Solubility of C_2BrH_5 in H_2O at saturated pressure	
(°C)	Wt %, W_2 [b]	Mole fraction, x_2
0	1.0557	0.001761
10	0.9557	0.001593
20	0.9057	0.001509
25	0.8939	0.001489
30	0.8880	0.001479

[a]M.wt. = 108.97; m.p. = -118.6°C; b.p. = 38.4°C; d_4^{20} = 1.4604 g cc^{-1}.

[b]W_2 = 1.0557 - 0.01309t + 3.3850 x 10^{-4}t^2 - 2.9500 x 10^{-6}t^3; standard error of estimate for W_2 = 0.000357.

CBrH=CBrH[a]

Ethene, cis-1,2-dibromo-

Temperature, t	Solubility of $C_2Br_2H_2$ in H_2O at saturated pressure	
($^\circ$C)	Wt %, W_2	Mole fraction, x_2
25	0.8906	0.0008702

[a]M.wt. = 185.86; m.p. = -53.0°C; b.p. = 112.5°C; d_4^{20} = 2.2464 g cc^{-1}.

CBrH$_2$-CBrH$_2$[a]

Ethane, 1,2-dibromo-

Temperature, t	Solubility of $C_2Br_2H_4$ in H_2O at saturated pressure	
($^\circ$C)	Wt %, W_2[b]	Mole fraction, x_2
0	0.3658	0.0003520
10	0.3817	0.0003673
20	0.4021	0.0003870
25	0.4152	0.0003997
30	0.4310	0.0004149
40	0.4722	0.0004547
50	0.5297	0.0005104
60	0.6073	0.0005856
70	0.7089	0.0006842
80	0.8384	0.0008101

[a]M.wt. = 187.87; m.p. = 9.79°C; b.p. = 131.36°C; d_4^{20} = 2.1792 g cc^{-1}.

[b]W_2 = 0.36583 + 1.4836 x 10^{-3}t + 3.48175 x 10^{-6}t^2 + 6.47685 x 10^{-7}t^3; standard error of estimate for W_2 = 0.0007336.

$CBr_2H-CBr_2H^a$

Ethane, 1,1,2,2-tetrabromo-

Temperature, t	Solubility of $C_2Br_4H_2$ in H_2O at saturated pressure	
(°C)	Wt %, W_2^b	Mole fraction, x_2
0	0.05042	0.00002629
10	0.05703	0.00002974
20	0.06424	0.00003350
25	0.06849	0.00003572
30	0.07339	0.00003828
40	0.08586	0.00004478
50	0.10300	0.00005373
60	0.1262	0.00006583
70	0.1567	0.00008180
80	0.1960	0.0001024
90	0.2455	0.0001282
100	0.3064	0.0001601

[a] M.wt. = 345.67; m.p. = 0°C; b.p. = 243.5°C; d_4^{20} = 2.9656 g cc^{-1}.

[b] W_2 = 5.0422 x 10^{-2} + 6.7693 x 10^{-4}t - 3.8430 x 10^{-6}t^2 + 2.2668 x 10^{-7}t^3; standard error of estimate for W_2 = 0.001686.

$CClH_2-CFH_2^a$

Ethane, 1-chloro-2-fluoro-   (R 151)

Temperature, t	Solubility of C_2ClFH_4 in H_2O at saturated pressure	
(°C)	Wt %, W_2	Mole fraction, x_2
25	2.4390	0.005429

[a] M.wt. = 82.51; m.p. = -50°C; b.p. = 57.0°C; d_4^{20} = 1.1747 g cc^{-1}.

$CClF_2-CH_3{}^a$
Ethane, 1-chloro-1,1-difluoro-   (R 142b)

Temperature, t	Solubility of $C_2ClF_2H_3$ in H_2O at 1 atm partial pressure	
(°C)	Wt %, W_2	Mole fraction, x_2
13	0.214	0.0003843
21	0.140	0.0002512

[a]M.wt. = 100.50; m.p. = −130.8°C; b.p. = −9.5°C; d^{30} = 1.096 g cc^{-1}.

$CClF_2-CF_3{}^a$
Ethane, chloropentafluoro-   (R 115)

Total pressure,	Solubility of C_2ClF_5 in H_2O (g/100 cc H_2O)		
(atm)	25°C	50°C	75°C
1.00	0.0058	0.0028	0.0013
1.70	0.0077	0.0046	0.0025
3.40	0.0123	0.0083	0.0051
5.10	0.0166	0.0115	0.0076
6.80	0.0210	0.0145	0.0100

[a]M.wt. = 154.47; m.p. = −106.0°C; b.p. = −38.0°C; d^{30} = 1.265 g cc^{-1}.

CClH=CH$_2$[a]

Ethene, chloro-   (R 1140)

Temperature, t	Solubility of C_2ClH_3 in H_2O at 1 atm partial pressure	
(°C)	Wt %, W_2	Mole fraction, x_2
0	0.5460	0.00158
25	0.2763	0.000798
50	0.1421	0.000410
75	0.0780	0.000225
100	0.0538	0.000155
150	0.0337	0.000097
175	0.0320	0.000092
200	0.03208	0.000092
230	0.0347	0.000100
260	0.0376	0.000108
300	0.0415	0.000120
330	0.0449	0.000129
350	0.0467	0.000135
374	0.04857	0.000140

Temperature, t	Saturated vapor pressure	Solubility of C_2ClH_3 in H_2O	
(°C)	(atm)	Wt %, W_2	Mole fraction, x_2
0.2	1.7208	0.9133	0.002650
25	3.9091	0.8787	0.002549
50	7.7899	0.9802	0.002845
75	13.9168	1.3985	0.004072

Temperature, t	Solubility of C_2ClH_3 in H_2O (wt %)			
	Total pressure (atm)			
(°C)	1.00	1.36	3.06	6.12
0.2	0.7488	0.8380	-	-
25	0.2697	-	0.71105	-
50	0.1277	-	0.3978	0.7625
75	0.0507	-	0.2120	0.4944

[a]M.wt. = 62.50; m.p. = -153.8°C; b.p. = -13.37°C; d_4^{20} = 0.9106 g cc^{-1}.

$CCIH_2-CH_3$ [a]
Ethane, chloro-   (R 160)

Temperature, t	Solubility of C_2ClH_5 in H_2O at saturated pressure	
($^\circ$C)	Wt %, W_2 [b]	Mole fraction, x_2
0	0.5842	0.001638
10	0.5742	0.001610
20	0.5678	0.001592

[a] M.wt. = 64.52; m.p. = -136.4°C; b.p. = 12.27°C; d_4^{20} = 0.8978 g cc^{-1}.

[b] $W_2 = 0.5842 - 1.6863 \times 10^{-3}t + 9.3949 \times 10^{-5}t^2 - 2.5316 \times 10^{-6}t^3$; standard error of estimate for W_2 = 0.

$CClF_2-CClF_2$ [a]
Ethane, 1,2-dichloro-1,1,2,2-tetrafluoro-   (R 114)

Total pressure	Solubility of $C_2Cl_2F_4$ in H_2O (g/100 cc H_2O)		
(atm)	25°C	50°C	75°C
1.00	0.0137	0.0063	0.0041
1.70	0.0233	0.0107	0.0070
3.40	-	0.0214	0.0141
5.10	-	-	0.0211

[a] M.wt. = 170.92; m.p. = -93.90°C; b.p. = 3.8°C; d^{25} = 1.456 g cc^{-1}.

$CCl_2=CH_2$ [a]

Ethene, 1,1-dichloro-   (R 1130a)

Temperature, t	Solubility of $C_2Cl_2H_2$ in H_2O at saturated pressure	
(°C)	Wt %, W_2	Mole fraction, x_2
20	0.040	0.00007436
25	0.021	0.00003903

Temperature, t	Solubility of $C_2Cl_2H_2$ in H_2O at 1 atm total pressure	
(°C)	Wt %, W_2	Mole fraction, x_2
10	0.6650	0.001242
20	0.4195	0.0007823
25	0.3344	0.0006232
30	0.2688	0.0005006
40	0.1814	0.0003375
50	0.1322	0.0002460
60	0.1028	0.0001912
70	0.0809	0.0001505
80	0.0610	0.0001134

[a]M.wt. = 96.94; m.p. = -122.1°C; b.p. = 37.0°C; d_4^{20} = 1.218 g cc^{-1}.

$CClH=CClH$ [a]

Ethene, cis-1,2-dichloro-   (R 1130 cis)

Temperature, t	Solubility of $C_2Cl_2H_2$ in H_2O at saturated pressure	
(°C)	Wt %, W_2	Mole fraction, x_2
10	0.400	0.0007457
25	0.350	0.0006522

[a]M.wt. = 96.94; m.p. = -80.5°C; b.p. = 60.3°C; d_4^{20} = 1.2837 g cc^{-1}.

CClH=CClH[a]

Ethene, trans-1,2-dichloro-   (R 1130 trans)

Temperature, t	Solubility of $C_2Cl_2H_2$ in H_2O at saturated pressure	
($^\circ$C)	Wt %, W_2	Mole fraction, x_2
25	0.626	0.001169

[a]M.wt. = 96.94; m.p. = -50.0°C; b.p. = 47.5°C; d_4^{20} = 1.2565 g cc^{-1}.

CCl_2H-CH_3[a]

Ethane, 1,1-dichloro-   (R 150a)

Temperature, t	Solubility of $C_2Cl_2H_4$ in H_2O at saturated pressure	
($^\circ$C)	Wt %, W_2[b]	Mole fraction, x_2
0	0.5826	0.001066
10	0.5193	0.0009494
20	0.4843	0.0008851
25	0.4767	0.0008713
30	0.4754	0.0008689
40	0.4905	0.0008966
50	0.5274	0.0009642
60	0.5838	0.001068
70	0.6575	0.001203
80	0.7465	0.001367

[a]M.wt. = 98.96; m.p. = -96.98°C; b.p. = 57.28°C; d_4^{20} = 1.1757 g cc^{-1}.

[b]W_2 = 0.5826 - 7.8236 x 10^{-3}t + 1.5268 x 10^{-4}t^2 - 3.6609 x 10^{-7}t^3; standard error of estimate for W_2 = 0.008973.

CClH$_2$-CClH$_2$[a]
Ethane, 1,2-dichloro-   (R 150)

| Temperature, t | Solubility of C$_2$Cl$_2$H$_4$ in H$_2$O at saturated pressure | |
(°C)	Wt %, W$_2$[b]	Mole fraction, x$_2$
0	0.8880	0.001628
10	0.8579	0.001572
20	0.8524	0.001563
25	0.8608	0.001578
30	0.8775	0.001609
40	0.9391	0.001723
50	1.0430	0.001915
60	1.1951	0.002197
70	1.4014	0.002581
80	1.6678	0.003078

| Temperature, t | Solubility of C$_2$Cl$_2$H$_4$ in H$_2$O at 1 atm total pressure | |
(°C)	Wt %, W$_2$	Mole fraction, x$_2$
72.0	0.87	0.001595
89.3	0.59	0.001079
92.3	0.43	0.000786
94.0	0.33	0.000602
98.0	0.13	0.000237

[a]M.wt. = 98.96; m.p. = -35.36°C; b.p. = 83.47°C; d$_4^{20}$ = 1.2351 g cc^{-1}.

[b]W$_2$ = 0.888 - 4.0468 x 10^{-3}t + 9.3738 x 10^{-5}t^2 + 9.8365 x 10^{-7}t^3; standard error of estimate for W$_2$ = 0.005337.

$CCl_2F-CClF_2$ [a]
Ethane, 1,1,2-trichloro-1,2,2-trifluoro-   (R 113)

Temperature, t	Solubility of $C_2Cl_3F_3$ in H_2O at saturated pressure	
(°C)	Wt %, W_2 [b]	Mole fraction, x_2
0	0.01065	0.00001024
10	0.01361	0.00001309
20	0.01575	0.00001515
25	0.01664	0.00001600
30	0.01747	0.00001680
40	0.01917	0.00001843
50	0.02125	0.00002043
60	0.02411	0.00002319
70	0.02816	0.00002708
80	0.03380	0.00003251

Total pressure	Solubility of $C_2Cl_3F_3$ in H_2O (g/100 cc H_2O)		
(atm)	30°C	50°C	75°C
0.34	0.011	0.007	0.004
0.68	-	0.013	0.009
1.02	-	0.020	0.013
2.04	-	-	0.026

[a] M.wt. = 187.38; m.p. = -36.4°C; b.p. = 47.7°C; d_4^{20} = 1.5635 g cc^{-1}.

[b] $W_2 = 1.0648 \times 10^{-2} + 3.51135 \times 10^{-4}t - 6.1331 \times 10^{-6}t^2 + 6.70235 \times 10^{-8}t^3$; standard error of estimate for W_2 = 0.0007634.

$CClH=CCl_2$[a]

Ethene, trichloro-   (R 1120)

Temperature, t	Solubility of C_2Cl_3H in H_2O at saturated pressure	
(°C)	Wt %, W_2[b]	Mole fraction, x_2
0	0.1049	0.0001440
10	0.1061	0.0001456
20	0.1083	0.0001487
25	0.1099	0.0001508
30	0.1117	0.0001533
40	0.1160	0.0001592
50	0.1212	0.0001663
60	0.1271	0.0001744
70	0.1335	0.0001833
80	0.1405	0.0001929

[a]M.wt. = 131.39; m.p. = -73.0°C; b.p. = 87.0°C; d_4^{20} = 1.4642 g cc^{-1}.

[b]W_2 = 0.10494 + 4.9038 x 10^{-5}t + 6.4541 x 10^{-6}t^2 - 1.8808 x 10^{-8}t^3; standard error of estimate for W_2 = 0.0002198.

CCl_3-CH_3[a]

Ethane, 1,1,1-trichloro-   (R 140a)

Temperature, t	Solubility of $C_2Cl_3H_3$ in H_2O at saturated pressure	
(°C)	Wt %, W_2[b]	Mole fraction, x_2
0	0.1910	0.0002584
10	0.1707	0.0002309
20	0.1554	0.0002101
25	0.1495	0.0002022
30	0.1449	0.0001959
40	0.1390	0.0001880
50	0.1377	0.0001861
60	0.1407	0.0001902
70	0.1479	0.0002000
80	0.1592	0.0002153

[a]M.wt. = 133.41; m.p. = -30.41°C; b.p. = 74.1°C; d_4^{20} = 1.3390 g cc^{-1}.

[b]W_2 = 0.1910 - 2.2811 x 10^{-3}t + 2.5529 x 10^{-5}t^2 - 2.4775 x 10^{-8}t^3; standard error of estimate for W_2 = 0.001131.

$CCl_2H-CClH_2$[a]

Ethane, 1,1,2-trichloro-   (R 140)

| Temperature, t | Solubility of $C_2Cl_3H_3$ in H_2O at saturated pressure | |
(°C)	Wt %, W_2[b]	Mole fraction, x_2
0	0.4814	0.0006527
10	0.4564	0.0006188
20	0.4420	0.0005992
25	0.4394	0.0005956
30	0.4401	0.0005966
40	0.4527	0.0006137
50	0.4815	0.0006530
60	0.5287	0.0007172
70	0.5961	0.0008092
80	0.6857	0.0009312

[a]M.wt. = 133.41; m.p. = -36.5°C; b.p. = 113.77°C; d_4^{20} = 1.4397 g cc^{-1}.

[b]W_2 = 0.48137 - 2.9594 x 10^{-3}t + 4.3162 x $10^{-5}t^2$ + 3.2190 x $10^{-7}t^3$; standard error of estimate for W_2 = 0.001943

$CCl_2=CCl_2$[a]

Ethene, tetrachloro-   (R 1110)

| Temperature, t | Solubility of C_2Cl_4 in H_2O at saturated pressure | |
(°C)	Wt %, W_2[b]	Mole fraction, x_2
0	0.01497	0.00001626
10	0.01488	0.00001616
20	0.01490	0.00001619
25	0.01503	0.00001633
30	0.01527	0.00001659
40	0.01620	0.00001760
50	0.01791	0.00001946
60	0.02063	0.00002242
70	0.02457	0.00002670
80	0.02997	0.00003257

[a]M.wt. = 165.83; m.p. = -19.0°C; b.p. = 121.0°C; d_4^{20} = 1.6227 g cc^{-1}.

[b]W_2 = 1.49696 x 10^{-2} - 7.8009 x 10^{-6}t - 5.1403 x $10^{-7}t^2$ + 3.6938 x $10^{-8}t^3$; standard error of estimate for W_2 = 0.0001881.

$CCl_2F-CCl_2F^a$

Ethane, 1,2-difluoro-1,1,2,2-tetrachloro-   (R 112)

Temperature, t	Solubility of $C_2Cl_4F_2$ in H_2O at saturated pressure	
($^\circ$C)	Wt %, W_2^b	Mole fraction, x_2
25	0.01200	0.00001061
30	0.01424	0.00001259
40	0.01570	0.00001388
50	0.01518	0.00001342
60	0.01517	0.00001341
70	0.01815	0.00001605
80	0.02660	0.00002352

Temperature, t	Solubility of $C_2Cl_4F_2$ in H_2O at 1 atm partial pressure	
($^\circ$C)	Wt %, W_2^c	Mole fraction, x_2
25	0.1792	0.0001586
30	0.1344	0.0001189
40	0.0941	0.0000800
50	0.07082	0.00006264
60	0.05583	0.00004937
70	0.04283	0.00003786
80	0.03487	0.00003083

Total pressure	Solubility of $C_2Cl_4F_2$ in H_2O (g/100 cc H_2O)		
(atm)	30°C	50°C	75°C
0.068	0.009	0.005	0.002
0.204	–	0.014	0.007
0.612	–	–	0.020

[a]M.wt. = 203.83; m.p. = 25.0°C; b.p. = 93.0°C; d_4^{25} = 1.6447 g cc^{-1}.

[b]W_2 = -2.6751 x 10^{-2} + 2.7781 x 10^{-3}t - 5.9457 x 10^{-5}t^2 + 4.1334 x 10^{-7}t^3; standard error of estimate for W_2 = 0.000896.

[c]W_2 = 1.11096 - 7.9458 x 10^{-2}t + 2.50225 x 10^{-3}t^2 - 3.9801 x 10^{-5}t^3 + 3.1254 x 10^{-7}t^4 - 9.6359 x 10^{-10}t^5; standard error of estimate for W_2 = 0.0007642.

$CCl_3-CClH_2^a$

Ethane, 1,1,1,2-tetrachloro-   (R 130a)

Temperature, t	Solubility of $C_2Cl_4H_2$ in H_2O at saturated pressure	
(°C)	Wt %, W_2^b	Mole fraction, x_2
0	0.1197	0.0001286
10	0.1133	0.0001217
20	0.1102	0.0001184
25	0.1100	0.0001182
30	0.1108	0.0001191
40	0.1156	0.0001242
50	0.1249	0.0001243
60	0.1393	0.0001497
70	0.1591	0.0001710
80	0.1848	0.0001987

[a] M.wt. = 167.85; m.p. = 170.2°C; b.p. = 130.5°C; d_4^{20} = 1.5406 g cc^{-1}.

[b] W_2 = 0.11968 - 7.87116 x 10^{-4}t + 1.42253 x $10^{-5}t^2$ + 7.24354 x $10^{-8}t^3$; standard error of estimate for W_2 = 0.0006981.

$CCl_2H-CCl_2H^a$

Ethane, 1,1,2,2-tetrachloro-   (R 130)

Temperature, t	Solubility of $C_2Cl_4H_2$ in H_2O at saturated pressure	
(°C)	Wt %, W_2^b	Mole fraction, x_2
0	0.3299	0.0003552
10	0.3116	0.0003354
20	0.2995	0.0003223
25	0.2962	0.0003187
30	0.2948	0.0003172
40	0.2984	0.0003212
50	0.3118	0.0003355
60	0.3356	0.0003613
70	0.3713	0.0003999
80	0.4199	0.0004524

[a] M.wt. = 167.85; m.p. = -36.0°C; b.p. = 146.2°C; d_4^{20} = 1.5953 g cc^{-1}.

[b] W_2 = 0.329934 - 2.10434 x 10^{-3}t + 2.54796 x $10^{-5}t^2$ + 1.86126 x $10^{-7}t^3$; standard error of estimate for W_2 = 0.0007877.

$CCl_3-CCl_2H^a$
Ethane, pentachloro-   (R 120)

| Temperature, t | Solubility of C_2Cl_5H in H_2O at saturated pressure | |
(°C)	Wt %, W_2 [b]	Mole fraction, x_2
0	0.05411	0.00004821
10	0.05168	0.00004605
20	0.05023	0.00004475
25	0.04995	0.00004450
30	0.05003	0.00004457
40	0.05136	0.00004576
50	0.05453	0.00004858
60	0.05980	0.00005329
70	0.06748	0.00006013
80	0.07783	0.00006936

[a] M.wt. = 202.30; m.p. = -29.0°C; b.p. = 162.0°C; d_4^{20} = 1.6796 g cc^{-1}.

[b] W_2 = 5.41068 x 10^{-2} - 2.81268 x 10^{-4}t + 3.41333 x 10^{-6}t^2 + 4.76167 x 10^{-8}t^3;
standard error of estimate for W_2 = 0.0003687.

$CCl_3-CCl_3^a$
Ethane, hexachloro-   (R 110)

| Temperature, t | Solubility of C_2Cl_6 in H_2O at saturated pressure | |
(°C)	Wt %, W_2	Mole fraction, x_2
22.3	0.0050	0.000003805

[a] M.wt. = 236.74; m.p. = 186.8°C; b.p. = 186.0°C at 777 mm Hg; d_4^{20} = 2.091 g cc^{-1}.

$CFH=CH_2^a$
Ethene, fluoro-   (R 1141)

| Temperature, t | Total pressure | Solubility of C_2FH_3 in H_2O | |
(°C)	(atm)	Wt %, W_2	Mole fraction, x_2
80	34	0.9312	0.003664
80	68	1.5166	0.005988

[a] M.wt. = 46.05; m.p. = -160.0°C; b.p. = -72.2°C; d^{25} = 0.615 g cc^{-1}.

CFH_2-CH_3[a]
Ethane, fluoro-   (R 161)

| Temperature, t | Solubility of C_2FH_5 in H_2O at 1 atm partial pressure | |
(°C)	Wt %, W_2	Mole fraction, x_2
14	0.4022	0.001511
25	0.2158	0.000810

[a]M.wt. = 48.06; m.p. = -143.2°C; b.p. = -37.7°C; d_4^{20} = 0.7182 g cc^{-1}.

$CF_2=CH_2$[a]
Ethene, 1,1-difluoro-   (R 1132a)

| Temperature, t | Solubility of $C_2F_2H_2$ in H_2O at 1 atm partial pressure | |
(°C)	Wt %, W_2	Mole fraction, x_2
25	0.01649	0.00004638

[a]M.wt. = 64.04; m.p. = -144.0°C; b.p. = -82.0°C; d^{24} = 0.617 g cc^{-1}.

CF_2H-CH_3[a]
Ethane, 1,1-difluoro-   (R 152a)

| Temperature, t | Solubility of $C_2F_2H_4$ in H_2O at 1 atm partial pressure | |
(°C)	Wt %, W_2	Mole fraction, x_2
0	0.54	0.001479
21	0.32	0.0008748
27.5	0.25	0.0006831

[a]M.wt. = 66.05; m.p. = -117.0°C; b.p. = -24.7°C; d^{20} = 0.950 g cc^{-1}.

$CF_2=CF_2$ [a]
Ethene, tetrafluoro-   (R 1114)

Temperature, t	Solubility of C_2F_4 in H_2O at 1 atm partial pressure	
(°C)	Wt %, W_2 [b]	Mole fraction, x_2
0	0.04062	0.00007319
10	0.02439	0.00004394
20	0.01773	0.00003194
25	0.01585	0.00002856
30	0.01435	0.00002586
40	0.01186	0.00002136
50	0.01005	0.00001811
60	0.009289	0.00001673
70	0.008782	0.00001582

Total pressure	Solubility of C_2F_4 in H_2O at 23°C	
(atm)	Wt %, W_2	Mole fraction, x_2
4.40	0.001	0.00000178
6.31	0.060	0.000108
7.80	0.120	0.000216
8.48	0.130	0.000234
9.91	0.160	0.000288
12.56	0.941	0.001708
13.45	1.274	0.002319

[a] M.wt. = 100.02; m.p. = -142.5°C; b.p. = -76.3°C; $d^{-76.3}$ = 1.519 g cc^{-1}.

[b] W_2 = 4.06185 x 10^{-2} - 2.44154 x 10^{-3}t + 1.04067 x 10^{-4}t^2 - 2.50745 x 10^{-6}t^3 + 3.01367 x 10^{-8}t^4 - 1.39456 x 10^{-10}t^5; standard error of estimate for W_2 = 0.0001045.

CF_3-CFH_2 [a]
Ethane, 1,1,1,2-tetrafluoro-   (R 134a)

Temperature, t	Solubility of $C_2F_4H_2$ in H_2O at 1 atm partial pressure	
(°C)	Wt %, W_2	Mole fraction, x_2
37	0.055	0.0000971

[a] M.wt. = 102.03; m.p. = -101.0°C; b.p. = -26.3°C; d_4^{20} = 1.10 g cc^{-1}.

CHI=CHI[a]

Ethene, cis-1,2-diiodo-

| Temperature, t | Solubility of $C_2H_2I_2$ in H_2O at saturated pressure | |
(°C)	Wt %, W_2	Mole fraction, x_2
25	0.04617	0.00002972

[a]M.wt. = 279.8472; m.p. = -14.0°C; b.p. = 188.0°C; d^{25} = 2.955 g cc^{-1}.

CHI=CHI[a]

Ethene, trans-1,2-diiodo-

| Temperature, t | Solubility of $C_2H_2I_2$ in H_2O at saturated pressure | |
(°C)	Wt %, W_2	Mole fraction, x_2
25	0.01475	0.000009495

[a]M.wt. = 279.8472; m.p. = 78.0°C; b.p. = 192.0°C; d^{83} = 2.826 g cc^{-1}.

CH_3-CH_2I[a]

Ethane, iodo-

| Temperature, t | Solubility of C_2H_5I in H_2O at saturated pressure | |
(°C)	Wt %, W_2[b]	Mole fraction, x_2
0	0.4391	0.0005092
10	0.4123	0.0004780
20	0.4014	0.0004653
25	0.4041	0.0004684
30	0.4133	0.0004791

[a]M.wt. = 155.97; m.p. = -108.0°C; b.p. = 72.3°C; d_4^{20} = 1.9358 g cc^{-1}.

[b]W_2 = 0.4391 - 3.24498 x 10^{-3}t + 4.49991 x $10^{-5}t^2$ + 1.15001 x $10^{-6}t^3$; standard error of estimate for W_2 = 0.001542.

$CBrH_2-CH_2-CClH_2$[a]
Propane, 1-bromo-3-chloro-

| Temperature, t | Solubility of C_3BrClH_6 in H_2O at saturated pressure | |
(°C)	Wt %, W_2	Mole fraction, x_2
25	1.8068	0.002101

[a]M.wt. = 157.44; m.p. = -57.87°C; b.p. = 143.36°C; d_4^{20} = 1.5969 g cc^{-1}.

$CH_2Br-CH_2-CH_3$[a]
Propane, 1-bromo-

| Temperature, t | Solubility of C_3BrH_7 in H_2O at saturated pressure | |
(°C)	Wt %, W_2[b]	Mole fraction, x_2
0	0.2971	0.0004363
10	0.2623	0.0003850
20	0.2444	0.0003587
25	0.2427	0.0003563
30	0.2464	0.0003617

[a]M.wt. = 123.00; m.p. = -109.85°C; b.p. = 71.0°C; d_4^{20} = 1.3537 g cc^{-1}.

[b]W_2 = 0.2971 - 0.004225t + 6.94998 x 10^{-5}t^2 + 5.00002 x 10^{-7}t^3; standard error of
estimate for W_2 = 0.000324.

$CH_3-CBrH-CH_3$[a]
Propane, 2-bromo-

| Temperature, t | Solubility of C_3BrH_7 in H_2O at saturated pressure | |
(°C)	Wt %, W_2[b]	Mole fraction, x_2
0	0.4163	0.0006119
10	0.3637	0.0005344
20	0.3170	0.0004656
25	0.3086	0.0004532
30	0.3170	0.0004656

[a]M.wt. = 123.00; m.p. = -89.0°C; b.p. = 59.38°C; d_4^{20} = 1.3140 g cc^{-1}.

[b]W_2 = 0.4163 - 4.19499 x 10^{-3}t - 1.7450 x 10^{-4}t^2 + 6.79999 x 10^{-6}t^3; standard error
of estimate for W_2 = 0.00008721.

$CBrH_2-CBrH-CClH_2$[a]

Propane, 1,2-dibromo-3-chloro-

| Temperature, t | Solubility of $C_3Br_2ClH_5$ in H_2O at saturated pressure | |
(°C)	Wt %, W_2	Mole fraction, x_2
25	0.10	0.00007629

[a]M.wt. = 236.35; b.p. = 78.0°C; d^{14} = 2.093 g cc^{-1}.

$CBrH_2-CBrH-CH_3$[a]

Propane, 1,2-dibromo-

| Temperature, t | Solubility of $C_3Br_2H_6$ in H_2O at saturated pressure | |
(°C)	Wt %, W_2	Mole fraction, x_2
25	0.1428	0.0001275

[a]M.wt. = 201.90; m.p. = -55.25°C; b.p. = 140.0°C; d_4^{20} = 1.9324 g cc^{-1}.

$CBrH_2-CH_2-CBrH_2$[a]

Propane, 1,3-dibromo-

| Temperature, t | Solubility of $C_3Br_2H_6$ in H_2O at saturated pressure | |
(°C)	Wt %, W_2	Mole fraction, x_2
25	0.1700	0.0001519
30	0.1677	0.0001498

[a]M.wt. = 201.90; m.p. = -34.2°C; b.p. = 167.3°C; d_4^{20} = 1.9822 g cc^{-1}.

$CClH_2-CH_2-CF_3$[a]

Propane, 1-chloro-3,3,3-trifluoro-

| Temperature, t | Solubility of $C_3ClF_3H_4$ in H_2O at saturated pressure | |
(°C)	Wt %, W_2	Mole fraction, x_2
20	0.1328	0.0001807

[a]M.wt. = 132.51; m.p. = -106.2°C; b.p. = 45.1°C; d_4^{20} = 1.3253 g cc^{-1}.

$CClH_2-CH=CH_2$ [a]
1-Propene, 3-chloro-

| Temperature, t | Solubility of C_3ClH_5 in H_2O at saturated pressure | |
| | Wt %, W_2 [b] | Mole fraction, x_2 |
(°C)		
20	0.3600	0.0008498
25	0.2858	0.0006743
30	0.2300	0.0005424
40	0.1613	0.0003802
50	0.1298	0.0003059
60	0.1114	0.0002624
70	0.0819	0.0001929

[a] M.wt. = 76.53; m.p. = -134.5°C; b.p. = 45.0°C; d_4^{20} = 0.9376 g cc^{-1}.

[b] W_2 = 0.90037 - 0.038776t + 6.68248 x $10^{-4}t^2$ - 4.01915 x $10^{-6}t^3$; standard error of estimate for W_2 = 0.0004567.

$CClH_2-CH_2-CH_3$ [a]
Propane, 1-chloro-

| Temperature, t | Solubility of C_3ClH_7 in H_2O at saturated pressure | |
| | Wt %, W_2 [b] | Mole fraction, x_2 |
(°C)		
0	0.3746	0.0008617
10	0.3090	0.0007105
20	0.2713	0.0006236
25	0.2651	0.0006093
30	0.2684	0.0006169

| Temperature, t | Solubility of C_3ClH_7 in H_2O at 0.0132 atm partial pressure | |
| | Wt %, W_2 | Mole fraction, x_2 |
(°C)		
20	0.0714	0.0001639

[a] M.wt. = 78.54; m.p. = -122.8°C; b.p. = 46.60°C; d_4^{20} = 0.8909 g cc^{-1}.

[b] W_2 = 0.3746 - 0.007725t + 1.05001 x $10^{-4}t^2$ + 1.14998 x $10^{-6}t^3$; standard error of estimate for W_2 = 0.

$CH_3-CC1H-CH_3$[a]
Propane, 2-chloro-

Temperature, t	Solubility of C_3ClH_7 in H_2O at saturated pressure	
(°C)	Wt %, W_2[b]	Mole fraction, x_2
0	0.4381	0.001008
10	0.3607	0.0008297
20	0.3041	0.0006992
25	0.2945	0.0006770
30	0.3031	0.0006969

[a]M.wt. = 78.54; m.p. = -117.18°C; b.p. = 35.74°C; d_4^{20} = 0.8617 g cc^{-1}.

[b]W_2 = 0.4381 - 7.61998 x 10^{-3}t - 7.00013 x $10^{-5}t^2$ + 5.80002 x $10^{-6}t^3$; standard error
of estimate for W_2 = 0.

$CC1H=CH-CC1H_2$[a]
1-Propene, cis-1,3-dichloro-

Temperature, t	Solubility of $C_3Cl_2H_4$ in H_2O at saturated pressure	
(°C)	Wt %, W_2	Mole fraction, x_2
20	0.0999	0.0001623

[a]M.wt. = 110.97; b.p. = 104.3°C; d_4^{20} = 1.217 g cc^{-1}.

$CC1H_2-CC1H-CH_3$[a]
Propane, 1,2-dichloro-

Temperature, t	Solubility of $C_3Cl_2H_6$ in H_2O at saturated pressure	
(°C)	Wt %, W_2[b]	Mole fraction, x_2
10	0.2300	0.0003674
20	0.2700	0.0004315
25	0.2800	0.0004475
30	0.2900	0.0004635

[a]M.wt. = 112.99; m.p. = -100.44°C; b.p. = 96.37°C; d_4^{20} = 1.1560 g cc^{-1}.

[b]W_2 = 0.1300 + 1.43332 x 10^{-2}t - 4.99992 x $10^{-4}t^2$ + 6.66654 x $10^{-6}t^3$; standard error
of estimate for W_2 = 0.

$CClH_2-CH_2-CClH_2$[a]

Propane, 1,3-dichloro-

Temperature, t	Solubility of $C_3Cl_2H_6$ in H_2O at saturated pressure	
(°C)	Wt %, W_2[b]	Mole fraction, x_2
10	0.2672	0.0004270
20	0.2693	0.0004304
25	0.2723	0.0004352
30	0.2862	0.0004574

[a]M.wt. = 112.99; m.p. = -99.5°C; b.p. = 120.4°C; d_4^{20} = 1.1878 g cc^{-1}.

[b]W_2 = 0.222301 + 8.54978 x 10^{-3}t - 5.01987 x 10^{-4}t^2 + 9.59978 x 10^{-6}t^3; standard
error of estimate for W_2 = 0.

$CClH_2-CClH-CClH_2$[a]

Propane, 1,2,3-trichloro-

Temperature, t	Solubility of $C_3Cl_3H_5$ in H_2O at saturated pressure	
(°C)	Wt %, W_2	Mole fraction, x_2
25	0.1896	0.0002321

[a]M.wt. = 147.43; m.p. = -14.7°C; b.p. = 156.85°C; d_4^{20} = 1.3889 g cc^{-1}.

$CCl_3-CCl=CCl_2$[a]

1-Propene, hexachloro-

Temperature, t	Solubility of C_3Cl_6 in H_2O at saturated pressure	
(°C)	Wt %, W_2[b]	Mole fraction, x_2
15	0.000947	0.00000072
20	0.001180	0.00000090
25	0.001700	0.00000129
30	0.002355	0.00000179
40	0.003460	0.00000263

[a]M.wt. = 236.74; m.p. = -72.9°C; b.p. = 141.0°C at 100 mm Hg; d_4^{25} = 1.7666 g cc^{-1}.

[b]W_2 = 3.49207 x 10^{-3} - 3.92755 x 10^{-4}t + 1.79164 x 10^{-5}t^2 - 2.02938 x 10^{-7}t^3;
standard error of estimate for W_2 = 0.

$CFH_2-CH=CH_2{}^a$

1-Propene, 3-fluoro-

Temperature, t	Solubility of C_3FH_5 in H_2O at 1 atm partial pressure	
(°C)	Wt %, W_2	Mole fraction, x_2
13	0.007167	0.00002149

[a]M.wt. = 60.07; b.p. = -3.0°C; d_4^{20} = 0.9379 g cc^{-1}.

$CFH_2-CH_2-CH_3{}^a$

Propane, 1-fluoro-

Temperature, t	Solubility of C_3FH_7 in H_2O at 1 atm partial pressure	
(°C)	Wt %, W_2	Mole fraction, x_2
14	0.3856	0.001122

[a]M.wt. = 62.09; m.p. = -159.0°C; b.p. = 2.5°C; d_4^{20} = 0.7956 g cc^{-1}.

$CH_3-CFH-CH_3{}^a$

Propane, 2-fluoro-

Temperature, t	Solubility of C_3FH_7 in H_2O at 1 atm partial pressure	
(°C)	Wt %, W_2	Mole fraction, x_2
15	0.3663	0.001066

[a]M.wt. = 62.09; m.p. = -133.4°C; b.p. = -9.4°C; d_4^{20} = 0.7238 g cc^{-1}.

$CF_2=CF-CF_3$ [a]

Propene, hexafluoro-

| Temperature, t | Solubility of C_3F_6 in H_2O at 1 atm partial pressure | |
(°C)	Wt %, W_2 [b]	Mole fraction, x_2
0	0.04164	0.00005002
10	0.03085	0.00003705
20	0.02248	0.00002700
25	0.01938	0.00002327
30	0.01691	0.00002030
40	0.01336	0.00001609
50	0.01076	0.00001292
60	0.008489	0.00001019
70	0.007132	0.00000856

[a] M.wt. = 150.03; m.p. = -156.2°C; b.p. = -29.4°C; d_4^{-40} = 1.583 g cc^{-1}.

[b] W_2 = 4.16377 x 10^{-2} - 1.14534 x 10^{-3}t + 2.04905 x 10^{-6}t^2 + 5.62581 x 10^{-7}t^3 - 1.10874 x 10^{-8}t^4 + 6.47770 x 10^{-11}t^5; standard error of estimate for W_2 = 0.00010372.

$CF_3-CF_2-CF_3$ [a]

Propane, octafluoro-

| Temperature, t | Solubility of C_3F_8 in H_2O at 1 atm partial pressure | |
(°C)	Wt %, W_2	Mole fraction, x_2
15	0.001495	0.000001432

[a] M.wt. = 188.02; m.p. = -183.0°C; b.p. = -38.0°C; $d^{0.2}$ = 1.450 g cc^{-1}.

$CH_3-CH_2-CH_2I$[a]
Propane, 1-iodo-

Temperature, t	Solubility of C_3H_7I in H_2O at saturated pressure	
(°C)	Wt %, W_2[b]	Mole fraction, x_2
0	0.1139	0.0001208
10	0.1100	0.0001167
20	0.1069	0.0001134
25	0.1051	0.0001115
30	0.1029	0.0001092

Temperature, t	Solubility of C_3H_7I in H_2O at 0.0132 atm partial pressure	
(°C)	Wt %, W_2	Mole fraction, x_2
20	0.02617	0.00002774

[a]M.wt. = 169.99; m.p. = -101.3°C; b.p. = 102.45°C; d_4^{20} = 1.7489 g cc^{-1}.

[b]W_2 = 0.1139 - 4.86669 x 10^{-4}t + 1.25003 x $10^{-5}t^2$ - 2.83340 x $10^{-7}t^3$; standard error of estimate for W_2 = 0.

$CH_3-CHI-CH_3$[a]
Propane, 2-iodo-

Temperature, t	Solubility of C_3H_7I in H_2O at saturated pressure	
(°C)	Wt %, W_2[b]	Mole fraction, x_2
0	0.1667	0.0001769
10	0.1428	0.0001515
20	0.1398	0.0001483
25	0.1387	0.0001471
30	0.1338	0.0001420

[a]M.wt. = 169.99; m.p. = -90.1°C; b.p. = 89.45°C; d_4^{20} = 1.7033 g cc^{-1}.

[b]W_2 = 0.1667 - 4.23167 x 10^{-3}t + 0.000224t^2 - 3.98334 x $10^{-6}t^3$; standard error of estimate for W_2 = 0.

$CBrH_2-CH_2-CH_2-CH_3$[a]

n-Butane, 1-bromo-

| Temperature, t | Solubility of C_4BrH_9 in H_2O at saturated pressure | |
(°C)	Wt %, W_2[b]	Mole fraction, x_2
0	0.07221	0.00009499
10	0.06416	0.00008440
20	0.05732	0.00007540
25	0.05483	0.00007212
30	0.05322	0.00007000

[a]M.wt. = 137.03; m.p. = -112.4°C; b.p. = 101.6°C; d_4^{20} = 1.2758 g cc^{-1}.

[b]W_2 = 0.07221 - 8.14496 x 10^{-4}t - 1.60030 x $10^{-6}t^2$ + 2.55005 x $10^{-7}t^3$; standard error of estimate for W_2 = 0.

$CBrH_2-CH(CH_3)-CH_3$[a]

Propane, 1-bromo-2-methyl-

| Temperature, t | Solubility of C_4BrH_9 in H_2O at saturated pressure | |
(°C)	Wt %, W_2	Mole fraction, x_2
18	0.0510	0.00006706
20	0.0508	0.00006681

[a]M.wt. = 137.03; m.p. = -111.9°C; b.p. = 91.20°C; d_4^{20} = 1.2585 g cc^{-1}.

$CClH_2-CH_2-CH_2-CH_3$[a]

n-Butane, 1-chloro-

| Temperature, t | Solubility of C_4ClH_9 in H_2O at saturated pressure | |
(°C)	Wt %, W_2[b]	Mole fraction, x_2
0	0.07703	0.0001500
10	0.07036	0.0001370
20	0.06420	0.0001250
25	0.06147	0.0001197
30	0.05906	0.0001150

[a]M.wt. = 92.57; m.p. = -123.1°C; b.p. = 78.44°C; d_4^{20} = 0.8862 g cc^{-1}.

[b]W_2 = 0.07703 - 6.75497 x 10^{-4}t - 1.00170 x $10^{-10}t^2$ + 8.50030 x $10^{-8}t^3$; standard error of estimate for W_2 = 0.

$CH_3-CClH-CH_2-CH_3$ [a]
n-Butane, 2-chloro-

| Temperature, t | Solubility of C_4ClH_9 in H_2O at saturated pressure | |
(°C)	Wt %, W_2	Mole fraction, x_2
25	0.0999	0.0001945

[a]M.wt. = 92.57; m.p. = -131.3°C; b.p. = 68.25°C; d_4^{20} = 0.8732 g cc^{-1}.

$CClH_2-CH(CH_3)-CH_3$ [a]
Propane, 1-chloro-2-methyl-

| Temperature, t | Solubility of C_4ClH_9 in H_2O at saturated pressure | |
(°C)	Wt %, W_2	Mole fraction, x_2
12.5	0.0920	0.0001791
20	0.0924	0.0001799

[a]M.wt. = 92.57; m.p. = -24.0°C; b.p. = 68.4°C; d_4^{25} = 0.8725 g cc^{-1}.

$CCl_2H-CH_2-CH_2-CH_3$ [a]
n-Butane, 1,1-dichloro-

| Temperature, t | Solubility of $C_4Cl_2H_8$ in H_2O at saturated pressure | |
(°C)	Wt %, W_2	Mole fraction, x_2
25	0.04997	0.00007088

[a]M.wt. = 127.03; m.p. = -81.0°C; b.p. = 113.8°C; d_4^{20} = 1.0963 g cc^{-1}.

$CH_3-CClH-CClH-CH_3$ [a]

n-Butane, 2,3-dichloro-

Temperature, t	Solubility of $C_4Cl_2H_8$ in H_2O at saturated pressure	
(°C)	Wt %, W_2 [b]	Mole fraction, x_2
0	0.1817	0.0002581
10	0.1167	0.0001656
20	0.05617	0.00007970
25	0.03337	0.00004734
30	0.01860	0.00002638
40	0.02230	0.00003163

[a] M.wt. = 127.03; m.p. = -80.0°C; b.p. = 116.0°C; d_4^{20} = 1.1134 g cc^{-1}.

[b] W_2 = 0.1817 - 6.12066 x 10^{-3}t - 6.89752 x $10^{-5}t^2$ + 3.05917 x $10^{-6}t^3$; standard error of estimate for W_2 = 0.

$CCl_2=C=CCl-CCl_2H$ [a]

1,2-Butadiene, 1,1,3,4,4-pentachloro-

Temperature, t	Solubility of C_4Cl_5H in H_2O at saturated pressure	
(°C)	Wt %, W_2	Mole fraction, x_2
20	0.001326	0.000001045

[a] M.wt. = 226.35; d_4^{20} = 1.6138 g cc^{-1}.

$CCl_2=CCl-CCl=CCl_2$ [a]

1,3-Butadiene, hexachloro-

Temperature, t	Solubility of C_4Cl_6 in H_2O at saturated pressure	
(°C)	Wt %, W_2 [b]	Mole fraction, x_2
15	0.000320	0.00000022
20	0.000380	0.00000026
25	0.000408	0.00000028
30	0.000432	0.00000030
40	0.000590	0.00000041

[a] M.wt. = 260.76; m.p. = -21.0°C; b.p. = 215.0°C; d_4^{20} = 1.6820 g cc^{-1}.

[b] W_2 = -3.46037 x 10^{-4} + 8.04054 x 10^{-5}t - 2.98524 x $10^{-6}t^2$ + 3.90033 x $10^{-8}t^3$; standard error of estimate for W_2 = 0.000367.

CF_2-CF_2 [a]
CF_2-CF_2

Cyclobutane, octafluoro-   (R C318)

Temperature, t	Solubility of C_4F_8 in H_2O at 1 atm partial pressure	
(°C)	Wt %, W_2	Mole fraction, x_2
21	0.014	0.00001259

Total pressure	Solubility of C_4F_8 in H_2O (g/100 g H_2O)		
(atm)	0°C	26°C	37.8°C
1.00	0.009	0.005	0.003
1.36	-	0.007	0.004
2.04	-	0.010	0.007
2.72	-	0.013	0.009

[a] M.wt. = 200.03; m.p. = -38.7°C; b.p. = -4.0°C; d_4^0 = 1.724 g cc^{-1}.

$CH_2I-CH_2-CH_2-CH_3$ [a]
n-Butane, 1-iodo-

Temperature, t	Solubility of C_4H_9I in H_2O at saturated pressure	
(°C)	Wt %, W_2 [b]	Mole fraction, x_2
0	0.02553	0.00002500
10	0.02247	0.00002200
20	0.01940	0.00001900
25	0.01819	0.00001781
30	0.01736	0.00001700

[a] M.wt. = 184.02; m.p. = -103.0°C; b.p. = 130.55°C; d_4^{20} = 1.6154 g cc^{-1}.

[b] W_2 = 0.02553 - 2.70832 x 10^{-4}t - 5.25005 x 10^{-6}t^2 + 1.73334 x 10^{-7}t^3; standard error of estimate for W_2 = 0.000783.

$CBrH_2-CH_2-CH(CH_3)-CH_3$ [a]
n-Butane, 1-bromo-3-methyl-

Temperature, t	Solubility of $C_5H_{11}Br$ in H_2O at saturated pressure	
(°C)	Wt %, W_2	Mole fraction, x_2
16.5	0.0200	0.00002386
20	0.0194	0.00002314

[a]M.wt. = 151.05; m.p. = 112.0°C; b.p. = 120.4°C; d_4^{20} = 1.2071 g cc^{-1}.

$CClH_2-CH_2-CH_2-CH_2-CH_3$ [a]
n-Pentane, 1-chloro-

Temperature, t	Solubility of C_5ClH_{11} in H_2O at saturated pressure	
(°C)	Wt %, W_2 [b]	Mole fraction, x_2
0	0.02426	0.00004101
10	0.02248	0.00003800
20	0.02071	0.00003501
25	0.01982	0.00003351
30	0.01893	0.00003200

[a]M.wt. = 106.60; m.p. = -99.0°C; b.p. = 107.8°C; d_4^{20} = 0.8818 g cc^{-1}.

[b]W_2 = 0.02426 - 1.79166 x 10^{-4}t + 1.49999 x $10^{-7}t^2$ - 3.33355 x $10^{-9}t^3$; standard error of estimate for W_2 = 0.

$CH_3-CClH-CH_2-CH_2-CH_3$ [a]
n-Pentane, 2-chloro-

Temperature, t	Solubility of C_5ClH_{11} in H_2O at saturated pressure	
(°C)	Wt %, W_2	Mole fraction, x_2
25	0.02499	0.00004224

[a]M.wt. = 106.60; m.p. = -137.0°C; b.p. = 96.86°C; d_4^{20} = 0.8698 g cc^{-1}.

$CH_3-CH_2-CClH-CH_2-CH_3$[a]

n-Pentane, 3-chloro-

Temperature, t	Solubility of C_5ClH_{11} in H_2O at saturated pressure	
(°C)	Wt %, W_2	Mole fraction, x_2
25	0.02499	0.00004224

[a]M.wt. = 106.60; m.p. = -105.0°C; b.p. = 97.8°C; d_4^{20} = 0.8731 g cc^{-1}.

$CH_3-CCl(CH_3)-CH_2-CH_3$[a]

n-Butane, 2-chloro-2-methyl-

Temperature, t	Solubility of C_5ClH_{11} in H_2O at saturated pressure	
(°C)	Wt %, W_2	Mole fraction, x_2
25	0.03322	0.00005615

[a]M.wt. = 106.60; m.p. = -73.5°C; b.p. = 85.6°C; d_4^{20} = 0.8653 g cc^{-1}.

$CH_3-CCl(CH_3)-CClH-CH_3$[a]

n-Butane, 2,3-dichloro-2-methyl-

Temperature, t	Solubility of $C_5Cl_2H_{10}$ in H_2O at saturated pressure	
(°C)	Wt %, W_2	Mole fraction, x_2
25	0.02856	0.00003647

[a]M.wt. = 141.04; b.p. = 138.0°C; d_4^{15} = 1.0696 g cc^{-1}.

$CCl \equiv C-CCl=CCl-CCl_3$ [a]

2-Butene-4-yne, hexachloro-

Temperature, t	Solubility of C_5Cl_6 in H_2O at saturated pressure	
(°C)	Wt %, W_2 [b]	Mole fraction, x_2
15	0.000199	0.00000013
20	0.000288	0.00000019
25	0.000315	0.00000021
30	0.000389	0.00000026
40	0.001100	0.00000073

[a] M.wt. = 272.77.

[b] $W_2 = -1.50856 \times 10^{-3} + 2.28720 \times 10^{-4}t - 9.80153 \times 10^{-6}t^2 + 1.42847 \times 10^{-7}t^3$;
    standard error of estimate for W_2 = 0.0006843.

$CCl_2-CCl=CCl-CCl=CCl$ [a]

1,3-Cyclopentadiene, hexachloro-

Temperature, t	Solubility of C_5Cl_6 in H_2O at saturated pressure	
(°C)	Wt %, W_2	Mole fraction, x_2
20	0.0003404	0.00000002248

[a] M.wt. = 272.81; m.p. = -9.0°C; b.p. = 239.0°C; d_4^{25} = 1.7019 g cc^{-1}.

$CCl_2=CCl-CCl_2-CCl=CCl_2$ [a]

1,4-Pentadiene, octachloro-

Temperature, t	Solubility of C_5Cl_8 in H_2O at saturated pressure	
(°C)	Wt %, W_2 [b]	Mole fraction, x_2
15	0.000014	0.00000001
20	0.000020	0.00000001
25	0.000031	0.00000002
30	0.000044	0.00000002
40	0.000060	0.00000003

[a] M.wt. = 343.716; m.p. = -78.0°C; b.p. = 72.4°C at 0.15 mm Hg; d_4^{25} = 1.7495 g cc^{-1}.

[b] $W_2 = 6.73352 \times 10^{-5} - 8.55514 \times 10^{-6}t + 4.10544 \times 10^{-7}t^2 - 5.03125 \times 10^{-9}t^3$;
    standard error of estimate for W_2 = 0.

C_6BrClH_4[a]

Benzene, 1-bromo-2-chloro-

Temperature, t	Solubility of C_6BrClH_4 in H_2O at saturated pressure	
(°C)	Wt %, W_2	Mole fraction, x_2
25	0.01240	0.00001167

[a]M.wt. = 191.46; m.p. = -12.3°C; b.p. = 204.0°C at 765 mm Hg; d_4^{25} = 1.6382 g cc^{-1}.

C_6BrClH_4[a]

Benzene, 1-bromo-3-chloro-

Temperature, t	Solubility of C_6BrClH_4 in H_2O at saturated pressure	
(°C)	Wt %, W_2	Mole fraction, x_2
25	0.01184	0.00001114

[a]M.wt. = 191.46; m.p. = -21.5°C; b.p. = 196.0°C; d_4^{20} = 1.6302 g cc^{-1}.

C_6BrClH_4[a]

Benzene, 1-bromo-4-chloro-

Temperature, t	Solubility of C_6BrClH_4 in H_2O at saturated pressure	
(°C)	Wt %, W_2	Mole fraction, x_2
25	0.004501	0.000004236

[a]M.wt. = 191.46; m.p. = 68.0°C; b.p. = 196.0°C at 756 mm Hg; d_4^{71} = 1.576 g cc^{-1}.

C_6BrH_4I[a]

Benzene, 1-bromo-4-iodo-

Temperature, t	Solubility of C_6BrH_4I in H_2O at saturated pressure	
(°C)	Wt %, W_2	Mole fraction, x_2
25	0.0007815	0.0000004976

[a]M.wt. = 282.91; m.p. = 92.0°C; b.p. = 252.0°C at 754 mm Hg; d_4^{25} = 2.235 g cc^{-1}.

C_6BrH_5 [a]

Benzene, bromo-

Temperature, t	Solubility of C_6BrH_5 in H_2O at saturated pressure	
(°C)	Wt %, W_2	Mole fraction, x_2
25	0.0410	0.00004706
30	0.0446	0.00005118

[a]M.wt. = 157.02; m.p. = -30.82°C; b.p. = 156.0°C; d_4^{20} = 1.4950 g cc^{-1}.

$C_6Br_2H_4$ [a]

Benzene, 1,2-dibromo-

Temperature, t	Solubility of $C_6Br_2H_4$ in H_2O at saturated pressure	
(°C)	Wt %, W_2	Mole fraction, x_2
25	0.007482	0.000005714

[a]M.wt. = 235.92; m.p. = 7.1°C; b.p. = 225.0°C; d_4^{20} = 1.9873 g cc^{-1}.

$C_6Br_2H_4$ [a]

Benzene, 1,3-dibromo-

Temperature, t	Solubility of $C_6Br_2H_4$ in H_2O at saturated pressure	
(°C)	Wt %, W_2	Mole fraction, x_2
35	0.00675	0.00000515

[a]M.wt. = 235.92; m.p. = -7.0°C; b.p. = 218.0°C; d_4^{20} = 1.9523 g cc^{-1}.

$C_6Br_2H_4$ [a]

Benzene, 1,4-dibromo-

Temperature, t	Solubility of $C_6Br_2H_4$ in H_2O at saturated pressure	
(°C)	Wt %, W_2	Mole fraction, x_2
25	0.002	0.000001527

[a]M.wt. = 235.92; m.p. = 87.33°C; b.p. = 218.5°C; d^{17} = 2.261 g cc^{-1}.

$C_6Br_3H_3$ [a]
Benzene, 1,2,3-tribromo-

Temperature, t	Solubility of $C_6Br_3H_3$ in H_2O at saturated pressure	
(°C)	Wt %, W_2	Mole fraction, x_2
25	0.0002947	0.0000001686

[a]M.wt. = 314.82; m.p. = 87.8°C; d^{20} = 2.658 g cc^{-1}.

$C_6Br_3H_3$ [a]
Benzene, 1,2,4-tribromo-

Temperature, t	Solubility of $C_6Br_3H_3$ in H_2O at saturated pressure	
(°C)	Wt %, W_2	Mole fraction, x_2
25	0.0009985	0.0000005714

[a]M.wt. = 314.82; m.p. = 44.5°C; b.p. = 275.0°C.

$C_6Br_3H_3$ [a]
Benzene, 1,3,5-tribromo-

Temperature, t	Solubility of $C_6Br_3H_3$ in H_2O at saturated pressure	
(°C)	Wt %, W_2	Mole fraction, x_2
25	0.020	0.00001145

[a]M.wt. = 314.82; m.p. = 122.0°C; b.p. = 271.0°C at 765 mm Hg.

$C_6Br_4H_2$ [a]
Benzene, 1,2,4,5-tetrabromo-

Temperature, t	Solubility of $C_6Br_4H_2$ in H_2O at saturated pressure	
(°C)	Wt %, W_2	Mole fraction, x_2
25	0.000004135	0.000000001892

[a]M.wt. = 393.72; m.p. = 182.0°C; d^{20} = 3.072 g cc^{-1}.

$C_6ClH_4I^a$

Benzene, 1-chloro-2-iodo-

Temperature, t	Solubility of C_6ClH_4I in H_2O at saturated pressure	
(°C)	Wt %, W_2	Mole fraction, x_2
25	0.006897	0.000005211

[a]M.wt. = 238.46; m.p. = 0.7°C; b.p. = 234.5°C; d^{25} = 1.9515 g cc^{-1}.

$C_6ClH_4I^a$

Benzene, 1-chloro-3-iodo-

Temperature, t	Solubility of C_6ClH_4I in H_2O at saturated pressure	
(°C)	Wt %, W_2	Mole fraction, x_2
25	0.006740	0.000005093

[a]M.wt. = 238.46; b.p. = 230.0°C; d_4^{20} = 1.9255 g cc^{-1}.

$C_6ClH_4I^a$

Benzene, 1-chloro-4-iodo-

Temperature, t	Solubility of C_6ClH_4I in H_2O at saturated pressure	
(°C)	Wt %, W_2	Mole fraction, x_2
25	0.002232	0.000001686

[a]M.wt. = 238.46; m.p. = 57.0°C; b.p. = 227.0°C; d_4^{57} = 1.886 g cc^{-1}.

C_6ClH_5[a]
Benzene, chloro-

Temperature, t	Solubility of C_6ClH_5 in H_2O at saturated pressure	
(°C)	Wt %, W_2[b]	Mole fraction, x_2
20	0.04663	0.00007467
25	0.04979	0.00007972
30	0.05439	0.00008709
40	0.06816	0.0001092
50	0.08845	0.0001417
60	0.1158	0.0001855
70	0.1506	0.0002414
80	0.1935	0.0003102
90	0.2450	0.0003929

[a]M.wt. = 112.56; m.p. = -45.6°C; b.p. = 132.0°C; d_4^{20} = 1.1058 g cc^{-1}.

[b]W_2 = 4.7153 x 10^{-2} - 5.09717 x 10^{-4}t + 2.25008 x $10^{-5}t^2$ + 8.42385 x $10^{-8}t^3$;
standard error of estimate for W_2 = 0.0055879.

$CClH_2-CH_2-CH_2-CH_2-CH_2-CH_3$[a]
n-Hexane, 1-chloro-

Temperature, t	Solubility of C_6ClH_{13} in H_2O at saturated pressure	
(°C)	Wt %, W_2[b]	Mole fraction, x_2
0	0.01172	0.00001751
10	0.01011	0.00001510
20	0.009708	0.00001450
25	0.009101	0.00001359
30	0.007766	0.00001160

[a]M.wt. = 120.62; m.p. = -94.0°C; b.p. = 134.5°C; d_4^{20} = 0.8785 g cc^{-1}.

[b]W_2 = 0.01172 - 0.000313t + 1.07800 x $10^{-5}t^2$ - 4.58000 x $10^{-7}t^3$; standard error of
estimate for W_2 = 0.

$C_6Cl_2H_4$ [a]
Benzene, 1,2-dichloro-

| Temperature, t | Solubility of $C_6Cl_2H_4$ in H_2O at saturated pressure | |
(°C)	Wt %, W_2 [b]	Mole fraction, x_2
0	0.004497	0.00000551
10	0.006639	0.00000814
20	0.008396	0.00001029
25	0.009261	0.00001135
30	0.01019	0.00001248
40	0.01243	0.00001523
50	0.01554	0.00001904
60	0.01993	0.00002443
70	0.02603	0.00003190
80	0.03425	0.00004198
90	0.04501	0.00005517
100	0.05872	0.00007200

[a] M.wt. = 147.01; m.p. = -17.0°C; b.p. = 180.5°C; d_4^{20} = 1.3048 g cc^{-1}.

[b] W_2 = 4.49686 x 10^{-3} + 2.47344 x 10^{-4}t - 4.01210 x 10^{-6}t^2 + 6.96106 x 10^{-8}t^3; standard error of estimate for W_2 = 0.0020008.

$C_6Cl_2H_4$ [a]
Benzene, 1,3-dichloro-

| Temperature, t | Solubility of $C_6Cl_2H_4$ in H_2O at saturated pressure | |
(°C)	Wt %, W_2 [b]	Mole fraction, x_2
20	0.01107	0.00001357
25	0.01241	0.00001521
30	0.01381	0.00001692
40	0.01654	0.00002028
50	0.01880	0.00002304
60	0.02010	0.00002463

[a] M.wt. = 147.01; m.p. = -24.7°C; b.p. = 173.0°C; d_4^{20} = 1.2884 g cc^{-1}.

[b] W_2 = 7.49041 x 10^{-3} + 6.82924 x 10^{-5}t + 7.12568 x 10^{-6}t^2 - 7.93629 x 10^{-8}t^3; standard error of estimate for W_2 = 0.0001735.

$C_6Cl_2H_4$ [a]

Benzene, 1,4-dichloro-

Temperature, t	Solubility of $C_6Cl_2H_4$ in H_2O at saturated pressure	
(°C)	Wt %, W_2 [b]	Mole fraction, x_2
Solid:		
0	0.004432	0.00000543
10	0.006316	0.00000774
20	0.008048	0.00000986
25	0.008967	0.00001099
30	0.009979	0.00001223
40	0.01246	0.00001527
50	0.01584	0.00001942
Liquid:		
60	0.02048	0.00002510
70	0.02672	0.00003275
80	0.03491	0.00004279
90	0.04540	0.00005566
100	0.05855	0.00007179

[a] M.wt. = 147.01; m.p. = 53.1°C; b.p. = 174.0°C; d_4^{20} = 1.2475 g cc^{-1}.

[b] W_2 = 4.43225 x 10^{-3} + 2.07605 x 10^{-4}t - 2.51077 x $10^{-6}$$t^2$ + 5.84656 x $10^{-8}$$t^3$;
standard error of estimate for W_2 = 0.0012086.

$C_6Cl_3H_3$ [a]

Benzene, 1,2,3-trichloro-

Temperature, t	Solubility of $C_6Cl_3H_3$ in H_2O at saturated pressure	
(°C)	Wt %, W_2	Mole fraction, x_2
25	0.003162	0.000003140

[a] M.wt. = 181.45; m.p. = 53.5°C; b.p. = 218.5°C; d_4^{40} = 1.4533 g cc^{-1}.

$C_6Cl_3H_3$[a]

Benzene, 1,2,4-trichloro-

Temperature, t	Solubility of $C_6Cl_3H_3$ in H_2O at saturated pressure	
(°C)	Wt %, W_2	Mole fraction, x_2
20	0.003	0.00000298

[a]M.wt. = 181.45; m.p. = 16.95°C; b.p. = 213.5°C; d_4^{20} = 1.4542 g cc^{-1}.

$C_6Cl_3H_3$[a]

Benzene, 1,3,5-trichloro-

Temperature, t	Solubility of $C_6Cl_3H_3$ in H_2O at saturated pressure	
(°C)	Wt %, W_2	Mole fraction, x_2
25	0.0006607	0.0000006560

[a]M.wt. = 181.45; m.p. = 63.5°C; b.p. = 208.0°C at 763 mm Hg; d_4^{64} = 1.3865 g cc^{-1}.

$C_6Cl_4H_2$[a]

Benzene, 1,2,3,4-tetrachloro-

Temperature, t	Solubility of $C_6Cl_4H_2$ in H_2O at saturated pressure	
(°C)	Wt %, W_2	Mole fraction, x_2
25	0.0004320	0.0000003605

[a]M.wt. = 215.90; m.p. = 47.5°C; b.p. = 254.0°C.

$C_6Cl_4H_2$[a]

Benzene, 1,2,3,5-tetrachloro-

Temperature, t	Solubility of $C_6Cl_4H_2$ in H_2O at saturated pressure	
(°C)	Wt %, W_2	Mole fraction, x_2
25	0.0003512	0.0000002930

[a]M.wt. = 215.90; m.p. = 54.5°C; b.p. = 246.0°C.

$C_6Cl_4H_2$[a]
Benzene, 1,2,4,5-tetrachloro-

| Temperature, t | Solubility of $C_6Cl_4H_2$ in H_2O at saturated pressure | |
(°C)	Wt %, W_2	Mole fraction, x_2
25	0.00005964	0.00000004976

[a]M.wt. = 215.90; m.p. = 140.0°C; b.p. = 244.5°C; d^{22} = 1.858 g cc^{-1}.

C_6Cl_5H[a]
Benzene, pentachloro-

| Temperature, t | Solubility of C_6Cl_5H in H_2O at saturated pressure | |
(°C)	Wt %, W_2	Mole fraction, x_2
25	0.00005621	0.00000004045

[a]M.wt. = 250.34; m.p. = 86.0°C; b.p. = 277.0°C; $d^{16.5}$ = 1.8342 g cc^{-1}.

C_6Cl_6[a]
Benzene, hexachloro-

| Temperature, t | Solubility of C_6Cl_6 in H_2O at saturated pressure | |
(°C)	Wt.%, W_2	Mole fraction, x_2
25	0.0000005	3.163×10^{-10}

[a]M.wt. = 284.79; m.p. = 230.0°C; b.p. = 322.0°C; $d^{23.6}$ = 1.5691 g cc^{-1}.

$C_6Cl_6H_6$[a]
Cyclohexane, α-1,2,3,4,5,6-hexachloro-

| Temperature, t | Maximum size of particles | Solubility of $C_6Cl_6H_6$ in H_2O | |
(°C)	(μm)	Wt %, W_2	Mole fraction, x_2
28	0.10	0.000203	1.257×10^{-7}
28	0.05	0.000177	1.096×10^{-7}
100	-	0.0068	4.212×10^{-6}

[a]M.wt. = 290.83; m.p. = 159.8°C; b.p. = 288.0°C; d_4^{20} = 1.870 g cc^{-1}.

$C_6Cl_6H_6$ [a]
Cyclohexane, β-1,2,3,4,5,6-hexachloro-

Temperature, t	Maximum size of particles	Solubility of $C_6Cl_6H_6$ in H_2O	
(°C)	(μm)	Wt %, W_2	Mole fraction, x_2
28	0.10	0.00002	1.239×10^{-8}
100	-	0.0017	1.053×10^{-6}

[a] M.wt. = 290.83; m.p. = 314.5°C; b.p. = 60.0°C at 0.58 mm Hg; d^{19} = 1.890 g cc^{-1}.

$C_6Cl_6H_6$ [a]
Cyclohexane, δ-1,2,3,4,5,6-hexachloro-

Temperature, t	Maximum size of particles	Solubility of $C_6Cl_6H_6$ in H_2O	
(°C)	(μm)	Wt %, W_2	Mole fraction, x_2
28	0.10	0.00157	9.725×10^{-7}
28	0.05	0.00116	7.186×10^{-7}
100	-	0.0130	8.036×10^{-6}

[a] M.wt. = 290.83; m.p. = 141.8°C; b.p. = 60.0°C at 0.34 mm Hg.

$C_6Cl_6H_6$ [a]

Cyclohexane, γ-1,2,3,4,5,6-hexachloro-

Temperature, t	Maximum size of particles	Solubility of $C_6Cl_6H_6$ in H_2O	
(°C)	(μm)	Wt %, W_2 [b]	Mole fraction, x_2
28	0.10	0.000740	0.000000458
28	0.05	0.000661	0.000000409
15	–	0.000186	0.00000012
20	–	0.000558	0.00000035
25	–	0.000779	0.00000048
30	–	0.000913	0.00000057
40	–	0.001181	0.00000073
50	–	0.001881	0.00000117
60	–	0.003531	0.00000219
70	–	0.006649	0.00000412
80	–	0.01175	0.00000728
90	–	0.01937	0.00001200
100	–	0.03000	0.00001859

[a] M.wt. = 290.83; m.p. = 112.8°C; b.p. = 323.4°C.

[b] $W_2 = -0.002487 + 2.8193 \times 10^{-4}t - 8.21198 \times 10^{-6}t^2 + 8.64132 \times 10^{-8}t^3$; standard error of estimate for W_2 = 0.00016031.

C_6FH_5 [a,b]

Benzene, fluoro-

Temperature, t	Solubility of C_6FH_5 in H_2O at saturated pressure	
(°C)	Wt %, W_2	Mole fraction, x_2
25	0.1550	0.0002908
30	0.1538	0.0002886

[a] M.wt. = 96.11; m.p. = -41.2°C; b.p. = 85.1°C; d_4^{20} = 1.0225 g cc^{-1}.

[b] Further data: Fig. 2.73 and Table 2.107.

$C_6F_2H_4$ [a]
Benzene, 1,2-difluoro-

Temperature, t	Solubility of $C_6F_2H_4$ in H_2O at saturated pressure	
(°C)	Wt %, W_2	Mole fraction, x_2
25	0.1144	0.0001809

[a]M.wt. = 114.09; m.p. = -34.0°C; b.p. = 91.5°C at 751 mm Hg; d_4^{18} = 1.1599 g cc^{-1}.

$C_6F_2H_4$ [a]
Benzene, 1,3-difluoro-

Temperature, t	Solubility of $C_6F_2H_4$ in H_2O at saturated pressure	
(°C)	Wt %, W_2	Mole fraction, x_2
25	0.1144	0.0001809

[a]M.wt. = 114.09; m.p. = -59.0°C; b.p. = 83.0°C; d_4^{18} = 1.1552 g cc^{-1}.

$C_6H_4I_2$ [a]
Benzene, 1,2-diiodo-

Temperature, t	Solubility of $C_6H_4I_2$ in H_2O at saturated pressure	
(°C)	Wt %, W_2	Mole fraction, x_2
25	0.00149	8.137 x 10^{-7}

[a]M.wt. = 329.91; m.p. = 27.0°C; b.p. = 286.0°C; d_4^{20} = 2.54 g cc^{-1}.

$C_6H_4I_2$ [a]
Benzene, 1,3-diiodo-

Temperature, t	Solubility of $C_6H_4I_2$ in H_2O at saturated pressure	
(°C)	Wt %, W_2	Mole fraction, x_2
25	0.000967	5.281 x 10^{-7}

[a]M.wt. = 329.91; m.p. = 40.4°C; b.p. = 285.0°C; d^{25} = 2.47 g cc^{-1}.

$C_6H_4I_2$[a]

Benzene, 1,4-diiodo-

Temperature, t	Solubility of $C_6H_4I_2$ in H_2O at saturated pressure	
(°C)	Wt %, W_2	Mole fraction, x_2
25	0.00014	7.645×10^{-8}

[a]M.wt. = 329.91; m.p. = 131.5°C; b.p. = 285.0°C subl.

C_6H_5I[a]

Benzene, iodo-

Temperature, t	Solubility of C_6H_5I in H_2O at saturated pressure	
(°C)	Wt %, W_2	Mole fraction, x_2
25	0.02284	0.00002016
30	0.03399	0.00003002

[a]M.wt. = 204.01; m.p. = -31.27°C; b.p. = 188.3°C; d_4^{20} = 1.8308 g cc^{-1}.

Formula Index: Solubility of H_2O in Halogenated Hydrocarbons

$CBrClF_2$	$CBrClF_2$	Methane, bromochlorodifluoro-
$CBrClH_2$	$CBrClH_2$	Methane, bromochloro-
$CBrCl_3$	$CBrCl_3$	Methane, bromotrichloro-
$CBrF_3$	$CBrF_3$	Methane, bromotrifluoro-
CBr_2H_2	CBr_2H_2	Methane, dibromo-
CBr_3F	CBr_3F	Methane, tribromofluoro-
$CClF_2H$	$CClF_2H$	Methane, chlorodifluoro-
$CClF_3$	$CClF_3$	Methane, chlorotrifluoro-
$CClH_3$	$CClH_3$	Methane, chloro-
CCl_2FH	CCl_2FH	Methane, dichlorofluoro-
CCl_2F_2	CCl_2F_2	Methane, dichlorodifluoro-
CCl_2H_2	CCl_2H_2	Methane, dichloro-
CCl_3F	CCl_3F	Methane, trichlorofluoro-
CCl_3H	CCl_3H	Methane, trichloro-
CCl_4	CCl_4	Methane, tetrachloro-
CH_2I_2	CH_2I_2	Methane, diiodo-
C_2BrClF_3H	$CBrClH-CF_3$	Ethane, 1-bromo-1-chloro-2,2,2-trifluoro-
C_2BrH_5	$CBrH_2-CH_3$	Ethane, bromo-
$C_2Br_2ClH_3$	$CBrClH-CBrH_2$	Ethane, 1,2-dibromo-1-chloro-
$C_2Br_2Cl_2H_2$	$CBrClH-CBrClH$	Ethane, 1,2-dibromo-1,2-dichloro-
$C_2Br_2H_4$	$CBrH_2-CBrH_2$	Ethane, 1,2-dibromo-
$C_2Br_3H_3$	$CBr_2H-CBrH_2$	Ethane, 1,1,2-tribromo-
$C_2Br_4H_2$	CBr_2H-CBr_2H	Ethane, 1,1,2,2-tetrabromo-
$C_2ClF_2H_3$	$CClH_2-CF_2H$	Ethane, 1-chloro-2,2-difluoro-
$C_2ClF_2H_3$	$CClF_2-CH_3$	Ethane, 1-chloro-1,1-difluoro-
C_2ClH_3	$CClH=CH_2$	Ethene, chloro-
C_2ClH_5	$CClH_2-CH_3$	Ethane, chloro-
$C_2Cl_2F_3H$	CCl_2H-CF_3	Ethane, 1,1-dichloro-2,2,2-trifluoro-
$C_2Cl_2F_4$	$CClF_2-CClF_2$	Ethane, 1,2-dichloro-1,1,2,2-tetrafluoro-
$C_2Cl_2F_4$	CCl_2F-CF_3	Ethane, 1,1-dichloro-1,2,2,2-tetrafluoro-
$C_2Cl_2H_2$	$CCl_2=CH_2$	Ethene, 1,1-dichloro-
$C_2Cl_2H_2$	$CClH=CClH$	Ethene, cis-1,2-dichloro-
$C_2Cl_2H_2$	$CClH=CClH$	Ethene, trans-1,2-dichloro-
$C_2Cl_2H_4$	CCl_2H-CH_3	Ethane, 1,1-dichloro-
$C_2Cl_2H_4$	$CClH_2-CClH_2$	Ethane, 1,2-dichloro-
$C_2Cl_3F_3$	$CCl_2F-CClF_2$	Ethane, 1,1,2-trichloro-1,2,2-trifluoro-
C_2Cl_3H	$CCl_2=CClH$	Ethene, trichloro-
$C_2Cl_3H_3$	CCl_3-CH_3	Ethane, 1,1,1-trichloro-
$C_2Cl_3H_3$	$CCl_2H-CClH_2$	Ethane, 1,1,2-trichloro-

C_2Cl_4	$CCl_2=CCl_2$	Ethene, tetrachloro-
$C_2Cl_4F_2$	CCl_2F-CCl_2F	Ethane, 1,1,2,2-tetrachloro-1,2-difluoro-
$C_2Cl_4H_2$	CCl_3-CClH_2	Ethane, 1,1,1,2-tetrachloro-
$C_2Cl_4H_2$	CCl_2H-CCl_2H	Ethane, 1,1,2,2-tetrachloro-
C_2Cl_5H	CCl_3-CCl_2H	Ethane, pentachloro-
C_3BrClH_6	$CBrH_2-CH_2-CClH_2$	Propane, 1-bromo-3-chloro-
$C_3Br_2F_6$	$CBrF_2-CBrF-CF_3$	Propane, 1,2-dibromo-1,1,2,3,3,3-hexafluoro-
$C_3Br_2H_6$	$CBrH_2-CBrH-CH_3$	Propane, 1,2-dibromo-
C_3ClH_5	$CClH_2-CH=CH_2$	1-Propene, 3-chloro-
$C_3Cl_2F_6$	$CClF_2-CClF-CF_3$	Propane, 1,2-dichloro-1,1,2,3,3,3-hexafluoro-
$C_3Cl_2H_6$	$CClH_2-CClH-CH_3$	Propane, 1,2-dichloro-
$C_3Cl_3F_5$	$CCl_3-CF_2-CF_3$	Propane, 1,1,1-trichloro-2,2,3,3,3-pentafluoro-
$C_3Cl_4F_4$	$CCl_3-CF_2-CClF_2$	Propane, 1,1,1,3-tetrachloro-2,2,3,3-tetrafluoro-
C_3Cl_6	$CCl_3-CCl=CCl_2$	1-Propene, hexachloro-
C_4ClH_9	$CClH_2-CH_2-CH_2-CH_3$	n-Butane, 1-chloro-
C_4ClH_9	$CH_3-CClH-CH_2-CH_3$	n-Butane, 2-chloro-
C_4Cl_6	$CCl_2=CCl-CCl=CCl_2$	1,3-Butadiene, hexachloro-
C_4F_8	$\overline{CF_2-CF_2-CF_2-CF_2}$	Cyclobutane, octafluoro-
C_5ClH_{11}	$CClH_2-CH_2-CH_2-CH_2-CH_3$	n-Pentane, 1-chloro-
C_5Cl_6	$CCl\ C-CCl=CCl-CCl_3$	2-Butene-4-yne, hexachloro-
C_5Cl_8	$CCl_2=CCl-CCl_2-CCl=CCl_2$	1,4-Pentadiene, octachloro-
C_6BrH_5	C_6BrH_5	Benzene, bromo-
C_6ClH_5	C_6ClH_5	Benzene, chloro-
$C_6Cl_2H_4$	$C_6Cl_2H_4$	Benzene, 1,2-dichloro-
$C_6Cl_2H_4$	$C_6Cl_2H_4$	Benzene, 1,4-dichloro-
$C_6Cl_3H_3$	$C_6Cl_3H_3$	Benzene, 1,2,4-trichloro-
C_6FH_5	C_6FH_5	Benzene, fluoro-
C_6F_{14}	$CF_3-(CF_2)_4-CF_3$	n-Hexane, perfluoro-
C_6H_5I	C_6H_5I	Benzene, iodo-

538

CBrClF$_2$[a]

Methane, bromochlorodifluoro-   (R 12B1)

Temperature, t	Solubility of H_2O in CBrClF$_2$ at saturated pressure	
(°C)	Wt %, W_2[b]	Mole fraction, x_2
4	0.007705	0.0007068
6	0.007707	0.0007070
8	0.007707	0.0007070
10	0.007709	0.0007071

[a]M.wt. = 165.3648; m.p. = -159.5°C; b.p. = -3.3°C; d^{15} = 1.850 g cc^{-1}.

[b]W_2 = 7.67620 x 10^{-3} + 1.29997 x 10^{-5}t - 1.78723 x 10^{-6}t^2 + 8.12270 x 10^{-8}t^3;
   standard error of estimate for W_2 = 0.

CBrClH$_2$[a]

Methane, bromochloro-   (R 30B1)

Temperature, t	Solubility of H_2O in CBrClH$_2$ at saturated pressure	
(°C)	Wt %, W_2	Mole fraction, x_2
25	0.0899	0.006421

[a]M.wt. = 129.39; m.p. = -86.5°C; b.p. = 68.11°C; d_4^{20} = 1.9344 g cc^{-1}.

CBrCl$_3$[a]

Methane, bromotrichloro-   (R 10B1)

Temperature, t	Solubility of H_2O in CBrCl$_3$ at saturated pressure	
(°C)	Wt %, W_2	Mole fraction, x_2
25	0.0059	0.000649

[a]M.wt. = 198.28; m.p. = -5.65°C; b.p. = 104.7°C; d_4^{20} = 2.0122 g cc^{-1}.

$CBrF_3$ [a]
Methane, bromotrifluoro-   (R 13B1)

| Temperature, t | Solubility of H_2O in $CBrF_3$ at saturated pressure | |
(°C)	Wt %, W_2	Mole fraction, x_2
21	0.0095	0.000785

[a]M.wt. = 148.92; m.p. = -168.0°C; b.p. = -59.0°C; d_4^{25} = 1.538 g cc^{-1}.

CBr_2H_2 [a]
Methane, dibromo-   (R 30B2)

| Temperature, t | Solubility of H_2O in CBr_2H_2 at saturated pressure | |
(°C)	Wt %, W_2	Mole fraction, x_2
25	0.02803	0.002698

[a]M.wt. = 173.85; m.p. = -52.55°C; b.p. = 97.0°C; d_4^{20} = 2.4970 g cc^{-1}.

CBr_3F [a]
Methane, tribromofluoro-   (R 11B3)

| Temperature, t | Solubility of H_2O in CBr_3F at saturated pressure | |
(°C)	Wt %, W_2	Mole fraction, x_2
25	0.007234	0.001086

[a]M.wt. = 270.74; m.p. = -74.5°C; b.p. = 106.0°C; d_4^{20} = 2.7648 g cc^{-1}.

$CClF_2H^a$

Methane, chlorodifluoro-   (R 22)

Temperature, t	Solubility of H_2O in $CClF_2H$ at saturated pressure	
(°C)	Wt %, W_2^b	Mole fraction, x_2
-50	0.007381	0.0003542
-40	0.01180	0.0005663
-30	0.01847	0.0008857
-20	0.02812	0.001348
-10	0.04151	0.001989
0	0.05939	0.002844
10	0.08252	0.003948
20	0.1116	0.005335
25	0.1287	0.006146
30	0.1475	0.007040
40	0.1908	0.009094

[a] M.wt. = 86.47; m.p. = -146.5°C; b.p. = -40.8°C; d^{-69} = 1.4906 g cc^{-1}.

[b] W_2 = 5.93923 x 10^{-2} + 2.03775 x $10^{-3}t$ + 2.62039 x $10^{-5}t^2$ + 1.25065 x $10^{-7}t^3$;
standard error of estimate for W_2 = 0.00061735.

$CClF_3^a$

Methane, chlorotrifluoro-   (R 13)

Temperature, t	Solubility of H_2O in $CClF_3$ at saturated pressure	
(°C)	Wt %, W_2	Mole fraction, x_2
-40	0.000135	0.00000783
-30	0.000229	0.00001329
-20	0.000546	0.00003168
-10	0.001055	0.00006119
0	0.001990	0.0001154
10	0.003750	0.0002174
20	0.006793	0.0003937
25	0.008930	0.0005176
30	0.01154	0.0006686
40	0.01826	0.001058
50	0.02699	0.001563

[a] M.wt. = 104.46; m.p. = -181.0°C; b.p. = -81.1°C; d^{-30} = 1.298 g cc^{-1}.

[b] W_2 = 1.99047 x 10^{-3} + 1.27238 x $10^{-4}t$ + 4.09520 x $10^{-6}t^2$ + 7.56958 x $10^{-8}t^3$ +
2.55793 x $10^{-10}t^4$ - 8.51039 x $10^{-12}t^5$; standard error of estimate for W_2 = 0.000147.

CClH$_3$[a]

Methane, chloro-   (R 40)

Temperature, t	Solubility of H$_2$O in CClH$_3$ at saturated pressure	
(°C)	Wt %, W$_2$[b]	Mole fraction, x$_2$
-40	0.03041	0.0008519
-30	0.04716	0.001321
-20	0.06702	0.001876
-10	0.09314	0.002606
0	0.1287	0.003598
10	0.1768	0.004938
20	0.2406	0.006713
25	0.2793	0.007789
30	0.3232	0.009006
40	0.4279	0.01190
50	0.5577	0.01547

[a]M.wt. = 50.49; m.p. = -97.73°C; b.p. = -24.2°C; d$_4^{20}$ = 0.9159 g cc^{-1}.

[b]W$_2$ = 0.128671 + 4.12859 x 10^{-3}t + 6.27975 x 10^{-5}t^2 + 5.24871 x 10^{-7}t^3; standard
error of estimate for W$_2$ = 0.005641.

CCl$_2$FH[a]

Methane, dichlorofluoro-   (R 21)

Temperature, t	Solubility of H$_2$O in CCl$_2$FH at saturated pressure	
(°C)	Wt %, W$_2$	Mole fraction, x$_2$
-50	0.005333	0.0003046
-40	0.009168	0.0005235
-30	0.01475	0.0008420
-20	0.02307	0.001316
-10	0.03546	0.002022
0	0.05347	0.003047
10	0.07869	0.004479
20	0.1127	0.006403
25	0.1334	0.007573
30	0.1568	0.008891
40	0.2120	0.01199

[a]M.wt. = 102.92; m.p. = -135.0°C; b.p. = 9.0°C; d^9 = 1.3724 g cc^{-1}.

[b]W$_2$ = 0.0534657 + 2.13446 x 10^{-3}t + 3.61152 x 10^{-5}t^2 + 2.68507 x 10^{-7}t^3 -
2.54527 x 10^{-10}t^4 - 1.10595 x 10^{-11}t^5; standard error of estimate for W$_2$ = 0.000984.

CCl_2F_2 [a]

Methane, dichlorodifluoro-   (R 12)

Temperature, t	Solubility of H_2O in CCl_2F_2 at saturated pressure	
(°C)	Wt %, W_2 [b]	Mole fraction, x_2
−50	0.000069	0.00000463
−40	0.000148	0.00000991
−30	0.000343	0.00002300
−20	0.000722	0.00004845
−10	0.001388	0.00009317
0	0.002504	0.0001681
10	0.004316	0.0002896
20	0.007177	0.0004815
25	0.009147	0.0006136
30	0.01157	0.0007763
40	0.01815	0.001217

[a] M.wt. = 120.91; m.p. = −158.0°C; b.p. = −29.8°C; d^{57} = 1.1834 g cc^{-1}.

[b] W_2 = 2.50441 x 10^{-3} + 1.41469 x 10^{-4}t + 3.43292 x 10^{-6}t^2 + 4.89761 x 10^{-8}t^3 + 4.49635 x 10^{-10}t^4 + 2.02411 x 10^{-12}t^5; standard error of estimate for W_2 = 0.000043034.

CCl_2H_2 [a]

Methane, dichloro-   (R 30)

Temperature, t	Solubility of H_2O in CCl_2H_2 at saturated pressure	
(°C)	Wt %, W_2 [b]	Mole fraction, x_2
-40	0.01896	0.0008933
-30	0.02505	0.001180
-20	0.03758	0.001769
-10	0.05695	0.002679
0	0.08354	0.003926
10	0.1177	0.005526
20	0.1599	0.007495
25	0.1842	0.008623
30	0.2105	0.009848
40	0.2699	0.01260
50	0.3384	0.01576

[a] M.wt. = 84.93; m.p. = -95.1°C; b.p. = 40.0°C; d_4^{20} = 1.3266 g cc^{-1}.

[b] W_2 = 8.35359 x 10^{-2} + 3.03307 x 10^{-3}t + 3.80578 x 10^{-5}t^2 + 6.47566 x 10^{-8}t^3;

standard error of estimate for W_2 = 0.010851.

CCl_3F^a

Methane, trichlorofluoro-   (R 11)

Temperature, t	Solubility of H_2O in CCl_3F at saturated pressure	
($^\circ$C)	Wt %, W_2^b	Mole fraction, x_2
-50	0.000215	0.00001641
-40	0.000446	0.00003401
-30	0.000783	0.00005969
-20	0.001313	0.0001001
-10	0.002174	0.0001657
0	0.003539	0.0002698
10	0.005609	0.0004276
20	0.008603	0.0006556
25	0.01052	0.0008013
30	0.01274	0.0009709
40	0.01825	0.001390

[a] M.wt. = 137.37; m.p. = -111.0°C; b.p. = 23.82°C; d_4^{25} = 1.467 g cc^{-1}.

[b] W_2 = 3.53856 x 10^{-3} + 1.68243 x 10^{-4}t + 3.52209 x 10^{-6}t^2 + 3.53778 x 10^{-8}t^3 + 6.78216 x 10^{-11}t^4 - 9.01921 x 10^{-13}t^5; standard error of estimate for W_2 = 0.000080894.

CCl_3H[a]

Methane, trichloro-   (R 20)

Temperature, t	Solubility of H_2O in CCl_3H at saturated pressure	
(°C)	Wt %, W_2[b]	Mole fraction, x_2
-40	0.007381	0.0004889
-30	0.01202	0.0007963
-20	0.01736	0.001149
-10	0.02460	0.001628
0	0.03496	0.002312
10	0.04965	0.003281
20	0.06990	0.004614
25	0.08248	0.005441
30	0.09691	0.006387
40	0.1319	0.008676
50	0.1761	0.01156
60	0.2307	0.01509
70	0.2970	0.01935
80	0.3760	0.02440
90	0.4692	0.03029

[a]M.wt. = 119.38; m.p. = -63.5°C; b.p. = 61.7°C; d_4^{20} = 1.4832 g cc^{-1}.

[b]W_2 = 3.49553 x 10^{-2} + 1.23247 x 10^{-3}t + 2.16812 x 10^{-5}t^2 + 2.02591 x 10^{-7}t^3;

standard error of estimate for W_2 = 0.0027898.

CCl$_4$[a]

Methane, tetrachloro-   (R 10)

Temperature, t ($^\circ$C)	Solubility of H$_2$O in CCl$_4$ at saturated pressure	
	Wt %, W$_2$	Mole fraction, x$_2$
-30	0.000983	0.00008391
-20	0.001670	0.0001426
-10	0.002817	0.0002404
0	0.004510	0.0003850
10	0.006837	0.0005835
20	0.009884	0.0008433
25	0.01170	0.0009985
30	0.01374	0.001172
40	0.01848	0.001576
50	0.02421	0.002063
60	0.03100	0.002641
70	0.03895	0.003316

[a]M.wt. = 153.82; m.p. = -22.99°C; b.p. = 76.54°C; d$_4^{20}$ = 1.5940 g cc^{-1}.

[b]W$_2$ = 4.51021 x 10^{-3} + 1.99578 x 10^{-4}t + 3.16636 x 10^{-6}t^2 + 1.44357 x 10^{-8}t^3; standard error of estimate for W$_2$ = 0.00071324.

CH$_2$I$_2$[a]

Methane, diiodo-

Temperature, t ($^\circ$C)	Solubility of H$_2$O in CH$_2$I$_2$ at saturated pressure	
	Wt %, W$_2$	Mole fraction, x$_2$
25	0.0217	0.003217

[a]M.wt. = 267.84; m.p. = 6.1°C; b.p. = 182.0°C; d$_4^{20}$ = 3.3254 g cc^{-1}.

CBrClH-CF$_3$[a]
Ethane, 1-bromo-1-chloro-2,2,2-trifluoro-   (R 123B1)

Temperature, t	Solubility of H$_2$O in C$_2$BrClF$_3$H at saturated pressure	
(°C)	Wt %, W$_2$	Mole fraction, x$_2$
25	0.035	0.003822

[a]M.wt. = 197.39; m.p. = -118.27°C; b.p. = 50.30°C; d_4^{20} = 1.776 g cc^{-1}.

CBrH$_2$-CH$_3$[a]
Ethane, bromo-

Temperature, t	Solubility of H$_2$O in C$_2$H$_5$Br at saturated pressure	
(°C)	Wt %, W$_2$	Mole fraction, x$_2$
25	0.07507	0.004524

[a]M.wt. = 108.97; m.p. = -118.6°C; b.p. = 38.4°C; d_4^{20} = 1.4604 g cc^{-1}.

CBrClH-CBrH$_2$[a]
Ethane, 1,2-dibromo-1-chloro-

Temperature, t	Solubility of H$_2$O in C$_2$Br$_2$ClH$_3$ at saturated pressure	
(°C)	Wt %, W$_2$	Mole fraction, x$_2$
25	0.05996	0.007349

[a]M.wt. = 222.31; b.p. = 163.0°C; d_4^{19} = 2.248 g cc^{-1}.

CBrClH-CBrClH[a]
Ethane, 1,2-dibromo-1,2-dichloro-

Temperature, t	Solubility of H$_2$O in C$_2$Br$_2$Cl$_2$H$_2$ at saturated pressure	
(°C)	Wt %, W$_2$	Mole fraction, x$_2$
25	0.06995	0.009878

[a]M.wt. = 256.76; m.p. = -26.0°C; b.p. = 195.0°C; d_4^{20} = 2.135 g cc^{-1}.

$CBrH_2-CBrH_2$[a]
Ethane, 1,2-dibromo-

Temperature, t	Solubility of H_2O in $C_2H_4Br_2$ at saturated pressure	
(°C)	Wt %, W_2[b]	Mole fraction, x_2
20	0.06200	0.006428
25	0.06535	0.006773
30	0.07172	0.007429
40	0.09153	0.009463
50	0.1175	0.01211
60	0.1455	0.01497
70	0.1718	0.01763
80	0.1923	0.01970

[a]M.wt. = 187.87; m.p. = 9.79°C; b.p. = 131.36°C; d_4^{20} = 2.1792 g cc^{-1}.

[b]W_2 = 8.86957 x 10^{-2} - 3.26907 x 10^{-3}t + 1.09939 x 10^{-4}t^2 - 6.61051 x 10^{-7}t^3;
standard error of estimate for W_2 = 0.0010586.

$CBr_2H-CBrH_2$[a]
Ethane, 1,1,2-tribromo-

Temperature, t	Solubility of H_2O in $C_2H_3Br_3$ at saturated pressure	
(°C)	Wt %, W_2	Mole fraction, x_2
25	0.05	0.00735

[a]M.wt. = 266.77; m.p. = 129.3°C; b.p. = 188.93°C; d_4^{20} = 2.6211 g cc^{-1}.

CBr_2H-CBr_2H[a]
Ethane, 1,1,2,2-tetrabromo-

Temperature, t	Solubility of H_2O in $C_2Br_4H_2$ at saturated pressure	
(°C)	Wt %, W_2	Mole fraction, x_2
25	0.03998	0.007616

[a]M.wt. = 345.67; m.p. = 0.0°C; b.p. = 243.5°C; d_4^{20} = 2.9656 g cc^{-1}.

$CClH_2-CF_2H$[a]

Ethane, 1-chloro-2,2-difluoro-   (R 142)

Temperature, t	Solubility of H_2O in $C_2ClF_2H_3$ at saturated pressure	
(°C)	Wt %, W_2[b]	Mole fraction, x_2
-40	0.007024	0.0003917
-30	0.01549	0.0008633
-20	0.03057	0.001703
-10	0.05184	0.002885
0	0.07886	0.004383
10	0.1112	0.006171
20	0.1484	0.008222
25	0.1687	0.009337
30	0.1900	0.01051
40	0.2356	0.01301
50	0.2848	0.01568
60	0.3371	0.01852
70	0.3921	0.02149
80	0.4494	0.02456

[a]M.wt. = 100.50; b.p. = 35.1°C; d^{15} = 1.312 g cc^{-1}.

[b]W_2 = 7.88571 x 10^{-2} + 2.97441 x $10^{-3}t$ + 2.65477 x $10^{-5}t^2$ - 7.29174 x $10^{-8}t^3$;
standard error of estimate for W_2 = 0.016428.

$CClF_2-CH_3$[a]

Ethane, 1-chloro-1,1-difluoro-   (R 142b)

Temperature, t	Solubility of H_2O in $C_2ClF_2H_3$ at saturated pressure	
(°C)	Wt %, W_2	Mole fraction, x_2
21	0.051	0.00284

[a]M.wt. = 100.50; m.p. = -130.8°C; b.p. = -9.5°C; d^{30} = 1.096 g cc^{-1}.

CClH=CH$_2$[a]

Ethene, chloro-   (R 1140)

| Temperature, t | Solubility of H_2O in C_2H_3Cl at saturated pressure | |
(°C)	Wt %, W_2[b]	Mole fraction, x_2
-30	0.009725	0.0003373
-20	0.01899	0.0006585
-10	0.02947	0.001022
0	0.04273	0.001481
10	0.06033	0.002090
20	0.08383	0.002902
25	0.09829	0.003402
30	0.1148	0.003972
40	0.1548	0.008351
50	0.2054	0.007092
60	0.2682	0.009245
70	0.3448	0.01186
80	0.4366	0.01498

[a]M.wt. = 62.5; m.p. = -153.8°C; b.p. = -13.37°C; d_4^{20} = 0.9106 g cc^{-1}.

[b]W_2 = 0.042726 + 1.51675 x 10^{-3}t + 2.17158 x 10^{-5}t^2 + 2.60846 x 10^{-7}t^3; standard error of estimate for W_2 = 0.0011988.

CClH$_2$-CH$_3$[a]

Ethane, chloro-   (R 160)

| Temperature, t | Solubility of H_2O in C_2H_5Cl at saturated pressure | |
(°C)	Wt %, W_2[b]	Mole fraction, x_2
0	0.06995	0.002501
10	0.1012	0.003616
20	0.1400	0.004996
25	0.1640	0.005849
30	0.1920	0.006844
40	0.2631	0.009360
50	0.3590	0.01274

[a]M.wt. = 64.52; m.p. = -136.4°C; b.p. = 12.27°C; d_4^{20} = 0.8978 g cc^{-1}.

[b]W_2 = 0.06995 + 2.94548 x 10^{-3}t + 8.61107 x 10^{-6}t^2 + 9.61985 x 10^{-7}t^3; standard error of estimate for W_2 = 0.

$CCl_2H\text{-}CF_3$ [a]

Ethane, 1,1-dichloro-2,2,2-trifluoro-   (R 123)

Temperature, t	Solubility of H_2O in $C_2Cl_2F_3H$ at saturated pressure	
(°C)	Wt %, W_2 [b]	Mole fraction, x_2
-30	0.008486	0.0007199
-20	0.01252	0.001061
-10	0.01799	0.001525
0	0.02519	0.002134
10	0.03439	0.002912
20	0.04586	0.003880
25	0.05254	0.004442
30	0.05989	0.005061
40	0.07674	0.006477
50	0.09669	0.008149

[a] M.wt. = 152.93; m.p. = -107.0°C; b.p. = 28.2°C; d^{15} = 1.475 g cc^{-1}.

[b] W_2 = 2.51909 x 10^{-2} + 8.15259 x $10^{-4}t$ + 9.99445 x $10^{-6}t^2$ + 4.59968 x $10^{-8}t^3$; standard error of estimate for W_2 = 0.00057433.

CClF$_2$-CClF$_2$[a]
Ethane, 1,2-dichloro-1,1,2,2-tetrafluoro-   (R 114)

| Temperature, t | Solubility of H$_2$O in C$_2$Cl$_2$F$_4$ at saturated pressure | |
(°C)	Wt %, W$_2$[b]	Mole fraction, x$_2$
-50	0.000109	0.00001033
-40	0.000237	0.00002250
-30	0.000450	0.00004269
-20	0.000829	0.00007864
-10	0.001498	0.0001421
0	0.002621	0.0002486
10	0.004398	0.0004171
20	0.007064	0.0006698
25	0.008812	0.0008354
30	0.01088	0.001032
40	0.01615	0.001530

[a]M.wt. = 170.92; m.p. = -93.90°C; b.p. = 3.8°C; d^{25} = 1.456 g cc^{-1}.

[b]W_2 = 2.62082 x 10^{-3} + 1.41368 x 10^{-4}t + 3.25780 x 10^{-6}t^2 + 3.63774 x 10^{-8}t^3 + 1.41407 x 10^{-10}t^4 - 2.41060 x 10^{-13}t^5;   standard error of estimate for W_2 = 0.000041493.

CCl$_2$F-CF$_3$[a]
Ethane, 1,1-dichloro-1,2,2,2-tetrafluoro-   (R 114a)

| Temperature, t | Solubility of H$_2$O in C$_2$Cl$_2$F$_4$ at saturated pressure | |
(°C)	Wt %, W$_2$	Mole fraction, x$_2$
21	0.0060	0.0000063

[a]M.wt. = 170.92; m.p. = -94.0°C; b.p. = 3.6°C; d_4^{25} = 1.455 g cc^{-1}.

$CCl_2=CH_2$[a]

Ethene, 1,1-dichloro-   (R 1130a)

Temperature, t	Solubility of H_2O in $C_2Cl_2H_2$ at saturated pressure	
(°C)	Wt %, W_2	Mole fraction, x_2
20	0.033	0.00177
25	0.035	0.00188

[a]M.wt. = 96.94; m.p. = -122.1°C; b.p. = 37.0°C; d_4^{20} = 1.218 g cc^{-1}.

$CClH=CClH$[a]

Ethene, cis-1,2-dichloro-   (R 1130 cis)

Temperature, t	Solubility of H_2O in $C_2Cl_2H_2$ at saturated pressure	
(°C)	Wt %, W_2[b]	Mole fraction, x_2
-40	0.02972	0.001597
-30	0.07178	0.003850
-20	0.08203	0.004398
-10	0.1142	0.006116
0	0.1877	0.01002
10	0.2995	0.01591
20	0.4365	0.02304
25	0.5105	0.02687
30	0.5875	0.03082
40	0.7555	0.03935
50	0.9696	0.05005
60	1.2975	0.06606

[a]M.wt. = 96.94; m.p. = -80.5°C; b.p. = 60.3°C; d_4^{20} = 1.2837 g cc^{-1}.

[b]W_2 = 0.187705 + 9.43795 x 10^{-3}t + 1.9580 x $10^{-4}t^2$ - 1.84619 x $10^{-6}t^3$ -
4.23377 x $10^{-8}t^4$ + 1.01092 x $10^{-9}t^5$; standard error of estimate for
W_2 = 0.052499.

CClH=CClH[a]

Ethene, trans-1,2-dichloro-   (R 1130 trans)

| Temperature, t | Solubility of H_2O in $C_2Cl_2H_2$ at saturated pressure | |
(°C)	Wt %, W_2[b]	Mole fraction, x_2
-40	0.02972	0.001597
-30	0.07171	0.003847
-20	0.08206	0.004400
-10	0.1143	0.006118
0	0.1876	0.01001
10	0.2992	0.01589
20	0.4360	0.02302
25	0.5100	0.02684
30	0.5870	0.03079
40	0.7552	0.03933
50	0.9696	0.05005
60	1.2975	0.06606

[a]M.wt. = 96.94; m.p. = -50.0°C; b.p. = 47.5°C; d_4^{20} = 1.2565 g cc^{-1}.

[b]W_2 = 0.187602 + 9.41860 x 10^{-3}t + 1.95394 x 10^{-4}t^2 - 1.82773 x 10^{-6}t^3 - 4.21059 x 10^{-8}t^4 + 1.00542 x 10^{-9}t^5; standard error of estimate for W_2 = 0.053240.

CCl$_2$H-CH$_3$[a]

Ethane, 1,1-dichloro-   (R 150a)

| Temperature, t | Solubility of H_2O in $C_2Cl_2H_4$ at saturated pressure | |
(°C)	Wt %, W_2[b]	Mole fraction, x_2
0	0.04620	0.002532
10	0.06200	0.003396
20	0.08250	0.004515
25	0.09670	0.005289
30	0.1147	0.006268

[a]M.wt. = 98.96; m.p. = -96.98°C; b.p. = 57.28°C; d_4^{20} = 1.1757 g cc^{-1}.

[b]W_2 = 0.0462 + 1.57834 x 10^{-3}t - 1.15008 x 10^{-5}t^2 + 1.16668 x 10^{-6}t^3; standard error of estimate for W_2 = 0.

$CClH_2-CClH_2$ [a]

Ethane, 1,2-dichloro- (R 150)

Temperature, t	Solubility of H_2O in $C_2Cl_2H_4$ at saturated pressure	
(°C)	Wt %, W_2 [b]	Mole fraction, x_2
-20	0.05035	0.002759
-10	0.07162	0.003921
0	0.09680	0.005294
10	0.1285	0.007017
20	0.1693	0.009227
25	0.1939	0.01056
30	0.2217	0.01206
40	0.2885	0.01565
50	0.3722	0.02011
60	0.4755	0.02557
70	0.6008	0.03213
80	0.7508	0.03990

[a] M.wt. = 98.96; m.p. = -35.36°C; b.p. = 83.47°C; d_4^{20} = 1.2351 g cc^{-1}.

[b] W_2 = 9.67945 x 10^{-2} + 2.79932 x $10^{-3}t$ + 3.25215 x $10^{-5}t^2$ + 4.33495 x $10^{-7}t^3$;
standard error of estimate for W_2 = 0.0028071.

$CCl_2F-CClF_2$ [a]

Ethane, 1,1,2-trichloro-1,2,2-trifluoro-   (R 113)

Temperature, t	Solubility of H_2O in $C_2Cl_3F_3$ at saturated pressure	
($^\circ$C)	Wt %, W_2 [b]	Mole fraction, x_2
-50	0.000168	0.00001748
-40	0.000439	0.00004568
-30	0.000798	0.00008296
-20	0.001327	0.0001381
-10	0.002174	0.0002261
0	0.003527	0.0003668
10	0.005599	0.0005820
20	0.008608	0.0008946
25	0.01053	0.001094
30	0.01276	0.001325
40	0.01822	0.001892

[a] M.wt. = 187.38; m.p. = -36.4°C; b.p. = 47.7°C; d_4^{20} = 1.5635 g cc^{-1}.

[b] W_2 = 3.52727 x 10^{-3} + 1.67576 x 10^{-4}t + 3.59204 x 10^{-6}t^2 + 3.66941 x 10^{-8}t^3 + 2.13630 x 10^{-11}t^4 - 1.57682 x 10^{-12}t^5; standard error of estimate for W_2 = 0.000091134.

$CCl_2=CClH^a$
Ethene, trichloro-   (R 1120)

Temperature, t	Solubility of H_2O in C_2Cl_3H at saturated pressure	
(°C)	Wt %, W_2	Mole fraction, x_2
-50	0.00010	0.0000729
-40	0.0014	0.000102
-30	0.0025	0.000182
-20	0.0040	0.000292
-10	0.0062	0.000452
0	0.0097	0.000707
10	0.0162	0.001180
20	0.0253	0.001842
25	0.03199	0.002328
30	0.0379	0.002758
40	0.0510	0.003708
50	0.0655	0.004757
60	0.0805	0.005841
70	0.1024	0.007420
80	0.1250	0.009045

[a]M.wt. = 131.39; m.p. = -73.0°C; b.p. = 87.0°C; d_4^{20} = 1.4642 g cc^{-1}.

$CCl_3-CH_3{}^a$
Ethane, 1,1,1-trichloro-   (R 140a)

Temperature, t	Solubility of H_2O in $C_2Cl_3H_3$ at saturated pressure	
(°C)	Wt %, $W_2{}^b$	Mole fraction, x_2
0	0.0161	0.001191
10	0.0212	0.001568
20	0.0289	0.002137
25	0.0347	0.002564
30	0.0423	0.003124

[a]M.wt. = 133.41; m.p. = -30.41°C; b.p. = 74.1°C; d_4^{20} = 1.3390 g cc^{-1}.

[b]W_2 = 0.0161 + 4.82336 x 10^{-4}t - 2.36690 x $10^{-6}t^2$ + 5.13338 x $10^{-7}t^3$; standard error of estimate for W_2 = 0.

$CCl_2H-CClH_2$ [a]

Ethane, 1,1,2-trichloro-   (R 140)

Temperature, t	Solubility of H_2O in $C_2Cl_3H_3$ at saturated pressure	
(°C)	Wt %, W_2 [b]	Mole fraction, x_2
-10	0.0180	0.00133
0	0.0624	0.00460
10	0.0749	0.00552
20	0.0947	0.00697
25	0.1196	0.00879
30	0.1610	0.01180

[a]M.wt. = 133.41; m.p. = -36.5°C; b.p. = 113.77°C; d_4^{20} = 1.4397 g cc^{-1}.

[b]W_2 = 0.06236 + 2.19112 x 10^{-3}t - 1.59178 x 10^{-4}t^2 + 6.52467 x 10^{-6}t^3; standard error of estimate for W_2 = 0.

$CCl_2=CCl_2$ [a]

Ethene, tetrachloro-   (R 1110)

Temperature, t	Solubility of H_2O in C_2Cl_4 at saturated pressure	
(°C)	Wt %, W_2 [b]	Mole fraction, x_2
-30	0.001090	0.0001003
-20	0.001258	0.0001158
-10	0.002071	0.0001906
0	0.003471	0.0003194
10	0.005446	0.0005011
20	0.008025	0.0007382
25	0.009564	0.0008797
30	0.01129	0.001038
40	0.01538	0.001414
50	0.02049	0.001883
60	0.02690	0.002471
70	0.03493	0.003206
80	0.04501	0.004128

[a]M.wt. = 165.83; m.p. = -19.0°C; b.p. = 121.0°C; d_4^{20} = 1.6227 g cc^{-1}.

[b]W_2 = 3.47135 x 10^{-3} + 1.68636 x 10^{-4}t + 2.84888 x 10^{-6}t^2 + 1.12482 x 10^{-9}t^3 + 1.90515 x 10^{-10}t^4 + 4.39421 x 10^{-13}t^5; standard error of estimate for W_2 = 0.000852.

CCl_2F-CCl_2F[a]

Ethane, 1,1,2,2-tetrachloro-1,2-difluoro-   (R 112)

Temperature, t	Solubility of H_2O in $C_2Cl_4F_2$ at saturated pressure	
(°C)	Wt %, W_2	Mole fraction, x_2
28	0.0099	0.00112

[a]M.wt. = 203.83; m.p. = 25.0°C; b.p. = 93.0°C; d_4^{25} = 1.6447 g cc^{-1}.

CCl_3-CClH_2[a]

Ethane, 1,1,1,2-tetrachloro-   (R 130a)

Temperature, t	Solubility of H_2O in $C_2Cl_4H_2$ at saturated pressure	
(°C)	Wt %, W_2[b]	Mole fraction, x_2
0	0.02301	0.002140
10	0.03660	0.003400
20	0.04931	0.004575
25	0.05531	0.005130
30	0.06106	0.005660

[a]M.wt. = 167.85; m.p. = -70.2°C; b.p. = 130.5°C; d_4^{20} = 1.5406 g cc^{-1}.

[b]W_2 = 0.02301 + 1.40034 x 10^{-3}t - 4.00059 x $10^{-6}t^2$ - 1.33212 x $10^{-8}t^3$; standard error of estimate for W_2 = 0.

CCl_2H-CCl_2H[a]

Ethane, 1,1,2,2-tetrachloro-   (R 130)

Temperature, t	Solubility of H_2O in $C_2Cl_4H_2$ at saturated pressure	
(°C)	Wt %, W_2[b]	Mole fraction, x_2
0	0.06182	0.005730
10	0.07630	0.007064
20	0.09468	0.008752
25	0.1105	0.01020
30	0.1335	0.01230

[a]M.wt. = 167.85; m.p. = -36.0°C; b.p. = 146.2°C; d_4^{20} = 1.5953 g cc^{-1}.

[b]W_2 = 0.06182 + 1.80454 x 10^{-3}t - 6.32274 x $10^{-5}t^2$ + 2.75735 x $10^{-6}t^3$; standard error of estimate for W_2 = 0.

$CCl_3-CCl_2H^a$

Ethane, pentachloro-   (R 120)

Temperature, t	Solubility of H_2O in C_2Cl_5H at saturated pressure	
(°C)	Wt %, $W_2{}^b$	Mole fraction, x_2
0	0.01623	0.001820
10	0.0220	0.002465
20	0.02966	0.003321
25	0.03494	0.003910
30	0.04158	0.004649

[a] M.wt. = 202.30; m.p. = -29.0°C; b.p. = 162.0°C; d_4^{20} = 1.6796 g cc^{-1}.

[b] W_2 = 0.01623 + 5.61404 x 10^{-4}t - 2.38698 x 10^{-6}t^2 + 3.94673 x 10^{-7}t^3; standard error of estimate for W_2 = 0.

$CBrH_2-CH_2-CClH_2{}^a$

Propane, 1-bromo-3-chloro-

Temperature, t	Solubility of H_2O in C_3BrClH_6 at saturated pressure	
(°C)	Wt %, W_2	Mole fraction, x_2
25	0.030	0.002616

[a] M.wt. = 157.44; m.p. = -58.87°C; b.p. = 143.36°C; d_4^{20} = 1.5969 g cc^{-1}.

$CBrF_2-CBrF-CF_3$ [a]

Propane, 1,2-dibromo-1,1,2,3,3,3-hexafluoro-   (R 216B2)

| Temperature, t | Solubility of H_2O in $C_3Br_2F_6$ at saturated pressure | |
(°C)	Wt %, W_2 [b]	Mole fraction, x_2
-30	0.001559	0.0002680
-20	0.002118	0.0003642
-10	0.002893	0.0004974
0	0.003904	0.0006711
10	0.005168	0.0008882
20	0.006705	0.001152
25	0.007582	0.001302
30	0.008533	0.001466
40	0.01067	0.001832
50	0.01314	0.002255

[a] M.wt. = 309.87; m.p. = -94.0°C; b.p. = 72.6°C; d_4^{25} = 2.157 g cc^{-1}.

[b] W_2 = 3.90387 x 10^{-3} + 1.13428 x 10^{-4}t + 1.26894 x $10^{-6}t^2$ + 3.12680 x $10^{-9}t^3$;

standard error of estimate for W_2 = 0.000040712.

$CBrH_2-CBrH-CH_3$ [a]

Propane, 1,2-dibromo-

| Temperature, t | Solubility of H_2O in $C_3Br_2H_6$ at saturated pressure | |
(°C)	Wt %, W_2	Mole fraction, x_2
25	0.05197	0.005794

[a] M.wt. = 201.90; m.p. = -55.25°C; b.p. = 140.0°C; d_4^{20} = 1.9324 g cc^{-1}.

$CClH_2-CH=CH_2$ [a]

1-Propene, 3-chloro-

| Temperature, t | Solubility of H_2O in C_3ClH_5 at saturated pressure | |
(°C)	Wt %, W_2	Mole fraction, x_2
20	0.08	0.003390

[a] M.wt. = 76.53; m.p. = -134.5°C; b.p. = 45.0°C; d_4^{20} = 0.9376 g cc^{-1}.

$CClF_2-CClF-CF_3$[a]
Propane, 1,2-dichloro-1,1,2,3,3,3-hexafluoro-   (R 216)

Temperature, t	Solubility of H_2O in $C_3Cl_2F_6$ at saturated pressure	
(°C)	Wt %, W_2[b]	Mole fraction, x_2
-30	0.002110	0.0002587
-20	0.003011	0.0003692
-10	0.004169	0.0005111
0	0.005599	0.0006862
10	0.007317	0.0008966
20	0.009339	0.001144
25	0.01047	0.001283
30	0.01168	0.001430
40	0.01436	0.001758
50	0.01739	0.002129

[a]M.wt. = 220.95; m.p. = -136.0°C; b.p. = 34.7°C; d_4^{25} = 1.577 g cc^{-1}.

[b]W_2 = 5.59881 x 10^{-3} + 1.57122 x 10^{-4}t + 1.44086 x $10^{-6}t^2$ + 2.68197 x $10^{-9}t^3$;
   standard error of estimate for W_2 = 0.000049368.

$CClH_2-CClH-CH_3$[a]
Propane, 1,2-dichloro-

Temperature, t	Solubility of H_2O in $C_3Cl_2H_6$ at saturated pressure	
(°C)	Wt %, W_2	Mole fraction, x_2
20	0.040	0.002503
25	0.060	0.003751

[a]M.wt. = 112.99; m.p. = -100.44°C; b.p. = 96.37°C; d_4^{20} = 1.1560 g cc^{-1}.

$CCl_3-CF_2-CF_3$ [a]

Propane, 1,1,1-trichloro-2,2,3,3,3-pentafluoro- (R 215)

Temperature, t	Solubility of H_2O in $C_3Cl_3F_5$ at saturated pressure	
(°C)	Wt %, W_2 [b]	Mole fraction, x_2
-30	0.001353	0.0001783
-20	0.001876	0.0002471
-10	0.002551	0.0003361
0	0.003401	0.0004480
10	0.004449	0.0005859
20	0.005715	0.0007526
25	0.006437	0.0008476
30	0.007223	0.0009510
40	0.008994	0.001184
50	0.01105	0.001454

[a] M.wt. = 237.40; m.p. = -80.0°C; b.p. = 74.39°C; d_4^{25} = 1.646 g cc^{-1}.

[b] W_2 = 3.40140 x 10^{-3} + 9.45026 x $10^{-5}t$ + 9.85045 x $10^{-7}t^2$ + 3.69293 x $10^{-9}t^3$;
standard error of estimate for W_2 = 0.00003894.

$CCl_3-CF_2-CClF_2$ [a]

Propane, 1,1,1,3-tetrachloro-2,2,3,3-tetrafluoro-   (R 214)

Temperature, t	Solubility of H_2O in $C_3Cl_4F_4$ at saturated pressure	
(°C)	Wt %, W_2 [b]	Mole fraction, x_2
-30	0.001160	0.0001635
-20	0.001657	0.0002335
-10	0.002273	0.0003202
0	0.003025	0.0004261
10	0.003934	0.0005540
20	0.005018	0.0007067
25	0.005632	0.0007931
30	0.006298	0.0008867
40	0.007792	0.001097
50	0.009520	0.001340

[a] M.wt. = 253.86; m.p. = -92.78°C; b.p. = 113.95°C; d_4^{25} = 1.695 g cc^{-1}.

[b] W_2 = 3.02484 x 10^{-3} + 8.27246 x 10^{-5}t + 7.82313 x $10^{-7}t^2$ + 3.22365 x $10^{-9}t^3$;

standard error of estimate for W_2 = 0.000041882.

CCl_3-CCl=CCl_2 [a]

1-Propene, hexachloro-

Temperature, t	Solubility of H_2O in C_3Cl_6 at saturated pressure	
(°C)	Wt %, W_2 [b]	Mole fraction, x_2
0	0.001756	0.0002307
10	0.003267	0.0004291
20	0.004140	0.0005437
25	0.004828	0.0006340
30	0.005844	0.0007674
40	0.009160	0.001202
50	0.01460	0.001916
60	0.02284	0.002993
70	0.03510	0.004593
80	0.05363	0.007002
90	0.08205	0.01068
100	0.1259	0.01629

[a] M.wt. = 236.74; m.p. = -72.9°C; b.p. = 141.0°C at 100 mm Hg; d_4^{25} = 1.7666 g cc^{-1}.

[b] W_2 = 1.75626 x 10^{-3} + 2.57445 x 10^{-4}t - 1.54300 x $10^{-5}t^2$ + 5.39060 x $10^{-7}t^3$ - 6.36269 x $10^{-9}t^4$ + 3.49866 x $10^{-11}t^5$; standard error of estimate for W_2 = 0.0002375.

$CClH_2$-CH_2-CH_2-CH_3 [a]

n-Butane, 1-chloro-

Temperature, t	Solubility of H_2O in C_4ClH_9 at saturated pressure	
(°C)	Wt %, W_2	Mole fraction, x_2
20	0.080	0.004097

[a] M.wt. = 92.57; m.p. = -123.1°C; b.p. = 78.44°C; d_4^{20} = 0.8862 g cc^{-1}.

CH_3-$CClH$-CH_2-CH_3 [a]

n-Butane, 2-chloro-

Temperature, t	Solubility of H_2O in C_4ClH_9 at saturated pressure	
(°C)	Wt %, W_2	Mole fraction, x_2
25	0.10	0.00512

[a] M.wt. = 92.57; m.p. = -131.3°C; b.p. = 68.25°C; d_4^{20} = 0.8732 g cc^{-1}.

$CCl_2=CCl-CCl=CCl_2$ [a]

1,3-Butadiene, hexachloro-

Temperature, t	Solubility of H_2O in C_4Cl_6 at saturated pressure	
(°C)	Wt %, W_2 [b]	Mole fraction, x_2
0	0.001803	0.0002609
10	0.003360	0.0004861
20	0.004528	0.0006550
25	0.005453	0.0007887
30	0.006757	0.0009772
40	0.01079	0.001559
50	0.01717	0.002479
60	0.02682	0.003868
70	0.04154	0.005979
80	0.06453	0.009260
90	0.1010	0.01442
100	0.1585	0.02247

[a] M.wt. = 260.76; m.p. = -21.0°C; b.p. = 215.0°C; d_4^{20} = 1.6820 g cc^{-1}.

[b] W_2 = 1.80320 x 10^{-3} + 2.51961 x 10^{-4}t - 1.46873 x 10^{-5}t^2 + 5.75564 x 10^{-7}t^3 - 7.41537 x 10^{-9}t^4 + 4.44390 x 10^{-11}t^5; standard error of estimate for W_2 = 0.00029499.

$$CF_2-CF_2$$
$$CF_2-CF_2$$

Cyclobutane, octafluoro- (R C318)

Temperature, t	Solubility of H_2O in C_4F_8 at saturated pressure	
(°C)	Wt %, W_2[b]	Mole fraction, x_2
-40	0.006028	0.0006689
-30	0.008263	0.0009167
-20	0.01454	0.001612
-10	0.02444	0.002707
0	0.03756	0.004155
10	0.05348	0.005907
20	0.07179	0.007914
25	0.08172	0.008999
30	0.09208	0.01013
40	0.1139	0.01251
50	0.1369	0.01500
60	0.1607	0.01755

[a] M.wt. = 200.03; m.p. = -38.7°C; b.p. = -4.0°C; d_4^0 = 1.724 g cc^{-1}.

[b] W_2 = 3.75635 x 10^{-2} + 1.45890 x 10^{-3}t + 1.40079 x 10^{-5}t^2 - 6.88660 x 10^{-8}t^3; standard error of estimate for W_2 = 0.0023105.

$$CClH_2-CH_2-CH_2-CH_2-CH_3$$[a]
n-Pentane, 1-chloro-

Temperature, t	Solubility of H_2O in C_5ClH_{11} at saturated pressure	
(°C)	Wt %, W_2	Mole fraction, x_2
25	0.2697	0.01575

[a] M.wt. = 106.60; m.p. = -99.0°C; b.p. = 107.8°C; d_4^{20} = 0.8818 g cc^{-1}.

CCl≡C-CCl=CCl-CCl$_3$ [a]

2-Butene-4-yne, hexachloro-

Temperature, t	Solubility of H_2O in C_5Cl_6 at saturated pressure	
(°C)	Wt %, W_2 [b]	Mole fraction, x_2
15	0.000354	0.00005362
20	0.000691	0.0001046
25	0.001161	0.0001758
30	0.001978	0.0002994
40	0.005500	0.0008321

[a]M.wt. = 272.77.

[b]W_2 = $-1.97853 \times 10^{-3} + 3.06663 \times 10^{-4}t - 1.43272 \times 10^{-5}t^2 + 2.83367 \times 10^{-7}t^3$; standard error of estimate for W_2 = 0.

CCl$_2$=CCl-CCl$_2$-CCl=CCl$_2$ [a]

1,4-Pentadiene, octachloro-

Temperature, t	Solubility of H_2O in C_5Cl_8 at saturated pressure	
(°C)	Wt %, W_2 [b]	Mole fraction, x_2
15	0.000017	0.00000333
20	0.000020	0.00000391
25	0.000033	0.00000642
30	0.000053	0.00001024
40	0.000100	0.00001925

[a]M.wt. = 343.716; b.p. = 72.4°C at 0.15 mm Hg; d_4^{25} = 1.7495 g cc^{-1}.

[b]W_2 = $1.01716 \times 10^{-4} - 1.16202 \times 10^{-5}t + 4.65482 \times 10^{-7}t^2 - 4.40123 \times 10^{-9}t^3$; standard error of estimate for W_2 = 0.

C_6BrH_5[a]

Benzene, bromo-

Temperature, t	Solubility of H_2O in C_6BrH_5 at saturated pressure	
(°C)	Wt %, W_2[b]	Mole fraction, x_2
25	0.02740	0.002383
30	0.03140	0.002730
40	0.03930	0.003415

[a]M.wt. = 157.02; m.p. = -30.82°C; b.p. = 156.0°C; d_4^{20} = 1.4950 g cc^{-1}.

[b]W_2 = 1.09002 x 10^{-2} + 4.43314 x 10^{-4}t + 1.20007 x 10^{-5}t^2 - 1.33342 x 10^{-7}t^3;
standard error of estimate for W_2 = 0.

C_6ClH_5[a]

Benzene, chloro-

Temperature, t	Solubility of H_2O in C_6ClH_5 at saturated pressure	
(°C)	Wt %, W_2	Mole fraction, x_2
-10	0.007423	0.0004636
0	0.01145	0.0007151
10	0.01933	0.001207
20	0.03182	0.001985
25	0.04002	0.002495
30	0.04966	0.003094
40	0.07359	0.004580
50	0.1044	0.006486
60	0.1428	0.008854
70	0.1895	0.01172
80	0.2453	0.01513

[a]M.wt. = 112.56; m.p. = -45.6°C; b.p. = 132.0°C; d_4^{20} = 1.1058 g cc^{-1}.

[b]W_2 = 1.14517 x 10^{-2} + 5.83054 x 10^{-4}t + 1.92658 x 10^{-5}t^2 + 1.24888 x 10^{-7}t^3;
standard error of estimate for W_2 = 0.00098413.

$C_6Cl_2H_4$ [a]
Benzene, 1,2-dichloro-

Temperature, t	Solubility of H_2O in $C_6Cl_2H_4$ at saturated pressure	
(°C)	Wt %, W_2 [b]	Mole fraction, x_2
25	0.03340	0.002719
30	0.03870	0.003149
40	0.05230	0.004252
50	0.07390	0.005999

[a]M.wt. = 147.01; m.p. = -17.0°C; b.p. = 180.5°C; d_4^{20} = 1.3048 g cc^{-1}.

[b]W_2 = -2.09925 x 10^{-3} + 2.31993 x 10^{-3}t - 5.59979 x $10^{-5}t^2$ + 7.99979 x $10^{-7}t^3$;
standard error of estimate for W_2 = 0.

$C_6Cl_2H_4$ [a]
Benzene, 1,4-dichloro-

Temperature, t	Solubility of H_2O in $C_6Cl_2H_4$ at saturated pressure	
(°C)	Wt %, W_2 [b]	Mole fraction, x_2
60	0.1370	0.01107
70	0.1480	0.01195
80	0.1590	0.01283
90	0.1700	0.01371

[a]M.wt. = 147.01; m.p. = 53.1°C; b.p. = 174.0°C; d_4^{20} = 1.2475 g cc^{-1}.

[b]W_2 = 7.10038 x 10^{-2} + 1.09984 x 10^{-3}t + 2.27359 x $10^{-9}t^2$ - 1.04774 x $10^{-11}t^3$;
standard error of estimate for W_2 = 0.

$C_6Cl_3H_3$ [a]
Benzene, 1,2,4-trichloro-

Temperature, t	Solubility of H_2O in $C_6Cl_3H_3$ at saturated pressure	
(°C)	Wt %, W_2	Mole fraction, x_2
25	0.02023	0.002034

[a]M.wt. = 181.45; m.p. = 16.95°C; b.p. = 213.5°C; d_4^{20} = 1.4542 g cc^{-1}.

C_6FH_5[a]
Benzene, fluoro-

Temperature, t	Solubility of H_2O in C_6FH_5 at saturated pressure	
(°C)	Wt %, W_2	Mole fraction, x_2
25	0.0310	0.001652

[a]M.wt. = 96.11; m.p. = -41.2°C; b.p. = 85.1°C; d_4^{20} = 1.0225 g cc^{-1}.

$CF_3-CF_2-CF_2-CF_2-CF_2-CF_3$[a]
n-Hexane, perfluoro-

Temperature, t	Solubility of H_2O in C_6F_{14} at saturated pressure	
(°C)	Wt %, W_2[b]	Mole fraction, x_2
1	0.000630	0.000136
10	0.000893	0.000192
20	0.001297	0.000270
25	0.001554	0.000335
30	0.001857	0.000400
40	0.002626	0.000565
50	0.003665	0.000789

[a]M.wt. = 388.05; m.p. = -78.0°C; b.p. = 82.43°C; d_4^{20} = 1.6697 g cc^{-1}.

[b]$W_2 = 6.00497 \times 10^{-4} + 2.62116 \times 10^{-5}t + 2.44435 \times 10^{-7}t^2 + 9.16842 \times 10^{-9}t^3$;
standard error of estimate for W_2 = 0.0000034017.

C_6H_5I[a]

Benzene, iodo-

Temperature, t	Solubility of H_2O in C_6H_5I at saturated pressure	
($^\circ$C)	Wt %, W_2[b]	Mole fraction, x_2
25	0.02676	0.003022
30	0.03059	0.003453
40	0.03961	0.004467
50	0.05331	0.006004

[a]M.wt. = 204.01; m.p. = -31.27°C; b.p. = 188.3°C; d_4^{20} = 1.8308 g cc^{-1}.

[b]W_2 = $-2.78956 \times 10^{-3} + 1.95862 \times 10^{-3}t - 4.53984 \times 10^{-5}t^2 + 5.73315 \times 10^{-7}t^3$;
standard error of estimate for W_2 = 0.

Chapter 9

Expressions, Units, and Conversions
Of Solubility Data

There is no other physical property in the literature that has so many different
sorts of units for expressing its value as does solubility. As a consequence of the
wide variety of units, there is often confusion regarding the interpretation of published solubility data (Friend and Adler, 1957). Two problems are especially common:
one involves the unit "grams per liter" and the other occurs when the gas solubility
is published at 1 atm pressure. In the first case, there should be an indication
regarding grams per liter solvent or solution, and in the second case, whether the
1 atm refers to partial pressure of the solute gas or total pressure of the system.

Regarding the solubility of a solid in water, Heslop (1974) carried out an examination of Seidell's "Solubilities of Inorganic and Organic Compounds" (Seidell and
Linke, 1952), for guidance. He found six methods for expressing solubility in water:

1. g solute/100 g water
2. g solute/100 g saturated solution
3. g solute/1000 g saturated solution
4. g solute/1000 cc saturated solution
5. g-mole solute/1000 g water
6. millimol solute/100 cc saturated solution

According to his suggestion, the solubilities are most simply expressed as mass of
solute per mass of solvent in saturated solution at the specified temperature. Expressing solubility this way has many advantages, particularly for its conversion
into other units.

Particular attention should be paid to concentration when it is given in molality
or molarity. A molar solution contains 1 mole or g-molecular weight of the solute in
1 liter of solution. A molal solution contains 1 g-mole per 1000 g of solvent.

In general, the unit of solubility means the same thing as the concentration of
a saturated solution. The conversion formulas for solutions having concentrations expressed in various ways are given in, for example, the "Chemical Rubber Handbook"
(Weast, 1976), "Experimental Thermodynamics" (Kertes et al., 1975), and Horvath
(1976b). In addition to these publications, several conversion tables and formulas
appear in various books: Landolt-Börnstein (1962, 1964, 1976a, b), International
Critical Tables (Washburn, 1928, 1929), Lange (1967). Tables 9.1 and 9.2 give the

Table 9.1 Conversion Formulas Between Various Concentration Units[a]

From \ To	W_2	X_2	m_2	c_2	G_2
Weight percent (W_2)	1	$\dfrac{100X_2M_2}{X_2M_2 + (1-X_2)M_1}$	$\dfrac{100m_2M_2}{1000 + m_2M_2}$	$\dfrac{c_2M_2}{10d_{12}}$	$\dfrac{G_2}{10d_{12}}$
Mole fraction (X_2)	$\dfrac{\frac{W_2}{M_2}}{\frac{W_2}{M_2} + \frac{100 - W_2}{M_1}}$	1	$\dfrac{m_2M_1}{1000 + m_2M_1}$	$\dfrac{c_2M_1}{c_2(M_1-M_2)+1000d_{12}}$	$\dfrac{G_2M_1}{G_2(M_1-M_2)+1000M_2d_{12}}$
Molality (m_2)	$\dfrac{1000W_2}{M_2(100 - W_2)}$	$\dfrac{1000X_2}{M_1 - X_2M_1}$	1	$\dfrac{1000c_2}{1000d_{12} - (c_2M_2)}$	$\dfrac{1000G_2}{M_2(1000d_{12} - G_2)}$
Molarity (c_2)	$\dfrac{10d_{12}W_2}{M_2}$	$\dfrac{1000d_{12}X_2}{X_2M_2 + (1 - X_2)M_1}$	$\dfrac{1000d_{12}m_2}{1000 + m_2M_2}$	1	$\dfrac{G_2}{M_2}$
Grams per liter of solution (G_2)	$10W_2d_{12}$	$\dfrac{1000d_{12}X_2M_2}{X_2M_2 + (1 - X_2)M_1}$	$\dfrac{1000d_{12}m_2M_2}{m_2M_2 + 1000}$	c_2M_2	1

[a] M = molecular weight; d_{12} = density of solution (g cc^{-1}); subscript 1 = solvent; subscript 2 = solute.

Table 9.2 Conversion Formulas Between Various Concentration Units[a]

From \ To	X_2	w_2	θ_2	m_2	c_2
Mole fraction (X_2)	1	$\dfrac{X_2 M_2}{X_1 M_1 + X_2 M_2}$	$\dfrac{X_2 V^*_2}{X_1 V^*_1 + X_2 V^*_2}$	$\dfrac{1000 X_2}{X_1 M_1}$	$\dfrac{1000 X_2 d_{12}}{X_1 M_1 + X_2 M_2}$
Weight fraction (w_2)	$\dfrac{w_2 M_1}{w_1 M_2 + w_2 M_1}$	1	$\dfrac{w_2 V^*_2 M_1}{w_2 V^*_2 M_1 + w_1 V^*_1 M_2}$	$\dfrac{1000 w_2}{w_1 M_2}$	$\dfrac{1000 w_2 d_{12}}{M_2}$
Volume fraction (θ_2)	$\dfrac{\theta_2 V^*_1}{\theta_1 V^*_2 + \theta_2 V^*_1}$	$\dfrac{\theta_2 V^*_1 M_2}{\theta_2 V^*_1 M_2 + \theta_1 V^*_2 M_1}$	1	$\dfrac{1000 \theta_2 V^*_1}{\theta_1 V^*_2 M_1}$	$\dfrac{1000 \theta_2 V^*_1 d_{12}}{\theta_1 V^*_2 M_1 + \theta_2 V^*_1 M_2}$
Molality (m_2)	$\dfrac{m_2 M_1}{m_2 M_1 + 1000}$	$\dfrac{m_2 M_2}{m_2 M_2 + 1000}$	$\dfrac{m_2 V^*_2 M_1}{m_2 V^*_2 M_1 + 1000 V^*_1}$	1	$\dfrac{m_2 d_{12}}{1 + 0.001 M_2 m_2}$
Molarity (c_2)	$\dfrac{c_2 M_1}{c_2 M_1 - c_2 M_2 + 1000 d_{12}}$	$\dfrac{c_2 M_2}{1000 d_{12}}$	$\dfrac{c_2 V^*_2 M_1}{c_2 V^*_2 M_1 + (1000 d_{12} - c_2 M_2) V^*_1}$	$\dfrac{c_2}{d_{12} - 0.001 c_2 M_2}$	1

[a] M = molecular weight; d_{12} = density of the solution (g cc^{-1}); subscript 1 = solvent; subscript 2 = solute.

$$V^*_2 = \frac{1}{m_2}\left[\frac{1000 + m_2 M_2}{d_{12}} - \frac{1000}{d_2}\right]$$

various conversion formulas between concentration units.

The most common conversion needed in this book is probably the conversion from weight percentage to mole fraction, or vice versa. Therefore, for the reader who has no quick access to a packet calculator, a nomogram will be useful (see Fig. 9.1).

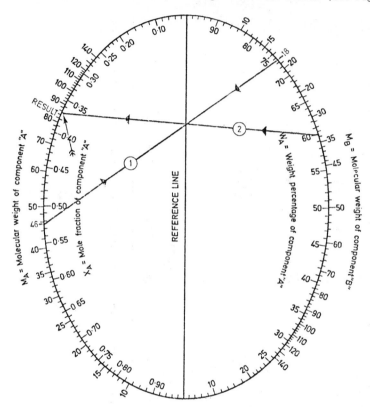

Fig. 9.1 Conversion between weight percentage and mole fraction. The example is shown for a binary mixture containing 60% ethyl alcohol (M_A = 46.07) and 40% water (M_B = 18.015). From the nomogram, the mole fraction of ethyl alcohol in the mixture is X_A = 0.367.

The difference between solubility and concentration units is that solubility always refer to the equilibrium condition between the solute and solvent (Friend and Adler, 1975; IUPAC, 1974). In other words, solubility gives the maximum amount of solute possible to dissolve in the solvent at the specified condition, whereas concentration, expressed for example in weight percentage, does not necessarily mean a saturated solution at the condition in question.

Until now, the solubility units discussed have been those connected to solid-liquid or liquid-liquid solubilities. Gas-liquid solubility is usually expressed, in

addition to the above-mentioned units, in different ways. Probably the oldest, but
still widely used, is Ostwald's solubility coefficient (Ostwald, 1894). Several other
solubility coefficients can be found in the literature. In some cases the weight of
water in the organic phase is given in relation to the weight of water in the vapor
phase (du Pont, 1963a). That is, the distribution of water between the vapor and
liquid phases. Another form used to express the concentration of water is the ratio
of the densities between the saturated water vapor and saturated organic vapor (du
Pont, 1956).

Kirshenbaum (1951) reported the relative solubilities of water and heavy water
in organic liquids at 25°C. The solubility was expressed as the ratio of the mole
fraction of H_2O in saturated solution to the mole fraction of D_2O in saturated solution. The definitions for these coefficients are summarized in the Glossary. The
conversion formulas between different units for gas solubility are given in Table
9.3. A computer program is also presented for conversion between various units (see
Table 9.4). Regarding letter symbols, signs, and abbreviations, see British Standards
Institution (1967). The recommended key values for thermodynamics and fundamental
physical constants are available in CODATA publications (CODATA, 1973a and b).

In connection with air pollution, it is quite common to express the impurities
of chemicals in air or gas samples as part per million (ppm) or parts per billion
(ppb). These units are also used in water pollution studies. However, it is essential
to state that ppm and ppb refer to mass or volume or both, similarly to percentages,
that is, wether weight or volume percent or both in ratio. For example, 10% can have
several meanings:

10% w/w   (by weight)
10% v/v   (by volume)
10% w/v   (by weight/volume)
10% v/w   (by volume/weight)

The conversion formula between percentage, ppm, and ppb is

$$1\% \quad = \quad 10,000 \text{ ppm} \quad = \quad 10,000,000 \text{ ppb}$$

One gram-mole of an ideal gas or vapor has a volume of 24.45 liters at 25°C
and a pressure of 760 mm Hg. Therefore, the conversion of mg liter^{-1} becomes

$$1 \text{ ppm (w/v)} \quad = \quad \frac{24,450}{\text{mol. wt.}} \text{ mg liter}^{-1}$$

For example, a 1-liter air sample of chloroform contains chloroform expressed in
ppm (w/v):

$$= \quad \frac{24,450}{119.38} \text{ mg liter}^{-1} \quad = \quad 204.81 \text{ mg in sample}$$

Table 9.3　Conversion Factors Between Various Solubility Units[a]

From ＼ To	Bunsen coeff., B_2	Ostwald coeff., O_2	Technical coeff., T_2	Kuenen coeff., K_2	Weight percent, W_2	Mole fraction, X_2
Bunsen coefficient, B_2	1	$\dfrac{273.15\,O_2}{T}$	$\dfrac{dT_2}{0.9678}$	dK_2	$\dfrac{22{,}415zdW_2}{(100 - W_2)MP}$	$\dfrac{22{,}415zdX_2}{(1 - X_2)M_LP}$
Ostwald coefficient, O_2	$\dfrac{TB_2}{273.15}$	1	$\dfrac{dTT_2}{264.35}$	$\dfrac{dTK_2}{273.15}$	$\dfrac{22{,}415zdTW_2}{273.15(100 - W_2)MP}$	$\dfrac{22{,}415zdTX_2}{273.15(1 - X_2)M_LP}$
Technical coefficient, T_2	$\dfrac{0.9678B_2}{d}$	$\dfrac{264.35\,O_2}{dT}$	1	$0.9678K_2$	$0.9678\,\dfrac{22{,}416zW_2}{(100 - W_2)MP}$	$0.9678\,\dfrac{22{,}415zX_2}{(1 - X_2)M_LP}$
Kuenen coefficient, K_2	$\dfrac{B_2}{d}$	$\dfrac{273.15\,O_2}{dT}$	$\dfrac{T_2}{0.9678}$	1	$\dfrac{22{,}415zW_2}{(100 - W_2)MP}$	$0.9678\,\dfrac{22{,}415zX_2}{(1 - X_2)M_LP}$
Weight percent, W_2	$\dfrac{100}{1 + \dfrac{22{,}415zd}{MPB_2}}$	$\dfrac{100}{1 + \dfrac{22{,}415zd}{273.15MPO_2}}$	$\dfrac{100}{1 + \dfrac{22{,}415z\;0.9678}{MPT_2}}$	$\dfrac{100}{1 + \dfrac{22{,}415z}{MPK_2}}$	1	$\dfrac{100}{1 + \dfrac{(1 - X_2)M_L}{MX_2}}$
Mole fraction, X_2	$\dfrac{1}{1 + \dfrac{22{,}415zd}{M_LPB_2}}$	$\dfrac{1}{1 + \dfrac{22{,}415zd}{273.15M_LPO_2}}$	$\dfrac{1}{1 + \dfrac{22{,}415z\;0.9678}{M_LPT_2}}$	$\dfrac{1}{1 + \dfrac{22{,}415z}{M_LPK_2}}$	$\dfrac{W_2/M}{\dfrac{W_2}{M} + \dfrac{100 - W_2}{M_L}}$	1

[a] d = density of liquid solvent (g cc^{-1}); z = compressibility factor of gaseous solute; T = absolute temperature (K); P = partial pressure of solute (atm); M = molecular weight of solute; M_L = molecular weight of solvent.

Table 9.4 Conversion and Expression of Solubility data in Various Units

```
100   PRINT "CONVERSION AND EXPRESSION OF SOLUBILITY DATA IN VARIOUS UNITS."
110   PRINT "****************************************************************"
120   PRINT
130   PRINT "S$ = CODE FOR THE AVAILABLE (OR INPUT) SOLUBILITY DATA."
140   PRINT "B2 = BUNSEN COEFFICIENT, N ML/ML,ATM."
150   PRINT "O2 = OSTWALD COEFFICIENT, N ML/ML."
160   PRINT "T2 = TECHNICAL SOLUBILITY COEFFICIENT, N ML/G, AT."
170   PRINT "K2 = KUENENE COEFFICIENT, N ML/G, ATM."
180   PRINT "W2 = SOLUBILITY, WT.%."
190   PRINT "X2 = SOLUBILITY, MOLE FRACTION."
200   PRINT
210   DIM A$(70),S$(2)
220   READ L
230   REM L = NUMBER OF SOLUBILITY SYATEMS.
240   FOR J=1 to L
250   READ A$
260   REM A$ = DESCRIPTION AND NAME OF THE SYSTEM. (REFERENCE.)
270   PRINT A$
280   PRINT ":::::::::::::::::::::::::::::::::::::::::::::::::::::::::::"
290   PRINT
300   PRINT "TEMP.  BUNSEN,  KUENEN,  OSTWALD, TECHNICAL,   WT.%.  MOLE FRACTION,"
310   PRINT " *C. NML/ML,ATM NML/G,ATM NML/ML   NML/G,AT     W2.       X2."
320   PRINT
330   IMAGE 3D.D,1X,2D.5D,1X,2D.5D,1X,2D.5D,1X,2D.6D,3X,2D.5D,3X,D.8D
340   READ M1,M2
350   REM M1 = MOLECULAR WEIGHT OF SOLVENT.
360   REM M2 = MOLECULAR WEIGHT OF SOLUTE.
370   READ N
380   REM N = NUMBER OF SOLUBILITY DATA.
390   FOR I=1 TO N
400   READ S$
410   REM S$ = CODE FOR THE AVAILABLE (OR INPUT) SOLUBILITY DATA.
420   IF S$≠"B2" THEN 440
430   GOTO 880
440   IF S$≠"O2" THEN 460
450   GOTO 810
460   IF S$≠"T2" THEN 480
470   GOTO 740
```

Table 9.4 (Continued)

```
480   IF S$≠"K2" THEN 500
490   GOTO 670
500   IF S$≠"W2" THEN 520
510   GOTO 600
520   IF S$≠"X2" THEN 9999
530   READ X2,T1,P,D
532   REM T1 = TEMPERATURE, *C.
534   REM P = PARTIAL PRESSURE OF SOLUTE, ATM.
536   REM D = DENSITY OF LIQUID SOLVENT, G/CC.
540   T2=22415*0.9678*X2/((1-X2)*M1*P)
550   B2=22415*X2*D/((1-X2)*M1*P)
560   O2=22415*X2*D*(T1+27315)/((1-X2)*273.15*M1*P)
570   K2=22415*X2/((1-X2)*M1*P)
580   W2=100*X2*M2/(X2*M2+(1-X2)*M1)
590   GOTO 900
600   READ W2,T1,P,D
610   T2=22415*W2*0.9678/((100-W2)*M2*P)
620   B2=22415*W2*D/((100-W2)*M2*P)
630   O2=22415*W2*D*(T1+273.15)/((100-W2)*273.15*M2*P)
640   K2=22415*W2/((100-W2)*M2*P)
650   X2=(W2/M2)/((W2/M2)+(100-W2)/M1)
660   GOTO 900
670   READ K2,T1,P,D
680   T2=0.9678*K2
690   B2=K2*D
700   O2=K2*D*(T1+273.15)/273.15
710   W2=100*K2*M2*P/(22415+K2*M2*P)
720   X2=K2*M1*P/(22415+K2*M1*P)
730   GOTO 900
740   READ T2,T1,P,D
750   B2=D*T2/0.9678
760   O2=D*T2*(T1+273.15)/264.35
770   K2=T2/0.9678
780   W2=100*T2*M2*P/(22415*0.9678+M2*P*T2)
790   X2=T2*M1*P/(22415*0.9678+M1*P*T2)
800   GOTO 900
810   READ O2,T1,P,D
820   T2=264.35*O2/(D*(T1+273.15))
830   B2=273.15*O2/(T1+273.15)
```

Table 9.4 (Continued)

```
840   K2=273.15*O2/(D*(T1+273.15))
850   W2=27315*M2*O2*P/(22415*D*(T1+273.15)+273.15*M2*P*O2)
860   X2=273.15*O2*M1*P/(22415*D*(T1+273.15)+273.15*O2*M1*P)
870   GOTO 900
880   READ B2,T1,P,D
890   T2=0.9678*B2/D
900   O2=B2*(T1+273.15)/273.15
910   K2=B2/D
920   W2=100*B2*M2*P/(22415*D*B2*M2*P)
930   X2=B2*M1*P/(22415*D*B2*M1*P)
940   PRINT USING 330;T1,B2,K2,O2,T2,W2,X2
950   NEXT I
960   PRINT
970   PRINT
980   NEXT J
1000  DATA 8
1010  DATA "SOLUBILITY OF HCL(G) IN HEXANE."
1020  DATA 86.18,36.46
1030  DATA 1
1040  DATA "T2"
1050  DATA 5.06,20,1,0.65937
1060  DATA "SOLUBILITY OF CL2(G) IN CCL4."
1070  DATA 153.82,70.906
1080  DATA 1
1090  DATA "B2"
1100  DATA 57.5,15,1,1.5942
1110  DATA "SOLUBILITY OF HELIUM IN WATER."
1120  DATA 18.0153,4.0026
1130  DATA 2
1140  DATA "B2",0.0097,0,1,0.9999
1150  DATA "B2",0.0086,20,1,0.9982
1160  DATA "SOLUBILITY OF CO(G) IN BENZENE."
1165  DATA 78.11,28.01
1170  DATA 2
1180  DATA "X2",0.00065,12,1,0.899
1190  DATA "X2",0.00066,20,1,0.87
1200  DATA "SOLUBILITY OF CH4(G) IN ACETONE."
1205  DATA 58.08,16.04
1210  DATA 2
```

Table 9.4 (Continued)

```
1220   DATA "B2",0.62,0,1,0.8
1230   DATA "B2",0.58,20,1,0.7908
1240   DATA "SOLUBILITY OF C2H2(G) IN CCL4."
1250   DATA 153.82,26.04
1260   DATA 2
1270   DATA "B2",3.97,0,1,1.61
1280   DATA "B2",2.88,20,1,1.5942
1290   DATA "SOLUBILITY OF CO2(G) IN ETHANOL."
1300   DATA 46.07,44.01
1310   DATA 1
1320   DATA "X2",0.007,20,1,0.7893
1330   DATA "SOLUBILITY OF C2H4(G) IN BENZENE."
1340   DATA 78.11,28.05
1350   DATA 2
1360   DATA "B2",4.2,5,1,0.899
1370   DATA "X2", 0.0133,20,1,0.87
9999   END
RUN
```

CONVERSION AND EXPRESSION OF SOLUBILITY DATA IN VARIOUS UNITS.
****

S$ = CODE FOR THE AVAILABLE (OR INPUT) SOLUBILITY DATA.

B2 = BUNSEN COEFFICIENT, N ML/ML,ATM.

O2 = OSTWALD COEFFICIENT, N ML/ML.

T2 = TECHNICAL SOLUBILITY COEFFICIENT, N ML/G, AT.

K2 = KUENENE COEFFICIENT, N ML/G, ATM.

W2 = SOLUBILITY, WT.%.

X2 = SOLUBILITY, MOLE FRACTION.

SOLUBILITY OF HCL(G) IN HEXANE.
::

TEMP.	BUNSEN,	KUENEN,	OSTWALD,	TECHNICAL,	WT.%.	MOLE FRACTION,
*C.	NML/ML,ATM	NML/G,ATM	NML/ML	NML/G,AT	W2.	X2.
20.0	3.44742	5.22835	3.69984	5.060000	0.84327	0.01970558

SOLUBILITY OF CL2(G) IN CCL4.
::

TEMP.	BUNSEN,	KUENEN,	OSTWALD,	TECHNICAL,	WT.%.	MOLE FRACTION,
*C.	NML/ML,ATM	NML/G,ATM	NML/ML	NML/G,AT	W2.	X2.
15.0	57.50000	36.06825	60.65760	34.906853	10.24111	0.19840556

Table 9.4 (Continued)

SOLUBILITY OF HELIUM IN WATER.

::

TEMP.	BUNSEN,	KUENEN,	OSTWALD,	TECHNICAL,	WT.%.	MOLE FRACTION,
*C.	NML/ML,ATM	NML/G,ATM	NML/ML	NML/G,AT	W2.	X2.
0.0	0.00970	0.00970	0.00970	0.009389	0.00017	0.00000780
20.0	0.00860	0.00862	0.00923	0.008338	0.00015	0.00000692

SOLUBILITY OF CO(G) IN BENZENE.

::

TEMP.	BUNSEN,	KUENEN,	OSTWALD,	TECHNICAL,	WT.%.	MOLE FRACTION,
*C.	NML/ML,ATM	NML/G,ATM	NML/ML	NML/G,AT	W2.	X2.
12.0	0.16780	0.18665	0.17517	0.180640	0.02332	0.00065000
20.0	0.16489	0.18952	0.17696	0.183421	0.02368	0.00066000

SOLUBILITY OF CH4(G) IN ACETONE.

::

TEMP.	BUNSEN,	KUENEN,	OSTWALD,	TECHNICAL,	WT.%.	MOLE FRACTION,
*C.	NML/ML,ATM	NML/G,ATM	NML/ML	NML/G,AT	W2.	X2.
0.0	0.62000	0.77500	0.62000	0.750045	0.05543	0.00200410
20.0	0.58000	0.73343	0.62247	0.709818	0.05246	0.00189681

SOLUBILITY OF C2H2(G) IN CCL4.

::

TEMP.	BUNSEN,	KUENEN,	OSTWALD,	TECHNICAL,	WT.%.	MOLE FRACTION,
*C.	NML/ML,ATM	NML/G,ATM	NML/ML	NML/G,AT	W2.	X2.
0.0	3.97000	2.46584	3.97000	2.386439	0.28564	0.01663992
20.0	2.88000	1.80655	3.09087	1.748378	0.20943	0.01224540

SOLUBILITY OF CO2(G) IN ETHANOL.

::

TEMP.	BUNSEN,	KUENEN,	OSTWALD,	TECHNICAL,	WT.%.	MOLE FRACTION,
*C.	NML/ML,ATM	NML/G,ATM	NML/ML	NML/G,AT	W2.	X2.
20.0	2.70714	3.42980	2.90536	3.319364	0.66891	0.00700000

SOLUBILITY OF C2H4(G) IN BENZENE.

::

TEMP.	BUNSEN,	KUENEN,	OSTWALD,	TECHNICAL,	WT.%.	MOLE FRACTION,
*C.	NML/ML,ATM	NML/G,ATM	NML/ML	NML/G,AT	W2.	X2.
5.0	4.20000	4.67186	4.27688	4.521423	0.58124	0.01601932
20.0	3.36525	3.86811	3.61166	3.743556	0.48172	0.01330000

DONE

In the case of 1 mg liter^{-1} trichloroethylene, the concentration corresponds to

$$\frac{24,450}{131.39} = 186.1 \text{ ppm (w/v)}$$

Gas concentrations might be expressed by the partial pressure method, that is:

$$1 \text{ ppm of constituent} = \frac{\text{partial pressure of one constituent}}{\text{total barometric pressure}}$$

The gas solubilities are often expressed by the Henry's law constant (H),

$$H = \frac{p_2}{x_2}$$

where p_2 = partial pressure of the solute, mm Hg

x_2 = mole fraction of the solute in the solvent

Although this form is most common, there are other units for partial pressure and concentration (e.g., partial pressure in lb in^{-2} and concentration in g g-mole^{-1} solution), so that the reader should be careful not to use the numerical values, quoted as a Henry's law constant, without giving its unit.

The Henry's law constant is often expressed according to the Henry–Dalton laws:

$$\log_{10} H = \frac{A}{2.3026 \, R \, T} + B$$

where A is the heat absorbed when 1 g-mole of gas is vaporized reversibly from an infinite amount of the solution, and B is a constant. The value of A is usually negative.

The Henry's law coefficient can be calculated from the Bunsen absorption coefficient (β) according to the relationship

$$H = \frac{17.0324 \times 10^6 \, d \left(1 + \frac{M \, p_2 \, \beta}{17.0324 \times 10^6 \, d} \right)}{M \, \beta}$$

The solubility of gases in liquids are both temperature- and pressure-dependent. Because of the various units used for temperature and pressure in the literature, the reader will find use for conversion tables between different temperature and pressure units. Tables 9.5a and b show the relationships between the units of temperature values and the temperature interval, and Table 9.6 gives the $\log_{10} T$, $\log_e T$, and 1000/T values between 0 and 100°C.

The conversion factors between various units are presented in Table 9.7.

Table 9.5a   Conversion Formulas Between Various Units of Temperature

From \ To	$^\circ$C	$^\circ$F	K	$^\circ$R	$^\circ$r
$^\circ$C	1	$\dfrac{9}{5}\,^\circ\text{C} + 32$	$^\circ\text{C} + 273.15$	$\dfrac{9}{5}\,^\circ\text{C} + 491.67$	$\dfrac{4}{5}\,^\circ\text{C}$
$^\circ$F	$\dfrac{5}{9}(^\circ\text{F} - 32)$	1	$\dfrac{5}{9}(^\circ\text{F} + 459.67)$	$^\circ\text{F} + 459.67$	$\dfrac{4}{9}(^\circ\text{F} - 32)$
K	$\text{K} - 273.15$	$\dfrac{9}{5}\,\text{K} - 459.67$	1	$\dfrac{9}{5}\,\text{K}$	$\dfrac{4}{5}(\text{K} - 273.15)$
$^\circ$R	$\dfrac{5}{9}(^\circ\text{R} - 491.67)$	$^\circ\text{R} - 459.67$	$\dfrac{5}{9}\,^\circ\text{R}$	1	$\dfrac{4}{9}(^\circ\text{R} - 491.67)$
$^\circ$r	$\dfrac{5}{4}\,^\circ\text{r}$	$\dfrac{9}{4}\,^\circ\text{r} + 32$	$\dfrac{5}{4}\,^\circ\text{r} + 273.15$	$\dfrac{9}{4}\,^\circ\text{r} + 491.67$	1

Table 9.5b Conversion Factors Between Various Temperature Intervals or Differences[a]

From \ To	°C or K	°F or °R	°r
1°C = 1 K	1	$\frac{9}{5}$	$\frac{4}{5}$
1°F = 1°R	$\frac{5}{9}$	1	$\frac{4}{9}$
1°r	$\frac{5}{4}$	$\frac{9}{4}$	1

[a] C = Celsius; K = Kelvin; F = Fahrenheit; R = Rankine; r = Reaumur.

Table 9.6 Calculation of log T Values Between 0 and 100°C

| Temperature | | log T | ln T | 1000/T |
°C	K	(K)	(K)	(K^{-1})
0	273.15	2.436401	5.610022	3.660992
1	274.15	2.437988	5.613675	3.647638
2	275.15	2.439569	5.617316	3.634381
3	276.15	2.441145	5.620944	3.621220
4	277.15	2.442715	5.624558	3.608154
5	278.15	2.444279	5.628160	3.595182
6	279.15	2.445837	5.631749	3.582303
7	280.15	2.447391	5.635325	3.569516
8	281.15	2.448938	5.638888	3.556820
9	282.15	2.450480	5.642439	3.544214
10	283.15	2.452016	5.645976	3.531697
11	284.15	2.453547	5.649502	3.519268
12	285.15	2.455073	5.653015	3.506926
13	286.15	2.456594	5.656516	3.494670
14	287.15	2.458109	5.660005	3.482500
15	288.15	2.459619	5.663481	3.470415
16	289.15	2.461123	5.666945	3.458412
17	290.15	2.462623	5.670398	3.446493
18	291.15	2.464117	5.673839	3.434655
19	292.15	2.465606	5.677267	3.422899
20	293.15	2.467090	5.680685	3.411222
21	294.15	2.468569	5.684091	3.399626
22	295.15	2.470043	5.687484	3.388107
23	296.15	2.471511	5.690866	3.376667
24	297.15	2.472976	5.694238	3.365304
25	298.15	2.474435	5.697597	3.354016
26	299.15	2.475889	5.700946	3.342804
27	300.15	2.477338	5.704283	3.331667
28	301.15	2.478783	5.707608	3.320604
29	302.15	2.480222	5.710923	3.309614
30	303.15	2.481658	5.714229	3.298697
31	304.15	2.483088	5.717522	3.287851
32	305.15	2.484513	5.720803	3.277077
33	306.15	2.485934	5.724075	3.266372
34	307.15	2.487351	5.727337	3.255738
35	308.15	2.488762	5.730586	3.245172
36	309.15	2.490169	5.733827	3.234675
37	310.15	2.491572	5.737057	3.224246
38	311.15	2.492970	5.740274	3.213884
39	312.15	2.494364	5.743484	3.203588
40	313.15	2.495752	5.746682	3.193357
41	314.15	2.497137	5.749870	3.183192
42	315.15	2.498517	5.753048	3.173092
43	316.15	2.499893	5.756217	3.163055
44	317.15	2.501265	5.759375	3.153082
45	318.15	2.502632	5.762523	3.143171
46	319.15	2.503995	5.765661	3.133323
47	320.15	2.505354	5.768790	3.123536
48	321.15	2.506708	5.771908	3.113810
49	322.15	2.508058	5.775017	3.104144
50	323.15	2.509404	5.778116	3.094538
51	324.15	2.510746	5.781206	3.084991
52	325.15	2.512084	5.784286	3.075503

Table 9.6 (Continued)

53	326.15	2.513417	5.787357	3.066074
54	327.15	2.514747	5.790420	3.056702
55	328.15	2.516072	5.793470	3.047387
56	329.15	2.517394	5.796514	3.038128
57	330.15	2.518711	5.799547	3.028926
58	331.15	2.520025	5.802572	3.019779
59	332.15	2.521334	5.805586	3.010688
60	333.15	2.522640	5.808593	3.001651
61	334.15	2.523942	5.811590	2.992668
62	335.15	2.525239	5.814578	2.983738
63	336.15	2.526533	5.817557	2.974862
64	337.15	2.527823	5.820528	2.966039
65	338.15	2.529109	5.823490	2.957267
66	339.15	2.530392	5.826442	2.948548
67	340.15	2.531670	5.829386	2.939879
68	341.15	2.532945	5.832322	2.931262
69	342.15	2.534217	5.835250	2.922695
70	343.15	2.535484	5.838167	2.914177
71	344.15	2.536747	5.841077	2.905710
72	345.15	2.538008	5.843979	2.897291
73	346.15	2.539264	5.846872	2.888921
74	347.15	2.540517	5.849756	2.880599
75	348.15	2.541766	5.852633	2.872325
76	349.15	2.543012	5.855501	2.864099
77	350.15	2.544254	5.858362	2.855919
78	351.15	2.545493	5.861214	2.847785
79	352.15	2.546728	5.864058	2.839699
80	353.15	2.547959	5.866894	2.831658
81	354.15	2.549187	5.869720	2.823662
82	355.15	2.550412	5.872540	2.815711
83	356.15	2.551633	5.875352	2.807806
84	357.15	2.552851	5.878156	2.799944
85	358.15	2.554065	5.880952	2.792126
86	359.15	2.555276	5.883740	2.784352
87	360.15	2.556484	5.886521	2.776621
88	361.15	2.557688	5.889294	2.768932
89	362.15	2.558889	5.892059	2.761287
90	363.15	2.560086	5.894815	2.753683
91	364.15	2.561280	5.897566	2.746121
92	365.15	2.562471	5.900309	2.738600
93	366.15	2.563659	5.903043	2.731121
94	367.15	2.564843	5.905770	2.723682
95	368.15	2.566025	5.908491	2.716284
96	369.15	2.567203	5.911202	2.708926
97	370.15	2.568378	5.913909	2.701607
98	371.15	2.569549	5.916606	2.694328
99	372.15	2.570718	5.919296	2.687088
100	373.15	2.571883	5.921980	2.679887

Table 9.7 Conversion Factors Between Various Units of Pressure

From \ To	Pa Nm^{-2}	bar	kgf/cm^2 at	atm	mm Hg torr	psi lbf/in^2	mbar
Pa	1	1.000×10^{-5}	1.020×10^{-5}	9.869×10^{-6}	7.501×10^{-3}	1.450×10^{-4}	1.000×10^{-2}
bar	1.000×10^{5}	1	1.020	9.869×10^{-1}	7.501×10^{2}	1.450×10^{1}	1.000×10^{3}
kgf/cm^2	9.807×10^{4}	9.807×10^{-1}	1	9.678×10^{-1}	7.356×10^{2}	1.422×10^{1}	9.807×10^{2}
atm	1.013×10^{5}	1.01325	1.033	1	7.600×10^{2}	1.470×10^{1}	1.01325×10^{3}
mm Hg	1.333×10^{2}	1.333×10^{-3}	1.360×10^{-3}	1.316×10^{-3}	1	1.934×10^{-2}	1.3332
psi	6.895×10^{3}	6.895×10^{-2}	7.031×10^{-2}	6.805×10^{-2}	5.171×10^{-1}	1	6.895×10^{1}
mbar	1.000×10^{2}	1.000×10^{-3}	1.020×10^{-3}	9.869×10^{-4}	7.501×10^{-1}	1.450×10^{-2}	1

The different values of the gas constant (R) are given in Table 9.8.

The Solubility parameter values are usually given in $(cal/cc)^{\frac{1}{2}}$ in the literature; however, the SI unit is $(J/m^3)^{\frac{1}{2}}$. The conversion factor is

$$1 \ (cal/cc)^{\frac{1}{2}} \ = \ 4045.5 \ (J/m^3)^{\frac{1}{2}}$$

Table 9.8 Gas Constant in Various Units

Volume	Temperature	moles	atm	psia	mm Hg	cm Hg	in Hg	in H_2O	ft H_2O
ft^3	K	g	0.00290	0.0426	2.20	0.220	0.0867	1.18	0.0982
		lb	1.31	19.31	999	99.9	39.3	535	44.6
	$°R$	g	0.00161	0.02366	1.22	0.122	0.0482	0.655	0.0546
		lb	0.730	10.73	555	55.5	21.8	297	24.8
cm^3	K	g	82.05	1,206	62,400	6,240	2,450	33,400	2,780
		lb	37,200	547,000	2.83×10^7	2.83×10^6	1.11×10^6	1.51×10^7	1.26×10^6
	$°R$	g	45.6	670	34,600	3,460	1,360	18,500	1,550
		lb	20,700	304,000	1.57×10^7	1.57×10^6	6.19×10^5	8.41×10^6	7.01×10^5
liters	K	g	0.08205	1.206	62.4	6.24	2.45	33.4	2.78
		lb	37.2	547	2.83×10^4	2,830	1,113	1.514×10^4	1,262
	$°R$	g	0.0456	0.670	34.6	3.46	1.36	18.5	1.55
		lb	20.7	304	1.57×10^4	1,570	619	8,410	701

Table 9.8   (Continued)

Energy	Temperature	Mass unit	R
cal	K	g-mole	1.9872
abs. J	K	g-mole	8.3143
internal. J	K	g-mole	8.312
Btu	$^{\circ}$R	lb-mole	1.9859
hp-hr	$^{\circ}$R	lb-mole	0.000780
kWhr	$^{\circ}$R	lb-mole	0.000583

Glossary

Acentric factor (ω): expresses the effect of finite core size, regardless of the shape of the core, upon intermolecular forces:

$$\omega = -\log_{10}\left(\frac{P^o_{0.7T_{red}}}{P_{cr}}\right) - 1.0$$

where $P^o_{0.7T_{red}}$ is the vapor pressure of the compound when the reduced temperature is equal to 0.7. P_{cr} is the critical pressure. The acentric factor represents the extra entropy of vaporization in addition to that of a perfect fluid.

Activity coefficient (γ): a correction factor, indicating a departure from ideal solutions:

$$p_i = \gamma_i x_i P^o_i$$

where p_i = partial pressure of component i

x_i = concentration of component i in the liquid phase, mole fraction

P^o_i = vapor pressure of pure component i

The activity coefficient becomes unity as the mole fraction approaches unity.

Amagat's law: expresses the molar volume of a liquid mixture (V^{liq}_{mix}) when its components are similar at the same temperature:

$$V^{liq}_{mix} = \sum_i x_i V^{liq}_i$$

where x_i = mole fraction of component i

V^{liq}_i = molar volume of component i

Ångström's law: states that when a gas dissolves in a liquid, the volume of the liquid increases proportionally with the gas volume, taken at standard condition ($0^\circ C$ and 760 mm Hg).

Antoine equation: represents the vapor pressure data (P^o, mm Hg) as a function of temperature (t, $^\circ C$);

$$\log_{10} P^o = A + \frac{B}{C + t}$$

where A, B, and C are constants. This equation is also used to express the vapor pressure of mixtures.

Athermal solution: defined by the assumption that the change of enthalpy is zero upon mixing its components at constant temperature. However, the heat of mixing is zero only for solutions of molecules with equal size. It is sometimes called a semi-ideal mixture.

Azeotropes: also called constant-boiling mixtures, these are characterized by the ratio of the activity coefficients (γ_i/γ_j), which equal the reciprocal ratio of the vapor pressures (P_2^o/P_1^o). In a binary system the relationship becomes

$$\frac{\gamma_1}{\gamma_2} = \frac{P_2^o}{P_1^o}$$

At the azeotrope the vapor and liquid compositions are equal. There are maximum and minimum compositions in addition to the homogeneous and heterogeneous ones.

Boyle's law: states that the product of the gas pressure (P) and the gas volume (V) are constant at constant temperature:

$$(PV)_T = \text{constant} \quad \text{or} \quad (P_1V_1 = P_2V_2)_T$$

In other words, at a constant temperature, the volume (V) of a given quantity of any gas varies inversely with the pressure (P) to which the gas is subjected.

Bubble point: in a liquid-vapor mixture, the bubble point is represented by the temperature at which the last bubble of vapor dissapears and the system becomes wholly liquid.

Bunsen absorption coefficient (β): defined as the volume of gas (V_{STD}) reduced from the temperature of the experiment to its volume at zero degrees Celsius, which dissolves in a given volume of solvent when the partial pressure (p_i) of the test gas (i.e., total pressure minus solvent pressure) is 1 atm (760 mm Hg).

Charles' law: states that the volume (V_{gas}^o) of a gas at constant pressure increases with temperature:

$$\frac{V_{gas}^t}{V_{gas}^o} = \frac{273.15 + t(^oC)}{273.15}$$

where V_{gas}^o and V_{gas}^t are the gas volumes at 0 and t^oC, respectively.

Chebyshev polynominals: polynominals applied particularly for expressing the pressure (P^o) of substances as a function of temperature (T, K):

$$T \log_e P^o = \frac{A_o}{2} + \sum_{i=1}^{n} A_i E_i(x)$$

where A_o and A_i = coefficients
$E_i(x)$ = Chebyshev polynominal $\quad x = \dfrac{2T - (T_{max} + T_{min})}{T_{max} - T_{min}}$

Clapeyron equation: a relationship between the vapor pressure (P^O) and temperature (T, K):

$$\frac{\partial P}{\partial T} = \frac{\Delta H_{vap}}{T(V_{gas} - V_{liq})}$$

where ΔH_{vap} = latent heat of vaporization
V_{gas} = molar volume of gas
V_{liq} = molar volume of liquid

In other words, the relationship expresses the dynamic equilibrium between the vapor and the condensed phase of a pure substance.

Clausius-Clapeyron equation: derived from the Clapeyron's equation by assuming that at low pressure the volume of the condensed phase (V_{liq}) is negligible in relation to the gas volume (V_{gas}), and the ideal-gas law is adoptable:

$$\frac{\partial P^O}{\partial T} = \frac{P^O H_{vap}}{RT^2}$$

where R = gas constant. This expression is often rearranged to

$$\frac{d(\log_e P^O)}{d(1/T)} = - \frac{\Delta H_{vap}}{R}$$

Compressibility factor (z): represents the nonideality of a gas. Its value depends upon the temperature (T) and pressure (P):

$$z = \frac{nPV}{RT}$$

where n = number of moles
P = pressure
V = molar volume
T = absolute temperature
R = gas constant

The compressibility factor is usually less than unity for real gases at low pressures; however, its value becomes greater than unity at high temperatures and pressures.

Corresponding states: used to express the generalized properties of substances (e.g., compressibility factor, fugacity, heat capacity, enthalpy, viscosity, thermal conductivity) through their relation to the critical properties. The assumption is that the reduced properties of the substances (T/T_{cr}, P/P_{cr}, V/V_{cr}; etc.) are the same.

Cox chart: a graphical correlation or illustration of vapor-pressure data. The ordinate represents the logarithm of the vapor pressure (to base 10) and the abscissa gives the corresponding temperature expressed in reciprocal absolute temperature. The Antoine vapor-pressure equation closely resembles the Cox chart. The vapor pressure is a straight line except close to the critical and triple points of the substances.

Critical point: represents a thermodynamic condition when the meniscus disappears between the liquid and vapor phases. The critical-point condition is characterized by the critical temperature, pressure, and volume.

Dew point: the dew point of a vapor-liquid mixture is characterized by a temperature where the condensation of the vapor takes place in the system.

Dulong and Petit's law: states that the atomic heat capacity of most solid elements is about 26 J g-atom^{-1} K^{-1}.

Einstein heat-capacity equation: characterizes the monoatomic crystalline solids at constant volume (C_v):

$$C_v = \frac{3Ru^2e^u}{(e^u - 1)^2}$$

where $u = \dfrac{h\nu}{kT}$

ν = harmonic frequency

h = Planck's constant

k = Boltzmann's constant

R = gas constant

Eötvös constant (k_E): used in the empirical Eötvös' equation for surface tension (γ):

$$k_E = \frac{\gamma V^{2/3}}{T_{cr} - T}$$

where V =   liquid molar volume

T_{cr} = critical temperature, K

T =   absolute temperature, K

The Eötvös constant (k_E) is about 2.12 for low-molecular-weight polar and nonpolar substances.

Extensive properties: properties such as volume and entropy that are defined by internal properties; therefore, they are representative for the sample or system.

Fugacity: the fugacity of a component in a mixture is related to the escaping tendency of the component from the mixture. At equilibrium condition, the fugacity of each component is the same in each phase.

Gibbs-Duhem relation: represents the activity coefficient (γ_i) of component i in the mixture as function of composition expressed as a mole fraction (x_i) at constant temperature and pressure:

$$\left(\sum_i x_i \, d \log_e \gamma_i = 0\right)_{T,P}$$

Guldberg's rule: expresses the ratio between the normal boiling point and the critical temperature, both expressed in K, as a constant for a pure substance:

$$\frac{T_{b.p.}}{T_{cr}} = 0.613$$

Haggenmacher equation: gives a more accurate expression than the Clausius-Clapeyron's equation for latent heat of vaporization (ΔH_{vap}):

$$\Delta H_{vap} = \frac{RT^2}{P}\left(1 - \frac{T_{cr}^3 P}{T^3 P_{cr}}\right)^{\frac{1}{2}}\left(\frac{dP}{dT}\right)$$

where R =   gas constant

   T =   absolute temperature

   P =   pressure

   T_{cr} = critical temperature

   P_{cr} = critical pressure

Henry's law constant (H): defined by the expression

$$H = \frac{p_i}{x_i}$$

where p_i = partial pressure of component i

   x_i = mole fraction of component i in the solution

The Henry's law constant (H) depends on temperature and the chemical nature of both solute and solvent. The theory of Henry's law assumes ideal-gas behavior in the gas phase ($\emptyset_2$ = 1) and ideal dilute solution behavior in the liquid phase (γ_2 = 1). A large difference in the chemical properties of solute and solvent gives large H values and consequently small solubility values.

Herrington equal-area test: applied for checking the consistency of isothermal binary-phase equilibrium data. By plotting $\log_e(\gamma_1/\gamma_2)$ versus the mole fraction (x_2) over the entire concentration range, two equal areas should be obtained for consistent data (γ_i = activity coefficient).

Hess's law: states that the enthalpies of reactions are additive and depend only on the initial and final states of the reaction, not on the particular sequence of steps by which the overall reaction is achieved.

Hildebrand's rules: expresses the latent heat of vaporization (ΔH_{vap}) of organic substances at 25°C and at their normal boiling points ($T_{b.p.}$, K):

$$\Delta H_{vap}^{25^o} \text{(cal g-mole}^{-1}) = -2950 + 23.7T_{b.p.} + 0.02T_{b.p.}^2$$

and

$$\Delta H_{vap}^{b.p.} \text{(cal g-mole}^{-1}) = 17.0T_{b.p.} + 0.009T_{b.p.}^2$$

Ideal gas: has no intermolecular forces, or the interaction between the molecules is so weak that it can be neglected. It is characterized by the gas law:

$$PV = nRT$$

where P = pressure

V = molar volume

n = number of moles

R = gas constant

T = absolute temperature

Ideal solutions: characterized by no heat or volume changes taking place upon mixing, and obey Raoult's law. If the structure, size, and chemical nature of the molecules are very similar, the solution is approximately ideal.

Intensive properties: properties such as temperature, pressure, and composition that do not need to be specified in relation to the system.

Joule-Thomson effect: an effect observed during cooling, when a compressed gas is allowed to expand without external work.

Kay's rule: expresses the pseudocritical temperature of a mixture on the mole fraction average:

$$T_{pseudo} = \sum_i x_i T_{cr_i}$$

Kirkhoff equation: evaluates the temperature variation of enthalpies (ΔH_2):

$$\Delta H_2 = \Delta H_1 + \int_{T_1}^{T_2} \Delta C_p \, dT$$

where ΔH_2 = enthalpy at T_2

ΔH_1 = enthalpy at T_1

ΔC_p = heat capacity at constant pressure

Kopp's rule: states that the heat capacity (C_p) of a solid substance is approximately equal to the sum of the constituent elements:

$$C_p \text{(solid molecule)} = n_1 C_{p_1} + n_2 C_{p_2} + n_3 C_{p_3} + \cdots$$

where n = number of atoms of an element in the molecule

C_{p_1}, C_{p_2}, C_{p_3}, etc. = heat capacity of the elements, cal g-mole^{-1} K^{-1}

__Kuenen solubility coefficient__: defined as the volume of gas, expressed in cc, corrected to 0°C and 1 atm at a partial pressure of 1 atm, which will be dissolved in 1 g of solvent. The usual unit for Kuenen's solubility coefficient is cc g^{-1}.

__Le Châtelier's principle__: states that if an external pressure displaces the equilibrium in a system, the direction of the effect will tend to release the stress.

__Lewis fugacity rule__: gives the second virial coefficient of a binary mixture by taking the arithmetical average of the pure components:

$$B_{12} = \frac{B_1 + B_2}{2}$$

The condition for this assumption is that the components are chemically similar and that there is little difference in the size of molecules.

__Lorenz-Lorentz equation__: gives the specific refraction (n_{sp}) when the refractive index (n) and density (d) are known:

$$n_{sp} = \frac{n^2 - 1}{(n^2 + 2)d}$$

__Margules equation__: expresses the activity coefficient (γ_i) as a function of concentration, expressed as a mole fraction (x_i) in a power series. For a binary system the expression becomes

$$\log_e \gamma_1 = A_1 x_2 + B_1 x_2^2 + C_1 x_2^3 + D_1 x_2^4 + \cdots$$

A similar expression can be written for γ_2.

__Mixing rules__: rules proposed to represent physical properties of mixtures, calculated from physical properties of the pure constituents. The most common expressions are based upon the arithmetic average:

$$M_{AB} = \frac{A + B}{2}$$

or the geometric mean:

$$M_{AB} = (A\ B)^{\frac{1}{2}}$$

However, the interaction parameter (IP) plays a significant role in numerous mixing rules: for example,

$$M_{AB} = IP \ \frac{A + B}{2}$$

Nernst-Bingham rule: relates the latent heat of vaporization (ΔH_{vap}, cal g-mole^{-1}) of a compound to its normal boiling point ($T_{b.p.}$, K):

$$\Delta H_{vap} = 17.0 T_{b.p.} + 0.011 T_{b.p.}^2$$

NRTL (nonrandom, two-liquid) equation: a three-parameter expression of the activity coefficient (γ_i) as a function of composition, expressed as a mole fraction (x_i). For a binary system, γ_1 becomes

$$\log_e \gamma_1 = x_2^2 \left[\tau_{21} \left(\frac{G_{21}}{x_1 + x_2 G_{21}} \right)^2 + \frac{\tau_{12} G_{12}}{(x_2 + x_1 G_{12})^2} \right]$$

where τ_{21}, τ_{12}, G_{12}, and G_{21} are the parameters of the equation. A similar expression can be written for γ_2. The NRTL equation is applicable to both vapor-liquid and liquid-liquid equilibria data.

Ostwald solubility coefficient: defined as the volume of gas dissolved by a unit volume of solvent at a given temperature. In the calculations there are three assumptions:

Boyle's law holds (at a constant temperature the volume of a given quantity of any gas varies inversely as the pressure to which the gas is subjected; i.e., PV = constant).

Dalton's law holds ($P_{tot} = p_1 + p_2 + \cdots$: the total pressure of a mixture of gases is equal to the sum of the separate pressures that each gas would exert if it alone occupied the whole volume).

Henry's law holds (the mass of a slightly soluble gas that dissolves in a definite mass of liquids at a given temperature is very nearly directly proportional to the partial pressure of the gas; i.e., $p_2 = Hx_2$).

Because of the three assumptions (restrictions), Ostwald's solubility coefficient is independent of the partial pressure of the gas above the solution. If the above-mentioned laws do not hold, such data should not be reported as Ostwald solubility coefficients. In other words, liquid-liquid solubility should not be expressed as an Ostwald solubility coefficient.

Parachor: a comparative volume between two liquids at equal surface tensions. It is an additive function of the atoms and groups in the molecule (group contribution).

Poynting factor: used in connection with the calculation of the fugacity of pure liquids. It indicates the deviation from unity:

$$\exp \left[\int_{P_i^0}^{P} \frac{V_i^{liq}(T,P)}{RT} \, dP \right]$$

where V_i^{liq} = liquid molar volume

    R =   gas constant

    P = system pressure

    P_i^0 = vapor pressure of component i

    T =   absolute temperature

Pseudocritical properties of mixtures: calculated as the simple mole fraction (x_i) average of the pure-component critical properties. For example, the pseudo-critical temperature (T_{pseudo}) of the binary system is

$$T_{pseudo} = x_1 T_{cr_1} + x_2 T_{cr_2}$$

where T_{cr_1} and T_{cr_2} are the critical temperature of components 1 and 2, respectively.

Raoult absorption coefficient (C_R): expressed by

$$C_R = \frac{c}{P}$$

where c = dissolved concentration, g/1000 cc liquid

    P = standard atmospheric pressure

Raoult's law: applied to ideal solutions in which the activity (a_i) is equal to the mole fraction (x_i) over the entire composition range at constant temperature:

$$p_i = x_i P_i^0$$

where p_i = partial pressure of component i in the system

    P_i^0 = vapor pressure of pure component i

Redlich-Kister equation: expresses the activity coefficient (γ_i) in a binary system as function of composition, expressed as a mole fraction (x_i):

$$\log_{10} \gamma_1 = x_2^2 \, B + C(3x_1 - x_2) + D(x_1 - x_2)(5x_1 - x_2) +$$
$$E(x_1 - x_2)^2 (7x_1 - x_2) + \cdots$$

The coefficients B, C, D, E, $\cdots$, are characteristic for the system. A similar equation can be written for γ_2.

Redlich-Kwong equation of state: expresses the temperature (T), pressure (P), and volume (V) relationship for a single or multicomponent system:

$$\left[P + \frac{a_i}{T^{\frac{1}{2}}(V + b_i)V} \right](V - b_i) = RT$$

where a_i and b_i = coefficients of the equation, can be calculated from critical
        properties

Reduced property: a ratio between the actual and the critical properties, that is,
a fraction of the critical property. Common expressions are: reduced temperature,
T/T_{cr}; reduced pressure, P/P_{cr}; and reduced volume, V/V_{cr}.

Regular solution: does not involve an entropy change when a small amount of one
of its components is transferred to it from an ideal solution of the same composition, the total volume remaining unchanged.

Solubility parameter (δ): represents the square root of the cohesive energy density:

$$\delta = \left(\frac{\Delta E}{V_{liq}} \right)^{\frac{1}{2}}$$

where ΔE = energy of vaporization (i.e., energy required to evaporate a liquid
        isothermally from the saturated liquid to the ideal gas)

    V_{liq} = liquid molar volume

The energy of vaporization (ΔE) is often approximated from the latent heat of vaporization (ΔH_{vap}):

$$\Delta E \approx \Delta H_{vap} - RT$$

Standard conditions (STP): arbitrarily chosen for gases at $0^\circ C$ and 760 mm Hg
pressure.

Stiell polar factor (χ): the logarithm of the ratio between the reduced vapor
pressure (P^o_{red}) and the reduced vapor pressure calculated from the acentric factor
($P^o_{red, normal}$):

$$\chi = \log_{10} \left(\frac{P^o_{red}}{P^o_{red, normal}} \right) \text{ at } T_{red} = 0.6$$

where T_{red} is the reduced temperature.

Sugden's parachor (P_{Sgd}): related to the molecular volume (V_{mol}) and surface tension (γ) of the substance at the same temperature:

$$P_{Sgd} = \frac{M\gamma^{\frac{1}{4}}}{d_{liq} - d_{vap}}$$

where M = molecular weight

    d_{liq} and d_{vap} = density of liquid and vapor, respectively

Tielines: lines drawn between points on the equilibrium curve in a ternary system when two mutually insoluble liquids are present and the two activities $\gamma_1 x_1$ and $\gamma_2 x_2$ are equal.

Triple point: represents a thermodynamic condition when the melted crystal is in equilibrium with the liquid and vapor phase of the pure compound at saturated pressure. The equilibrium temperature is the triple-point temperature.

Trouton's rule: gives the approximate relationship between the normal boiling point, expressed degrees Kelvin ($T_{b.p.}$) and the latent heat of vaporization (ΔH_{vap}) at the same temperature for a pure nonpolar liquid:

$$\frac{\Delta H_{vap}}{T_{b.p.}} = 21 \text{ cal g-mole}^{-1} \text{ K}^{-1}$$

UNIFAC (UNIQUAC functional-group activity coefficients): the group-contribution method provides interaction parameters between pairs of structural groups in nonelectrolyte systems for the calculation of activity coefficients. The parameters can be used for systems that have not been studied experimentally previously.

UNIQUAC (universal quasi chemical) **equation**: an expression for the activity coefficient (γ_i) as function of concentration (x_i = mole fraction). For example, for a binary system,

$$\log_e \gamma_1 = \log_e\left(\frac{\phi_1}{x_1}\right) + \frac{z}{2} q_1 \log_e\left(\frac{Q_1}{\phi_1}\right) + Q_2\left(l_1 - \frac{\tau_1}{\tau_2} l_2\right) -$$

$$- q_1 \log_e(Q_1 + Q_2 \tau_{12}) + Q_2 q_1 \frac{\tau_{21}}{Q_1 + Q_2 \tau_{21}} - \frac{\tau_{12}}{Q_2 + Q_1 \tau_{12}}$$

where $l_1 = \dfrac{z}{2} (\tau_1 - q_1) - (\tau_1 - 1)$

$l_2 = \dfrac{z}{2} (\tau_2 - q_2) - (\tau_2 - 1)$

ϕ = component volume fraction

z = lattice coordination number

q = pure-component area parameter

τ = pure-component volume parameter

Van Laar equation: expresses the logarithm of the activity coefficient (γ_i) as function of composition, expressed as a mole fraction (x_i). For example, for a binary system

$$\log_{10} \gamma_1 = \frac{A_{12}}{\left(1 + \dfrac{A_{12} x_1}{A_{21} x_2}\right)^2}$$

where A_{12} and A_{21} are parameters of the equation. A similar expression can be written for γ_2. The van Laar's equation cannot represent activity coefficient curves that show a minimum or maximum.

Van't Hoff equation: relates the solubility, expressed as a mole fraction (x_2) at various temperatures to the heat of solution (ΔH_{soln}):

$$\left(\frac{\partial \log_e x_2}{\partial T} \right)_P = \frac{\Delta H_{soln}}{RT^2}$$

where T = absolute temperature

R = gas constant

P = pressure

Virial coefficients: represent the constants in the virial equation

$$\frac{PV}{RT} = 1 + \frac{B}{V} + \frac{C}{V^2} + \frac{D}{V^3} + \cdots$$

where B, C, D, $\cdots$ = second, third, fourth, $\cdots$ virial coefficients, respectively

P =   pressure

V =   volume

T =   absolute temperature

R =   gas constant

Wilson equation: used to correlate vapor-liquid equilibria data for binary and multicomponent homogeneous systems. The equation expresses the logarithm of the activity coefficient (γ_i) as a function of composition, expressed as a mole fraction (x_i). For example, for a binary system,

$$\log_e \gamma_1 = - \log_e(x_1 + A_{12}x_2) + x_2 \left(\frac{A_{12}}{x_1 + A_{12}x_2} - \frac{A_{21}}{x_2 + A_{21}x_1} \right)$$

where A_{12} and A_{21} are Wilson's parameters. A similar expression can be written for γ_2.

Appendix

Table A.1. Ionization Potential of Halogenated Hydrocarbons

Halogenated hydrocarbon	Ionization potential (eV)	Reference
CH_3Br	10.5	Schuster et al. (1976)
	10.53	Gordon & Ford (1972)
	10.69	Gibson (1977)
CH_3Cl	11.28	Schuster et al. (1976)
	11.26	Bingham et al. (1975)
	11.33	Gibson (1977)
	11.26	Gordon & Ford (1972)
CH_3F	12.54	Gibson (1977)
	12.54	Bingham et al. (1975)
	13.31	Dewar & Rzepa (1978)
CH_3I	9.1	Schuster et al. (1976)
	9.54	Gordon & Ford (1972)
	9.86	Gibson (1977)
CH_2Cl_2	11.35	Schuster et al. (1976)
	11.33	Bingham et al. (1975)
CH_2F_2	13.17	Dewar & Rzepa (1978)
	12.72	Bingham et al. (1975)
$CHCl_3$	11.42	Schuster et al. (1976)
	11.50	Bingham et al. (1975)
CHF_3	14.67	Dewar & Rzepa (1978)
	13.8	Bingham et al. (1975)
CCl_4	11.47	Schuster et al. (1976)
	11.47	Bingham et al. (1975)
CF_4	16.23	Dewar & Rzepa (1978)
	15.35	Bengham et al. (1975)
C_2H_5Br	10.46	Gibson (1977)
C_2H_5Cl	11.01	Gibson (1977)
	10.97	Bingham et al. (1975)
C_2H_5F	12.00	Gibson (1977)
	11.50	Bingham et al. (1975)
	12.43	Dewar & Rzepa (1978)
C_2H_5I	9.64	Gibson (1977)
CH_3-CHF_2	12.8	Dewar & Rzepa (1978)
	12.68	Bingham et al. (1975)
CH_3-CF_3	13.8	Dewar & Rzepa (1978)
	12.14	Bingham et al. (1975)

Table A.1. (Continued)

CF_3-CF_3	14.6	Dewar & Rzepa (1978)
	12.62	Bingham et al. (1975)
$CH_2=CHF$	10.58	Dewar & Rzepa (1978)
	10.31	Bingham et al. (1975)
$CH_2=CF_2$	10.72	Dewar & Rzepa (1978)
	10.15	Bingham et al. (1975)
$CHF=CHF$ (cis)	10.43	Dewar & Rzepa (1978)
$CHF=CHF$ (trans)	10.38	Dewar & Rzepa (1978)
$CHF=CF_2$	10.53	Dewar & Rzepa (1978)
	10.14	Bingham et al. (1975)
$CH\equiv CF$	11.26	Bingham et al. (1975)
$CF_2=CF_2$	10.54	Dewar & Rzepa (1978)
$CF\equiv CF$	11.30	Dewar & Rzepa (1978)
$CH_3-CHBr-CH_3$	10.26	Gibson (1977)
$CH_3-CHCl-CH_3$	10.78	Gibson (1977)
$CH_3-CHI-CH_3$	9.44	Gibson (1977)
$(CH_3)_2CHF$	11.12	Bingham et al. (1975)
$(CH_3)_3CF$	10.94	Bingham et al. (1975)
$CH_2=CHCl$	9.99	Bingham et al. (1975)
$CH_2=CCl_2$	9.46	Bingham et al. (1975)
$CHCl=CHCl$ (cis)	9.66	Bingham et al. (1975)
$CHCl=CHCl$ (trans)	9.95	Bingham et al. (1975)
$CCl_2=CCl_2$	9.65	Bingham et al. (1975)
CH_3-CCl_3	10.82	Bingham et al. (1975)
CCl_3-CCl_3	12.11	Bingham et al. (1975)
$(CH_3)_2CHCl$	11.2	Bingham et al. (1975)
$(CH_3)_3CCl$	10.61	Bingham et al. (1975)
$CH_2-CHCl-CH_2$	9.99	Bingham et al. (1975)
C_6H_6	9.2	Schuster et al. (1976)
C_6F_6	10.12	Dewar & Rzepa (1978)
	9.97	Bingham et al. (1975)
C_6H_5F	9.19	Dewar & Rzepa (1978)
	9.20	Bingham et al. (1975)
	9.19	Schuster et al. (1976)
$C_6H_4F_2$ (1,2-)	9.68	Dewar & Rzepa (1978)
$C_6H_4F_2$ (1,3-)	9.68	Dewar & Rzepa (1978)
$C_6H_4F_2$ (1,4-)	9.15	Dewar & Rzepa (1978)
$C_6H_4Cl_2$ (1,4-)	8.95	Bingham et al. (1975)

Table A.1. (Continued)

C_6H_5Cl	9.07	Bingham et al. (1975)
	9.07	Gordon & Ford (1972)
C_6H_5Br	8.98	Gordon & Ford (1972)
C_6H_5I	8.62	Gordon & Ford (1972)

Table A.2. Temperature Dependence of Dielectric Constant of Liquid Halocarbons

Halocarbon	Temperature (°C)			
	20	45	70	95
CCl_4	2.2443	2.1938	2.4315	–
CH_2Br-CH_2Br	4.791	4.650	4.497	–
$CH_2Br-CH_2-CH_2Br$	9.482	8.524	7.720	6.994
$CH_2Br-(CH_2)_2-CH_2Br$	8.829	8.162	7.561	–
$CH_2Br-(CH_2)_3-CH_2Br$	9.183	8.287	7.453	–
$CH_2Br-(CH_2)_4-CH_2Br$	8.436	7.756	7.155	6.587
$CH_2Br-(CH_2)_7-CH_2Br$	7.153	6.626	6.174	–
$CH_2Br-(CH_2)_8-CH_2Br$	–	6.261	5.805	5.415

Source: Ketelaar & van Meurs, 1957a, b.

Table A.3. Dielectric Constant of Liquid Halogenated Hydrocarbons

Halogenated hydrocarbon	Temperature (°C)	Dielectric constant	Reference
CCl_4	25	2.228	Finsy & van Loon (1976)
	25	2.23	Davis (1968)
	20	2.238	Gray (1963)
$CHCl_3$	20	4.81	Davis (1968)
	20	4.806	Gray (1963)
CH_2Cl_2	20	9.08	Gray (1963)
	20	9.08	Davis (1968)
CH_3Cl	-20	12.6	Gray (1963)
$CHBr_3$	20	4.39	Gray (1963)
CH_2Br_2	10	7.77	Gray (1963)
CH_3Br	0	9.82	Gray (1963)
CH_2I_2	25	5.32	Gray (1963)
CH_3I	20	7.00	Gray (1963)
$CCl_2=CCl_2$	25	2.30	Gray (1963)
	25	2.30	Davis (1968)
$CHCl=CCl_2$	16	3.42	Gray (1963)
$CCl_2=CH_2$	16	4.67	Gray (1963)
$CHCl=CHCl$ (cis)	25	9.20	Gray (1963)
$CHCl=CHCl$ (trans)	25	2.14	Gray (1963)
CCl_3-CH_3	0	7.949	Finsy & van Loon (1976)
	20	7.52	Gray (1963)
CCl_3-CHCl_2	20	3.73	Gray (1963)
$CHCl_2-CHCl_2$	20	8.20	Gray (1963)
CH_2Cl-CH_2Cl	25	10.23	Davis (1968)
	25	10.36	Gray (1963)
	20	10.63	Gray (1963)
$CHCl_2-CH_3$	18	10.0	Gray (1963)
	25	9.90	Davis (1968)
CH_3Br	0	10.6	Vuks (1969)
C_2H_5Br	25	9.2	Vuks (1969)
C_3H_7Br	25	8.09	Vuks (1969)
$n-C_4H_9Br$	25	6.93	Vuks (1969)
$n-C_5H_{11}Br$	25	6.31	Vuks (1969)
$n-C_6H_{13}Br$	25	5.82	Vuks (1969)
$n-C_7H_{15}Br$	25	5.33	Vuks (1969)

Table A.3. (Continued)

n-C_8H_{17}Br	25	5.00	Vuks (1969)
n-C_9H_{19}Br	25	4.74	Vuks (1969)
n-$C_{10}H_{21}$Br	25	4.44	Vuks (1969)
n-$C_{11}H_{23}$Br	-9	4.73	Vuks (1969)
n-$C_{12}H_{25}$Br	25	4.07	Vuks (1969)
n-$C_{13}H_{27}$Br	10	4.20	Vuks (1969)
n-$C_{14}H_{29}$Br	25	3.84	Vuks (1969)
n-$C_{15}H_{31}$Br	20	3.89	Vuks (1969)
n-$C_{16}H_{33}$Br	25	3.68	Vuks (1969)
n-$C_{18}H_{37}$Br	30	3.53	Vuks (1969)
n-$C_{22}H_{45}$Br	55	3.12	Vuks (1969)
CH_3Cl	-20	12.6	Vuks (1969)
C_3H_7Cl	20	7.7	Vuks (1969)
n-C_4H_9Cl	25	7.54	Vuks (1969)
n-$C_5H_{11}Cl$	11	6.5	Vuks (1969)
n-$C_7H_{15}Cl$	22	5.48	Vuks (1969)
n-$C_8H_{17}Cl$	25	5.75	Vuks (1969)
n-$C_{10}H_{21}Cl$	24.5	5.57	Vuks (1969)
n-$C_{12}H_{25}Cl$	25	4.17	Vuks (1969)
n-$C_{16}H_{33}Cl$	24.5	3.70	Vuks (1969)
CCl_3F	25	2.5	Döring (1977)
CCl_2F_2	25	2.1	Döring (1977)
$CHClF_2$	25	6.6	Döring (1977)
$CCl_2F-CClF_2$	25	2.6	Döring (1977)
$CClF_2-CClF_2$	25	2.2	Döring (1977)
$CBrF_2-CBrF_2$	25	2.7	Döring (1977)
C_6H_5Cl	25	5.62	Davis (1968)
	25	5.621	Gray (1963)
	20	5.708	Gray (1963)
C_6H_5Br	25	5.4	Gray (1963)
C_6H_5F	25	5.42	Gray (1963)
C_6H_5I	20	4.63	Gray (1963)
$C_6H_4Cl_2$ (1,2-)	25	9.93	Davis (1968)
	25	9.93	Gray (1963)
$C_6H_4Cl_2$ (1,3-)	25	5.04	Gray (1963)
$C_6H_4Cl_2$ (1,4-)	50	2.41	Gray (1963)
$C_6H_4Br_2$ (1,2-)	20	7.35	Gray (1963)
$C_6H_4Br_2$ (1,3-)	20	4.80	Gray (1963)

Table A.3. (Continued)

$C_6H_4Br_2$ (1,4-)	95	2.57	Gray (1963)
$C_6H_4I_2$ (1,2-)	20	5.70	Gray (1963)
$C_6H_4I_2$ (1,3-)	25	4.25	Gray (1963)
$C_6H_4I_2$ (1,4-)	120	2.88	Gray (1963)
C_6H_4BrCl (1,2-)	20	6.80	Gray (1963)
C_6H_4BrCl (1,3-)	20	4.58	Gray (1963)
$C_6H_3F_3$ (1,3,4-)	30	9.18	Davis (1968)
$C_6H_{11}Br$	25	7.92	Gray (1963)
$C_6H_{13}Br$	25	5.82	Gray (1963)

Table A.4. Dipole Moment of Halogenated Hydrocarbons

Halogenated hydrocarbon	Dipole moment (debyes)	Reference
CH_3Br	1.84	McClellan (1974)
CH_3Cl	1.87	Boggs & Buck (1958)
	1.892	McClellan (1974)
	1.87	Bingham et al. (1975)
CH_2Cl_2	1.60	Bingham et al. (1975)
	1.58	Boggs & Buck (1958)
	1.62	Dowell & Stewart (1976)
	1.59	McClellan (1974)
$CHCl_3$	1.06	Dowell & Stewart (1976)
	1.22	McClellan (1974)
	1.01	Bingham et al. (1975)
CCl_4	0.1	McClellan (1974)
	0.0	Bingham et al. (1975)
$CClF_3$	0.40	Boggs & Buck (1958)
	0.50	McClellan (1974)
$CHClF_2$	1.41	Boggs & Buck (1958)
	1.43	McClellan (1974)
CH_2ClF	1.82	Boggs & Buck (1958)
CH_2F_2	1.93	Boggs & Buck (1958)
	1.96	Dewar & Rzepa (1978)
	1.97	Bingham et al. (1975)
CH_3F	1.85	Boggs & Buck (1958)
	1.8471	McClellan (1974)
	1.86	Dewar & Rzepa (1978)
	1.71	Bingham et al. (1975)
CHF_3	1.649	McClellan (1974)
	1.65	Bingham et al. (1975)
	1.65	Dewar & Rzepa (1978)
CCl_2F_2	0.50	McClellan (1974)
CCl_3F	0.46	McClellan (1974)
$CBrCl_3$	0.59	McClellan (1974)
CBr_4	0.57	McClellan (1974)
$CHBrCl_2$	1.31	McClellan (1974)
$CHBr_3$	1.02	McClellan (1974)
	1.00	Dowell & Stewart (1976)

Table A.4. (Continued)

CH_2BrCl	1.66	McClellan (1974)
CH_2Br_2	1.50	McClellan (1974)
CH_3I	1.618	McClellan (1974)
CH_2I_2	1.09	Dowell & Stewart (1976)
	1.08	McClellan (1974)
CHI_3	0.86	McClellan (1974)
$CH_2{=}CHF$	1.427	Ellis et al. (1975)
	1.46	McClellan (1974)
	1.43	Bingham et al. (1975)
	1.43	Dewar & Rzepa (1978)
$CH_2{=}CF_2$	1.385	Ellis et al. (1975)
	1.3843	McClellan (1974)
	1.38	Bingham et al. (1975)
	1.39	Dewar & Rzepa (1978)
$CHF{=}CHF$ (cis)	2.42	Ellis et al. (1975)
	2.42	Dewar & Rzepa (1978)
$CH{\equiv}CF$	0.73	McClellan (1974)
	0.73	Dewar & Rzepa (1978)
	0.60	Bingham et al. (1975)
$CHF{=}CF_2$	1.30	Dewar & Rzepa (1978)
	1.39	Bingham et al. (1975)
	1.32	McClellan (1974)
$CF_3{-}CF_3$	0.0	Bingham et al. (1975)
$CBrF{=}CF_2$	0.76	McClellan (1974)
$CBrF{=}CHBr$ (cis)	1.20	McClellan (1974)
$CBrF{=}CHBr$ (trans)	1.36	McClellan (1974)
$CHBr{=}CBr_2$	0.82	McClellan (1974)
$CHBr{=}CFH$ (cis)	1.94	McClellan (1974)
$CHBr{=}CFH$ (trans)	0.39	McClellan (1974)
$CHBr{=}CHI$ (cis)	1.30	McClellan (1974)
$CHBr{=}CHI$ (trans)	0.39	McClellan (1974)
$CHBr_2{-}CHBr_2$	1.41	McClellan (1974)
$CH_2{=}CHBr$	1.36	McClellan (1974)
$CHBr_2{-}CH_2Br$	1.41	McClellan (1974)
$CHBr_2{-}CH_3$	1.94	McClellan (1974)
$CH_2Br{-}CH_2Br$	1.28	McClellan (1974)
$CH_3{-}CH_2Br$	2.069	McClellan (1974)
$CClF{=}CF_2$	0.58	McClellan (1974)

Table A.4. (Continued)

CHCl=CHF (cis)	1.6	McClellan (1974)
CClH$_2$-CFH$_2$	2.72	McClellan (1974)
CH$_3$-CH$_2$F	1.94	Bingham et al. (1975)
	1.96	Dewar & Rzepa (1978)
CH$_3$-CHF$_2$	2.27	Bingham et al. (1975)
	2.30	Dewar & Rzepa (1978)
CH$_2$F-CH$_2$F	2.67	McClellan (1974)
	2.67	Dewar & Rzepa (1978)
CHF$_2$-CH$_2$F	1.58	McClellan (1974)
CH$_3$-CF$_3$	2.32	Dewar & Rzepa (1978)
CF$_2$=CFI	1.04	McClellan (1974)
CH$_2$=CHCl	1.45	Bingham et al. (1975)
	0.52	McClellan (1974)
CH$_2$=CCl$_2$	1.34	Bingham et al. (1975)
	1.28	McClellan (1974)
CHCl=CHCl (cis)	1.90	Bingham et al. (1975)
	1.84	McClellan (1974)
CHCl=CHCl (trans)	0.0	Bingham et al. (1975)
CCl$_2$=CCl$_2$	0.0	Bingham et al. (1975)
CHCl=CCl$_2$	0.99	McClellan (1974)
CH$_3$-CH$_2$Cl	1.78	Bingham et al. (1975)
	1.90	McClellan (1974)
CCl$_2$-CCl$_3$	0.0	Bingham et al. (1975)
CCl$_3$-CHCl$_2$	1.07	McClellan (1974)
CCl$_3$-CH$_2$Cl	1.45	McClellan (1974)
CHCl$_2$-CHCl$_2$	1.70	McClellan (1974)
CCl$_3$-CH$_3$	1.755	McClellan (1974)
CCl$_2$H-CH$_3$	1.88	McClellan (1974)
CHCl=CHI (trans)	0.549	McClellan (1974)
CHI=CHI (cis)	1.14	McClellan (1974)
CCl$_2$H-CClH$_2$	1.57	McClellan (1974)
CH$_2$I-CH$_2$I	0.53	McClellan (1974)
CH$_3$-CH$_2$I	1.75	McClellan (1974)
CH$_2$=C=CHF	1.97	Ellis et al. (1975)
CH$_2$=C=CF$_2$	2.07	Ellis et al. (1975)
CHF=C=CHF	1.77	Ellis et al. (1975)
CH$_2$=C=CHBr	1.45	McClellan (1974)
CH$_2$=C=CHCl	1.57	McClellan (1974)

Table A.4. (Continued)

$CH_2=CH-CCl_3$	1.56	McClellan (1974)
$CH_2=CH-CHF_2$	0.889	McClellan (1974)
$CH_2=CH-CH_2Br$	1.73	McClellan (1974)
$CH_2=CH-CH_2Cl$ (cis)	1.64	McClellan (1974)
$CHCl=CH-CH_3$	1.75	McClellan (1974)
$CH_2=CH-CH_2F$ (cis)	1.46	McClellan (1974)
$CHF=CH-CH_3$ (cis)	1.765	McClellan (1974)
$CH_2=CI-CH_3$	1.57	McClellan (1974)
C_6H_5Br	1.71	Dowell & Stewart (1976)
	1.55	McClellan (1974)
C_6H_5Cl	1.72	Dowell & Stewart (1976)
	1.55	McClellan (1974)
	1.69	Bingham et al. (1975)
C_6H_5F	1.60	Dowell & Stewart (1976)
	1.43	McClellan (1974)
	1.66	Dewar & Rzepa (1978)
	1.60	Bingham et al. (1975)
C_6H_5I	1.72	Dowell & Stewart (1976)
	1.36	McClellan (1974)
$C_6H_4Cl_2$ (1,4-)	0.0	Bingham et al. (1975)
	0.0	McClellan (1974)
$C_6H_4Cl_2$ (1,2-)	2.05	McClellan (1974)
$C_6H_4Cl_2$ (1,3-)	1.36	McClellan (1974)
$C_6H_4F_2$ (1,2-)	2.59	Dewar & Rzepa (1978)
	2.46	McClellan (1974)
$C_6H_4F_2$ (1,4-)	1.51	Dewar & Rzepa (1978)
	0.0	McClellan (1974)
C_6H_4BrF (1,4-)	0.0	McClellan (1974)
C_6H_4ClF (1,2-)	2.41	McClellan (1974)
C_6H_4ClF (1,3-)	1.47	McClellan (1975)
C_6H_4ClF (1,4-)	0.0	McClellan (1974)
$C_6H_4F_2$ (1,3-)	1.51	McClellan (1974)
C_6H_4BrCl (1,2-)	2.26	McClellan (1974)
C_6H_4BrCl (1,4-)	0.0	McClellan (1974)
$C_6H_4Br_2$ (1,4-)	0.0	McClellan (1974)
$C_6H_4Br_2$ (1,3-)	1.24	McClellan (1974)

Table A.5.   Partial Molar Heat Capacities of Compounds in Aqueous Solutions at 25°C

Solute	Partial molar heat capacity (cal g-mole^{-1} K^{-1})
Methyl fluoride	35.0
Methyl chloride	43.0
Methyl bromide	44.0
Methyl iodide	86.0
Methane, tetrafluoro-	98.0
Ethane, hexafluoro-	173.0
Propane, octafluoro-	3000.0 (at 10°C)
Cyclobutane, octafluoro-	189.0
Methane	62.62
Ethane	72.66
Propane	87.48
n-Butane	95.36
n-Pentane	136.71
n-Hexane	151.77
Benzene	86.28
Toluene	102.77
Ethylbenzene	120.46
Propylbenzene	144.84
Ethylene	58.08
Formic acid	22.94
Acetic acid	39.44
Cyclohexane	123.33
Acetone	57.60
Aniline	73.37
Methanol	37.81
Ethanol	62.21
n-Propanol	84.35
n-Butanol	104.45
n-Pentanol	125.19

Source: Alexander et al., 1971; D'Orazio & Wood, 1963; Gill et al., 1976; Guthrie, 1977; Jolicoeur & Lacroix, 1976; Muccitelli, 1978; Nemethy & Scherega, 1962a, b; Nichols et al., 1976; Roux et al., 1978; Wauchope & Haque, 1972; Wen & Muccitelli, 1978.

Table A.6. Heat of Solution of Halocarbons and Other Compounds in Water

Solute	Temperature ($^\circ$C)	Heat of solution (cal g-mole^{-1})	Reference
CBrH$_3$	35	-5,175	Swain & Thornton (1962)
	0	-7,390	Glew (1962)
	20	-6,845	Carey et al. (1966)
	25	-6,400	Robertson (1967)
	25	-6,282	Klausutis (1961)
	25	-6,275	Glew & Moelwyn-Hughes(1953)
CBr$_3$H	25	-177	Larsen & Magid (1974)
CClH$_3$	45	-4,865	Svetlanov et al. (1971)
	35	-5,212	Swain & Thornton (1962)
	25	-6,300	Butler (1937)
	4	-7,000	Glew & Moelwyn-Hughes(1953)
	25	-5,670	Glew & Moelwyn-Hughes(1953)
	25	-5,700	Robertson (1967)
	25	-6,900	Valentiner (1927)
CCl$_2$H$_2$	45	-6,110	Svetlanov et al. (1971)
	25	-349	Larsen & Magid (1974)
CCl$_3$H	45	-4,910	Svetlanov et al. (1971)
	25	-167	Larsen & Magid (1974)
	25	-9,800	Butler (1937)
	15	-7,261	Jones et al. (1957)
CCl$_4$	45	-6,980	Svetlanov et al. (1971)
	25	-1,020	Larsen & Magid (1974)
	25	-6,900	Butler (1937)
	25	-3,700	Valentiner (1927)
	15	-5,957	Jones et al. (1957)
CFH$_3$	35	-4,143	Swain & Thornton (1962)
	25	-4,404	Glew & Moelwyn-Hughes(1953)
CF$_3$H	25	-8,565	Stepakoff & Modica (1973)
CF$_4$	5	-5,400	Ashton et al. (1968)
	15	-4,600	Ashton et al. (1968)
	25	-3,700	Ashton et al. (1968)
	35	-2,800	Ashton et al. (1968)
	45	-1,900	Ashton et al. (1968)
	25	-7,608	Stepakoff & Modica (1973)
	5	-5,254	Muccitelli (1978)

Table A.6. (Continued)

CF_4	10	-4,754	Muccitelli (1978)
	15	-4,245	Muccitelli (1978)
	20	-3,727	Muccitelli (1978)
	25	-3,201	Muccitelli (1978)
	30	-2,665	Muccitelli (1978)
	5	-5,250	Wen & Muccitelli (1979)
	10	-4,750	Wen & Muccitelli (1979)
	15	-4,250	Wen & Muccitelli (1979)
	20	-3,730	Wen & Muccitelli (1979)
	25	-3,200	Wen & Muccitelli (1979)
	30	-2,670	Wen & Muccitelli (1979)
CH_3I	35	-5,916	Swain & Thornton (1962)
	0	-8,500	Glew (1962)
	25	-6,400	Robertson (1967)
	25	-6,375	Glew & Moelwyn-Hughes (1953)
$CClF_3$	25	-4,042	Stepakoff & Modica (1973)
$CClF_2H$	25	-4,868	Stepakoff & Modica (1973)
$CClFH_2$	10	-6,475	Stepakoff & Modica (1973)
	20	-6,475	Carey et al. (1966)
	25	-6,244	Klausutis (1961)
CCl_2FH	10	-8,608	Stepakoff & Modica (1973)
	10	-7,880	Carey et al. (1966)
	20	-8,365	Carey et al. (1966)
	25	-8,429.5	Klausutis (1961)
CCl_2F_2	25	-4,720	Stepakoff & Modica (1973)
$CBrH_2-CH_3$	25	-7,000	Robertson (1967)
$CClH_2-CH_3$	0	-8,400	Glew (1962)
	4	-700	Lucas (1970)
	25	-6,700	Robertson (1967)
$CClH_2-CClH_2$	25	-143	De Lisi et al. (1980)
$CClF_2-CClF_2$	0	-6,597	Stepakoff & Modica (1973)
$CClF_2-CH_3$	10	-6,970	Stepakoff & Modica (1973)
	25	-6,970	Carey et al. (1966)
$CClF_2-CF_3$	25	-6,199	Stepakoff & Modica (1973)
$CF_2=CF_2$	0	-7,200	Volokhonovich et al. (1966)
	20	-5,100	Volokhonovich et al. (1966)
	40	-3,400	Volokhonovich et al. (1966)
	60	-2,000	Volokhonovich et al. (1966)

Table A.6. (Continued)

CF_3-CF_3	5	-7,360	Wen & Muccitelli (1979)
	10	-6,460	Wen & Muccitelli (1979)
	15	-5,530	Wen & Muccitelli (1979)
	20	-4,590	Wen & Muccitelli (1979)
	25	-3,640	Wen & Muccitelli (1979)
	30	-2,670	Wen & Muccitelli (1979)
	5	-7,360	Muccitelli (1978)
	10	-6,457	Muccitelli (1978)
	15	-5,534	Muccitelli (1978)
	20	-4,594	Muccitelli (1978)
	25	-3,639	Muccitelli (1978)
	30	-2,667	Muccitelli (1978)
CH_3-CH_2I	25	-7,000	Robertson (1967)
$CF_3-CF_2-CF_3$	5	-28,000	Muccitelli (1978)
	10	-13,000	Muccitelli (1978)
	15	2,315	Muccitelli (1978)
$CBrH_2-CH_2-CH_3$	25	-8,600	Robertson (1967)
$CClH_2-CH=CH_2$	25	-8,560	Treger et al. (1964)
	50	-4,500	Treger et al. (1964)
	70	-1,200	Treger et al. (1964)
$CClH_2-CH_2-CH_3$	25	-8,100	Robertson (1967)
$CH_3-CH_2-CH_2I$	25	-8,600	Robertson (1967)
$CBrH_2-(CH_2)_2-CH_3$	25	-9,700	Robertson (1967)
$CClH_2-(CH_2)_2-CH_3$	25	-9,100	Robertson (1967)
$CH_3-(CH_2)_2-CH_2I$	25	-9,700	Robertson (1967)
$CF_2-CF_2-CF_2-CF_2$	0	-4,050	Stepakoff & Modica (1973)
	5	-9,269	Muccitelli (1978)
	10	-8,299	Muccitelli (1978)
	15	-7,312	Muccitelli (1978)
	20	-6,307	Muccitelli (1978)
	25	-5,286	Muccitelli (1978)
	30	-4,247	Muccitelli (1978)
	5	-9,270	Wen & Muccitelli (1979)
	10	-8,300	Wen & Muccitelli (1979)
	15	-7,310	Wen & Muccitelli (1979)
	20	-6,310	Wen & Muccitelli (1979)
	25	-5,290	Wen & Muccitelli (1979)
	30	-4,250	Wen & Muccitelli (1979)

Table A.6. (Continued)

$CBrH_2-(CH_2)_3-CH_3$	25	-10,800	Robertson (1967)
$CClH_2-(CH_2)_3-CH_3$	25	-10,300	Robertson (1967)
$CH_3-(CH_2)_3-CH_2I$	25	-10,800	Robertson (1967)
C_6BrH_5	25	1,635	Vesala (1973)
C_6ClH_5	25	-600	Lucas (1970)
	25	-597	De Lisi et al. (1980)
	0	1,520	Katayama et al. (1966)
	5	953	Katayama et al. (1966)
	10	473	Katayama et al. (1966)
	15	53	Katayama et al. (1966)
	20	-450	Katayama et al. (1966)
	25	-930	Katayama et al. (1966)
	30	-1,560	Katayama et al. (1966)
	35	-2,500	Katayama et al. (1966)
$1,3-C_6Cl_2H_4$	25	2,063	Vesala (1973)
$1,4-C_6Cl_2H_4$	25	4,486	Vesala (1973)
	25	-11,360	Wauchope (1970)
C_6FH_5	15	-6,281	Jones et al. (1957)
C_6H_5I	25	1,840	Vesala (1973)
CH_4	25	-2,800	Lucas (1970)
	25	-1,900	Shinoda & Fujihira (1968)
	25	-2,600	Ben-Naim et al. (1973)
	25	-3,170	Zahradnik et al. (1975)
	25	-2,560	Nemethy & Scheraga (1962b)
	25	-3,850	Frank & Franks (1968)
	25	-3,180	Frank & Evans (1945)
	25	-2,960	Wauchope (1970)
CH_3-CH_3	25	-4,300	Lucas (1970)
	25	-2,000	Shinoda & Fujihira (1968)
	25	-4,200	Ben-Naim et al. (1973)
	25	-1,820	Nemethy & Scheraga (1962b)
	25	-470	Frank & Franks (1968)
	25	-4,430	Frank & Evans (1945)
	25	-4,620	Wauchope (1970)
$CH_3-CH_2-CH_3$	25	-5,300	Lucas (1970)
	25	-1,880	Shinoda & Fujihira (1968)
	25	-1,770	Nemethy & Scheraga (1962b)
	25	-2,580	Frank & Franks (1968)
	25	-5,700	Wauchope (1970)

Table A.6. (Continued)

$CH_3-CH_2-CH_2-CH_3$	25	-820	Shinoda & Fujihira (1968)
	25	-5,500	Ben-Naim et al. (1973)
	25	-840	Nemethy & Scheraga (1962b)
	25	-2,890	Frank & Franks (1968)
	25	-6,000	Wauchope (1970)
$C_6H_5-CH_3$	25	-530	Bohon & Claussen (1951)
	25	-640	Shinoda & Fujihira (1968)
	25	-200	Krishnan & Friedman (1969)
	25	-640	Nemethy & Scheraga (1962b)
	25	-430	De Lisi et al. (1980)
	25	-413	Gill et al. (1976)
	25	-8,640	Wauchope (1970)
	25	469	Vesala (1973)
$C_6H_5-CH_2-CH_3$	25	-9,580	Wauchope (1970)
	25	588	Vesala (1973)
Naphthalene	25	-11,010	Wauchope (1970)
	25	6,231	Vesala (1973)
Biphenyl	25	-12,040	Wauchope (1970)
	25	6,822	Vesala (1973)

Table A.7.   Heat of Solution of Benzene in Water --- Compilation

Temperature (°C)	Heat of solution (cal g-mole^{-1})	Reference
25	191.04	Reid et al. (1969)
25	110	Krishnan & Friedman (1969)
25	580	Shinoda & Fujihira (1968)
25	580	Nemethy & Scheraga (1962b)
25	570	Bohon & Claussen (1951)
25	452.2	Robertson (1967)
25	430	Franks et al. (1963)
30	889	Lyashchenko & Kalinowska (1977)
25	559	De Lisi et al. (1980)
25	497	Gill et al. (1975)
25	566	Vesala (1973)
25	-7,640	Wauchope (1970)

Table A.8. Heat of Solution of Water in Halocarbons and Other Liquids

Solvent	Temperature (°C)	Heat of solution (cal g-mole^{-1})	Reference
CCl_2H_2	25	2,870	Staverman (1941b)
	40	5,900	Prosyanov et al. (1974)
CCl_3H	25	3,010	Staverman (1941b)
CCl_4	20	5,160	Bell (1932)
	25	4,090	Staverman (1941b)
	25	5,400	Glasoe & Schultz (1972)
	76.54	5,880	Prosyanov et al. (1974)
CCl_2F_2	25	9,200	Zhukoborskii (1973)
$CClH=CH_2$	-13.37	6,500	Prosyanov et al. (1974)
$CCl_2=CCl_2$	20	3,760	Bell (1932)
	37	7,100	Prosyanov et al. (1974)
CCl_2H-CH_3	25	3,100	Staverman (1941b)
$CBrH_2-CBrH_2$	25	2,950	Staverman (1941b)
$CClH_2-CClH_2$	25	2,710	Staverman (1941b)
	25	4,036	De Lisi et al. (1980)
$CCl_2H-CClH_2$	25	2,800	Staverman (1941b)
CCl_2H-CCl_2H	25	2,710	Staverman (1941b)
	146.2	6,300	Prosyanov et al. (1974)
CCl_3-CClH_2	25	3,100	Staverman (1941b)
CCl_3-CCl_2H	25	3,280	Staverman (1941b)
C_6ClH_5	20	6,580	Bell (1932)
	25	5,770	Goldman (1969)
	25	10,170	Jones & Monk (1963)
	132	6,300	Prosyanov et al. (1974)
	25	9,888	De Lisi et al. (1980)
C_6BrH_5	20	5,050	Bell (1932)
	25	10,870	Jones & Monk (1963)
C_6H_5I	25	11,850	Jones & Monk (1963)
$C_6Cl_2H_4$ (1,2-)	25	5,620	Goldman (1969)
	25	12,000	Jones & Monk (1963)
C_6F_{14}	25	6,352	Shields (1976a, b)
	50	6,783	Shields (1976a, b)
C_6H_6	20	3,770	Bell (1932)
	25	5,574	Karlsson (1972)
	25	6,100	Goldman (1969)

Table A.8. (Continued)

C_6H_6	25	8,090	Jones & Monk (1963)
	25	4,944	De Lisi et al. (1980)
$C_6H_5-CH_3$	20	4,580	Bell (1932)
	25	4,400	Glasoe & Schultz (1972)
	25	9,080	Jones & Monk (1963)
	25	7,380	De Lisi et al. (1980)
$C_6H_5-CH_2-CH_3$	25	10,100	Jones & Monk (1963)
CH_3-CH_3	25	3,540	Staverman (1941b)
$CH_2=CHCl$	25	5,682	Clarke et al. (1981)

Table A.9. Properties of Water

Temperature ($^\circ$C)	Vapor pressure (mm Hg)	Sat. liquid density (g cc^{-1})	Thermal expansion coeff. ($^\circ$C^{-1} x 10^{-3})	Compressibility coeff. (atm^{-1} x 10^{-6})	Refractive index Na 0.5893
0	4.581	0.99984	-0.07	50.6	1.33464
10	9.204	0.99970	0.088	48.6	1.33389
20	17.53	0.99820	0.207	47.0	1.33299
25	23.75	0.99705	0.255	46.5	1.33287
30	31.81	0.99565	0.303	46.0	1.33192
40	55.32	0.99222	0.385	45.3	1.33051
50	92.52	0.98805	0.457	45.0	1.32894
60	149.39	0.98321	0.523	45.0	1.32718
70	233.7	0.97779	0.585	45.2	1.32511
80	355.2	0.97183	0.643	45.7	1.32287
90	525.9	0.96532	0.698	46.5	1.32050
100	760.0	0.95835	0.752	48.0	1.31783

Table A.9. (Continued)

Temperature (°C)	Heat capacity ($J\ g^{-1}\ {}^{\circ}C^{-1}$)	Latent heat of vap. ($kJ\ kg^{-1}$)	Enthalpy ($kJ\ kg^{-1}$)	Entropy ($kJ\ kg^{-1}\ {}^{\circ}C^{-1}$)	Free energy ($kJ\ kg^{-1}$)	Velocity of sound ($m\ s^{-1}$)	Dielectric constant
0	4.2174	2500.5	0.00	0.0000	0.00	1402.74	87.69
10	4.1919	2476.9	42.03	0.1511	42.03	1447.59	83.82
20	4.1816	2453.4	83.86	0.2963	83.86	1482.66	80.08
25	4.1793	2441.7	104.74	0.3669	104.74	1497.00	78.25
30	4.1782	2429.9	125.61	0.4364	146.46	1509.44	76.49
40	4.1783	2406.2	167.34	0.5718	167.33	1529.18	73.02
50	4.1804	2382.2	209.11	0.7031	209.10	1542.87	69.70
60	4.1841	2357.9	250.91	0.8304	250.89	1551.30	66.51
70	4.1893	2333.1	292.78	0.9542	292.75	1555.12	63.45
80	4.1961	2307.8	334.72	1.0747	334.67	1554.81	60.54
90	4.2048	2281.9	376.75	1.1920	376.68	1550.79	57.77
100	4.2156	2255.5	418.88	1.3063	418.77	1543.41	55.15

Table A.9. (Continued)

Temperature (°C)	Viscosity (cP)	Thermal conductivity (W m^{-1} °C^{-1})	Surface tension (dyn cm^{-1})	Electrical conductivity[a] (ohm^{-1} cm^{-1} x 10^{-8})	Enthalpy of ionization (kJ g-mole^{-1})
0	1.788	0.550	75.62	1.61	62.81
10	1.305	0.576	74.20	2.85	59.64
20	1.004	0.598	72.75	4.94	57.00
25	0.903	0.608	71.96	6.34	55.84
30	0.801	0.617	71.15	8.04	54.75
40	0.653	0.633	69.55	12.53	52.75
50	0.550	0.647	67.90	18.90	50.90
60	0.470	0.658	66.17	27.58	49.13
70	0.406	0.667	64.41	38.93	47.39
80	0.355	0.675	62.60	53.03	45.64
90	0.315	0.680	60.74	69.65	43.86
100	0.282	0.683	58.84	88.10	42.05

[a]Source: Holzapfel, 1969.

Table A.10. Physical Properties of Refrigerants - Methane Series

Refrigerant number	Formula	Name	M.wt.	M.P. (°C)	B.P. (°C)	Liquid density (g cc^{-1})	Crit. temp. (°C)
10	CCl_4	Methane, tetrachloro-	153.82	-22.99	76.54	1.5940(20°C)	283.2
11	CCl_3F	Methane, fluorotrichloro-	137.37	-111.0	23.82	1.467(25°C)	198.0
12	CCl_2F_2	Methane, dichlorodifluoro-	120.91	-158.0	-29.79	1.311(25°C)	111.8
13	$CClF_3$	Methane, chlorotrifluoro-	104.46	-181.0	-81.4	1.298(-30°C)	28.81
14	CF_4	Methane, tetrafluoro-	88.01	-184.0	-127.96	1.317(-80°C)	-45.55
20	CCl_3H	Methane, trichloro-	119.38	-63.5	61.7	1.4832(20°C)	263.2
21	CCl_2FH	Methane, dichlorofluoro-	102.92	-135.0	8.92	1.366(25°C)	178.43
22	$CClF_2H$	Methane, chlorodifluoro-	80.47	-160.0	-40.75	1.194(25°C)	96.10
23	CF_3H	Methane, trifluoro-	70.01	-155.2	-82.03	0.670(25°C)	26.15
30	CCl_2H_2	Methane, dichloro-	84.93	-95.1	40.0	1.3266(20°C)	237.0
31	$CClFH_2$	Methane, chlorofluoro-	68.48	-133.0	-9.1	1.271(20°C)	153.4[a]
32	CF_2H_2	Methane, difluoro-	52.02	-136.0	-51.6	0.909(20°C)	78.40
40	$CClH_3$	Methane, chloro-	50.49	-97.73	-24.2	0.9159(20°C)	143.10
41	CFH_3	Methane, fluoro-	34.03	-141.8	-78.4	0.8428(-60°C)	44.60
10B1	$CBrCl_3$	Methane, bromotrichloro-	198.28	-5.65	104.7	2.0122(20°C)	333.1[a]
10B2	CBr_2Cl_2	Methane, dibromodichloro-	242.74	38.0	150.2	2.42(25°C)	411.8[a]
10B3	CBr_3Cl	Methane, chlorotribromo-	287.19	55.0	158.5	2.71(15°C)	432.4[a]
11B1	$CBrCl_2F$	Methane, bromodichlorofluoro-	181.90	—	52.0	—	246.9[a]
11B2	CBr_2ClF	Methane, chlorodibromofluoro-	226.30	—	80.0	—	297.4[a]
11B3	CBr_3F	Methane, fluorotribromo-	270.74	-74.5	106.0	2.7648(20°C)	345.7[a]
12B1	$CBrClF_2$	Methane, bromochlorodifluoro-	165.40	-160.5	-4.0	1.850(15°C)	153.73
12B2	CBr_2F_2	Methane, dibromodifluoro-	209.83	-110.0	24.5	2.3063(15°C)	198.15
13B1	$CBrF_3$	Methane, bromotrifluoro-	148.92	-168.0	-57.75	1.538(25°C)	67.0

Table A.10. (Continued)

20B1	CBrCl$_2$H	Methane, bromodichloro-	163.83	-57.1	90.0	1.980(20°C)	312.7[a]
20B2	CBr$_2$ClH	Methane, chlorodibromo-	208.29	-22.0	125.0	2.451(20°C)	375.7[a]
20B3	CBr$_3$H	Methane, tribromo-	252.75	8.3	149.5	2.8899(20°C)	422.9[a]
21B1	CBrClFH	Methane, bromochlorofluoro-	147.38	-115.0	36.1	1.9771(0°C)	225.0[a]
21B2	CBr$_2$FH	Methane, dibromofluoro-	191.84	26.5	64.9	2.421(20°C)	276.9[a]
22B1	CBrF$_2$H	Methane, bromodifluoro-	130.93	-14.5	-	1.55(16°C)	142.9[a]
30B1	CBrClH$_2$	Methane, bromochloro-	129.39	-86.5	68.11	1.9344(20°C)	284.6[a]
30B2	CBr$_2$H$_2$	Methane, dibromo-	173.85	-52.55	97.0	2.4970(20°C)	308.5[a]
31B1	CBrFH$_2$	Methane, bromofluoro-	112.94	-	19.0	-	203.7
40B1	CBrH$_3$	Methane, bromo-	94.94	-93.6	3.56	1.6755(20°C)	191.0

[a]Estimated values (Lydersen, 1955).

Table A.11 Literature Data for the Mutual Solubility Between Halocarbons and Water
Formula Index: Substances

$CBrClF_2$	Methane, bromochlorodifluoro-	R 12B1
$CBrClH_2$	Methane, bromochloro-	R 30B1
$CBrCl_3$	Methane, bromotrichloro-	R 10B1
$CBrF_3$	Methane, bromotrifluoro-	R 13B1
$CBrH_3$	Methane, bromo-	R 40B1
CBr_2H_2	Methane, dibromo-	R 30B2
CBr_3F	Methane, tribromofluoro-	R 11B3
CBr_3H	Methane, tribromo-	R 20B3
CBr_4	Methane, tetrabromo-	R 10B4
$CClFH_2$	Methane, chlorofluoro-	R 31
$CClF_2H$	Methane, chlorodifluoro-	R 22
$CClF_3$	Methane, chlorotrifluoro-	R 13
$CClH_3$	Methane, chloro-	R 40
CCl_2FH	Methane, dichlorofluoro-	R 21
CCl_2F_2	Methane, dichlorodifluoro-	R 12
CCl_2H_2	Methane, dichloro-	R 30
CCl_3F	Methane, trichlorofluoro-	R 11
CCl_3H	Methane, trichloro-	R 20
CCl_4	Methane, tetrachloro-	R 10
CFH_3	Methane, fluoro-	R 41
CF_2H_2	Methane, difluoro-	R 32
CF_3H	Methane, trifluoro-	R 23
CF_4	Methane, tetrafluoro-	R 14
CHI_2	Methane, triiodo-	
CH_2I_2	Methane, diiodo-	
CH_3I	Methane, iodo-	
CI_4	Methane, tetraiodo-	
C_2BrClF_3H	Ethane, 1-bromo-1-chloro-2,2,2-trifluoro-	R 123B1
C_2BrClH_4	Ethane, 1-bromo-2-chloro-	
C_2BrF_4H	Ethane, 1-bromo-1,2,2,2-tetrafluoro-	
C_2BrH_5	Ethane, bromo-	
$C_2Br_2ClH_3$	Ethane, 1,2-dibromo-1-chloro-	
$C_2Br_2Cl_2H_2$	Ethane, 1,2-dibromo-1,2-dichloro-	
$C_2Br_2F_4$	Ethane, 1,2-dibromo-1,1,2,2-tetrafluoro-	R 114B2
$C_2Br_2H_2$	Ethene, 1,2-dibromo- (cis-)	
$C_2Br_2H_4$	Ethane, 1,2-dibromo-	

Table A.11 (Continued)

$C_2Br_3H_3$	Ethane, 1,1,2-tribromo-	
$C_2Br_4H_2$	Ethane, 1,1,2,2-tetrabromo-	
C_2ClFH_4	Ethane, 1-chloro-2-fluoro-	R 151
$C_2ClF_2H_3$	Ethane, 1-chloro-2,2-difluoro-	R 142
$C_2ClF_2H_3$	Ethane, 1-chloro-1,1-difluoro-	R 142b
$C_2ClF_3H_2$	Ethane, 1-chloro-2,2,2-trifluoro-	R 133a
C_2ClF_4H	Ethane, 1-chloro-1,2,2,2-tetrafluoro-	R 124
C_2ClF_5	Ethane, chloropentafluoro-	R 115
C_2ClH_3	Ethene, chloro-	R 1140
C_2ClH_5	Ethane, chloro-	R 160
$C_2Cl_2F_2H_2$	Ethane, 1,2-dichloro-2,2-difluoro-	R 132b
$C_2Cl_2F_3H$	Ethane, 1,1-dichloro-2,2,2-trifluoro-	R 123
$C_2Cl_2F_4$	Ethane, 1,2-dichloro-1,1,2,2-tetrafluoro-	R 114
$C_2Cl_2F_4$	Ethane, 1,1-dichloro-1,2,2,2-tetrafluoro-	R 114a
$C_2Cl_2H_2$	Ethene, 1,1-dichloro-	R 1130a
$C_2Cl_2H_2$	Ethene, 1,2-dichloro- (cis-)	R 1130(c)
$C_2Cl_2H_2$	Ethene, 1,2-dichloro- (trans-)	R 1130(t)
$C_2Cl_2H_4$	Ethane, 1,1-dichloro-	R 150a
$C_2Cl_2H_4$	Ethane, 1,2-dichloro-	R 150
$C_2Cl_3F_3$	Ethane, 1,1,2-trichloro-1,2,2-trifluoro-	R 113
C_2Cl_3H	Ethene, trichloro-	R 1120
$C_2Cl_3H_3$	Ethane, 1,1,1-trichloro-	R 140a
$C_2Cl_3H_3$	Ethane, 1,1,2-trichloro-	R 140
C_2Cl_4	Ethene, tetrachloro-	R 1110
$C_2Cl_4F_2$	Ethane, 1,2-difluoro-1,1,2,2-tetrachloro-	R 112
$C_2Cl_4H_2$	Ethane, 1,1,1,2-tetrachloro-	R 130a
$C_2Cl_4H_2$	Ethane, 1,1,2,2-tetrachloro-	R 130
C_2Cl_5H	Ethane, pentachloro-	R 120
C_2Cl_6	Ethane, hexachloro-	R 110
C_2FH_3	Ethene, fluoro-	R 1141
C_2FH_5	Ethane, fluoro-	R 161
$C_2F_2H_2$	Ethene, 1,1-difluoro-	R 1132a
$C_2F_2H_4$	Ethane, 1,1-difluoro-	R 152a
C_2F_4	Ethene, tetrafluoro-	R 1114
$C_2F_4H_2$	Ethane, 1,1,1,2-tetrafluoro-	R 134a
C_2F_5H	Ethane, pentafluoro-	R 125
C_2F_6	Ethane, hexafluoro-	R 116
$C_2H_2I_2$	Ethene, 1,2-diiodo- (cis-)	
$C_2H_2I_2$	Ethene, 1,2-diiodo- (trans-)	

Table A.11 (Continued)

C_2H_5I	Ethane, iodo-	
C_3BrClH_6	Propane, 1-bromo-3-chloro-	
C_3BrH_7	Propane, 1-bromo-	
C_3BrH_7	Propane, 2-bromo-	
$C_3Br_2ClH_5$	Propane, 1,2-dibromo-3-chloro-	
$C_3Br_2F_6$	Propane, 1,2-dibromohexafluoro-	R 216B2
$C_3Br_2H_6$	Propane, 1,2-dibromo-	
$C_3Br_2H_6$	Propane, 1,3-dibromo-	
$C_3ClF_3H_4$	Propane, 1-chloro-3,3,3-trifluoro-	
C_3ClH_5	1-Propene, 3-chloro-	
C_3ClH_7	Propane, 1-chloro-	
C_3ClH_7	Propane, 2-chloro-	
$C_3Cl_2F_6$	Propane, 1,2-dichloro-1,1,2,3,3,3-hexafluoro-	R 216
$C_3Cl_2H_4$	Propene, 1,3-dichloro- (cis-)	
$C_3Cl_2H_4$	Propene, 1,3-dichloro- (trans-)	
$C_3Cl_2H_4$	2-Propene, 1,2-dichloro-	
$C_3Cl_2H_6$	Propane, 1,2-dichloro-	
$C_3Cl_2H_6$	Propane, 1,3-dichloro-	
$C_3Cl_3F_5$	Propane, 1,1,1-trichloro-2,2,3,3,3-pentafluoro-	R 215
$C_3Cl_3H_5$	Propane, 1,2,3-trichloro-	
$C_3Cl_4F_4$	Propane, 1,1,1,3-tetrachloro-2,2,3,3-tetrafluoro-	R 214
C_3Cl_6	Propane, hexachloro-	
C_3FH_5	1-Propene, 3-fluoro-	
C_3FH_7	Propane, 1-fluoro-	
C_3FH_7	Propane, 2-fluoro-	
C_3F_8	Propane, octafluoro-	
C_3H_7I	Propane, 1-iodo-	
C_3H_7I	Propane, 2-iodo-	
C_4BrH_9	n-Butane, 1-bromo-	
C_4BrH_9	Propane, 1-bromo-2-methyl-	
C_4ClH_9	n-Butane, 1-chloro-	
C_4ClH_9	n-Butane, 2-chloro-	
C_4ClH_9	Propane, 1-chloro-2-methyl-	
$C_4Cl_2H_8$	n-Butane, 1,1-dichloro-	
$C_4Cl_2H_8$	n-Butane, 2,3-dichloro-	
C_4Cl_5H	1,2-Butadiene, 1,1,3,4,4-pentachloro-	
C_4Cl_6	1,3-Butadiene, hexachloro-	
C_4F_8	Cyclobutane, octafluoro-	R C318
C_4H_9I	n-Butane, 1-iodo-	

Table A.11 (Continued)

C_5BrH_{11}	n-Butane, 1-bromo-3-methyl-
C_5ClH_{11}	n-Pentane, 1-chloro-
C_5ClH_{11}	n-Pentane, 2-chloro-
C_5ClH_{11}	n-Pentane, 3-chloro-
C_5ClH_{11}	n-Butane, 2-chloro-2-methyl-
$C_5Cl_2H_{10}$	n-Butane, 2,3-dichloro-2-methyl-
C_5Cl_6	2-Penten-4-yne, hexachloro-
C_5Cl_6	1,3-Cyclopentadiene, hexachloro-
C_5Cl_8	1,4-Pentadiene, octachloro-
C_6BrClH_4	Benzene, 1-bromo-2-chloro-
C_6BrClH_4	Benzene, 1-bromo-3-chloro-
C_6BrClH_4	Benzene, 1-bromo-4-chloro-
C_6BrH_4I	Benzene, 1-bromo-2-iodo-
C_6BrH_4I	Benzene, 1-bromo-3-iodo-
C_6BrH_4I	Benzene, 1-bromo-4-iodo-
C_6BrH_5	Benzene, bromo-
$C_6Br_2H_4$	Benzene, 1,2-dibromo-
$C_6Br_2H_4$	Benzene, 1,3-dibromo-
$C_6Br_2H_4$	Benzene, 1,4-dibromo-
$C_6Br_3H_3$	Benzene, 1,2,3-tribromo-
$C_6Br_3H_3$	Benzene, 1,2,4-tribromo-
$C_6Br_3H_3$	Benzene, 1,3,5-tribromo-
$C_6Br_4H_2$	Benzene, 1,2,4,5-tetrabromo-
C_6ClH_4I	Benzene, 1-chloro-2-iodo-
C_6ClH_4I	Benzene, 1-chloro-3-iodo-
C_6ClH_4I	Benzene, 1-chloro-4-iodo-
C_6ClH_5	Benzene, chloro-
C_6ClH_{13}	n-Hexane, 1-chloro-
$C_6Cl_2H_4$	Benzene, 1,2-dichloro-
$C_6Cl_2H_4$	Benzene, 1,3-dichloro-
$C_6Cl_2H_4$	Benzene, 1,4-dichloro-
$C_6Cl_3H_3$	Benzene, 1,2,3-trichloro-
$C_6Cl_3H_3$	Benzene, 1,2,4-trichloro-
$C_6Cl_3H_3$	Benzene, 1,3,5-trichloro-
$C_6Cl_4H_2$	Benzene, 1,2,3,4-tetrachloro-
$C_6Cl_4H_2$	Benzene, 1,2,3,5-tetrachloro-
$C_6Cl_4H_2$	Benzene, 1,2,4,5-tetrachloro-
C_6Cl_5H	Benzene, pentachloro-
C_6Cl_6	Benzene, hexachloro-

Table A.11 (Continued)

$C_6Cl_6H_6$	Cyclohexane, α-hexachloro-
$C_6Cl_6H_6$	Cyclohexane, β-hexachloro-
$C_6Cl_6H_6$	Cyclohexane, δ-hexachloro-
$C_6Cl_6H_6$	Cyclohexane, γ-hexachloro-
$C_6Cl_6H_6$	Cyclohexane, hexachloro-(α-, β-, δ-, γ-mixture)
C_6FH_5	Benzene, fluoro-
$C_6F_2H_4$	Benzene, 1,2-difluoro-
$C_6F_2H_4$	Benzene, 1,3-difluoro-
$C_6F_2H_4$	Benzene, 1,4-difluoro-
C_6F_{14}	n-Hexane, perfluoro-
$C_6H_3I_3$	Benzene, 1,2,3-triiodo-
$C_6H_3I_3$	Benzene, 1,2,4-triiodo-
$C_6H_3I_3$	Benzene, 1,3,5-triiodo-
$C_6H_4I_2$	Benzene, 1,2-diiodo-
$C_6H_4I_2$	Benzene, 1,3-diiodo-
$C_6H_4I_2$	Benzene, 1,4-diiodo-
C_6H_5I	Benzene, iodo-

CBrClF$_2$[a]
Methane, bromochlorodifluoro- (R 12B1)

Temperature ($^{\circ}$C)	Solubility in H$_2$O	Solubility of H$_2$O in	Reference
-20	0.1347 wt %	–	Glew (1960)
-15	0.1108	–	"
-10	0.0942	–	"
-5	0.0827	–	"
0	0.0747	–	"
2	0.0991	–	"
4	0.1353	0.0077048 wt %	"
6	0.1845	0.0077074	"
8	0.2531	0.0077074	"
10	0.3495	0.0077087	"
5.8 (1-4.5 bars)	Graph only	–	Filatkin et al.
.		–	(1976)
.		–	"
.		–	"
35		–	"

[a]M.wt. = 165.3648; m.p. = -159.5°C; b.p. = -3.3°C.

CBrClH$_2$[a]
Methane, bromochloro-

Temperature ($^{\circ}$C)	Solubility in H$_2$O	Solubility of H$_2$O in	Reference
25	1.5 g/100 g H$_2$O	0.09 g/100 g CBrClH$_2$	O'Connell (1963)
25	0.9	–	Kirk-Othmer (1964a)
25	1.5	0.09	Jolles (1966)

[a]M.wt. = 129.39; m.p. = -86.5°C; b.p. = 68.11°C; d_4^{20} = 1.9344 g cc^{-1}.

CBrCl$_3$[a]
Methane, bromotrichloro-

Temperature ($^{\circ}$C)	Solubility in H$_2$O	Solubility of H$_2$O in	Reference
25	–	0.006 g/100 g CBrCl$_3$	O'Connell (1963)
25	–	0.06	Jolles (1966)

[a]M.wt. = 198.28; m.p. = -5.65°C; b.p. = 104.7°C; d_4^{20} = 2.0122 g cc^{-1}.

CBrF$_3$[a]
Methane, bromotrifluoro- (R 13B1)

Temperature ($^{\circ}$C)	Solubility in H$_2$O	Solubility of H$_2$O in	Reference
21	–	0.0095 wt %	du Pont (1969a)
25	0.03 wt %	–	"
25 (0-225 psia total pressure)	Graphs only	–	du Pont (1952c)
.		–	"
:		–	"
75		–	"
21	–	0.0095 wt %	du Pont (1966c)
25 (14.7 psia)	0.032 wt %	–	"
25 (14.7-200 psia total pressure)	Table	–	"
50		–	"
75		–	"
21	–	95 ppm by weight	Elchardus & Maestre
25	0.03 wt %	–	(1965)
21	–	0.0095 wt %	Matheson Company
25	0.03 wt %	–	(1966, 1971)
25 (1 atm)	0.03 wt %	–	Kirk-Othmer (1966a)
20	–	0.0092 g/100 g CBrF$_3$	Kali-Chemie (1972)
25	0.03 g/100 g H$_2$O	–	"

[a]M.wt. = 148.92; m.p. = -168.0°C; b.p. = -59.0°C; d_4^{25} = 1.538 g cc^{-1}.

CBrH$_3^a$

Methane, bromo-

Temperature (^{o}C)	Solubility in H$_2$O	Solubility of H$_2$O in	Reference
5.01	2,070 mm Hg mol^{-1} l^{-1} soln.	–	Glew & Moelwyn-
10.01	2,550	–	Hughes (1953)
14.99	3,200	–	"
20.01	3,990	–	"
24.98	4,730	–	"
30.01	5,580	–	"
40.06	7,540	–	"
50.03	9,760	–	"
60.04	12,400	–	"
70.01	14,900	–	"
80.01	18,000	–	"
10	2.679 g/100 g H$_2$O	–	Haight (1951)
17	1.830	–	"
25	1.341	–	"
32	1.149	–	"
20	1.2933 wt %	–	Korenman et al. (1971)
20(748 mm Hg)	1.75 g/100 g soln.	–	Matheson Company (1971)
20	47,600 mm Hg mol^{-1} l^{-1} soln.	–	Moelwyn-Hughes (1938)
25(760 mm Hg)	0.002946 mole fraction	–	Moelwyn-Hughes (1964)
20	0.09 g/100 ml H$_2$O	–	Patty (1962)
29.4	5,510 mm Hg mol^{-1} l^{-1} soln.	–	Swain & Thornton
40.3	7,443	–	(1962)
49.6	9,400	–	"
25	1.34 g/100 g H$_2$O	–	Gunther et al. (1968)
20	900 ppm by weight	–	Nex & Swezey (1954)
20(748 mm Hg)	1.75 g/100 g soln.	–	Kirk-Othmer (1964a)
20	>0.1 g/100 g H$_2$O	–	Manufacturing Chemists Assoc. (1956-1971:SD-35)
25	1.34 g/100 cc H$_2$O	–	Jolles (1966)

$CBrH_3$ (Continued)

14.73 (1151 mm Hg) 2.84 wt %			
14.73 (1151 mm Hg) 2.84 wt %		–	Colten et al. (1972)
20	16000 ppm by weight	–	Metcalf (1962)
20	0.09 g/100 cc H_2O	–	"
20	1.75 g/100 g H_2O	–	Andelman (1978)

[a]M.wt. = 94.94; m.p. = -93.6°C; b.p. = 3.56°C; d_4^{20} = 1.6755 g cc^{-1}.

CBr_2H_2[a]
Methane, dibromo-

Temperature (°C)	Solubility in H_2O	Solubility of H_2O in	Reference
25	0.70 cc/100 cc H_2O	–	Booth & Everson (1948)
30	11.93 g/1000 g H_2O	–	Gross & Saylor
15	11.70	–	(1931)
25	1.10 g/100 g H_2O	0.07 g/100 g CBr_2H_2	O'Connell (1963)
0	1.173 g/100 g H_2O	–	Rex (1906)
10	1.146	–	"
20	1.148	–	"
30	1.176	–	"
30	0.00686 mol/100 g H_2O	–	Van Arkel & Vles (1936)
0	1.15 wt %	–	Washburn (1928)
10	1.13	–	"
20	1.13	–	"
30	1.16	–	"
0	1.173 g/100 g H_2O	–	Lange (1967)
20	1.150	–	"
25	1.1 g/100 g H_2O	0.7 g/100 g CBr_2H_2	Jolles (1966)
15	1.15 wt %	–	Stephen & Stephen
30	1.18	–	(1963)
0	1.173 g/100 g H_2O	–	Andelman (1978)
10	1.146	–	"
20	1.148	–	"
30	1.176	–	"

[a]M.wt. = 173.85; m.p. = -52.55°C; b.p. = 97.0°C; d_4^{20} = 2.4970 g cc^{-1}.

CBr_3F[a]

Methane tribromofluoro- (R 11B3)

Temperature, ($^\circ$C)	Solubility in H_2O	Solubility of H_2O in	Reference
25	0.04 g/100 g H_2O	0.02 g/100 g CBr_3F	O'Connell (1963)
20	Insoluble	–	Gmelin (1974)
25	0.4 g/100 g H_2O	0.2 g/100 g CBr_3F	Jolles (1966)

[a]M.wt. = 270.74; b.p. = 106.0°C; d_4^{20} = 2.7648 g cc^{-1}.

CBr_3H[a]

Methane, tribromo- (R 20B3)

Temperature ($^\circ$C)	Solubility in H_2O	Solubility of H_2O in	Reference
25	<0.02 cc/100 cc H_2O	–	Booth & Everson (1948)
15	3.01 g/1000 g H_2O	–	Gross & Saylor
30	3.19	–	(1931)
25	0.311 g/100 g H_2O	–	O'Connell (1963)
30	0.318 wt %	–	Riddick & Bunger (1970)
20	1.0 g/800 g H_2O	–	Squire & Caines (1905)
30	0.00126 mol/100 g H_2O	–	Van Arkel & Vles (1936)
30	0.3 g/100 cc H_2O	–	Patty (1962)
0 (cold)	0.1 g/100 g H_2O	–	Lange (1967)
30	0.3 g/100 g H_2O	–	Jolles (1966)
15	0.301 g/100 g H_2O	–	Andelman (1978)
30	0.319	–	"

[a]M.wt. = 252.75; m.p. = 8.30°C; b.p. = 149.5°C; d_4^{20} = 2.8899 g cc^{-1}.

CBr_4[a]

Methane, tetrabromo-

Temperature ($^\circ$C)	Solubility in H_2O	Solubility of H_2O in	Reference
30	0.24 g/1000 g H_2O	–	Gross & Saylor (1931)
30	0.00007 mol/100 g H_2O	–	Van Arkel & Vles (1936)
30	0.024 g/100 cc H_2O	–	Patty (1962)
30	0.24 g/1000 g H_2O	–	Gmelin (1974)
30	0.024 wt %	–	Stephen & Stephen (1963)
30	0.024 g/100 g H_2O	–	Andelman (1978)

[a]M.wt. = 331.65; m.p. = 90.5°C; b.p. = 189.5°C; d_4^{20} = 2.9609 g cc^{-1}.

$CClFH_2$[a]

Methane, chlorofluoro- (R 31)

Temperature ($^\circ$C)	Solubility in H_2O	Solubility of H_2O in	Reference
10.25	0.244 mol l^{-1} atm^{-1}	–	Boggs & Buck
34.45	0.116	–	(1958)
59.25	0.0666	–	"
79.15	0.0488	–	"
17.88 (2147 mm Hg)	4.12 wt %	–	Colten et al. (1972)
25 (1 atm)	1.27 wt %	0.12 wt % H_2O	Sanders (1979)

[a]M.wt. = 68.48; b.p. = -9.1°C.

CClF$_2$H[a]
Methane, chlorodifluoro- (R 22)

Temperature (°C)	Solubility in H$_2$O	Solubility of H$_2$O in	Reference
10.25	0.0606 mol l^{-1} atm^{-1}	–	Boggs & Buck
23.55	0.0354	–	(1958)
40.95	0.0201	–	"
59.25	0.0138	–	"
79.15	0.0105	–	"
-30	–	205.0 mg/kg CClF$_2$H	Chernyshev (1969)
-10	–	426.0	"
10	–	830.0	"
-50	–	Monograph	"
.	–	"	"
.	–	"	"
.			
30	–	"	"
-30	–	200.0 mg/kg CClF$_2$H	Badyl'kes (1962)
-10	–	424.0	"
10	–	815.0	"
0	–	0.060 wt % H$_2$O	du Pont (1969a)
25	0.30 wt %	–	"
27 (137 psia)	22.9 g/kg H$_2$O	–	du Pont (1961b)
25 (0-137 psia)	0.1885 g l^{-1} psia^{-1}	–	du Pont (1952a)
50 (0-174 psia)	0.0923	–	"
75 (0-171 psia)	0.0592	–	"
25 (152.7 psia)	28.0 g l^{-1} H$_2$O	–	du Pont (1952a)
50 (284.0 psia)	24.0	–	"
75 (490.0 psia)	21.0	–	"
25	0.30 wt %	0.13 wt % H$_2$O	Weast (1976)
-17.7	–	0.031 wt % H$_2$O	du Pont (1966a)
21.1	–	0.114	"
25 (1 atm)	0.30 g/100 g H$_2$O	–	"
25 (0-137 psia)	Graph	–	"
50 (0-160 psia)	"	–	"
70 (0-165 psia)	"	–	"
-73	–	Graph	"
:	–	"	"
:	–	⌄	"
16	–	"	"

CClF$_2$H (Continued)

25 (14.7 psia)	0.026 lb/gal H$_2$O	–	du Pont (1966b)
25 (50 psia)	0.084	–	"
25 (100 psia)	0.158	–	"
25 (0–137 psia)	Graph	–	"
50 (0–172 psia)	"	–	"
75 (0–172 psia)	"	–	"
-73.33	–	19 ppm by weight H$_2$O	du Pont (1966d)
-62.22	–	37	"
-51.11	–	68	"
-40.00	–	120	"
-28.89	–	195	"
-17.78	–	308	"
-6.67	–	472	"
10.00	–	830	"
37.78	–	1800	"
26	–	1300 ppm by weight H$_2$O	du Pont (1969b)
21	–	0.114 wt % H$_2$O	du Pont (1966c)
-73.3	–	0.0019	"
-62.2	–	0.0037	"
-51.1	–	0.0068	"
-40.0	–	0.0120	"
-28.9	–	0.0195	"
-17.8	–	0.0308	"
-6.7	–	0.0472	"
4.4	–	0.0690	"
15.6	–	0.0970	"
26.7	–	0.1350	"
37.8	–	0.1800	"
25 (14.7 psia)	0.288 wt %	–	"
21.1 (14.7–100 psia)	Table + graph	–	"
48.9 (14.7–100 psia)	"	–	"
76.7 (14.7–100 psia)	"	–	"
-17.7	–	0.031 wt % H$_2$O	Elchardus &
21.1	–	0.114	Maestre (1965)
-50	–	Graph	"
⋮	–	"	"
	–	"	"
50	–	"	"

$CClF_2H$ (Continued)

26	–	1300 ppm by weight H_2O	Komedera (1966)
21	2.68 wt %	0.114 wt % H_2O	Mannheimer (1956)
-1	–	573 ppm by weight H_2O	Richards (1966)
26	–	1300	"
25 (14.7 psia)	0.026 lb/gal H_2O	–	Parmelee (1953-
25 (50 psia)	0.084	–	1954:61)
25 (100 psia)	0.158	–	"
25 (0-137 psia)	Graph	–	"
50 (0-172 psia)	"	–	"
75 (0-172 psia)	"	–	"
25	5630 mol/million	8150 mol/million	"
-40.8	0.000708 g cc^{-1} $psia^{-1}$	–	Wittstruck et al. (1961)
0	–	0.06 wt % H_2O	Matheson Company
30 (760 mm Hg)	o.12 wt % $CClF_2H$	0.15	(1966, 1971)
25 (760 mm Hg)	0.30	–	Matheson Company (1966)
25 (760 mm Hg)	0.30 wt % $CClF_2H$	–	Kirk-Othmer (1966a)
21 (1 atm)	0.33 wt % $CClF_2H$	0.114 wt % H_2O	Allied Chem. Corp. (1974)
0	–	0.060 g/100 g $CClF_2H$	Kali-Chemie (1972)
30	–	0.150	"
20 (1 atm)	0.38 wt % $CClF_2H$	–	Saito & Tanaka
5 (0-2300 mm Hg)	Graph	–	(1965)
⋮	"	–	"
30	"	–	"
-100	–	Graph	Petrak (1975)
⋮	–	"	"
96	–	"	"
21	–	0.12 wt % H_2O	Shepherd (1961)
-40	–	0.120 g/kg $CClF_2H$	Hantzschel &
-20	–	0.300	Kittner (1969)
0	–	0.570	"
20	–	1.000	"
40	–	1.600	"
-40	–	120 ppm by weight H_2O	Ackroyd (1978)
-20	–	280	"
-10	–	410	"

$CClF_2H$ (Continued)

0	–	590 ppm by weight H_2O	Ackroyd (1978)
15	–	950	"
30	–	1450	"
0	–	0.060 g/100 g $CClF_2H$	Imperial Smelting Corp. (1957)
25 (1 atm)	0.30 wt % $CClF_2H$	0.13 wt % H_2O	Sanders (1979)
25 (1 bar)	0.30 wt % $CClF_2H$	–	Grant (1978)
20 (saturated)	2.8 wt % $CClF_2H$	0.11 wt % H_2O	du Pont (1966e)

[a]M.wt. = 86.47; m.p. = -146.5°C; b.p. = -40.8°C; d^{-69} = 1.4906 g cc^{-1}.

$CClF_3$[a]
Methane, chlorotrifluoro- (R 13)

Temperature (°C)	Solubility in H_2O	Solubility of H_2O in	Reference
10.25	0.0005 mol l^{-1} atm^{-1}	–	Boggs & Buck
59.25	0.0001	–	(1958)
-40	–	Monograph	Chernyshev
⋮	–	"	(1969)
·	–	"	"
50	–	"	"
25	0.009 wt % $CClF_3$	–	du Pont (1969a)
25 (0-366 psia)	Graph	–	du Pont (1952b)
50 (0-360 psia)	"	–	"
75 (0-360 psia)	"	–	"
25 (14.7 psia)	0.0008 lb/gal H_2O	–	du Pont (1966b)
25 (50 psia)	0.0026	–	"
25 (100 psia)	0.0046	–	"
25 (0-366 psia)	Graph	–	"
50 (0-360 psia)	"	–	"
75 (0-360 psia)	"	–	"
25 (14.7 psia)	0.009 wt % $CClF_3$	–	du Pont (1966c)
25 (14.7-350 psia)	Table	–	"
50 (14.7-350 psia)	"	–	"
75 (14.7-350 psia)	"	–	"
-40	–	Graph	Elchardus &
⋮	–	"	Maestre (1965)
30	–	"	"

CClF$_3$ (Continued)

25 (14.7 psia)	0.0008 lb/gal H$_2$O	–	Parmelee (1953)
25 (50 psia)	0.0026	–	"
25 (100 psia)	0.0046	–	"
25 (0–366 psia)	Graph	–	"
50 (0–360 psia)	"	–	"
75 (0–360 psia)	"	–	"
25 (760 mm Hg)	0.096 g l^{-1} H$_2$O	–	Gmelin (1974)
25 (2580 mm Hg)	0.312	–	"
25 (5170 mm Hg)	0.551	–	"
25	0.009 wt % CClF$_3$	–	Weast (1976)
25 (760 mm Hg)	0.009 wt % CClF$_3$	–	Matheson Company (1971)
25 (1 atm)	0.009 wt % CClF$_3$	–	Kirk-Othmer (1966a)
–54	–	Graph	Petrak (1975)
.	–	"	"
.	–	"	"
.	–	"	"
28	–	"	"
–40	–	1.5 ppm by weight H$_2$O	Ackroyd (1978)
–20	–	6.0	"
–10	–	11	"
0	–	19	"
15	–	42	"
30	–	82	"

[a] M.wt. = 104.46; m.p. = –181.0°C; b.p. = –81.1°C.

CClH$_3^a$

Methane, chloro- (R 40)

Temperature (^oC)	Solubility in H_2O	Solubility of H_2O in	Reference
-20	Graph	Graph	Allied Chemical
⋮	"	"	Corp. (1966)
50	"	"	"
10.25	0.1670 mol l^{-1} atm H_2O	–	Boggs & Buck
23.55	0.0979	–	(1958)
36.95	0.0714	–	"
37.45	0.0691	–	"
59.25	0.0452	–	"
-50	–	Monograph	Chernyshev (1969)
⋮	–	"	"
	–	"	"
20	–	"	"
-50	–	Graph	Elchardus &
⋮	–	"	Maestre (1965)
50	–	"	"
20 (1 bar)	0.0019 mol/mol H_2O	–	Gerrard (1971)
25	–	0.29 wt % H_2O	Gladis (1960)
30	0.74 wt % CClH$_3$	–	"
4.09	3.28 × 10^5 mm Hg mol^{-1} l^{-1}	–	Glew & Moelwyn-
10.17	4.26 × 10^5	–	Hughes (1953)
15.06	5.11 × 10^5	–	"
20.03	6.12 × 10^5	–	"
25.01	7.18 × 10^5	–	"
39.91	11.06 × 10^5	–	"
49.78	13.95 × 10^5	–	"
59.99	16.95 × 10^5	–	"
70.06	20.60 × 10^5	–	"
80.08	23.80 × 10^5	–	"
10	3.73 cc(STP)/cc H_2O/atm	–	Mamedaliev &
15	3.25	–	Musachanly
20	2.88	–	(1940)
30	2.29	–	"
25	2.05 wt % CClH$_3$	0.25 wt % H_2O	Mannheimer (1956)
-12	–	0.026 wt % H_2O	Mellan (1957)
20(760 mm Hg)	303 cc/100 cc H_2O	–	"
25(760 mm Hg)	0.48 g/100 g H_2O	–	"

$CClH_3$ (Continued)

20	10.4 mm Hg/mol/l H_2O	–	Moelwyn-Hughes (1938)
25 (760 mm Hg)	19.13×10^{-4} mole fraction	–	Moelwyn-Hughes (1964)
30 (1 atm)	0.74 g/100 g H_2O	–	McGovern (1943)
25	–	0.285 g/100 g $CClH_3$	"
-40	–	Graph	"
⋮	–	"	"
50	–	"	"
20	400 cc/100 cc H_2O	–	Patty (1962)
15 (1 atm)	9.00 g/1000 g H_2O	–	Svetlanov et al.
30 (1 atm)	6.52	–	(1971)
45 (1 atm)	4.36	–	"
60 (1 atm)	2.64	–	"
29.4	8,106 mm Hg/mol/l soln.	–	Swain &
40.3	10,927	–	Thornthon
49.6	13,430	–	(1962)
30	4.1 g/1000 g H_2O	–	Van Arkel & Vles (1936)
30	4.1 g/1000 g H_2O	–	Van Arkel (1946)
0 (311 mm Hg)	0.02 mol/108 g H_2O	–	von Stackelberg (1949)
25	–	0.13 wt % H_2O	Lees & Sarram (1971)
20	4 cc/cc H_2O	–	Gunther et al. (1968)
20	0.6 wt % H_2O	–	Union Carbide Corp. (1973)
0 (760 mm Hg)	3.4 cc/cc H_2O	–	du Pont (1947)
10 (760 mm Hg)	2.7	–	"
20 (760 mm Hg)	2.2	–	"
30 (760 mm Hg)	1.7	–	"
40 (760 mm Hg)	1.3	–	"
-24 (760 mm Hg)	–	0.026 wt % H_2O	"
1 (760 mm Hg)	–	0.30	"
25	2.5 cc(STP)/g H_2O	–	Lax (1967)
0 (760 mm Hg)	340 cc/100 cc H_2O	–	Matheson Company
20 (760 mm Hg)	220	–	(1966, 1971)
16	280 cc/100 g H_2O	–	Lange (1967)
25	0.48 g/100 g H_2O	0.0725 g/100 g $CClH_3$	Kirk-Othmer (1964b)

CClH$_3$ (Continued)

0 (1 atm)	3.4 cc/cc H$_2$O	–	Manufacturing
10 (1 atm)	2.7	–	Chemists Assoc.
20 (1 atm)	2.2	–	(1956-1971:SD-
25 (1 atm)	–	0.284 wt % H$_2$O	40)
30 (1 atm)	1.7	–	"
40 (1 atm)	1.3	–	"
20	2.86 cc/cc H$_2$O	–	Plank (1921)
30	1.95	–	"
40	1.40	–	"
50	1.05	–	"
60	0.76	–	"
70	0.61	–	"
80	0.52	–	"
20 (1 atm)	7250 ppm by weight	–	Pearson & McConnell (1975)
20.4 (3640 mm Hg)	2.95 wt % CClH$_3$	–	Colten et al. (1972)
20	0.46 wt % CClH$_3$	–	Nathan (1978)
25 (1 atm)	5380 ppm by weight	–	Dilling (1977)
20 (1 atm)	6270	–	"
0 (1 atm)	7600 ppm by volume	–	Andelman (1978)
20 (1 atm)	6300	–	"
25 (1 atm)	5900	–	"
30 (1 atm)	7300	–	"
0 (1 bar)	0.00521 mole fraction	–	Horvath (1975c)
10 (1 bar)	0.00314	–	"
20 (1 bar)	0.00218	–	"
30 (1 bar)	0.00159	–	"
40 (1 bar)	0.00122	–	"
50 (1 bar)	0.00097	–	"
60 (1 bar)	0.00080	–	"
70 (1 bar)	0.00068	–	"
80 (1 bar)	0.00059	–	"

[a]M.wt. = 50.49; m.p. = -97.73°C; b.p. = -24.2°C; d$_4^{20}$ = 0.9159 g cc^{-1}.

CCl_2FH^a
Methane, dichlorofluoro- (R 21)

Temperature (^oC)	Solubility in H_2O	Solubility of H_2O in	Reference
-50	-	58.0 mg/kg CCl_2FH	Chernyshev (1969)
-30	-	162.0	"
-10	-	420.0	"
10	-	940.0	"
-50	-	Monograph	"
⋮	-	"	"
30	-	"	"
-50	-	56.0 mg/kg CCl_2FH	Badyl'kes (1962)
-30	-	168.0	"
-10	-	409.0	"
10	-	920.0	"
0	-	0.055 wt % H_2O	du Pont (1969a)
25 (1 atm)	0.95 wt % CCl_2FH	0.13	"
31	0.521 g/psia/kg H_2O	-	du Pont (1961b)
28	0.606	-	"
52	0.263	-	"
52	0.264	-	"
75	0.145	-	"
75	0.135	-	"
7	Graph	-	"
⋮	"	-	"
76	"	-	"
-17.7	-	0.026 wt % H_2O	du Pont (1966a)
21.1	-	0.115	"
30 (1 atm)	0.69 g/100 g H_2O	-	"
-73	-	Graph	"
⋮	-	"	"
16	-	"	"
-73.3	-	12 ppm by weight H_2O	du Pont (1966d)
-62.2	-	25	"
-51.1	-	49	"
-40.0	-	90	"
-28.9	-	158	"
-17.8	-	260	"
-6.7	-	398	"

CCl$_2$FH (Continued)

4.4	–	645 ppm by weight H$_2$O	du Pont (1966d)
15.6	–	960	"
26.7	–	1410	"
37.8	–	1980	"
-73.3	–	Graph	"
$\vdots$	–	"	"
	–	"	"
18.2	–	"	"
21	–	0.115 wt % H$_2$O	du Pont (1966c)
25 (14.7-50 psia)	Table + graph	–	"
50 (14.7-50 psia)	"	–	"
75 (14.7-50 psia)	"	–	"
-73.3	–	0.0012 wt % H$_2$O	"
-62.2	–	0.0025	"
-51.1	–	0.0049	"
-40.0	–	0.0090	"
-28.9	–	0.0158	"
-17.8	–	0.0260	"
-6.7	–	0.0398	"
4.4	–	0.0645	"
15.6	–	0.0960	"
26.7	–	0.1410	"
37.8	–	0.1980	"
-17.7	–	0.026 wt % H$_2$O	Elchardus &
21.1	–	0.115	Maestre (1965)
-50	–	Graph	"
$\vdots$	–	"	"
	–	"	"
10	–	"	"
25	0.95 wt % CCl$_2$FH	0.13 wt % H$_2$O	Weast (1976)
0	–	0.055 wt % H$_2$O	Matheson Company
30 (1 atm)	0.69 wt % CCl$_2$FH	0.16	(1966, 1971)
25	0.95 wt % CCl$_2$FH	–	Kirk-Othmer (1966a)
20 (1 atm)	1.20 wt % CCl$_2$FH	–	Saito & Tanaka
5 (250-400 mm Hg)	Graph	–	(1965)
10 (310-780 mm Hg)	"	–	"
20 (400-1120 mm Hg)	"	–	"
30 (450-1400 mm Hg)	"	–	"

CCl$_2$FH (Continued)

25 (1 atm)	0.81 g/100 g H$_2$O	–	Grewer (1962)
50 (1 atm)	0.30	–	"
70 (1 atm)	0.205	–	"
81 (1 atm)	0.183	–	"
45 (760 mm Hg)	0.31 wt % CCl$_2$FH	–	Colten et al.
54 (760 mm Hg)	0.25	–	(1972)
63 (760 mm Hg)	0.22	–	"
72.5 (760 mm Hg)	0.18	–	"
8.7 (761 mm Hg)	1.80	–	"
–50	–	Graph	Petrak (1975)
⋮	–	"	"
	–	"	"
180	–	"	"
21	–	0.13 wt % H$_2$O	Shepherd (1961)
0	–	0.055 g/100 g CCl$_2$FH	Imperial Smelting Corp. (1957)
25 (1 atm)	0.95 wt % CCl$_2$FH	0.13 wt % H$_2$O	Sanders (1979)
20 (saturated)	3.0 wt % CCl$_2$FH	0.11 wt % H$_2$O	du Pont (1966e)

[a]M.wt. = 102.92; m.p. = –135.0°C; b.p. = 9.0°C; d^9 = 1.3724 g cc^{-1}.

$CCl_2F_2^a$
Methane, dichlorodifluoro- (R 12)

Temperature (°C)	Solubility in H_2O	Solubility of H_2O in	Reference
-50	-	Monograph	Chernyshev (1961)
⋮	-	"	"
30	-	"	"
0	-	0.0026 wt % H_2O	du Pont (1969a)
25 (1 atm)	0.028 wt % CCl_2F_2	0.009	"
-17.7	-	0.008 wt % H_2O	du Pont (1966a)
21.1	-	0.0076	"
25 (1 atm)	0.028 g/100 g H_2O	-	"
25 (0-95 psia)	Graph	-	"
50 (0-125 psia)	"	-	"
75 (0-120 psia)	"	-	"
-72	-	Graph	"
⋮	-	"	"
49	-	"	"
25 (14.7 psia)	0.0023 lb/gal H_2O	-	du Pont (1966b)
25 (50 psia)	0.0082	-	"
25 (0-110 psia)	Graph	-	"
50 (0-122 psia)	"	-	"
75 (0-122 psia)	"	-	"
-73.3	-	0.1 ppm by weight H_2O	du Pont (1966d)
-62.2	-	0.3	"
-51.1	-	0.7	"
-40.0	-	1.7	"
-28.9	-	3.8	"
-17.8	-	8.3	"
-6.7	-	16.6	"
4.4	-	32	"
15.6	-	58	"
26.7	-	98	"
37.8	-	165	"
-73.3	-	Graph	"
	-	"	"
⋮	-	"	"
37.8	-	"	"

CCl$_2$F$_2$ (Continued)

21	–	0.0076 wt % H$_2$O	du Pont (1966c)
-73.3	–	0.00001	"
-62.2	–	0.00003	"
-51.1	–	0.00007	"
-40.0	–	0.00017	"
-28.9	–	0.00038	"
-17.8	–	0.00083	"
-6.7	–	0.00166	"
4.4	–	0.0032	"
15.6	–	0.0058	"
26.7	–	0.0098	"
37.8	–	0.0165	"
25 (14.7-75 psia)	Table	–	du Pont (1966c)
54 (14.7-100 psia)	"	–	"
76.7 (14.7-100 psia)	"	–	"
20 (0-80 psia)	Graph	–	"
25 (0-96 psia)	"	–	"
35 (0-116 psia)	"	–	"
40 (0-118 psia)	"	–	"
50 (0-125 psia)	"	–	"
60 (0-126 psia)	"	–	"
75 (0-126 psia)	"	–	"
26	–	93 ppm by weight H$_2$O	du Pont (1969b)
-17.7	–	0.0008 wt % H$_2$O	Elchardus &
21.1	–	0.0076	Maestre (1965)
-42	–	Graph	"
⋮	–	"	"
50	–	"	"
20	5.7 cc/100 cc H$_2$O	–	Patty (1962)
-40	–	0.00017 wt % H$_2$O	Elsey & Flowers
-28.9	–	0.00038	(1949)
-17.8	–	0.00083	"
-6.7	–	0.00166	"
-1.1	–	0.00233	"
4.4	–	0.0032	"
10.0	–	0.0044	"
15.6	–	0.0058	"
21.1	–	0.0076	"

CCl_2F_2 (Continued)

26.7	–	0.0098 wt % H_2O	Elsey & Flowers
32.2	–	0.0128	(1949)
37.8	–	0.0165	"
-40	–	Graph	"
⋮	–	"	"
38	–	"	"
26	–	93 ppm by weight H_2O	Komedera (1966)
21	0.187 wt % CCl_2F_2	0.0076 wt % H_2O	Mannheimer (1956)
26	0.028 wt % CCl_2F_2	–	Mellan (1950)
25 (1 atm)	0.0023 lb/gal H_2O	–	Parmelee (1953-1954:
25 (1 atm)	270 mol/10^6 H_2O	625 mol/10^6 CCl_2F_2	"
25 (0-100 psia)	Graph	–	"
50 (0-120 psia)	"	–	"
75 (0-120 psia)	"	–	"
25 (0-80 psia)	"	–	"
-1	–	23 ppm by weight H_2O	Richards (1966)
26	–	93	"
-29.8 (14.7 psia)	1.06×10^{-5} g/cc H_2O	–	Wittstruck et al. (1961)
-40	–	0.00017 wt % H_2O	Gmelin (1974)
-29	–	0.00038	"
-18	–	0.00083	"
-7	–	0.00166	"
-1	–	0.00232	"
4	–	0.00320	"
10	–	0.00440	"
16	–	0.00580	"
21	–	0.0076	"
27	–	0.0098	"
32	–	0.0128	"
38	–	0.0165	"
25 (760 mm Hg)	0.276 g l^{-1} H_2O	–	"
25 (2580 mm Hg)	0.984	–	"
25 (5170 mm Hg)	1.75	–	"
0	–	0.0024 wt % H_2O	Union Carbide Corp.
21	–	0.0080	(1973)
0	–	0.006 wt % H_2O	du Pont (1947)
16	–	0.0114	"
25	0.028 wt % CCl_2F_2	0.009 wt % H_2O	Weast (1976)

CCl_2F_2 (Continued)

Temperature	Solubility		Reference
0	–	0.0026 wt % H_2O	Matheson Company
25 (760 mm Hg)	0.028 wt % CCl_2F_2	–	(1966, 1971)
30	–	0.0120 wt % H_2O	"
25 (1 atm)	0.028 wt % CCl_2F_2	–	Kirk-Othmer (1966a)
21 (1 atm)	0.040 wt % CCl_2F_2	0.008 wt % H_2O	Allied Chemical Corp. (1974)
0	–	0.0026 g/100 g CCl_2F_2	Imperial Chemical
30	–	0.012	Industries (1953)
0	–	0.0026 g/100 g CCl_2F_2	Kali-Chemie (1972)
30	–	0.012	"
25	0.028 wt % CCl_2F_2	0.0098 wt % H_2O	Rhône Progil (1973)
20 (1 atm)	0.03 wt % CCl_2F_2	–	Saito & Tanaka (1965)
5 (300-2200 mm Hg)	Graph	–	"
10 (300-2400 mm Hg)	"	–	"
20 (300-2400 mm Hg)	"	–	"
30 (300-2400 mm Hg)	"	–	"
50 (380-2400 mm Hg)	"	–	"
7.5 (3.83 atm)	0.0316 g l^{-1} H_2O	–	Chepstov (1974)
7.7 (3.93 atm)	0.0468	–	"
8.0 (3.92 atm)	0.0761	–	"
8.0 (2.88 atm)	0.0560	–	"
8.0 (3.93 atm)	0.0763	–	"
8.2 (3.92 atm)	0.1020	–	"
8.7 (3.98 atm)	0.1921	–	"
8.8 (3.09 atm)	0.1660	–	"
10.3 (3.58 atm)	0.6740	–	"
11.0 (3.76 atm)	1.0609	–	"
11.0 (4.31 atm)	1.2168	–	"
25 (1 atm)	0.028 wt % CCl_2F_2	–	Rauws et al. (1973)
4 (1 bar)	0.0860 wt % CCl_2F_2	–	Zeininger (1975)
12 (1 bar)	0.073	–	"
20 (1 bar)	0.052	–	"
40 (1 bar)	0.0405	–	"
25 (1 atm)	280 ppm by weight CCl_2F_2	–	Pearson & McConnell (1975)
6 (1-5 bars)	Graph	–	Filatkin et al.
⋮	"	–	"
35	"	–	"
12.1 (3435 mm Hg)	0.19 wt % CCl_2F_2	–	Colten et al. (1972)

CCl_2F_2 (Continued)

Temp			Reference
−30	−	0.0000202 mole fraction H_2O	Zhukoborskii
−20	−	0.0000410	(1973)
−10	−	0.0000796	"
0	−	0.0001593	"
10	−	0.0002903	"
20	−	0.0004710	"
30	−	0.0007930	"
40	−	0.0012400	"
−60	−	0.31 ppm by weight H_2O	Cremer (1969)
−50	−	0.76	"
−40	−	1.73	"
−30	−	3.65	"
−20	−	7.3	"
−10	−	13.8	"
0	−	24.5	"
10	−	43.0	"
20	−	73	"
30	−	118	"
40	−	185	"
−100	−	Graph	Petrak (1975)
:	−	"	"
112	−	"	"
21	−	0.008 wt % H_2O	Shepherd (1961)
−40	−	0.0016 g/kg CCl_2F_2	Häntzschel &
−20	−	0.0070	Kittner
0	−	0.0250	(1969)
20	−	0.0750	"
40	−	0.205	"
6	Graph	−	Filatkin et al.
:	"	−	(1975)
35	"	−	"
−40	−	2 ppm by weight H_2O	Ackroyd (1978)
−20	−	7	"
−10	−	14	"
0	−	25	"
15	−	57	"
30	−	115	"
25	0.028 wt % CCl_2F_2	0.009 wt % H_2O	Sanders (1979)

CCl_2F_2 (Continued)

0	–	0.0026 g/100 g CCl_2F_2	Imperial Smelting
30	–	0.012	Corp. (1957)
25	0.028 wt % CCl_2F_2	–	Grant (1978)
20 (saturated)	0.16 wt % CCl_2F_2	0.0072 wt % H_2O	du Pont (1966e)
0	–	0.0026 wt % H_2O	Banks (1979)
30	–	0.0120	"
25 (1 atm)	0.028 wt % CCl_2F_2	–	"

[a]M.wt. = 120.91; m.p. = -158.0°C; b.p. = -29.8°C; d^{57} = 1.1834 g cc^{-1}.

$CCl_2H_2^a$

Methane, dichloro- (R 30)

Temperature (^oC)	Solubility in H_2O	Solubility of H_2O in	Reference
-20	-	Graph	Allied Chemical
0	Graph	"	Corp. (1966)
⋮	"	"	"
50	"	"	"
18	-	0.14 wt % H_2O	Antropov et al.
20	2.0 wt % CCl_2H_2	-	(1972)
22	-	0.16	"
26	-	0.18	"
20	2.0 g/100 cc H_2O	0.5 g/100 cc CCl_2H_2	Bakowski & Treszczanowicz (1937)
0	2.363 g/100 g H_2O	-	Bakowski &
10	2.122	-	Treszczanowicz (1938)
20	2.000	-	"
30	1.969	-	"
25	2.60 cc/100 cc H_2O	-	Booth & Everson (1948)
20	2.00 g/100 g H_2O	-	Carlisle & Levine (1932)
25	6.1 cc (STP)/g H_2O	-	Lax (1967)
20	2.0 wt % CCl_2H_2	-	"
-50	-	Monograph	Chernyshev (1969)
⋮	-	"	"
0	-	"	"
-20	-	0.03 wt % H_2O	Davies et al. (1949)
-10	-	0.04	"
0	-	0.07	"
10	-	0.10	"
20	-	0.14	"
30	-	0.19	"
40	-	0.26	"
-50	-	Graph	Elchardus & Maestre
⋮	-	"	(1965)
50	-	"	"
25	1.32 wt % CCl_2H_2	0.20 wt % H_2O	Gladis (1960)
25	1.40 wt % CCl_2H_2	0.20 wt % H_2O	Maretic & Sirocic (1962)

CCl$_2$H$_2$ (Continued)

25	1.32 wt % CCl$_2$H$_2$	0.17 g/100 g CCl$_2$H$_2$	Mellan (1957)
25	1.32 g/100 g H$_2$O	0.198 g/100 g CCl$_2$H$_2$	McGovern (1943)
-40	-	Graph	"
-20	-	"	"
0	Graph	"	"
.	"	"	"
.			
50	"	"	"
20	1.629 wt % CCl$_2$H$_2$	0.239 wt % H$_2$O	Niini (1938)
20	2 g/100 cc H$_2$O	-	Patty (1962)
25	-	0.19 wt % H$_2$O	Pechiney-Saint-Gobain (1971: 113CM)
0	2.363 g/100 g H$_2$O	-	Rex (1906)
10	2.122	-	"
20	2.000	-	"
30	1.969	-	"
25	1.30 wt % CCl$_2$H$_2$	0.198 wt % H$_2$O	Riddick & Bunger (1970)
20	0.6 wt % CCl$_2$H$_2$	0.15 wt % H$_2$O	Sabinin et al.
25	1.31	0.197	(1970)
30	2.6	0.19	"
0	-	0.00400 mole fraction	Staverman (1941b)
20	0.00413 mole fraction	-	"
25	-	0.00788 mole fraction	"
30	-	0.00925	"
15 (1 atm)	25.0 g/1000 g H$_2$O	-	Svetlanov et al.
30 (1 atm)	15.6	-	(1971)
45 (1 atm)	8.8	-	"
60 (1 atm)	5.30	-	"
30	0.0232 mol/100 g H$_2$O	-	Van Arkel & Vles (1936)
0	2.30 wt % CCl$_2$H$_2$	-	Washburn (1928)
10	2.08	-	"
20	1.96	-	"
30	1.93	-	"
20	12.24 g l^{-1} H$_2$O	-	Salkowski (1920)
0	2.363 wt % CCl$_2$H$_2$	-	Marsden (1963)
10	2.122	-	"
20	2.00	0.14 wt % H$_2$O	"
30	1.969	-	"

CCl_2H_2 (Continued)

26	5.7 cc/100 g H_2O	–	Lange (1967)
20	2.0 g/100 g H_2O	–	"
0	2.36 g/100 g H_2O	–	Kirk-Othmer
10	2.12	–	(1964b)
20	2.00	–	"
25	–	0.17 g/100 g CCl_2H_2	"
30	1.97	–	"
25	1.3 wt % CCl_2H_2	0.198 wt % H_2O	Interdyne Inc. (1976)
20	2.0 g/100 g H_2O	–	Pittsburgh Plate
25	–	0.2 g/100 g CCl_2H_2	Industries (1975)
25 (1 atm)	13200 ppm by weight CCl_2H_2	–	Pearson & McConnell (1975)
25	–	0.17 wt % H_2O	Shepherd (1961)
25	2.0 g/100 g H_2O	0.20 g/100 g CCl_2H_2	Archer & Stevens (1977)
20	1.96 wt % CCl_2H_2	–	Nathan (1978)
25	–	0.167 wt % H_2O	"
7	–	0.0344 wt % H_2O	Prosyanov et al.
⋮	–	⋮	(1974)
37	–	0.205 wt % H_2O	"
25 (1 atm)	13200 ppm by weight CCl_2H_2	–	Selenka & Bauer (1978)
25 (1 atm)	1.32 wt % CCl_2H_2	–	Sanders (1979)
20 (1 atm)	1.38 wt % CCl_2H_2	0.15 wt % H_2O	du Pont (1966e)
25 (1 atm)	19400 ppm by weight CCl_2H_2	–	Dilling (1977)
1.5 (1 atm)	22700	–	"
0 (1 atm)	23100 ppm by volume CCl_2H_2	–	Andelman (1978)
10 (1 atm)	20800	–	"
20 (1 atm)	19600	–	"
25 (1 atm)	20000	–	"
30 (1 atm)	19000	–	"
20 (1 atm)	2.0 wt % CCl_2H_2	–	Matthews (1975, 1979)
25 (1 atm)	19400 g m^{-3} H_2O	–	Afghan & Mackay (1980)

[a] M.wt. = 84.93; m.p. = $-95.1^{\circ}C$; b.p. = $40.0^{\circ}C$; d_4^{20} = 1.3266 g cc^{-1}.

CCl_3F[a]

Methane, trichlorofluoro- (R 11)

Temperature (°C)	Solubility in H_2O	Solubility of H_2O in	Reference
-30	–	5.65 mg/kg CCl_3F	Chernyshev (1969)
-10	–	17.9	"
10	–	50.0	"
-50	–	Monograph	"
⋮	–	"	"
20	–	"	"
-30	–	5.60 mg/kg CCl_3F	Badyl'kes (1962)
-10	–	18.0	"
10	–	48.5	"
0	–	0.0036 wt % H_2O	du Pont (1969a)
25 (1 atm)	0.11 wt % CCl_3F	0.011	"
31	0.0720 g/psia/kg H_2O	–	du Pont (1961b)
50	0.0397	–	"
75	0.0210	–	"
7	Graph	–	"
⋮	"	–	"
76	"	–	"
-17.7	–	0.0015 wt % H_2O	du Pont (1966a)
21.1	–	0.0090	"
27 (1 atm)	0.11 g/100 g H_2O	–	"
-73	–	Graph	"
⋮	–	"	"
49	–	"	"
-73.3	–	0.3 ppm by weight H_2O	du Pont (1966d)
-62.2	–	0.8	"
-51.1	–	2	"
-40.0	–	4	"
-28.9	–	8	"
-17.8	–	15	"
-6.7	–	26	"
4.4	–	44	"
15.6	–	70	"
26.7	–	113	"
37.8	–	168	"

CCl$_3$F (Continued)

-73.3	-	Graph	du Pont (1966d)
⋮	-	"	"
48	-	"	"
21.1	-	0.0090 wt % H$_2$O	du Pont (1966c)
31 (14.7 psia)	0.106 wt % CCl$_3$F	-	"
30 (14.7 psia)	1.08 g l^{-1} H$_2$O	-	"
50 (14.7-30 psia)	Table	-	"
75 (14.7-50 psia)	"	-	"
-73.3	-	0.00003 wt % H$_2$O	"
-62.2	-	0.00006	"
-51.1	-	0.0002	"
-40.0	-	0.0004	"
-28.9	-	0.0008	"
-17.8	-	0.0015	"
-6.7	-	0.0026	"
4.4	-	0.0044	"
15.6	-	0.0070	"
26.7	-	0.0113	"
37.8	-	0.0168	"
-17.7	-	0.0015 wt % H$_2$O	Elchardus &
21.1	-	0.0090	Maestre (1965)
-50	-	Graph	"
⋮	-	"	"
50	-	"	"
0	-	0.0036 wt % H$_2$O	Union Carbide
21	-	0.0090	Corp. (1973)
20	0.14 wt % CCl$_3$F	0.0086	"
25	0.11 wt % CCl$_3$F	0.011 wt % H$_2$O	Weast (1976)
0	-	0.0036 wt % H$_2$O	Matheson Company
25 (1 atm)	0.11 wt % CCl$_3$F	-	(1966, 1971)
30 (1 atm)	-	0.0130 wt % H$_2$O	"
25 (1 atm)	0.11 wt % CCl$_3$F	-	Kirk-Othmer (1966a)
21 (1 atm)	0.14 wt % CCl$_3$F	0.009 wt % H$_2$O	Allied Chemical Corp. (1974a)
0	-	0.0036 g/100 g CCl$_3$F	Imperial Chemical
30	-	0.013	Industries (1953)
0	-	0.0036 g/100 g CCl$_3$F	Kali-Chemie (1972)
30	-	0.013	"

CCl$_3$F (Continued)

25	0.11 wt % CCl$_3$F	–	Rhône Progil
30	–	0.013 wt % H$_2$O	(1973)
25 (1 atm)	0.11 wt % CCl$_3$F	–	Rauws et al. (1973)
0.5 (1 bar)	0.360 wt % CCl$_3$F	–	Zeininger (1975)
10 (1 bar)	0.248	–	"
20 (1 bar)	0.152	–	"
40 (1 bar)	0.068	–	"
20 (1 atm)	1100 ppm by weight CCl$_3$F	–	Pearson & McConnell (1975)
-50	–	Graph	Petrak (1975)
⋮	–	"	"
198	–	"	"
21	–	0.009 wt % H$_2$O	Shepherd (1961)
-40	–	4 ppm by weight H$_2$O	Ackroyd (1978)
-20	–	13	"
-10	–	22	"
0	–	36	"
15	–	68	"
30	–	120	"
0	–	0.0036 g/100 g CCl$_3$F	Imperial Smelting
30	–	0.013	Corp. (1957)
25 (1 atm)	0.11 wt % CCl$_3$F	0.011 wt % H$_2$O	Sanders (1979)
25 (1 bar)	0.11 wt % CCl$_3$F	–	Grant (1978)
20 (saturated)	0.035 wt % CCl$_3$F	0.0086 wt % H$_2$O	du Pont (1966e)
0	–	0.0036 wt % H$_2$O	Banks (1979)
30	–	0.013	"
25 (1 atm)	0.11 wt % CCl$_3$F	–	"

[a]M.wt. = 137.37; m.p. = -111.0°C; b.p. = 23.82°C; d_4^{25} = 1.467 g cc^{-1}.

CCl_3H^a

Methane, trichloro- (R 20)

Temperature (^{o}C)	Solubility in H_2O	Solubility of H_2O in	Reference
-20	-	Graph	Allied Chemical
0	Graph	"	Corp. (1966a)
:	"	"	"
50	"	"	"
37	4.0 cc/cc H_2O (Ostwald)	-	Allott et al. (1973)
20	0.79 wt % CCl_3H	-	Antonov (1907)
15	1.0 wt % CCl_3H	-	Antropov et al.
18	-	0.062 wt % H_2O	(1972)
22	-	0.065	"
26	-	0.079	"
25	0.9 cc/100 cc H_2O	-	Booth & Everson (1948)
0	9.87 g l^{-1} H_2O	-	Chancel &
3.2	8.90	-	Parmentier
17.4	7.12	-	(1885a, b)
29.4	7.05	-	"
41.6	7.12	-	"
54.9	7.75	-	"
-10	-	Monograph	Chernyshev (1970)
:	-	"	"
80	-	"	"
18	-	0.062 wt % H_2O	Chistyakov (1965)
22	-	0.065	"
24	-	0.079	"
31	-	0.109	"
24.5	-	0.084 g/100 g soln.	Clifford (1921b)
26.7	-	0.107	"
27.8	-	0.116	"
25.0	0.75 wt % CCl_3H	0.08 wt % H_2O	Conti et al. (1960)
56.1	0.80	0.17	"
20	-	0.063 wt % H_2O	Chistyakov & Shapurova (1964)
15	-	0.084 wt % H_2O	Eberius (1954)
25	-	0.00488 mol/100 g soln.	Eidinoff (1955)

CCl_3H (Continued)

20	0.80 wt % CCl_3H	0.20 wt % H_2O	Evans (1936)
15	0.84	-	"
30	0.77	-	"
-25	-	0.006 wt % H_2O	Gibby & Hall
-15	-	0.009	(1931)
-1	-	0.014	"
3	-	0.019	"
11	-	0.043	"
17	-	0.061	"
22	-	0.065	"
23	-	0.072	"
31	-	0.100	"
45	-	0.118	"
54	-	0.165	"
25	0.80 wt % CCl_3H	0.10 wt % H_2O	Gladis (1960)
15	8.52 g/1000 g H_2O	-	Gross & Saylor
30	7.71	-	(1931)
22	-	0.152 g/147.6 g CCl_3H	Herz (1898)
40	4.6 cc/cc H_2O (Ostwald)	-	Larson et al.
37	3.8	-	(1962b)
30	4.17	-	"
20	7.7	-	"
10	5.9	-	"
10	-	0.06 g/100 g CCl_3H	Mellan (1957)
20	0.8 wt % CCl_3H	-	"
25	0.792 wt % CCl_3H	0.080 wt % H_2O	Miller (1969)
25	0.79 g/100 g H_2O	0.079 g/100 g CCl_3H	McGovern (1943)
-40	-	Graph	"
0	Graph	"	"
:	"	"	"
:			
60	"	"	"
25	-	0.073 mol l^{-1} soln.	Ödberg et al. (1972)
37	3.8 cc/cc H_2O (Ostwald)	-	Papper & Kitz (1963)
0	1.062 g/100 g H_2O	-	Rex (1906)
10	0.895	-	"
20	0.822	-	"
30	0.776	-	"

CCl$_3$H (Continued)

20	0.815 wt % CCl$_3$H	–	Riddick & Bunger
23	–	0.072 wt % H$_2$O	(1970)
20	0.666 g/100 g H$_2$O	–	Secher (1971)
20	0.80 g/100 g H$_2$O	–	Smith (1932)
0	0.00148 mole fraction	0.00352 mole fraction	Staverman (1941b)
20	0.00123	–	"
25	–	0.00623 mole fraction	"
30	0.00116 mole fraction	0.00750	"
37	4.0 cc/cc H$_2$O (Ostwald)	–	Steward et al. (1973)
15 (1 atm)	4.82 g/1000 g H$_2$O	–	Svetlanov et al.
30 (1 atm)	3.20	–	(1971)
45 (1 atm)	2.05	–	"
60 (1 atm)	1.30	–	"
30	0.00646 mol/100 g H$_2$O	–	Van Arkel & Vles (1936)
30	0.82 g/l soln.	–	Van Arkel (1946)
0	0.98 wt % CCl$_3$H	–	Washburn (1928)
10	0.86	–	"
20	0.80	–	"
30	0.76	–	"
40	0.735	–	"
50	0.745	–	"
55	0.770	–	"
25	1 g/125 cc H$_2$O	–	Wright & Schaffer (1932)
25	–	1.44 g l^{-1} CCl$_3$H	Zielinski (1959)
20	7.5 g l^{-1} H$_2$O	–	Salkowski (1920)
15	1.0 g/100 cc H$_2$O	–	Patty (1962)
15	1.0 wt % CCl$_3$H	–	Gunther et al. (1968)
25	5.6 cc(STP)/g H$_2$O	–	Lax (1967)
20	0.82 wt % CCl$_3$H	–	"
15	1.00 g/100 cc H$_2$O	–	Scheflan & Jacobs
20	0.82	–	(1953)
25	–	0.097 g/100 g CCl$_3$H	"
10	–	0.06 wt % H$_2$O	Marsden (1963)
20	0.80 wt % CCl$_3$H	0.097	"
20	0.82 g/100 g H$_2$O	–	Lange (1967)
0	–	1.062 g/100 g CCl$_3$H	Kirk-Othmer
10	–	0.895	(1964b)

CCl$_3$H (Continued)

20	–	0.822 g/100 g CCl$_3$H	Kirk-Othmer (1964b)
22	0.0806 g/100 g CCl$_3$H	–	"
30	–	0.776 g/100 g CCl$_3$H	"
0	0.98 wt % CCl$_3$H	0.1 wt % H$_2$O	Reinders & de Minjer
25	0.90	–	(1947)
42	0.71	–	"
60	0.75	0.17 wt % H$_2$O	"
20	0.75 g/100 cc H$_2$O	–	Macintosh et al.
37	405 cc (STP)/100 cc H$_2$O	–	(1963)
20	0.815 wt % CCl$_3$H	–	Interdyne Inc. (1976)
15	0.84 wt % CCl$_3$H	–	Stephen & Stephen
20	0.80	0.2 wt % H$_2$O	"
30	0.76	–	"
20	820 ppm by weight CCl$_3$H	–	Pearson & McConnell (1975)
25	–	0.0738 mol l^{-1} CCl$_3$H	Masterton & Gendrano (1966)
25	0.82 wt % CCl$_3$H	0.10 wt % H$_2$O	Mitchell & Smith (1977)
20	0.8 wt % CCl$_3$H	–	Nathan (1978)
25	–	0.932 wt % H$_2$O	"
15	0.859 g/100 g H$_2$O	–	Jones et al. (1957)
20	0.0675 mol/l H$_2$O	–	Pavlovskaya et al. (1977)
20	8200 ppm by weight CCl$_3$H	–	Selenka & Baner (1978)
25	–	0.00596 mole fraction	Kirchnerova (1975)
38	3.8 cc/cc H$_2$O (Ostwald)	–	Grant (1978)
20	0.82 wt % CCl$_3$H	0.07 wt % H$_2$O	du Pont (1966e)
25	7840 ppm by weight CCl$_3$H	–	Dilling (1977)
1.5	10300	–	"
25	7950 ppm by weight CCl$_3$H	–	Chiou et al. (1977)
25	1270 mg/100 cc H$_2$O	–	Aref'eva et al. (1979)
0	10510 ppm by volume CCl$_3$H	–	Anderman (1978)
10	8870	–	"
20	8100	–	"
25	7800	–	"
30	7600	–	"
20	0.80 wt % CCl$_3$H	–	Matthews (1975, 1979)

CCl$_3$H (Continued)

25	7950 **g m**$^{-3}$ H$_2$O	–	Afghan & Mackay (1980)
25	0.0605 mol l^{-1} soln.	–	Banerjee et al. (1980)

---

[a]M.wt. = 119.38; m.p. = -63.5°C; b.p. = 61.7°C; d$_4^{20}$ = 1.4832 g cc^{-1}.

CCl_4^a

Methane, tetrachloro- (R 10)

Temperature (°C)	Solubility in H_2O	Solubility of H_2O in	Reference
-20	-	Graph	Allied Chemical
0	Graph	"	Corp. (1966a)
⋮	"	"	"
50	"	"	"
15	0.00676 wt % CCl_4	-	Antropov et al.
18	-	0.0071 wt % H_2O	(1972)
20	0.00795 wt % CCl_4	-	"
22	-	0.0083 wt % H_2O	"
26	0.00976 wt % CCl_4	0.0120	"
40	0.0253	0.0196	"
25	0.10 cc/100 cc H_2O	-	Booth & Everson (1948)
25	0.10 cc/100 cc H_2O	-	Booth & Everson
60	0.10	-	(1949)
-10	-	Monograph	Chernyshev (1970)
⋮	-	"	"
80	-	"	"
18	-	0.007 wt % H_2O	Chistyakov (1965)
22	-	0.009	"
24	-	0.010	"
29	-	0.013	"
24	-	0.010 g/100 g soln.	Clifford (1921b)
28.5	0.013 g/100 g H_2O	-	"
30	0.80 g kg^{-1} H_2O	-	du Pont (1961b)
20	-	0.003 wt % H_2O	Chistyakov & Balezin (1961)
15	-	0.012 wt % H_2O	Eberius (1954)
25	-	0.000543 mol/100 g soln.	Eidinoff (1955)
15	-	0.0062 mol l^{-1} CCl_4	Glasoe & Schultz
25	-	0.0086	(1972)
30	-	0.0093	"
35	-	0.0114	"
45	-	0.0156	"
18	-	0.007 wt % H_2O	Chistyakov & Balezin
20	-	0.008	(1961)

CCl_4 (Continued)

25	-	0.0105 wt % H_2O	Chistyakov & Balezin
30	-	0.0130	(1961)
40	-	0.0180	"
50	-	0.0230	"
55	-	0.0255	"
25	0.08 wt % CCl_4	0.013 wt % H_2O	Gladis (1960)
10	-	0.00523 mol l^{-1} CCl_4	Goldman (1974)
15	-	0.00605	"
20	-	0.00747	"
25	-	0.00887	"
30	-	0.01035	"
40	-	0.01433	"
25	0.077 g/100 g H_2O	-	Gross (1929a)
15	0.77 g/1000 g H_2O	-	Gross & Saylor
30	0.81	-	(1931)
25	-	0.010 mol l^{-1} CCl_4	Högfeldt & Fredlund (1970)
10	-	57 ppm by volume H_2O	Luft (1957)
20	-	Monograph	"
⋮	-	"	"
120	-	"	"
20	0.08 wt % CCl_4	0.008 wt % H_2O	Mellan (1957)
25	0.08 wt % CCl_4	0.013 wt % H_2O	Miller (1969)
25	0.0006184 mole fraction	-	Moelwyn-Hughes (1964)
25	0.77 g/1000 g H_2O	-	Gross (1929b)
25	0.08 g/100 g H_2O	0.013 g/100 g CCl_4	McGovern (1943)
-30	-	Graph	"
0	Graph	"	"
⋮	"	"	"
70	"	"	"
20	0.071 wt % CCl_4	0.005 wt % H_2O	Niini (1938)
25	0.157 $\times 10^4$ atm/mole frac.	-	Pierotti (1965)
0	0.097 g/100 g H_2O	-	Rex (1906)
10	0.083	-	"
20	0.080	-	"
30	0.085	-	"
25	0.077 wt % CCl_4	-	Riddick & Bunger
24	-	0.010 wt % H_2O	(1970)

CCl$_4$ (Continued)

Temp (°C)			Reference
10	–	0.00711 g/100 g CCl$_4$	Rosenbaum & Walton
20	–	0.00844	(1930)
30	–	0.0109	"
40	–	0.0152	"
50	–	0.0237	"
0	–	0.00302 wt % H$_2$O	Simonov et al.
30	–	0.0135	(1970)
50	–	0.0305	"
60	–	0.0512	"
80	–	0.1318	"
100	–	0.3388	"
112	–	0.5970	"
20	0.08 g/100 g H$_2$O	–	Smith (1932)
0	–	0.000428 mole fraction	Staverman (1941b)
25	0.000090 mole fraction	0.000991	"
30	–	0.001350	"
15	0.81 g/1000 g H$_2$O	–	Svetlanov et al.
30	0.45	–	(1971)
45	0.25	–	"
60	0.14	–	"
30	0.00053 mol/100 g H$_2$O	–	Van Arkel & Vles (1936)
30	0.08 g l^{-1} H$_2$O	–	Van Arkel (1946)
0	0.097 wt % CCl$_4$	–	Washburn (1928)
10	0.083	–	"
20	0.080	–	"
24	–	0.010 wt % H$_2$O	"
28.5	–	0.013	"
30	0.085 wt % CCl$_4$	–	"
25	1 g/1250 cc H$_2$O	–	Wright & Schaffer (1932)
25	–	1.54 g l^{-1} CCl$_4$	Zielinski (1959)
20	0.08 g/100 g H$_2$O	–	Patty (1962)
25	0.08 g/100 g H$_2$O	–	Gunther et al. (1968)
15	0.755 g/1000 g H$_2$O	–	Liu & Huang (1961)
20	0.748	–	"
25	0.762	–	"
20	0.0058 mol l^{-1} soln.	–	Fitzgerald et al. (1956)

CCl$_4$ (Continued)

15	0.755 g/1000 g H$_2$O	-	Gmelin (1974)
15	0.81	-	"
20	0.748	-	"
25	0.762	-	"
30	0.45	-	"
45	0.25	-	"
60	0.14	-	"
20	0.08 g/100 g H$_2$O	0.013 g/100 g CCl$_4$	Hancock (1973)
20	-	0.0075 mol l^{-1} CCl$_4$	Greinacher et al. (1955)
25	0.08 g/100 g H$_2$O	0.013 g/100 g CCl$_4$	Kirk-Othmer (1964b)
0	-	34 ppm by weight H$_2$O	Lax (1967)
20	0.077 wt % CCl$_4$	85	"
25	0.75 cc (STP)/g H$_2$O	-	"
40	-	175 ppm by weight H$_2$O	"
10	-	0.0071 wt % H$_2$O	Marsden (1963)
20	0.08 wt % CCl$_4$	0.0084	"
30	-	0.0109	"
40	-	0.0152	"
50	-	0.0237	"
0	0.097 g/100 g H$_2$O	-	Lange (1967)
20	0.080	-	"
40	0.025	0.0197 wt % H$_2$O	Simonov et al. (1974)
0	0.097 wt % CCl$_4$	-	Stephen & Stephen
10	0.083	-	(1963)
20	0.080	-	"
30	0.085	-	"
25	0.077 wt % CCl$_4$	0.01 wt % H$_2$O	Interdyne Inc. (1976)
20	785 ppm by weight CCl$_4$	-	Pearson & McConnell (1975)
25	-	0.00085 mole fraction	Kirchnerova & Cave (1976)
25	0.077 wt % CCl$_4$	0.01 wt % H$_2$O	Mitchell & Smith (1977)
25	0.16 wt % CCl$_4$	0.0116 wt % H$_2$O	Nathan (1978)
15	0.077 g/100 g H$_2$O	-	Jones et al. (1957)
25	-	0.00025 wt % H$_2$O	Prosyanov et al.
$\vdots$	-	.	(1974)
70	-	0.00313 wt % H$_2$O	"

CCl$_4$ (Continued)

20	-	0.0072 wt % H$_2$O	Zakuskina et al.
80	-	0.089	(1975)
20	788 ppm by weight CCl$_4$	-	Selenka & Baner (1978)
20	0.08 g/100 g H$_2$O	-	Metcalf (1962)
25	-	0.00085 mole fraction	Kirchnerova (1975)
25	-	0.0087 mol l^{-1} soln.	Johnston et al. (1966)
20	0.08 wt % CCl$_4$	0.01 wt % H$_2$O	du Pont (1966e)
25	800 ppm by weight CCl$_4$	-	Dilling (1977)
25	-	0.0087 mol l^{-1} soln.	Christian et al. (1968)
20	800 ppm by weight CCl$_4$	-	Chiou et al. (1977)
25	87.0 mg/100 cc H$_2$O	-	Aref'eva et al. (1979)
20	793 ppm by volume CCl$_4$	-	Andelman (1978)
25	800	-	"
30	814	-	"
20	0.08 wt % CCl$_4$	-	Matthews (1975, 1979)
25	0.00492 mol l^{-1} soln.	-	Banerjee et al. (1980)
50	-	0.0021 mole fraction	Ksiazczak & Buchowski (1980)

^{a}M.wt. = 153.82; m.p. = -22.99°C; b.p. = 76.54°C; d$_4^{20}$ = 1.5940 g cc^{-1}.

CFH$_3$[a]

Methane, fluoro-   (R 41)

Temperature ($^\circ$C)	Solubility in H$_2$O	Solubility of H$_2$O in	Reference
0.06	6,130 mm Hg mol^{-1} l^{-1} soln.	–	Glew & Moelwyn-
10.73	8,830	–	Hughes (1953)
20.19	11,410	–	"
25.06	12,920	–	"
30.03	14,650	–	"
39.90	17,940	–	"
49.49	21,600	–	"
60.08	25,400	–	"
80.25	32,500	–	"
29.4	14,280 mm Hg mol^{-1} l^{-1} soln.	–	Swain & Thornton
40.3	18,150	–	(1962)
15	166 cc/100 g H$_2$O	–	Lange (1967)

[a]M.wt. = 34.03; m.p. = -141.8°C; b.p. = -78.4°C; d_4^{20} = 0.5786 g cc^{-1}.

CF$_2$H$_2$[a]

Methane, difluoro-   (R 32)

Temperature ($^\circ$C)	Solubility in H$_2$O	Solubility of H$_2$O in	Reference
25 (1 atm)	0.44 wt % CF$_2$H$_2$	–	Sanders (1979)

[a]M.wt. = 52.02; b.p. = -51.6°C; d^{20} = 0.909 g cc^{-1}.

CF_3H[a]

Methane, trifluoro- (R 23)

Temperature ($^{\circ}$C)	Solubility in H_2O	Solubility of H_2O in	Reference
25 (1 atm)	0.10 wt % CF_3H	–	du Pont (1969a)
25 (0-306 psia)	Graph	–	du Pont (1952b)
50 (0-347 psia)	"	–	"
75 (0-342 psia)	"	–	"
25 (14.7 psia)	0.008 lb/gal H_2O	–	du Pont (1966b)
25 (50 psia)	0.027	–	"
25 (100 psia)	0.050	–	"
25 (0-320 psia)	Graph	–	"
50 (0-347 psia)	"	–	"
75 (0-342 psia)	"	–	"
25 (14.7 psia)	0.090 wt % CF_3H	–	du Pont (1966c)
25 (14.7-300 psia)	Table	–	"
50 (14.7-300 psia)	"	–	"
75 (14.7-300 psia)	"	–	"
25 (760 mm Hg)	0.001056 mole fraction CF_3H	–	Moelwyn-Hughes (1964)
25 (14.7 psia)	0.008 lb/gal H_2O	–	Parmelee (1953-
25 (50 psia)	0.027	–	1954:61)
25 (100 psia)	0.050	–	"
25 (0-320 psia)	Graph	–	"
50 (0-347 psia)	"	–	"
75 (0-342 psia)	"	–	"
20	75 cc/100 cc H_2O	–	Meslans (1894)
25	0.10 wt % CF_3H	–	Weast (1976)
25 (1 atm)	0.10 wt % CF_3H	–	Matheson Company (1971)
20	75 cc/100 g H_2O	–	Lange (1967)
25 (1 atm)	0.10 wt % CF_3H	–	Kirk-Othmer (1966a)

[a]M.wt. = 70.01; m.p. = -160.0°C; b.p. = -82.2°C; d^{-100} = 1.52 g cc^{-1}.

CF_4^a

Methane, tetrafluoro- (R 14)

Temperature (°C)	Solubility in H_2O	Solubility of H_2O in	Reference
2.4	0.00900 cc (STP)/g H_2O	–	Ashton et al.
5.0	0.00826	–	(1968)
10.0	0.00700	–	"
15.0	0.00594	–	"
20.0	0.00530	–	"
25.0	0.00474	–	"
30.0	0.00430	–	"
35.0	0.00395	–	"
40.0	0.00375	–	"
45.0	0.00352	–	"
50.0	0.00339	–	"
25 (1 atm)	0.0015 wt % CF_4	–	du Pont (1969a)
25 (0-174 psia)	Graph	–	du Pont (1953)
50 (0-115 psia)	"	–	"
75 (0-135 psia)	"	–	"
25 (14.7 psia)	0.0002 lb/gal H_2O	–	du Pont (1966b)
25 (50 psia)	0.0005	–	"
25 (100 psia)	0.0010	–	"
25 (0-140 psia)	Graph	–	"
50 (0-115 psia)	"	–	"
75 (0-135 psia)	"	–	"
25 (1 atm)	0.0015 wt % CF_4	–	Weast (1976)
25 (14.7 psia)	0.002 wt % CF_4	–	du Pont (1966c)
25 (14.7-150 psia)	Table	–	"
50 (14.7-150 psia)	"	–	"
75 (14.7-150 psia)	"	–	"
25 (0-140 psia)	Graph	–	"
35 (0-128 psia)	"	–	"
50 (0-125 psia)	"	–	"
60 (0-126 psia)	"	–	"
75 (0-124 psia)	"	–	"
5	0.00000663 mole fraction	–	Miller (1966)
10	0.00000560	–	"
15	0.00000483	–	"
20	0.00000425	–	"

CF_4 (Continued)

25	0.00000379 mole fraction	–	Miller (1966)
30	0.00000348	–	"
35	0.00000324	–	"
40	0.00000299	–	"
45	0.00000281	–	"
25	0.0000038 mole fraction	–	Miller & Hildebrand (1968)
6.7 (1 atm)	6.5 cc (STP)/1000 g H_2O	–	Morrison & Johnston
13.1 (1 atm)	5.6	–	(1954)
19.0 (1 atm)	5.0	–	"
24.9 (1 atm)	4.5	–	"
38.5 (1 atm)	3.7	–	"
25	276,000 atm/mole fraction	–	Pierotti (1965)
25	0.0015 wt % CF_4	–	Matheson Company (1971)
25 (14.7 psia)	0.0002 lb/gal H_2O	–	Parmelee (1953-1954:
25 (50 psia)	0.0005	–	61)
25 (100 psia)	0.0010	–	"
25 (0-140 psia)	Graph	–	"
50 (0-115 psia)	"	–	"
75 (0-135 psia)	"	–	"
16	7.45 cc/1000 cc H_2O	–	Gmelin (1974)
17	7.25	–	"
20	6.50	–	"
25 (760 mm Hg)	0.0002 g 1^{-1} H_2O	–	"
25 (2580 mm Hg)	0.060	–	"
25 (5170 mm Hg)	18.94	–	"
16	7.45 cc/100 cc H_2O	–	Treadvell & Köhle
17	7.25	–	(1926)
20	6.50	–	"
25 (1 atm)	0.0015 wt % CF_4	–	Kirk-Othmer (1966a)
15	0.00623 cc/cc H_2O (Ostwald)	–	Muccitelli & Wen (1978)
5 (1 atm)	0.00000668 mole fraction	–	Wen & Muccitelli
10 (1 atm)	0.00000562	–	(1979)
15 (1 atm)	0.00000477	–	"
20 (1 atm)	0.00000416	–	"
25 (1 atm)	0.00000374	–	"
30 (1 atm)	0.00000337	–	"

CF_4 (Continued)

5 (1 atm)	0.00000662 mole fraction	–	Muccitelli (1978)
10 (1 atm)	0.00000555	–	"
15 (1 atm)	0.00000469	–	"
20 (1 atm)	0.00000406	–	"
25 (1 atm)	0.00000362	–	"
30 (1 atm)	0.00000323	–	"
25	0.00493 cc/cc H_2O (Bunsen)	–	Maharajh (1973)

[a]M.wt. = 88.01; m.p. = -150.0°C; b.p. = -129.0°C; d^0 = 3.034 g cc^{-1}.

CHI_3[a]
Methane, triiodo-

Temperature (°C)	Solubility in H_2O	Solubility of H_2O in	Reference
25	0.02 g/100 cc H_2O	–	Booth & Everson (1948)
22	0.01 g/100 g H_2O	–	Dehn (1917)
25	0.00003 mol/100 g H_2O	–	Van Arkel & Vles (1936)
25	0.01 g/100 cc H_2O	–	Patty (1962)
25	0.01 g/100 cc H_2O	–	Kirk-Othmer (1966b)
25	0.01 g/100 g H_2O	–	Andelman (1978)

[a]M.wt. = 393.73; m.p. = 123.0°C; b.p. = 218.0°C; d_4^{20} = 4.008 g cc^{-1}.

CH_2I_2[a]
Methane, diiodo-

Temperature (°C)	Solubility in H_2O	Solubility of H_2O in	Reference
25	0.00311 mol l^{-1} soln.	–	Andrews & Keefer (1951)
25	–	0.001207 mol/100 g soln.	Eidinoff (1955)
30	1.24 g/1000 g H_2O	–	Gross & Saylor (1931)
25	0.124 g/100 g H_2O	–	O'Connell (1963)
30	0.124 wt % CH_2I_2	–	Riddick & Bunger (1970)
30	0.00046 mol/100 g H_2O	–	Van Arkel & Vles (1936)
0	1.6 g/100 g H_2O	–	Lange (1967)
20	1.4	–	"
20	1.42 g/100 cc H_2O	–	Kirk-Othmer (1966b)
30	1.24 wt % CH_2I_2	–	Stephen & Stephen (1963)
30	0.124 g/100 g H_2O	–	Andelman (1976)

[a] M.wt. = 267.84; m.p. = 6.1°C; b.p. = 182.0°C; d_4^{20} = 3.3254 g cc^{-1}.

CH_3I^a
Methane, iodo-

Temperature (^oC)	Solubility in H_2O	Solubility of H_2O in	Reference
22	1.362 wt % CH_3I	–	Führner (1924)
0.01	1274 mm Hg mol^{-1} l^{-1} soln.	–	Glew & Moelwyn-
5.04	1686	–	Hughes (1953)
10.20	2160	–	"
15.12	2730	–	"
19.95	3380	–	"
24.93	4060	–	"
29.93	4730	–	"
39.87	5540	–	"
44.88	6380	–	"
49.76	8040	–	"
20	1.4 wt % CH_3I	–	Harms (1939)
20	1.399 wt % CH_3I	–	Korenman et al. (1971)
20	0.0960 mol l^{-1} H_2O	–	Merckel (1937)
25 (760 mm Hg)	0.003383 mole fraction	–	Moelwyn-Hughes (1964)
15	1 cc/125 cc H_2O	–	Patty (1962)
0	1.565 g/100 g H_2O	–	Rex (1906)
10	1.446	–	"
20	1.419	–	"
30	1.429	–	"
20	1.4 wt % CH_3I	–	Riddick & Bunger (1970)
22	0.0959 mol/1000 g H_2O	–	Saracco & Spaccamela Marchetti (1958)
29.4	4890 mm Hg mol^{-1} l^{-1} soln.	–	Swain & Thornthon
40.3	6885	–	(1962)
30	0.0101 mol/100 g H_2O	–	Van Arkel & Vles (1936)
0	1.54 wt % CH_3I	–	Washburn (1928)
10	1.42	–	"
20	1.40	–	"
30	1.41	–	"
15	1.8 g/100 g H_2O	–	Lange (1967)
20	1.4 g/100 cc H_2O	–	Kirk-Othmer (1966b)

CH$_3$I (Continued)

22	1.362 wt % CH$_3$I	–	Stephen & Stephen
30	1.413	–	(1963)
0	1.565 g/100 g H$_2$O	–	Andelman (1978)
10	1.446	–	"
20	1.419	–	"
30	1.429	–	"

[a]M.wt. = 141.94; m.p. = -66.45°C; b.p. = 42.4°C; d_4^{20} = 2.279 g cc^{-1}.

CI$_4^{a}$

Temperature (°C)	Solubility in H$_2$O	Solubility of H$_2$O in	Reference
20	None	–	Gmelin (1974)
20	None	–	Bellino (1932)

[a]M.wt. = 519.63; m.p. = 171.0°C; b.p. = subl. at 130°C; d^{20} = 4.23 g cc^{-1}.

CBrClH-CF$_3$[a]

Ethane, 1-bromo-1-chloro-2,2,2-trifluoro- (R 123B1)

Temperature ($^\circ$C)	Solubility in H$_2$O	Solubility of H$_2$O in	Reference
37	0.80 cc/cc H$_2$O (Ostwald)	-	Allott et al. (1973)
20	1.0 g/400 g H$_2$O	-	Berndt (1964)
37	0.74 cc/cc H$_2$O (Ostwald)	-	Eger & Larson (1964)
10	0.53 wt % C$_2$BrClF$_3$H	-	Imperial Chemical
20	0.45	-	Industries (1968)
30	0.42	-	"
37	0.40	-	"
37	0.74 cc/cc H$_2$O (Ostwald)	-	Larson et al. (1962a)
37	0.74 cc/cc H$_2$O (Ostwald)	-	Papper & Kitz (1963)
37	1.04 cc/cc H$_2$O (Ostwald)	-	Saidman et al. (1966)
20	0.345 g/100 g H$_2$O	-	Secher (1971)
37	0.80 cc/cc H$_2$O (Ostwald)	-	Steward et al. (1973)
23	0.345 g/100 cc H$_2$O	-	Sadove & Wellace (1962)
25	-	0.03 wt % H$_2$O	Lees & Sarram (1971)
24	0.372 wt % C$_2$BrClF$_3$H	0.035 wt % H$_2$O	Horvath (1975c)
20	0.345 g/100 g H$_2$O	-	Stephen & Little (1961)
20	0.34 g/100 cc H$_2$O	-	Macintosh et al. (1963)
37	0.345 g/100 g H$_2$O	-	Tarrant (1969)
36.85	0.80 cc/cc H$_2$O (Ostwald)	-	Halliday et al. (1977)
24	1.07 cc/cc H$_2$O (Ostwald)	-	Lauven et al.
37	0.77	-	(1979)
37	0.74 cc/cc H$_2$O (Ostwald)	-	Grant (1978)

[a]M.wt. = 197.39; m.p. = -118.27°C; b.p. = 50.30°C; d$_4^{20}$ = 1.776 g cc^{-1}.

CBrH$_2$-CClH$_2^a$
Ethane, 1-bromo-2-chloro-

Temperature (°C)	Solubility in H$_2$O	Solubility of H$_2$O in	Reference
30	6.88 g/1000 g H$_2$O	-	Gross et al. (1933)
30	0.688 g/100 cc H$_2$O	-	Gunther et al. (1968)
30	0.69 g/100 g H$_2$O	-	Lange (1967)
30	0.683 wt % C$_2$BrClH$_4$	-	Stephen & Stephen (1963)

[a]M.wt. = 143.42; m.p. = -16.7°C; b.p. = 107.0°C; d$_4^{20}$ = 1.7392 g cc^{-1}.

CBrFH-CF$_3^a$
Ethane, 1-bromo-1,2,2,2-tetrafluoro-

Temperature (°C)	Solubility in H$_2$O	Solubility of H$_2$O in	Reference
37	0.04 g/100 g H$_2$O	-	Tarrant (1969)

$CBrH_2-CH_3^a$

Ethane, bromo-

Temperature ($^\circ$C)	Solubility in H_2O	Solubility of H_2O in	Reference
25	–	0.00417 mol/100 g soln.	Eidinoff (1955)
17.5	0.952 wt % C_2BrH_5	–	Fühner (1924)
20	0.90 wt % C_2BrH_5	–	Harms (1939)
20	0.9615 wt % C_2BrH_5	–	Korenman et al. (1971)
0	1.067 g/100 g H_2O	–	Rex (1906)
10	0.965	–	"
20	0.914	–	"
30	0.896	–	"
20	0.91 wt % C_2BrH_5	–	Riddick & Bunger (1970)
17.5	0.088 mol/1000 g H_2O	–	Saracco & Spacca-mela Marchetti (1958)
30	0.0082 mol/100 g H_2O	–	Van Arkel & Vles (1936)
0	1.055 wt % C_2BrH_5	–	Washburn (1928)
10	0.956	–	"
20	0.906	–	"
30	0.888	–	"
20	0.91 g/100 cc H_2O	–	Patty (1962)
0	1.06 g/100 g H_2O	–	Lange (1967)
30	0.90	–	"
25	0.90 g/100 g H_2O	–	Kirk-Othmer (1964a)

[a] M.wt. = 108.97; m.p. = -118.6°C; b.p. = 38.4°C; d_4^{20} = 1.4604 g cc^{-1}.

$CBrClH-CBrH_2^a$

Ethane, 1,2-dibromo-1-chloro-

Temperature ($^\circ$C)	Solubility in H_2O	Solubility of H_2O in	Reference
25	–	0.06 g/100 g $C_2Br_2ClH_3$	O'Connell (1963)
25	–	0.6 g/1000 g $C_2Br_2ClH_3$	Jolles (1966)

[a] M.wt. = 222.31; m.p. = -25.0°C; b.p. = 163.4°C; d^{25} = 2.246 g cc^{-1}.

CBrClH-CBrClH[a]

Ethane, 1,2-dibromo-1,2-dichloro-

Temperature ($^{\circ}$C)	Solubility in H_2O	Solubility of H_2O in	Reference
25	-	0.07 g/100 g $C_2Br_2Cl_2H_2$	O'Connell (1963)
25	-	0.7 g/1000 g $C_2Br_2Cl_2H_2$	Jolles (1966)

[a]M.wt. = 256.76; m.p. = -26.0°C; b.p. = 195.0°C; d_4^{20} = 2.135 g cc^{-1}.

CBrF$_2$-CBrF$_2$[a]

Ethane, 1,2-dibromo-1,1,2,2-tetrafluoro- (R 114B2)

Temperature ($^{\circ}$C)	Solubility in H_2O	Solubility of H_2O in	Reference
25	-	0.003 wt % H_2O	Semerikova et al. (1976)

[a]M.wt. = 259.82; m.p. = -112.0°C; b.p. = 46.4°C; d_4^{25} = 2.149 g cc^{-1}.

CBrH=CBrH[a]

Ethene, 1,2-dibromo- (cis-)

Temperature ($^{\circ}$C)	Solubility in H_2O	Solubility of H_2O in	Reference
25	0.40 cc/100 cc H_2O	-	Booth & Everson (1948)
25	0.41 wt % $C_2Br_2H_2$	-	Gunther et al. (1968)

[a]M.wt. = 185.86; m.p. = -53.0°C; b.p. = 112.5°C; d_4^{20} = 2.2464 g cc^{-1}.

$CBrH_2-CBrH_2^a$
Ethane, 1,2-dibromo-

Temperature ($^\circ$C)	Solubility in H_2O	Solubility of H_2O in	Reference
25	–	0.00324 mol/100 g soln.	Eidinoff (1955)
25	–	0.06 wt % H_2O	Ethyl Corp. (1972)
15	3.92 g/1000 g H_2O	–	Gross & Saylor
30	4.31	–	(1931)
25	0.404 wt % $C_2Br_2H_4$	–	Miller (1969)
25	0.417 g/100 g H_2O	0.071 g/100 g $C_2Br_2H_4$	O'Connell (1963)
30	0.429 wt % $C_2Br_2H_4$	–	Riddick & Bunger
25	–	0.071 wt % H_2O	(1970)
0	0.25 wt % $C_2Br_2H_4$	0.35 wt % H_2O	Shostakovskii &
25	0.36	0.54	Druzhinin (1942)
35	0.42	0.63	"
50	0.54	1.18	"
75	0.77	1.83	"
0	0.000322 mole fraction	–	Staverman (1941b)
20	0.000387	–	"
25	–	0.00685 mole fraction	"
30	–	0.00746	"
35	0.000432 mole fraction	–	"
30	0.43 g/100 cc H_2O	–	Patty (1962)
0	0.335 g/100 g H_2O	–	Van Arkel & Vles
20	0.404	–	(1936)
35	0.451	–	"
50	0.532	–	"
30	0.43 g/100 g H_2O	–	Gunther et al. (1968)
25	–	0.071 g/100 g $C_2Br_2H_4$	Dreisbach (1955- 1961: Ser. 22)
30	0.43 g/100 g H_2O	–	Lange (1967)
20	0.404 g/100 g soln.	–	Kirk-Othmer (1964a)
25	0.417 g/100 g H_2O	0.71 g/1000 g $C_2Br_2H_4$	Jolles (1966)
30	0.431 g/100 cc H_2O	–	"
20	3370 ppm by weight $C_2Br_2H_4$	–	Metcalf (1962)
30	0.43 g/100 cc H_2O	–	"
20	3.37 mg/cc soln.	–	Call (1957)

[a]M.wt. = 187.87; m.p. = 9.79°C; b.p. = 131.36°C; d_4^{20} = 2.1792 g cc^{-1}.

$CBr_2H-CBrH_2$[a]
Ethane, 1,1,2-tribromo-

Temperature (°C)	Solubility in H_2O	Solubility of H_2O in	Reference
25	–	0.05 g/100 g $C_2Br_3H_3$	O'Connell (1963)
25	–	0.5 g/1000 g $C_2Br_3H_3$	Jolles (1966)

[a]M.wt. = 266.77; m.p. = -29.3°C; b.p. = 188.93°C; d_4^{20} = 2.6211 g cc^{-1}.

CBr_2H-CBr_2H[a]
Ethane, 1,1,2,2-tetrabromo-

Temperature (°C)	Solubility in H_2O	Solubility of H_2O in	Reference
25	<0.02 cc/100 cc H_2O	–	Booth & Everson (1948)
25	–	0.002113 mol/100 g soln.	Eidinoff (1955)
30	0.0651 g/100 g H_2O	–	Gross & Saylor (1931)
30	0.0651 g/100 g H_2O	–	Gross et al. (1933)
51	0.30 g/100 g H_2O	–	Blumberg & Melzer (1960)
80	0.28 g/100 g H_2O	–	Dow Chemical Co. (1963)
1	0.0000272 mole fraction	–	Gooch et al.
10	0.0000295	–	(1972)
20	0.0000328	–	"
25	0.0000352	–	"
30	0.0000377	–	"
40	0.0000456	–	"
50	0.0000552	–	"
60	0.0000657	–	"
70	0.0000813	–	"
80	0.000101	–	"
90	0.000128	–	"
97	0.000150	–	"
30	0.065 g/100 g H_2O	–	O'Connell (1963)
25	–	0.04 g/100 g $C_2Br_4H_2$	"
30	0.0651 wt % $C_2Br_4H_2$	–	Riddick & Bunger (1970)

CBr_2H-CBr_2H (Continued)

30	0.065 g/100 cc H_2O	–	Patty (1962)
1	0.0000272 mole fraction	–	Gooch (1971)
10	0.0000295	–	"
20	0.0000328	–	"
25	0.0000352	–	"
30	0.0000377	–	"
40	0.0000456	–	"
50	0.0000552	–	"
60	0.0000657	–	"
70	0.0000813	–	"
80	0.000101	–	"
90	0.000128	–	"
97	0.000150	–	"
20	0.56 g l^{-1} H_2O	–	Jolles (1966)
30	0.065 g/100 g H_2O	0.04 g/100 g $C_2Br_4H_2$	"
30	0.651 g l^{-1} H_2O	–	Stephen & Stephen (1963)

[a]M.wt. = 345.67; m.p. = 0°C; b.p. = 243.5°C; d_4^{20} = 2.9656 g cc^{-1}.

$CClH_2-CFH_2$[a]
Ethane, 1-chloro-2-fluoro-

Temperature (°C)	Solubility in H_2O	Solubility of H_2O in	Reference
25	25 g/1000 cc H_2O	–	Rasumovskii & Friedenberg (1949)

[a]M.wt. = 82.51; m.p. = -50.0°C; b.p. = 57.0°C; d_4^{20} = 1.1747 g cc^{-1}.

$CClH_2-CF_2H$[a]

Ethane, 1-chloro-2,2-difluoro- (R 142)

Temperature (^oC)	Solubility in H_2O	Solubility of H_2O in	Reference
-40	–	Graph	Petrak (1975)
⋮	–	"	"
80	–	"	"

[a]M.wt. = 100.50; b.p. = 35.1^oC; d^{20} = 1.312 g cc^{-1}.

$CClF_2-CH_3$[a]

Ethane, 1-chloro-1,1-difluoro- (R 142b)

Temperature (^oC)	Solubility in H_2O	Solubility of H_2O in	Reference
21 (1 atm)	0.14 wt % $C_2ClF_2H_3$	0.054 wt % H_2O	Allied Chemical Corp. (1960a)
21 (1 atm)	0.19 wt % $C_2ClF_2H_3$	0.048 wt % H_2O	du Pont (1961a)
21 (14.7 psia)	0.140 wt % $C_2ClF_2H_3$	0.054 wt % H_2O	du Pont (1966c)
21 (1 atm)	0.19 wt % $C_2ClF_2H_3$	0.048 wt % H_2O	Matheson Company (1966, 1971)
13.09 (1743 mm Hg)	0.49 wt % $C_2ClF_2H_3$	–	Colten et al. (1972)
21 (1 atm)	–	0.054 wt % H_2O	Shephard (1961)
13.22 (512.7 mm Hg)	0.1458 wt % $C_2ClF_2H_3$	–	Fernandez et al.
13.22 (787.3 mm Hg)	0.2216	–	(1967)
13.22 (908.6 mm Hg)	0.2529	–	"
13.22 (1046.4 mm Hg)	0.2963	–	"
13.22 (1253.7 mm Hg)	0.3535	–	"
13.22 (1570.8 mm Hg)	0.4435	–	"
10.03 (1401.6 mm Hg)	0.4973	–	"
11.52 (1506.5 mm Hg)	0.4574	–	"
13.14 (1520.6 mm Hg)	0.4230	–	"
14.66 (1566.6 mm Hg)	0.3981	–	"
16.15 (1687.1 mm Hg)	0.3746	–	"
18.74 (1538.2 mm Hg)	0.3462	–	"
21.1 (1 atm)	0.054 wt % $C_2ClF_2H_3$	–	Sanders (1979)

[a]M.wt. = 100.50; m.p. = -130.8^oC; b.p. = -9.5^oC; d^{30} = 1.096 g cc^{-1}.

$CClH_2-CF_3^a$

Ethane, 1-chloro-2,2,2-trifluoro- (R 133a)

Temperature (^{o}C)	Solubility in H_2O	Solubility of H_2O in	Reference
25 (1 atm)	0.43 wt % $C_2ClF_3H_2$	–	Sanders (1979)

[a] M.wt. = 118.49; m.p. = -105.5^{o}C; b.p. = 6.93^{o}C; d_4^0 = 1.389 g cc^{-1}

$CClFH-CF_3^a$

Ethane, 1-chloro-1,2,2,2-tetrafluoro- (R 124)

Temperature (^{o}C)	Solubility in H_2O	Solubility of H_2O in	Reference
25 (1 atm)	0.15 wt % C_2ClF_4H	–	Sanders (1979)

[a] M.wt. = 136.5; b.p. = -11.0^{o}C; d^{21} = 1.38 g cc^{-1}.

$CClF_2-CF_3^a$

Ethane, chloropentafluoro- (R 115)

Temperature (^{o}C)	Solubility in H_2O	Solubility of H_2O in	Reference
25 (1 atm)	0.006 wt % C_2ClF_5	–	du Pont (1969a)
25 (0-125 psia)	Graph	–	du Pont (1952c)
50 (0-115 psia)	"	–	"
75 (0-120 psia)	"	–	"
25 (14.7 psia)	0.0004 lb/gal H_2O	–	du Pont (1966b)
25 (50 psia)	0.0010	–	"
25 (100 psia)	0.0018	–	"
25 (0-120 psia)	Graph	–	"
50 (0-120 psia)	"	–	"
75 (0-120 psia)	"	–	"
25 (14.7 psia)	0.006 wt % C_2ClF_5	–	du Pont (1966c)
25 (14.7-100 psia)	Table	–	"
50 (14.7-100 psia)	"	–	"
75 (14.7-100 psia)	"	–	"

$CClF_2-CF_3$ (Continued)

25 (0-120 psia)	Graph	–	du Pont (1966c)
30 (0-120 psia)	"	–	"
40 (0-120 psia)	"	–	"
50 (0-120 psia)	"	–	"
60 (0-120 psia)	"	–	"
75 (0-120 psia)	*	–	"
25	0.006 wt % C_2ClF_5	–	Weast (1976)
25 (14.7 psia)	0.0004 lb/gal H_2O	–	Parmelee (1953-
25 (50 psia)	0.0010	–	1954:61)
25 (100 psia)	0.0018	–	"
25 (0-120 psia)	Graph	–	"
50 (0-120 psia)	"	–	"
75 (0-120 psia)	"	–	"
25	0.006 wt % C_2ClF_5	–	Riddick & Bunger (1970)
25 (760 mm Hg)	0.006 wt % C_2ClF_5	–	Matheson Company (1966, 1971)
25 (1 atm)	0.006 g/100 g H_2O	–	Kirk-Othmer (1966a)
5 (1-5 bars)	Graph	–	Filatkin et al.
⋮	"	–	(1970)
40	"	–	"
25 (1 atm)	0.006 wt % C_2ClF_5	–	Sanders (1979)

[a]M.wt. = 154.47; m.p. = -106.0°C; b.p. = -38.0°C; d^{30} = 1.265 g cc^{-1}.

$CClH=CH_2^a$

Ethene, chloro- (R 1140)

Temperature (°C)	Solubility in H_2O	Solubility of H_2O in	Reference
25	–	0.11 g/100 g C_2ClH_3	Dreisbach (1955–1961: Ser. 22)
5.0 (30.2 psia)	–	0.00172 mole fraction	Ethyl Corp. (1968)
23.0 (55.0 psia)	–	0.00320	"
30.0 (67.6 psia)	–	0.00397	"
40.0 (90.0 psia)	–	0.00540	"
60.0 (148.0 psia)	–	0.00920	"
80.0 (231.8 psia)	–	0.0150	"
–25 (saturated)	–	Graph	"
⋮	–	"	"
80	–	"	"
50 (saturated)	1 g/100 g H_2O	–	Gerrens et al. (1967)
0 (1 atm)	Graph	–	Hayduk & Laudie
⋮	"	–	(1973)
374	0.000140 mole fraction	–	"
25 (1 atm)	0.000798 mole fraction	–	Hayduk & Laudie
50 (1 atm)	0.000410	–	(1974a)
75 (1 atm)	0.000225	–	"
0.2 (1.0–2.15 atm)	Table	–	Hayduk & Laudie
25 (1.0–3.06 atm)	"	–	(1974b)
50 (1.0–6.12 atm)	"	–	"
75 (1.0–6.12 atm)	"	–	"
0.2 (0.6–1.6 atm)	Graph	–	"
25 (0.6–3.8 atm)	"	–	"
50 (0.6–8.0 atm)	"	–	"
75 (0.6–13 atm)	"	–	"
10	–	0.11 wt % H_2O	Mannheimer (1956)
50	0.042 wt % C_2ClH_3	–	"
25	–	0.11 wt % H_2O	Miller (1969)
21	–	0.09 wt % H_2O	Union Carbide Corp. (1973)
25	–	0.11 g/100 g C_2ClH_3	Dreisbach (1955–1961: Ser. 22)
–15	–	0.03 g/100 g C_2ClH_3	Stamford Research Inst. (1965)

CClH=CH$_2$ (Continued)

25	0.11 g/100 g H$_2$O	-	Kirk-Othmer (1964b)
-15	-	0.03 g/100 g C$_2$ClH$_3$	"
0	-	0.05 wt % H$_2$O	Imperial Chemical
25	-	0.09	Industries (1959)
40	-	0.12	"
25 (760 mm Hg)	0.11 g/100 g H$_2$O	-	Matheson Company (1971)
0	-	0.05 wt % H$_2$O	Imperial Chemical
25	54 cc (STP)/100 cc H$_2$O	0.09	Industries (1965)
40	-	0.12	"
25 (saturated)	179 cc (STP)/100 cc H$_2$O	-	"
0 (1 atm total)	-	290 ppm by weight C$_2$ClH$_3$	Imperial Chemical
5 (1 atm total)	0.49 wt % C$_2$ClH$_3$	-	Industries (1951)
10 (1 atm total)	0.40	-	"
20 (1 atm total)	0.27	1120 ppm by weight C$_2$ClH$_3$	"
30 (1 atm total)	0.18	1720	"
40 (1 atm total)	0.11	2320	"
50 (1 atm total)	0.08	3260	"
60 (1 atm total)	0.06	4170	"
3	149 cc (STP)/cc H$_2$O	-	Mizutani & Yamas-
6	160	-	hita (1950)
10	107	-	"
20	99	-	"
23	83	-	"
25	79.5	-	"
30	70.5	-	"
38	46.7	-	"
40	42.8	-	"
45	30	-	"
50	23	-	"
10 (1 atm)	60 ppm by weight C$_2$ClH$_3$	-	Péarson & McConnell (1975)
20 (1 atm)	0.68 wt % C$_2$ClH$_3$	0.9 wt % H$_2$O	Nathan (1978)
20 (0-3.4 atm)	Graph	-	Zampachova (1962)
30 (0-4.5 atm)	"	-	"
40 (0-6.1 atm)	"	-	"
50 (0-7.8 atm)	"	-	"
20 (1 atm)	60 ppm by weight C$_2$ClH$_3$	-	Selenka & Baner (1978)

CClH=CH$_2$ (Continued)

10	60 ppm by weight C$_2$ClH$_3$	–	Dilling (1977)
25	2700	–	"
25	0.11 wt % C$_2$ClH$_3$	–	Pittsburgh Plate Glass Ind. (1975)

[a]M.wt. = 62.50; m.p. = -153.8°C; b.p. = -13.37°C; d_4^{20} = 0.9106 g cc^{-1}.

CClH$_2$-CH$_3^{\text{a}}$
Ethane, chloro-

Temperature (°C)	Solubility in H$_2$O	Solubility of H$_2$O in	Reference
37	1.2 cc/cc H$_2$O (Ostwald)	–	Allott et al. (1973)
12.5	0.57 wt % C$_2$ClH$_5$	–	Fühner (1924)
20	0.51 wt % C$_2$ClH$_5$	0.14 wt % H$_2$O	Gladis (1960)
20	–	0.2 wt % H$_2$O	Jenkin & Shorthose (1923)
20	0.9028 wt % C$_2$ClH$_5$	–	Korenman et al. (1971)
20	0.574 g/100 g H$_2$O	–	Mellan (1957)
21	–	48.3 mg/100 g C$_2$ClH$_5$	"
20	0.109 wt % C$_2$ClH$_5$	–	Miller (1969)
0	0.45 g/100 g H$_2$O	–	McGovern (1943)
20	2.12 cc/cc H$_2$O (Ostwald)	–	Nicloux & Scotti
25	1.62	–	Foglieni (1929)
30	1.37	–	"
40	0.89	–	"
0	0.447 g/100 g H$_2$O	0.07 g/100 g C$_2$ClH$_5$	Pittsburgh Plate Glass Ind. (1971d)
0	0.447 wt % C$_2$ClH$_5$	–	Riddick & Bunger (1970)
17.5	0.089 mol/1000 g H$_2$O	–	Saracco & Marchetti Spaccamela (1958)
30	1.31 cc/cc H$_2$O (Ostwald)	–	Scotti-Foglieni (1930a)
30	1.31	–	Scotti-Foglieni (1930b)
30	1.31	–	Scotti-Foglieni (1931a)

$CClH_2$-CH_3 (Continued)

30	1.31 cc/cc H_2O (Ostwald)	–	Scotti-Foglieni (1931b)
30	1.31	–	Scotti-Foglieni (1931c)
30	1.31	–	Scotti-Foglieni (1931d)
20	0.574-0.62 g/100 g H_2O	–	Secher (1971)
37	1.20 cc/cc H_2O (Ostwald)	–	Steward et al. (1973)
0	0.447 g/100 g H_2O	–	Van Arkel & Vles (1936)
11	0.2 wt % C_2ClH_5	–	Washburn (1928)
20	0.57 g/100 cc H_2O	–	Patty (1962)
20	0.57 wt % C_2ClH_5	–	Marsden (1963)
0 (760 mm Hg)	0.447 g/100 g H_2O	–	Matheson Company (1966, 1971)
0 (760 mm Hg)	0.45 g/100 g H_2O	–	Lange (1967)
0 (1 atm)	0.447 g/100 g H_2O	0.07 g/100 g C_2ClH_5	Kirk-Othmer (1964b)
50 (1 atm)	–	0.36	"
20	2.12 cc/cc H_2O (Ostwald)	–	Imperial Chemical
25	1.62	–	Industries (1959)
30	1.37	–	"
0	0.45 g/100 cc H_2O	–	Macintosh et al.
20	160 cc/100 cc H_2O	–	(1963)
0	0.447 wt % C_2ClH_5	–	Stephen & Stephen
17.5	0.57	–	(1963)
20	0.574 wt % C_2ClH_5	–	Nathan (1978)
20 (1 atm)	0.57 wt % C_2ClH_5	–	Sanders (1979)
20 (1 atm)	5700 ppm by weight C_2ClH_5	–	Dilling (1977)

[a]M.wt. = 64.52; m.p. = -136.4°C; b.p. = 12.27°C; d_4^{20} = 0.8978 g cc^{-1}.

$CClF_2$-$CClH_2$[a]
Ethane, 1,2-dichloro-2,2-difluoro- (R 132b)

Temperature (°C)	Solubility in H_2O	Solubility of H_2O in	Reference
23.9 (1 atm)	0.49 wt % $C_2Cl_2F_2H_2$	0.085 wt % H_2O	Sanders (1979)

[a]M.wt. = 134.97; m.p. = -101.2°C; b.p. = 46.8°C; d_4^{20} = 1.4163 g cc^{-1}.

$CCl_2H-CF_3^a$
Ethane, 1,1-dichloro-2,2,2-trifluoro- (R 123)

Temperature (^oC)	Solubility in H_2O	Solubility of H_2O in	Reference
25	–	504 ppm by weight H_2O	du Pont (1963b)
21	–	0.047 wt % H_2O	du Pont (1966c)
-28.9	–	0.0084	"
-23.3	–	0.0118	"
-17.8	–	0.0135	"
-12.2	–	0.0165	"
-6.7	–	0.0200	"
-1.1	–	0.0245	"
4.4	–	0.0290	"
10.0	–	0.0345	"
15.6	–	0.040	"
21.1	–	0.047	"
26.7	–	0.055	"
32.2	–	0.064	"
37.8	–	0.073	"
43.3	–	0.083	"
48.9	–	0.094	"
25 (1 atm)	0.46 wt % $C_2Cl_2F_3H$	–	Sanders (1979)

[a]M.wt. = 152.93; m.p. = -107.0^oC; b.p. = 27.6^oC; d_4^{25} = 1.464 g cc^{-1}.

$CClF_2-CClF_2^a$
Ethane, 1,2-dichloro-1,1,2,2-tetrafluoro- (R 114)

Temperature (^oC)	Solubility in H_2O	Solubility of H_2O in	Reference
-50	–	Monograph	Chernyshev (1969)
:	–	"	"
50	–	"	"
0	–	0.0026 wt % H_2O	du Pont (1969a)
25 (1 atm)	0.013 wt % $C_2Cl_2F_4$	0.009	"
26.5	0.00883 g/psia/kg H_2O	–	du Pont (1961b)
53	0.00429	–	"

$CClF_2$-$CClF_2$ (Continued)

75	0.00281 g/psia/kg H_2O	–	du Pont (1961b)
7	Graph	–	"
.	"	–	"
76	"	–	"
-17.7	–	0.001 wt % H_2O	du Pont (1966a)
21.1	–	0.007	"
25 (1 atm)	0.024 g/100 g H_2O	–	"
-73	–	Graph	"
.	–	"	"
49	–	"	"
-73.33	–	0.1 ppm by weight H_2O	du Pont (1966d)
-62.22	–	0.4	"
-51.11	–	1.0	"
-40.00	–	2.0	"
-28.89	–	5.0	"
-17.78	–	10.0	"
-6.67	–	18	"
4.44	–	33	"
15.56	–	57	"
26.67	–	95	"
37.78	–	148	"
21.1	–	0.0074 wt % H_2O	du Pont (1966c)
-73.3	–	0.00001	"
-62.2	–	0.00004	"
-51.1	–	0.0001	"
-40.0	–	0.0002	"
-28.9	–	0.0005	"
-17.8	–	0.0010	"
-6.7	–	0.0018	"
4.4	–	0.0033	"
15.6	–	0.0057	"
26.7	–	0.0095	"
37.8	–	0.0148	"
26.5 (14.7 psia)	0.013 wt % $C_2Cl_2F_4$	–	"
25.0 (14.7-25 psia)	Table	–	"
50.0 (14.7-50 psia)	"	–	"
75.0 (14.7-75 psia)	"	–	"
20 (1 atm)	0.14 g/kg H_2O	–	• Pennsalt Chemical Corp. (1958)

$CClF_2-CClF_2$ (Continued)

-17.7	–	0.001 wt % H_2O	Elchardus &
21.1	–	0.007	Maestre (1965)
-47.0	–	Graph	"
⋮	–	"	"
50	–	"	"
25	0.013 wt % $C_2Cl_2F_4$	0.009 wt % H_2O	Weast (1976)
0 (760 mm Hg)	–	0.0026 wt % H_2O	Matheson Company
25 (760 mm Hg)	0.013 wt % $C_2Cl_2F_4$	–	(1966, 1971)
30 (760 mm Hg)	–	0.011 wt % H_2O	"
25 (1 atm)	0.013 wt % $C_2Cl_2F_4$	–	Kirk-Othmer (1966a)
21 (1 atm)	0.013 wt % $C_2Cl_2F_4$	0.007 wt % H_2O	Allied Chemical Corp. (1974a)
0	–	0.0026 g/100 g $C_2Cl_2F_4$	Imperial Chemical
30	–	0.011	Industries (1953)
0	–	0.0025 g/100 g $C_2Cl_2F_4$	Kali-Chemie (1972)
30	–	0.011	"
25	0.013 wt % $C_2Cl_2F_4$	–	Rhône Progil
30	–	0.011 wt % H_2O	(1973)
25	0.013 wt % $C_2Cl_2F_4$	–	Rauws et al. (1973)
6 (1-3.2 atm)	Graph	–	Filatkin et al.
⋮	"	–	(1976)
35	"	–	"
0 (0-25 psia)	Graph	–	Stepakoff &
⋮	"	–	Modica (1972)
50	"	–	"
0 (0-7 psia)	Graph	–	Stepakoff &
25 (0-20 psia)	"	–	Modica (1973)
50 (0-25 psia)	"	–	"
0 (5-10 psia)	Table	–	"
25 (5-20 psia)	"	–	"
50 (5-20 psia)	"	–	"
-3.88 (1 atm)	347 ppm by weight $C_2Cl_2F_4$	–	"
0 (1 atm)	295	–	"
25 (1 atm)	104	–	"
51 (1 atm)	46	–	"
21	–	0.008 wt % H_2O	Shepherd (1961)
0	–	0.0026 g/100 g $C_2Cl_2F_4$	Imperial Smelting Corp. (1957)

$CClF_2$-$CClF_2$ (Continued)

-40	-	2 ppm by weight H_2O	Ackroyd (1978)
-20	-	9	"
-10	-	16	"
0	-	27	"
15	-	55	"
30	-	105	"
33	97.2 ppm by weight $C_2Cl_2F_4$/atm	-	Hellström et al.
37.4	85.8	-	(1976)
56	60.1	-	"
63	62.9	-	"
69	55.2	-	"
86.6	43.0	-	"
90.3	44.2	-	"
25 (1 atm)	0.013 wt % $C_2Cl_2F_4$	0.009 wt % H_2O	Sanders (1979)
20 (saturated)	0.027 wt % $C_2Cl_2F_4$	0.0071 wt % H_2O	du Pont (1966e)
0	-	0.0026 wt % H_2O	Banks (1979)
30	-	0.0110	"
25 (1 atm)	0.013 wt % $C_2Cl_2F_4$	-	"

[a]M.wt. = 170.92; m.p. = -93.90°C; b.p. = 3.8°C; d^{25} = 1.456 g cc^{-1}.

CCl_2F-CF_3[a]
Ethane, 1,1-dichloro-1,2,2,2-terafluoro- (R 114a)

Temperature (°C)	Solubility in H_2O	Solubility of H_2O in	Reference
21	-	0.006 wt % H_2O	Shepherd (1961)

[a]M.wt. = 170.92; m.p. = -94.0°C; b.p. = 3.6°C; d_4^{25} = 1.455 g cc^{-1}.

$CCl_2=CH_2^a$
Ethene, 1,1-dichloro-

Temperature (°C)	Solubility in H_2O	Solubility of H_2O in	Reference
20	–	0.035 wt % H_2O	Pittsburgh Plate Glass Ind. (1970d)
25	0.021 wt % $C_2Cl_2H_2$	0.035 wt % H_2O	Riddick & Bunger (1970)
20	0.05 g/100 g H_2O	0.33 g/100 g $C_2Cl_2H_2$	Stamford Research Inst. (1965)
25	0.021 wt % $C_2Cl_2H_2$	0.035 wt % H_2O	Dow Chemical Co. (1966)
20	0.04 g/100 cc H_2O	0.04 g/100 cc $C_2Cl_2H_2$	Kirk-Othmer (1964b)
20	400 ppm by weight $C_2Cl_2H_2$	–	Pearson & McConnell (1975)
10 (1 atm total)	0.001242 mole fraction	–	Horvath (1975c)
20 (1 atm total)	0.0007823	–	"
25 (1 atm total)	0.0006232	–	"
30 (1 atm total)	0.0005006	–	"
40 (1 atm total)	0.0003375	–	"
50 (1 atm total)	0.0002460	–	"
60 (1 atm total)	0.0001912	–	"
70 (1 atm total)	0.0001505	–	"
80 (1 atm total)	0.0001134	–	"
20	400 ppm by weight $C_2Cl_2H_2$	–	Dilling (1977)

[a]M.wt. = 96.94; m.p. = -122.1°C; b.p. = 37.0°C; d_4^{20} = 1.218 g cc^{-1}.

CClH=CClH (cis-)[a]

Ethene, 1,2-dichloro- (cis-)

Temperature ($^{\circ}$C)	Solubility in H_2O	Solubility of H_2O in	Reference
25	0.35 wt % $C_2Cl_2H_2$	0.55 wt % H_2O	Miller (1969)
25	0.35 g/100 g H_2O	0.55 g/100 g $C_2Cl_2H_2$	McGovern (1943)
-40	-	Graph	"
$\vdots$	-	"	"
60	-	"	"
25	0.35 wt % $C_2Cl_2H_2$	0.55 wt % H_2O	Riddick & Bunger (1970)
20	0.35 g/100 cc H_2O	-	Patty (1962)
10	0.04 wt % $C_2Cl_2H_2$	-	Marsden (1963)
25	-	0.77 wt % H_2O	"
25	0.35 g/100 g H_2O	0.55 g/100 g $C_2Cl_2H_2$	Kirk-Othmer (1964b)
25	8 cc 1^{-1} H_2O	-	Vallaud et al. (1957)
25	-	0.4 wt % H_2O	Mitchell & Smith (1977)
25	3500 ppm by weight $C_2Cl_2H_2$	-	Dilling (1977)
10	0.04 wt % $C_2Cl_2H_4$	-	Matthews (1979)

[a]M.wt. = 96.94; m.p. = -80.5°C; b.p. = 60.3°C; d_4^{20} = 1.2837 g cc^{-1}.

CClH=CClH (trans-)[a]

Ethene, 1,2-dichloro- (trans-)

Temperature (^{o}C)	Solubility in H_2O	Solubility of H_2O in	Reference
20	0.8 cc/100 cc H_2O	–	Salkowski (1920)
25	0.63 wt % $C_2Cl_2H_2$	0.55 wt % H_2O	Miller (1969)
25	0.63 g/100 g H_2O	0.55 g/100 g $C_2Cl_2H_2$	McGovern (1943)
-40	–	Graph	"
⋮	–	"	"
60	–	"	"
25	0.63 wt % $C_2Cl_2H_2$	0.55 wt % H_2O	Riddick & Bunger (1970)
20	0.63 g/100 cc H_2O	–	Patty (1962)
10	0.03 wt % $C_2Cl_2H_2$	–	Marsden (1963)
25	–	0.63 wt % H_2O	"
25	0.63 g/100 g H_2O	0.55 g/100 g $C_2Cl_2H_2$	Kirk-Othmer (1964b)
25	8 cc l^{-1} H_2O	–	Vallaud et al. (1957)
25	6300 ppm by weight $C_2Cl_2H_2$	–	Dilling (1977)
10	0.03 wt % $C_2Cl_2H_2$	–	Matthews (1979)

[a]M.wt. = 96.94; m.p. = -50.0^{o}C; b.p. = 47.5^{o}C; d_4^{20} = 1.2565 g cc^{-1}.

$CCl_2H-CH_3^a$
Ethane, 1,1-dichloro-

Temperature (^oC)	Solubility in H_2O	Solubility of H_2O in	Reference
25	0.506 g/100 g H_2O	–	Gross (1929a)
25	0.503 wt % $C_2Cl_2H_4$	–	Miller (1969)
25	5.06 g/1000 g H_2O	–	Gross (1929b)
0	0.656 g/100 g H_2O	–	Rex (1906)
10	0.595	–	"
20	0.550	–	"
30	0.540	–	"
20	5.03 wt % $C_2Cl_2H_4$	–	Riddick & Bunger
25	–	0.2 wt % H_2O	(1970)
0	0.594 g/100 g H_2O	–	Van Arkel & Vles
20	0.506	–	(1936)
35	0.482	–	"
50	0.519	–	"
0	0.00108 mole fraction	0.00254 mole fraction	Staverman (1941b)
20	0.00092	–	"
25	–	0.00531 mole fraction	"
30	–	0.00630	"
35	0.00088 mole fraction	–	"
20	0.5 g/100 cc H_2O	–	Patty (1962)
0	0.652 wt % $C_2Cl_2H_4$	–	Washburn (1928)
10	0.591	–	"
20	0.547	–	"
30	0.537	–	"
25	1 g/180 cc H_2O	–	Wright & Schaffer (1932)
0	0.70 g/100 g H_2O	–	Lange (1967)
30	0.50	–	"
20	0.55 g/100 g H_2O	0.0009 g/100 g $C_2Cl_2H_4$	Kirk-Othmer (1964b)
20	1810 ppm by volume $C_2Cl_2H_4$	–	Imperial Chemical Industries (1972)
10	0.503 wt % $C_2Cl_2H_4$	–	Walraevens et al.
20	0.483	–	(1974)
25	0.480	–	"
30	0.481	–	"

CCl_2H-CH_3 (Continued)

40	0.496 wt % $C_2Cl_2H_4$	–	Walraevens et al.
50	0.528	–	(1974)
60	0.579	–	"
70	0.652	–	"
80	0.751	–	"
30	0.537 wt % $C_2Cl_2H_4$	0.115 wt % H_2O	Nathan (1978)
25	5100 ppm by weight $C_2Cl_2H_4$	–	Dilling (1977)

[a]M.wt. = 98.96; m.p. = -96.98°C; b.p. = 57.28°C; d_4^{20} = 1.1757 g cc^{-1}.

$CClH_2-CClH_2$[a]
Ethane, 1,2-dichloro- (R 150)

Temperature (°C)	Solubility in H_2O	Solubility of H_2O in	Reference
72.0	0.87 wt % $C_2Cl_2H_4$	–	Baranaev et al.
89.3	0.59	–	(1954)
92.3	0.43	–	"
94.0	0.33	–	"
98.0	0.13	–	"
20	0.87 wt % $C_2Cl_2H_4$	0.16 wt % H_2O	BP Chemicals (1970)
20	0.87 wt % $C_2Cl_2H_4$	0.16 wt % H_2O	BP Chemicals (1968)
18	–	0.082 wt % H_2O	Chistyakov (1965)
26	–	0.114	"
-20	–	0.05 wt % H_2O	Davies et al.
-10	–	0.07	(1948)
0	–	0.10	"
10	–	0.13	"
20	–	0.17	"
30	–	0.22	"
40	–	0.29	"
50	–	0.37	"
60	–	0.48	"
70	–	0.60	"
80	–	0.75	"
25	1.06 wt % $C_2Cl_2H_4$	0.18 wt % H_2O	Coca & Diaz (1980)

$CClH_2-CClH_2$ (Continued)

20	0.87 wt % $C_2Cl_2H_4$	0.16 wt % H_2O	Doolittle (1935)
20	–	0.09 wt % H_2O	Chistyakov & Shapurova (1964)
4	0.91 wt % $C_2Cl_2H_4$	0.12 wt % H_2O	Ethyl Corp. (1964)
16	0.87	0.16	"
20	0.84	0.20	"
38	–	0.27	"
49	–	0.34	"
60	–	0.45	"
71	–	0.60	"
20	0.87 wt % $C_2Cl_2H_4$	0.16 wt % H_2O	Gladis (1960)
25	0.865 g/100 g H_2O	–	Gross (1929a)
15	8.72 g/1000 g H_2O	–	Gross & Saylor
30	9.00	–	(1931)
10	–	0.0812 mol l^{-1} soln.	Johnson et al.
25	–	0.1262	(1966)
25	0.857 wt % $C_2Cl_2H_4$	–	Miller (1969)
25	8.65 g/1000 g H_2O	–	Gross (1929b)
20	–	0.16 g/100 g $C_2Cl_2H_4$	McGovern (1943)
25	0.84 g/100 g H_2O	–	"
0	Graph	–	"
⋮	"	–	"
70	"	–	"
20	0.87 g/100 g H_2O	–	O'Connell (1963)
25	–	0.15 g/100 g $C_2Cl_2H_4$	"
25	–	123 mmol l^{-1} $C_2Cl_2H_4$	Ödberg & Högfeldt (1969)
20	–	0.16 g/100 g $C_2Cl_2H_4$	Pittsburgh Plate
25	0.84 g/100 g H_2O	–	Glass Ind. (1970c)
20	0.81 wt % $C_2Cl_2H_4$	0.15 wt % H_2O	Riddick & Bunger (1970)
0	0.922 g/100 g H_2O	–	Rex (1906)
10	0.885	–	"
20	0.869	–	"
30	0.894	–	"
35	0.895 g/100 g H_2O	–	Seidell (1941)
56	1.030	–	"
25	–	0.129 vol % H_2O	Sellers (1971)
0	0.00159 mole fraction	0.00495 mole fraction	Staverman (1941b)

$CClH_2-CClH_2$ (Continued)

Temp			Reference
20	0.00155 mole fraction	–	Staverman (1941b)
25	–	0.01025 mole fraction	"
30	–	0.01190	"
35	0.00163 mole fraction	–	"
30 (760 mm Hg)	8.00 g/1000 g H_2O	–	Svetlanov et al. (1971)
25	–	1.5 g l^{-1} $C_2Cl_2H_4$	Zielinski (1959)
19	–	0.0857 wt % H_2O	Udovenko &
23	0.8774 wt % $C_2Cl_2H_4$	–	Fakulina
25.5	–	0.1136 wt % H_2O	(1952a, b)
30.0	0.9098 wt % $C_2Cl_2H_4$	–	"
33.5	0.9239	0.1509 wt % H_2O	"
42.0	–	0.2079	"
44.5	1.0001 wt % $C_2Cl_2H_4$	–	"
46.5	1.0242	–	"
47.5	–	0.2448 wt % H_2O	"
53.0	–	0.2994	"
56.5	1.1278 wt % $C_2Cl_2H_4$	–	"
57.0	–	0.3494 wt % H_2O	"
58.0	–	0.3689	"
67.0	1.2906 wt % $C_2Cl_2H_4$	–	"
69.0	–	0.5148 wt % H_2O	"
72.5	1.3886 wt % $C_2Cl_2H_4$	–	"
0	0.873 g/100 g H_2O	–	Van Arkel & Vles
20	0.849	–	(1936)
35	0.895	–	"
56	1.030	–	"
0	0.914 wt % $C_2Cl_2H_4$	–	Washburn (1928)
10	0.875	–	"
20	0.861	–	"
30	0.885	–	"
25	1 g/115 cc H_2O	–	Wright & Schaffer (1932)
20	4 cc/500 cc H_2O	–	Salkowski (1920)
20	0.9 g/100 cc H_2O	–	Patty (1962)
20	0.87 wt % $C_2Cl_2H_4$	–	Gunther et al. (1968)
20	0.81 wt % $C_2Cl_2H_4$	0.15 wt % H_2O	Union Carbide Corp. (1962, 1973)
25	0.865 wt % $C_2Cl_2H_4$	–	Lax (1967)

$CClH_2-CClH_2$ (Continued)

26.7	-	0.222 wt % H_2O	Ethyl Corp. (1964)
32.2	-	0.245	"
37.8	-	0.268	"
43.3	-	0.295	"
48.9	-	0.326	"
54.4	-	0.365	"
60.0	-	0.413	"
65.6	-	0.480	"
71.1	-	0.582	"
0	0.90 wt % $C_2Cl_2H_4$	0.08 wt % H_2O	Marsden (1963)
10	0.83	0.11	"
20	0.80	0.16	"
30	0.85	0.20	"
40	0.96	0.26	"
50	-	0.34	"
0	0.90 g/100 g H_2O	-	Lange (1967)
30	0.90	-	"
20	0.869 g/100 g H_2O	0.160 g/100 g $C_2Cl_2H_4$	Kirk-Othmer (1964b)
20	0.81 wt % $C_2Cl_2H_4$	0.15 wt % H_2O	Manufacturing Chemists Assoc. (1956-1971: SD-18)
20	0.78 g/100 g H_2O	-	Imperial Chemical
71	0.97	-	Industries (1966)
0	0.922 g/100 g H_2O	-	Lichosherstov et
20	0.869	-	al. (1935)
30	0.894	-	"
15	8720 ppm wt/vol H_2O	-	Imperial Chemical
30	9000	-	Industries (1972)
10	0.852 wt % $C_2Cl_2H_4$	-	Walraevens et al.
20	0.849	-	(1974)
25	0.861	-	"
30	0.880	-	"
40	0.944	-	"
50	1.046	-	"
60	1.192 wt % $C_2Cl_2H_4$	-	"
70	1.395	-	"
80	1.672	-	"
20	0.81 wt % $C_2Cl_2H_4$	0.15 wt % H_2O	Interdyne Inc. (1976)

$CClH_2$-$CClH_2$ (Continued)

20	8800 ppm by weight $C_2Cl_2H_4$	–	Pearson & McConnell (1975)
10	–	0.118 g/100 g $C_2Cl_2H_4$	Imperial Chemical
15	–	0.127	Industries (1966)
33	–	0.191	"
65.5	–	0.384	"
5	–	0.0696 mol l^{-1} $C_2Cl_2H_4$	Masterton &
25	–	0.1264	Gendrano (1966)
20	0.81 wt % $C_2Cl_2H_4$	–	Earhart et al. (1977)
25	0.51 wt % $C_2Cl_2H_4$	0.9 wt % H_2O	Mitchell & Smith (1977)
25	–	0.187 wt % H_2O	Nathan (1978)
20	0.87 g/100 cc H_2O	–	Metcalf (1962)
25	–	0.0101 mole fraction	Kirchnerova (1975)
20	0.90 wt % $C_2Cl_2H_4$	0.15 wt % H_2O	du Pont (1966e)
25	8700 ppm by weight $C_2Cl_2H_4$	–	Dilling (1977)
10	–	0.0812 mol l^{-1} soln.	Christian et al.
25	–	0.1262	(1968)
20	0.80 wt % $C_2Cl_2H_4$	–	Matthews (1975, 1979)
25	–	0.126 mol l^{-1} soln.	Czapkiewicz & Czapkiewicz-Tutaj (1980)
25	0.0807 mol l^{-1} soln.	–	Banerjee et al. (1980)

[a] M.wt. = 98.96; m.p. = -35.36°C; b.p. = 83.47°C; d_4^{20} = 1.2351 g cc^{-1}.

CCl$_2$F-CClF$_2$[a]
Ethane, 1,1,2-trichloro-1,2,2-trifluoro- (R 113)

Temperature (^oC)	Solubility in H$_2$O	Solubility of H$_2$O in	Reference
21.1	0.017 wt % C$_2$Cl$_3$F$_3$	0.009 wt % H$_2$O	Bacquies (1971)
-50	-	Monograph	Chernyshev (1969)
⋮	-	"	"
50	-	"	"
0	-	0.0036 wt % H$_2$O	du Pont (1969a)
25	0.017 wt % C$_2$Cl$_3$F$_3$	0.011	"
27.5	0.0238 g/psia/kg H$_2$O	-	du Pont (1961b)
50	0.0131	-	"
75	0.0087	-	"
7	Graph	-	"
⋮	"	-	"
76	"	-	"
-17.7	-	0.002 wt % H$_2$O	du Pont (1966a)
21.1	-	0.009	"
-73	-	Graph	"
⋮	-	"	"
49	-	"	"
-34.4	-	6 ppm by weight H$_2$O	du Pont (1966d)
-23.3	-	11	"
-12.2	-	20	"
-1.1	-	34	"
10.0	-	55	"
21.1	-	90	♥
32.2	-	140	"
37.8	-	168	"
-73.3	-	Graph	"
⋮	-	"	"
48.9	-	"	"
21.1	-	0.0090 wt % H$_2$O	du Pont (1966c)
-34.4	-	0.0006	"
-23.3	-	0.0011	"
-12.2	-	0.0020	"
-1.1	-	0.0034	"
10.0	-	0.0055	"
21.1	-	0.0090	"

$CCl_2F-CClF_2$ (Continued)

32.2	–	0.0140	du Pont (1966c)
37.8	–	0.0168	"
30 (5 psia)	0.11 g l^{-1} $C_2Cl_3F_3$	–	"
50 (5–15 psia)	Table	–	"
75 (5–30 psia)	"	–	"
−17.7	–	0.002 wt % H_2O	Elchardus &
21.1	–	0.009	Maestre (1965)
−20	–	Graph	"
⋮	–	"	"
10	–	"	"
25	0.01 g/100 g H_2O	0.009 g/100 g $C_2Cl_3F_3$	Imperial Chemical Industries (1965)
0	–	0.0036 wt % H_2O	Pechiney–Saint–
25	0.017 wt % $C_2Cl_3F_3$	–	Gobain (1971:
30	–	0.013 wt % H_2O	113)
25	–	0.010 wt % H_2O	Pechiney–Saint– Gobain (1971: 113CM)
25	–	0.010 wt % H_2O	Pechiney–Saint– Gobain (1971: 113M)
25	0.017 wt % $C_2Cl_3F_3$	0.011 wt % H_2O	Riddick & Bunger (1970)
21	0.03 wt % $C_2Cl_3F_3$	0.009 wt % H_2O	Union Carbide Corp. (1973)
25	0.017 wt % $C_2Cl_3F_3$	0.011 wt % H_2O	Weast (1976)
0	–	0.013 wt % H_2O	Matheson Company
25	0.017 wt % $C_2Cl_3F_3$	–	(1971)
30	–	0.0036 wt % H_2O	"
25	0.017 wt % $C_2Cl_3F_3$	–	Kirk–Othmer (1966a)
21 (1 atm)	0.035 wt % $C_2Cl_3F_3$	0.009 wt % H_2O	Allied Chemical Corp. (1974)
0	–	0.0036 g/100 g $C_2Cl_3F_3$	Kali–Chemie (1972)
30	–	0.0130	"
20	235 ppm by volume $C_2Cl_3F_3$	–	Imperial Chemical
25	170	–	Industries (1972)
25 (1 atm)	0.017 wt % $C_2Cl_3F_3$	–	Rauws et al. (1973)
0	–	0.0036 g/100 g $C_2Cl_3F_3$	Imperial Smelting Corp. (1957)

CCl$_2$F-CClF$_2$ (Continued)

-30	–	Graph	Petrak (1975)
⋮	–	"	"
200	–	"	"
-20	–	13 ppm by weight H$_2$O	Ackroyd (1978)
-10	–	22	"
0	–	36	"
15	–	68	"
30	–	120	"
30	407.3 ppm by weight atm^{-1}	–	Hellström et al.
34.2	395.1	–	(1976)
57.5	211.9	–	"
73.0	154.7	–	"
90.0	109.5	–	"
93.2	98.1	–	"
25	0.017 wt % C$_2$Cl$_3$F$_3$	0.011 wt % H$_2$O	Sanders (1979)
20 (saturated)	0.013 wt % C$_2$Cl$_3$F$_3$	0.0086 wt % H$_2$O	du Pont (1966e)
25	0.01 wt % C$_2$Cl$_3$F$_3$	–	Matthews (1979)

[a]M.wt. = 187.38; m.p. = -36.4°C; b.p. = 47.7°C; d$_4^{20}$ = 1.5635 g cc^{-1}.

$CCl_2=CClH^a$

Ethene, trichloro-

Temperature (^oC)	Solubility in H_2O	Solubility of H_2O in	Reference
37	1.7 cc/cc H_2O (Ostwald)	-	Allott et al. (1973)
15	0.101 wt % C_2Cl_3H	-	Antropov et al.
18	' -	0.021 wt % H_2O	(1972)
20	0.109 wt % C_2Cl_3H	-	"
22	-	0.024 wt % H_2O	"
26	0.122 wt % C_2Cl_3H	0.032	"
-50	-	Graph	Carlisle & Levine
⋮	-	"	(1932)
28	-	"	"
-10		Monograph	Chernyshev (1970)
⋮	-	"	"
80	-	"	"
15	-	0.011 wt % H_2O	Eberius (1954)
25	-	0.0013 mol/100 g soln.	Eidinoff (1955)
25	0.1 wt % C_2Cl_3H	-	Ethyl Corp. (1958a)
25	0.11 wt % C_2Cl_3H	0.03 wt % H_2O	Gladis (1960)
18	0.18 cc/100 cc H_2O	-	Salkowski (1920)
20	3.0 cc/cc H_2O (Ostwald)	-	Larson et al.
37	1.55	-	(1962b)
25	0.111 wt % C_2Cl_3H	0.033 wt % H_2O	Miller (1969)
25	0.11 g/100 g H_2O	0.032 g/100 g C_2Cl_3H	McGovern (1943)
-40	-	Graph	"
0	Graph	"	"
⋮	"	"	"
80	"	"	"
25	0.11 g/100 g H_2O	0.02 g/100 g C_2Cl_3H	O'Connell (1963)
37	1.55 cc/cc H_2O (Ostwald)	-	Papper & Kitz (1963)
25	0.11 g/100 g H_2O	0.032 g/100 g C_2Cl_3H	Pittsburgh Plate Glass (1971b)
16	0.081 wt % C_2Cl_3H	-	Reilly et al.
18	-	0.0250 wt % H_2O	"
25	0.11 wt % C_2Cl_3H	0.32 wt % H_2O	Riddick & Bunger (1970)
25	0.11 g/100 g H_2O	0.027 g/100 g C_2Cl_3H	Sconce (1962)
25	0.01 wt % C_2Cl_3H	0.03 wt % H_2O	Coca & Diaz (1980)

$CCl_2=CClH$ (Continued)

20	0.07 g/100 g H_2O	–	Secher (1971)
37	1.7 cc/cc H_2O (Ostwald)	–	Steward et al. (1973)
25	1 g/550 cc H_2O	–	Wright & Schaffer (1932)
20	1.8 cc 1^{-1} H_2O	–	Salkowski (1920)
20	0.1 g/100 cc H_2O	–	Patty (1962)
25	–	0.02 wt % H_2O	Lees & Sarram (1971)
25	1.9 cc (STP)/g H_2O	–	Lax (1967)
25	0.1 wt % C_2Cl_3H	0.02 wt % H_2O	Marsden (1963)
25	0.1 g/100 g H_2O	–	Lange (1967)
25	0.1 g/100 g H_2O	0.04 g/100 g C_2Cl_3H	Archer & Stevens (1977)
0	–	0.01 g/100 g C_2Cl_3H	Kirk-Othmer
25	0.11 g/100 g H_2O	0.033	(1964b)
60	0.125	0.080	"
20	0.1 g/100 cc H_2O	–	Manufacturing Chemists Assoc. (1956-1971: SD-14)
10	–	0.017 g/100 g C_2Cl_3H	Marius (1937)
18	–	0.026	"
28	–	0.036	"
25	1100 ppm w/v H_2O	–	Imperial Chemical Industries (1972)
20	230 cc/100 cc H_2O	–	Macintosh et al.
25	0.11 cc/100 cc H_2O	–	(1963)
25	0.1 g/100 g H_2O	–	Key (1973)
25	0.11 wt % C_2Cl_3H	0.32 wt % H_2O	Interdyne Inc. (1976)
20	1100 ppm by weight C_2Cl_3H	–	Pearson & McConnell (1975)
25	0.1 g/100 g H_2O	–	Aviado et al. (1976)
25	0.11 wt % C_2Cl_3H	–	Nathan (1978)
25	–	0.000884 wt % H_2O	Prosyanov et al.
⋮	–	·	(1974)
82	–	0.0136 wt % H_2O	"
36.85	1.64 cc/cc H_2O (Ostwald)	–	Halliday et al. (1977)
20	1100 ppm by weight C_2Cl_3H	–	Selenka & Baner (1978)

$CCl_2=CClH$ (Continued)

20	0.11 wt % C_2Cl_3H	0.02 wt % H_2O	du Pont (1966e)
10	1.28 g/liter H_2O	–	Vallaud et al.
20	1.285	–	(1957)
25	1.288	–	"
30	1.29	–	"
40	1.305	–	"
50	1.33	–	"
37	1.6 cc/cc H_2O (Ostwald)	–	Grant (1978)
25	1100 ppm by weight C_2Cl_3H	–	Dilling (1977)
25	0.10 wt % C_2Cl_3H	–	Matthews (1975, 1979)
25	0.0112 mol l^{-1} soln.	–	Banerjee et al. (1980)

[a]M.wt. = 131.39; m.p. = -73.0°C; b.p. = 87.0°C; d_4^{20} = 1.4642 g cc^{-1}.

CCl_3-CH_3[a]

Temperature (^oC)	Solubility in H_2O	Solubility of H_2O in	Reference
25	–	0.1 g/100 g $C_2Cl_3H_3$	Dow Chemical Europe (1967)
25	0.07 g/100 g H_2O	0.05 g/100 g $C_2Cl_3H_3$	Dow Chemical Co. (1972)
25	–	0.1 g/100 g $C_2Cl_3H_3$	Dow Chemical Co. (1973)
25	0.1 wt % $C_2Cl_3H_3$	–	Ethyl Corp. (1958a)
20	0.132 wt % $C_2Cl_3H_3$	–	Miller (1969)
25	0.130 g/100 g H_2O	0.03 g/100 g $C_2Cl_3H_3$	O'Connell (1963)
25	0.07 g/100 g H_2O	0.05 g/100 g $C_2Cl_3H_3$	Pittsburgh Plate Glass Ind. (1971a)
20	0.132 wt % $C_2Cl_3H_3$	–	Riddick & Bunger
25	–	0.034 wt % H_2O	(1970)
0	0.000214 mole fraction	0.00119 mole fraction	Staverman (1941b)
20	0.000178	–	"
25	–	0.00257 mole fraction	"
30	–	0.00313	"
35	0.000169 mole fraction	–	"
0	0.159 g/100 g H_2O	–	Van Arkel & Vles (1936)
20	0.132	–	"
35	0.126	–	"
50	0.128	–	"
25	–	0.05 wt % H_2O	Lees & Sarram (1971)
20	0.44 g/100 g H_2O	0.05 g/100 g $C_2Cl_3H_3$	Kirk-Othmer (1964b)
20	1320 ppm w/v H_2O	–	Imperial Chemical Industries (1972)
25	1600	–	
10	0.173 wt % $C_2Cl_3H_3$	–	Walraevens et al. (1974)
20	0.155	–	
25	0.149	–	"
30	0.144	–	"
40	0.139	–	"
50	0.138	–	"
60	0.141	–	"
70	0.148	–	"
80	0.159	–	"
25	0.01 wt % $C_2Cl_3H_3$	0.02 wt % H_2O	Coca & Diaz (1980)

CCl_3-CH_3 (Continued)

25	0.132 wt % $C_2Cl_3H_3$	0.034 wt % H_2O	Interdyne Inc. (1976)
20	480 ppm by weight $C_2Cl_3H_3$	–	Pearson & McConnell (1975)
20	–	0.0125 wt % H_2O	Imperaial Chemical Industries (1965)
25	260 ppm by weight $C_2Cl_3H_3$	–	Aviado et al. (1976)
25	0.07 g/100 g H_2O	0.05 g/100 g $C_2Cl_3H_3$	Archer & Stevens (1977)
20	–	0.0339 wt % H_2O	Nathan (1978)
25	0.44 g/100 g H_2O	–	National Institute for Occupational Safety and Health (1976)
20	480 ppm by weight $C_2Cl_3H_3$	–	Selenka & Baner (1978)
23.5	0.118 wt % $C_2Cl_3H_3$	–	Schwarz (1980)
25	0.10 wt % $C_2Cl_3H_3$	–	du Pont (1966e)
25	720 ppm by weight $C_2Cl_3H_3$	–	Dilling (1977)
1.5	880	–	"
20	730	–	"
25	720 g m^{-3} H_2O	–	Afghan & Mackay (1980)
10	0.180 wt % $C_2Cl_3H_3$	–	Schwarz & Miller
20	0.185	–	(1980)
30	0.159	–	"
25	0.0100 mol l^{-1} soln.	–	Banerjee et al. (1980)

[a]M.wt. = 133.41; m.p. = -30.41°C; b.p. = 74.1°C; d_4^{20} = 1.3390 g cc^{-1}.

$CCl_2H-CClH_2$[a]

Ethane, 1,1,2-trichloro-

Temperature (°C)	Solubility in H_2O	Solubility of H_2O in	Reference
20	0.45 wt % $C_2Cl_3H_3$	0.05 wt % H_2O	Gladis (1960)
20	0.434 wt % $C_2Cl_3H_3$	-	Miller (1969)
25	0.44 g/100 g H_2O	-	McGovern (1943)
-9	-	0.025 g/100 g $C_2Cl_3H_3$	"
0	Graph	-	"
⋮	"	-	"
70	"	-	"
0	0.00063 mole fraction	0.00460 mole fraction	Staverman (1941b)
20	0.00059	-	"
25	-	0.00879 mole fraction	"
30	-	0.01180	"
35	0.00062 mole fraction	-	"
0	0.466 g/100 g H_2O	-	Van Arkel & Vles
20	0.436	-	(1936)
25	0.458	-	"
55	0.532	-	"
25	1 g/270 cc H_2O	-	Wright & Schaffer (1932)
20	0.44 g/100 g H_2O	-	Patty (1962)
20	0.45 wt % $C_2Cl_3H_3$	0.05 wt % H_2O	Marsden (1963)
20	0.44 g/100 g H_2O	-	Lange (1967)
20	0.45 g/100 g H_2O	0.05 g/100 g $C_2Cl_3H_3$	Kirk-Othmer (1964b)
20	4360 ppm w/v H_2O	-	Imperial Chemical Industries (1972)
20	0.45 wt % $C_2Cl_3H_3$	0.05 wt % H_2O	McClure (1944)
10	0.460 wt % $C_2Cl_3H_3$	-	Walraevens et al.
20	0.441	-	(1974)
25	0.438	-	"
30	0.439	-	"
40	0.453	-	"
50	0.483	-	"
60	0.529	-	"
70	0.595	-	"
80	0.686	-	"
20	0.45 wt % $C_2Cl_3H_3$	0.05 wt % H_2O	Union Carbide Chemicals Co. (1962)
25	0.51 wt % $C_2Cl_3H_3$	0.08 wt % H_2O	Coca & Diaz (1980)

$CCl_2H-CClH_2$ (Continued)

20	0.44 wt % $C_2Cl_3H_3$	0.05 wt % H_2O	du Pont (1966e)
25	4420 ppm by weight $C_2Cl_3H_3$	–	Dilling (1977)

[a]M.wt. = 133.41; m.p. = -36.5°C; b.p. = 113.77°C; d_4^{20} = 1.4397 g cc^{-1}.

$CCl_2=CCl_2$[a]
Ethene, tetrachloro-

Temperature (°C)	Solubility in H_2O	Solubility of H_2O in	Reference
15	0.0075 wt % C_2Cl_4	–	Antropov et al.
18	–	0.00562 wt % H_2O	(1972)
20	0.00874 wt % C_2Cl_4	–	"
22	–	0.00691 wt % H_2O	"
26	0.0151 wt % C_2Cl_4	0.0085	"
25	0.1 wt % C_2Cl_4	–	Ethyl Corp. (1958a)
25	0.015 wt % C_2Cl_4	0.011 wt % H_2O	Gladis (1960)
25	0.015 wt % C_2Cl_4	0.008 wt % H_2O	Miller (1969)
25	0.015 g/100 g H_2O	0.0105 g/100 g C_2Cl_4	McGovern (1943)
-25	–	Graph	"
⋮	–	"	"
0	Graph	"	"
⋮	"	"	"
80	"	"	"
25	0.015 g/100 g H_2O	–	O'Connell (1963)
25	0.015 g/100 g H_2O	0.0105 g/100 g C_2Cl_4	Pittsburgh Plate Glase Ind. (1971c)
25	0.015 wt % C_2Cl_4	0.0105 wt % H_2O	Riddick & Bunger (1970)
25	0.015 g/100 g H_2O	0.0105 g/100 g C_2Cl_4	Sconce (1962)
0	–	0.00282 wt % H_2O	Simonov et al.
30	–	0.0104	(1970)
50	–	0.0282	"
60	–	0.0465	"
80	–	0.1185	"
100	–	0.3019	"
112	–	0.5291	"
25	0.02 wt % C_2Cl_4	0.01 wt % H_2O	Coca & Diaz (1980)

$CCl_2=CCl_2$ (Continued)

25	1 g/5300 cc H_2O	–	Wright & Schaffer (1932)
25	0.02 wt % C_2Cl_4	–	Gunther et al. (1968)
25	0.015 g/100 g H_2O	0.0105 g/100 g C_2Cl_4	Hancock (1973)
20	0.02 wt % C_2Cl_4	–	Marsden (1963)
25	–	0.04 wt % H_2O	"
25	0.015 g/100 g H_2O	0.008 g/100 g C_2Cl_4	Kirk-Othmer (1964b)
25	150 ppm w/v H_2O	–	Imperial Chemical Industries (1972)
25	0.015 g/100 g H_2O	0.0105 g/100 g C_2Cl_4	Hooker Chemical Corp. (1966)
40	0.0166 wt % C_2Cl_4	0.0166 wt % H_2O	Simonov et al. (1974)
25	0.015 wt % C_2Cl_4	0.0105 wt % H_2O	Interdyne Inc. (1976)
25	1 g/10,000 g H_2O	0.01 wt % H_2O	Valluad et al. (1957)
20	150 ppm by weight C_2Cl_4	–	Pearson & McConnell (1975)
20	206.5 mg/liter H_2O	–	Moiseeva et al. (1977)
25	0.015 wt % C_2Cl_4	0.02 wt % H_2O	Mitchell & Smith (1977)
25	0.015 g/100 g H_2O	0.01 g/100 g C_2Cl_4	Archer & Stevens (1977)
25	0.015 wt % C_2Cl_4	0.0105 wt % H_2O	Nathan (1978)
25	–	0.000219 wt % H_2O	Prosyanov et al.
$\vdots$	–	.	(1974)
85	–	0.00246 wt % H_2O	"
20	150 ppm by weight C_2Cl_4	–	Selenka & Baner (1978)
25	150 mg l^{-1} H_2O	–	Schwarzenbach et al. (1979)
20	0.02 wt % C_2Cl_4	0.01 wt % H_2O	du Pont (1966e)
25	140 ppm by weight C_2Cl_4	–	Dilling (1977)
20	120	–	"
1.5	130	–	"
25	400 ppm by weight C_2Cl_4	–	Chiou et al. (1977)
25	0.04 wt % C_2Cl_4	–	Matthews (1975, 1979)
25	0.00292 mol l^{-1} soln.	–	Banerjee et al. (1980)

[a]M.wt. = 165.83; m.p. = -19.0°C; b.p. = 121.0°C; d_4^{20} = 1.6227 g cc^{-1}.

$CCl_2F-CCl_2F^a$
Ethane, 1,2-difluoro-1,1,2,2-tetrachloro- (R 112)

Temperature (°C)	Solubility in H_2O	Solubility of H_2O in	Reference
25 (saturated)	0.012 wt % $C_2Cl_4F_2$	–	du Pont (1969a)
28	–	99 ppm by weight H_2O	du Pont (1957)
27	0.1073 g/psia/kg H_2O	–	du Pont (1961b)
50	0.0477	–	"
75	0.0253	–	"
7	Graph & equation	–	"
⋮	"	–	"
76	"	–	"
28	–	0.0099 wt % H_2O	du Pont (1966c)
27 (14.7 psia)	0.158 wt % $C_2Cl_4F_2$	–	"
30 (1 psia)	0.09 g/liter H_2O	–	"
50 (1-9 psia)	Table	–	"
75 (1-9 psia)	"	–	"
25	0.012 wt % $C_2Cl_4F_2$	–	Riddick & Bunger
27.8	–	0.0099 wt % H_2O	(1970)
20	0.013 wt % $C_2Cl_4F_2$	–	Union Carbide Corp.
28	–	0.0099 wt % H_2O	(1973)
25 (saturated)	0.12 wt % $C_2Cl_4F_2$	–	Weast (1976)
25 (saturated)	0.012 wt % $C_2Cl_4F_2$	–	Kirk-Othmer (1966a)

[a]M.wt. = 203.83; m.p. = 25.0°C; b.p. = 93.0°C; d_4^{25} = 1.6447 g cc^{-1}.

$CCl_3-CClH_2^a$
Ethane, 1,1,1,2-tetrachloro-

Temperature (°C)	Solubility in H_2O	Solubility of H_2O in	Reference
0	0.000129 mole fraction	0.00214 mole fraction	Staverman (1941b)
20	0.000117	-	"
25	-	0.00513 mole fraction	"
30	-	0.00566	"
35	0.000124 mole fraction	-	"
0	0.120 g/100 g H_2O	-	Van Arkel & Vles
20	0.109	-	(1936)
35	0.115	-	"
50	0.125	-	"
20	0.02 g/100 g H_2O	-	Lange (1967)
20	1100 ppm vy volume $C_2Cl_4H_2$	-	Imperial Chemical Industries (1972)
20	-	0.0308 wt % H_2O	Imperial Chemical Industries (1977)
10	0.113 wt % $C_2Cl_4H_2$	-	Walraevens et al.
20	0.110	-	(1974)
25	0.110	-	"
30	0.111	-	"
40	0.116	-	"
50	0.125	-	"
60	0.139	-	"
70	0.159	-	"
80	0.185	-	"
20	0.02 wt % C_2Cl_4	-	du Pont (1966e)
25	1100 ppm by weight $C_2Cl_4H_2$	-	Dilling (1977)

[a] M.wt. = 167.85; m.p. = -70.2°C; b.p. = 130.5°C; d_4^{20} = 1.5406 g cc^{-1}.

$CCl_2H-CCl_2H^a$
Ethane, 1,1,2,2-tetrachloro-

Temperature (°C)	Solubility in H_2O	Solubility of H_2O in	Reference
25	0.3 wt % $C_2Cl_4H_2$	–	Anonymous (1969)
18	–	0.082 wt % H_2O	Antropov et al.
20	0.811 wt % $C_2Cl_4H_2$	–	(1972)
26	–	0.114 wt % H_2O	"
25	–	0.00641 mol/100 g soln.	Eidinoff (1955)
25	0.29 wt % $C_2Cl_4H_2$	1.13 wt % H_2O	Gladis (1960)
25	–	0.101 mol l^{-1} soln.	Johnson et al. (1966)
25	0.29 g/100 g H_2O	1.13 g/100 g $C_2Cl_4H_2$	McGovern (1943)
0	Graph	–	"
⋮	"	–	"
70	"	–	"
25	0.28 wt % $C_2Cl_4H_2$	0.13 wt % H_2O	Othmer et al. (1941)
20	0.287 wt % $C_2Cl_4H_2$	–	Riddick & Bunger (1970)
25	0.288 g/100 g H_2O	–	Van Arkel & Vles
55.6	0.336	–	(1936)
0	–	0.00573 mole fraction	Staverman (1941b)
20	0.00031 mole fraction	–	"
25	–	0.01020 mole fraction	"
30	–	0.01230	"
20	0.288 g/100 g H_2O	–	Stephen & Stephen
55	0.536	–	(1963)
25	1 g/350 cc H_2O	–	Wright & Schaffer (1932)
25	–	2.08 g l^{-1} $C_2Cl_4H_2$	Zielinski (1959)
25	–	0.04 wt % H_2O	Lees & Sarram (1971)
20	–	0.03 wt % H_2O	Marsden (1963)
25	0.32 wt % $C_2Cl_4H_2$	–	"
20	0.29 g/100 g H_2O	–	Lange (1967)
25	0.29 g/100 g H_2O	1.13 g/100 g $C_2Cl_4H_2$	Kirk-Othmer (1964b)
20	2880 ppm w/v H_2O	–	Imperial Chemical Industries (1972)
20	–	0.0844 wt % H_2O	Imperial Chemical Industries (1977)
25	–	0.11 wt % H_2O	Mitchell & Smith (1977)

CCl_2H-CCl_2H (Continued)

Temp			Reference
10	0.320 wt % $C_2Cl_4H_2$	-	Walraevens et al.
20	0.301	-	(1974)
25	0.296	-	"
30	0.294	-	"
40	0.298	-	"
50	0.312	-	"
60	0.336	-	"
70	0.371	-	"
80	0.420	-	"
97.9 (1 atm)	1.0 mol % $C_2Cl_4H_2$	-	Hollo & Lengyel
94.5 (1 atm)	5.1	-	(1960)
94.0 (1 atm)	17.79	-	"
93.4 (1 atm)	37.0	-	"
94.1 (1 atm)	53.1	-	"
94.1 (1 atm)	85.87	-	"
94.0 (1 atm)	91.8	-	"
95.3 (1 atm)	96.85	-	"
116.0 (1 atm)	96.9	-	"
131.5 (1 atm)	99.01	-	"
146.2 (1 atm)	100.0	-	"
25	-	0.00333 wt % H_2O	Prosyanov et al.
⋮	-	.	(1974)
90	-	0.05 wt % H_2O	"
23.5	0.296 wt % $C_2Cl_4H_2$	-	Schwarz (1980)
25	-	0.0107 mole fraction	Kirchnerova (1975)
25	0.29 wt % $C_2Cl_4H_2$	0.13 wt % H_2O	du Pont (1966e)
25	3000 ppm by weight $C_2Cl_4H_2$	-	Dilling (1977)
25	-	0.101 mol l^{-1} soln.	Christian et al. (1968)
25	480.0 g/m^3 H_2O	-	Afghan & Mackay (1980)
10	0.372 wt % $C_2Cl_4H_2$	-	Schwarz & Miller
20	0.385	-	(1980)
30	0.367	-	"
25	0.0177 mol l^{-1} soln.	-	Banerjee et al. (1980)

[a] M.wt. = 167.85; m.p. = -36.0°C; b.p. = 146.2°C; d_4^{20} = 1.5953 g cc^{-1}.

$CCl_3-CCl_2H^a$
Ethane, pentachloro-

Temperature (^oC)	Solubility in H_2O	Solubility of H_2O in	Reference
25	–	0.00186 mol/100 g soln.	Eidinoff (1955)
25	0.05 wt % C_2Cl_5H	0.03 wt % H_2O	Gladis (1960)
25	0.05 g/100 g H_2O	0.03 g/100 g C_2Cl_5H	McGovern (1943)
0	Graph	–	"
⋮	"	–	"
70	"	–	"
25	0.047 g/100 g H_2O	0.24 g/100 g C_2Cl_5H	O'Connell (1963)
25	0.05 wt % C_2Cl_5H	0.03 wt % H_2O	Riddick & Bunger (1970)
20	0.047 g/100 g H_2O	–	Van Arkel & Vles (1936)
0	–	0.00182 mole fraction	Staverman (1941b)
20	0.000041 mole fraction	–	"
25	–	0.00391 mole fraction	"
30	–	0.00465	"
20	0.047 g/100 g H_2O	–	Stephen & Stephen (1963)
25	1 g/2900 cc H_2O	–	Wright & Schaffer (1932)
20	0.05 wt % C_2Cl_5H	0.25 wt % H_2O	Marsden (1963)
20	0.05 g/100 g H_2O	–	Lange (1967)
20	0.05 g/100 g H_2O	0.03 g/100 g C_2Cl_5H	Kirk-Othmer (1964b)
20	0.047 g/100 g H_2O	–	Imperial Chemical Industries (1956)
20	470 ppm w/v H_2O	–	Imperial Chemical Industries (1972)
10	0.052 wt % C_2Cl_5H	–	Walraevans et al.
20	0.050	–	(1974)
25	0.050	–	"
30	0.050	–	"
40	0.051	–	"
50	0.055	–	"
60	0.060	–	"
70	0.067	–	"
80	0.078	–	"
25	480 ppm by weight	–	Dilling (1977)

CCl_3-CCl_2H (Continued)

20	-	0.0205 wt % H_2O	Imperial Chemical Industries (1977)

[a]M.wt. = 202.30; m.p. = -29.0°C; b.p. = 162.0°C; d_4^{20} = 1.6796 g cc^{-1}.

CCl_3-CCl_3[a]

Temperature (°C)	Solubility in H_2O	Solubility of H_2O in	Reference
22.3	0.005 g/100 g H_2O	-	McGovern (1943)
22.3	0.005 g/100 g H_2O	-	Van Arkel & Vles (1936)
22.3	0.005 g/100 g H_2O	-	Kirk-Othmer (1964b)
22.3	0.005 wt % C_2Cl_6	-	Stephen & Stephen (1963)
22	0.005 g/100 g H_2O	-	Patty (1962)
22	0.005 g/100 g H_2O	-	Lange (1967)
20	2.1 ppm by volume C_2Cl_6	-	Imperial Chemical Industries (1972)
25	8 ppm by weight C_2Cl_6	-	Dilling (1977)

[a]M.wt. = 236.74; m.p. = 186.8°C; b.p. = 186.0 at 777 mm Hg; d_4^{20} = 2.091 g cc^{-1}.

CFH=CH_2[a]
Ethene, fluoro-

Temperature (°C)	Solubility in H_2O	Solubility of H_2O in	Reference
80 (500 psia)	0.94 g/100 g H_2O	-	Kirk-Othmer
80 (1000 psia)	1.54	-	(1966a)

[a]M.wt. = 46.05; m.p. = -160.0°C; b.p. = -72.2°C; d_4^{25} = 0.615 g cc^{-1}.

$CFH_2-CH_3^a$
Ethane, fluoro-

Temperature ($^{\circ}$C)	Solubility in H_2O	Solubility of H_2O in	Reference
14	198 cc/100 cc H_2O	-	Moissan (1890)
14	198 cc/100 g H_2O	-	Lange (1967)

[a]M.wt. = 48.06; m.p. = -143.2°C; b.p. = -37.7°C; d_4^{20} = 0.7182 g cc^{-1}.

$CF_2=CH_2^a$
Ethene, 1,1-difluoro-

Temperature ($^{\circ}$C)	Solubility in H_2O	Solubility of H_2O in	Reference
25 (1 atm)	6.3 cc/100 g H_2O	-	Matheson Company (1966, 1971)

[a]M.wt. = 64.04; m.p. = -144.0°C; b.p. = -82.0°C; d^{-82} = 0.617 g cc^{-1}.

$CF_2H-CH_3^a$
Ethane, 1,1-difluoro-

Temperature ($^{\circ}$C)	Solubility in	Solubility of H_2O in	Reference
21 (1 atm part.)	0.32 wt % $C_2F_2H_4$	0.17 wt % H_2O	Allied Chemical Co. (1960a)
21 (1 atm part.)	0.32 wt % $C_2F_2H_4$	0.17 wt % H_2O	du Pont (1961a)
21	-	0.17 wt % H_2O	du Pont (1966c)
0	0.54 wt % $C_2F_2H_4$	-	Mannheimer (1956)
15	-	0.159 wt % H_2O	"
0	0.54 wt % $C_2F_2H_4$	-	Mellan (1950)
27.5	0.25	-	"
0	0.54 wt % $C_2F_2H_4$	-	Matheson Company
27.5	0.25	-	(1966, 1971)
21 (1 atm)	0.32 wt % $C_2F_2H_4$	0.17 wt % H_2O	Allied Chemical Co. (1974a)
21	-	0.17 wt % H_2O	Shepherd (1961)
25	0.28 wt % $C_2F_2H_4$	-	Sanders (1979)

[a]M.wt. = 66.05; m.p. = -117.0°C; b.p. = -24.7°C; $d_{sat.}^{20}$ = 0.950 g cc^{-1}.

$CF_2=CF_2^a$

Ethene, tetrafluoro-

Temperature ($^\circ$C)	Solubility in H_2O	Solubility of H_2O in	Reference
0 (1 atm)	0.004062 mol l^{-1} H_2O	–	Volokhonovich et
5 (1 atm)	0.002991	–	al. (1966)
10 (1 atm)	0.002433	–	"
15 (1 atm)	0.002053	–	"
20 (1 atm)	0.001785	–	"
25 (1 atm)	0.001584	–	"
30 (1 atm)	0.001428	–	"
35 (1 atm)	0.001294	–	"
40 (1 atm)	0.001183	–	"
45 (1 atm)	0.001084	–	"
50 (1 atm)	0.001013	–	"
55 (1 atm)	0.000959	–	"
60 (1 atm)	0.000924	–	"
65 (1 atm)	0.000897	–	"
70 (1 atm)	0.000879	–	"
20 (115 psia)	0.1 g/100 g H_2O	–	Kirk-Othmer (1966a)
20 (100 psig)	0.083 wt % C_2F_4	–	Imperial Chemical
20 (300 psig)	0.21	–	Industries (1977)
23 (50 psig)	0.001 g/100 cc H_2O	–	"
23 (78 psig)	0.06	–	"
23 (100 psig)	0.12	–	"
23 (170 psig)	0.95	–	"
23 (183 psig)	1.29	–	"
0 (10 kg cm^{-2})	0.0000033 mol cc^{-1} H_2O	–	Watanabe & Okamoto
20 (10 kg cm^{-2})	0.0000017	–	(1978)
40 (10 kg cm^{-2})	0.0000006	–	"
60 (10 kg cm^{-2})	0.0000005	–	"
40 (10 kg cm^{-2})	0.0000006	–	"
40 (20 kg cm^{-2})	0.0000011	–	"
40 (30 kg cm^{-2})	0.0000016	–	"

[a]M.wt. = 100.02; m.p. = -142.5°C; b.p. = -76.3°C; $d^{-76.3}$ = 1.519 g cc^{-1}.

CF_3-CFH_2[a]
Ethane, 1,1,1,2-tetrafluoro-

Temperature (^oC)	Solubility in H_2O	Solubility of H_2O in	Reference
37	0.055 g/100 g H_2O	-	Tarrant (1969)

[a]M.wt. = 102.03; m.p. = -101.0^oC; b.p. = -26.3^oC; d_4^{20} = 1.1 g cc^{-1}.

CF_3-CF_2H[a]
Ethane, pentafluoro- (R 125)

Temperature (^oC)	Solubility in H_2O	Solubility of H_2O in	Reference
25 (1 atm)	0.11 wt % C_2F_5H	-	Sanders (1979)

[a]M.wt. = 120.0; m.p. = -103.0^oC; b.p. = -48.5^oC; d^{21}= 1.23 g cc^{-1}.

CF_3-CF_3[a]
Ethane, hexafluoro- (R 116)

Temperature (^oC)	Solubility in H_2O	Solubility of H_2O in	Reference
15	0.00188 cc/cc H_2O (Ostwald)	-	Muccitelli & Wen (1978)
5 (1 atm)	0.00000215 mole fraction	-	Wen & Muccitelli
10 (1 atm)	0.00000173	-	(1979)
15 (1 atm)	0.00000145	-	"
20 (1 atm)	0.00000115	-	"
25 (1 atm)	0.00000103	-	"
30 (1 atm)	0.000000942	-	"
5 (1 atm)	0.00000213	-	Muccitelli (1978)
10 (1 atm)	0.00000171	-	"
15 (1 atm)	0.00000143	-	"
20 (1 atm)	0.00000112	-	"
25 (1 atm)	0.000000997	-	"
30 (1 atm)	0.000000903	-	"

[a]M.wt. = 138.01; m.p. = -94.0^oC; b.p. = -79.0^oC; d^{-78} = 1.590 g cc^{-1}.

CHI=CHI (cis-)[a]

Ethene, cis-1,2-diiodo-

Temperature (°C)	Solubility in H_2O	Solubility of H_2O in	Reference
25	0.00165 mol l^{-1} soln.	–	Andrews & Keefer (1951)

[a]M.wt. = 279.8472; m.p. = -14.0°C; b.p. = 188.0°C (decomposes); d_4^{25} = 2.955 g cc^{-1}.

CHI=CHI (trans-)[a]

Ethene, trans-1,2-diiodo-

Temperature (°C)	Solubility in H_2O	Solubility of H_2O in	Reference
25	0.000527 mol l^{-1} soln.	–	Andrews & Keefer (1951)

[a]M.wt. = 279.8472; m.p. = 78.0°C; b.p. = 192.0°C; d^{83} = 2.826 g cc^{-1}.

$CH_3-CH_2I^a$

Ethane, iodo-

Temperature ($^\circ$C)	Solubility in H_2O	Solubility of H_2O in	Reference
22.5	0.391 wt % C_2H_5I	–	Fühner (1924)
30	4.04 g/1000 g H_2O	–	Gross & Saylor (1931)
20	0.40 wt % C_2H_5I	–	Harms (1939)
20	0.390 wt % C_2H_5I	–	Korenman et al. (1971)
20	0.0251 mol 1^{-1} H_2O	–	Mercel (1937)
0	0.441 g/100 g H_2O	–	Rex (1906)
10	0.414	–	"
20	0.403	–	"
30	0.415	–	"
30	3.88 wt % C_2H_5I	–	Riddick & Bunger (1970)
22.5	0.0251 mol/1000 g H_2O	–	Saracco & Spaccamela Marchetti (1958)
30	0.00259 mol/100 g H_2O	–	Van Arkel & Vles (1936)
0	0.339 wt % C_2H_5I	–	Washburn (1928)
10	0.412	–	"
20	0.401	–	"
30	0.413	–	"
20	0.4 g/100 cc H_2O	–	Patty (1962)
20	0.40 g/100 g H_2O	–	Lange (1967)
20	0.40 g/100 cc H_2O	–	Kirk-Othmer (1966b)
20	0.40 wt % C_2H_5I	–	Hildebrand (1936a)

[a]M.wt. = 155.97; m.p. = -108.0°C; b.p. = 72.3°C; d_4^{20} = 1.9358 g cc^{-1}.

$CBrH_2-CH_2-CClH_2^a$
Propane, 1-bromo-3-chloro-

Temperature (°C)	Solubility in H_2O	Solubility of H_2O in	Reference
25	1.84 g/100 g H_2O	–	Dreisbach (1955–1961: Ser. 22)
25	–	300 ppm by weight H_2O	Kirk-Othmer (1964a)
20	1400 ppm by volume C_3BrClH_6	–	Imperial Chemical Industries (1972)

[a]M.wt. = 157.44; m.p. = -58.87°C; b.p. = 143.36°C; d_4^{20} = 1.5969 g cc^{-1}.

$CBrH_2-CH_2-CH_3^a$
Propane, 1-bromo-

Temperature (°C)	Solubility in H_2O	Solubility of H_2O in	Reference
19.5	0.226 wt % C_3BrH_7	–	Führner (1924)
30	2.31 g/1000 g H_2O	–	Gross & Saylor (1931)
20	0.24 wt % C_3BrH_7	–	Harmes (1939)
20	0.2446 wt % C_3BrH_7	–	Korenman et al. (1971)
0	0.298 g/100 g H_2O	–	Rex (1906)
10	0.263	–	"
20	0.245	–	"
30	0.247	–	"
30	0.230 wt % C_3BrH_7	–	Riddick & Bunger (1970)
19.5	0.0185 mol/1000 g H_2O	–	Saracco & Spaccamela Marchetti (1958)
30	0.00188 mol/100 g H_2O	–	Van Arkel & Vles (1936)
0	0.297 wt % C_3BrH_7	–	Washburn (1928)
10	0.262	–	"
20	0.244	–	"
30	0.246	–	"
20	0.25 g/100 cc H_2O	–	Patty (1962)
20	0.25 g/100 g H_2O	–	Lange (1967)
19.5	0.226 wt % C_3BrH_7	–	Stephen & Stephen
30	0.230	–	(1963)

[a]M.wt. = 123.00; m.p. = -109.85°C; b.p. = 71.0°C; d_4^{20} = 1.3537 g cc^{-1}.

$CH_3-CBrH-CH_3$[a]
Propane, 2-bromo-

Temperature (°C)	Solubility in H_2O	Solubility of H_2O in	Reference
18	0.286 wt % C_3BrH_7	–	Fühner (1924)
20	0.2873 wt % C_3BrH_7	–	Korenman et al. (1971)
0	0.418 g/100 g H_2O	–	Rex (1906)
10	0.365	–	"
20	0.318	–	"
30	0.318	–	"
30	0.0026 mol/100 g H_2O	–	Van Arkel & Vles (1936)
0	0.416 wt % C_3BrH_7	–	Washburn (1928)
10	0.364	–	"
20	0.317	–	"
30	0.317	–	"
20	0.32 g/100 g H_2O	–	Lange (1967)
18	0.286 wt % C_3BrH_7	–	Riddick & Bunger (1970)
18	0.286 wt % C_3BrH_7	–	Stephen & Stephen (1963)

[a]M.wt. = 123.00; m.p. = -89.0°C; b.p. = 59.38°C; d_4^{20} = 1.3140 g cc^{-1}.

$CBrH_2-CBrH-CClH_2$[a]
Propane, 1,2-dibromo-3-chloro-

Temperature (°C)	Solubility in H_2O	Solubility of H_2O in	Reference
20	0.1 wt % $C_3Br_2ClH_5$	–	Gunther et al. (1968)
25	0.1 wt % $C_3Br_2ClH_5$	–	Jolles (1966)
20	1230 ppm by weight $C_3Br_2ClH_5$	–	Metcalf (1962)

[a]M.wt. = 236.35; b.p. = 78.0°C at 16 mm Hg; d_4^{14} = 2.093 g cc^{-1}.

CBrF$_2$-CBrF-CF$_3^a$
Propane, 1,2-dibromohexafluoro-   (R 216B2)

Temperature ($^{\circ}$C)	Solubility in H$_2$O	Solubility of H$_2$O in	Reference
25	–	75.0 ppm by weight H$_2$O	du Pont (1963d)
2	–	40.0	"
-29	–	Graph	"
⋮	–	"	"
50	–	"	"
21	–	0.0068 wt % H$_2$O	du Pont (1966c)
-28.9	–	0.0016	"
-23.3	–	0.0019	"
-17.8	–	0.0023	"
-12.2	–	0.0027	"
-6.7	–	0.0032	"
-1.1	–	0.0038	"
4.4	–	0.0044	
10.0	–	0.0052	"
15.6	–	0.0060	"
21.1	–	0.0068	"
26.7	–	0.0079	"
32.2	–	0.0090	"
37.8	–	0.0102	"
43.3	–	0.0115	"
48.9	–	0.0128	"

[a]M.wt. = 309.87; m.p. = -94.0°C; b.p. = 72.6°C; d_4^{25} = 2.157 g cc^{-1}.

CBrH$_2$-CBrH-CH$_3^a$
Propane, 1,2-dibromo-

Temperature ($^{\circ}$C)	Solubility in H$_2$O	Solubility of H$_2$O in	Reference
25	0.143 g/100 g H$_2$O	0.052 g/100 g C$_3$Br$_2$H$_6$	Dreisbach (1955-1961: Ser. 22)
20	0.25 g/100 g H$_2$O	–	Lange (1967)

[a]M.wt. = 201.90; m.p. = -55.25°C; b.p. = 140.0°C; d_4^{20} = 1.9324 g cc^{-1}.

$CBrH_2-CH_2-CBrH_2^a$
Propane, 1,3-dibromo-

Temperature ($^{\circ}$C)	Solubility in H_2O	Solubility of H_2O in	Reference
30	1.68 g/1000 g H_2O	–	Gross et al. (1933)
25	0.17 wt % $C_3Br_2H_6$	–	Gunther et al. (1968)
30	0.168 g/100 g H_2O	–	Lange (1967)
30	0.168 wt % $C_3Br_2H_6$	–	Stephen & Stephen (1963)

[a] M.wt. = 201.90; m.p. = -34.2°C; b.p. = 167.3°C; d_4^{20} = 1.9822 g cc^{-1}.

$CClH_2-CH_2-CF_3^a$
Propane, 1-chloro-3,3,3-trifluoro-

Temperature ($^{\circ}$C)	Solubility in H_2O	Solubility of H_2O in	Reference
20	1.33 g l^{-1} H_2O	–	Selyuzhitskii (1967)

[a] M.wt. = 132.51; m.p. = -106.2°C; b.p. = 45.1°C; d_4^{20} = 1.3253 g cc^{-1}.

$CClH_2-CH=CH_2^a$
1-Propene, 3-chloro-

Temperature ($^\circ$C)	Solubility in H_2O	Solubility of H_2O in	Reference
20	0.36 wt % C_3ClH_5	0.08 wt % H_2O	Riddick & Bunger (1970)
20	0.30 g/100 g H_2O	-	Manufacturing Chemists Assoc. (1956-1971: SD-99)
20	0.33 g/100 g H_2O	-	Kirk-Othmer (1964b)
20	0.36 wt % C_3ClH_5	0.08 wt % H_2O	Shell Chemical Corp. (1949)
25	4.0 g/1000 g H_2O	-	Treger et al.
50	1.3	-	(1964)
70	0.82	-	"
20	0.36 wt % C_3ClH_5	0.08 wt % H_2O	Gladis (1960)
20	0.1 wt % C_3ClH_5	0.1 wt % H_2O	Nathan (1978)
25	3370 ppm by weight C_3ClH_5	-	Dilling (1977)

[a]M.wt. = 76.53; m.p. = -134.5°C; b.p. = 45.0°C; d_4^{20} = 0.9376 g cc^{-1}.

$CClH_2-CH_2-CH_3^a$
Propane, 1-chloro-

Temperature (°C)	Solubility in H_2O	Solubility of H_2O in	Reference
12.5	0.232 wt % C_3ClH_7	–	Fühner (1924)
20	0.27 wt % C_3ClH_7	–	Harms (1939)
20 (10 mm Hg)	0.00091 mol/1000 g H_2O	–	Hildebrand (1949c)
20 (saturated)	0.0255	–	"
20	0.2842 wt % C_3ClH_7	–	Korenman et al. (1971)
0	0.376 g/100 g H_2O	–	Rex (1906)
10	0.323	–	"
20	0.272	–	"
30	0.277	–	"
20	0.271 wt % C_3ClH_7	–	Riddick & Bunger (1970)
20	0.0297 mol/1000 g H_2O	–	Saracco Spaccamela Marchetti (1958)
30	0.00353 mol/100 g H_2O	–	Van Arkel & Vles (1936)
0	0.375 wt % C_3ClH_7	–	Washburn (1928)
10	0.322	–	"
20	0.271	–	"
30	0.276	–	"
25	1 g/400 cc H_2O	–	Wright & Schaffer (1932)
20	0.27 g/100 cc H_2O	–	Patty (1962)
20	0.27 g/100 g H_2O	–	Lange (1967)
20	0.27 wt % C_3ClH_7	–	Hildebrand (1936a)
12.5	0.232 wt % C_3ClH_7	–	Stephen & Stephen (1963)
20	0.27 wt % C_3ClH_7	–	Nathan (1978)
20	0.27 wt % C_3ClH_7	–	du Pont (1966e)

[a]M.wt. = 78.54; m.p. = -122.8°C; b.p. = 46.60°C; d_4^{20} = 0.8909 g cc^{-1}.

CH$_3$-CClH-CH$_3^a$
Propane, 2-chloro-

Temperature ($^{\circ}$C)	Solubility in H$_2$O	Solubility of H$_2$O in	Reference
12.5	0.342 wt % C$_3$ClH$_7$	–	Fühner (1924)
20	0.3415 wt % C$_3$ClH$_7$	–	Korenman et al. (1971)
0	0.440 g/100 g H$_2$O	–	Rex (1906)
10	0.362	–	"
20	0.305	–	"
30	0.304	–	"
12.5	0.342 wt % C$_3$ClH$_7$	–	Riddick & Bunger (1970)
30	0.0039 mol/100 g H$_2$O	–	Van Arkel & Vles (1936)
0	0.438 wt % C$_3$ClH$_7$	–	Washburn (1928)
10	0.361	–	"
20	0.304	–	"
30	0.303	–	"
20	0.31 g/100 cc H$_2$O	–	Patty (1962)
20	0.31 g/100 g H$_2$O	–	Lange (1967)
12.5	0.342 wt % C$_3$ClH$_7$	–	Stephen & Stephen (1963)
20	0.31 wt % C$_3$ClH$_7$	0.33 wt % H$_2$O	Nathan (1978)
20	0.31 wt % C$_3$ClH$_7$	–	du Pont (1966e)

[a] M.wt. = 78.54; m.p. = -117.18°C; b.p. = 35.74°C; d$_4^{20}$ = 0.8617 g cc^{-1}.

$CClF_2-CClF-CF_3$[a]
Propane, 1,2-dichloro-1,1,2,3,3,3-hexafluoro- (R 216)

Temperature (^{o}C)	Solubility in H_2O	Solubility of H_2O in	Reference
25	–	103.0 ppm by weight H_2O	du Pont (1963c)
1.7	–	59.0	"
-29	–	Graph	"
⋮	–	"	"
49	–	"	"
21	–	0.0096 wt % H_2O	du Pont (1966c)
-28.9	–	0.0022	"
-23.3	–	0.0027	"
-17.8	–	0.0032	"
-12.2	–	0.0039	"
-6.7	–	0.0046	"
-1.1	–	0.0055	"
4.4	–	0.0063	"
10.0	–	0.0073	"
15.6	–	0.0084	"
21.1	–	0.0096	"
26.7	–	0.0108	"
32.2	–	0.0123	"
37.8	–	0.0137	"
43.3	–	0.0154	"
48.9	–	0.0170	"

[a]M.wt. = 220.95; m.p. = -212.8^{o}C; b.p. = 34.5^{o}C; d_4^{25} = 1.577 g cc^{-1}.

CClH=CH-CClH$_2^a$
Propene, cis-1,3-dichloro-

Temperature (°C)	Solubility in H_2O	Solubility of H_2O in	Reference
20	0.1 g/100 g H_2O	–	Gunther et al. (1968)
20	2700 ppm by weight $C_3Cl_2H_4$	–	Metcalf (1962)
25	2700 ppm by weight $C_3Cl_2H_4$	–	Dilling (1977)

[a]M.wt. = 110.97; b.p. = 104.3°C; d_4^{20} = 1.217 g cc^{-1}.

CClH=CH-CClH$_2^a$
Propene, trans-1,3-dichloro-

Temperature (°C)	Solubility in H_2O	Solubility of H_2O in	Reference
20	2800 ppm by weight $C_3Cl_2H_4$	–	Metcalf (1962)
25	2800 ppm by weight $C_3Cl_2H_4$	–	Dilling (1977)

[a]M.wt. = 110.97; b.p. = 112.0°C; d_4^{20} = 1.224 g cc^{-1}.

CClH$_2$-CCl=CH$_2^a$
2-Propene, 1,2-dichloro-

Temperature (°C)	Solubility in H_2O	Solubility of H_2O in	Reference
25	2150 ppm by weight $C_3Cl_2H_4$	–	Dilling (1977)

[a]M.wt. = 110.97; b.p. = 94.0°C; d_4^{20} = 1.211 g cc^{-1}.

$CClH_2-CClH-CH_3^a$

Propane, 1,2-dichloro-

Temperature ($^{\circ}$C)	Solubility in H_2O	Solubility of H_2O in	Reference
25	–	0.006481 mol/100 g soln.	Eidinoff (1955)
25	0.26 wt % $C_3Cl_2H_6$	0.06 wt % H_2O	Gladis (1960)
25	0.28 g/100 g H_2O	–	Gross (1929a)
25	2.80 g/1000 g H_2O	–	Gross (1929b)
20	0.27 g/100 g H_2O	0.04 g/100 g $C_3Cl_2H_6$	McGovern (1943)
25	1 g/340 cc H_2O	–	Wright & Schaffer (1932)
20	0.27 g/100 cc H_2O	–	Patty (1962)
20	0.27 g/100 g H_2O	–	Gunther et al. (1968)
27	0.32 wt % $C_3Cl_2H_6$	0.14 wt % H_2O	Imperial Chemical Industries (1977)
25	0.275 g/100 g H_2O	0.132 g/100 g $C_3Cl_2H_6$	Dreisbach (1955-1961: Ser. 22)
10	0.23 wt % $C_3Cl_2H_6$	–	Marsden (1963)
20	0.27	0.040 wt % H_2O	"
30	0.29	–	"
20	0.27 g/100 g H_2O	–	Lange (1967)
20	2700 ppm w/v H_2O	–	Imperial Chemical Industries (1972)
25	0.280 wt % $C_3Cl_2H_6$	–	Stephen & Stephen (1963)
20	0.27 wt % $C_3Cl_2H_6$	0.06 wt % H_2O	Nathan (1978)
20	0.26 wt % $C_3Cl_2H_6$	0.06 wt % H_2O	Union Carbide Chemicals Co. (1962)
20	0.27 wt % $C_3Cl_2H_6$	0.06 wt % H_2O	du Pont (1966e)

[a]M.wt. = 112.99; m.p. = -100.44°C; b.p. = 96.37°C; d_4^{20} = 1.1560 g cc^{-1}.

$CClH_2-CH_2-CClH_2^a$
Propane, 1,3-dichloro-

Temperature (°C)	Solubility in H_2O	Solubility of H_2O in	Reference
25	0.273 g/100 g H_2O	−	Gross (1929a)
30	2.87 g/1000 g H_2O	−	Gross et al. (1933)
25	2.73 g/1000 g H_2O	−	Gross (1929b)
25	1 g/350 cc H_2O	−	Wright & Schaffer (1932)
20	0.27 g/100 g H_2O	−	Gunther et al. (1968)
25	0.27 g/100 g H_2O	−	Lange (1967)
25	0.273 wt % $C_3Cl_2H_6$	−	Stephen & Stephen
30	0.287	−	(1963)

[a]M.wt. = 112.99; m.p. = -99.5°C; b.p. = 120.4°C; d_4^{20} = 1.1878 g cc^{-1}.

$CCl_3-CF_2-CF_3^a$
Propane, 1,1,1-trichloro-2,2,3,3,3-pentafluoro- (R 215)

Temperature (°C)	Solubility in H_2O	Solubility of H_2O in	Reference
25	−	66.0 ppm by weight H_2O	du Pont (1963c)
1.7	−	35.0	"
-29	−	Graph	"
⋮	−	"	"
49	−	"	"
21	−	0.0058 wt % H_2O	du Pont (1966c)
-28.9	−	0.0014	"
-23.3	−	0.0017	"
-17.8	−	0.0020	"
-12.2	−	0.0024	"
-6.7	−	0.0028	"
-1.1	−	0.0033	"
4.4	−	0.0038	"
10.0	−	0.0045	"
15.6	−	0.0051	"
21.1	−	0.0059	"

$CClH_2-CH_2-CClH_2$ (Continued)

26.7	–	0.0067	du Pont (1966c)
32.2	–	0.0076	"
37.8	–	0.0085	"
43.3	–	0.0097	"
48.9	–	0.0108	"

[a]M.wt. = 237.40; m.p. = -80.0°C; b.p. = 74.39°C; d_4^{25} = 1.646 g cc^{-1}.

$CClH_2-CClH-CClH_2$[a]
Propane, 1,2,3-trichloro-

Temperature (°C)	Solubility in H_2O	Solubility of H_2O in	Reference
20	2.5 g/1000 g H_2O	–	Shell Chemical Co. (1948)
25	1.9 g/1000 g H_2O	0.05 wt % H_2O	Union Carbide Chemical Co. (1959, 1962)
25	1900 ppm by weight $C_3Cl_3H_5$	–	Dilling (1977)
25	1900 g/m^3 H_2O	–	Afghan & Mackay (1980)

[a]M.wt. = 147.43; m.p. = -14.7°C; b.p. = 156.85°C; d_4^{20} = 1.3889 g cc^{-1}.

$CCl_3-CF_2-CClF_2^a$
Propane, 1,1,1,3-tetrachloro-2,2,3,3-tetrafluoro- (R 214)

Temperature ($^\circ$C)	Solubility in H_2O	Solubility of H_2O in	Reference
25	-	56.0 ppm by weight H_2O	du Pont (1963c)
1.7	-	32.0	"
-29	-	Graph	"
⋮	-	"	"
49	-	"	"
21	-	0.0052 wt % H_2O	du Pont (1966c)
-28.9	-	0.0012	"
-23.3	-	0.0015	"
-17.8	-	0.0018	"
-12.2	-	0.0021	"
-6.7	-	0.0025	"
-1.1	-	0.0029	"
4.4	-	0.0034	"
10.0	-	0.0040	"
15.6	-	0.0045	"
21.1	-	0.0052	"
26.7	-	0.0058	"
32.2	-	0.0066	"
37.8	-	0.0074	"
43.3	-	0.0084	"
48.9	-	0.0093	"

[a]M.wt. = 253.86; m.p. = -92.8°C; b.p. = 114.0°C; d_4^{25} = 1.659 g cc^{-1}.

$CCl_3-CCl=CCl_2^a$

Propene, hexachloro-

Temperature (°C)	Solubility in H_2O	Solubility of H_2O in	Reference
15	0.000947 wt % C_3Cl_6	–	Antropov et al.
18	–	0.00371 wt % H_2O	(1972)
20	0.00118 wt % C_3Cl_6	–	"
22	–	0.00438 wt % H_2O	"
26	0.0017 wt % C_3Cl_6	0.00525	"
40	0.0034	0.00907	"
25	0.015 g/100 g H_2O	0.014 g/100 g C_3Cl_6	Dreisbach (1955-1961: Ser. 22)
0	–	0.00178 wt % H_2O	Simonov et al.
30	–	0.00590	(1970)
50	–	0.0144	"
60	–	0.02289	"
80	–	0.0537	"
100	–	0.1258	"
112	–	0.2098	"
40	0.005 wt % C_3Cl_6	0.0091 wt % H_2O	Simonov et al. (1974)

[a]M.wt. = 236.74; m.p. = -72.9°C; b.p. = 141.4°C at 100 mm Hg; d_4^{25} = 1.7666 g cc^{-1}.

$CFH_2-CH=CH_2^a$

1-Propene, 3-fluoro-

Temperature (°C)	Solubility in H_2O	Solubility of H_2O in	Reference
13	2.8 cc/100 cc H_2O	–	Meslans (1894)

[a]M.wt. = 60.07; b.p. = -3.0°C; d_4^{20} = 0.9379 g cc^{-1}.

CFH$_2$-CH$_2$-CH$_3^a$
Propane, 1-fluoro-

Temperature (°C)	Solubility in H$_2$O	Solubility of H$_2$O in	Reference
14	150 cc/100 cc H$_2$O	–	Meslans (1894)

[a]M.wt. = 62.09; m.p. = -159.0°C; b.p. = 2.5°C; d_4^{20} = 0.7956 g cc^{-1}.

CH$_3$-CFH-CH$_3^a$
Propane, 2-fluoro-

Temperature (°C)	Solubility in H$_2$O	Solubility of H$_2$O in	Reference
15	140 cc/100 cc H$_2$O	–	Meslans (1894)

[a]M.wt. = 62.09; m.p. = -133.4°C; b.p. = -10.0°C; d_4^{20} = 0.7238 g cc^{-1}.

CF$_3$-CF$_2$-CF$_3^a$
Propane, octafluoro-

Temperature (°C)	Solubility in H$_2$O	Solubility of H$_2$O in	Reference
15	0.00072 cc/cc H$_2$O (Ostwald)	–	Muccitelli & Wen (1978)
5 (1 atm)	0.00000132 mole fraction	–	Wen & Muccitelli
10 (1 atm)	0.000000674	–	(1979)
15 (1 atm)	0.000000560	–	"
5 (1 atm)	0.00000131 mole fraction	–	Muccitelli (1978)
10 (1 atm)	0.000000666	–	"
15 (1 atm)	0.000000551	–	"

[a]M.wt. = 188.02; m.p. = -183.0°C; b.p. = -38.0°C; $d^{0.2}$ = 1.450 g cc^{-1}.

$CH_3-CH_2-CH_2I^a$
Propane, 1-iodo-

Temperature (°C)	Solubility in H_2O	Solubility of H_2O in	Reference
20	0.087 wt % C_3H_7I	–	Fühner (1924)
30	1.04 g/1000 g H_2O	–	Gross & Saylor (1931)
20	0.11 wt % C_3H_7I	–	Harms (1939)
20 (10 mm Hg)	0.00154 mol/1000 g H_2O	–	Hildebrand (1949c)
20 (saturated)	0.0054	–	"
20	0.1070 wt % C_3H_7I	–	Korenman et al. (1971)
20	0.0051 mol l^{-1} H_2O	–	Mercel (1937)
0	0.114 g/100 g H_2O	–	Rex (1906)
10	0.103	–	"
20	0.107	–	"
30	0.103	–	"
30	0.104 wt % C_3H_7I	–	Riddick & Bunger (1970)
20	0.0051 mol/1000 g H_2O	–	Saracco & Spaccamela Marchetti (1958)
30	0.00061 mol/100 g H_2O	–	Van Arkel & Vles (1936)
0	0.114 wt % C_3H_7I	–	Washburn (1928)
10	0.103.	–	"
20	0.107	–	"
30	0.103	–	"
20	0.11 g/100 g H_2O	–	Lange (1967)
20	0.11 wt % C_3H_7I	–	Hildebrand (1936a)
20	0.087 wt % C_3H_7I	–	Stephen & Stephen
30	0.104	–	(1963)
23.5	0.107 wt % C_3H_7I	–	Schwarz (1980)

[a]M.wt. = 169.99; m.p. = -101.3°C; b.p. = 102.45°C; d_4^{20} = 1.7489 g cc^{-1}.

$CH_3-CHI-CH_3$[a]
Propane, 2-iodo-

Temperature (^{o}C)	Solubility in H_2O	Solubility of H_2O in	Reference
0	0.167 g/100 g H_2O	–	Rex (1906)
10	0.143	–	"
20	0.140	–	"
30	0.134	–	"
20	0.140 wt % C_3H_7I	–	Riddick & Bunger (1970)
30	0.00079 mol/100 g H_2O	–	Van Arkel & Vles (1936)
0	0.167 wt % C_3H_7I	–	Washburn (1928)
10	0.143	–	"
20	0.140	–	"
30	0.134	–	"
20	0.14 g/100 g H_2O	–	Lange (1967)

[a]M.wt. = 169.99; m.p. = -90.1^{o}C; b.p. = 89.45^{o}C; d_4^{20} = 1.7033 g cc^{-1}.

$CBrH_2-CH_2-CH_2-CH_3^a$
n-Butane, 1-bromo-

Temperature (^oC)	Solubility in H_2O	Solubility of H_2O in	References
25	0.02 cc/100 cc H_2O	–	Booth & Everson (1948)
16	0.058 wt % C_4BrH_9	–	Fühner (1924)
30	0.608 g/1000 g H_2O	–	Gross & Saylor (1931)
20	0.0508 wt % C_4BrH_9	–	Korenman et al. (1971)
17	0.0043 mol/1000 g H_2O	–	Saracco & Spaccamela Marchetti (1958)
16	0.06 g/100 g H_2O	–	Lange (1967)
16	0.058 wt % C_4BrH_9	–	Stephen & Stephen
18	0.051	–	(1963)
25	0.0045 mol/1000 g H_2O	–	Kakovsky (1957)

[a] M.wt. = 137.03; m.p. = -112.4^oC; b.p. = 101.6^oC; d_4^{20} = 1.2758 g cc^{-1}.

$CBrH_2-CH(CH_3)-CH_3^a$
Propane, 1-bromo-2-methyl-

Temperature (^oC)	Solubility in H_2O	Solubility of H_2O in	Reference
18	0.051 wt % C_4BrH_9	–	Fühner (1924)
20	0.0508 wt % C_4BrH_9	–	Korenman et al. (1971)

[a] M.wt. = 137.03; m.p. = -111.9^oC; b.p. = 91.20^oC; d_4^{20} = 1.2585 g cc^{-1}.

$CClH_2-CH_2-CH_2-CH_3^a$
n-Butane, 1-chloro-

Temperature (^{o}C)	Solubility in H_2O	Solubility of H_2O in	Reference
12.5	0.066 wt % C_4ClH_9	–	Führner (1924)
20	0.0669 wt % C_4ClH_9	–	Korenman et al. (1971)
20	0.11 wt % C_4ClH_9	0.08 wt % H_2O	Riddick & Bunger (1970)
17.5	0.0072 mol/1000 g H_2O	–	Saracco & Spaccamela Marchetti (1958)
25	1 g/1700 cc H_2O	–	Wright & Schaffer (1932)
12.5	0.070 g/100 g H_2O	–	Lange (1967)
12.5	0.066 wt % C_4ClH_9	–	Stephen & Stephen (1963)
25	0.008 mol/1000 g H_2O	–	Kakovsky (1957)
12.5	0.07 wt % C_4ClH_9	–	Nathan (1978)
20	0.11 wt % C_4ClH_9	0.05 wt % H_2O	Union Carbide Chemical Co. (1962)

[a]M.wt. = 92.57; m.p. = -123.1^{o}C; b.p. = 78.44^{o}C; d_4^{20} = 0.8862 g cc^{-1}.

$CH_3-CClH-CH_2-CH_3^a$
n-Butane, 2-chloro-

Temperature (^{o}C)	Solubility in H_2O	Solubility of H_2O in	Reference
25	0.1 wt % C_4ClH_9	< 0.1 wt % H_2O	Riddick & Bunger (1970)
25	1 g /100 cc H_2O	–	Wright & Schaffer (1932)
25	0.1 wt % C_4ClH_9	–	Nathan (1978)

[a]M.wt. = 92.57; m.p. = -131.3^{o}C; b.p. = 68.25^{o}C; d_4^{20} = 0.8732 g cc^{-1}.

$CClH_2-CH(CH_3)-CH_3^a$
Propane, 1-chloro-2-methyl-

Temperature (^oC)	Solubility in H_2O	Solubility of H_2O in	Reference
12.5	0.092 wt % C_4ClH_9	–	Führner (1924)
20	0.0924 wt % C_4ClH_9	–	Korenman et al. (1971)
12.5	0.092 wt % C_4ClH_9	–	Riddick & Bunger (1970)
12.5	0.92 wt % C_4ClH_9	–	Stephen & Stephen (1963)

[a]M.wt. = 92.57; m.p. = -131.2^oC; b.p. = 68.4^oC; d_4^{25} = 0.8725 g cc^{-1}.

$CCl_2H-CH_2-CH_2-CH_3^a$
n-Butane, 1,1-dichloro-

Temperature (^oC)	Solubility in H_2O	Solubility of H_2O in	Reference
25	1 g/2000 cc H_2O	–	Wright & Schaffer (1932)

[a]M.wt. = 127.03; m.p. = -81.0^oC; b.p. = 113.8^oC; d_4^{20} = 1.0863 g cc^{-1}.

$CH_3-CClH-CClH-CH_3^a$
n-Butane, 2,3-dichloro-

Temperature (^oC)	Solubility in H_2O	Solubility of H_2O in	Reference
0	0.182 g/100 g H_2O	–	Lichosherstov
20	0.0562	–	et al. (1935)
30	0.0186	–	"
40	0.0223	–	"

[a]M.wt. = 127.03; m.p. = -80.0^oC; b.p. = 116.0^oC; d_4^{20} = 1.1134 g cc^{-1}.

$CCl_2=C=CCl-CCl_2H$[a]

1,2-Butadiene, 1,1,3,4,4-pentachloro-

Temperature (°C)	Solubility in H_2O	Solubility of H_2O in	Reference
20	7.8 ppm by volume C_4Cl_5H	–	Imperial Chemical Industries (1972)

[a]M.wt. = 226.35; d_4^{20} = 1.6138 g cc^{-1}.

$CCl_2=CCl-CCl=CCl_2$[a]

1,3-Butadiene, hexachloro-

Temperature (°C)	Solubility in H_2O	Solubility of H_2O in	Reference
15	0.00032 wt % C_4Cl_6	–	Antropov et al.
18	–	0.000834 wt % H_2O	(1972)
20	0.00038 wt % C_4Cl_6	–	"
22	–	0.00115 wt % H_2O	"
26	0.000412 wt % C_4Cl_6	0.00565	"
20	0.0005 wt % C_4Cl_6	–	McBee & Hatton (1949)
0	–	0.00182 wt % H_2O	Simonov et al.
30	–	0.00676	(1970)
50	–	0.0212	"
60	–	0.0265	"
80	–	0.0648	"
100	–	0.1584	"
112	–	0.2708	"
20	5.0 ppm w/v H_2O	–	Imperial Chemical Industries (1972)
40	0.0059 wt % C_4Cl_6	0.0110 wt % H_2O	Simonov et al. (1974)
20	2.0 ppm by weight C_4Cl_6	–	Pearson & McConnell (1975)
25	0.0000124 mol/liter soln.	–	Banerjee et al. (1980)

[a]M.wt. = 260.76; m.p. = -21.0°C; b.p. = 215.0°C; d_4^{20} = 1.6820 g cc^{-1}.

$$CF_2-CF_2$$
$$\mid \quad \mid$$
$$CF_2-CF_2$$

Cyclobutane, octafluoro- (R C318)

Temperature ($^{\circ}$C)	Solubility in H_2O	Solubility of H_2O in	Reference
25 (14.7 psia)	0.0005 wt % C_4F_8	–	du Pont (1966c)
0 (14.7 psia)	Table	–	"
26 (14.7-40 psia)	"	–	"
37.8 (14.7-40 psia)	"	–	"
21 (760 mm Hg)	0.014 wt % C_4F_8	–	Matheson Company (1971)
6 (1-4.2 bars)	Equation	–	Filatkin et al.
$\vdots$	"	–	(1970)
38 (1-4.2 bars)	"	–	"
-40	–	0.000055 g/g C_4F_8	Petrak (1975)
-20	–	0.00016	"
0	–	0.00037	"
20	–	0.00070	"
40	–	0.00116	"
60	–	0.0016	"
21 (14.7 psia)	0.0024 wt % C_4F_8	–	Sandell & Johnson
23.5 (14.7 psia)	0.0027	–	(1967)
-12 (0-20 psia)	Graph	–	"
$\vdots$	"	–	"
38 (0-50 psia)	"	–	"
21	0.014 wt % C_4F_8	–	Shepherd (1961)
15	0.00414 cc/cc H_2O (Ostwald)	–	Muccitelli & Wen (1978)
5 (1 atm)	0.00000573 mole fraction	–	Wen & Muccitelli
10 (1 atm)	0.00000429	–	(1979)
15 (1 atm)	0.00000321	–	"
20 (1 atm)	0.00000262	–	"
25 (1 atm)	0.00000220	–	"
30 (1 atm)	0.00000188	–	"
5 (1 atm)	0.00000568 mole fraction	–	Muccitelli (1978)
10 (1 atm)	0.00000424	–	"
15 (1 atm)	0.00000321	–	"
20 (1 atm)	0.00000256	–	"
25 (1 atm)	0.00000213	–	"
30 (1 atm)	0.00000180	–	"

[a]M.wt. = 200.03; m.p. = -38.7°C; b.p. = -4.0°C; d_4^0 = 1.724 g cc^{-1}.

CH$_3$-CH$_2$-CH$_2$-CH$_2$I[a]
n-Butane, 1-iodo-

Temperature ($^\circ$C)	Solubility in H$_2$O	Solubility of H$_2$O in	Reference
17.5	0.021 wt % C$_4$H$_9$I	–	Fühner (1924)
20	0.0211 wt % C$_4$H$_9$I	–	Korenman et al. (1971)
20	0.0011 mol/liter H$_2$O	–	Mercel (1937)
17.5	0.0011 mol/1000 g H$_2$O	– Saracco	Saracco & Spaccamela Marchetti (1958)
17.5	0.021 wt % C$_4$H$_9$I	–	Stephen & Stephen (1963)
25	0.0017 mol/1000 g H$_2$O	–	Kakovsky (1957)

[a]M.wt. = 184.02; m.p. = -103.0°C; b.p. = 130.53°C; d$_4^{20}$ = 1.6154 g cc^{-1}.

CBrH$_2$-CH$_2$-CH(CH$_3$)-CH$_3$[a]
n-Butane, 1-bromo-3-methyl-

Temperature ($^\circ$C)	Solubility in H$_2$O	Solubility of H$_2$O in	Reference
16.5	0.020 wt % C$_5$BrH$_{11}$	–	Fühner (1924)
20	0.0194 wt % C$_5$BrH$_{11}$	–	Korenman et al. (1971)
0	16.5 wt % C$_5$BrH$_{11}$	–	Stephen & Stephen (1963)

[a]M.wt. = 151.05; m.p. = -112.0°C; b.p. = 120.4°C; d$_4^{20}$ = 1.2071 g cc^{-1}.

$CCIH_2-CH_2-CH_2-CH_2-CH_3^a$
n-Pentane, 1-chloro-

Temperature (°C)	Solubility in H_2O	Solubility of H_2O in	Reference
25	0.09 cc/100 cc H_2O	-	Booth & Everson (1948)
25	-	0.00253 mol/100 g soln.	Eidinoff (1955)
25	0.02 wt % C_5ClH_{11}	-	Riddick & Bunger (1970)
25	1 g/5000 cc H_2O	-	Wright & Schaffer (1932)

[a]M.wt. = 106.60; m.p. = -99.0°C; b.p. = 107.8°C; d_4^{20} = 0.8818 g cc^{-1}.

$CH_3-CCIH-CH_2-CH_2-CH_3^a$
n-Pentane, 2-chloro-

Temperature (°C)	Solubility in H_2O	Solubility of H_2O in	Reference
25	1 g/4000 cc H_2O	-	Wright & Schaffer (1932)

[a]M.wt. = 106.60; m.p. = -137.0°C; b.p. = 96.86°C; d_4^{20} = 0.8698 g cc^{-1}.

$CH_3-CH_2-CCIH-CH_2-CH_3^a$
n-Pentane, 3-chloro-

Temperature (°C)	Solubility in H_2O	Solubility of H_2O in	Reference
25	1 g/4000 cc H_2O	-	Wright & Schaffer (1932)

[a]M.wt. = 106.60; m.p. = -105.0°C; b.p. = 97.8°C; d_4^{20} = 0.8731 g cc^{-1}.

$CH_3-CCl(CH_3)-CH_2-CH_3^a$

n-Butane, 2-chloro-2-methyl-

Temperature (^oC)	Solubility in H_2O	Solubility of H_2O in	Reference
25	1 g/3oo cc H_2O	–	Wright & Schaffer (1932)

[a] M.wt. = 106.60; m.p. = -73.5^oC; b.p. = 85.6^oC; d_4^{20} = 0.8653 g cc^{-1}.

$CH_3-CCl(CH_3)-CClH-CH_3^a$

n-Butane, 2,3-dichloro-2-methyl-

Temperature (^oC)	Solubility in H_2O	Solubility of H_2O in	Reference
25	1 g/3500 cc H_2O	–	Wright & Schaffer (1932)

[a] M.wt. = 141.04; b.p. = 138.0^oC; d_4^{15} = 1.0696 g cc^{-1}.

$CCl\equiv C-CCl=CCl-CCl_3^a$

2-Pentene-4-yne, hexachloro-

Temperature (^oC)	Solubility in H_2O	Solubility of H_2O in	Reference
15	0.000199 wt % C_5Cl_6	–	Antropov et al.
18	–	0.000552 wt % H_2O	(1972)
20	0.000288 wt % C_5Cl_6	–	"
22	–	0.000851 wt % H_2O	"
26	0.000323 wt % C_5Cl_6	0.00129	"
40	0.0011 wt % C_5Cl_6	0.0055 wt % H_2O	Simonov et al. (1974)

[a] M.wt. = 272.77.

$CCl_2-CCl=CCl-CCl=CCl$[a]

1,3-Cyclopentadiene, hexachloro-

Temperature (°C)	Solubility in H_2O	Solubility of H_2O in	Reference
20	2.0 ppm vy volume C_5Cl_6	–	Imperial Chemical Industries (1972)

[a]M.wt. = 272.81; m.p. = -9.0°C; b.p. = 239.0°C; d_4^{25} = 1.7019 g cc^{-1}.

$CCl_2=CCl-CCl_2-CCl=CCl_2$[a]
1,4-Pentadiene, octachloro-

Temperature (°C)	Solubility in H_2O	Solubility of H_2O in	Reference
15	0.0000144 wt % C_5Cl_8	–	Antropov et al.
18	–	0.0000177 wt % H_2O	(1972)
20	0.0000202 wt % C_5Cl_8	–	"
22	–	0.0000245 wt % H_2O	"
26	0.000034 wt % C_5Cl_8	0.0000369	"
40	0.0006 wt % C_5Cl_8	0.0001 wt % H_2O	Simonov et al. (1974)

[a]M.wt. = 343.716; m.p. = -78.0°C.

C_6BrClH_4[a]
Benzene, 1-bromo-2-chloro-

Temperature (°C)	Solubility in H_2O	Solubility of H_2O in	Reference
25	0.0006456 mol l^{-1} soln.	–	Yalkowsky et al. (1979)
25	0.0006456	–	Yalkowsky & Valvani (1980)

[a]M.wt. = 191.46; m.p. = -12.3°C; b.p. = 204.0°C at 765 mm Hg; d_4^{25} = 1.6382 g cc^{-1}.

C_6BrClH_4[a]

Benzene, 1-bromo-3-chloro-

Temperature (^{O}C)	Solubility in H_2O	Solubility of H_2O in	Reference
25	0.0006166 mol l^{-1} soln.	-	Yalkowsky et al. (1979)
25	0.0006166	-	Yalkowsky & Valvani (1980)

[a]M.wt. = 191.46; m.p. = -21.5^{O}C; b.p. = 196.0^{O}C; d_4^{20} = 1.6302 g cc^{-1}.

C_6BrClH_4[a]

Benzene, 1-bromo-4-chloro-

Temperature (^{O}C)	Solubility in H_2O	Solubility of H_2O in	Reference
25	0.0002344 mol l^{-1} soln.	-	Yalkowsky et al. (1979)
25	0.0002344	-	Yalkowsky & Valvani (1980)

[a]M.wt. = 191.46; m.p. = 68.0^{O}C; b.p. = 196.0^{O}C at 756 mm Hg; d_4^{71} = 1.576 g cc^{-1}.

C_6BrH_4I[a]

Benzene, 1-bromo-2-iodo-

Temperature (^{O}C)	Solubility in H_2O	Solubility of H_2O in	Reference
25	0.000148 mol l^{-1} soln.	-	Yalkowsky & Valvani (1980)

[a]M.wt. = 282.91; m.p. = 9.5^{O}C; b.p. = 257.0^{O}C at 754 mm Hg; d_4^{25} = 2.2571 g cc^{-1}.

C_6BrH_4I[a]
Benzene, 1-bromo-3-iodo-

Temperature ($^\circ$C)	Solubility in H_2O	Solubility of H_2O in	Reference
25	0.000148 mol l^{-1} soln.	–	Yalkowsky & Valvani (1980)

[a]M.wt. = 282.91; m.p. = -9.3°C; b.p. = 252.0°C at 754 mm Hg.

C_6BrH_4I[a]
Benzene, 1-bromo-4-iodo-

Temperature ($^\circ$C)	Solubility in H_2O	Solubility of H_2O in	Reference
25	0.00002754 mol l^{-1} soln.	–	Yalkowsky & Valvani (1980)
25	0.00002754	–	Yalkowsky et al. (1979)

[a]M.wt. = 282.91; m.p. = 92.0°C; b.p. = 252.0°C at 754 mm Hg; d^{25} = 2.235 g cc^{-1}.

C_6BrH_5[a]
Benzene, bromo-

Temperature ($^\circ$C)	Solubility in H_2O	Solubility of H_2O in	Reference
25	0.041 g/100 cc soln.	–	Andrew & Keefer (1950)
30	0.0446 g/100 g H_2O	–	Seidell (1941)
30	0.446 g/1000 g H_2O	–	Gross & Saylor (1931)
35	2.92 mmol/1000 cc soln.	–	Hine et al. (1963)
25	–	0.00041 cc/cc C_6BrH_5	Jones & Monk
30	–	0.00047	(1963)
35	–	0.00053	"

C_6BrH_5 (Continued)

25	–	0.0424 wt % H_2O	Riddick & Bunger
30	0.0446 wt % C_6BrH_5	–	(1970)
30	0.00028 mol/100 g H_2O	–	Van Arkel & Vles (1936)
25	–	0.0284 wt % H_2O	Wing (1956)
25	–	0.00246 mole fraction	Wing & Johnston (1957)
10.0	0.000002482 mol g^{-1} H_2O	–	Vesala (1973)
15.2	0.000002587	–	"
19.6	0.000002619	–	"
25.0	0.000002913	–	"
30.0	0.000002960	–	"
35.0	0.000003117	–	"
25	0.00000284 mol g^{-1} H_2O	–	Vesala (1974)
25	0.0424 vol % C_6BrH_5	–	Stephen & Stephen
30	0.0446	–	(1963)
25	0.04 wt % C_6BrH_5	0.03 wt % H_2O	Mitchell & Smith
25	–	0.041 cc/100 cc C_6BrH_5	(1977)
30	–	0.047	"
35	–	0.053	"
5	0.0000134 mole fraction	–	Nelson & Smit
25	0.000017	–	(1978)
39	0.000028	–	"
25	0.002291 mol l^{-1} soln.	–	Yalkowsky et al. (1979)
25	0.002291 mol l^{-1} soln.	–	Yalkowsky & Valvani (1980)
30	446 ppm by weight C_6BrH_5	–	Chiou et al. (1977)

[a]M.wt. = 157.02; m.p. = -30.82°C; b.p. = 156.0°C; d_4^{20} = 1.4950 g cc^{-1}.

$C_6Br_2H_4$[a]

Benzene, 1,2-dibromo-

Temperature (°C)	Solubility in H_2O	Solubility of H_2O in	Reference
25	0.0003162 mol 1^{-1} soln.	–	Yalkowsky et al. (1979)
25	0.0003162	–	Yalkowsky & Valvani (1980)

[a]M.wt. = 235.92; m.p. = 7.1°C; b.p. = 225.0°C; d_4^{20} = 1.9873 g cc^{-1}.

$C_6Br_2H_4$[a]

Benzene, 1,3-dibromo-

Temperature (°C)	Solubility in H_2O	Solubility of H_2O in	Reference
35	0.286 mmol/1000 cc soln.	–	Hine et al. (1963)
25	0.0004169 mol 1^{-1} soln.	–	Yalkowsky et al. (1979)
25	0.0004169	–	Yalkowsky & Valvani (1980)

[a]M.wt. = 235.92; m.p. = -7.0°C; b.p. = 218.0°C; d_4^{20} = 1.9523 g cc^{-1}.

$C_6Br_2H_4$[a]

Benzene, 1,4-dibromo-

Temperature (°C)	Solubility in H_2O	Solubility of H_2O in	Reference
25	0.0020 g/100 cc soln.	–	Andrews & Keefer (1950)
35	0.112 mmol/1000 cc soln.	–	Hine et al. (1963)
25	0.00008511 mol 1^{-1} soln.	–	Yalkowsky et al. (1979)
25	0.00008511 mol 1^{-1} soln.	–	Yalkowsky & Valvani (1980)

[a]M.wt. = 235.92; m.p. = 87.33°C; b.p. = 218.5°C; d_4^{17} = 2.261 g cc^{-1}.

$C_6Br_3H_3^a$
Benzene, 1,2,3-tribromo-

Temperature (°C)	Solubility in H_2O	Solubility of H_2O in	Reference
25	0.000009332 mol l^{-1} soln.	–	Yalkowsky et al. (1979)
25	0.000009332 mol l^{-1} soln.	–	Yalkowsky & Valvani (1980)

[a]M.wt. = 314.82; m.p. = 87.8°C; d_4^{20} = 2.658 g cc^{-1}.

$C_6Br_3H_3^a$
Benzene, 1,2,4-tribromo-

Temperature (°C)	Solubility in H_2O	Solubility of H_2O in	Reference
25	0.00003162 mol l^{-1} soln.	–	Yalkowsky et al. (1979)
25	0.00003162	–	Yalkowsky & Valvani (1980)

[a]M.wt. = 314.82; m.p. = 44.5°C; b.p. = 275.0°C.

$C_6Br_3H_3^a$
Benzene, 1,3,5-tribromo-

Temperature (°C)	Solubility in H_2O	Solubility of H_2O in	Reference
25	0.02 g/100 cc H_2O	–	Booth & Everson (1948)
20	0.004 wt % $C_6Br_3H_3$	–	Stephen & Stephen (1963)
25	0.000002512 mol l^{-1} soln.	–	Yalkowsky et al. (1979)
25	0.000002512 mol l^{-1} soln.	–	Yalkowsky & Valvani (1980)

[a]M.wt. = 314.82; m.p. = 122.0°C; b.p. = 271.0°C at 765 mm Hg.

$C_6Br_4H_2^a$
Benzene, 1,2,4,5-tetrabromo-

Temperature (°C)	Solubility in H_2O	Solubility of H_2O in	Reference
25	0.0000001047 mol l^{-1} soln.	–	Yalkowsky et al. (1979)
25	0.0000001047	–	Yalkowsky & Valvani (1980)

[a] M.wt. = 393.72; m.p. = 182.0°C; d^{20} = 3.072 g cc^{-1}.

$C_6ClH_4I^a$
Benzene, 1-chloro-2-iodo-

Temperature (°C)	Solubility in H_2O	Solubility of H_2O in	Reference
25	0.0002884 mol l^{-1} soln.	–	Yalkowsky et al. (1979)
25	0.0002884	–	Yalkowsky & Valvani (1980)

[a] M.wt. = 238.46; m.p. = 0.7°C; b.p. = 234.5°C; d^{25} = 1.9515 g cc^{-1}.

$C_6ClH_4I^a$
Benzene, 1-chloro-3-iodo-

Temperature (°C)	Solubility in H_2O	Solubility of H_2O in	Reference
25	0.0002818 mol l^{-1} soln.	–	Yalkowsky et al. (1979)
25	0.0002818	–	Yalkowsky & Valvani (1980)

[a] M.wt. = 238.46; b.p. = 230.0°C; d_4^{20} = 1.9255 g cc^{-1}.

$C_6ClH_4I^a$

Benzene, 1-chloro-4-iodo-

Temperature (^{o}C)	Solubility in H_2O	Solubility of H_2O in	Reference
25	0.00009332 mol l^{-1} soln.	–	Yalkowsky et al. (1979)
25	0.00009332	–	Yalkowsky & Valvani (1980)

[a]M.wt. = 238.46; m.p. = 57.0^{o}C; b.p. = 227.0^{o}C; d_4^{57} = 1.886 g cc^{-1}.

$C_6ClH_5^a$

Benzene, chloro-

Temperature (^{o}C)	Solubility in H_2O	Solubility of H_2O in	Reference
25	0.05 g/100 cc soln.	–	Andrews & Keefer (1950)
30	0.0488 g/100 g H_2O	–	Seidell (1941)
25	0.02 cc/100 cc H_2O	–	Booth & Everson (1948)
-10	–	Monograph	Chernyshev
⋮	–	"	(1970)
80	–	"	"
21	0.0534 g/100 g H_2O	–	Chey & Calder (1972)
25	0.050 g/100 g H_2O	4.4 g/100 g C_6ClH_5	Dreisbach (1955-1961: Ser. 15)
25	–	0.0025 mol/100 g soln.	Eidinoff (1955)
25	–	0.0290 mol l^{-1} soln.	Goldman (1969)
35.1	–	0.0411	"
45.06	–	0.0535	"
30	0.488 g/1000 g H_2O	–	Gross & Saylor (1931)
25	–	0.00034 cc/cc C_6ClH_5	Jones & Monk
30	–	0.000395	(1963)
35	–	0.00046	"
25	0.11 wt % C_6ClH_5	0.18 wt % H_2O	Othmer et al. (1941)

C_6ClH_5 (Continued)

30	0.0490 wt % C_6ClH_5	–	Kisarov (1962)
40	0.0705	–	"
50	0.0960	–	"
60	0.1100	–	"
70	0.1605	–	"
80	0.1805	–	"
90	0.2500	–	"
17.7	–	0.0275 wt % H_2O	Filippov & Furman
22.9	–	0.0333	(1952)
25.2	–	0.0382	"
28.0	–	0.0470	"
30.4	–	0.0512	"
32.4	–	0.0560	"
35.3	–	0.0621	"
36.6	–	0.0660	"
39.8	–	0.0731	"
42.7	–	0.0825	"
47.9	–	0.0955	"
49.0	–	0.0990	"
20	0.049 g/100 cc H_2O	–	Patty (1962)
25	–	0.04 g/100 g C_6ClH_5	Pittsburgh Plate
30	0.0488 g/100 g H_2O	–	Glass Ind. (1970a)
25	–	0.0327 wt % H_2O	Riddick & Bunger
30	0.0488 wt % C_6ClH_5	–	(1970)
25	0.1 g/100 g H_2O	–	Sconce (1962)
30	0.00043 mol/100 g H_2O	–	Van Arkel & Vles (1936)
25	–	0.0326 wt % H_2O	Wing (1956)
25	–	0.00203 mole fraction	Wing & Johnston (1957)
-27	–	0.005 wt % H_2O	Imperial Chemical
-14	–	0.010	Industries
0	–	0.020	(1964)
1.5	–	0.021	"
3	0.01 wt % C_6ClH_5	–	"
7	0.02	–	"
8.3	–	0.0283 wt % H_2O	"
11	–	0.03	"
12	0.03 wt % C_6ClH_5	–	"

C_6ClH_5 (Continued)

16.4	–	0.034 wt % H_2O	Imperial Chemical
20	0.05 wt % C_6ClH_5	–	Industries
21.3	–	0.04 wt % H_2O	(1964)
25	–	0.042	"
29	–	0.050	"
31.2	–	0.058	"
32	0.07 wt % C_6ClH_5	–	"
33	–	0.060 wt % H_2O	"
36	–	0.070	"
36.5	–	0.070	"
40.5	–	0.080	"
43.5	–	0.090	"
45.0	0.09 wt % C_6ClH_5	–	"
20	0.049 wt % C_6ClH_5	–	Perry (1963)
30	0.049 wt % C_6ClH_5	–	Lax (1967)
25	0.049 wt % C_6ClH_5	–	Marsden (1963)
20	0.049 g/100 g H_2O	–	Lange (1967)
240	Graph	–	Vorozcov &
$\vdots$	"	–	Kobelev (1938)
260	"	–	"
25	0.00000411 mol/g H_2O	–	Vesala (1973, 1974)
30	488 ppm w/v H_2O	–	Imperaial Chemical Industries (1972)
25	0.036 vol % C_6ClH_5	–	Stephen & Stephen
30	0.0488 wt % C_6ClH_5	–	(1963)
25	–	0.030 mol l^{-1} C_6ClH_5	Högfeldt & Bolander (1963)
25	–	0.00297 mole fraction	Kirchnerova & Cave (1976)
25	0.05 wt % C_6ClH_5	0.05 wt % H_2O	Mitchell & Smith
25	–	0.034 cc/100 cc C_6ClH_5	(1977)
30	–	0.040	"
35	–	0.046	"
25	0.0512 wt % C_6ClH_5	–	Nathan (1978)
30	–	0.049 wt % H_2O	"
25	–	0.00141 wt % H_2O	Prosyanov et al.
$\vdots$	–	.	(1974)
88	–	.	"

C_6ClH_5 (Continued)

5	0.0000064 mole fraction	–	Nelson & Smit
25	0.0000171	–	(1978)
35	0.0000428	–	"
45	0.0000641	–	"
25	0.004467 mol l^{-1} soln.	–	Yalkowsky et al. (1979)
25	0.004467	–	Yalkowsky & Valvani (1980)
25	472 g m^{-3} H$_2$O	–	Mackay et al. (1979)
25	0.10 ppb by weight C_6ClH_5	–	Lu & Metcalf (1975)
25	–	0.00297 mole fraction	Kirchnerova (1975)
20	0.05 wt % C_6ClH_5	–	du Pont (1966e)
25	–	0.04 wt % H$_2$O	"
30	448 ppm by weight C_6ClH_5	–	Chiou et al. (1977)
25	471.7 mg l^{-1} soln.	–	Aquan-Yuen et al. (1979)
25	0.049 wt % C_6ClH_5	–	Matthews (1979)
25	471.7 g m^{-3} H$_2$O	–	Afghan & Mackay (1980)
10	0.046 wt % C_6ClH_5	–	Schwarz & Miller
20	0.045	–	(1980)
30	0.050	–	"

[a]M.wt. = 112.56; m.p. = -45.6°C; b.p. = 132.0°C; d_4^{20} = 1.1058 g cc^{-1}.

$CClH_2-CH_2-CH_2-CH_2-CH_2-CH_3$[a]
n-Hexane, 1-chloro-

Temperature (°C)	Solubility in H$_2$O	Solubility of H$_2$O in	Reference
25	1 g/12,000 cc H$_2$O	–	Wright & Schaffer (1932)

[a]M.wt. = 120.62; m.p. = -94.0°C; b.p. = 134.5°C; d_4^{20} = 0.8785 g cc^{-1}.

$C_6Cl_2H_4^a$
Benzene, 1,2-dichloro-

Temperature ($^{\circ}$C)	Solubility in H_2O	Solubility of H_2O in	Reference
25	0.02 cc/100 cc H_2O	–	Booth & Everson (1948)
25	–	2.1 g/100 g $C_6Cl_2H_4$	Dreisbach (1955–1961: Ser. 15)
25	–	0.0241 mol l^{-1} soln.	Goldman (1969)
35.1	–	0.0321	"
45.06	–	0.0439	"
25	–	0.00032 cc/cc $C_6Cl_2H_4$	Jones & Monk
30	–	0.000375	(1963)
35	–	0.000425	"
20	0.134 g/1000 g H_2O	–	Klemenc & Löw
25	0.145	–	(1930)
30	0.171	–	"
35	0.183	–	"
40	0.194	–	"
45	0.203	–	"
55	0.223	–	"
60	0.232	–	"
20	0.0134 g/100 g H_2O	–	Pittsburgh Plate
25	–	0.03 g/100 g $C_6Cl_2H_4$	Glass Ind. (1970b)
25	0.026 wt % $C_6Cl_2H_4$	0.309 wt % H_2O	Riddick & Bunger (1970)
25	0.008 g/100 g H_2O	–	Sconce (1962)
25	–	0.0239 wt % H_2O	Wing (1956)
25	–	0.00193 mole fraction	Wing & Johnston (1957)
25	145 ppm by weight $C_6Cl_2H_4$	–	Gunther et al. (1968)
0	47.36 mg/1000 g H_2O	–	Wauchope (1970)
25	83.05	–	"
50	169.7	–	"
75	289.0	–	"
100	589.9	–	"
25	0.013 wt % $C_6Cl_2H_4$	0.031 wt % H_2O	Marsden (1963)
25	0.0309 wt % $C_6Cl_2H_4$	–	Stephen & Stephen (1963)

$C_6Cl_2H_4$ (Continued)

25	–	0.00271 mole fraction	Kirchnerova & Cave (1976)
25	0.07 wt % $C_6Cl_2H_4$	0.03 wt % H_2O	Mitchell & Smith
25	–	0.032 cc/100 cc $C_6Cl_2H_4$	(1977)
30	–	0.038	"
35	–	0.043	"
25	0.0145 wt % $C_6Cl_2H_4$	–	Nathan (1978)
25	0.0006310 mol 1^{-1} soln.	–	Yalkowsky et al. (1979)
25	0.0006310 mol 1^{-1} soln.	–	Yalkowsky & Valvani (1980)
25	–	0.00271 mole fraction	Kirchnerova (1975)
25	0.01 wt % $C_6Cl_2H_4$	–	du Pont (1966e)
20	0.0134 wt % $C_6Cl_2H_4$	–	Matthews (1979)
10	0.0162 wt % $C_6Cl_2H_4$	–	Schwarz & Miller
20	0.0126	–	(1980)
25	0.00106 mol 1^{-1} soln.	–	Banerlee et al. (1980)

[a]M.wt. = 147.01; m.p. = -17.0°C; b.p. = 180.5°C; d_4^{20} = 1.3048 g cc^{-1}.

$C_6Cl_2H_4{}^a$
Benzene, 1,3-dichloro-

Temperature (^oC)	Solubility in H_2O	Solubility of H_2O in	Reference
20	0.111 g/1000 g H_2O	–	Klemenc & Löw
25	0.123	–	(1930)
30	0.140	–	"
35	0.150	–	"
40	0.167	–	"
45	0.177	–	"
55	0.196	–	"
60	0.201	–	"
20	0.0111 wt % $C_6Cl_2H_4$	–	Riddick & Bunger (1970)
25	0.00000070 mol g^{-1} H_2O	–	Vesala (1974)
25	0.0123 wt % $C_6Cl_2H_4$	–	Nathan (1978)
25	0.0008128 mol l^{-1} soln.	–	Yalkowsky et al. (1979)
25	0.0008128 mol l^{-1} soln.	–	Yalkowsky & Valvani (1980)
23.5	0.0147 wt % $C_6Cl_2H_4$	–	Schwarz (1980)
20	0.011 wt % $C_6Cl_2H_4$	–	Matthews (1979)
10	0.0118 wt % $C_6Cl_2H_4$	–	Schwarz & Miller
20	0.0101	–	(1980)
30	0.0135	–	"
25	0.000908 mol l^{-1} soln.	–	Banerjee et al. (1980)
10.0	0.0000006642 mol g^{-1} H_2O		Vesala (1973)
15.2	0.0000006651	–	"
19.6	0.0000006689	–	"
25.1	0.0000006889	–	"
30.0	0.0000007670	–	"
35.0	0.0000008241	–	"

[a]M.wt. = 147.01; m.p. = -24.7^oC; b.p. = 173.0^oC; d_4^{20} = 1.2884 g cc^{-1}.

$C_6Cl_2H_4$[a]

Benzene, 1,4-dichloro-

Temperature (°C)	Solubility in H_2O	Solubility of H_2O in	Reference
25	0.0076 g/100 cc soln.	–	Andrews & Keefer (1950)
25	0.00791 g/100 g H_2O	–	Seidell (1941)
25	0.05 g/100 cc H_2O	–	Booth & Everson (1948)
30	0.077 g/1000 g H_2O	–	Gross & Saylor (1931)
20	0.0688 g/1000 g H_2O	–	Klemenc & Löw
25	0.0792	–	(1930)
30	0.0933	–	"
35	0.0973	–	"
40	0.1008	–	"
45	0.1219	–	"
55	0.156	–	"
60	0.163	–	"
35	0.010 wt % $C_6Cl_2H_4$	–	Riddick & Bunger
35 (α form)	0.008	–	(1970)
55 (β form)	0.016	–	"
25	0.008 g/100 g H_2O	–	Sconce (1962)
25	0.00000058 mol g^{-1} H_2O	–	Vesala (1974)
25	0.008 g/100 g H_2O	–	Gunther et al. (1968)
30	77 ppm by volume $C_6Cl_2H_4$	–	Imperial Chemical (1972)
30	0.077 wt % $C_6Cl_2H_4$,	–	Stephen & Stephen (1963)
0	47.0 ppm by weight $C_6Cl_2H_4$	–	Wauchope &
22.2	77.5	–	Getzen (1972)
24.6	82.2	–	"
25	83.1	–	"
25.5	84.1	–	"
30	94.6	–	"
34.5	106.8	–	"
38.4	119.3	–	"
47.5	156.0	–	"
50.0	170.0	–	"
20	0.007 wt % $C_6Cl_2H_4$	–	Matthews (1979)

$C_6Cl_2H_4$ (Continued)

50.1	170.0 ppm by weight $C_6Cl_2H_4$	–	Wauchope & Getzen
59.2	210.0	–	(1972)
60.7	216	–	"
65.1	234	–	"
65.2	236	–	"
73.4	279	–	"
75.0	289	–	"
100	589.9	–	"
25	0.0006166 mol 1^{-1} soln.	–	Yalkowsky et al. (1979)
25	0.0006166 mol 1^{-1} soln.	–	Yalkowsky & Valvani (1980)
25	80 mg 1^{-1} H_2O	–	Schwarzenbach et al. (1979)
25	79 ppm by weight $C_6Cl_2H_4$	–	Chiou et al. (1977)
25	87.15 mg 1^{-1} soln.	–	Aquan-Yuen et al. (1979)
25	0.000502 mol 1^{-1} soln.	–	Banerjee et al. (1980)
10	0.00000388 mol/10 g H_2O	–	Vesala (1973)
15.2	0.000004673	–	"
20.0	0.000004850	–	"
25.1	0.000005702	–	"
30.0	0.000006598	–	"
35.0	0.000007540	–	"

[a]M.wt. = 147.01; m.p. = 53.1°C; b.p. = 174.0°C; d_4^{20} = 1.2475 g cc^{-1}.

$C_6Cl_3H_3^a$
Benzene, 1,2,3-trichloro-

Temperature (°C)	Solubility in H_2O	Solubility of H_2O in	Reference
25	0.0001738 mol 1^{-1} soln.	–	Yalkowsky et al. (1979)
25	0.0001738 mol 1^{-1} soln.	–	Yalkowsky & Valvani (1980)
25	16.6 g m^{-3} H_2O	–	Afghan & Mackay (1980)

[a]M.wt. = 181.45; m.p. = 53.5°C; b.p. = 218.5°C; d_4^{40} = 1.4533 g cc^{-1}.

$C_6Cl_3H_3$[a]
Benzene, 1,2,4-trichloro-

Temperature ($^{\circ}$C)	Solubility in H_2O	Solubility of H_2O in	Reference
25	–	0.001123 mol/100 g soln.	Eidinoff (1955)
20	30 mg l^{-1} H_2O	–	Meleshchenko (1960a, b)
25	0.1 g/100 g H_2O	–	Sconce (1962)
25	0.22 g/100 cc H_2O	0.0025 g/100 cc $C_6Cl_3H_3$	Marsden (1963)
25	0.0001905 mol l^{-1} soln.	–	Yalkowsky et al. (1979)
25	0.0001905 mol l^{-1} soln.	–	Yalkowsky & Valvani (1980)

[a]M.wt. = 181.45; m.p. = 16.95°C; b.p. = 213.5°C; d_4^{20} = 1.4542 g cc^{-1}

$C_6Cl_3H_3$[a]
Benzene, 1,3,5-trichloro-

Temperature ($^{\circ}$C)	Solubility in H_2O	Solubility of H_2O in	Reference
25	0.00003631 mol l^{-1} soln.	–	Yalkowsky et al. (1979)
25	0.00003631	–	Yalkowsky & Valvani (1980)

[a]M.wt. = 181.45; m.p. = 63.5°C; b.p. = 208.0°C at 763 mm Hg; d_4^{64} = 1.3865 g cc^{-1}.

$C_6Cl_4H_2$[a]
Benzene, 1,2,3,4-tetrachloro-

Temperature ($^{\circ}$C)	Solubility in H_2O	Solubility of H_2O in	Reference
25	0.00001995 mol l^{-1} soln.	–	Yalkowsky et al. (1979)
25	0.00001995	–	Yalkowsky & Valvani (1980)

[a]M.wt. = 215.90; m.p. = 47.5°C; b.p. = 254.0°C.

$C_6Cl_4H_2^a$

Benzene, 1,2,3,5-tetrachloro-

Temperature ($^{\circ}$C)	Solubility in H_2O	Solubility of H_2O in	Reference
25	0.00001622 mol l^{-1} soln.	–	Yalkowsky et al. (1979)
25	0.00001622	–	Yalkowsky & Valvani (1980)
25	3.57 g m^{-3} H_2O	–	Afghan & Mackay (1980)
25	0.0000186 mol l^{-1} soln.	–	Banerjee et al. (1980)

[a]M.wt. = 215.90; m.p. = 54.5°C; b.p. = 246.0°C.

$C_6Cl_4H_2^a$

Benzene, 1,2,4,5-tetrachloro-

Temperature ($^{\circ}$C)	Solubility in H_2O	Solubility of H_2O in	Reference
25	0.000002754 mol l^{-1} soln.	–	Yalkowsky et al. (1979)
25	0.000002754	–	Yalkowsky & Valvani (1980)

[a]M.wt. = 215.90; m.p. = 140.0°C; b.p. = 244.5°C; d^{22} = 1.858 g cc^{-1}.

$C_6Cl_5H^a$

Benzene, pentachloro-

Temperature ($^{\circ}$C)	Solubility in H_2O	Solubility of H_2O in	Reference
25	0.000002239 mol l^{-1} soln.	–	Yalkowsky et al. (1979)
25	0.000002239	–	Yalkowsky & Valvani (1980)
25	0.00000532 mol l^{-1} soln.	–	Banerjee et al. (1980)

[a]M.wt. = 250.34; m.p. = 86.0°C; b.p. = 277.0°C; $d^{16.5}$ = 1.8342 g cc^{-1}.

$C_6Cl_6^a$

Benzene, hexachloro-

Temperature (°C)	Solubility in H_2O	Solubility of H_2O in	Reference
25	5.0 µg/1000 cc H_2O	–	Weil et al. (1974)
80	0.001 g/100 cc H_2O	–	Sharov (1975)
100	0.004	–	"
120	0.007	–	"
140	0.011	–	"
160	0.0165	–	"
180	0.0250	–	"
200	0.0400	–	"
210	0.0480	–	"
220	0.0575	–	"
230	0.0725	–	"
25	0.00000001738 mol l^{-1} soln.	–	Yalkowsky et al. (1979)
25	0.00000001738	–	Yalkowsky & Valvani (1980)
25	0.000006 ppb by weight C_6Cl_6	–	Lu & Metcalf (1975)
24	100 µg l^{-1} H_2O	–	Hollifield (1979)

[a] M.wt. = 284.79; m.p. = 230.0°C; b.p. = 322°C subl.; $d^{23.6}$ = 1.5691 g cc^{-1}.

$C_6Cl_6H_6^a$
Cyclohexane, α -hexachloro-

Temperature (°C)	Solubility in H_2O	Solubility of H_2O in	Reference
25	1.63 ppm by weight $C_6Cl_6H_6$	–	Kanazawa et al. (1971)
20	10 ppm by weight $C_6Cl_6H_6$	–	Sconce (1962)
20	10 ppm by weight $C_6Cl_6H_6$	–	Slade (1945)
20	10 ppm by weight $C_6Cl_6H_6$	–	Gunther et al. (1968)
20	1.5 ppm by weight $C_6Cl_6H_6$	–	Imperial Chemical
100	68.0	–	Industries (1955)
20	10 ppm by weight $C_6Cl_6H_6$	–	Patty (1962)
25	2000 μg l^{-1} H_2O	–	Weil et al. (1974)
25	1.63 ppm by weight $C_6Cl_6H_6$	–	Brooks (1974)
28 (0.1 μm)	2.03 ppm by weight $C_6Cl_6H_6$	–	Kurihara et al.
28 (0.1 μm)	1.21	–	(1973)
28 (0.05 μm)	1.77	–	"
28 (0.05 μm)	1.48	–	"

[a]M.wt. = 290.83; m.p. = 159.8°C; b.p. = 288.0°C; d_4^{20} = 1.87 g cc^{-1}.

$C_6Cl_6H_6^a$
Cyclohexane, β-hexachloro-

Temperature (°C)	Solubility in H_2O	Solubility of H_2O in	Reference
25	0.70 ppm by weight $C_6Cl_6H_6$	–	Kanazawa et al. (1971)
20	5 ppm by weight $C_6Cl_6H_6$	–	Sconce (1962)
20	5 ppm by weight $C_6Cl_6H_6$	–	Slade (1945)
20	5 ppm by weight $C_6Cl_6H_6$	–	Gunther et al. (1968)
20	0.2 ppm by weight $C_6Cl_6H_6$	–	Imperial Chemical
100	17.0	–	Industries (1955)
20	5.0 ppm by weight $C_6Cl_6H_6$	–	Patty (1962)
25	240 µg l^{-1} H_2O	–	Weil et al. (1974)
25	0.70 ppm by weight $C_6Cl_6H_6$	–	Brooks (1974)
28 (0.1 µm)	0.20 ppm by weight $C_6Cl_6H_6$	–	Kurihara et al.
28 (0.1 µm)	0.13	–	(1973)

[a] M.wt. = 290.83; m.p. = 314.5°C subl.; b.p. = 60.0°C at 0.58 mm Hg; d^{19} = 1.89 g cc^{-1}.

$C_6Cl_6H_6$[a]
Cyclohexane, δ-hexachloro-

Temperature (°C)	Solubility in H_2O	Solubility of H_2O in	Reference
25	21.3 ppm by weight $C_6Cl_6H_6$	–	Kanazawa et al. (1971)
20	10 ppm by weight $C_6Cl_6H_6$	–	Sconce (1962)
20	10 ppm by weight $C_6Cl_6H_6$	–	Slade (1945)
20	10 ppm by weight $C_6Cl_6H_6$	–	Gunther et al. (1968)
20	10 ppm by weight $C_6Cl_6H_6$	–	Patty (1962)
20	4.0 ppm by weight $C_6Cl_6H_6$	–	Imperial Chemical
100	130.0	–	Industries (1955)
25	31,400 µg l^{-1} H_2O	–	Weil et al. (1974)
25	21.3 ppm by weight $C_6Cl_6H_6$	–	Brooks (1974)
28 (0.1 µm)	15.7 ppm by weight $C_6Cl_6H_6$	–	Kurihara et al.
28 (0.1 µm)	10.7	–	(1973)
28 (0.05 µm)	11.6	–	"
28 (0.05 µm)	8.64	–	"

[a]M.wt. = 290.83; m.p. = 141.8°C; b.p. = 60.0°C at 0.34 mm Hg.

$C_6Cl_6H_6^a$

Cyclohexane, γ-hexachloro-

Temperature (°C)	Solubility in H_2O	Solubility of H_2O in	Reference
25	7.90 ppm by weight $C_6Cl_6H_6$	–	Kanazawa et al. (1971)
25	8.7 ppm by weight $C_6Cl_6H_6$	–	Lipke & Kearns (1960)
28 (0.1 μm)	7.40 ppm by weight $C_6Cl_6H_6$	–	Kurihara et al.
28 (0.1 μm)	5.75	–	(1973)
28 (0.05 μm)	6.61	–	"
28 (0.05 μm)	6.24	–	"
20	10.0 ppm by weight $C_6Cl_6H_6$	–	Sconce (1962)
20	10 ppm by weight $C_6Cl_6H_6$	–	Slade (1945)
20	16.0 ppm by weight $C_6Cl_6H_6$	–	Imperial Chemical
100	300.0	–	Industries (1955)
20	10 ppm by weight $C_6Cl_6H_6$	–	Gunther et al.
25	7.3	–	(1968)
35	12.0	–	"
45	14.0	–	"
20	10 ppm by weight $C_6Cl_6H_6$	–	Patty (1962)
20	7 μg/cc H_2O	–	Hancock & Laws (1955)
25	7800 μg l^{-1} H_2O	–	Weil et al. (1974)
25	7.90 ppm by weight $C_6Cl_6H_6$	–	Brooks (1974)
25 (0.05 μm)	0.6 ppm by weight $C_6Cl_6H_6$	–	Biggar et al.
25 (5.0 μm)	6.8	–	(1966)
15 (5.0 μm)	2.15	–	"
35 (5.0 μm)	11.4	–	"
25 (0.04 μm)	0.5 ppm by weight $C_6Cl_6H_6$	–	Robeck et al.
25 (5.0 μm)	6.6	–	(1965)

[a]M.wt. = 290.83; m.p. = 112.8°C; b.p. = 323.4°C.

$C_6Cl_6H_6^a$

Cyclohexane, hexachloro- (α-, β-, δ-, γ-mixture)

Temperature (^oC)	Solubility in H_2O	Solubility of H_2O in	Reference
25	0.02 g/100 cc H_2O	–	Booth & Everson (1948)
20	8.25 mg l^{-1} H_2O	–	Ivanov (1956)

[a]M.wt. = 290.83.

$C_6FH_5^a$
Benzene, fluoro-

Temperature (°C)	Solubility in H_2O	Solubility of H_2O in	Reference
25	0.155 g/100 cc soln.	–	Andrews & Keefer (1950)
30	0.154 g/100 g H_2O	–	Seidell (1941)
30	1.54 g/1000 g H_2O	–	Gross et al. (1933)
25	–	0.0316 wt % H_2O	Riddick & Bunger
30	0.153 wt % C_6FH_5	–	(1970)
30	0.0016 mol/100 g H_2O	–	Van Arkel & Vles (1936)
25	–	0.0308 wt % H_2O	Wing (1956)
25	–	0.00196 mole fraction	Wing & Johnston (1957)
25	–	0.322 g/100 g C_6FH_5	Kirk-Othmer
30	0.154 g/100 g H_2O	–	(1966a)
25	0.0316 vol % C_6FH_5	–	Stephen & Stephen (1963)
25	–	0.032 cc/100 cc C_6FH_5	Mitchell & Smith (1977)
300 (0-3500 bars)	Graph	–	Götze et al.
⋮	"	–	(1975); Jockers
⋮	"	–	& Schneider
360	"	–	(1978)
30	0.154 g/100 g H_2O	–	Jones et al. (1957)
5	0.0000222 mole fraction	–	Nelson & Smit
25	0.0000355	–	(1978)
35	0.000063	–	"
45	0.000096	–	"
25	0.01622 mol l^{-1} soln.	–	Yalkowsky et al. (1979)
25	0.01622 mol l^{-1} soln.	–	Yalkowsky & Valvani (1980)
30	1540 ppm by weight C_6FH_5	–	Chiou et al. (1977)

[a] M.wt. = 96.11; m.p. = -41.2°C; b.p. = 85.1°C; d_4^{20} = 1.0225 g cc^{-1}.

$C_6F_2H_4^a$
Benzene, 1,2-difluoro-

Temperature (°C)	Solubility in H_2O	Solubility of H_2O in	Reference
25	0.010 mol l^{-1} soln.	–	Yalkowsky et al. (1979)
25	0.010 mol l^{-1} soln.	–	Yalkowsky & Valvani (1980)

[a]M.wt. = 114.09; m.p. = -34.0°C; b.p. = 91.5°C at 751 mm Hg; d_4^{18} = 1.1599 g cc^{-1}.

$C_6F_2H_4^a$
Benzene, 1,3-difluoro-

Temperature (°C)	Solubility in H_2O	Solubility of H_2O in	Reference
25	0.010 mol l^{-1} soln.	–	Yalkowsky et al. (1979)
25	0.010	–	Yalkowsky & Valvani (1980)

[a]M.wt. = 114.09; m.p. = -59.0°C; b.p. = 83.0°C; d_4^{18} = 1.1552 g cc^{-1}.

$C_6F_2H_4^a$
Benzene, 1,4-difluoro-

Temperature (°C)	Solubility in H_2O	Solubility of H_2O in	Reference
300 (0-3500 bars)	Graph	–	Jockers &
⋮	"	–	Schneider (1978)
340	"	–	"
25	0.01072 mol l^{-1} soln.	–	Yalkowsky et al. (1979)
25	0.01072	–	Yalkowsky & Valvani (1980)

[a]M.wt. = 114.10; m.p. = -13.0°C; b.p. = 88.5°C; d_4^{20} = 1.1701 g cc^{-1}.

$C_6F_{14}^a$
n-Hexane, perfluoro-

Temperature (°C)	Solubility in H_2O	Solubility of H_2O in	Reference
2	–	6.56×10^{-6} µg/g C_6F_{14}	Shields (1976a,
5	–	7.37	b)
10	–	8.93	"
15	–	10.78	"
20	–	12.97	"
25	–	15.55	"
30	–	18.58	"
35	–	22.12	"
40	–	26.26	"
45	–	31.07	"
50	–	36.65	"
53	–	40.40	"

[a]M.wt. = 338.04; m.p. = -82.26°C; b.p. = 57.15°C; d_4^{25} = 1.6717 g cc^{-1}.

$C_6H_3I_3^a$
Benzene, 1,2,3-triiodo-

Temperature (°C)	Solubility in H_2O	Solubility of H_2O in	Reference
25	0.000000661 mol l^{-1} soln.	–	Yalkowsky & Valvani (1980)

[a]M.wt. = 455.80; m.p. = 116.0°C; b.p. = subl.

$C_6H_3I_3^a$
Benzene, 1,2,4-triiodo-

Temperature (°C)	Solubility in H_2O	Solubility of H_2O in	Reference
25	0.00000118 mol l^{-1} soln.	–	Yalkowsky & Valvani (1980)

[a]M.wt. = 455.80; m.p. = 91.5°C; b.p. = subl.

$C_6H_3I_3$[a]
Benzene, 1,3,5-triiodo-

Temperature (°C)	Solubility in H_2O	Solubility of H_2O in	Reference
25	0.000000155 mol l^{-1} soln.	–	Yalkowsky & Valvani (1980)

[a]M.wt. = 455.80; m.p. = 184.2°C; b.p. = subl.

$C_6H_4I_2$[a]
Benzene, 1,2-diiodo-

Temperature (°C)	Solubility in H_2O	Solubility of H_2O in	Reference
25	0.0000452 mol l^{-1} soln.	–	Andrews & Keefer (1951)
25	0.00005754 mol l^{-1} soln.	–	Yalkowsky et al. (1979)
25	0.00005754	–	Yalkowsky & Valvani (1980)

[a]M.wt. = 329.91; m.p. = 27.0°C; b.p. = 286.0°C; d_4^{20} = 2.54 g cc^{-1}.

$C_6H_4I_2$[a]
Benzene, 1,3-diiodo-

Temperature (°C)	Solubility in H_2O	Solubility of H_2O in	Reference
25	0.0000293 mol l^{-1} soln.	–	Andrews & Keefer (1951)
25	0.00002692 mol l^{-1} soln.	–	Yalkowsky et al. (1979)
25	0.00002692	–	Yalkowsky & Valvani (1980)

[a]M.wt. = 329.91; m.p. = 40.4°C; b.p. = 285.0°C; d_4^{25} = 2.47 g cc^{-1}.

$C_6H_4I_2^a$
Benzene, 1,4-diiodo-

Temperature (^{o}C)	Solubility in H_2O	Solubility of H_2O in	Reference
25	0.00014 g/100 cc soln.	–	Andrews & Keefer (1950)
25	0.000005623 mol l^{-1} soln.	–	Yalkowsky et al. (1979)
25	0.000005623	–	Yalkowsky & Valvani (1980)

[a]M.wt. = 329.91; m.p. = 131.5^{o}C; b.p. = 285.0^{o}C subl.

$C_6H_5I^a$
Benzene, iodo-

Temperature (°C)	Solubility in H_2O	Solubility of H_2O in	Reference
25	0.018 g/100 cc soln.	–	Andrews & Keefer (1950)
30	0.034 g/100 g H_2O	–	Seidell (1941)
30	0.34 g/1000 g H_2O	–	Gross et al. (1933)
25	–	0.00049 cc/cc C_6H_5I	Jones & Monk
30	–	0.00056	(1963)
35	–	0.000635	"
30	0.00017 mol/100 g H_2O	–	Van Arkel & Vles (1936)
25	–	0.0276 cc/100 g C_6H_5I	Wing (1956)
25	–	0.00312 mole fraction	Wing & Johnston (1957)
25	0.00000112 mol/g H_2O	–	Vesala (1974)
25	0.0503 vol % C_6H_5I	–	Stephen & Stephen
30	0.034 wt % C_6H_5I	–	(1963)
25	–	0.049 cc/100 cc C_6H_5I	Mitchell & Smith
30	–	0.056	(1977)
35	–	0.064	"
5	0.00000485 mole fraction	–	Nelson & Smit
25	0.00000839	–	(1978)
35	0.0000127	–	"
45.5	0.0000183	–	"
25	0.001122 mol l^{-1} soln.	–	Yalkowsky et al. (1979)
25	0.001122	–	Yalkowsky & Valvani (1980)
30	340 ppm by weight C_6H_5I	–	Chiou et al. (1977)
10.0	0.0000009574 mol/g H_2O		Vesala (1973)
15.2	0.0000009636	–	"
19.6	0.000001061	–	"
25.1	0.000001120	–	"
30.0	0.000001158	–	"
35.0	0.000001231	–	"

[a] M.wt. = 204.01; m.p. = -31.27°C; b.p. = 188.3°C; d_4^{20} = 1.8308 g cc^{-1}.

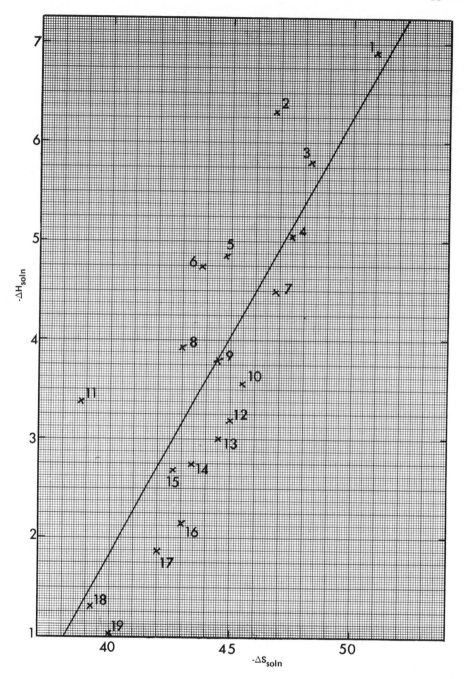

Fig. A.1. Correlation of heat of solution with entropy of solution for gases in water at 25°C. (From Butler, 1937.) ΔH_{soln} = heat of solution (kcal g-mole^{-1}); ΔS_{soln} = entropy of solution (cal g-mole^{-1} K^{-1}). The correlated equation was: $-\Delta H_{soln}$ = -15.7793 + 0.4407($-\Delta S_{soln}$). 1 = CCl_4; 2 = $CHCl_3$; 3 = COS; 4 = Rn; 5 = N_2O; 6 = CO_2; 7 = Xe; 8 = CO; 9 = C_2H_4; 10 = Kr; 11 = C_2H_2; 12 = CH_4; 13 = O_2; 14 = Ar; 15 = NO; 16 = N_2; 17 = Ne; 18 = H_2; 19 = He.

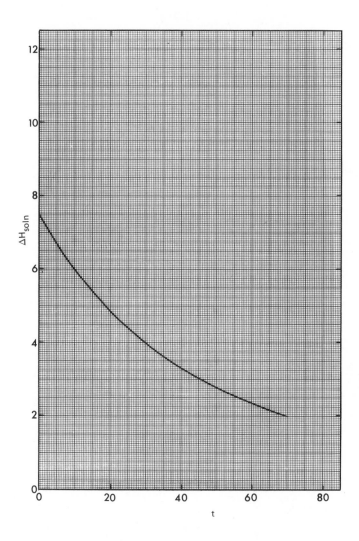

Fig. A.2. Heat of solution of $CF_2=CF_2$ in H_2O. ΔH_{soln} = heat of solution (kcal g-mole^{-1}); t = temperature ($^{\circ}C$). (Source: Volokhonovich et al., 1966.)

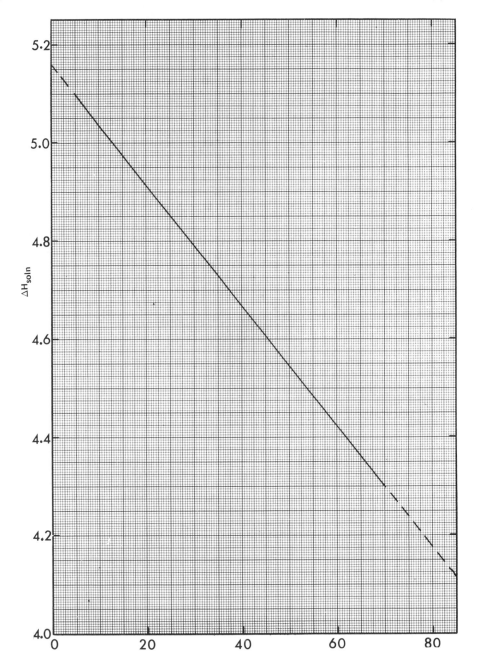

Fig. A.3. Heat of solution of hexafluoropropene in water. ΔH_{soln} = heat of solution
(kcal g-mole^{-1}); t = temperature ($^{\circ}$C). (Source: Nosov & Barlyaev, 1968.)

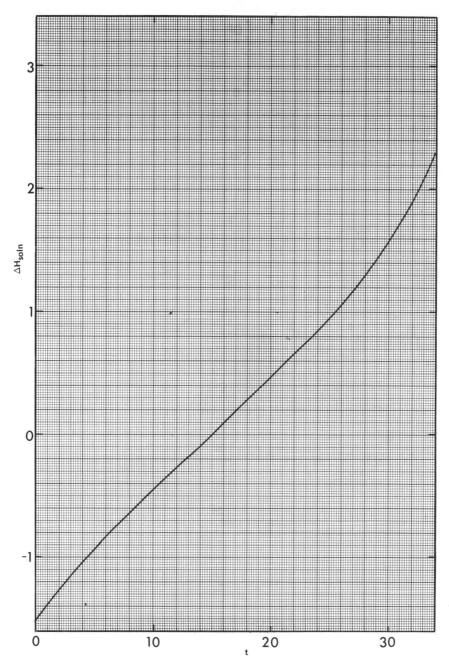

Fig. A.4. Heat of solution of chlorobenzene in water. ΔH_{soln} = heat of solution (kcal g-mole^{-1}); t = temperature (^{o}C). (Source: Katayama et al., 1966.)

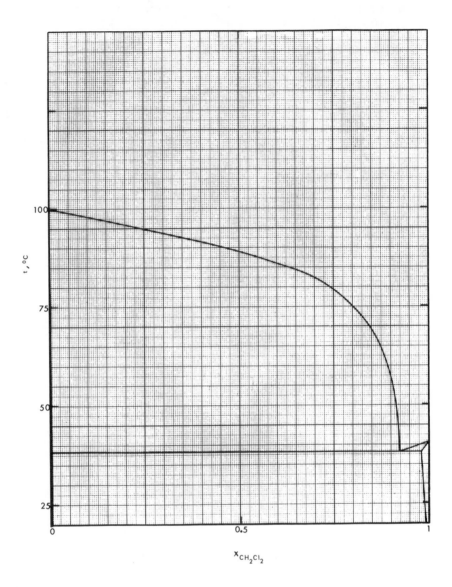

Fig. A.5. Vapor-Liquid equilibrium in water - dichloromethane system at 1 bar pressure. (From Maretic and Sirocic, 1962.)

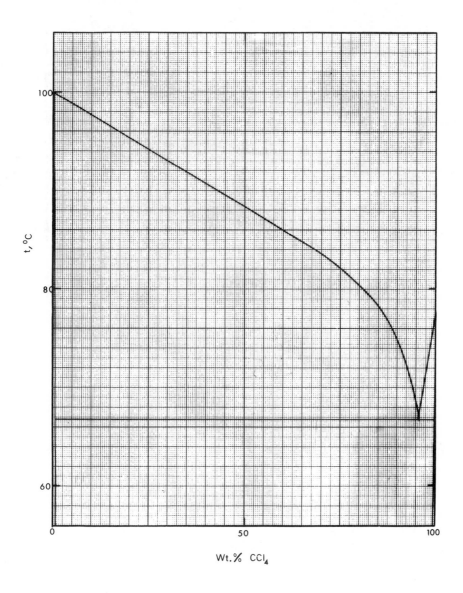

Fig. A.6. Vapor–Liquid equilibrium in H_2O – CCl_4 system at 1 atm pressure.

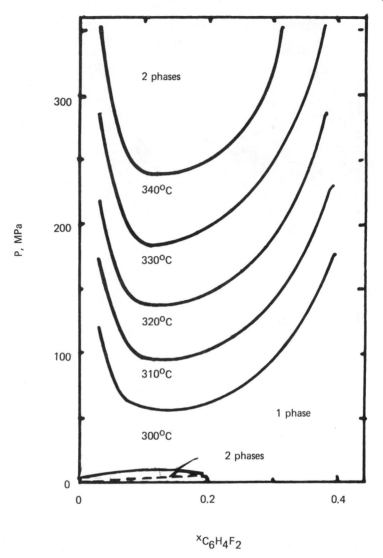

Fig. A.7. P(x) isotherms of $H_2O - 1,4-C_6H_4F_2$ system.
(From Jockers and Schneider, 1978.)

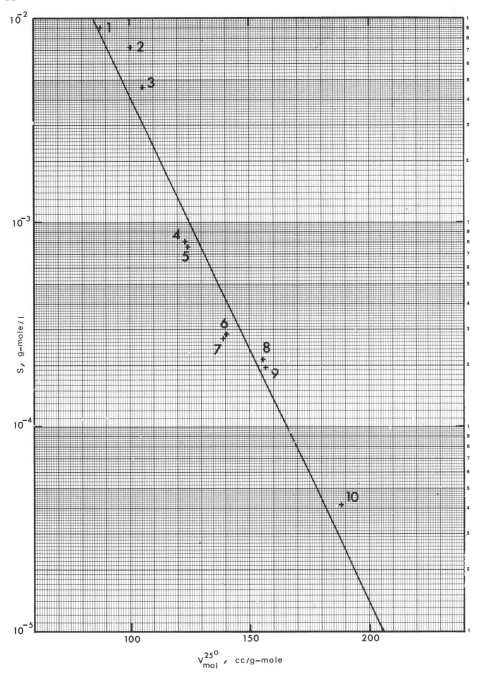

Fig. A.8. Correlation of solubility in water with molar volume of solute at 25°C.
1 = Cyclopentane; 2 = cyclohexane; 3 = 2-methyl-2-butene; 4 = 2-hexene; 5 = 1-hexene;
6 = 2-heptene; 7 = 1-heptene; 8 = 2-octene; 9 = 1-octene; 10 = 1-docene. The correlated
equation was: $\log_{10} S$ (g-mole liter^{-1}) = 0.132454 − 0.0248044 $V_{mol}^{25°}$ (cc g-mole^{-1}).
(Source: Natarajan & Venkatachalam, 1972.)

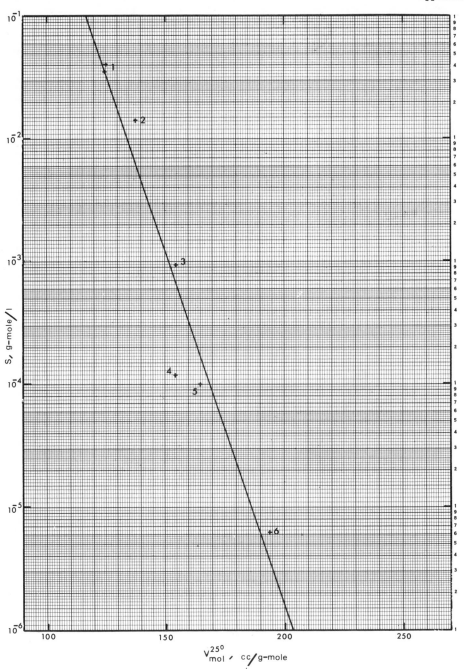

Fig. A.9. Correlation of solubility in water with molar volume of solute at 25°C.
1 = 2,2'-Bipyridyl (two measurements); 2 = 1,10-phenanthroline; 3 = 5-bromo-1,10-
phenanthroline; 4 = 5-nitro-1,10-phenanthroline; 5 = 4,7-dimethyl-1,10-phenanthroline;
6 = 3,4,7,8-tetramethyl-1,10-phenanthroline. The correlated equation was: $\log_{10} S$
(g-mole liter^{-1}) = 5.75668 − 0.057814 $V_{mol}^{25°}$ (cc g-mole^{-1}). (Source: Burgess & Haines,
1978.)

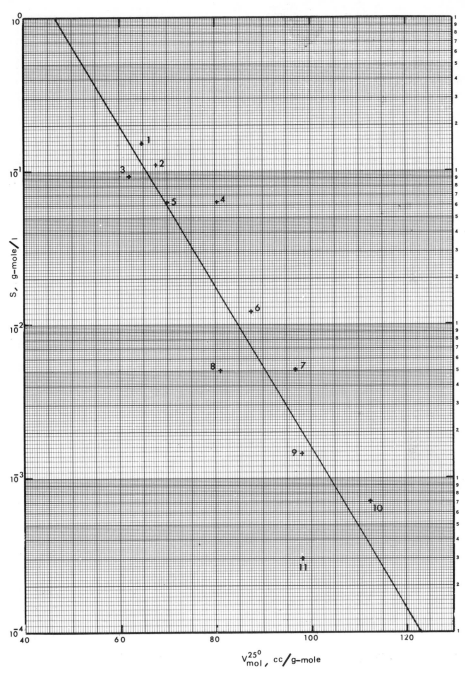

Fig. A.10. Correlation of the solubility of halogenated methanes in water with molar volume of solutes at 25°C. 1 = CCl_2H_2; 2 = $CBrClH_2$; 3 = CH_3I; 4 = CCl_3H; 5 = CBr_2H_2; 6 = CBr_3H; 7 = CCl_4; 8 = CH_2I_2; 9 = CBr_3F; 10 = CBr_4; 11 = CHI_3. The correlated equation was: $\log_{10} S$ (g-mole liter^{-1}) = 2.48078 - 0.0528392 $V_{mol}^{25°}$ (cc g-mole^{-1}).

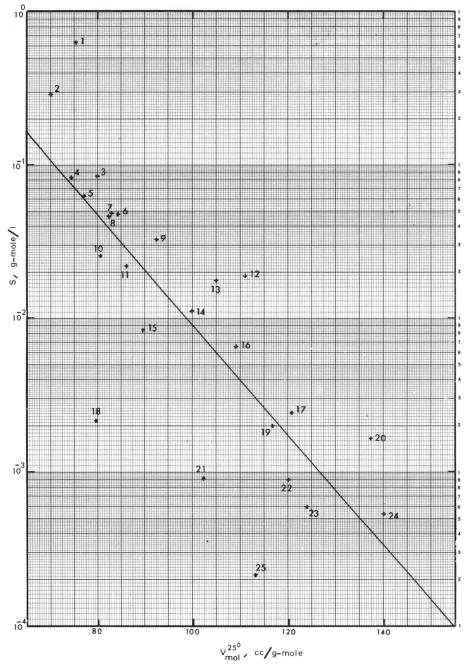

Fig. A.11. Correlation of solubility of halogenated ethanes in water with molar volume of solutes at 25°C. 1 = cis-CClH=CClH; 2 = CClH$_2$CFH$_2$; 3 = CClH$_2$CClH$_2$; 4 = CBrH$_2$CH$_3$; 5 = trans-CClH=CClH; 6 = CCl$_2$HCH$_3$; 7 = CBrH=CBrH; 8 = CBrH$_2$CClH$_2$; 9 = CCl$_2$HCClH$_2$; 10 = CH$_3$CH$_2$I; 11 = CBrH$_2$CBrH$_2$; 12 = CBrClHCF$_3$; 13 = CCl$_2$HCCl$_2$H; 14 = CClH=CCl$_2$; 16 = CCl$_3$CClH$_2$; 17 = CCl$_3$CCl$_2$H; 18 = CCl$_2$=CH$_2$; 19 = CBr$_2$HCBr$_2$H; 20 = cis-CHI=CHI; 21 = CCl$_2$=CCl$_2$; 22 = CCl$_2$FCClF$_2$; 23 = CCl$_2$FCCl$_2$F; 24 = trans-CHI=CHI; 25 = CCl$_3$CCl$_3$. The correlated equation was: $\log_{10}$ S (g-mole liter^{-1}) = 1.55484 - 0.0359923 V$_{mol}^{25°}$ (cc g-mole^{-1}).

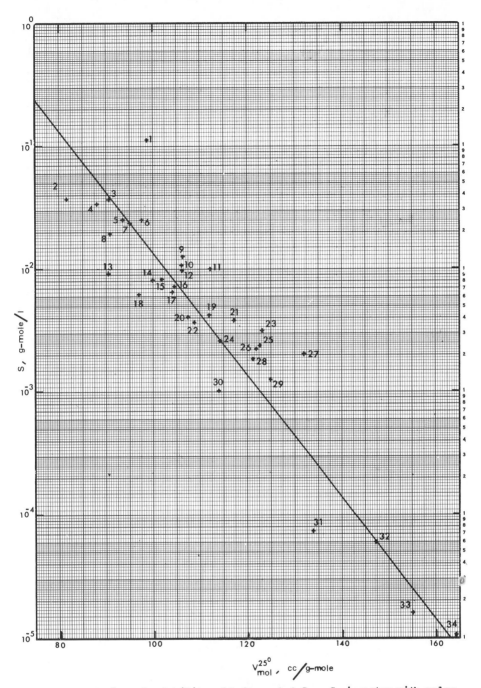

Fig. A.12. Correlation of solubility of halogenated C_3 - C_5 in water with molar volume of solutes at 25°C. 1 = $CBrH_2CH_2CClH_2$; 2 = $CClH_2CH=CH_2$; 3 = $CH_3CClHCH_3$; 4 = $CClH_2CH_2CH_3$; 5 = $CH_3CBrHCH_3$; 6 = $CClH_2CClHCH_3$; 7 = $CClH_2CH_2CClH_2$; 8 = $CBrH_2CH_2CH_3$; 9 = $CClH_2CClHCClH_2$; 10 = $CH_3CClHCH_2CH_3$; 11 = $CClH_2CH_2CF_3$; 12 = $CClH_2CH(CH_3)CH_3$; 13 = $CClH=CHClH_2$; 14 = CH_3CHICH_3; 15 = $CBrH_2CH_2CBrH_2$; 16 = $CBrH_2CBrHCH_3$; 17 = $CClH_2CH_2CH_2CH_3$; 18 = $CH_3CH_2CH_2I$; 19 = $CBrH_2CBrHCClH_2$; 20 = $CBrH_2CH_2CH_2CH_3$; 21 = $CCl_2HCH_2CH_2CH_3$; 22 = $CBrH_2CH(CH_3)CH_3$; 23 = $CH_3CCl(CH_3)CH_2CH_3$; 24 = $CH_3CClHCClHCH_3$; 25 = $CH_3CClHCH_2CH_2CH_3$; 26 = $CH_3CH_2CClHCH_2CH_3$; 27 = $CH_3CCl(CH_3)CClHCH_3$; 28 = $CClH_2CH_2CH_2CH_2CH_3$; 29 = $CBrH_2CH_2CH(CH_3)CH_3$; 30 = $CH_2ICH_2CH_2CH_3$; 31 = $CCl_3CCl=CCl_2$; 32 = $CCl_2=C=CClCCl_2H$; 33 = $CCl_2=CCl=CCl=CCl_2$; 34 = $CCl\equiv CCCl=CClCCl_3$; 35 = $CCl_2CCl=CClCCl=CCl$ (out of scale). The correlated equation was:
$$\log_{10} S \text{ (g-mole liter}^{-1}) = 3.07427 - 0.0494859 \, V_{mol}^{25°} \text{ (cc g-mole}^{-1}).$$

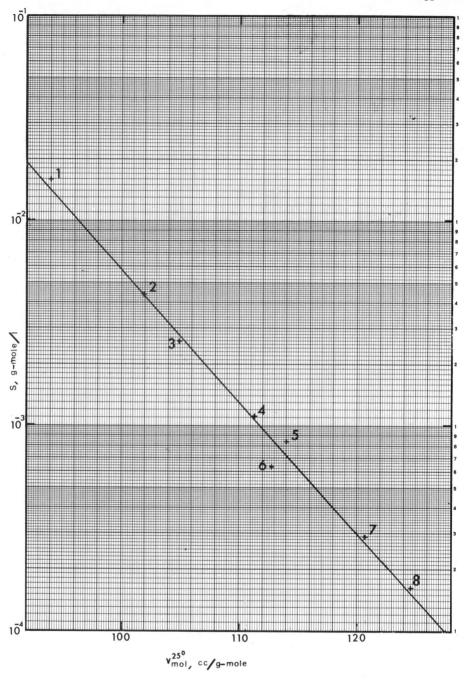

Fig. A.13. Correlation of solubility of halogenated benzenes in water with molar volume of solutes at 25°C. 1 = C_6FH_5; 2 = C_6ClH_5; 3 = C_6BrH_5; 4 = C_6H_5I; 5 = 1,3-$C_6Cl_2H_4$; 6 = 1,2-$C_6Cl_2H_4$; 7 = 1,3-$C_6Br_2H_4$; 8 = 1,2,4-$C_6Cl_3H_3$. The correlated equation was: $\log_{10} S$ (g-mole liter^{-1}) = 4.17442 - 6.40668 × 10^{-2} $V_{mol}^{25°}$ (cc g-mole^{-1}).

References

Abbound, J-L. M. and Taft, R. W. (1979). J. Phys. Chem. $\underline{83}$(3), 412-419.

Abraham, M. H. (1979). J. Am. Chem. Soc. $\underline{101}$(19), 5477-5484.

Abrahamson, A. A. (1963). Phys. Rev. $\underline{130}$(2), 693-707.

Abrams, D. S. and Prausnitz, J. M. (1975). AIChE J. $\underline{21}$(1), 116-128.

Ackermann, T. (1964). Z. Phys. Chem. Neue Folge $\underline{42}$(1/2), 119-122.

Ackroyd, K. (1978). Chem. Eng. (London) No. 332, 366-370.

Adams, D. J. and Matheson, A. J. (1972). J. Chem. Soc. Faraday Trans. II $\underline{68}$, 1536-1546.

Adams, W. A., Greer, G., Desnoyers, J. E., Atkinson, G., Kell, G. S., Oldham, K. B., and Walkley, J., Eds. (1975). Chemistry and Physics of Aqueous Gas Solutions. Symposium, Toronto. Electrochem. Soc., Princeton, N. J. 521 pp.

Afghan, B. K. and Mackay, D., Eds. (1980). Hydrocarbons and Halogenated Hydrocarbons in the Aquatic Environment. Environ. Sci. Res. Vol. 16. Plenum, New York. 602 pp.

Afshar, R. and Saxena, S. C. (1980). Internat. J. Thermophys. $\underline{1}$(1), 51-59.

Ageno, M. (1967). Proc. Natl. Acad. Sci. U. S. A. $\underline{57}$, 567-572.

Ahmad, H. and Yaseen, M. (1974). J. Colour Soc. $\underline{13}$(3), 7-11.

Ahmad, H. and Yaseen, M. (1977). J. Oil. Colour Chem. Assoc. $\underline{60}$(3), 99-103.

Ahmad, H. and Yaseen, M. (1978). Paint Manuf. $\underline{48}$(5), 28, 30, 32-34.

Ahmad, H. and Yaseen, M. (1979). Polym. Eng. Sci. $\underline{19}$(12), 858-863.

Ahmad, H. and Yaseen, M. (1980). Paint India $\underline{30}$(4), 3-6.

Ahmed, W. (1979). J. Chem. Educ. $\underline{56}$(12), 795-798.

Albert, A. (1973). Selective Toxicity. The Physico-chemical Basis of Therapy (5th ed.). Chapman & Hall, London. 597 pp.

Alder, B. J. and Wainwright, T. E. (1957). J. Chem. Phys. $\underline{27}$(5), 1208-1209.

Alder, B. J. and Wainwright, T. E. (1959). Sci. Am. $\underline{201}$(4), 113-126.

Alexander, D. M. (1959). J. Chem. Eng. Data $\underline{4}$(3), 252-254.

Alexander, D. M. and Hill, D. J. T. (1969). Aust. J. Chem. $\underline{22}$, 347-356.

Alexander, D. M., Hill, D. J. T., and White, L. R. (1971). Aust. J. Chem. $\underline{24}$, 1143-1155.

Alexander, R., Ko, E. C. F., Parker, A. J., and Broxton, T. J. (1968). J. Am. Chem. Soc. $\underline{90}$(19), 5049-5069.

Alexejew, W. (1886). Ann. Phys. Chem. $\underline{28}$, 305-338.

Allen, L. (1973). New Sci. $\underline{59}$(859), 376-380.

Allied Chemical Corporation (1960a). Genetron 142b, CH_3-$CClF_2$. Gen. Chem. Div. Tech. Data Bull. 142-7-60. New York. 10 pp.

Allied Chemical Corporation (1960b). Genetron 152a, CH_3-CHF_2. Uses and Properties. Gen. Chem. Div. Tech. Data Bull. 152a-7-60. New York. 9 pp.

Allied Chemical Corporation (1963a). Genetron 23, CHF_3. Gen. Chem. Div. Prod. Data Sheet PD-G23563. New York. 2 pp.

Allied Chemical Corporation (1963b). Genetron 32, CH_2F_2. Gen. Chem. Div. Prod. Data Sheet PD-G32-763. New York. 9 pp.

Allied Chemical Corporation (1966a). Chloromethanes. Tech. Eng. Sales Bull. No. 21. New York.

Allied Chemical Corporation (1966b). Genetron 500. Ind. Chem. Div., Morristown, N.J. 11 pp.

Allied Chemical Corporation (1974a). Genetron Aerosol Propellants. Spec. Chem. Div. GAPB-1. Morristown, N.J. 12 pp.

Allied Chemical Corporation (1974b). Genetron Super-dry Refrigerants. Spec. Chem. Div., Morristown, N.J. 5 pp.

Allott, P. R., Steward, A., Flook, V., and Mapleson, W. W. (1973). Br. J. Anaesth. $\underline{45}$, 294-300.

Almgren, M., Grieser, F., Powell, J. R., and Thomas, J. K. (1979). J. Chem. Eng. Data $\underline{24}$(4), 285-287.

Ambrose, D. (1977). Pure Appl. Chem. $\underline{49}$(8), 1437-1464.

American Chemical Society Committee on Environmental Improvement (1980). Anal. Chem. $\underline{52}$(14), 2242-2249.

American Institute of Chemical Engineers (1979). Nonideal Vapor-Liquid and Liquid-Liquid Equilibria in Theory and in the Real Word. Parts 1 to 3. 87th Nat. Meeting. Boston, Mass. (Aug. 19-22) Sessions 1, 12, and 32.

American Mutual Insurance Alliance (1972). Handbook of Organic Industrial Solvents (4th ed.). Tech. Guide No. 6. Chicago, Ill. 75 pp.

American Petroleum Institute (1970). Technical Data Book -- Petroleum Refining (2nd ed.). American Petroleum Institute, Washington, D.C. 14 chaps.

Amidon, G. L. (1974). J. Pharm. Sci. $\underline{63}$(10), 1520-1523.

Amidon, G. L. (1976). Personal communication, April 2.

Amidon, G. L. and Anik, S. T. (1976). J. Pharm. Sci. $\underline{65}$(6), 801-806.

Amidon, G. L. and Anik, S. T. (1980). J. Pharm. Sci. $\underline{84}$(9), 970-974.

Amidon, G. L., Yalkowsky, S. H., and Leung, S. (1974). J. Pharm. Sci. $\underline{63}$(12), 1858-1866.

Amidon, G. L., Yalkowsky, S. H., Anik, S. T., and Valvani, S. C. (1975). J. Phys. Chem. $\underline{79}$(21), 2239-2246.

Amis, E. S. and Hinton, J. F. (1973). Solvent Effects on Chemical Phenomena. Academic, New York. 474 pp.

Amos, A. T. and Yoffe, J. A. (1976). Theor. Chim. Acta 42(3), 247-260.

Anantaraman, A. V. and Goldman, S. (1980). Can. J. Chem. 58(12), 1183-1187.

Andelman, J. B. (1978). Chloroform, Carbon Tetrachloride, and Other Halomethanes: An Environmental Assessment. National Academy of Sciences, Washington. 294 pp.

Andersen, H. C. (1975). Annu. Rev. Phys. Chem. 26, 145-167.

Anderson, E. V. (1975). Chem. Eng. News 53(33), 8-9.

Andon, R. J. L. and Cox, J. D. (1952). J. Chem. Soc. 4601-4606.

Andrews, L. J. and Keefer, R. M. (1950). J. Am. Chem. Soc. 72, 3113-3116.

Andrews, L. J. and Keefer, R. M. (1951). J. Am. Chem. Soc. 73, 5733-5736.

Angell, C. A. and Tucker, J. C. (1973). Science 181(4097), 342-344.

Anik, S. T. (1978). A Thermodynamic Analysis of the Aqueous Solubility and Hydrophobicity of Hydrocarbons Using Molecular Surface Area. Ph. D. thesis. Univ. of Wisconsin. Madison. 140 pp. (Univ. Microfilm No. 78-14,251).

Anonymous (1953). Disc. Farady Soc. No. 15. 292 pp.

Anonymous (1969). Tetrachloroethane. Peint. Pigm. Vernis 45(143), 824-825.

Anonymous (1973). Book of ASTM Standards, Part 18. Am. Soc. Testing and Materials, Philadelphia, pp. 756-763.

Anonymous (1974). Chem. Ind. No. 4, 123.

Anonymous (1975a). Chem. Eng. 82(17), 41.

Anonymous (1975b). Chem. Eng. News 53(16), 21-25.

Anonymous (1975c). Chem. Eng. News 53(17), 18-19.

Anonymous (1975d). Chem. Eng. News 53(25), 5-6.

Anonymous (1975e). Chem. Eng. News 53(31), 7-8.

Anonymous (1975f). Chem. Eng. News 53(35), 15-16.

Anonymous (1975g). Chem. Eng. News 53(51), 5.

Anonymous (1975h). Fluorocarbons and the Environment. Counc. Environ. Qual. and Fed. Counc. for Sci. and Technol., Washington, D.C., June. 109 pp.

Anonymous (1975i). J. Chem. Educ. 52(6), 413.

Anonymous (1975j). Water Quality Parameters. ASTM Spec. Tech. Publ. 573. Am. Soc. Testing and Materials, Philadelphia. 580 pp.

Anonymous (1976a). Chem. Eng. News 54(10), 6.

Anonymous (1976b). New Sci. 70(995), 61.

Anonymous (1976c). Chem. Eng. News 54(14), 4.

Anonymous (1976d). Chem. Eng. News 54(16), 7.

Anonymous (1976e). Chem. Eng. News 54(22), 4.

Anonymous (1976f). Halocarbons: Environmental Effects. Natl. Acad. Sci., Washington, D.C.

Anonymous (1976g). J. Water Pollut. Control Fed. 48(1), 188-197.

Anonymous (1976h). Mar. Pollut. Bull. 7(3), 400.

Anonymous (1976i). Sci. Am. 234(4), 57-58.

Anonymous (1976j). Water Pollution Research 1973. Dept. of Environ. HMS Office, London.

Anonymous (1977a). Chem. Eng. $\underline{84}$(12), 75.

Anonymous (1977b). Chem. Eng. News $\underline{55}$(5), 5-6.

Anonymous (1977c). Chem. Eng. News $\underline{55}$(20), 4.

Anonymous (1977d). Chem. Eng. News $\underline{55}$(43), 14.

Anonymous (1977e). J. Am. Water Works Assoc. $\underline{69}$(4), 67.

Anonymous (1977f). Science $\underline{196}$(4290), 632-636.

Anonymous (1978a). J. Chem. Educ. $\underline{55}$(7), 467.

Anonymous (1978b). Chem. Eng. News $\underline{56}$(46), 17.

Anonymous (1978c). Chem. Eng. News $\underline{56}$(46), 17.

Anonymous (1981). Nature $\underline{289}$(5797), 431-432.

Antonov, G. N. (1907). J. Chim. Phys. $\underline{5}$, 372-385.

Antropov, L. I., Pogulyai, V. E., Simonov, V. D., and Shamsutdinov, T. M. (1972). Russ. J. Phys. Chem. $\underline{46}$(2), 311-312. (VINITI No. 3739-71.)

Aquan-Yuen, M., Mackay, D., and Shiu, W. Y. (1979). J. Chem. Eng. Data $\underline{24}$(1), 30-34.

Arakawa, K. (1974). Kagaku No Ryoiki Zokan $\underline{106}$, 15-29. (CA 82:129362s.)

Arakawa, K., Tokiwano, K., and Kojima, K. (1977). Bull. Chem. Soc. Japan $\underline{50}$(1), 65.

Archer, G. and Hildebrand, J. H. (1963). J. Phys. Chem. $\underline{67}$, 1830-1833.

Archer, W. L. and Stevens, V. L. (1977). Ind. Eng. Chem. Prod. Res. Dev. $\underline{16}$(4), 319.

Aref'eva, R. P., Korenman, I. M., and Gorokhov, A. A. (1979). Determination of the Solubility of Liquid and Solid Organic Substances in Water, Using Diphenylthiocarbazone. USSR Patent 672548 (July 5). 3 pp. (CA 91:113256k.)

Arich, G., Kikic, I., and Alessi, P. (1975). Chem. Eng. Sci. $\underline{30}$(2), 187-191.

Armitage, J. W. and Gray, P. (1963). J. Chem. Soc. 1807-1813.

Armitage, J. W., Gray, P., and Wright, P. G. (1963). J. Chem. Soc. 1796-1806.

Arnett, E. M., Kover, W. B., and Carter, J. V. (1969). J. Am. Chem. Soc. $\underline{91}$(15), 4028-4034.

Arnold, D. S., Plank, C. A., Erickson, E. E., and Pike, F. P. (1958). J. Chem. Eng. Data $\underline{3}$(2), 253-256.

ASHRAE (1959). Refrigerant Tables, Charts, and Characteristics. Am. Soc. Heat. Refrig. Air-Cond. Eng. Co., New York. (Revised.)

ASHRAE Tables (1969). ASHRAE Thermodynamic Properties of Refrigerants (2nd ed.). Am. Heat. Refrig. Air-Cond. Eng. Co., New York. 329 pp.

Ashton, J. T., Dawe, R. A., Miller, K. W., Smith, E. B., and Stickings, B. J. (1968). J. Chem. Soc., No. 8, 1793-1796.

Ashwell, G. E., Eggett, P. A., Emery, R., and Gobbie, H. A. (1974). Nature $\underline{247}$(5438), 196-197.

Asinger, F. (1968). Paraffins. Chemistry and Technology. Pergamon, Oxford. 896 pp.

Atack, D. and Schneider, W. G. (1950). J. Phys. Chem. $\underline{54}$(9), 1323-1336.

Atoji, M. (1956). J. Chem. Phys. $\underline{25}$, 174.

Aviado, D. M., Zakhari, S., Simaan, J. A., and Ulsamer, A. G. (1976). Methyl Chloroform and Trichloroethylenen in the Environment. Chemical Rubber Co., Cleveland,
   Ohio. 102 pp.

Aziz, R. A. (1974). AIChE J. $\underline{20}$(4), 817-818.

Baba, K. and Kamiyoshi, K. (1977). J. Phys. Chem. $\underline{81}$(19), 1872-1876.

Babb, S. E. and Christian, S. D. (1977). J. Chem. Phys. $\underline{66}$(10), 4713-4714.

Bacquies, G. (1971). Galvano $\underline{40}$(413-414), 677-681, 767-770.

Badyl'kes, I. S. (1962). Working Fluids in Process Refrigeration Equipment. Gostor-
   gizdat, Moscow. 280 pp.

Badyl'kes, I. S. (1974). Properties of Refrigerants. Pichevaya Promychlennosti,
   Moscow. 175 pp.

Bae, J. H. and Reed, T. M. (1967). Ind. Eng. Chem. $\underline{6}$(1), 67-72.

Bae, J. H. and Reed, T. M. (1971). Ind. Eng. Chem. Fundam. $\underline{10}$(2), 269-272.

Baker, E. G. (1959). Science $\underline{129}$(3353), 871-874.

Baker, W. O. and Smyth, C. P. (1939a). J. Am. Chem. Soc. $\underline{61}$, 2063-2071.

Baker, W. O. and Smyth, C. P. (1939b). J. Am. Chem. Soc. $\underline{61}$, 2798-2805.

Bakin, V. M. (1971). Russ. J. Phys. Chem. $\underline{45}$(7), 1068-1069.

Bakowski, S. and Treszczanowicz, E. (1937). Przem. Chem. $\underline{21}$, 204-208.

Bakowski, S. and Treszczanowicz, E. (1938). Przem. Chem. $\underline{22}$, 211-227.

Baldwin, R. R. and Daniel, S. G. (1952). J. Appl. Chem. $\underline{2}$, 161-165.

Ball, T. M. D. (1977). Solvents -- The Neglected Parameters. 2nd Solvents Symp.
   UMIST. Manchester.

Banerjee, S., Yalkowsky, S. H., and Valvani, S. C. (1980). Envir. Sci. Technol. $\underline{14}$
   (10), 1227-1229.

Banks, R. E., Ed. (1979). Organofluorine Chemicals and Their Industrial Applications.
   Ellis Horwood, Chichester. 255 pp.

Banks, W. P., Heston, B. O., and Blankenship, F. F. (1954). J. Phys. Chem. $\underline{58}$, 962-
   965.

Baranaev, M. K., Gilman, I. S., Kogan, L. M., and Rodinova, N. P. (1954). J. Appl.
   Chem. USSR $\underline{27}$(10), 1031-1036.

Barboi, V. M., Garber, Yu. I., and Shashkov, Yu. I. (1971). Theor. Osn. Khim. Tekhnol.
   $\underline{5}$(2), 308-312.

Barclay, I. M. and Butler, J. A. V. (1938). Trans. Faraday Soc. $\underline{34}$, 1445-1454.

Barduhn, A. J., Towlson, H. E., and Hu, Y-C. (1960). The Properties of Gas Hydrates
   and Their Use in Demineralizing Sea Water. Res. Dev. Prog. Rep. No. 44. Office of
   Saline Water, Syracuse Univ., Syracuse, N.Y., September. 69 pp.

Barduhn, A. J., Towlson, H. E., and Hue, Y-C. (1962). AIChE J. $\underline{8}$(2), 176-183.

Barduhn, A. J., Roux, G. M., Richard, H. A., Giuliano, G. B., and Stern, S. A. (1976).
   5th In. Symp. Fresh Water from the Sea, Vol. 3, p. 253-261.

Barfield, M. and Johnston, M. D. (1973). Chem. Rev. $\underline{73}$(1), 53-73.

Barker, J. A. (1963a). Annu. Rev. Phys. $\underline{14}$, 229-250.

Barker, J. A. (1963b). Lattice Theories of the Liquid State. Pergamon, Oxford. 133 pp.

Barker, J. A. and Fock, W. (1953). Disc. Faraday Soc. $\underline{15}$, 188-195.

Barker, J. A. and Henderson, D. (1967). J. Chem. Phys. $\underline{47}$(8), 2856-2861, (11), 4714-4721.

Barker, W. and Mossman, A. L. (1971). Matheson Gas Data Book (5th ed.). Matheson Co., East Rutherford, N.J. 574 pp.

Barton, A. F. M. (1974). The Dynamic Liquid State. Longman, London. 159 pp.

Barton, A. F. M. (1975). Chem. Rev. $\underline{75}$(6), 731-753.

Bathe, P. (1976). Manuf. Chem. Aerosol News $\underline{47}$(2), 49-51.

Batsanov, S. S. and Pakhomov, V. I. (1956). Zh. Fiz. Khim. $\underline{30}$, 133-141.

Battino, R. (1971). Chem. Rev. $\underline{71}$(1), 5-45.

Battino, R. and Clever, H. L. (1966). Chem. Rev. $\underline{66}$, 395-463.

Baumgaertner, M., Moorwood, R. A. S., and Wenzel, H. (1980). Am. Chem. Soc. Symp. Ser. $\underline{133}$, 415-434.

Baumgartner, E. K. and Atkinson, G. (1971). J. Phys. Chem. $\underline{75}$(15), 2336-2340.

Bean, H. S., Beckett, A. H., and Carless, J. E., Eds. (1974). Advances in Pharmaceutical Sciences (4th ed.). Academic, London. 444 pp.

Beattie, J. A. and Bridgeman, O. C. (1927). J. Am. Chem. Soc. $\underline{49}$, 1665-1667.

Becker, F., Kiefer, M., and Rhensius, P. (1972). Z. Naturforsch. Teil A $\underline{27}$, 1611-1616; $\underline{28}$, 772-784.

Beilsteins Handbuch der Organischen Chemie (4th ed.)(1958-1964). Beilstein Institut für Literatur der Organischen Chemie. Springer-Verlag, Berlin. Vol. I, Part 1 (1958); Vol. V, Part 1 (1963); Vol. V, Part 2 (1964).

Bell, G. H. (1973). Chem. Phys. Lipids $\underline{10}$, 1-10.

Bell, R. P. (1932). J. Chem. Soc. 2905-2911.

Bell, R. P. (1937). Trans. Faraday Soc. $\underline{33}$, 496-501.

Bell, R. P. (1941). Acid-Base Catalysis. Oxford University Press, Oxford. 211 pp.

Bellino, F. (1932). Gazz. Chim. Ital. $\underline{62}$, 795-798.

Belzile, J. L., Kaliaguine, S., and Ramalho, R. S. (1976). Can. J. Chem. Eng. $\underline{54}$(5), 446-450.

Benedict, M., Webb, G. B., and Rubin, L. C. (1940). J. Chem. Phys. $\underline{8}$(4), 334-345.

Ben-Naim, A. (1971a). J. Chem. Phys. $\underline{54}$(3), 1387-1404.

Ben-Naim, A. (1971b). J. Chem. Phys. $\underline{54}$(9), 3696-3711.

Ben-Naim, A. (1972). Mol. Phys. $\underline{24}$, 705-733.

Ben-Naim, A. (1974). Water and Aqueous Solution. Introduction to a Molecular Theory. Plenum, New York. 474 pp.

Ben-Naim, A. (1974). Proc. 8th Int. Conf. on the Properties of Water and Steam. Giens (France). September. p. 911-917.

Ben-Naim, A. (1975). J. Phys. Chem. $\underline{79}$(13), 1268-1274.

Ben-Naim, A. (1978a). Phys. Chem. Liq. $\underline{7}$, 375-386.

Ben-Naim, A. (1978b). Energetics and Structure of Halophilic Microorganisms. Caplan,
    S. R. and Ginzburg, M., Eds., Elesevier, Amsterdam. p. 1-12.

Ben-Naim, A. (1980). Hydrophobic Interactions. Plenum, New York. 311 pp.

Ben-Naim, A. and Baer, S. (1963). Trans. Faraday Soc. $\underline{59}$, 2735-2738.

Ben-Naim, A. and Egel-Thal, M. (1965). J. Phys. Chem. $\underline{69}$(10), 3240-3253.

Ben-Naim, A. and Tenne, R. (1977). J. Chem. Phys. $\underline{67}$(2), 627-635.

Ben-Naim, A. and Wilf, J. (1980). J. Phys. Chem. $\underline{84}$, 583-586.

Ben-Naim, A., Wiff, J., and Yaacobi, M. (1973). J. Phys. Chem. $\underline{77}$(1), 95-102.

Benson, B. B. and Krause, D. (1976). J. Chem. Phys. $\underline{64}$(2), 689-709.

Benson, S. W. (1978). J. Am. Chem. Soc. $\underline{100}$(18), 5640-5644.

Benson, S. W., Cruickshank, F. R., Golden, D. M., Haugen, G. R., O'Neal, H. E.,
    Rodgers, A. S., Shaw, R., and Walsh, R. (1969). Chem. Rev. $\underline{69}$(1), 279-324.

Berens, A. R. (1974). Am. Chem. Soc. Polym. Repr. $\underline{15}$(2), 197-202.

Berg, C. and McKinnis, A. C. (1948). Ind. Eng. Chem. $\underline{40}$(7), 1309-1311.

Bergström, S. and Olofsson, G. (1975). J. Solution Chem. $\underline{4}$(7), 535-555.

Bernal, J. D. (1959). Nature $\underline{183}$(4655), 141-147.

Bernal, J. D. (1964). Proc. Roy. Soc. Lond. Ser. A $\underline{280}$, 299-322.

Bernal, J. D. and Fowler, R. H. (1933). J. Chem. Phys. $\underline{1}$(8), 515-548.

Berndt, P. (1964). Pharm. Prax. No. 12, 277-278.

Berthelot, M. D. (1899). J. Phys. $\underline{8}$, 263-274.

Bett, K. E. and Cappi, J. B. (1965). Nature $\underline{207}$(4997), 620-621.

Beutier, D. and Renon, H. (1978). AIChE J. $\underline{24}$(6), 1122-1125.

Bhattacharyya, B. (1976). Indian J. Chem. $\underline{14A}$(10), 796-798.

Biddulph, M. W. and Meachin, A. J. (1978). Can. J. Chem. Eng. $\underline{56}$(3), 382-388.

Bidleman, T. F. and Olney, C. E. (1974). Science $\underline{183}$(4124), 516-518.

Bier, K., Maurer, G., and Sand, H. (1975). 4th Inter. Conf. Thermodyn. Chim. Vol. 2
    August. p. 95-102.

Bierenbaum, M. L. (1976). Chem. Tech. $\underline{6}$(5), 314-315.

Biggar, J. W. and Riggs, R. L. (1974). Hilgardia $\underline{42}$(10), 383-391.

Biggar, J. W., Donnen, L. D., and Riggs, R. L. (1966). Soil Interaction with Organi-
    cally Polluted Water. Summary Rep. Dept. of Water Sci. and Eng., Univ. of
    California, Davis, Calif.

Binder, K. (1975). J. Chem. Phys. $\underline{63}$(5), 2265-2266.

Bingham, R. C., Dewar, M. J. S., and Lo, D. H. (1975). J. Am. Chem. Soc. $\underline{97}$(6), 1294-
    1311.

Birks, J. W. and Leck, T. J. (1976). Nature $\underline{260}$(5546), 8.

Bittrich, H.-J., Gedan, H., and Feix, G. (1979). Z. phys. Chem. $\underline{260}$(5), 1009-1013.

Black, C., Derr, E. L., and Papadoulos, M. N. (1963). Ind. Eng. Chem. $\underline{55}$(8), 40-47.

Black, C., Derr, E. L., and Taylor, H. S. (1948). J. Chem. Phys. $\underline{16}$(5), 537-543.

Blackburn, G. M., Lilley, T. H., and Walmsey, E. (1980). J. Chem. Soc. Comm. No. 22, 1091-1093.

Blandamer, M. J. (1970). Quart. Rev. $\underline{24}$(2), 169-184.

Blandamer, M. J. (1977). Solvents -- The Neglected Parameter. 2nd Solvents Symp. UMIST. Manchester.

Blandamer, M. J. and Burgess, J. (1975). Chem. Soc. Rev. $\underline{4}$(1), 55-75.

Blandamer, M. J., Godfrey, E., and Membrey, J. R. (1974). J. Solution Chem. $\underline{3}$(12), 881-887.

Blank, Yu. I. (1975). Russ. J. Phys. Chem. $\underline{49}$(7), 1087-1088. (VINITI No. 1008-75.)

Blank, Yu. I. and Kinshova, L. A. (1975a). Azerb. Khim. Zh. No. 4, 81-86.

Blank, Yu. I. and Kinshova, L. A. (1975b). Azerb. Khim. Zh. No. 5, 121-124.

Blank, Yu. I. and Kinshova, L. A. (1976). Zh. Fiz. Khim. $\underline{50}$(3), 810. (VINITI No. 3522-75.)

Blanks, R. F. and Prausnitz, J. M. (1964). Ind. Eng. Chem. Fundam. $\underline{3}$(1), 1-8.

Blumberg, R. and Melzer, P. (1960). TBE Bull. $\underline{2}$, 5.

Bodor, E., Bor, G., Mohai, B., and Siposs, G. (1957). Veszpremi Vegyip. Egy. Közl. $\underline{1}$, 55-62. (CA 55:3175h.)

Boehn, P. D. and Quinn, J. G. (1975). Environ. Sci. Technol. $\underline{9}$(4), 365-366.

Bogaard, M. P. and Orr, B. J. (1975). Electric Dipole Polarisabilities of Atoms and Molecules. In Physical Chemistry, Series 2, Vol. 2, Molecular Structure and Properties (A. D. Buckingham, Ed.). Butterworths, London, p. 149-194.

Boggs, J. E. and Buck, A. E. (1958). J. Phys. Chem. $\underline{62}$, 1459-1462.

Boggs, J. E. and Mysher, H. P. (1960). J. Am. Chem. Soc. $\underline{82}$, 3517-3519.

Bohon, R. L. and Claussen, W. F. (1951). J. Am. Chem. Soc. $\underline{73}$, 1571-1578.

Bolz, R. E. and Tuve, G. L., Eds. (1970). Handbook of Tables for Applied Engineering Science. Chemical Rubbber Co., Cleveland, Ohio.

Bondi, A. (1964). J. Phys. Chem. $\underline{68}$(3), 441-451.

Bondi, A. (1968). Physical Properties of Molecular Crystals, Liquids, and Glasses. Wiley, New York. 502 pp.

Bondi, A. and Simkin, D. J. (1957). AIChE J. $\underline{3}$(4), 473-479.

Bonner, O. D. (1971). Rec. Chem. Prog. $\underline{32}$(1), 1-15.

Bonner, O. D. and Choi, Y. S. (1974). J. Phys. Chem. $\underline{78}$(17), 1723-1731.

Bonner, O. D. and Woolsey, G. B. (1968). J. Phys. Chem. $\underline{72}$, 899-905.

Booth, H. S. and Everson, H. E. (1948). Ind. Eng. Chem. $\underline{40}$(8), 1491-1493.

Booth, H. S. and Everson, H. E. (1949). Ind. Eng. Chem. $\underline{41}$, 2627-2628.

Borchardt, H. J. (1976). ASHRAE J. $\underline{18}$(8), 38-39.

Borina, A. F. (1977). Russ. J. Phys. Chem. $\underline{51}$(2), 235-237.

Böttcher, C. J. F. (1946a). Recl. trav. chim. $\underline{65}$, 14-18.

Böttcher, C. J. F. (1946b). Recl. trav. chim. $\underline{65}$, 19-38.

Böttcher, C. J. F. (1946c). Recl. trav. chim. $\underline{65}$, 39-49.

Boublik, T. (1972). Coll. Czech. Chem. Commun. 37, 2499-2506.

Boublik, T. and Aim, K. (1972). Coll. Czech. Chem. Commun. 37, 3513-3521.

Boublik, T., Fried, V., and Hala, E. (1973). The Vapour Pressure of Pure Substances. Elsevier, Amsterdam. 626 pp.

Boublik, T., Nezbeda, I., and Hlavaty, K. (1980). Statistical Thermodynamics of Simple Liquids and Their Mixtures. Elsevier, Amsterdam. 146 pp.

Boucher, R. E. and Skau, E. L. (1954). Solubility Charts for Homologous Long Chain Organic Compounds. A Comprehensive Graphical Correlation of Literature Data for 138 Systems Involving 11 Homologous Series and 17 Solvents. U.S. Dept. of Agriculture, Southern Regional Res. Lab., New Orleans, La. 77 pp.

Boulegue, J. (1978). Phosphorus and Sulfur 5, 127-128.

Bowman, M. C., Acree, F., and Corbett, M. K. (1960). J. Agric. Food Chem. 8, 406-408.

Boyer, F. L. (1960). The Solubility of Several Gases in Some Alcohols. (Ph. D. thesis), Vanderbilt Univ., Nashville, Tenn. 105 pp.

BP Chemicals (U.K.) (1968). Bisol. Ethylene Dichloride. Technigram 391. London, December. 4 pp.

BP Chemicals (U.K.) (1970). Organic Chemicals Data Book (2nd ed.). London, January. 37 pp.

Bradley, R. S., Dew, M. J., and Munro, D. C. (1973). High Temp. High Press. 5(3), 169-176.

Brandani, V. (1974). Ind. Eng. Chem. Fundam. 13(2), 154-156.

Brandani, V. (1975). Ind. Eng. Chem. Fundam. 14(1), 73.

Breon, T. L., Mauger, J. W., and Paruta, A. N. (1980). Drug Dev. Ind. Pharm. 6(1), 87-98.

Breusch, F. L. and Kirkali, A. (1968). Fette-Seifen-Anstrichm. 70(11), 864-871.

Brian, P. L. T. (1965). Ind. Eng. Chem. Fundam. 4(1), 100-101.

Bridges, B. A. (1976). Nature 261(5557), 195-200.

Briegleb, G. (1961). Elektronen-Donator-Acceptor-Komplexe. Springer-Verlag, Berlin.

Briggs, F. A. and Barduhn, A. J. (1963). Saline Water Conversion--II. Adv. Chem. Ser. No. 38. Am. Chem. Soc., Washington, D.C., p. 190-199.

Briggs, F. A., Hu, Y. C., and Barduhn, A. J. (1962). New Agents for Use in the Hydrate Process for Demineralizing Sea Water. Res. Dev. Prog. Rep. No. 59. Office of Saline Water, U.S. Dept. of the Interior, Washington, D.C. March. 41 pp.

British Standards Institution (1967). Recommendations for Letter Symbols, Signs, and Abbreviations, Part 1. BS 1991. London. 18 pp.

Brocks, W. B. (1952). Phase Equilibria in the 1-Butene - Water System and in the n-Butane - Water System (Ph. D. thesis). Univ. of Texas, Austin, Tex. May. 136 pp.

Brockway, L. O. (1937). J. Phys. Chem. 41(2), 185-195.

Brönsted, J. N. (1928). Chem. Ber. 61, 2049-2063.

Brooker, P. J. and Ellison, M. (1974). Chem. Ind., No. 19, 785-788.

Brooks, G. T. (1974). Chlorinated Insecticides, Vol. 1, Technology and Application. Chemical Rubber Co., Cleveland, Ohio. 249 pp.

Broul, M., Nyvlt, J., and Söhnel, O. (1980). Solubility in Inorganic Two-Component Systems. Physical Sciences Data 6. Elsevier, Amsterdam. 576 pp.

Brown, R. L. and Wasik, S. P. (1974). J. Res. Natl. Bur. Std. 78A(4), 453-460.

Buckingham, A. D., Lippert, E., and Bratos, S. (1978). Organic Liquids. Structure, Dynamics, and Chemical Properties. Wiley, New York. 352 pp.

Buffington, C. and Turndorf, H. (1976). Bull. N. Y. Acad. Med. 52(7), 838-841.

Buijs, K. and Choppin, G. R. (1963-1964). J. Chem. Phys. 39, 2035-2050(1963); 40, 3120(1964).

Bukovskii, M. I., Zhukov, V. I., Kozhukhova, T. V., Sanotsky, I. V., and Sidorov, K. K. (1977). Maximum Allowable Concentrations for Harmful Substances in the Environment. All-Union Sci. Res. Inst. for Safety in the Chem. Ind. and Sci. Res. Inst. for Ind. Hyg. and Occupational Diseases of the Acad. Med. Sci. USSR. Severodonetsk. 85 pp.

Bunnett, J. F. (1963). Annu. Rev. Phys. Chem. 14, 271-290.

Burd, S. D. (1968). Phase Equilibria of Partially Miscible Mixtures of Hydrocarbons and Water (Ph. D. thesis). Pennsylvania State Univ., University Park, Pa. 209 pp.

Burgess, J. and Haines, R. I. (1978). J. Chem. Eng. Data 23(3), 196-197.

Burgin, J., Hearne, G., and Rust, F. (1941). Ind. Eng. Chem. 33(3), 385-388.

Burgos, E. and Bonadeo, H. (1977). Chem. Phys. Lett. 49(3), 475-478.

Burrell, H. (1975). Solubility Parameter Values. In Polymer Handbook (2nd ed.; J. Brandrup and E. H. Immergut, Eds.). Wiley-Interscience, New York. p. IV 337 - IV 359.

Burrell, H. (1955). Interchem. Rev. 14(1), 3-16; (2), 31-46.

Burrows, G. and Preece, F. H. (1953). J. Appl. Chem. 3, 451-462.

Burshtein, A. I. (1972). Dokl. Phys. Chem. 205(1/2), 628-631.

Bushinskii, V. I., Freidlin, G. N., and Kolomiets, V. A. (1974). Zh. Prikl. Khim. 47(5), 1054-1058.

Butler, J. A. V. (1937). Trans. Faraday Soc. 33, 229-238.

Butler, J. A. V. (1962). Chemical Thermodynamics (5th ed.). Macmillan, London. 601 pp.

Butler, J. A. V. and Harrower, P. (1937). Trans. Faraday Soc. 33, 171-178.

Butler, J. A. V. and Reid, W. S. (1936). J. Chem. Soc. 1171-1173.

Byk, S. Sh. and Fomina, V. I. (1968). Russ. Chem. Rev. 37(6), 469-491.

Cabral, J. R. P., Shubik, P., Mollner, T., and Raitano, F. (1977). Nature 269(5628), 510-511.

Caddock, B. D. and Davies, P. L. (1959). Nature 184(4704), 2011.

Cairns, J. (1981). Nature 289(5796), 353-357.

Cali, J. P. (1976). Pure Appl. Chem. 48, 503-515.

Call, F. (1957). J. Sci. Food Agric. $\underline{8}$(11), 630-639.

Cammarata, A. (1979). J. Pharm. Sci. $\underline{68}$(7), 839-842.

Camp, T. R. and Meserve, R. L. (1974). Water and Its Impurities (2nd ed.). Dowden, Hutchinson & Ross, Stroudsburg, Pa. 384 pp.

Carey, W. W. (1965). Solubility of CH_3CClF_2 in Water and Aqueous Sodium Chloride Solutions (M. Sc. thesis). Syracuse Univ., Syracuse, N. Y. 55 pp.

Carey, W. W., Klausutis, N. A., and Barduhn, A. J. (1966). Desalination $\underline{1}$(4), 342-358.

Carless, J. E. and Swarbrick, J. (1964). J. Pharm. Pharmacol. $\underline{16}$, 633-634.

Carlisle, P. J. and Levine, A. A. (1932a). Ind. Eng. Chem. $\underline{24}$(2), 146-147.

Carlisle, P. J. and Levine, A. A. (1932b). Ind. Eng. Chem. $\underline{24}$(10), 1164-1168.

Carlson, H. C. and Colburn, A. P. (1942). Ind. Eng. Chem. $\underline{34}$(5), 581-589.

Carothers, W. H., Berchet, G. J., and Collins, A. M. (1932). J. Am. Chem. Soc. $\underline{54}$, 4066-4070.

Carothers, W. H., Williams, I., Collins, A. M., and Kirby, J. E. (1931). J. Am. Chem. Soc. $\underline{53}$, 4203-4225.

Carra, S., Santacesaria, E., Ferrario, P., and Bovone, P. (1980). Fluid Phase Equil. $\underline{4}$(1), 89-104.

Carstensen, J. T. and Anik, S. T. (1976). J. Pharm. Sci. $\underline{63}$(1), 158-160.

Carter, G. C. (1980). J. Chem. Inf. Comput. Sci. $\underline{20}$(3), 146-152.

Casparian, A. S. and Cole, R. H. (1974). J. Chem. Phys. $\underline{60}$(3), 1106-1109.

Casteel, J. F. and Sears, P. G. (1974). J. Chem. Eng. Data $\underline{19}$(3), 196-200.

Cave, G., Kothari, R., Puisieux, F., et al. (1980). Int. J. Pharm. $\underline{5}$(4), 267-272.

Certain, P. R. and Bruch, L. W. (1972). Intermolecular Forces. In Physical Chemistry, Series 1, Vol. 1, Theoretical Chemistry (W. Byers Brown, Ed.). Butterworths, London. p. 113-165.

Cesaro, A. and Russo, E. (1978). J. Chem. Educ. $\underline{55}$(2), 133-134.

Chancel, G. and Parmentier, F. (1885a). C. R. Acad. Sci. Ser. A $\underline{100}$, 27-30.

Chancel, G. and Parmentier, F. (1885b). C. R. Acad. Sci. Ser. A $\underline{100}$, 773-776.

Chandrasekaran, R. S., Govindarajan, G., and Andiappan, A. N. (1976). Chem. Era $\underline{12}$(9), 327-328.

Chao, K. C. and Greenkorn, R. A. (1975). Thermodynamics of Fluids. An Introduction to Equilibrium Theory. Dekker, New York. 553 pp.

Chao, K. C. Robinson, R. L., Smith, M. L., and Kuo, C. M. (1967). Chem. Eng. Prog. Symp. Ser. $\underline{63}$(81), 121-127.

Charykov, A. K. and Tal'nikova, T. V. (1973). Determination of the Solubility of Difficultly Soluble Organic Liquids in Aqueous and Water - Salt Systems. USSR Patent No. 399659, August 10.

Cheesman, G. H. and Whitaker, A. M. B. (1952). Proc. Roy. Soc. London Ser. A $\underline{212}$, 406-425.

Chen, S.-H. (1971). Structure of Liquids. In Physical Chemistry: An Advanced Treatise, Vol. 8A, Liquid State (D. Henderson, Ed.). Academic, New York. p. 85-156.

Cheng, W. K. and Pinder, K. L. (1976). Can. J. Chem. Eng. 54(5), 377-381.

Chepstov, A. S. (1974). Poverkhnost. Yavleniya Dispersn. Sistemakh., No. 3, 231-234. (CA 83:85738h.)

Chernyshev, A. K. (1969). Kholod. Tekh. 46(1), 62.

Chernyshev, A. K. (1970). Chem. Technol. Fuels Oils 15(4), 308-310.

Chey, W. and Calder, G. V. (1972). J. Chem. Eng. Data 17(2), 199-200.

Chinworth, H. E. and Katz, D. L. (1947). Refrig. Eng. 54(10), 359-363.

Chiou, C. T., Freed, V. H., Schmedding, D. W., and Kohnert, R. L. (1977). Envir. Sci. Techn. 11(5), 475-478.

Chistyakov, V. M. (1965). J. Appl. Chem. USSR 38(5), 1007-1010.

Chistyakov, V. M. and Balezin, S. A. (1961). Izv. Vyssh. Uchebn. Zaved., Khim. Khim. Tekhnol. 4(6), 955-967.

Chistyakov, V. M. and Shapurova, V. V. (1964). Izv. Vyssh. Uchebn. Zaved., Khim. Khim. Tekhnol. 7(2), 349-350.

Christian, S. D. and Tucker, E. E. (1977). Science 198(4313), 210.

Christian, S. D., Affsprung, H. E., and Johnson, J. R. (1963). J. Chem. Soc. 1896-1898.

Christian, S. D., Affsprung, H. E., Johnson, J. R., and Worsley, J. D. (1963). J. Chem. Educ. 40(8), 419-421.

Christian, S. D., Johnson, J. R., Affsprung, H. E., and Kilpatrick, P. J. (1966). J. Phys. Chem. 70, 3376-3377.

Christian, S. D., Taba, A. A., and Gash, B. W. (1970). Quart. Rev. 24(1), 20-36.

Christian, S. D., Affsprung, H. E., Hunter, J. A., Gillam, W. S., and McCoy, W. H. (1968). U. S. Office Saline Water Res. Dev. Progr. Rep. 301. January. 107 pp. (CA 69:100003w.)

Christie, A. D. and Crisp, D. J. (1967). J. Appl. Chem. 17(1), 11-14.

Chu, B. (1967). Molecular Forces Based on the Baker Lectures of P. J. W. Debye. Wiley-Interscience, New York. 176 pp.

Chu, J. C., Getty, R. J., Brennecke, L. F., and Paul, R. F. (1950). Distillation Equilibrium Data. Reinhold, New York.

Chu, J. C., Wang, S. L., Levy, S. L., and Paul, R. F. (1956). Vapor-Liquid Equilibrium Data. Edwards, Ann Arbor, Mich.

Ciaccio, L. L., Ed. (1971-1973). Water and Water Pollution Handbook, Vols. 1-4. Dekker, New York. 1945 pp.

Clark, H. M. (1972). The Measurement of Radioactivity. In Techniques of Chemistry, Vol. 1, Physical Methods of Chemistry, Part 3B (A. Weissberger, Ed.). Wiley-Interscience, New York. p. 587-659.

Clarke, E. C. W. and Glew, D. N. (1966). Trans. Faraday Soc. 62, 539-547.

Clarke, E. C. W., Glew, D. N., and Maisonneuve, D. J. (1981). Can. J. Chem. 59(5), 768-771.

Claussen, W. F. (1951). J. Chem. Phys. 19, 259-260, 1425-1426.

Claussen, W. F. and Polglase, M. F. (1952). J. Am. Chem. Soc. 74, 4817-4819.

Claxton, G. (1958). Physical and Azeotropic Data. National Benzole and Allied Prod. Assoc., London. 146 pp.

Claxton, G., Ed. (1961). Benzoles Production and Uses. National Benzole and Allied Prod. Assoc., London. 979 pp.

Clayton, J. W. (1967). Fluorocarbon Toxicity and Biological Action. In Fluorine Chemistry Reviews, Vol. 1 (P. Tarrant, Ed.). Dekker, New York. p. 192-252.

Clever, H. L. and Battino, R. (1975). The Solubility of Gases in Liquids. In Techniques of Chemistry, Vol. 8, Solutions and Solubilities, Part 1 (M. R. J. Dack, Ed.). Wiley, New York. p. 379-441.

Clever, H. L. and Han, C. H. (1980). Am. Chem. Soc. Symp. Ser. 133, 513-536.

Clifford, A. A., Gray, P., and Scott, A. C. (1979). J. Chem. Soc. Faraday Trans. I 75(7), 1752-1756.

Clifford, C. W. (1921a). J. Ind. Eng. Chem. 13(7), 628-631.

Clifford, C. W. (1921b). J. Ind. Eng. Chem. 13(7), 631-632.

Coca, J. and Diaz, R. (1980). J. Chem. Eng. Data 25(1), 80-83.

CODATA (1969). International Compendium of Numerical Data Projects. A Survey and Analysis. Springer-Verlag, Berlin. 295 pp.

CODATA (1972). Proc. 3rd Bienn. Inter. CODATA Conf. Le Creusot, France. CODATA Secretariat, Paris. 100 pp.

CODATA Bulletin 10 (1973a). CODATA Recommended Key Values for Thermodynamics, 1973. Frankfurt am Main. December. 12 pp.

CODATA Bulletin 11 (1973b). Recommended Consistent Values of the Fundamental Physical Constants, 1973. Frankfurt am Main. December. 8 pp.

CODATA (1975). Proc. 4th Bienn. Inter. CODATA Conf. Tsakhcadzor, USSR. Pergamon, Oxford. 171 pp.

CODATA Bulletin 18 (1976). Abstr. 5th Inter. CODATA Conf. Boulder, Color. June 28 - July 1, 1976; Paris. Ayril. 43 pp.

CODATA Bulletin 21 (1976). Proc. Plenetary Sess. 5th Inter. CODATA Conf. Boulder, Color. CODATA Secretariat, Paris. October. 122 pp.

CODATA Newsletter 16 (1976). Program -- 5th Inter. CODATA Conf., Paris. March. 15 pp.

CODATA (1977). Proc. 5th Bienn. Inter. CODATA Conf. Boulder, U.S.A. Pergamon, Oxford. 642 pp.

CODATA (1979). Proc. 6th Inter. CODATA Conf. Santa Flavia, Italy. Pergamon, Oxford. 433 pp.

CODATA (1981). Proc. 7th Inter. CODATA Conf. Kyoto, Japan. Pergamon, Oxford. 524 pp.

Coetzee, J. F. and Ritchie, C. D., Eds. (1969-1976). Solute--Solvent Interactions. Dekker, New York. Vol. 1 (1969), 672 pp., Vol. 2 (1976), 480 pp.

Cohen, E. N., Brown, B. W., Bruce, D. L., Cascorbi, H. F., Corbett, T. H., Jones, T. W., and Whitcher, C. E. (1975). J. Am. Dent. Assoc. 90, 1291-1296.

Cohen, J. L. and Connor, K. A. (1970). J. Pharm. Sci. 59(9), 1271-1276.

Cohen, N. C. (1971). Tetrahedron 27, 789-797.

Cohen, N. C. and Regnier, G. (1976). Bull. Soc. Chim. Fr. No. 11, 2034-2038.

Cole, G. H. A. (1967). An Introduction to the Statistical Theory of Classical Simple Dense Fluids. Pergamon, Oxford.

Cole, K. S. and Cole, R. H. (1941). J. Chem. Phys. 9(4), 341-351.

Colten, S. L., Lin, F. S., Tsao, T. C., Stern, S. A., and Barduhn, A. J. (1972). Hydrolysis Losses in the Hydrate Process for Desalination: Rate Measurement and Economic Analysis. Res. Dev. Prog. Rep. No. 753. U. S. Office of Saline Water, Syracuse Univ., Syracuse, N. Y., February. 72 pp.

Connolly, J. F. (1966). J. Chem. Eng. Data 11(1), 13-16.

Conti, J. J., Othmer, D. F., and Gilmont, R. (1960). J. Chem. Eng. Data 5(3), 301-307.

Cook, M. W. and Hanson, D. N. (1957). Rev. Sci. Instr. 28(5), 370-374.

Coolidge, A. S. (1928). J. Am. Chem. Soc. 50, 2166-2178.

Copley, M. J., Zellhoefer, G. F., and Marvel, C. S. (1938a). J. Am. Chem. Soc. 60, 2666-2673.

Copley, M. J., Zellhoefer, G. F., and Marvel, C. S. (1938b). J. Am. Chem. Soc. 60, 2714-2716.

Copp, J. L. (1953). Disc. Faraday Soc., No. 15, 265-267.

Copp, J. L. and Everett, D. H. (1953). Disc. Faraday Soc., No. 15, 174-188, 268-269.

Cosaert, E. (1971). Chim. Peint. 34(5), 169-178.

Cotton, F. A. (1976). Nature 260(5549), 280.

Cottrell, D. W. (1976). Chem. Ind., No. 7, 319-320.

Covington, A. K. and Dickinson, T., Eds. (1973). Physical Chemistry of Organic Solvent Systems. Plenum, London. 823 pp.

Covington, A. K. and Jones, P., Eds. (1968). Hydrogen Bonded Solvent Systems. Taylor & Francis, London. 354 pp.

Cowles, A. L. (1970). The Uptake and Distribution of Inhalation Anesthetics (Ph. D. thesis). Univ. of Rochester, Rochester, N.Y. 368 pp.

Cox, J. D. and Head, A. J. (1962). Trans. Faraday Soc. 58, 1839-1845.

Cramer, F. D. (1955). Rev. Pure Appl. Chem. 5, 143-164.

Craubner, H. (1976). Z. phys. Chem. Neue Folge 103(1/4), 45-74.

Craver, J. K. (1970). J. Appl. Polym. Sci. 14, 1755-1765.

Cremaschi, P., Gamba, A., and Simonetta, M. (1972). Theor. Chim. Acta 25, 237-247.

Cremaschi, P., Gamba, A., and Simonetta, M. (1977). J. Chem. Soc. Perkin Trans. II, No. 2, 162-166.

Cremer, J. G. (1969). Kältetech.-Klim. 21(4), 97-99.

Criss, C. M. and Salomon, M. (1973). Thermodynamic Measurements, Parts 1 and 2. In Physical Chemistry of Organic Solvent Systems (A. K. Covington and T. Dickinson, Eds.). Plenum, London. p. 23-135 and 253-329.

Crossland, J. and Brodine, V. (1973). Environment 15(3), 11-19.

Crowley, J. D., Teague, G. S., and Lowe, J. W. (1966). J. Paint Technol. 38(496), 269-280.

Croxton, C. A. (1975). Endeavour 34(122), 79-83.

Cysewski, G. R. and Prausnitz, J. M. (1976). Ind. Eng. Chem. Fundam. 15(4), 404-309.

Czapkiewicz, J. and Czapkiewicz-Tutaj, B. (1980). J. Chem. Soc. Faraday I, 76(8), 1663-1668.

Czolbe, P. (1975). Beitrag zur Bestimmung der Löslichkeit von Gasen und ihrer Sättigungsdrucke als Voraussetzung für die Lösung erdölgeologischer Erkundungsaufgaben. VEB Deutscher Verlag für Grundstoffindustrie, Leipzig, East Germany. 86 pp.

Dack, M. R. J. (1974). J. Chem. Educ. 51(4), 231-234.

Dack, M. R. J. (1975a). Chem. Soc. Rev. 4(2), 211-229.

Dack, M. R. J., Ed. (1975b). Solutions and Solubilities, Part 1. Techniques of Chemistry, Vol. 8., Wiley, New York. 475 pp.

Dack, M. R. J., Ed. (1976). Solutions and Solubilities, Part 2. Techniques of Chemistry, Vol. 8, Wiley, New York. 499 pp.

Dagani, R. (1981). Chem. Eng. News 59(10), 26-29.

Dahler, J. S. and Cohen, E. G. D. (1960). Physics 26, 81-102.

Dainton, F. S. (1965). Disc. Faraday Soc. 40, 1-291.

Dalgarno, A. and Lynn, N. (1956). Proc. Phys. Soc. (London) 69, 821-832.

Danforth, W. E. (1931). Phys. Rev. 38, 1224-1235.

Dankleff, M. A. P., Curci, R., Edwards, J. O., and Pyun, H.-Y. (1968). J. Am. Chem. Soc. 90(12), 3209-3218.

David, H. G., Hamann, S. D., and Pearse, J. F. (1952). J. Chem. Phys. 20(6), 969-972.

Davidson, D. W. and Cole, R. H. (1950). J. Chem. Phys. 18, 1417.

Davies, M. (1946). Annu. Rep. Prog. Chem. Chem. Soc., London, p. 5-30.

Davies, T. M. C. (1977). German Patent No. 2,639,594 (to British Vinegars Ltd., London) March 17. 24 pp.

Davies, W., Jagger, J. B., and Whalley, H. K. (1949). J. Soc. Chem. Ind. 68(1), 26-31.

Davis, C. M. and Litovitz, T. A. (1965). J. Chem. Phys. 42, 2563-2576.

Davis, J. C. (1975). Chem. Eng. 82(18), 60-64.

Davis, M. M. (1968). Acid-Base Behaviour in Aprotic Organic Solvents. Natl. Bur. Std., Monogr. 105. Washington, D. C., August. 151 pp.

Davis, S. S. (1970). Experientia 26(6), 671-672.

Davis, S. S. (1973a). J. Pharm. Pharmacol. 25, 1-12.

Davis, S. S. (1973b). J. Pharm. Pharmacol. 25, 293-296.

Davis, S. S. (1973c). J. Pharm. Pharmacol. 25, 769-778.

Davis, S. S. (1973d). J. Pharm. Pharmacol. 25, 982-992.

Davis, S. S. (1973e). Pharm. J. 210, 205-208.

Davis, S. S. and Mukhayer, G. I. (1972). J. Pharm. Pharmacol. 24(Suppl.), 129-130.

Davis, S. S., Higuchi, T., and Rytting, J. H. (1972). J. Pharm. Pharmacol. 24, 30-46.

Davis, S. S., Higuchi, T., and Rytting, J. H. (1974). Adv. Pharm. Sci. 4, 73-261.

Dayantis, J. (1977). Plast. Mod. Elastomeres 29(2), 58-62.

Deak, Gy. (1973). Hung. J. Ind. Chem. 1, 343-350.

Deal, C. H. and Derr, E. L. (1968). Ind. Eng. Chem. 60(4), 28-38.

Deal, C. H., Derr, E. L., and Papadopoulos, M. N. (1962). Ind. Eng. Chem. Fundam. 1(1), 17-19.

De Boer, J. (1954). Physics 20, 655-664.

De Boer, J. and Uhlenbeck, G. E., Eds. (1964). Studies in Statistical Mechanics, Vol. 2, The Cell Theory of Liquids (J. M. H. Levelt and E. G. D. Cohen, Eds.). North-Holland, Amsterdam.

Debye, P. (1912). Phys. Z. 13(3), 97-100 and 295.

De Groot, S. R. (1949). Rev. Opt. 28, 627-634.

Dehn, W. M. (1917). J. Am. Chem. Soc. 39, 1399-1404.

De La Mare, P. B. D. (1976). Electrophilic Halogenation. Cambridge University Press, Cambridge. 231 pp.

Delfino, J. J. (1976). Chem. Eng. News 54(10), 2.

De Ligny, C. L. and van der Veen, N. G. (1972). Chem. Eng. Sci. 27, 391-401.

De Ligny, C. L. and van der Veen, N. G. (1975). J. Solution Chem. 4(10), 841-851.

De Ligny, C. L., van der Veen, N. G., and van Houvelingen, J. C. (1976). Ind. Eng. Chem. Fundam. 15(4), 336-341.

De Lisi, R., Goffredi, M., and Livery, V. T. (1980). J. Chem. Soc. Faraday I, 76(8), 1660-1662.

De Mateo, A. and Kurata, F. (1975). Ind. Eng. Chem. Prod. Res. Dev. 14(2), 137-140.

De Minjer, C. H. (1939). Damp - Vloeistofevenwichten in eenige Ternaire Stelsels (Ph. D. thesis). Univ. of Delft, Drukkerij Waltman, Delft, Holland. 165 pp.

Denbigh, K. G. (1940). Trans. Faraday Soc. 36, 936-948.

Deno, N. C. and Berkheimer, H. E. (1960). J. Chem. Eng. Data 5(1), 1-5.

Derjaguin, B. V. and Churaev, N. V. (1973). Nature 244(5416), 430-431.

Derr, E. L. and Deal, C. H. (1969). Inst. Chem. Eng. (Lond.) Symp. Ser. No. 32. Paper 3:40. London.

De Santis, R., Breedveld, G. J. F., and Prausnitz, J. M. (1974). Ind. Eng. Chem. Prod. Res. Dev. 13(4), 374-377.

Desplanches, H., Chevalier, J. L., and Llinas, R. (1975). Rev. Inst. Fr. Pet. 30(6), 951-967.

Dewan, A. R., Tao, L. C., and Weber, J. H. (1978). Ind. Eng. Chem. Process Des. Dev. 17(3), 371-373.

Dewar, M. J. S. and Grisdale, P. J. (1962). J. Am. Chem. Soc. 84, 3548-3553.

Dewar, M. J. S. and Rzepa, H. S. (1978). J. Am. Chem. Soc. 100(1), 58-67.

Dexter, R. N. (1976). An Application of Equilibrium Adsorption Theory to the Chemical Dynamics of Organic Compounds in Marine Ecosystems. (Ph. D. thesis). Univ. of Washington, Seattle, Washington, D. C., 195 pp. (Univ. Microfilm No. 77:566.)

Dexter, R. N. and Pavlou, S. P. (1978). Mar. Chem. 6(1), 41-53.

Diamond, W. J. (1959). Appl. Specrosc. 13(3), 77-78.

Dickson, A. G. and Riley, J. P. (1976). Mar. Pollut. Bull. 7(9), 167-169.

Dieterici, K. H. (1899). Ann. Phys. 69, 685-705.

Dietz, E. A. and Singley, K. F. (1979). Anal. Chem. 51(11), 1809-1814.

Dilling, W. L. (1977). Environ. Sci. Technol. 11(4), 405-409.

Dilling, W. L., Tefartiller, N. B., and Kallos, G. J. (1975). Environ. Sci. Technol. 9(9), 833-838.

Di Giacomo, A. and Smyth, C. P. (1955). J. Am. Chem. Soc. 77, 774-777.

Di Paolo, T. Kier, L. B., and Hall, L. H. (1979). J. Pharm. Sci. 68(1), 39-42.

Di Paolo, T. and Sandorfy, C. (1974a). Chem. Phys. Lett. 26(4), 466-469.

Di Paolo, T. and Sandorfy, C. (1974b). Nature 252(5483), 471-472.

Di Paolo, T., Kier, L. B., and Hall, L. H. (1977). Mol. Pharmacol. 13(1), 31-37.

Djordjevic, B. D., Mihajlov-Dudukovic, A. N., and Tasik, A. Z. (1980). AIChE J. 26 (5), 858-862.

Djordjevic, B. D., Mihajlov-Dudukovic, A. N., Grondanic, D. K., Tasik, A. Z., and Horvath, A. L. (1977). Chem. Eng. Sci. 32(9), 1103-1107.

Donahue, D. J. and Bartell, F. E. (1952). J. Phys. Chem. 56, 480-484.

Dooms, I. L. (1975). Books on Water and Wastewaters. Natl. Cent. Sci. Technol. Docum., Brussels. 102 pp.

D'Orazio, L. A. and Wood, R. H. (1963). J. Phys. Chem. 67(7), 1435-1438.

Döring, R. (1977). Kälte Klimatech. 30(9), 348-354.

Döring, R., Knapp, H., Oellrich, L., Plöcker, U., and Prausnitz, J. M. (1979). Vapor-Liquid Equilibria for Mixtures of Low Boiling Fluids. Deutsche Gesellschaft für Chemisches Apparatewesen, Chemistry Data Series 6(1).

Dorsey, N. E., Ed. (1968). Properties of Ordinary Water Substances in All Its Phases: Water-Vapor, Water and All the Ices. Reinhold, New York (reprinting of 1940 ed.); Am. Chem. Soc. Monogr. Ser. No. 81. 673 pp.

Douabul, A. and Riley, J. (1979). J. Chem. Eng. Data 24(4), 274-276.

Douglas, F. (1964). J. Phys. Chem. 68(1), 169-174.

Dow Chemical Company (1963). Bromine and Brominated Products Handbook. Tech. Serv. Dev., Midland, Mich., p. 55-56.

Dow Chemical Company (1966). Vinylidene Chloride Monomer. Tech. Serv. Dev., Midland, Mich.

Dow Chemical Company (1972). Chlorothene VG Cleaning Solvent. Bull. No. 100-5343-72.
    Midland, Mich. 11 pp.

Dow Chemical Company (1973). Chlorothene NU Solvent. Bull. No. 100-71-73. Midland,
    Mich. 38 pp.

Dow Chemical Europe (1967). Chlorothene NU. Bull. EU 8513-E-867. Zurich. 21 pp.

Dowell, F. and Stewart, G. H. (1976). Phosphorus Sulfur 1(2/3), 135-142.

Dowty, B. J., Carlisle, D. R., and Laseter, J. L. (1975). Environ. Sci. Technol. 9
    (8), 762-765.

Dowty, B. J., Carlisle, D. R., Laseter, J. L., and Storer, J. L. (1975). Science
    187(4171), 75-77.

Draiko, L. I. (1969). Razrab. Usoversk. Tekhnol. Khim. Proizvod., 116-118.

Dreisbach, D. (1966). Liquids and Solutions. Houghton Miffin, Boston. 194 pp.

Dreisbach, R. R. (1955-1961). Physical Properties of Chemical Compounds. Am. Chem.
    Soc. Adv. Chem. Ser. 15 (1955, 536 pp.); 22 (1959, 491 pp.); and 29 (1961, 489 pp.).
    Washington, D.C.

Drost-Hansen, W. (1975). Anomalous Temperature and Pressure Dependencies of Gas Sol-
    ubilities: Laboratory and Field Observations. Chem. Phys. Aqueous Gas Solutions,
    Proc. Symp., Eng. Proc., p. 233-256.

Drozdov, V. A., Kreshkov, A. P., and Petrov, S. I. (1969). Russ. Chem. Rev. 38(1),
    47-64.

Dummett, G. A. (1969). Int. Symp. Distill., Inst. Chem. Eng. (London), Brighton,
    England. September 8-10. (6 sessions.)

Dundon, M. L. and Mack, E. (1923). J. Am. Chem. Soc. 45(11), 2479-2485.

du Pont de Nemours & Company (1947). Methyl Chloride. Arctic the Refrigerant. Prac-
    tical Manual for Design and Service Engineers in the Refrigerating Industry. Elec-
    trochem. Dept., Wilmington, Del. 69 pp.

du Pont de Nemours & Company (1950). The Freon Family. Freon Prod. Div. Tech. Bull.
    RT-9. Wilmington, Del. 8 pp.

du Pont de Nemours & Company (1952a). The solubility of Freon Compounds in Water.
    II. Freon-22, Jackson Lab. Kinet. Tech. Rep. KSS-141. Wilmington, Del. April 10.

du Pont de Nemours & Company (1952b). The Solubility of Freon Compounds in Water.
    III. Freon-13 and Freon-23, Jackson Lab. Kinet. Tech. Rep. KSS-151. Wilmington,
    Del. April 22. 6 pp.

du Pont de Nemours & Company (1952c). The Solubility of Freon Compounds in Water.
    V. Freon-115 and Freon-13B1, Jackson Lab. Kinet. Tech. Rep. KSS-188. Wilmington,
    Del. October 9. 6 pp.

du Pont de Nemours & Company (1953). The Solubility of Freon Compounds in Water.
    VI. Freon-14, Tetrafluoromethane, Jackson Lab. Kinet. Tech. Rep. KSS-241. Wilming-
    ton, Del. May 14. 4 pp.

du Pont de Nemours & Company (1956). Concentration of Water in Saturated Freon Vapor
    at 100 Percent Humidity. Freon Prod. Div. Tech. Rep. D-15. Wilmington, Del. July 2.

References

817

du Pont de Nemours & Company (1957). Freon – BF Solvent, Freon-112, Stability and Solubility for Water. Kinet. Lab. Tech. Rep. KSS-1213. Wilmington, Del. January 29. 2 pp.

du Pont de Nemours & Company (1961a). Physical Properties of Fluorinated Hydrocarbons for B-2 Bulletin. Freon Prod. Div. Tech. Rep. KSS-3275. Wilmington, Del. July 3. 6 pp.

du Pont de Nemours & Company (1961b). The Solubility of Freon-11, Freon-21, Freon-112, Freon-113, and Freon-114 in Water. Freon Prod. Div. Tech. Rep. KSS-3560. Wilmington, Del. December 29. 11 pp.

du Pont de Nemours & Company (1963a). Distribution of Water Between the Vapor and Liquid Phase of the Freon Refrigerants. Tech. Bull. M-21. Wilmington, Del. 1 p.

du Pont de Nemours & Company (1963b). Technical Bulletin Data for Fluorocarbon 123-- Physical Properties. Solubilities of Compounds in the Liquid, and Compatabilities with Magnet Wire Coatings. Freon Prod. Div. Tech. Rep. KSS-4277. Wilmington, Del. July 12. 9 pp.

du Pont de Nemours & Company (1963c). Data for Technical Bulletin III--Miscellaneous Physical Property, Solubility, and Compability Data for Freon-214, Freon-215, and Freon-216 Fluorocarbons. Freon Prod. Div. Tech. Rep. KSS-4217C. Wilmington, Del. August 7. 15 pp.

du Pont de Nemours & Company (1963d). Technical Bulletin Data for Fluorocarbon 216B2. I. Physical Properties, Solubilities of Compounds in the Liquids, and Compatabilities with Magnet Wire Coatings. Freon Prod. Div. Tech. Rep. KSS-4599A. Wilmington, Del. December 11. 11 pp.

du Pont de Nemours & Company (1966a). Solubility Relationships of the Freon Fluorocarbon Compounds. Tech. Bull. B-7. Wilmington, Del. 16 pp.

du Pont de Nemours & Company (1966b). Water Solubility of Freon Refrigerants. Part 1. Compounds Boiling Below 32°F. Tech. Bull. B-17. Wilmington, Del. 6 pp.

du Pont de Nemours & Company (1966c). Solubility Relationships Between Fluorocarbons and Water. Tech. Bull. B-43. Wilmington, Del. 17 pp.

du Pont de Nemours & Company (1966d). Solubility of Water in the Liquid Phase of the Freon Products. Tech. Bull. M-5. Wilmington, Del. 4 pp.

du Pont de Nemours & Company (1966e). Solvent Properties Comparison Chart, Freon Aerosol Report FA-26. Wilmington, Del. 6 pp.

du Pont de Nemours & Company (1969a). Freon Fluorocarbons Properties and Applications. Tech. Bull. B-2. Wilmington, Del. 11 pp.

du Pont de Nemours & Company (1969b). Freon 502 Refrigerant and a Comparison with Refrigerants 12 and 22. Tech. Bull. RT-31. Wilmington, Del. 19 pp.

du Pont de Nemours & Company (1969c). Thermodynamic Properties of Freon 502 Refrigerant. Org. Chem. Dept. Bull.-T502. Wilmington, Del. 35 pp.

Durrans, T. H. (1950). Solvents (6th ed.). Chapman & Hall, London.

Dyke, T. R., Mack, K. M., and Muenter, J. S. (1977). J. Chem. Phys. $\underline{66}$(2), 498-510.

Dymond, J. and Hildebrand, J. H. (1967). Ind. Eng. Chem. Fundam. $\underline{6}$(1), 130-131.

Dymond, J. H. and Smith, E. B. (1979). The Second Virial Coefficients of Pure Gases and Mixtures (2nd ed.). Oxford University Press, Oxford.

Earhart, J. P., Won, K. W., Wong, H. Y., Prausnitz, J. M., and King, C. J. (1977). Chem. Eng. Prog. $\underline{73}$(5), 67-73.

Eberius, E. (1954). Wasserbestimmung mit Karl-Fischer Lösung. Verlag Chemie, Weinheim, West Germany. 138 pp.

Eberson, L. (1969). Acidity and Hydrogen Bonding of Carboxyl Groups. In Chemistry of Carboxylic Acids and Esters (S. Patai, Ed.). Interscience, London. p. 236.

Eckert, C. A., Renon, H., and Prausnitz, J. M. (1976). Ind. Eng. Chem. $\underline{6}$(1), 58-67.

Eduljee, G. H. and Tiwari, K. K. (1976). Chem. Eng. Sci. $\underline{31}$(7), 535-540.

Eduljee, G. H. and Tiwari, K. K. (1977). Chem. Eng. Sci. $\underline{32}$(6), 569-570.

Edward, J. T., Farrell, P. G., and Shahidi, F. (1977). J. Chem. Soc. Fraday Trans. I $\underline{73}$(5), 705-721.

Eganhouse, R. P. and Calder, J. A. (1976). Geochim. Cosmochim. Acta $\underline{40}$, 555-561.

Egelstaff, P. A. (1967). An Introduction to the Liquid State. Academic, New York.

Egelstaff, P. A. (1973). The Structure of Simple Liquids. Annu. Rev. Phys. Chem. $\underline{24}$, 159-187.

Egelstaff, P. A. (1979). Pure Appl. Chem. $\underline{51}$, 2131-2145.

Eger, E. I. and Larson, C. P. (1964). Br. J. Anaesth. $\underline{36}$, 140-149.

Eidinoff, M. L. (1955). Distillation of Water. In Production of Heavy Water (M. L. Eidinoff, G. G. Joris, E. Taylor, H. S. Taylor, and H. C. Urey, Eds.). McGraw-Hill, New York. p. 120-144.

Eiseman, B. J. (1955). Refrig. Eng. (April.) p. 61-66 and 133-139.

Eisenberg, D. and Kauzmann, W. (1969). The Structure and Properties of Water. Oxford University Press, Oxford. 296 pp.

Eisenschitz, R. and London, F. (1930). Z. phys. $\underline{60}$, 491-502.

El Khishen, S. A. (1948). Anal. Chem. $\underline{20}$(11), 1078-1081.

Elchardus, M. E. M. and Maestre, M. (1965). Genie Chem. $\underline{93}$(1), 6-24.

Eley, D. D. (1939a). Trans. Faraday Soc. $\underline{35}$, 1281-1293.

Eley, D. D. (1939b). Trans. Faraday Soc. $\underline{35}$, 1421-1432.

Eley, D. D. (1953). Disc. Faraday Soc., No. 15, 262-264.

Ellis, P. D., Li, Y. S., Tong, C. C., Zens, A. P., and During, J. R. (1975). J. Chem. Phys. $\underline{62}$(4), 1311-1313.

Elsey, H. M. and Flowers, L. C. (1949). Refrig. Eng. $\underline{57}$(2), 153-157.

Ember, L. (1975). Environ. Sci. Technol. $\underline{9}$(13), 1116-1121.

Emsley, J. (1980). Chem. Soc. Rev. $\underline{9}$(1), 91-124.

Engineering Science Data Units Item No. 75010 (1975). Vapour Pressure and Critical Points of Liquids. V. Halogenated Methanes. London. June. 31 pp.

Englin, B. A., Plate, A. F., Tugolukova, V. M., and Pryanishnikova, M. A. (1965).
    Khim. Tekhnol. Topl. Masel, No. 9, 42-46.

Environmental Protection Agency (1975). Fed. Regist. 40(51), 11990-11998.

Environmental Protection Agency (1978). Fed. Regist. 43(28), 5756-5780.

Environmental Studies Board (1973). Water Quality Criteria, 1972. Environmental
    Protection Agency EPA-R3-73-033. Washington, D. C., March. 594 pp.

Epley, T. D. and Drago, R. S. (1967). J. Am. Chem. Soc. 89(23), 5770-5773.

Epshtein, N. A. and Nizhnii, S. V. (1979a). Pharm Chem. J. 13(3), 268-273.

Epshtein, N. A. and Nizhnii, S. V. (1979b). Khim.-Farm. Zh. 13(4), 57-67.

Epshtein, N. A. and Nizhnii, S. V. (1979c). Khim.-Farm. Zh. 13(10), 57-62.

Epstein, S. S. and Swartz, J. B. (1981). Nature 285(5794), 127-130.

Erdey-Gruz, T. (1974). Transport Phenomena in Aqueous Solutions. Adam Hilger, London.
    512 pp.

Ethyl Corporation (1958a). Chlorinated Solvents. Rep. ICD-1006 (58). New York. 12 pp.

Ethyl Corporation (1958b). Physical Property Charts VCL - EDC Know-How. Rep. TDM 246,
    RWP. New York, February 2.

Ethyl Corporation (1964). Ethylene Dichloride. Tech. Bull. No. IC-556 (R/11/64). New
    York. 5 pp.

Ethyl Corporation (1968). Vapor-Liquid Equilibrium of Water - Vinyl Chloride - Hyd-
    rogen Chloride at Various Temperatures and Moderate Pressures. Rep. 68-2. New
    York. 71 pp.

Ethyl Corporation (1972). Ethylene Dibromide. Tech. Bull. IC-73 (12/72). New York.
    3 pp.

Eucken, A. (1946). Theory of the Condition of Water. Nachr. Akad. Wiss. Göttingen
    Math.-Phys. Kl. p. 38-48. (CA 43:7283d.)

Eucken, A. (1947). Effect of Dissolved Substances on the Constitution of Water. Nachr.
    Akad. Wiss. Göttingen Math.-Phys. Kl., Math.-Phys.-Chem. Abt., p. 33-36. (CA 43:
    8243g.)

Eucken, A. (1948-1949). Z. Elektrochem. 52, 255 (1948); 53, 102-105 (1949).

Evans, M. G. (1937). Trans. Faraday Soc. 33, 166-170.

Evans, M. G. and Polanyi, M. (1936). Trans. Faraday Soc. 32, 1333-1360.

Evans, T. W. (1936). Ind. Eng. Chem. Anal. Ed. 8(3), 206-208.

Evelein, K. A., Moore, R. G., and Heidemann, R. A. (1976). Ind. Eng. Chem. Process
    Des. Dev. 15(3), 423-428.

Everett, D. H. (1953). Disc. Faraday Soc., No. 15, 267-268.

Ewell, R. H. and Eyring, H. (1937). J. Chem. Phys. 5, 726-736.

Ewell, R. H., Harrison, J. M., and Berg, L. (1944). Ind. Eng. Chem. 36(10), 871-875.

Ewing, G. E. (1975). Acc. Chem. Res. 8(6), 185-192.

Exner, O. (1973). Prog. Phys. Org. Chem. 10, 411-482.

Eyring, H. (1936). J. Chem. Phys. 4(4), 283-291.

Eyring, H. and Hirschfelder, J. O. (1937). J. Phys. Chem. 41(2), 249-257.

Eyring, H. and Jhon, M. S. (1969). Significant Liquid Structures. Wiley, New York. 149 pp.

Eyring, H. and Marchi, R. P. (1963). J. Chem. Educ. 40(11), 562-572.

Eyring, H. and Ree, T. (1961). Proc. Natl. Acad. Sci. U.S.A. 47, 526.

Eyring, H., Hildebrand, J., and Rice, S. (1963). Int. Sci. Technol., No. 15, 56-66.

Fabuss, B. M. and Korosi, A. (1968). Properties of Sea Water and Solutions Containing Sodium Chloride, Potassium Chloride, Sodium Sulfate, and Magnesium Sulfate. Res. Dev. Prog. Rep. No. 384, U.S. Dept. of the Interior, Washington, D.C.

Fair, J. R. (1980). AIChE Monograph Ser. 76(13), 41 pp.

Farbwerke Hoechst (1969). Hoechst Solvents (4th ed.). Frankfurt. 231 pp.

Farkas, E. J. (1965). Anal. Chem. 37(9), 1173-1174.

Fedors, R. F. (1974). Polym. Eng. Sci. 14(2), 147-154; 14(6), 472.

Feingold, A. (1976). Anesth. Analg. Curr. Res. 55(4), 593-595.

Fenton, T. M. and Garner, W. E. (1930). J. Chem. Soc. 694-700.

Fernandez, R., Carey, W. W., Bozzo, A. T., and Barduhn, A. J. (1967). Thermodynamics and Kinetics in the Hydrate and Freezing Processes. Res. Dev. Prog. Rep. No. 229. Office of Saline Water, U.S. Dept. of the Interior, Washington, D.C., January. 70 pp.

Ferstanding, L. L. (1978). Anesth. Analg. 57(3), 328-345.

Fetsko, J. M. (1974). NPIRI Raw Materials Data Handbook, Vol. 1, Organic Solvents. Natl. Printing Ink Res. Inst., Lehigh Univ., Bethlehem, Pa. 130 pp.

Filatkin, V. N., Plotnikov, V. T., and Alishev, A. G. (1975). Kholod. Mash. Appar., p. 185-187.

Filatkin, V. N., Plotnikov, V. T., and Alishev, A. G. (1976). Kholod. Tekh., No. 2, 23-25.

Filippov, L. P. (1954). Vestn. Mosk. Univ. 9(12) Ser. Fiz.-Mat. Estestv. Nauk., No. 8, 45-48.

Filippov, T. S. and Furman, A. A. (1952). Zh. Prikl. Khim. 25, 895-897.

Filomenko, G. V. and Korol, A. N. (1976). J. Chromatogr. 119, 157-166.

Finney, J. L. (1970). Proc. Roy. Soc. Lond. Ser. A 319, 495-507.

Finsy, R. and van Loon, R. (1976). J. Phys. Chem. 80(25), 2783-2788.

Fisher, I. Z. (1962). Usp. Fiz. Nauk 76, 499-518.

Fisher, I. Z. (1964). Statistical Theory of Liquids. Univ. of Chicago Press, Chicago. 335 pp.

Fitzgerald, M. E., Griffing, V., and Sullivan, J. (1956). J. Chem. Phys. 25(5), 926-933.

Flemr, V. (1976). Coll. Czech. Chem. Commun. 41(11), 3347-3349.

Fletcher, N. H. (1970). The Chemical Physics of Ice. Cambridge University Press, Cambridge.

Fleury, D. and Hayduk, W. (1975). Can. J. Chem. Eng. 53(2), 195-199.

Flid, R. M. and Golymetz, Yu. F. (1959). Izv. Vyssh. Uchebn. Zaved., Khim. Khim. Tekhnol. 2, 173-179.

Flowers, B. H. and Mendoza, E. (1970). Properties of Matter. Wiley, New York. 318 pp.

Flynn, C. M. (1975). J. Chem. Educ. 52(10), 641.

Fonenko, V. N. (1965). Gig. Sanit. 30(1), 9-15.

Ford, C. L. (1975). An Overview of Halon 1301 Systems. In Halogenated Fire Suppressants. Am. Chem. Soc. Symp. Ser. 16 (R. G. Gann, Ed.). Washington, D.C. p. 1-11.

Forslind, E. (1952). Acta Polytech., No. 15, Chem. Incl. Metall. Ser. 3, No. 5.

Foster, R., Ed. (1975). Molecular Association, Vol. 1. Academic, London. 365 pp.

Fowler, R. D., Hamilton, J. M., Kasper, J. S., Weber, C. E., Buford, W. B., and Anderson, H. C. (1947). Ind. Eng. Chem. 39(3), 375-378.

Fox, J. J. and Martin, A. E. (1940). Proc. Roy. Soc. Lond. Ser. A 174, 234-262.

Francis, A. W. (1963). Liquid-Liquid Equilibriums. Interscience, New York.

Francis, A. W. (1972). Handbook for Components in Solvent Extraction. Gordon and Breach, New York.

Franck, E. U. and Mayer, F. (1959). Z. Elektrochem. 63(5), 571-582.

Franck, E. U., Harder, W., Hill, W., and Reuter, K. (1975). Measurement and Discussion of the Dielectric Constant of Methylcyanide, Methylfluoride, and Methyltrifluoride to 300°C and 2000 bar. Proc. 4th Int. Conf. on High Press., Phys.-Chem. Soc., Kyoto, Japan, 1974, p. 602-603.

Frank, H. S. (1945a). J. Chem. Phys. 13(11), 478-492.

Frank, H. S. (1945b). J. Chem. Phys. 13(11), 493-507.

Frank, H. S. (1958). Proc. Roy. Soc. Lond. Ser. A 481-492.

Frank, H. S. (1963). Questions About Water Structure. Natl. Acad. Sci., Natl. Res. Counc. Publ. No. 942, Washington, D.C., p. 141-155.

Frank, H. S. (1965). Fed. Proc., Suppl. No. 15, 24(2), Part 3, p. 1-11.

Frank, H. S. and Evans, M. W. (1945). J. Chem. Phys. 13(11), 507-532.

Frank, H. S. and Franks, F. (1968). J. Chem. Phys. 48(10), 4746-4757.

Frank, H. S. and Wen, W.-Y. (1957). Disc. Faraday Soc. 24, 133-140.

Franks, F. (1966). Nature 210(5031), 87-88.

Franks, F. (1967). Physical-Chemical Processes in Mixed Aqueous Solvents. Heinemann Educational Books, London. 152 pp.

Franks, F. Ed. (1972-1979). Water. A Comprehensive Treatise. Plenum, New York. Vol. 1, The Physics and Physical Chemistry of Water, 1972, 596 pp; Vol. 2, Water in Crystalline Hydrates: Aqueous Solutions of Simple Non-Electrolytes, 1973, 684 pp; Vol. 3, Aqueous Solutions of Simple Electrolytes, 1973, 472 pp; Vol. 4, Aqueous Solutions of Macromolecules: Water in Disperse Systems, 1973, 587 pp; Vol. 5, Water in Disperse Systems, 1975, 384 pp; Vol. 6, Recent Advances, 1979, 470 pp.

Franks, F. and Ives, D. J. G. (1960). J. Chem. Soc., No. 1, 741-754.

Franks, F., Gent, M., and Johnson, H. H. (1963). J. Chem. Soc., 2716-2723.

Fredenslund, A. and Grausø, L. (1975). Determination and Prediction of Henry's Constants. 4th Int. Conf. on Chem. Thermodyn., IUPAC, Montpellier, August 26-30, Part IV/6, p. 36-45.

Fredenslund, A., Gmehling, J., and Rasmussen, P. (1977). Vapor-Liquid Equilibria Using UNIFAC. A Group-Contribution Method. Elsevier, Amsterdam. 380 pp.

Fredenslund, A., Jones, R. L., and Prausnitz, J. M. (1975). AIChE J. 21(6), 1086-1099.

Fredenslund, A., Michelsen, M. L., and Prausnitz, J. M. (1976). Chem. Eng. Prog. 72(9), 67-69.

Freier, R. K. (1978). Aqueous Solutions. Walter de Gruyter, Berlin. p. 111-255.

Frenkel, J. (1955). Kinetic Theory of Liquids. Dover, New York.

Friedman, H. L. (1954). J. Am. Chem. Soc. 76, 3294-3297.

Friedman, H. L. (1973). Chem. Br. 9(7), 300-305.

Friend, L. and Adler, S. B. (1957). Chem. Eng. Prog. 53(9), 452-458.

Frisch, H. L. and Lebowitz, J. L., Eds. (1964). The Equilibrium Theory of Classical Fluids. Benjamin, New York.

Frisch, H. L. and Salsbury, F. W., Eds. (1968). Simple Dense Fluids. Academic, New York.

Fritz, F. (1957). Die wichtigsten Lösung- und Weichmachungs-mittel. VEB Verlag Technik, Berlin.

Fritz, J. S. (1976). Chem. Eng. News 54(16), 35-36.

Fritz, J. S. (1977). Acc. Chem. Res. 10(2), 67-72.

Fröhlich, H. (1946). Trans. Faraday Soc. 42A, 3-7.

Fröhlich, H. (1958). Theory of Dielectrics (2nd ed.). Oxford University Press, Oxford.

Frontas'ev, V. P. (1956). Dokl. Akad. Nauk SSSR 111, 1014-1016.

Fühner, H. (1924). Chem. Ber. 57B, 510-515.

Fuoss, R. M. and Kirkwood, J. G. (1941). J. Am. Chem. Soc. 63, 385-394.

Fürer, R. and Geiger, M. (1977). Pestic. Sci. 8(4), 337-344.

Furniss, B. S., Hannaford, A. J., Rogers, V., Smith, P. W. G., and Tatchell, A. R. (1978). Vogel's Textbook of Practical Organic Chemistry (4th ed.). Longman, London. 1371 pp.

Fürth, R. (1949). Sci. Prog. 146, 202-218.

Gallant, R. W. (1968-1970). Physical Properties of Hydrocarbons. Gulf Publ. Co., Houston, Tex. Vol. 1 (1968), 225 pp.; Vol. 2 (1970), 201 pp.

Gann, R. G., Ed. (1975). Halogenated Fire Suppressants. Am. Chem. Soc. Symp. Ser. No. 16, Washington, D.C. 453 pp.

Gardiner, G. E. (1970). Gas-Liquid Equilibriums at High Pressures with Special Reference to the Solubility of Helium in Water and Aqueous Sodium Chloride. (Ph. D. thesis), Fordham Univ., New York. 142 pp.

Gardon, J. L. (1966). J. Paint Technol. 38(492), 43-56.

Gardon, J. L. (1977a). Prog. Org. Coat. 5(1), 1-20.

Gardon, J. L. (1977b). J. Colloid Interface Sci. 59(3), 582-596.

Garner, R. C. and Hertzog, P. J. (1981). Nature 289(5799), 627.

Garrett, H. E. (1972). Surface Active Chemicals. Pergamon Press, Oxford. 167 pp.

Gaube, J. and Koenen, E. (1979). Chem.-Ing.-Techn. 51(5), 496-497.

Gebelein, C. G. (1978). Org. Coat. Plast. Chem. 39, 524-528.

Gentric, E., Le Narvon, A., and Saunague, P. (1970). C. R. Acad. Sci. Ser. C 270, 1053-1056.

Gerrard, W. (1971). Chem. Ind., July 31, 884-885.

Gerrard, W. (1972). Chem. Ind., October 21, 804-805.

Gerrard, W. (1973). J. Appl. Chem. Biotechnol. 23, 1-17.

Gerrard, W. (1976). Solubility of Gases and Liquids. A Graphical Approach. Plenum, London. 275 pp.

Gerrard, W. (1980). Gas Solubilities. Widespread Applications. Pergamon Press, Oxford. 497 pp.

Gerrens, H., Fink, W., and Köhnlein, E. (1967). J. Polym. Sci., Part C, No. 16, 2781-2793.

Getzen, F. W. (1970). The Use of Slightly Soluble Non-Polar Solutes as Probes for Obtaining Evidence of Water Structure. In Liquid Crystals and Ordered Fluids (J. F. Johnson and R. S. Porter, Eds.). Plenum, New York. p. 53-67.

Getzen, F. W. (1976). Structure of Water and Aqueous Solubility. In Techniques of Chemistry, Vol. 8, Solutions and Solubilities, Part 2 (M. R. J. Dack, Ed.). Wiley, New York. p. 363-436.

Getzen, F. W. and Ward, T. M. (1971). Ind. Eng. Chem. Prod. Res. Dev. 10(2), 122-132.

Ghosh, S. K. and Chopra, S. J. (1975). Ind. Eng. Chem. Process Des. Dev. 14(3), 304-308.

Gibbard, H. F. and Emptage, M. R. (1975). J. Chem. Educ. 52(10$\frac{1}{4}$, 673-676.

Gibbs, J. H. (1977). Ann. N. Y. Acad. Sci. 303, 20-29.

Gibbs, J. H., Cohen, C., Fleming, P. D., and Porosoff, H. (1973). J. Solution Chem. 2(2/3), 277-299.

Gibby, C. W. and Hall, J. (1931). J. Chem. Soc., 691-693.

Gibney, L.-Y. (1975a). Chem. Eng. News 53(10), 17.

Gibney, L.-Y. (1975b). Chem. Eng. News 53(37), 17.

Gibney, L.-Y. (1976). Chem. Eng. News 54(18), 14-15.

Gibson, H. W. (1977). Can. J. Chem. 55(14), 2637-2641.

Gill, S. J. and Wadsö, I. (1976). Proc. Natl. Acad. Sci. U.S.A. 73(9), 2955-2958.

Gill, S. J., Nichols, N. F., and Wadsö, I. (1975). J. Chem. Thermodyn. 7, 175-183.

Gill, S. J., Nichols, N. F., and Wadsö, I. (1976). J. Chem. Thermodyn. 8, 445-452.

Gilmour, J. B. (1965). Solubilities and Free Energies of Mixing of Fluorocarbons at
    Low Temperature (Ph. D. thesis). Univ. of California, Los Angeles, Calif. 165 pp.

Ginell, R. (1979). Chem. Tech. $\underline{9}$(7), 446-450.

Gingold, M. P. (1973). Bull. Soc. Chim. Fr., No. 5, 1629-1634.

Ginning, D. C. and Furukawa, G. T. (1953). J. Am. Chem. Soc. $\underline{75}$(3), 522-527.

Gjaldbaek, J. Ch. and Anderson, E. K. (1954). Acta Chem. Scand. $\underline{8}$, 1398-1413.

Gladis, G. P. (1960). Chem. Eng. Prog. $\underline{56}$(10), 43-51.

Glasoe, P. K. and Schultz, S. D. (1972). J. Chem. Eng. Data $\underline{17}$(1), 66-68.

Glasstone, S. (1937). Trans. Faraday Soc. $\underline{33}$, 200-214.

Glaze, W. H. and Henderson, J. E. (1975). J. Water Pollut. Control Fed. $\underline{47}$(10), 2511-
    2515.

Glew, D. N. (1952). Some Static and Dynamic Properties of Methyl Halides in Aqueous
    Solutions (Ph. D. dissertation 2123), Cambridge Univ., Cambridge. 245 pp.

Glew, D. N. (1960). Can. J. Chem. $\underline{38}$, 208-221.

Glew, D. N. (1962). J. Phys. Chem. $\underline{66}$, 605-609.

Glew, D. N. and Moelwyn-Hughes, E. A. (1953). Disc. Faraday Soc., No. 15, 150-161.

Glew, D. N. and Robertson, R. E. (1956). J. Phys. Chem. $\underline{60}$, 332-337.

Gmehling, J. and Onken, U. (1977). Vapor-Liquid Equilibrium Data Collection. Aqueous-
    Organic Systems. Deutsche Ges. Chem. Apparatewesen, Chem. Data Ser., Vol. 1, No.
    1. 698 pp.

Gmelins Handbuch der Anorganischen Chemie (1963). Sauerstoff, Lief. 5. Verlag Chemie,
    Weinheim, West Germany. p. 1191-1732.

Gmelins Handbuch der Anorganischen Chemie (1964). Sauerstoff, Lief. 6, Verlag Chemie,
    Weinheim, West Germany. p. 1733-1750; 1750-1759.

Gmelins Handbuch der Anorganischen Chemie (1974). Kohlenstoff, Kohlenstoff-Halogen-
    Verbindungen. Syst. No. 14, Part D2. Springer-Verlag, Berlin. 386 pp.

Gnamm, H. and Sommer, W. (1958). Lösungsmittel und Weichmachungsmittel (7th ed.).
    Wissenschaftliche Verlagsgesellschaft, Stuttgart, West Germany. 971 pp.

Gokcen, N. A. (1973). J. Chem. Soc. Faraday Trans. I $\underline{69}$(2), 438-443.

Gokcen, N. A. and Chang, E. T. (1973). Temperature Sependence of Solubilities of
    Electrolytes and Nonelectrolytes. U. S. Dept. of Commerce, Natl. Tech. Inf. Serv.,
    Springfield, Va., March 7. 18 pp.

Gokcen, N. A. and Chang, E. T. (1975). Denki Kagaku Oyobi Kogyo Butsuri Kagaku $\underline{43}$(5),
    232-237. (CA 83:137555t.)

Gokcen, N. A. and Chang, E. T. (1977). J. Chem. Educ. $\underline{54}$(6), 368.

Goldberg, E. D., Ed. (1975). The Nature of Seawater. Report of the Dahlem Workshop
    on the Nature of Seawater. Dahlem Konf. bei Abakon Verlagsges., Berlin. 719 pp.

Goldman, S. (1969). The Hydrates and Lower Aggregates of Tetra-N-Butylammonium
    Picrate in Organic Solvents: A Thermodynamic Study (Ph. D. thesis). McGill Univ.,
    Montreal, Quebec. July.

Goldman, S. (1974). Can. J. Chem. $\underline{52}$(9), 1668-1680.

Goldman, S. (1976). J. Phys. Chem. $\underline{80}$(15), 1697-1700.

Goldman, S. (1977). J. Solution Chem. $\underline{6}$(7), 461-474.

Goldman, S. and Krishnan, T. R. (1976). J. Solution Chem. $\underline{5}$(10), 693-707.

Goldman, S. (1979). Acc. Chem. Res. $\underline{12}$, 409-415.

Goldsmith, R. H. (1974). J. Chem. Educ. $\underline{51}$(4), 272-273.

Gooch, J. P. (1971). The Solubility of 1,1,2,2-Tetrabromoethane in Water as a Function of Temperature (Ph. D. thesis). Univ. of Alabama, Birmingham, Ala. 68 pp.

Gooch, J. P., Landis, E. K., and Browning, J. S. (1972). Solubility of 1,1,2,2-Tetrabromoethane in Water as a Function of Temperature. U. S. Natl. Tech. Inf. Serv. PB Rep. No. 211354. Washington, D.C. 24 pp.

Good, R. J. and Elbing, E. (1970). Ind. Eng. Chem. $\underline{62}$(3), 54-78.

Goodman, G. T. (1974). Proc. Roy. Soc. Lond. Ser. B $\underline{185}$, 127-148.

Gorbunov, A. N. (1980). Russ. J. Phys. Chem. $\underline{54}$(6), 805-807.

Gordon, A. J. and Ford, R. A. (1972). The Chemist's Companion. Wiley, New York.

Gordon, J. J. and Bright, A. W. (1939). Process for Azeotropic Distillation of Aliphatic Acids. U. S. Patent No. 2,171,549 (to Eastman Kodak Co.), September 5. 10 pp.

Gordon, M., Hope, C. S., Loan, L. D., and Roe, R. J. (1960). Proc. Roy. Soc. Lond. Ser. A $\underline{258}$(1293), 215-236.

Gordy, W. and Stanford, S. C. (1939-1941). J. Chem. Phys. $\underline{7}$, 93-99(1939); $\underline{8}$, 170-177 (1940); $\underline{9}$, 204-223(1941).

Gory, G. (1925). Ann. chim. applicata $\underline{15}$, 283-300.

Gorodyskii, V. A., Kardashina, L. F., and Bakhshiev, N. G. (1975). Russ. J. Phys. Chem. $\underline{49}$(5), 641-646.

Gothard, F. A. (1975). Echilibre lichid-vapori. Editura ARSR, Bucharest.

Gothard, F. A., Ciobanu, M. F. C., Breban, D. G., Bucur, C. I., and Soroscu, G. V. (1975). 4th Conf. Int. Thermodyn. Chim. (C. R.), Part IV, p. 53-63.

Gothard, F. A., Ciobanu, M. F. C., Breban, D. G., Bucur, C. I., and Sorescu, G. V. (1976). Ind. Eng. Chem. Process Des. Dev. $\underline{15}$(2), 333-337.

Gotoh, K. (1971). Nature Phys. Sci. $\underline{232}$, 64-65.

Gotoh, K. (1972). Nature Phys. Sci. $\underline{239}$, 154-156.

Gotoh, K. (1976). Ind. Eng. Chem. Fundam. $\underline{15}$(4), 269-274.

Götze, G., Jockers, R., and Schneider, G. M. (1975). PVT and Phase Equilibrium Data for Aqueous Mixtures at High Pressures. 4th Int. Conf. on Chem. Thermodyn. IUPAC, Montpellier. August 26-30. Part IV/9, p. 57-64.

Grant, W. J. (1978). Medical Gases - Their Properties and Uses. HM + M Publ., Aylesbury. 199 pp.

Gray, D. E., Ed. (1963). American Institute of Physics Handbook (2nd ed.). McGraw-Hill, New York. p. 5.121-5.134.

Green, H. S. (1970). The Molecular Theory of Fluids. North-Holland, Amsterdam.

Green, S. W. (1969). Chem. Ind., January 18, 63-67.

Green, W. J. and Frank, H. S. (1979). J. Solution Chem. $\underline{8}$(3), 187-196.

Greenacre, G. C. and Young, R. N. (1976). J. Chem. Soc. Perkin Trans. II, No. 8, 874-876.

Greer, E. J. (1930). J. Am. Chem. Soc. $\underline{52}$, 4191-4201.

Gregory, M. D., Affsprung, H. E., and Christian, S. D. (1968). J. Phys. Chem. $\underline{72}$(5), 1748-1751.

Greinacher, E., Lüttke, W., and Mecke, R. (1955). Z. Elektrochem. $\underline{59}$(1), 23-31.

Grewer, Th. (1962). Löslichkeit von Frigen 21 in Wasser und von Frigen 22 in Äthanol-Wasser-Gemisch. Hoechst Aktiengesell., 1079-III-1. Frankfurt, June 13. 2 pp.

Gribbin, J. (1979). New Scientist $\underline{81}$(1138), 164-167.

Grifalco, L. A. and Good, R. J. (1957). J. Phys. Chem. $\underline{61}$(7), 904-909.

Gross, P. M. (1929a). J. Am. Chem. Soc. $\underline{51}$, 2362-2366.

Gross, P. M. (1929b). Z. phys. Chem. $\underline{6B}$, 215-220.

Gross, P. M. (1931). Phys. Z. $\underline{32}$, 587-592.

Gross, P. M. and Saylor, J. H. (1931). J. Am. Chem. Soc. $\underline{53}$, 1744-1751.

Gross, P. M., Saylor, J. H., and Gorman, M. A. (1933). J. Am. Chem. Soc. $\underline{55}$, 650-652.

Grosse, A. V. and Cady, G. H. (1947). Ind. Eng. Chem. $\underline{39}$(3), 367-374.

Grundwald, E., Pan, K.-C., Anderson, S. P., Effio, A., and Gould, S. E. (1976). J. Phys. Chem. $\underline{80}$(27), 2929-2944.

Guerrant, R. P. (1964). Hydrocarbon - Water Solubilities at High Temperatures Under Vapor-Liquid-Liquid Equilibrium Conditions (M. Sc. thesis). Pennsylvania State Univ., University Park, Pa., December. 124 pp.

Guggenheim, E. A. (1935). Proc. Roy. Soc. Lond. Ser. A $\underline{148}$, 304-312.

Guggenheim, E. A. (1952). Mixtures. The Theory of the Equilibrium Properties of Some Simple Classes of Mixtures, Solutions, and Alloys. Clarendon Press, Oxford. 270 pp.

Guggenheim, E. A. (1953). Disc. Faraday Soc., No. 15, 271-272.

Guggenheim, E. A. (1966). Applications of Statistical Mechanics. Clarendon Press, Oxford.

Gunther, F. A., Westlake, W. E., and Jaglan, P. S. (1968). Residue Rev. $\underline{20}$, 1-148.

Gupta, S. P. and Singh, P. (1979). Bull. Chem. Soc. Japan $\underline{52}$(9), 2745-2746.

Gurfein, L. N. and Pavlova, Z. K. (1960). Sanit. Okhr. Vodoemov Zagryaz. Prom. Stochnymi Vodami, No. 4, 117-127. (CA 56:7060i.)

Gurikov, Yu. V. (1969). Russ. J. Phys. Chem. $\underline{43}$(1), 85-88.

Gurikov, Yu. V. (1980). Russ. J. Phys. Chem. $\underline{54}$(5), 697-700.

Gurnham, C. F., Ed. (1965). Industrial Wastewater Control. A Textbook and Reference Work. Academic, New York.

Guseva, A. N. and Parnov, E. I. (1964). Russ. J. Phys. Chem. $\underline{38}$(3), 439-440.

Guthrie, J. P. (1977). Can. J. Chem. $\underline{55}$(21), 3700-3706.

Gutmann, V. (1977). Chem. Tech. $\underline{7}$(4), 255-263.

Gutowsky, H. S. (1976). Removal of Halocarbons at the Surface. Solubility in and
Removal by the Oceans. In Halocarbons: Effects on Statospheric Ozone. Natl. Acad.
Sci., Washington, D. C., p. 179-185.

Haase, R. (1953). Disc. Faraday Soc., No. 15, 270-271.

Hadzi, D. and Thompson, H. W., Eds. (1959). Hydrogen Bonding. Proc. 1st Int. Conf. on
Hydrogen Bonding. Ljubljane, 1957. Pergamon, New York. 571 pp.

Hagan, M. (1962). Clathrate Inclusion Compounds. Reinhold, New York.

Hagen, A. P. and Elphingstone, E. A. (1974). J. Inorg. Nucl. Chem. $\underline{36}$(3), 509-511.

Hagler, A. T., Scheraga, H. A., and Nemethy, G. (1972). J. Phys. Chem. $\underline{76}$, 3229-3243.

Haight, G. P. (1951). Ind. Eng. Chem. $\underline{43}$(8), 1827-1828.

Hala, E., Pick, J., Fried, V., and Vilin, O. (1967). Vapour-Liquid Equilibrium (2nd
ed.). Pergamon, Oxford.

Hala, E., Wichterle, I., Polak, J., and Boublik, T. (1968). Vapour-Liquid Equilibrium
Data at Normal Pressures. Pergamon, Oxford.

Hales, J. L. (1980). Bibligraphy of Fluid Density: Experimental Determinations on
Single Substances Over a Temperature Range Exceeding 40 K., Natl. Phys. Lab. Rep.
CHEM 106, Teddington. February. 94 pp.

Halkiadakis, E. A. and Bowrey, R. G. (1975). Chem. Eng. Sci. $\underline{30}$(1), 53-60.

Hall, L. H., and Kier, L. B. (1977a). J. Pharm. Sci. $\underline{66}$(5), 642-644.

Hall, L. H. and Kier, L. B. (1977b). Tetrahedron $\underline{33}$(15), 1953-1957.

Hall, L. H. and Kier, L. B. (1978). Eur. J. Med. Chem. $\underline{13}$(1), 89-92.

Hall, L. H. and Kier, L. B. (1978b). J. Pharm. Sci. $\underline{67}$(12), 1743-1747.

Hall, L. H., Kier, L. B., and Murray, W. J. (1975). J. Pharm. Sci. $\underline{64}$(12), 1974-1977.

Halliday, M. M., MacDonald, I., and MacGregor, M. H. G. (1977). Br. J. Anaesth. $\underline{49}$(5),
413-417.

Hamilton, J. M. (1963). The Organic Fluorochemicals Industry. In Advances in Fluorine
Chemistry, Vol. 3 (M. Stacey, J. C. Tatlow, and A. G. Sharpe, Eds.). Butterworths,
London. p. 117-180.

Hamilton, W. C. and Ibers, J. A. (1968). Hydrogen Bonding in Solids. Methods of Molecular Structure Determination. W. A. Benjamin, New York. 284 pp.

Hammarstrand, K. (1976). Varian Instr. Appl. $\underline{10}$(2), 2-4.

Hammer, P. M., Hayes, J. M., Jenkins, W. J., and Gagosian, R. B. (1978). Geophys. Res.
Lett. $\underline{5}$(8), 645-648.

Hancock, E. G., Ed. (1973). Propylene and Its Industrial Derivatives. Ernest Benn,
London. 517 pp.

Hancock, W. and Laws, E. Q. (1955). Analyst $\underline{80}$, 665-674.

Hansch, C. (1-71). In Drug Design, Vol. 1 (E. J. Ariens, Ed.), Academic, New York.
p. 271-342.

Hansch, C. and Anderson, S. M. (1967). J. Org. Chem. $\underline{32}$, 2583-2586.

Hansch, C., Quinlan, J. E., and Lawrence, G. L. (1968). J. Org. Chem. 33(1), 347-350.

Hansch, C., Unger, S. H., and Forsythe, A. B. (1973). J. Med. Chem. 16(11), 1217-1222.

Hansch, C., Vittoria, A., Silipo, C., and Jow, P. Y. C. (1975). J. Med. Chem. 18(6), 546-548.

Hansen, C. M. (1967). The Three-Dimensional Solubility Parameter and Solvent Diffusion Coefficient. Danish Technical, Copenhagen. 106 pp.

Hansen, J.-P. and McDonald, I. R. (1976). Theory of Simple Liquids. Academic, New York. 395 pp.

Hanssens, I., Mullens, J., Deneuter, C., and Huyskens, P. (1968). Bull. Soc. Chim. Fr., No. 10, 3942-3945.

Häntzschel, H. and Kittner, M. (1969). Luft-Kältetech., No. 4, 184-187.

Hardy, C. J. (1959). J. Chromatogr. 2, 490-498.

Harms, H. (1939). Z. phys. Chem. Abt. B 43, 257-270.

Harms, L. L. and Looyenga, R. W. (1977). J. Am. Water Works Assoc. 69(5), 258-263.

Harnden, C. W. (1947). Bactericide and Algaecide. U. S. Patent No. 2,419,021 (to Shell Develop. Co., San Francisco), April 15. 2 pp.

Harris, M. J. (1971). An Evaluation of Thermodynamic Group Contributions from Ion Pair Extraction Equilibria for Use in the Prediction of Partition Coefficients (Ph. D. thesis). Univ. of Kansas, Lawrence, Kans. 149 pp.

Harris, M. J., Higuchi, T., and Rytting, J. H. (1973). J. Phys. Chem. 77(22), 2694-2703.

Harvey, H. H. (1975). Gas Disease in Fishes - A Review. In Chemistry and Physics of Aqueous Gas Solutions (W. A. Adams et al., Eds.). Electrochem. Soc., Princeton, N.J., p. 450-485.

Hashizume, M. (1964). Kogyo Kakaku Zasshi 67(4), 518-523.

Hasted, J. B. (1973). Aqueous Dielectrics, Chapman & Hall, London, 302 pp.

Haszeldine, R. N. and Sharpe, A. G. (1951). Fluorine and Its Compounds. Matheun, London.

Hawkins, D. T. (1975). J. Solution Chem. 4(8), 623-743.

Hawkins, D. T. (1976). Physical and Chemical Properties of Water. A Bibligraphy: 1957-1974. IFI/Plenum, New York. 556 pp.

Hawkins, D. T. (1980). J. Chem. Inf. Computer Sci. 20(3), 143-145.

Hay, A. (1978). Nature 274(5671), 533-534.

Hayashi, T. and Nakajima, T. (1975). Bull. Chem. Soc. Jap. 48(3), 980-984.

Hayashi, T. and Nakajima, T. (1976). Bull. Chem. Soc. Jap. 49(8), 2055-2066.

Hayashi, T. and Sasaki, T. (1956). Bull. Chem. Soc. Jap. 29(8), 857-859.

Hayden, J. G. and O'Connell, J. P. (1975). Ind. Eng. Chem. Process Des. Dev. 14(3), 209-216.

Hayduk, W. and Buckley, W. D. (1971). Can. J. Chem. Eng. 49, 667-671.

Hayduk, W. and Castaneda, R. (1973). Can. J. Chem. Eng. 51, 353-358.

Hayduk, W. and Cheng, S. C. (1970). Can. J. Chem. Eng. 48, 93-99.

Hayduk, W. and Laudie, H. (1973). AIChE J. 19(6), 1233-1238.

Hayduk, W. and Laudie, H. (1974a). AIChE J. 20(3), 611-615 and 1236.

Hayduk, W. and Laudie, H. (1974b). J. Chem. Eng. Data 19(3), 253-257.

Hayes, J. M. and Thompson, G. M. (1977). Trichlorofluoromethane in Ground Water --
    A Possible Indicator of Ground Water Age. Office of Water Res. Technol. Tech. Rep.
    No. 90., U. S. Dept. of the Interior, Washington, D.C. 25 pp.

Hayworth, K. E. (1969). Phase Behavior in a Multicomponent Heteroazeotropic System
    (Ph. D. thesis). Univ. of Southern California, Los Angeles, Calif. 143 pp.

Heidemann, R. A. (1974). AIChE J. 20(5), 847-855.

Heidemann, R. A. (1975). Ind. Eng. Chem. Fundam. 14(1), 72-73.

Heintz, A. and Lichtenthaler, R. N. (1976). Ber. Bunsen ges. Phys. Chem. 80(10), 962-
    965.

Hellstrom, G. W., Jacobs, H. R., and Boehm, R. F. (1976). The Solubility of Selected
    Secondary Fluids for Use in Direct Contact Geothermal Power Cycles. Univ. of Utah,
    Dept. of Mech. Eng., DGE/1549-6, Salt Lake City, Utah, December. 74 pp.

Helpinstill, J. G. and Van Winkle, M. (1968). Ind. Eng. Chem. Process Des. Dev. 7(2),
    213-220.

Henderson, D., Ed. (1971). Liquid State. In An Advanced Treatise in Physical Chemistry;
    Vol. 8A (H. Eyring, D. Henderson, and W. Jost, Eds.). Academic, New York.

Herrington, E. F. G. (1947). Nature 160(4070), 610-611.

Herrington, E. F. G. (1950). Research 3(Suppl.), 41-46.

Herrington, E. F. G. (1951a). J. Inst. Pet. 37, 457-470.

Herrington, E. F. G. (1951b). J. Am. Chem. Soc. 73, 5883-5884.

Herrington, E. F. G. (1953). Disc. Faraday Soc., No. 15, 273-274.

Herrington, E. F. G. (1977). Natl. Phys. Lab. Rep. Chem. 74 (July). 31 pp.

Herman, R. C. (1940). J. Chem. Phys. 8, 252-258.

Herman, R. C. and Hofstadter, R. (1938). J. Chem. Phys. 6, 534-540.

Herman, R. C. and Hofstadter, R. (1939). J. Chem. Phys. 7, 460-464.

Hermann, R. B. (1971). J. Phys. Chem. 75(3), 363-368.

Hermann, R. B. (1972). J. Phys. Chem. 76(19), 2754-2759.

Hermann, R. B. (1975). J. Phys. Chem. 79(2), 163-169.

Hermann, R. B. (1977). Proc. Natl. Acad. Sci. U.S.A. 74(10), 4144-4145.

Herz, W. (1898). Chem. Ber. 31, 2669-2672.

Heslop, R. B. (1974). Educ. Chem. 11(6), 216-217.

Heublein, G., Kuhmstedt, R., and Kadua, P. (1970). Tetrahedron 26(1), 81-90.

Hibbard, R. R. and Schalla, R. L. (1952). Solubility of Water in Hydrocarbons. Natl.
    Advisory Committee for Aeronautics, RM-E52-D24, Washington, D.C., July 10, 25 pp.

Hicks, C. P. (1978). Bibliography of Thermodynamic Quantities for Binary Fluid Mix-
    tures. In Chemical Thermodynamics, Vol. 2, Ch. 9 (M. L. McGlashan, Ed.). Chemical
    Society, London.

Higuchi, T. and Daviis, S. S. (1970). J. Pharm. Sci. $\underline{59}$, 1376-1383.

Hildebrand, J. H. (1924). Solubility (1st ed.). Chemical Catalog Company, New York.

Hildebrand, J. H. (1929). J. Am. Chem. Soc. $\underline{51}$(1), 66-80.

Hildebrand, J. H. (1935). J. Am. Chem. Soc. $\underline{57}$, 866-871.

Hildebrand, J. H. (1936a). Science $\underline{83}$(2141), 21-24.

Hildebrand, J. H. (1936b). Solubility of Nonelectrolytes (2nd ed.). Reinhold, New York.

Hildebrand, J. H. (1939). J. Chem. Phys. $\underline{7}$(4), 233-235.

Hildebrand, J. H. (1949a). Chem. Rev. $\underline{44}$, 37-45.

Hildebrand, J. H. (1949b). J. Chem. Phys. $\underline{17}$, 1346-1347.

Hildebrand, J. H. (1949c). J. Phys. Chem. $\underline{53}$, 973-974.

Hildebrand, J. H. (1952). Chem. Eng. Prog. Symp. Ser. $\underline{48}$(3), 3-9.

Hildebrand, J. H. (1953). Disc. Faraday Soc., No. 15, 9-23.

Hildebrand, J. H. (1965). Science $\underline{150}$, 441-450.

Hildebrand, J. H. (1969). Proc. Natl. Acad. Sci. U.S.A. $\underline{64}$, 1331-1334.

Hildebrand, J. H. (1974). Ind. Eng. Chem. Fundam. $\underline{13}$(2), 110-115.

Hildebrand, J. H. (1977a). Struct.-Solubility Relat. Polym., Proc. Symp., p. 1-10.

Hildebrand, J. H. (1977b). Viscosity and Diffusivity. Wiley, New York. 109 pp.

Hildebrand, J. H. (1978a). Ind. Eng. Chem. Fundam. $\underline{17}$(4), 365-366.

Hildebrand, J. H. (1978b). Faraday Disc. Chem. Soc., No. 66, 151-159.

Hildebrand, J. H. (1979a). Proc. Natl. Acad. Sci. U.S.A. $\underline{76}$(1), 194.

Hildebrand, J. H. (1979b). Proc. Natl. Acad. Sci. U.S.A. $\underline{76}$(12), 6040-6041.

Hildebrand, J. H. and Lamoreaux, R. H. (1974). Ind. Eng. Chem. Fundam. $\underline{13}$(2), 110-115.

Hildebrand, J. H. and Scott, R. L. (1950). The Solubility of Nonelectrolytes (3rd ed.). Reinhold, New York. 488 pp.

Hildebrand, J. H. and Scott, R. L. (1962). Regular Solutions. Prentice-Hall, Englewood Cliffs, N. J. 180 pp.

Hildebrand, J. H. and Wood, S. E. (1933). J. Chem. Phys. $\underline{1}$(12), 817-822.

Hildebrand, J. H., Prausnitz, J. M., and Scott, R. L. (1970). Regular and Related Solutions. The Solubility of Gases, Liquids, and Solids. Van Nostrand Reinhold, New York. 228 pp.

Hill, N. E., Vaughan, W. E., Price, A. H., and Davies, M. (1969). Dielectric Properties and Molecular Behavior. Van Nostrand, London.

Himmelblau, D. M. (1960). J. Chem. Eng. Data $\underline{5}$(1), 10-15.

Hine, J. (1975). Structural Effects on Equilibria in Organic Chemistry. Wiley, New York. 347 pp.

Hine, J. and Ehrenson, S. J. (1958). J. Am. Chem. Soc. $\underline{80}$, 824-830.

Hine, J. and Mookerjee, P. K. (1975). J. Org. Chem. $\underline{40}$(3), 292-298.

Hine, J. and Prosser, F. P. (1958). J. Am. Chem. Soc. $\underline{80}$, 4282-4285.

Hine, J., Butterworth, R., and Langford, P. B. (1958). J. Am. Chem. Soc. 80, 819-824.

Hine, J., Haworth, H. W., and Ramsay, O. B. (1963). J. Am. Chem. Soc. 85, 1473-1476.

Hiranuma, M. (1976). J. Chem. Eng. Japan 9(3), 231-233.

Hiranuma, M. and Honma, K. (1975). Ind. Eng. Chem. Process Des. Dev. 14(3), 221-226.

Hirata, M., Ore, S., and Nagahama, K. (1975). Computer Aided Data Book of Vapor-Liquid Equilibria. Kodansha, Tokyo. 941 pp.

Hirschfelder, J. O., Ed. (1967). Intermolecular Forces. Adv. Chem. Phys., Vol. 12. Interscience, New York. 643 pp.

Hirschfeleder, J. O., Curtiss, C. F., and Bird, R. B. (1964). Molecular Theory of Gases and Liquids (2nd ed.). Wiley, New York. 1249 pp.

Hirschfeleder, J. O., McClure, F. T., and Weeks, I. F. (1942). J. Chem. Phys. 10, 201-211.

Hirschfeleder, J. O., Buchler, R. J., McGee, H. A., and Sutton, J. R. (1958). Ind. Eng. Chem. 50(3), 375-385.

Hiza, M. J., Kidnay, A. J., and Miller, R. C. (1975). Equilibrium Properties of Fluid Mixtures. A Bibliography of Data on Fluids of Cryogenic Interest. NSRDS Bibliographic Series. Plenum, New York.

Ho, F. F.-L. and Kohler, R. R. (1974). Anal. Chem. 46(9), 1302-1304.

Hobza, P. and Zahradnik, R. (1974a). Coll. Czech. Chem. Commun. 39, 2857-2865.

Hobza, P. and Zahradnik, R. (1974b). Coll. Czech. Chem. Commun. 39, 2866-2876.

Hobza, P. and Zahradnik, R. (1975). Coll. Czech. Chem. Commun. 40(3), 809-814.

Hobza, P. and Zahradnik, R. (1976). Coll. Czech. Chem. Commun. 41(4), 1111-1120.

Hodgman, C. D., Weast, R. C., and Selby, S. M. (1956). Handbook of Chemistry and Physics (38th ed.). Chemical Rubber Co., Cleveland, Ohio.

Hoechst Aktiengesellschaft (1975). Hoechst Solvents. Manual for Laboratory and Industry (5th ed.). Verkauf Organische Chemikalien, Frankfurt am Main. 442 pp.

Hofmann, H.-J. and Birnstock, F. (1980). Z. phys. Chem. 261, 1212-1216.

Hoffman, J. D. (1952). J. Chem. Phys. 20(4), 740.

Högfeldt, E. (1976). Chem. Ind., No. 5, 184-189.

Högfeldt, E. and Bolander, B. (1963). Arkiv Kemi 21(16), 161-186.

Högfeldt, E. and Fredlund, F. (1970). Acta Chem. Scand. 24(5), 1858-1860.

Holland, C. D. (1963). Multicomponent Distillation. Prentice-Hall, Englewood Cliffs, N.J.

Holland, H. G. and Moelwyn-Hughes, E. A. (1956). Trans. Faraday Soc. 52, 297-299.

Hollenbeck, R. G. (1980). J. Pharm. Sci. 69(10), 1241-1242.

Hollifield, H. C. (1979). Bull. Environ. Contam. Toxicol. 23, 579-586.

Hollo, J. and Lengyel, T. (1960). Period. Polytech. 4, 125-140.

Holmes, C. F. (1973). J. Am. Chem. Soc. 95(4), 1014-1016.

Holmes, H. N. and Weiser, H. B., Eds. (1925). Colloid Symposium Monograph, 3rd International Symposium. Chemical Catalog Co., New York. 323 pp.

Holtzer, A. and Emerson, M. F. (1969). J. Phys. Chem. 73(1), 26-33.

Holzapfel, W. B. (1969). J. Chem. Phys. 50(10), 4424-4428.

Honeyborne, D. B., Cornell, J. B., and Arnold, L. (1971). Mater. Constr. 4(21), 181-192.

Hooker Chemical Corporation (1966). Hooker Perchloroethylene for Industrial Uses. Bull. No. 190. Niagara Falls, N. Y.

Hoot, W. F. (1957). Pet. Refiner 36(6), 198.

Hoot, W. F., Azarnoosh, A., and McKetta, J. J. (1957). Pet. Refiner 36(5), 255-256.

Hoppe, J. I. (1972). Educ. Chem. 9(4), 138-140.

Horiuti, J. (1931). Sci. Pap. Inst. Phys. Chem. Res. 17(341), 126-256.

Horne, R. A. (1969). Marine Chemistry. The Structure of Water and the Chemistry of the Hydrosphere. Wiley-Interscience, New York. 568 pp.

Horne, R. A. (1970). Water Properties. In Kirk-Othmer Encyclopedia of Chemical Technology (2nd ed.)., Vol. 21. Wiley-Interscience, New York. p. 668-688.

Horne, R. A. (1972a). Effect of Structure and Physical Characteristics of Water on Water Chemistry. In Water and Water Pollution Handbook, Vol. 3 (L. L. Ciaccio, Ed.). Dekker, New York. p. 915-947.

Horne, R. A., Ed. (1972b). Water and Aqueous Solutions. Structure, Thermodynamics, and Transport Processes. Wiley-Interscience, New York. 837 pp.

Hornig, D. F. (1964). J. Chem. Phys. 40, 3119-3120.

Horsley, L. H. (1973). Azeotropic Data--III. Adv. Chem. Ser. No. 116. Am. Chem. Soc., Washington, D.C. 628 pp.

Horvath, A. L. (1972a). Chem. Eng. Sci. 27, 1185-1189.

Horvath, A. L. (1972b). J. Chem. Doc. 12(3), 163-171.

Horvath, A. L. (1973). Chem. Eng. Sci. 28, 299-304.

Horvath, A. L. (1974). Chem. Eng. Sci. 29(5), 1334-1340.

Horvath, A. L. (1975a). Chem.-Ing.-Tech. 47(19), 815.

Horvath, A. L. (1975b). Physical Properties of Inorganic Compounds, SI Units. Edward Arnold, London. 480 pp.

Horvath, A. L. (1975c). Unpublished measurements.

Horvath, A. L. (1976a). Chem. Ind., No. 1, 26-27.

Horvath, A. L. (1976b). Chem.-Ing.-Tech. 48(2), 144-146.

Horvath, A. L. (1980). FIZDAT - A Computer Program for Physical Properties. Transport Properties of Polyatomic Fluids. Meeting of the Chemical Society, London, April 14-15.

Horvath, A. L. (1981). Physical Properties of Aqueous Electrolytes: Estimation Methods. Dekker, New York. (In preparation.)

Horvath, A. L., Rathbone, P., and Lines, D. (1973). Critical Appraisal and Presentation of Numerical Data for Scientists and Engineers. Paper presented at the Conf. of Inst. Inf. Sci., Northern Branch, Manchester, England. June 12.

Horvath, R. S. (1972). Bacteriol. Rev. $\underline{36}$(2), 146-155.

Howard, B. and Loscalzo, A. G. (1978). Functional Group Analysis and Solubility Data. Prentice-Hall, Tarrytown, N.Y., Filmstrip. (CA 92:100359s.)

Howard, B. B., Jumper, C. F., and Emerson, M. T. (1963). J. Mol. Spectrosc. $\underline{10}$, 117-130.

Howarth, O. (1974). Educ. Chem. $\underline{11}$(1), 15-16.

Howe, J. R. (1975). Lab. Practice $\underline{24}$(7), 457-467.

Hoy, K. L. (1970). J. Paint Technol. $\underline{42}$(541), 76-118.

Huang, C. P., Fennema, O., and Powrie, W. D. (1965). Cytobiologie $\underline{2}$(3), 109-115.

Huckel, W. (1958). Theoretical Principles of Organic Chemistry, Vol. 2. Elsevier, Amsterdam. p. 370-379.

Hudlicky, M. (1976). Chemistry of Organic Fluorine Compounds (2nd ed.). Ellis Horwood, Chichister, England.

Huggins, M. L. (1971). Angew. Chem. $\underline{83}$, 163-168.

Huggins, M. L. (1972). J. Paint Technol. $\underline{44}$(567), 55-66.

Huggins, M. L. (1980). Chem. Tech. $\underline{10}$(7), 422-429.

Hughel, T. L., Ed. (1965). Liquids: Structure, Properties, Solid Interactions. Proc. Symp., Warren, Mich., Elsevier, Amsterdam.

Hulett, C. A. (1901). Z. phys. Chem. $\underline{37}$, 385-406.

Hunter, C. H. and Honaker, C. B. (1979). Determination of Water Solubility in Organic Liquids. Rockwell Internal Rep. RHO-SA-113. Richland, Washington, D.C. June, 9 pp. (CA 93:113679s.)

Huntress, E. H. (1948). Organic Chlorine Compounds. Wiley, London.

Hutchison, C. A. and Lyon, A. M. (1943). Columbia Univ. Rep. A-745. July.

Hutchinson, T. C., Hellebust, J. A., Tam, D., Mackay, D., Mascarenhas, R. A., and Shiu, W. Y. (1980). The Correlation of the Toxicity to Algae of Hydrocarbons and Halogenated Hydrocarbons with their Physical-Chemical Properties. In Hydrocarbons and Halogenated Hydrocarbons in the Aquatic Environment (B. K. Afghan and D. MacKay, Eds.). Plenum, New York. p. 577-586.

Huttenlocker, D. F. (1976). ASHRAE J. $\underline{18}$(8), 33-37.

Hvidt, A. (1978). Acta Chem. Scand. $\underline{32A}$(8), 675-680.

Iguchi, A. (1974). Kagaku Sochi $\underline{16}$(8), 70-71.

Iguchi, A. (1977). Kagaku Sochi $\underline{19}$(4), 70-71; $\underline{19}$(7), 78-80.

Imperial Chemical Industries (1953). Arcton for Aerosols. Bull. No. 133/53/457. London. 13 pp.

Imperial Chemical Industries (1955). Unpublished data.

Imperial Chemical Industries (1959). General Chem. Div. Quart. Res. Rep., London, January-March. p. 53.

Imperial Chemical Industries (1964). Unpublished data.

Imperial Chemical Industries (1965). Arklone the Modern Solvent That Improves Cleaning and Lower Costs. Mond Div., Runcorn, England. 12 pp.

Imperial Chemical Industries (1966). Solubility of Ethylene Dichloride in Aqueous Solutions of NaOH, HCl, NaCl, and Na_2CO_3. Mond Div. Rep. MD 5153. Runcorn, England, December 15. 2 pp.

Imperial Chemical Industries (1968). Fluothane, Halothane B. P. (3rd ed.). Pharmaceutical Div., Macclesfield, England. 148 pp.

Imperial Chemical Industries (1972). Brixham Lab. Rep. BL/B/1417. Brixham, England. August.

Imperial Chemical Industries (1977). Unpublished data.

Imperial Smelting Corporation (1957). Isceon (1st ed.). Res. Dept. Publ., London, 133 pp.

Interdyne (1976). Solvent Data Sheet. SDS-104. Indianapolis, Ind. 2 pp.

Iogansen, A. V. (1971). Theor. Exp. Chem. USSR 7(3), 249-256.

Irmann, F. (1965). Chem.-Ing.-Tech. 37(8), 789-798.

Ishida, K. (1958). Bull. Chem. Soc. Japan 31(2), 143-154.

Ishizuka, I., Sarashina, E., Arai, Y., and Saito, S. (1980). J. Chem. Eng. Japan 13(2), 90-97.

IUPAC (1974). Pure Appl. Chem. 37(4), 463-468.

Ivanov, K. A. (1956). Gigi. Sanit. 21(9), 82-83.

Jacobson, B. (1950). Arkiv Kemi 2, 177-210.

Jaffe, H. H. and Orchin, M. (1962). Theory and Application of Ultraviolet Spectroscopy. Wiley, New York.

Jain, D. V. S., Singh, S., and Gombar, V. (1976). J. Chem. Soc. Faraday Trans. I 72(7), 1694-1696.

James, K. C. (1972). Educ. Chem. 9, 220-221, 224.

Janz, G. J. and Tomkins, R. P. T. (1972). Non-Aqueous Electrolytes Handbook, Vol. 1. Academic, New York.

Jaques, D. and Lee, D. A. (1966). Ind. Eng. Chem. Fundam. 5(1), 136-137.

Jarry, R. L. and Davis, W. (1953). J. Phys. Chem. 57, 600-604.

Jasper, J. J. (1972). J. Phys. Chem. Ref. Data 1(4), 841-1009.

Jayasri, A. and Yaseen, M, (1979). Progr. Org. Coatings 7, 167-207.

Jayasri, A. and Yaseen, M. (1980a). J. Coat. Technol. 52(667), 41-47.

Jayasri, A. and Yaseen, M. (1980b). J. Oil Col. Chem. Assoc. 63(2), 61-69.

Jeffrey, G. A. and McMullan, R. K. (1967). Prog. Inorg. Chem. 8, 43-108.

Jellinek, H. H. G., Ed. (1972). Water Structure at the Water – Polymer Interface. Proc. Symp. 161st Natl. Meet. Am. Chem. Soc., 1971. Plenum, New York. 182 pp.

Jenkin, C. F. and Shorthose, D. N. (1923). The Thermal Properties of Ethyl Chloride. Food Investigation Board Spec. Rep. No. 14. HMS Office, London. 35 pp.

Jensen, S., Lange, R., Berge, G., Pelmonk, K. H., and Renberg, L. (1975). Proc. Roy. Soc. Lond. Ser. B $\underline{189}$, 333-346.

Jhon, M. S. and Eyring, H. (1978). Theoretical Chemistry Vol. 3. Academic, New York p. 55.

Jhon, M. S., Eyring, H., and Sung, Y. K. (1972). Chem. Phys. Lett. $\underline{13}$(1), 36-39.

Jhon, M. S., Grosh, J., Ree, T., and Eyring, H. (1966). J. Chem. Phys. $\underline{44}$(4), 1465-1472.

Jockers, R. (1976). Phasengleichgewichte in binären und ternären wässrigen Systemen bei hohen Temperaturen und Drücken. (Ph. D. thesis). Univ. of Bochum, Bochum, West Germany. 123 pp.

Jockers, R. and Schneider, G. M. (1978). Ber. Bunsen ges. Phys. Chem. $\underline{82}$, 576-582.

Joesten, M. D. and Schaad, L. J. (1974). Hydrogen Bonding. Dekker, New York.

Joffe, J. (1977). J. Chem. Eng. Data $\underline{22}$(3), 348-350.

John, V. B. (1974). Understanding Phase Diagrams. Macmillan, London. 94 pp.

Johnson, A. I. and Modonis, J. A. (1959). Can. J. Chem. Eng. $\underline{37}$(2), 71-76.

Johnson, A. I., Huang, C.-J., and Kwei, T.-K. (1954). Can. J. Technol. $\underline{32}$, 127-132.

Johnson, E. W. and Nash, L. K. (1950). J. Am. Chem. Soc. $\underline{72}$, 547-556.

Johnson, J. F. and Porter, R. S., Eds. (1970). Liquid Crystals and Ordered Fluids. Plenum, New York.

Johnson, J. R., Christian, S. D., and Affsprung, H. E. (1965). J. Chem. Soc. p. 1-6.

Johnson, J. R., Christian, S. D., and Affsprung, H. E. (1966). J. Chem. Soc. Ser. A, 77-78.

Johnson, J. R., Christian, S. D., and Affsprung, H. E. (1967). J. Chem. Soc. Ser. A, 1924-1928.

Johnston, W. R., Ittihadieh, F. T., Craig, K. P., and Pillsbury, A. F. (1967). Water Resour. Res. $\underline{3}$(2), 525-537.

Jolicoeur, C. and Cabana, A. (1968). Can. J. Chem. $\underline{46}$, 567-570.

Jolicoeur, C. and Lacroix, G. (1976). Can. J. Chem. $\underline{54}$, 624-631.

Jolles, Z. E., Ed. (1966). Bromine and Its Compounds. Ernest Benn, London.

Jonah, D. A. and King, M. B. (1971). Proc. Roy. Soc. Lond. Ser. A $\underline{323}$, 361-375.

Jones, D. C., Ottewill, R. H., and Chater, A. P. J. (1957). Proc. 2nd Int. Congr. Surf. Activ., London, Vol. 1, p. 188-199.

Jones, J. R. and Monk, C. B. (1963). J. Chem. Soc., p. 2633-2636.

Jordan, O. (1937). The Technology of Solvents. Chemical Publ., New York. 329 pp.

Jordan, T. E. (1954). Vapor Pressure of Organic Compounds. Interscience, New York. 266 pp.

Joris, G. G. and Taylor, H. S. (1948). J. Chem. Phys. $\underline{16}$(1), 45-51.

Junge, C. (1976). Z. Naturforsch. Teil A $\underline{31}$, 482-487.

Juvala, A. (1930). Chem. Ber. $\underline{63}$, 1989-2009.

Kabadi, V. N. and Danner, R. P. (1979). Hydr. Proc. $\underline{58}$(5), 245-246.

Kafarov, V. V., Ed. (1961). Spravochnik po Rastvorimosti. Izdatel'stvo Akademii Nauk
    SSSR, Moscow.

Kafarov, V. V., Ed. (1967-1968). Reference Book on Solubility, Vol. 3. Izdatel'stvo
    Akademii Nauk SSSR, Moscow.

Kaiser, K. L. E. and Lawrence, J. (1977). Science 196(4295), 1205-1206.

Kajino, M. (1977). Osaka-Shi Suidokyoko Konubu Suishitsu Shikensho Ch. 27, 1-4.

Kakovsky, I. A. (1957). Proc. 2nd Int. Congr. Surf. Activ., London, Vol. 4, p. 225-
    237.

Kali-Chemie (1972). Taschenbuch für Kältetechniker. Kaltron (5th ed.). Anlage 208,
    Hannover, West Germany. 134 pp.

Kalman, E. (1974). Experimental Studies on the Structure of Liquid Water. In Proc.
    8th Int. Conf. Prop. of Water and Steam. Giens, France. September. p. 1063-1073.

Kamb, B. (1965). Science 150, 205-209.

Kanazawa, J., Yushima, T., and Kiritani, K. (1971). Kagaku 41, 384-391.

Kaneniwa, N. and Watari, N. (1978). Chem. Pharm. Bull. 26(3), 813-826.

Kaneniwa, N., Watari, N., and Iijima, H. (1978). Chem. Pharm. Bull. 26(9), 2603-2614.

Kanno, H., Speedy, R. J., and Angell, C. A. (1975). Science 189(4206), 880-881.

Karapet'yants, M. Kh. (1973). Approximate Method of Calculating Solubility. Thermodyn.
    Str. Rastvarov, No. 1, 12-27. (CA 83:121606z.)

Karapet'yants, M. Kh. (1976). Russ. J. Phys. Chem. 50(4), 649-653.

Karasz, F. E. and Halsey, G. D. (1958). J. Chem. Phys. 29(1), 173-179.

Karger, B. L., Castells, R. C., Sewell, P. A., and Hartkopf, A. (1971). J. Phys. Chem.
    75(25), 3870-3879.

Karger, B. L., Chatterjee, A. K., and King, J. W. (1971). Adsorption and Solution of
    Weakly Polar Vapors with Thin Layers of Water as Studied by Gas-Liquid Chromatog-
    raphy. Tech. Rep. No. 3. Dept. Chem., Northeastern Univ., Boston. May 10. 13 pp.

Karger, B. L., Snyder, L. R., and Eon, C. (1976). J. Chromatogr. 125(1), 71-88.

Karger, B. L., Snyder, L. R., and Eon, C. (1978). Anal. Chem. 50(14), 2126-2136.

Karger, B. L., Sewell, P. A., Castells, R. C., and Hartkopf, A. (1971). J. Colloid
    Interface Sci. 35(2), 328-339.

Karle, J. and Brockway, L. O. (1944). J. Am. Chem. Soc. 66, 574-584.

Karlsson, R. (1972). Talanta 19, 1639-1644.

Karlsson, R. and Karrman, K. J. (1971). Talanta 18, 459-465.

Karpov, B. D. (1964). Toxicological Research. Izdatel'stvo Akademii Nauk SSSR,
    Moscow.

Karyakin, A. V. and Muradov, G. A. (1971). Russ. J. Phys. Chem. 45(5), 591-594.

Karyakin, A. V., Maisuradze, G. V., and Muradova, G. A. (1970). Russ. J. Phys. Chem.
    44(8), 1179-1181.

Karyakin, A. V., Tokhadze, V. L., and Maisuradze, G. V. (1960). J. Anal. Chem. USSR
    25(2), 266-268.

Kass, J. R., Fernandez-Martin, R., Empie, H. L., and Barduhn, A. J. (1965). Thermodynamics and Kinetics of Hydrate Systems. Syracuse Univ. Res. Inst. Rep. No. 130.
Syracuse, N. Y. 40 pp.

Katayama, A., Takagishi, T., Konishi, K., and Kuroki, N. (1966). Kolloid Z. Z. Polym.
210(2), 126-132.

Katz, D. L., Cornell, D., Kobayashi, R., Poettmann, F. H., Vary, J. A., Elenbaas, J.
R., and Weinaug, C. F. (1959). Handbook of Natural Gas Engineering. McGraw-Hill,
New York. 802 pp.

Kavanau, J. L. (1964). Water and Solute - Water Interactions. Holden-Day, San Francisco. 101 pp.

Kay, R. L., Ed. (1973). The Physical Chemistry of Aqueous Systems. Plenum, New York.
258 pp.

Kaye, G. W. C. and Laby, T. H. (1973). Tables of Physical Chemical Constants (14th
ed.). Longman, London. 386 pp.

Keevil, T. A., Taylor, D. R., and Streitwieser, A. (1978). J. Chem. Eng. Data 23(3),
237-239.

Kemp, M. K., Thompson, R. E., and Zigrang, D. J. (1974). J. Chem. Educ. 52(12), 802.

Kendall, C. E. (1944). Chem. Ind., p. 211.

Kertes, A. S., Project Leader (1974-1976). Solubility Data Project. Information Sheets
A101 (March 1974), 3 pp.; S102 (March 1974), 2 pp.; A104 (June 1974), 4 pp.; A105
(July 1974), 14 pp.; S106 (November 1974), 19 pp.; IS107 (September 1975), 10 pp.;
ISS110 (March 1976), 14 pp.; ISS111 (May 1976), 4 pp.; Circulars 101 (May 1976),
1 pp.; 102 (October 1975), 2 pp.; 206 (March 1976), 1 pp.; 208 (June 1976), 1 pp.
IUPAC - CODATA - Gmelin Institut, Jerusalem, Israel.

Kertes, A. S. (1977). IUPAC Solubility Data Projects -- Sample Booklet. Pergamon,
Oxford, June. 48 pp.

Kertes, A. S., Battino, R., and Clever, H. L. (1977). Proc. 5th Int. CODATA Conf.,
pp. 217-225.

Kertes, A. S., Levy, O., and Markovits, G. Y. (1975). Solubility. In Experimental
Thermodynamics, Vol. 2, IUPAC (B. Le Neidre and B. Vodar, Eds.). Butterworths,
London. p. 725-748.

Ketelaar, J. A. A. (1958). Chemical Constitutions. An Introduction to the Theory of
the Chemical Bond (2nd ed.). Elsevier, Amsterdam. 448 pp.

Ketelaar, J. A. A. and van Meurs, N. (1957a). Recl. Trav. chim. 76, 437-479.

Ketelaar, J. A. A. and van Meurs, N. (1957b). Recl. Trav. chim. 76, 495-505.

Key, M. M. (1973). Criteria for a Recommended Standard Occupational Explosure to
Trichloroethylene. U. S. Dept. of Health, Education and Welfare, Natl. Inst. for
Occupational Safety and Health, Washington, D. C. 102 pp.

Khachat-Ryan, M. K. (1962). Sanit. Okhr. Vodoemov Zagryaz. Prom. Stochnymi Vodami,
No. 5, 44-61. (CA 61:442b.)

Khalil, S. A., Moustafa, M. A., and Abdallah, O. Y. (1976). Can. J. Pharm. Sci. 11(4), 121-126.

Khanina, E. P., Pavlenko, T. G., Malysheva, O. A., and Timofeev, V. S. (1978). Russ. J. Phys. Chem. 52(6), 904.

Khmara, Yu. I. (1976). Russ. J. Phys. Chem. 50(1), 88-89.

Khodeeva, S. M., Subbotina, L. A., and Gubochkina, I. V. (1977). Russ. J. Phys. Chem. 51(2), 266-268.

Kier, L. B. (1976). Private communication.

Kier, L. B. and Hall, L. H. (1976a). J. Pharm. Sci. 65(12), 1806-1809.

Kier, L. B. and Hall, L. H., Eds. (1976b). Molecular Connectivity in Chemistry and Drug Research. Medical Chemistry, Vol. 14 (G. de Stevens, Ed.). Academic, New York. 257 pp.

Kier, L. B. and Hall, L. H. (1977a). Eur. J. Med. Chem. Chim. Ther. 12(4), 307-312.

Kier, L. B. and Hall, L. H. (1977b). J. Med. Chem. 20(12), 1631-1636.

Kier, L. B. and Hall, L. H. (1978a). J. Pharm. Sci. 67(10$\frac{1}{4}$, 1408-1412.

Kier, L. B. and Hall, L. H. (1978b). J. Pharm. Sci. 68(1), 120-122.

Kier, L. B., Di Paolo, T., and Hall, L. H. (1977). J. Theor. Biol. 67(3), 585-595.

Kier, L. B., Murray, W. J., and Hall, J. H. (1975). J. Med. Chem. 18(12), 272-274.

Kier, L. B., Hall, L. H., Murray, W. J., and Randic, M. (1975). J. Pharm. Sci. 64(12), 1971-1981.

Kier, L. B., Murray, W. J., Randic, M., and Hall, L. H. (1976). J. Pharm. Sci. 65(8), 1226-1230.

Kihara, T. (1970). Intermolecular Forces. In Physical Chemistry: An Advanced Treatise, Vol. 5 (H. Eyring, Ed.). Academic, New York. p. 663-716.

Kihara, T. (1978). Intermolecular Forces. Wiley, London.

Kihara, T. and Jhon, M. S. (1970). Chem. Phys. Lett. 7(6), 559-562.

Kikik, I. and Alessi, P. (1977). Can. J. Chem. 55(1), 78-81.

Kim, J. I. and Brückl, N. (1978). Z. Phys. Chem. Neue Folge 110(2), 197-208.

Kinetic Chemicals (1956). New Kinetic Products--Freon-112. A Safe Solvent. Bull. B-1. Wilmington, Del. 15 pp.

King, E. J. (1973). Acid-Base Behavior. In Physical Chemistry of Organic Solvent Systems (A. K. Covington and T. Dickinson, Eds.). Plenum, London, p. 331-403.

King, M. B. (1969). Phase Equilibrium in Mixtures. Pergamon, Oxford. 585 pp.

King, M. B., Al-Najjar, H., and Ali, J. K. (1979). Chem. Eng. Sci. 34, 1080-1082.

Kirchnerova, J. (1975). Aggregates and Hydrates of Some Alcohols in Low-Dielectric Solvents: A Thermodynamic Study. (Ph. D. thesis). McGill Univ., Montreal. April. 280 pp.

Kirchnerova, J. and Cave, G. C. B. (1976). Can. J. Chem. 54(24), 3909-3916.

Kirk-Othmer Encyclopedie of Chemical Technology (1964a). Bromine Compounds, Organic Vol. 3 (2nd ed.). Wiley, New York. p. 770-783.

Kirk-Othmer Encyclopedia of Chemical Technology (1964b). Chlorocarbons and Chloro-
hydrocarbons, Vol. 5 (2nd ed.). Wiley, New York. p. 100-363.

Kirk-Othmer Encyclopedia of Chemical Technology (1966a). Fluorine Compounds, Organic,
Vol. 9 (2nd ed.). Wiley, New York. p. 686-847.

Kirk-Othmer Encyclopedia of Chemical Technology (1966b). Iodine Compounds, Organic,
Vol. 11 (2nd ed.). Wiley, New York. p. 861-865.

Kirk-Othmer Encyclopedia of Chemical Technology (1969). Solvents, Industrial, Vol. 18
(2nd ed.). Wiley, New York. p. 564-588.

Kirk-Othmer Encyclopedia of Chemical Technology (1970). Water Properties, Vol. 21
(2nd ed.). Wiley, New York. p. 668-707.

Kirk-Othmer Encyclopedia of Chemical Technology (1978). Azeotropic and Extractive
Distillation. Vol 3 (3rd ed.). Wiley, New York. p. 352-377.

Kirkwood, J. G. (1936). J. Chem. Phys. 4, 592-601.

Kirkwood, J. G. (1939a). J. Chem. Phys. 7, 911-919.

Kirkwood, J. G. (1939b). J. Chem. Phys. 7(10), 919-925.

Kirkwood, J. G. (1946). Trans. Faraday Soc. 42A, 7-12.

Kirschenbaum, D. M. (1980). J. Chem. Inf. Comput. Sci. 20(3), 152-153.

Kirshenbaum, I. (1951). Physical Properties and Analysis of Heavy Water. McGraw-Hill,
New York.

Kisarov, V. M. (1962). J. Appl. Chem. USSR 35, 2252-2253.

Kistenmacher, H., Popkie, H., and Clementi, E. (1974). J. Chem. Phys. 61(3), 799-
815.

Klausutis, N. A. (1961). The Solubilities of Some Halogenated Methanes and Ethanes
in Water and NaCl Solutions (M. Sc. thesis). Syracuse Univ., Syracuse, N. Y.
101 pp.

Klein, M., Hanley, H. J. M., Smith, F. J., and Holland, P. (1974). Tables of Collision Integrals and Second Virial Coefficients for the (m,6,8) Intermolecular
Potential Functions. Natl. Bur. Std. NSRDS NBS-47. Washington, D. C., June. 151 pp.

Klemenc, A. and Löw, M. (1930). Recl. Trav. chim. 49(4), 629-640.

Klevens, H. (1950). J. Phys. Colloid Chem. 54, 283-298.

Klots, C. E. (1963). J. Phys. Chem. 67, 933-934.

Klotz, I. M. and Rosenberg, R. M. (1972). Chemical Thermodynamics. Basic Theory and
Methods (3rd ed.). W. A. Benjamin, Menlo Park, Calif. 444 pp.

Knox, W. G., Hess, G. E., Jones, G. E., and Smith, H. B. (1961). Chem. Eng. Prog.
57(2), 66-71.

Kobatake, Y. and Hildebrand, J. H. (1961). J. Phys. Chem. 65, 331-335.

Kobayashi, R. and Katz, D. L. (1953). Ind. Eng. Chem. 45(2), 440-451.

Koenhen, D. M. and Smolders, C. A. (1975). J. Appl. Polym. Sci. 19(4), 1163-1179.

Kogan, L. M., Kol'tsov, N. S., and Litvinov, N. D. (1963). Russ. J. Phys. Chem. 37
(8), 1040-1042.

Kogan, V. B. and Friedman, V. M. (1961). Handbuch der Dampf-Flüssigkeits-gleichgewichte. VEB Deutscher Verlag der Wisserschaften, Berlin.

Kogan, V. B., Friedman, V. M., and Kafarov, V. V. (1961-1970). Spravochnik po Rastvorimosti. Izdatel'stvo Akademii Nauk SSSR, Moscow.

Kohler, F. (1972). The Liquid State. Monogr. Mod. Chem., Vol. 1, Verlag Chemie, Weinheim, West Germany. 256 pp.

Kohler, F. and Wilhelm, E. (1976). Adv. Mol. Relaxation Processes $\underline{8}$(3), 195-239.

Kojima, K. and Tochigi, K. (1979). Prediction of Vapor-Liquid Equilibria by the ASOG Method. Physical Science Data 3. Elsevier, Amsterdam. 264 pp.

Kölle, W. (1974). Determination of Organic Chlorine Compounds in the Rhein, in Water Works Activated Charcoal, and in Organisms. Ber. Kernforschungsanlage Juelich (Conf.) $\underline{11}$(3), 3-12.

Kollman, P. A. and Allen, L. C. (1972). Chem. Rev. $\underline{72}$(3), 283-303.

Komedera, M. (1966). Mod. Refrig. Air. Cond. $\underline{69}$(816), 218-220, 246.

Konda, C. and Yamamoto, T. (1977). Chem. Phys. Lett. $\underline{50}$(2), 324-326.

Konicek, J. and Wadsö, I. (1971). Acta Chem. Scand. $\underline{25}$, 1541-1551.

Konstam, A. H. and Feairheller, W. R. (1970). AIChE J. $\underline{16}$(5), 837-840.

Koonce, K. T. and Kobayashi, R. (1964). J. Chem. Eng. Data $\underline{9}$(4), 490-494.

Kopsel, R. (1974). Ausgewahlte rechnerische Methoden der Verfahrenstechnik. Akademie-Verlag, Berlin.

Koradia, P. B. and Kiovsky, J. R. (1977). Chem. Eng. Prog. $\underline{73}$(4), 105-106.

Korenman, I. M. (1975). USSR Patent No. 553,524, October 6.

Korenman, I. M. and Aref'eva, R. P. (1978). J. Appl. Chem. USSR $\underline{51}$(4), 923-924.

Korenman, I. M., Gur'ev, I. A., and Gur'eva, Z. M. (1971). Russ. J. Phys. Chem. $\underline{45}$(7), 1065-1066.(VINITI No. 2885-71.)

Körösi, F. (1937). Trans. Faraday Soc. $\underline{33}$, 416-425.

Koskas, A. and Durandet, J. (1967). Chim. Ind. Genie Chim. $\underline{98}$(8), 1386-1397.

Kostovetskii, Ya. I., Lisovskaya, E. V., Dyatlovitskaya, F. G., and Surkina, R. M. (1962). Sanit. Okhr. Vodoemov Zagryaz. Prom. Stochnymi Vodami, No. 5, 94-106.

Kozlova, V. S. and Korol, A. N. (1971). Russ. J. Phys. Chem. $\underline{45}$(11), 1667.

Krasnec, L. (1967). Chem. Zvesti $\underline{21}$(5), 370-382.

Krasnoshchenkova, R. Ya., Pahapill, Yu. A., and Gubergri, M. Ya. (1977). Khim. Tverd. Topl. No. 2, 133-136.

Krasovskii, I. V. and Gritsan, L. D. (1976). Russ. J. Phys. Chem. $\underline{50}$(9), 1441-1443.

Kreglewski, A., Wilhoit, R. C., and Zwolinski, B. J. (1973). J. Chem. Eng. Data $\underline{18}$(4), 432-435.

Kretschmer, C. B. and Wieber, R. (1954). J. Am. Chem. Soc. $\underline{76}$, 2579-2583.

Krichevskii, I. R. and Efremova, G. D. (1948). Zh. Fiz. Khim. $\underline{22}$(9), 1116-1125.

Krichevskii, I. R. and Kasarnovsky, Ya. S. (1935). Zh. Fiz. Khim. $\underline{6}$(10), 1320-1324.

Krichevskii, I. R. and Sorina, G. A. (1958). Zh. Fiz. Khim. $\underline{32}$, 2080-2086.

Krichevskii, I. R., Khodeeva, S. M., and Sominskaya, E. E. (1966). Dokl. Phys. Chem. 169(2), 468-470.

Krichevsky, I. R. and Kasarnovsky, J. S. (1935). J. Am. Chem. Soc. 57, 2168-2171.

Krishnan, C. V. and Friedman, H. I. (1969). J. Phys. Chem. 73(5), 1572-1580.

Kritchevsky, I. and Iliinskaya, A. (1945). Acta Physicochim. URSS 20(3), 327-348.

Kroon, J. and Kanters, J. A. (1974). Nature 248(5450), 667-669.

Kruus, P. (1977). Liquids and Solutions. Structure and Dynamics. Dekker, New York. 582 pp.

Ksiazczak, A. and Buchowski, H. (1980). Fluid Phase Equil. 5(1/2), 131-140.

Kudchadker, A. P., Kudchadker, S. A., Patnaik, P. R., and Mishra, P. P. (1978). J. Phys. Chem. Ref. Data 7(2), 425-439.

Kudchadker, S. A. and Kudchadker, A. P. (1978). J. Phys. Chem. Ref. Data 7(4), 1285-1278.

Kudrjawzewa, L., Toome, M., and Eisen, O. (1977). Chem. Tech. 29(11), 622-624.

Kurant, R. A., Rei, B. D., and Khorn, R. A. (1972). Elektrokhim. 8, 581-589.

Kurihara, N., Uchida, M., Fujita, T., and Nakajima, M. (1973). Pestic. Biochem. Physiol. 2(4), 383-390.

Kurtyka, Z. M. (1975). J. Chem. Ed. 52(6), 366.

Kurtyka, Z. M. and Kurtyka, E. A. (1979). J. Chem. Eng. Data 24(1), 15-16.

Kurtyka, Z. M. and Kurtyka, E. A. (1980). Ind. Eng. Chem. Fundam. 19(2), 225-227.

Kusano, K., Suurkuusk, J., and Wadsö, I. (1973). J. Chem. Thermodyn. 5, 757-767.

Kutepov, E. N. (1968). Gig. Sanit. 33(1), 32-37.

Kuznetsova, E. M. (1975). Zh. Fiz. Khim. 49(5), 1316-1317.

Kuznetsova, E. M. and Rashidov, D. S. (1977). Zh. Fiz. Khim. 51(2), 472-473.

Lachowicz, S. K. and Weale, K. E. (1958). J. Chem. Eng. Data 3(1), 162-166.

Lachowicz, S. K., Newitt, D. M., and Weale, K. E. (1955). Trans. Faraday Soc. 51, 1198-1205.

Lai, T. T., Doan-Nguyen, T. H., Vera, J. H., and Ratcliff, G. A. (1978). Can. J. Chem. Eng. 56, 358.

L'Air liquide (1976). Gas Encyclopedia. Elsevier, Amsterdam. 1150 pp.

Lamb, B. K. and Shair, F. H. (1976). Anal. Chem. 48(3), 473-475.

Lambert, J. D. (1953). Disc. Faraday Soc., No. 15, 226-233.

Landolt-Börnstein (1951). Zahlenwerte und Funktionen aus Physik, Chemie, Astronomie, Geophysik, und Technik (6th ed.). Vol. 1, Atom- und Molekularphysik, Part 3, Molekeln II. Springer-Verlag, Berlin. p. 509-517.

Landolt-Börnstein (1960). Zahlenwerte und Funktionen aus Physik, Chemie, Astronomie, Geophysik, und Technik (6th ed.). Vol. 2, Eigenschaften der Materie in ihren Aggregat-Zustäden, Part 2a, Gleichgewichte Dampf - Kondensat und Osmotische Phänomene. Springer-Verlag, Berlin. 974 pp.

Landolt-Börnstein (1962). Zahlenwerte und Funktionen aus Physik, Chemie, Astronomie, Geophysik, und Technik (6th ed.), Vol. 1, Eigenschaften der Materie in Ihren Aggregatzustäden, Part 2b, Lösungsgleichgewichte I. Springer-Verlag, Berlin.

Landolt-Börnstein (1964). Zahlenwerte und Funktionen aus Physik, Chemie, Astronomie, Geophysik, und Technik (6th ed.), Vol. 2, Eigenschaften der Materie in Ihren Aggregatzuständen, Part 2c, Lösungsgleichgewichte II. Springer- Verlag, Berlin.

Landolt- Börnstein (1975). Zahlenwerte und Funktionen aus Naturwissenschaften und Technik. Neue Serie. Thermodynamic Equilibria of Boiling Mixtures. Vol. 3. Springer-Verlag, Berlin. 376 pp.

Landolt-Börnstein (1976a). Zahlenwerte und Funktionen aus Physik, Chemie, Astronomie, Geophysik, und Technik (6th ed.). Vol. 4, Gleichgewicht der Absorption von Gases in Flüssigkeiten, Part 4c, Absorption in Flüssigkeiten von Neidrigem Dampfdruck. Wärmetechnik. Springer-Verlag, Berlin. 479 pp.

Landolt-Börnstein (1976b). Zahlenwerte und Funktionen aus Naturwissenschaften und Technik, Neue Serie, Gruppe IV, Makroskopische und technische Eigenschaften der Materie, Vol. 2, Heats of Mixing and Solution. Springer-Verlag, Berlin. 695 pp.

Landolt-Börnstein (1977). Zahlenwerte und Funktionen aus Naturwissenschaften und Technik. Neue Serie, Gruppe IV, Makroscopische und technische Eigenschaften der Materie, Vol. 1, Densities of Liquid Systems. Springer-Verlag, Berlin. 523 pp.

Landolt-Börnstein (1980a). Zahlenwerte und Funktionen aus Physik, Chemie, Astronomie, Geophysik, und Technik (6th ed.). Vol. 4, Gleichgewicht der Absorption von Gases in Flüssigkeiten, Part 4c, 2, Absorption in Flüssigkeiten von Hohen Dampfdruck. Springer-Verlag, Berlin.

Landolt-Börnstein (1980b). Zahlenwerte und Funktionen aus Naturwissenschaften und Technik. Neue Serie, Vol. 4, High-Pressure Properties of Matter. Springer-Verlag, Berlin. 427 pp.

Lange, N. A., Ed. (1973). Handbook of Chemistry (11th ed.). McGraw-Hill, New York. 2001 pp. (12th ed., 1979)

Langmuir, I. (1925). The Distribution and Orientation of Molecules. 3rd Colloid Symp. Monogr. Chemical Catalog Co., New York. p. 48-75.

Lannung, A. (1930). J. Chem. Soc. 52, 68-80.

Laprade, B., Mauger, J. W., Petersen, H., Lausier, J. M., and Paruta, A. N. (1977). Drug. Dev. Ind. Pharm. 3(1), 73-85.

Larsen, E. R. (1969). Fluorine Compounds in Anesthesiology. In Fluorine Chemistry Reviews, Vol. 3 (P. Tarrant, Ed.), Dekker, New York. p. 1-44.

Larsen, J. W. and Magid, L. J. (1974). J. Phys. Chem. 78(8), 834-839.

Larson, C. P., Eger, E. I., and Severinghaus, J. W. (1962a). Anesthesiology 23, 349-355.

Larson, C. P., Eger, E. I., and Severinghaus, J. W. (1962b). Anesthesiology 23, 686-689.

Larson, R. A. and Rockwell, A. L. (1978). Naturwiss. $\underline{65}$(9), 490.

Laszlo, P., Speert, A., and Raynes, W. T. (1969). J. Chem. Phys. $\underline{51}$, 1677-1678.

Lauven, P. M., Hack, G., and Stoeckel, H. (1979). Anaesthesist $\underline{28}$(3), 104-106.

Lavergne, M. and Drost-Hansen, W. (1956). Naturwiss. $\underline{43}$(22), 511-512.

Lawson, D. D. (1978). Applications of Solubility Parameters. Proc. Doe Chem. Hydrogen Energy Contract Rev. Conf.-771131. p. 109-114. (CA 92:47955e.)

Lawson, D. D. (1979). Methods for Calculation of Engineering Parameters for Gas Separation. Jet Propulsion Laboratory Rep. 79-46, Calif. Inst. of Technology, Pasadena, Calif. 26 pp. (CA 93:49240n and 113726e.)

Lawson, D. D., Moacanin, J., Scherer, K. V., Terrenova, T. F., and Ingham, J. D. (1978). J. Fluorine Chem. $\underline{12}$, 221-236.

Lax, E., Ed. (1967). Taschenbuch für Chemiker und Physiker (3rd ed.). Vol. 1. Springer-Verlag, Berlin.

Leach, M. J. (1977). Chem. Eng. $\underline{84}$(11), 137-140.

Lee, B. I. and Kester, M. G. (1975). AIChE J. $\underline{21}$(5), 510-527, 1040.

Lees, F. P. and Sarram, P. (1971). J. Chem. Eng. Data $\underline{16}$(1), 41-44.

Leffler, J. E. (1955). J. Org. Chem. $\underline{20}$(10), 1202-1231.

Leinonen, P. J. and Mackay, D. (1973). Can. J. Chem. Eng. $\underline{51}$(2), 230-233.

Leinonen, P. J., Mackay, D., and Phillips, C. R. (1971). Can. J. Chem. Eng. $\underline{49}$, 288-290.

Leites, I. L. and Sergeeva, L. E. (1973). Teor. Osn. Khim. Tekhnol. $\underline{7}$(5), 691-697.

Leland, T. W., McKetta, J. J., and Kobe, K. A. (1954). Phase Equilibrium in the 1-Butene – Water System and Correlation of Hydrocarbon – Water Solubility Data. Univ. of Texas, ADI 4493. Paper Presented before the Ind. Eng. Chem. Div. at the 125th Meet. Am. Chem. Soc., Kansa City, Mo./Austin, Tex.

Leland, T. W., McKetta, J. J., and Kobe, K. A. (1955). Ind. Eng. Chem. $\underline{47}$(6), 1265-1271.

Lennard-Jones, J. E. and Devonshire, A. F. (1937). Proc. Roy. Soc. Lond. Ser. A $\underline{163}$, 53-70.

Le Noble, W. J. (1965). J. Am. Chem. Soc. $\underline{87}$(11), 2434-2438.

Le Noble, W. J. and Duffy, M. (1964). J. Am. Chem. Soc. $\underline{86}$(20), 4512.

Lenoir, J. M. and Sakata, M. (1978). Ind. Eng. Chem. Fundam. $\underline{17}$(2), 71-84.

Lentz, B. R., Hagler, A. T., and Scheraga, H. A. (1974). J. Phys. Chem. $\underline{76}$(15), 1531-1550.

Leo, A., Hansch, C., and Elkins, D. (1971). Chem. Rev. $\underline{71}$(6), 525-616.

Leroy, G., Mengoni, H., Reviart, A., and Wilants, C. (1978). Bull. Soc. Chim. Belg. $\underline{87}$(3), 171-178.

Letcher, T. M. (1975). J. Chem. Thermodyn. $\underline{7}$(10), 969-972.

Lewis, G. N. and Randal, M. (1961). Thermodynamics (2nd ed.). rev. by K. S. Pitzer and L. Brewer. McGraw-Hill, New York.

Li, C. C. (1977). AIChE J. $\underline{23}$(2), 210-211.

Liabastre, A. A. (1974). Experimental Determination of the Solubility of Small Organic Molecules in H_2O and D_2O and the Application of the Scaled Particle Theory to Aqueous and Nonaqueous Solutions (Ph. D. thesis). Georgia Institute of Technology, Atlanta, Ga. 240 pp.

Lichosherstov, M. V., Alekseev, S. V., and Shalaeva, T. V. (1935). Zh. Chim. Prom. $\underline{12}$, 705-709.

Liddel, U. and Becker, E. D. (1956). J. Chem. Phys. $\underline{25}$, 173-174.

Lieberman, E. P. (1962). Official Dig. $\underline{34}$(444), 30-50.

Lieberman, E. and Wilhelm, E. (1976). Monatsh. Chem. $\underline{107}$(2), 367-369.

Lienert, E. (1975). Wien. Tierärztl. Monatsschr. $\underline{62}$(6-8), 235-240.

Lilich, L. S. and Mogilëv, M. E. (1964). Fiziko-Khimicheskie Svoistva Ratvorov. Symposium. Izd. Leningrad Univ., Leningrad.

Lillian, D., Singh, H. B., Appleby, A., Lobban, L., Arnts, R., Gumpert, R., Hague, R., Toomey, J., Kazazis, J., Antell, M., Hansen, D., and Scott, B. (1975). Environ. Sci. Technol. $\underline{9}$(12), 1042-1048.

Lin, S. H. (1970). Hydrogen Bonding. In Physical Chemistry: An Advanced Treatise, Vol. 5, Valency (H. Eyring, Ed.). Academic, New York. p. 439-482.

Lin, T. F., Christian, S. D., and Affsprung, H. E. (1965). J. Phys. Chem. $\underline{69}$(9), 2980-2983.

Lindenberg, A. B. (1956). C. R. Acad. Sci. $\underline{243}$, 2057-2060.

Line, R. A. and Hoftiezer, H. (1956). Improvements in or Relating to Method of Determination of Water in Liquids of Low Water Solubility. Br. Patent No. 805,012 (to Ansul Chemical Company, Wis.), March 20, 4 pp.

Linford, R. G. and Thornhill, D. G. T. (1977). J. Appl. Chem. Biotechnol. $\underline{27}$(9), 479-497.

Linford, R. G. and Thornhill, D. G. T. (1980). J. Chem. Tech. Biotechnol. $\underline{30}$(10), 547-555.

Linke, W. F. (1958-1965). Solubilities of Inorganic and Metal-Organic Compounds $(4^{th}$ ed.). D. Van Nostrand, Princeton, N. J., Vol. 1 (1958); Vol. 2 (1965).

Lipke, H. and Kearns, C. W. (1960). J. Econ. Entomol. $\underline{53}$, 31-35.

Liptak, B. G., Ed. (1974). Environmental Engineers' Handbook, Vol. 1, Water Pollution. Chilton, Radnor, Pa. 2018 pp.

Lisovskaya, E. V., Kostovetskii, Ya. I., and Dyatlovskaya, F. G. (1964). Sanit. Okhr. Vodoemov Zagryaz. Prom. Stochnymi Vodami, No. 6, 273-279.

Liss, P. S. and Slater, P. G. (1974). Nature $\underline{247}$, 181-184.

Liu, J.-L. and Huang, T.-C. (1961). Sci. Sin. (Peking) $\underline{10}$(6), 700-710.

Llor, J. and Cortijo, M. (1977). J. Chem. Soc. Perkin Trans. II, No. 9, 1111-1113.

Lombardo, S. R. and Missen, R. W. (1977). Can. J. Chem. Eng. $\underline{55}$(6), 753-754.

London, F. (1937). Trans. Faraday Soc. $\underline{33}$, 8-26.

Long, R. W., Hildebrand, J. H., and Morrell, W. E. (1943). J. Am. Chem. Soc. 65, 182-187.

Longuet-Higgins, H. C. (1951). Proc. Roy. Soc. Lond. Ser. A 205, 247-269.

Longuet-Higgins, H. C. and Salem, L. (1961). Proc. Roy. Soc. Lond. Ser. A 259, 433-441.

Loprest, F. J. (1957). J. Phys. Chem. 61, 1128-1130.

Lotter, Yu. G., Skripka, V. G., and Namiot, A. Yu. (1978). Russ. J. Phys. Chem. 52 (9), 1276-1279.

Lovelace, A. M., Rausch, D. A., and Postelnek, W. (1958). Aliphatic Fluorine Compounds. Reinhold, New York.

Lovelock, J. E. (1975). Nature 265(5514), 193-194.

Lovelock, J. E., Maggs, R. J., and Wade, R. J. (1973). Nature 241, 194-196.

Lown, D. A. and Thirsk, H. R. (1972). Trans. Faraday Soc. 68, 1982-1986.

Lozadac, M., Monfort, J. P., and del Río, F. (1977). Chem. Phys. Lett. 45(1), 130-133.

Lu, P.-Y. and Metcalf, R. L. (1975). Environ. Health Perspectives 10(4), 269-284.

Lucas, M. (1969). Bull. Soc. Chim. Fr., No. 9, 2994-3000.

Lucas, M. (1970). Bull. Soc. Chim. Fr., No. 9, 2902-2904.

Lucas, M. (1972). J. Phys. Chem. 76(26), 4030-4032.

Lucas, M. (1973). J. Phys. Chem. 77(20), 2479-2483.

Lucas, M. (1976). J. Phys. Chem. 80(4), 359-362.

Lucas, M. and Bury, R. (1976). J. Phys. Chem. 80(9), 999-1002.

Lucas, M. and Feillolay, A. (1970). Bull. Soc. Chim. Fr., No. 4, 1267-1270.

Luck, W. A. P. (1965). Ber. Bunsenges. 69(7), 626-637.

Luck, W. A. P. (1967). Disc. Faraday Soc. 43, 115-127.

Luck, W. A. P. (1970). Med. Welt, No. 3, 87-101.

Luck, W. A. P., Ed. (1974). Structure of Water and Aqueous Solutions. Verlag Chemie Physik Verlag, Weinheim, West Germany. 590 pp.

Luck, W. A. P. (1976). Hydrogen Bonds in Liquid Water. In Hydrogen Bond, Vol. 8, Ch. 28 (P. Schuster, G. Zundel, and C. Sandorfy, Eds.). North-Holland, Amsterdam. p. 1367-1423.

Luckhurst, G. R. and Gray, G. W., Eds. (1979). The Molecular Physics of Liquid Crystals. Academic, London. 494 pp.

Luft, N. W. (1957). Ind. Chem. 33(9), 446-447.

Luke, C. L. (1971). Anal. Chim. Acta 54, 447-459.

Luneau, J. (1965). Bull. Union Physiciens 59(483), 565-593.

Lyashchenko, A. K. and Kalinowska, B. (1977). Russ. J. Phys. Chem. 51(2), 182-183.

Lyashchenko, A. K. and Stunzas, P. (1980). Zh. Strukt. Khim. 21(3), 106-111.

Lydersen, A. L. (1955). Estimation of Critical Properties of Organic Compounds. College of Eng., Univ. of Wisconsin, Eng. Exp. Stat. Rep. 3, Madison, Wis., April.

Lykos, P., Ed. (1978). Computer Modeling of Matter. Am. Chem. Soc. Symp. Ser. No. 86. 271 pp.

Lyle, S. J. and Smith, D. B. (1975). Chem. Ind., No. 24, 1055-1057.

Mabey, W. and Mill, T. (1978). J. Phys. Chem. Ref. Data $\underline{7}$(2), 383-415.

McAulif, K. M. (1970). Org. Geokhim., No. 2, 168-182.

McAuliffe, C. A. (1963). Nature $\underline{200}$(4911), 1092-1093.

McAuliffe, C. A. (1966). J. Phys. Chem. $\underline{70}$(4), 1267-1275.

McAuliffe, C. A. (1969). Science $\underline{163}$, 478-479.

McAuliffe, C. A. (1980). The Multiple Gas-Phase Equilibration Method and Its Application to Environmental Studies. In Petroleum in the Marine Environment (L. Petrakis and F. T. Weiss, Eds.). Adv. in Chem. Ser. 185, Am. Chem. Soc., Washington, D. C. p. 193-218.

McBee, E. T. and Hatton, R. E. (1949). Ind. Eng. Chem. $\underline{41}$(4), 809-812.

McBee, E. T., Lindgren, V. V., and Ligett, W. B. (1947). Ind. Eng. Chem. $\underline{39}$, 378-379.

McClellan, A. L. (1963). Tables of Experimental Dipole Moments, Vol. 1. W. H. Freeman, San Francisco. 700 pp.

McClellan, A. L. (1974). Tables of Experimental Dipole Moments, Vol. 2, Rahara Enterprises, El Cerrito, Calif. 999 pp.

McClure, H. B. (1944). Chem. Eng. News $\underline{22}$(6), 416-421.

McConnell, G., Ferguson, D. M., and Pearson, C. R. (1975). Endeavour $\underline{34}$(121), 13-18.

McDaniel, A. S. (1911). J. Phys. Chem. $\underline{15}$, 587-610.

McDermott, C. and Ashton, N. (1977). Fluid Phase Equilib. $\underline{1}$(1), 33-35.

McDevit, W. F. and Long, F. A. (1952). J. Am. Chem. Soc. $\underline{74}$, 1773-1777.

Macdonald, D. D., Estep, M. E., Smith, M. D., and Hyne, J. B. (1974). J. Chem. Soc. $\underline{3}$(9), 713-725.

McDonald, I. R. and Singer, K. (1970). Quart. Rev. $\underline{24}$(2), 238-262.

McDonald, I. R. and Singer, K. (1973). Chem. Br. $\underline{9}$(2), 53-60 and 65.

MacDougall, F. H. (1936). J. Am. Chem. Soc. $\underline{58}$, 2585-2591.

MacDougall, F. H. (1941). J. Am. Chem. Soc. $\underline{63}$, 3420-3424.

McGlashan, M. L. (1962). Experimental Thermochemistry, Vol. 2 (H. A. Skinner, Ed.). Interscience, New York. p. 321-342.

McGovern, E. W. (1943). Ind. Eng. Chem. $\underline{35}$(12), 1230-1239.

McGowan, J. C. (1952). J. Appl. Chem. $\underline{2}$(6), 323-328.

McGowan, J. C. (1954). J. Appl. Chem. $\underline{4}$(1), 41-47.

McGowan, J. C. (1956). Rec. Trav. Chim. Pays-Bas $\underline{75}$, 193-208.

McGowan, J. C. (1971). Chem. Commun., No. 10, 514-515.

Nacintosh, R., Mushin, W. W., and Epstein, H. G. (1963). Physics for the Anaesthetist, Including a Section on Explosion. Blackwell Science, Oxford. 439 pp.

Mackay, D. (1979). Environ. Sci. Technol. $\underline{13}$(10), 1218-1223.

Mackay, D. and Leinonen, P. J. (1975). Environ. Sci. Technol. $\underline{9}$(13), 1178-1180.

Mackay, D. and Shiu, W.-Y. (1975a). Can. J. Chem. Eng. 53(2), 239-242.

Mackay, D. and Shiu, W.-Y. (1975b). The Aquous Solubility and Air - Water Exchange Characteristics of Hydrocarbons under Environmental Conditions. Chem. Phys. Aqueous Gas Solutions, p. 93-110.

Mackay, D. and Shiu, W.-Y. (1977). J. Chem. Eng. Data 22(4), 399-402.

Mackay, D. and Wolkoff, A. W. (1973). Environ. Sci. Technol. 7(7), 611-614.

Mackay, D., Mascarenhas, R., and Shiu, W. Y. (1980). Chemosphere 9(5-6), 257-264.

Mackay, D., Shiu, W.-Y., Sutherland, R. P. (1979). Environ. Sci. Technol. 13(3), 333-337.

Mackay, S., Shiu, W.-Y., and Wolkoff, A. W. (1975). Gas Chromatographic Determination of Low Concentration of Hydrocarbon in Water by Vapor Phase Extraction. In Water Quality Parameters. ASTM Spec. Tech. Publ. 573. Am. Soc. Testing and Materials, Philadelphia, p. 251-258.

McKee, J. E. and Wolf, H. W. (1963). Water Quality Criteria (2nd ed.). The Resources Agency of California. State Water Qual. Control Board Publ. No. 3-A. Pasadena, Calif. 548 pp.

Maclean, J. N., Rossotti, F. J. C., and Rossotti, H. S. (1962). J. Inorg. Nucl. Chem. 24, 1549-1554.

McRae, E. G. (1957). J. Phys. Chem. 61, 562-572.

Maczynski, A. (1976). Verified Vapor-Liquid Equilibrium Data. PWN-- Polish Scientific Publishers, Warsaw.

Mader, W. J. and Grady, L. T. (1971). Determination of Solubility. In Techniques of Chemistry, Vol. 1, Physical Methods of Chemistry, Part 5, Determination of Thermodynamic and Surface Properties (A. Weissberger and B. W. Rossiter, Eds.). Wiley-Interscience, New York. p. 257-308.

Mader, W. J., Vold, R. D., and Vold, M. J. (1963). Determination of Solubility. In Techniques of Chemistry, Vol. 1, Part 1, Physical Methods of Organic Chemistry (A. Weissberger, Ed.). Interscience, New York. p. 655-688.

Madgin, W. M. and Briscoe, H. V. A. (1927). J. Soc. Chem. Ind. 46, 107T-108T.

Magnusson, L. B. (1970). J. Phys. Chem. 74(24), 4221-4228.

Maharajh, D. (1973). Solubility and Diffusion of Gases in Water. (Ph. D. thesis). Simon Fraser Univ., British Columbia, Canada. May. 174 pp.

Maier, D. and Mackle, H. (1976). Vom Wasser 47, 379-397.

Mair, B. J., Glascow, A. R., and Rossini, F. D. (1941). J. Res. Natl. Bur. Std. 27 (1), 39-63.

Majer, V., Svoboda, V., Pick, J., and Holub, R. (1974). Coll. Czech. Chem. Commun. 39(1), 11-19.

Malanowski, S. (1974). Rownowaga ciecz-para. PWN--Polish Science Publishers, Warsaw.

Malek, K. R. and Stiel, L. J. (1972). Can. J. Chem. Eng. 50, 491-495.

Malesinski, W. (1965). Azeotropy and Other Theoretical Problems of Vapor-Liquid Equilibrium. Interscience, London. 222 pp.

Malijevska, I. (1979). Coll. Czech. Chem. Commun. $\underline{44}$(4), 1187-1196.

Malinowski, E. R. and Garg, S. K. (1977). J. Phys. Chem. $\underline{81}$(7), 685-686.

Malone, W. F. (1975). J. Chem. Educ. $\underline{52}$(10), A468-A470.

Mamedaliev, Yu. G. and Musachanly, S. (1940). Zh. Prikl. Khim. $\underline{13}$(5), 735-737.

Mandelcorn, I. (1959). Chem. Rev. $\underline{59}$, 827-839.

Mandelcorn, I. (1964). Non-Stoichiometric Compounds. Academic, New York.

Manja, K. S., Krishnan, B., Nambinarayauan, T. K., and Rao, A. S. (1974). Bull. Soc. Chim. Belg. $\underline{83}$(5-6), 197-200.

Mannheimer, M. (1956). Chem. Anal. $\underline{45}$(1), 8-10.

Manufacturing Chemists Association (1956-1971). Chemical Safety Data Sheets SD-14 (Trichloroethylene) 1956; SD-18 (Ethylene dichloride) 1971; SD-35 (Methyl Bromide) 1968; SD-40 (Methyl Chloride) 1970; SD-99 (Allyl Chloride) 1973. Washington, D. C.

Mapstone, G. E. (1952). Chem. Proc. $\underline{15}$(12), 176.

Marcelja, S., Mitchell, D. J., Ninham, B. W., and Sculley, M. J. (1977). J. Chem. Soc. Faraday Trans. II $\underline{73}$(5), 630-648.

Marcus, Y. (1977). Introduction to Liquid State Chemistry. Wiley, London. 357 pp.

Maretic, M. and Sirocic, V. (1962). Nafta (Zagreb). $\underline{13}$, 126-131.

Margenau, H. and Kester, N. R. (1969). Theory of Intermolecular Forces. Pergamon, Oxford. 360 pp.

Margules, M. (1895). Sitzber. Akad. Wiss. Math. Naturwiss. Kl. II $\underline{104}$, 1243-1278.

Marius, G. (1937). Riv. Ital. Essenze Prof. $\underline{19}$, 263-265 and XXX-XXXI.

Mrkham, A. E. and Kobe, K. A. (1941). Chem. Rev. $\underline{28}$, 519-588.

Marsden, C., Ed. (1963). Solvents Guide (2nd ed.). Cleaver-Hume, London.

Martin, E., Yalkowsky, S. H., and Wells, J. E. (1979). J. Pharm. Sci. $\underline{68}$(5), 565-568.

Martin, J. J. and Hou, Y.-C. (1955). AIChE J. $\underline{1}$(2), 142-151.

Martire, D. S. (1966). Anal. Chem. $\underline{38}$(2), 244-250.

Marvel, C. S., Copley, M. J., and Grinsberg, E. (1940). J. Am. Chem. Soc. $\underline{62}$, 3263-3264.

Marx, J. L. (1974). Science $\underline{186}$, 809-811.

Marx, J. L. (1977). Science $\underline{196}$(4290), 632-636.

Maryott, A. A. and Smith, E. R. (1951). Table of Dielectric Constants of Pure Liquids. Natl. Bur. Std. Circ. No. 514. Washington, D. C., August 10.

Mash, C. J. and Pemberton, R. C. (1980). Activity Coefficients at Very Low Concentrations for Organic Solutes in Water Determined by an Automatic Chromatographic Method. Natl. Phys. Lab. Rep. CHEM 111. Teddington, England. July. 24 pp.

Mashiko, Y., Iizuka, K., Saeki, S., and Kondo, S. (1978). Precision of Numerical and Graphical Data Presentation. 6th CODATA Conf., Santa Flavia, Italy. May.

Maslennikova, V. Ya., Goryunova, N. P., Subbotina, L. A., and Tsiklis, D. S. (1976). Russ. J. Phys. Chem. $\underline{50}$(2), 240-243.

Mason, E. A. and Spurling, T. H. (1969). The Virial Equation of State. Pergamon, Oxford. 297 pp.

Masterton, W. L. and Gendrano, M. C. (1966). J. Phys. Chem. 70(9), 2895-2898.

Masterton, W. L. and Seiler, H. K. (1968). J. Phys. Chem. 72(12), 4257-4262.

Masuoka, H., Tawaraya, R., Saito, S. (1979). J. Chem. Eng. Japan 12(4), 257-263.

Matheson Company (1966-1971). Matheson Gas Data Book (4th ed. 1966; 5th ed. 1971). East Rutherford, N. J. (500 and 574 pp. resp.)

Mattelin, A. C. and Verhoeye, L. A. J. (1975). Chem. Eng. Sci. 30, 193-200.

Matthews, P. J. (1975). Effl. Water Treat. J. 15(11), 565-567 and 626-627.

Matthews, P. J. (1979). Use of Vapor-Liquid Equilibrium Data for Estimating Trade Effluent Limits. Proc. Natl. Phys. Lab. Conf. on Chemical Thermodynamic Data on Fluids and Fluid Mixtures. Teddington, Sept. 11-12, 1978. IPC Science and Technology Press, Guildford, England. p. 53-61.

Maugh, T. H. (1978). Science 202(4363), 37-41.

Maurer, G. and Prausnitz, J. M. (1978). Fluid Phase Equilib. 2(2), 91-99.

May, W. E. (1980). The Solubility Behavior of Polycyclic Aromatic Hydrocarbons in Aqueous Systems. In Petroleum in the Marine Environment (L. Petrakis and F. T. Weiss, Rds.). Adv. in Chem. Ser. 185, Am. Chem. Soc., Washington, D. C., p. 143-192.

May, W. E., Wasik, S. P., and Freeman, D. H. (1978a). Anal. Chem. 50(1), 175-179.

May, W. E., Wasik, S. P., and Freeman, D. H. (1978b). Anal. Chem. 50(7), 997-1000.

Medvedev, V. A. (1963). Russ. J. Phys. Chem. 37(6), 751-753.

Megaw, W. J. (1976). Nature 261(5555), 10.

Meissner, H. P. and Greenfield, S. H. (1948). Ind. Eng. Chem. 40(3), 438-442.

Meleshchenko, K. F. (1960a). Gig. Sanit. 25(3), 13-18.

Meleshchenko, K. F. (1960b). Gig. Sanit. 25(5), 54-57.

Mellan, I. (1950). Industrial Solvents (2nd ed.). Reinhold, New York. 758 pp.

Mellan, I. (1957). Source Book of Industrial Solvents, Vol. 2, Reinhold, New York.

Mellan, I. (1977). Industrial Solvents Handbook (2nd ed.). Noyes Data Corp., Park Ridge, N. J. 567 pp.

Mel'tser, L. Z. and Smirnov, L. F. (1968). Kholod. Tekh. 45(5), 21-25.

Mercel, J. H. C. (1937). Recl. Trav. Chim. 56, 811-814.

Merriman, J. R. (1977). Solubility of Gases in CCl_2F_2: A Critical Review. Union Carbide Corp., Nucl. Div. Rep. No. KY-G-400. Gaseous Diffusion Plant, Peducah, Ky., March. 169 pp.

Meslans, M. M. (1894). Ann. Chim. Phys. 1, 346-423.

Metanomski, W. V. (1980). J. Chem. Inf. Comput. Sci. 20(3), 131.

Metcalf, R. L., Ed. (1962). Advances in Pest Control Research, Vol. 5. Interscience, New York. 329 pp.

Metcalf, R. L., Sanborn, J. R., Lu, P.-Y., and Nye, D. (1975). Arch. Environ. Contam. Toxicol. 3(2), 151-165.

Metropolis, N., Rosenbluth, A. W., Rosenbluth, M. N., Teller, A. H., and Teller, E. (1953). J. Chem. Phys. $\underline{21}$(6), 1087-1092.

Michels, A., Gerver, J., and Bijl, A. (1936). Physica $\underline{3}$(8), 797-808.

Miklashevskii, V. E., Tugarinova, V. N., Yakovleva, G. P., Aleseeva, N. P., and Rakhmanina, N. L. (1962). Sanit. Okhr. Vodoemov Zagryaz. Prom. Stochnymi Vodami, No. 5, 308-325.

Miller, A. A. (1963). J. Chem. Phys. $\underline{38}$, 1568-1571.

Miller, K. J. and Savchik, J. A. (1979). J. Am. Chem. Soc. $\underline{101}$(24), 7206-7213.

Miller, K. W. (1966). Molecular Interactions (Ph. D. thesis). Oxford Univ., Oxford.

Miller, K. W. and Hildebrand, J. H. (1968). J. Am. Chem. Soc. $\underline{90}$(12), 3001-3004.

Miller, R. C. and Smyth, C. P. (1956). J. Chem. Phys. $\underline{24}$(4), 814-817.

Miller, R. C. and Smyth, C. P. (1957). J. Am. Chem. Soc. $\underline{79}$, 20-24.

Miller, S. A., Ed. (1969). Ethylene and Its Industrial Derivatives. Ernest Benn, London. 1321 pp.

Millero, F. J. (1971). Chem. Rev. $\underline{71}$(2), 147-176.

Minkin, V. I., Osipov, O. A., and Zhdanov, Y. A. (1970). Dipole Moments in Organic Chemistry, Plenum, New York.

Minto, M. A. (1975). Lab. Pract. $\underline{24}$(2), 74.

Mishchenko, K. P. (1972). Thermodynamics of Electrolyte Solutions. In Physical Chemistry, Series 1, Vol. 10, Thermochemistry and Thermodynamics (A. D. Buckingham and H. A. Skinner, Eds.). MTP Int. Rev. Sci., Butterworths, London. p. 177-207.

Missen, R. W. (1978). Can. J. Chem. Eng. $\underline{56}$(1), 126-127.

Mistura, L. (1973). J. Chem. Phys. $\underline{59}$(8), 4563-4564.

Mitchell, A. G., Wan, L. S. C., and Bjaastad, S. G. (1964). J. Pharm. Pharmacol. $\underline{16}$, 632-633.

Mitchell, J. and Smith, D. M. (1977). Aquametry, Part 1: A Treatise on Methods for the Determination of Water (2nd ed.). Wiley, New York. 632 pp.

Mitomo, T. and Teshirogi, T. (1979-1980). Kagaku Kyoiku $\underline{27}$(6), 449-453 (1979); $\underline{28}$(1), 70-74 (1980).

Miyahara, K., Sadotomo, H., and Kitamura, K. (1970). J. Chem. Eng. Jap. $\underline{3}$(2), 157-160.

Mizutani, K. and Yamashita, K. (1950). Rep. Gov. Chem. Ind. Res. Inst., Tokyo, $\underline{45}$, 49-55.

Moelwyn-Hughes, E. A. (1938). Proc. Roy. Soc. Lond. Ser. A $\underline{164}$, 295-306.

Moelwyn-Hughes, E. A. (1964). Physical Chemistry (2nd ed.), Pergamon, Oxford. 1334 pp.

Moelwyn-Hughes, E. A. (1971). The Chemical Statistics and Kinetics of Solutions. Academic, London. 507 pp.

Moiseeva, L. M., Stepanov, G. G., and Pukhonto, A. N. (1977). Khim. Prom., No. 6, 435-437.

Moissan, H. (1890). Ann. Chim. Phys. $\underline{19}$(6), 266-280.

Molina, M. J. and Rowland, F. S. (1974). Nature 249(5460), 810-812.

Monfort, J. P., Varela-Ham, J.-R., and Perez-Meseguer, J.-L. (1977). J. Chim. Phys. 74(4), 409-423.

Moore, J. W. and Moore, E. A. (1976). J. Chem. Educ. 53(4), 240-243.

Morild, E. (1980). Acta Chem. Scand. 34A(10), 777-779.

Moriyoshi, T. and Aoki, Y. (1978). J. Chem. Eng. Jap. 11(5), 341-345.

Moriyoshi, T., Kaneshina, S., Aihara, K., and Yabumoto, K. (1975). J. Chem. Thermodyn. 7(6), 537-545.

Moriyoshi, T., Aoki, Y., and Kamiyama, H. (1977). J. Chem. Thermodyn. 9(5), 495-502.

Morokhov, I. D., Chizhik, S. P., Gladkikh, N. T., Grigor'eva, L. K., and Stepanov, S. V. (1979). Dokl. Akad. Nauk SSSR. 247(6), 1376-1380.

Morrell, W. E. and Hildebrand, J. H. (1936). J. Chem. Phys. 4(3), 224-227.

Morrison, T. J. and Billett, F. (1948). J. Chem. Soc. 2033-2035.

Morrison, T. J. and Johnstone, N. B. (1954). J. Chem. Soc. 3441-3446.

Moudgil, B. M., Somasundran, P., and Lin, I. J. (1974). Rev. Sci. Inst. 45(3), 406-409.

Mountain, R. D. (1970). Crit. Rev. Solid State Sci. 1, 5-46.

Mourits, F. M. and Rummens, F. H. (1977). Can. J. Chem. 55(16), 3007-3020.

Mruzik, M. R. (1977). Chem. Phys. Lett. 48(1), 171-175.

Muccitelli, J. A. (1978). Solubilities of Some Perfluorocarbon Gases in Water, Heavy Water and Ethanol-Water Mixtures and of Methane in Water and Aqueous Solutions of Triethylenediamine. (Ph. D. thesis). Clark Univ., Worcester, N. Y., 371 pp.

Muccitelli, J. and Wen, W.-Y. (1978). Am. Chem. Soc. Abstr. Pap. (Eng. Proc.), 175, (60).

Mukerjee, P. and Yang, A. Y. S. (1976). J. Phys. Chem. 80(12), 1388-1390.

Mukhopadhyay, M. (1975). Ind. Eng. Chem. Process Des. Dev. 14(2), 195-196.

Müller, E. and Hüther, F. (1931). Chem. Ber. 64, 589-600.

Müller, F. H. (1933). Phys. Z. 34, 689-710.

Müller, F. H. (1934). Phys. Z. 35, 346-349.

Müller, F. H. (1937). Phys. Z. 38, 283-292.

Müller, H.-D. (1975). Wiss. Z. Tech. Hochsch. Chem. "Carl Schorlemmer" Leuna-Merseburg 17(1), 72-82.

Müller, N. and Simon, P. (1967). J. Phys. Chem. 71(3), 568-572.

Munson, A., Sanders, V., Borzelle, J., and Barnes, D. (1977). Pharmacologist 19(2), 200.

Murgulescu, I. G. and Demetrescu, I. (1970). Stud. Cercet. Chim. 18(6), 545-566.

Murray, A. J. and Riley, J. P. (1973). Nature 242(5392), 37-38.

Murray, C. (1975). Chem. Eng. News 53(7), 22.

Murray, F. E. and Mason, S. G. (1952). Can. J. Chem. 30(7), 550-561.

Murray, W. J., Kier, L. B., and Hall, L. H. (1976). J. Med. Chem. 19(5), 573-578.

Murthy, A. S. N. and Rao, C. N. R. (1970). J. Mol. Struct. $\underline{6}$, 253-282.

Musgrave, W. K. R. and Smith, F. (1949). J. Chem. Soc., 3021-3026.

Myers, R. T. (1977). Inorg. Chem. $\underline{16}$(10), 2671-2674.

Nagahama, K., Suzuki, I., and Hirata, M. (1971). J. Chem. Eng. Jap. $\underline{4}$(1), 1-5.

Nagata, I., Ogura, M., and Nagashima, M. (1975). Ind. Eng. Chem. Process Des. Dev. $\underline{14}$(4), 500-502.

Nagata, I., Yamada, T., Gotoh, K., and Kazuma, K. (1975). J. Chem. Eng. Jap. $\underline{8}$(1), 71-72.

Nakanishi, K. (1968). J. Chem. Eng. Jap. $\underline{1}$(2), 104-109.

Namiot, A. Yu. (1967). Zh. Strukt. Khim. $\underline{8}$(3), 363-366.

Namiot, A. Yu. (1979). Russ. J. Phys. Chem. $\underline{53}$(12), 1740-1742.

Namiot, A. Yu. and Gorodetskaya, L. E. (1970). Dokl. Chem. $\underline{190}$(1-3), 86-88.

Namiot, A. Yu., Skripka, V. G., Gubkina, G. F., and Boksha, O. A. (1976). Russ. J. Phys. Chem. $\underline{50}$(4), 510-513.

Nango, M., Yamamoto, H., Joukou, K., Ueda, M., Katayama, A., and Kuroki, N. (1980). J. Chem. Soc. Chem. Commun., No. 3, 104-105.

Natarajan, G. S. and Venkatachalam, K. A. (1972). J. Chem. Eng. Data $\underline{17}$(3), 328-329.

Nath, J. (1977). J. Chem. Phys. $\underline{67}$(10), 4776-4777.

Nath, J., Das, S. S., and Yadava, M. L. (1976). Ind. Eng. Chem. Fundam. $\underline{15}$(3), 223-225.

Nathan, M. F. (1978). Chem. Eng. $\underline{85}$(3), 93-100.

National Institute for Occupational Safety and Health (1976). Criteria for a Recommended Standard Occupational Explosure to 1,1,1-Trichloroethane (Methyl Chloroform), Publication PB-267 069. Cincinnati, Ohio, July. 180 pp.

National Physical Laboratory (1979). Proc. of the NPL Conf. Chemical Thermodynamic Data on Fluids and Fluid Mixtures - Their Estimation, Correlation, and Use. IPC Sciences and Technology Press, Guildford, England. (Sept. 11-12), 215 pp.

National Technical Information Service (1977). Toxicity of Gaseous Halogenated Organic Compounds. A Bibliography with Abstracts. Publication NTIS/PS-77/0521. Springfield, Va., June. 178 pp.

Nauruzov, M. Kh. (1975). Tr. Inst. Khim. Nefti Prir. Solei Akad. Nauk Kaz., No. 8, 17-18.

Neff, R. O. and McQuarrie, D. A. (1973). J. Phys. Chem. $\underline{77}$(3), 413-418.

Nelson, H. D. and de Ligny, C. L. (1968a). Recl. Trav. Chim. $\underline{87}$, 528-544.

Nelson, H. D. and de Ligny, C. L. (1968b). Recl. Trav. Chim. $\underline{87}$, 623-640.

Nelson, H. D. and Smit, J. H. (1978). South Afr. Tydskr. Chem. $\underline{31}$(2), 76.

Nelson, R. D., Lide, D. R., and Maryott, A. A. (1967). Selected Values of Electric Dipole Moments for Molecules in the Gas Phase. Natl. Bur. Std. NSRDS-NBS 10. Washington, D. C., September. 49 pp.

Nemethy, G. (1974). Recent Structural Models for Liquid Water. In Structure of Water and Aqueous Solutions (A. P. Luck, Ed.). Verlag-Chemie, Weinheim, West Germany, p. 73-91.

Nemethy, G. and Scheraga, H. A. (1962a). J. Chem. Phys. $\underline{36}$(12), 3382-3400.

Nemethy, G. and Scheraga, H. A. (1962b). J. Chem. Phys. $\underline{36}$(12), 3401-3417.

Neumann, H. M. (1977). J. Solution Chem. $\underline{6}$(1), 33-38.

Nex, R. W. and Swezey, A. W. (1954). Weeds $\underline{3}$, 241-253.

Ng, H.-J. and Robinson, D. B. (1976). Ind. Eng. Chem. $\underline{15}$(4), 293-298.

Nichols, N., Sköld, R., Spink, C., Suurkuusk, J., and Wadsö, I. (1976). J. Chem. Thermodyn. $\underline{8}$(11), 1081-1093.

Nicloux, M. and Scotti-Foglieni, L. (1929). Ann. Physiol. Physicochim. Biol. $\underline{5}$, 434-482.

Nicolaides, G. L. and Eckert, C. A. (1978). Ind. Eng. Chem. Fundam. $\underline{17}$(4), 331-340.

Niini, A. (1938). Suomen Kemistil. $\underline{11A}$, 19-20.

Nikul'shin, R. K. and Petriman, E. F. (1976). Russ. J. Phys. Chem. $\underline{50}$(6), 853-854.

Nishino, N. and Nakamura, M. (1978). Bull. Chem. Soc. Jap. $\underline{51}$(6), 1617-1620.

Nitta, T., Takeuchi, S., and Katayama, T. (1974). Chem. Eng. Sci. $\underline{29}$(11), 2213-2218.

Nitta, T., Turek, E. A., Greenkorn, R. A., and Chao, K. C. (1977a). AIChE J. $\underline{23}$(2), 144-160.

Nitta, T., Turek, E. A., Greenkorn, R. A., and Chao, K. C. (1977b). A Group Contribution Molecular Model for Liquids and Solutions Composed of the Groups CH_3, CH_2, OH, and CO. In Phase Equilibria and Fluid Properties in Chemical Industry. Am. Chem. Soc. Symp. Ser. No. 60, p. 421-428.

Norman, C. and Sherwell, C. (1976). Nature $\underline{263}$(5575), 268-269.

Nosov, E. F. and Barlyaev, E. V. (1968). J. Gen. Chem. USSR $\underline{38}$(2), 215.

Novak, D. M. and Conway, B. E. (1973). Chem. Instr. $\underline{5}$(2), 79-90.

Null, H. R. (1970). Phase Equilibrium in Process Design. Wiley, New York.

Nutting, H. S. and Horsley, L. H. (1973). Graphical Method for Predicting Effect of Pressure on Azeotropic Systems. In Azeotropic Data--III (L. H. Horsley, Ed.). Adv. Chem. Ser. No. 116. Am. Chem. Soc., Washington, D. C., p. 626-628.

Nys, G. G. and Rekker, R. F. (1973). Chim. Ther. $\underline{8}$(5), 521-535. (CA 81:20777j.)

Obraztsov, V. I. and Khrustaleva, A. A. (1973). Russ. J. Phys. Chem. $\underline{47}$(4), 461-463.

O'Brien, R. N. (1972). Interferometry. In Techniques of Chemistry, Vol. 1, Physical Methods of Chemistry, Part 3A (A. Weissberger, Ed.). Wiley-Interscience, New York. p. 1-73.

O'Brien, R. N. and Hyslop, W. F. (1975). A Fabry-Perot Interferometer for Monitoring Gas-Liquid Exchange. In Chemistry and Physics of Aqueous Gas Solutions (W. A. Adams et al., Eds.). Electrochem. Soc., Princeton, N. J., p. 326-336.

O'Connell, W. L. (1963). Trans. Am. Inst. Mech. Eng. $\underline{226}$(2), 126-132.

O'Connor, J. T., Badorec, D., Thiem, L., and Popalisky, J. R. (1977). Design, Construction and Operation of a Pilot Plant for the Removal of Organic Substances from Missouri River Water. Proc. 97th AWWA Annu. Conf., Vol. 2, No. 33-2. 11 pp.

Ödberg, L. and Högfeldt, E. (1969). Acta Chem. Scand. $\underline{23}$, 1330-1342.

Ödberg, L., Löfvenberg, A., Högfeldt, E., and Fredlund, F. (1972). J. Inorg. Nucl. Chem. $\underline{34}$(8), 2605-2616.

Oellrich, L. R., Plocker, J., and Knapp, H. (1973). Vapor-Liquid Equilibria. Institute of Thermodynamics, Technical Univ. of Berlin, Berlin.

Oellrich, L., Plocker, U., Prausnitz, J. M., and Knapp, H. (1977). Chem.-Ing.-Tech. $\underline{49}$(12), 955-965.

Oh, Y., Jhon, M. S., and Eyring, H. (1977). Proc. Natl. Acad. Sci. U. S. A. $\underline{74}$(11), 4739-4743.

Ohe, S. (1976). Computer Aided Data Book of Vapor Pressure. Data Book, Tokyo. 2035 pp.

Ohnishi, R. and Tanabe, K. (1971). Bull. Chem. Soc. Jap. $\underline{41}$, 2647-2649.

Ohtaki, H. (1975). Kagaku Kyoiku $\underline{23}$(1), 65-69.

Olbregts, J. and Walgraeve, J. P. (1976). J. Chem. Educ. $\underline{53}$(9), 602-604.

Olofsson, G. and Olofsson, I. (1977). J. Chem. Thermodyn. $\underline{9}$(1), 65-69.

Onnes, H. K. (1901). Arch. Neerl. $\underline{6}$, 874-888.

Onsager, L. (1936). J. Am. Chem. Soc. $\underline{58}$, 1486-1493.

O'Reilly, M. G. and Edmonds, B. (1977). Chem. Eng. Lond., No. 328, 61-63.

Orentlicher, M. and Prausnitz, J. M. (1964). Chem. Eng. Sci. $\underline{19}$, 775-782.

Orlovskii, V. M. (1963). Vopr. Gig. Naselennykh Mest Sb. $\underline{4}$, 199-203.

Osipov, O. A. and Minkin, V. I. (1965). Handbook of Dipole Moments. Izdatel'stvo "Vyshaya Shkola," Moscow.

Osipov, O. A., Ismailov, K. M., Garnovskii, A. D., Orlova, L. V., and Kashireninova, O. E. (1965). J. Gen. Chem. USSR $\underline{35}$, 267-269.

Ostrenga, J. A. (1969). J. Pharm. Sci. $\underline{58}$(10), 1281-1282.

Ostwald, W. (1894). Manual of Physico-Chemical Measurements. Macmillan, London. 255 pp.

Othmer, D. F. (1963). Azeotropy and Azeotropic Distillation. In Kirk-Othmer Encyclopedia of Chemical Technology (2nd ed.). Vol. 2. Wiley-Interscience, New York. p. 839-859.

Othmer, D. F. (1965). Data--Interpretation and Correlation. In Kirk-Othmer Encyclopedia of Chemical Technology (2nd ed.). Vol. 6. Wiley-Interscience, New York. p. 705-755.

Othmer, D. F. and Chen, H.-T. (1968). Ind. Eng. Chem. $\underline{60}$(4), 39-61.

Othmer, D. F. and Roszkowski, E. S. (1949). Data--Interpretation and Correlation. In Kirk-Othmer Encyclopedia of Chemical Technology (1st ed.). Vol. 4. Wiley, New York. p. 846-873.

Othmer, D. F., White, R. E., and Trueger, E. (1941). Ind. Eng. Chem. $\underline{33}$(12), 1513.

Otson, R., Williams, D. T., and Bothwell, P. D. (1979). Environ. Sci. Technol. $\underline{13}$(8), 936-939.

Owicki, J. C., Shipman, L. L., and Scheraga, H. A. (1975). J. Phys. Chem. $\underline{79}$(17), 1794-1811.

Owicki, J. C., Lentz, B. R., Hagler, A. T., and Scheraga, H. A. (1975). J. Phys. Chem. $\underline{79}$(22), 2352-2361.

Palit, S. R. (1947). J. Phys. Colloid Chem. $\underline{51}$, 837-857.

Palit, S. R. and McBain, J. W. (1947). J. Soc. Chem. Ind. $\underline{66}$, 3-5.

Palmer, D. A. (1975). Chem. Eng. $\underline{82}$(12), 80-85.

Palmer, D. A. and Smith, B. D. (1972). Ind. Eng. Chem. Process Des. Dev. $\underline{11}$(1), 114-119.

Palmer, G. (1948). Ind. Eng. Chem. $\underline{40}$(1), 89-92.

Palmer, H. A. (1950). Characterization of Hydrocarbontype Hydrates (Ph. D. thesis). Univ. of Oklahoma, Norman, Okla.

Papazian, H. A. (1971). J. Am. Chem. Soc. $\underline{93}$(22), 5634-5636.

Papper, E. M. and Kitz, R. J. (1963). Uptake and Distribution of Anesthetic Agents. McGraw-Hill, New York. p. 10.

Parker, A. J. (1962). Quart. Rev. $\underline{16}$, 163-187.

Parmelee, H. M. (1953-1954). Refrig. Eng. $\underline{61}$(12), 1341-1345 (1953); $\underline{62}$(1), 61, 68 (1954); $\underline{62}$(2), 53 (1954).

Parnov, E. I. (1969). Russ. J. Phys. Chem. $\underline{43}$(4), 572.

Parshad, R. (1942). Indian J. Phys. $\underline{16}$, 1-11.

Partington, J. R. (1953). An Advanced Treatise on Physical Chemistry, Vol. 4, Physico-Chemical Optics, Longmans, London. 688 pp.

Partington, J. R. (1954). An Advanced Treatise on Physical Chemistry, Vol. 5, Molecular Spectra and Structure Dielectrics and Dipole Moments. Longmans, London. 565 pp.

Paruta, A. N. (1963). A Dielectric Constant Approach to Solubility Phenomena (Ph. D. thesis). Rutgers, The State Univ., New Brunswick, N. J. 135 pp.

Paruta, A. N., Sciarrone, B. J., and Lordi, N. G. (1962). J. Pharm. Sci. $\underline{51}$(7), 704-705.

Patrick, C. R. (1969). Chem. Ind., p. 940-942.

Patrick, C. R. (1971). Chem. Br. $\underline{7}$(4), 154-156.

Patsatsiya, K. M. and Krestov, G. A. (1970). Russ. J. Phys. Chem. $\underline{44}$(7), 1036-1037.

Patterson, D. (1976). Pure Appl. Chem. $\underline{47}$(4), 305-314.

Patterson, D. and Barbe, M. (1976). J. Phys. Chem. $\underline{80}$(21), 2435-2436.

Patterson, D., Tewari, Y. C., and Schreiber, H. P. (1972). Trans. Faraday Soc. $\underline{68}$, 885-894.

Patty, F. A., Ed. (1962). Industrial Hygiene and Technology (2nd rev. ed.). Vol. 2. Interscience, New York. p. 831-2377.

Pauling, L. (1960). The Nature of the Chemical Bond (3rd ed.). Cornell University Press, Ithaca, N. Y.

Paull, J. D. and Hewett, E. B. (1976). Anaesth. Intensive Care $\underline{4}$(1), 68-69.

Pavlovskaya, E. M., Charykov, A. K., and Tikhomirov, V. I. (1977). Zh. Obshch. Khim. $\underline{47}$(11), 2439-2444.

Peake, E. and Hodgson, G. W. (1966). J. Am. Oil Chem. Soc. $\underline{43}$(4), 215-222.

Peake, E. and Hodgson, G. W. (1967). J. Am. Oil Chem. Soc. 44(12), 696-702.

Pearson, C. R. and McConnell, G. (1975). Proc. Roy. Soc. Lond. Ser. B 189, 305-332.

Pechiney-Saint-Gobain (1971). Flugène 113, 113CM, 113M, 113MA. Solvant de Précision. Seuilly sur Seine, France.

Pecsar, R. E. and Martin, J. J. (1966). Anal. Chem. 38(12), 1661-1669.

Pedersen, P. V. and Brown, K. F. (1976a). J. Pharm. Sci. 65(10), 1437-1442.

Pedersen, P. V. and Brown, K. F. (1976b). J. Pharm. Sci. 65(10), 1442-1447.

Pegler, C. L. and Muir, W. R., Chairmen (1975). Fluorocarbons and the Environment: Report of Federal Task Force on Inadvertent Modification of the Statosphere (IMOS). Counc. on Environ. Qual. Fed. Counc. for Sci. Technol., Natl. Sci. Found. NSF-75-403. Washington, D. D., June. 109 pp.

Peiffer, D. G. (1980). J. Appl. Polym. Sci. 25(3), 369-380.

Peng, D. Y. and Robinson, D. B. (1976). Ind. Eng. Chem. Fundam. 15(1), 59-64.

Pennsalt Chemicals Corporation (1958). Data Sheet Trifluoroethylchloride. New York.

Perona, M. J. (1979). J. Chem. Educ. 56(11), 726-727.

Perram, J. W. and Anastasiou, N. (1981). J. Chem. Soc. Faraday Trans II 77(1), 101-108.

Perrin, D. D., Armarega, W. L. F., and Perrin, D. R. (1980). Purification of Laboratory Chemicals (2nd ed.). Pergamon, Oxford. 568 pp.

Perron, G. and Desnoyers, J. E. (1979). Fluid Phase Equil. 2(3), 239-262.

Perry, J. H. and Chilton, C. H., Eds. (1973). Chemical Engineers' Handbook (5th ed.). McGraw-Hill, New York.

Pesuit, D. R. (1978). Ind. Eng. Chem. Fundam. 17(4), 235-242.

Peterson, J. M. and Rodebush, W. H. (1928). J. Phys. Chem. 32, 709-718.

Peto, R. (1980). Nature 284, 297-300.

Petrak, J. (1975). Potravin. Chladici Tech. 6(1), 23-26.

Petrakis, L. and Weiss, F. T., Eds. (1980). Petroleum in the Marine Environment. Am. Chem. Soc., Adv. Chem. Ser. No. 185, Washington, D. C. 371 pp.

Petrov, A. N., Pankov, A. G., and Bogdanov, M. I. (1970). Uch. Zap. Yarosl. Tekhnol. Inst., No. 13, 186-190.

Petukhov, P. S. (1975). Russ. J. Phys. Chem. 49(10), 1520-1521.

Pfaender, F. K., Jonas, R. B., Stevens, A. A., and Moore, L. (1978). Envoron. Sci. Technol. 12(4), 438-441.

Phillips Petroleum Company (1966). 1,2-dichloro-1,1,2-trifluoroethane (PF-123a), 1,2-dichloro-1-fluoroethane (PF-141), 1,1,1-trifluoroethane (PF-143a), 1,1,2-trifluoroethane (PF-143), pentafluoroethane (PF-125), 1,2-dichloro-1,2-difluoroethane (PF-132), 1,1,2,2-tetrafluoroethane (PF-134). Chem.Dept., Commercial Dev. Div., Bartlesville, Okla. 7 pp.

Piel, E. (1979). J. Chem. Educ. 56(10), 695.

Pierotti, G. J., Deal, C. H., and Derr, E. L. (1959). Ind. Eng. Chem. 51(1), 95-102.

Pierotti, G. J., Deal, C. H., Derr, E. L., and Porter, P. E. (1956). J. Am. Chem. Soc. 78, 2989-2998.

Pierotti, R. A. (1963). J. Phys. Chem. 67. 1840-1845.

Pierotti, R. A. (1965). J. Phys. Chem. 69, 281-288.

Pierotti, R. A. (1967). J. Phys. Chem. 71(7), 2366-2367.

Pierotti, R. A. (1976). Chem. Rev. 76(6), 717-726.

Pierotti, R. A. and Liabastre, A. A. (1972). The S tructure and Properties of Water Solutions. Georgia Inst. Technol.; Environ. Resour. Cent., PB 211163. Atlanta, Ga, June. 102 pp.

Pimentel, G. C. and McClellan, A. L. (1960). The Hydrogen Bond. W. H. Freeman, San Francisco. 475 pp.

Pinder, K. L. (1973). J. Chem. Eng. Data 18(3), 275-277.

Pings, C. J. (1968). Physics of Simple Liquids. North-Holland, Amsterdam.

Pinsker, G. Z. (1972). Zh. Strukt. Khim. 13(6), 985-988.

Pittsburgh Plate Glass Industries (1970a). Monochlorobenzene. Bull. 30A (A-1007, 2.5M, 570), Pittsburgh, Pa. 2 pp.

Pittsburgh Plate Glass Industries (1970b). Orthodichlorobenzene. Bull. 30B (A-1005, 2.5M, 570). Pittsburgh, Pa. 2 pp.

Pittsburgh Plate Glass Industries (1970c). Ethylene Dichloride, CH_2Cl-CH_2Cl. Bull. 55A (A-1009, 2.5M, 570). Pittsburgh, Pa. 2 pp.

Pittsburgh Plate Glass Industries (1970d). Vinylidene Chloride. Bull. 120A (A-1001, 2.5M, 470). Pittsburgh, Pa. 2 pp.

Pittsburgh Plate Glass Industries (1971a). Tri-Ethane (1,1,1-trichloroethane). Bull. 35A (A-1019, 5M, 471). Pittsburgh, Pa. 2 pp.

Pittsburgh Plate Glass Indistries (1971b). Trichloroethylene. Bull. 35B (A-1021, 5M, 471). Pittsburgh, Pa. 2 pp.

Pittsburgh Plate Glass Industries (1971c). Perchloroethylene. Bull. 35C (A-1020, 5M, 471). Pittsburgh, Pa. 2 pp.

Pittsburgh Plate Glass Industries (1971d). Ethyl Chloride. Bull. 50A (A-1027, 2.5M, 471). Pittsburgh, Pa. 2 pp.

Pittsburgh Plate Glass Industries (1975a). Methylene Chloride. Bull. 35E (A-1042, 2M, 775). Pittsburgh, Pa. 2 pp.

Pittsburgh Plate Glass Industries (1975b). Vinyl Chloride Monomer. Handling and Properties. Bull. A-994-115R. Pittsburgh, Pa. 41 pp.

Pitzer, K. S. (1959). Adv. Chem. Phys. 2, 59-83.

Plank, J. (1921). Magy. Kem. Foly. 27, 7-11.

Platford, R. F. (1977). J. Chem. Soc. Faraday Trans. I 73(2), 267-271.

Polak, J. and Lu, B. C.-Y. (1973). Can. J. Chem. 51, 4018-4023.

Pople, J. A. (1951). Proc. Roy. Soc. Lond. Ser. A 205, 163-178.

Powell, R. E. and Latimer, W. M. (1951). J. Chem. Phys. 19(9), 1139-1141.

Prahl, W. and Mathes, W. (1934). Angew. Chem. $\underline{47}$(1), 11-13.

Pratt, L. R. and Chandler, D. (1980). J. Chem. Phys. $\underline{73}$(7), 3434-3441.

Prausnitz, J. M. (1958). AIChE J. $\underline{4}$(3), 269-272.

Prausnitz, J. M. (1965). Solubility of Solids in Dense Gases. Natl. Bur. Std. Tech. Note No. 316. Washington, D. C., July. 45 pp.

Prausnitz, J. M. (1969). Molecular Thermodynamics of Fluid-Phase Equilibria. Prentice-Hall, Englewood Cliffs, N. J. 523 pp.

Prausnitz, J. M. (1977). Recent Developments in the UNIFAC Method for Calculating Activity Coefficients from Group Contributions. In Phase Equilibria and Fluid Properties in Chemical Industry. Estimation and Correlation. Am. Chem. Soc. Symp. Ser. No. 60, p. 453-454.

Prausnitz, J. M. and Chueh, P. J. (1968). Computer Calculations for High-Pressure Vapor-Liquid Equilibria, Prentice-Hall, Englewood Cliffs, N. J.

Prausnitz, J. M. and Shair, F. H. (1961). AIChE J. $\underline{7}$(4), 682-687.

Prausnitz, J. M., Eckert, C. A., Orye, R. V., and O'Connell, J. P. (1967). Computer Calculations for Multicomponent Vapor-Liquid Equilibria. Prentice-Hall, Englewood Cliffs, N. J.

Prausnitz, J. M., Anderson, T. F., Grens, E. A., Eckert, C. A., Hsich, R., and O'Connell, J. P. (1980). Computer Calculations for Multicomponent Vapor-Liquid Equilibria. Prentice-Hall, Englewood Cliffs, N. J. 353 pp.

Pray, H. A., Schweickert, C. E., and Minnich, B. H. (1952). Ind. Eng. Chem. $\underline{44}$(5), 1146-1151.

Prelog, V. and Cerkovnikov, E. (1936). Ann. Chem. $\underline{525}$, 292-296.

Price, L. C. (1973). The Solubility of Hydrocarbons and Petroleum in Water as Applied to the Primary Migration of Petroleum. (Ph. D. thesis). Univ. of California, Riverside, Calif., 298 pp.

Prigogine, I. and Defay, R. (1967). Chemical Thermodynamics (4th impression). Longmars, London. 543 pp.

Prigogine, I. and Rice, S. A., Eds. (1975). Non-Simple Liquids. Adv. Chem. Phys., Vol. 31, Wiley, New York. 496 pp.

Prigogine, I., Bellemans, A., and Mathot, V. (1957). The Molecular Theory of Solutions. North Holland, Amsterdam. 448 pp.

Prosyanov, N. N., Shalygin, V. A., and Zel'venskii, Ya. D. (1973a). Tr. Mosk. Khim.-Tekhnol. Inst. $\underline{73}$, 183-186.

Prosyanov, N. N., Shalygin, V. A., and Zel'venskii, Ya. D. (1973b). Tr. Mosk. Khim.-Tekhnol. Inst. $\underline{75}$, 100-102.

Prosyanov, N. N., Shalygin, V. A., and Zel'venskii, Ya. D. (1974). Tr. Mosk. Khim.-Rekhnol. Inst. $\underline{81}$, 55-56.

Pruett, H. D. (1971). Prediction of Binary Azeotropes (Ph. D. thesis). Louisiana State Univ., Baton Rouge, La. 243 pp.

Pryde, J. A. (1969). The Liquid State (reprint). Hutchinson, London, 179 pp.

Prydz, R. and Straty, G. C. (1970). The Thermodynamic Properties of Compressed Gaseous and Liquid Fluorine. Natl. Bur. Std. Tech. Note No. 392. Washington, D. C., October. 182 pp.

Purchase, I. F. H., Longstaff, E., Ashby, J., Styles, J. A., Anderson, D., Lefevre, P. A., and Westwood, F. R. (1976). Nature $\underline{264}$(5587), 624-627.

Quayle, O. R. (1953). Chem. Rev. $\underline{53}$, 439-589.

Rabinovich, I. B. (1970). Influence of Isotropy on the Physicochemical Properties of Liquids. Consultants Bureau, New York. 304 pp.

Rademacher, P. (1976). J. Chem. Educ. $\underline{53}$(12), 757-761.

Radukov, E. S., Zamashchikov, V. V., Belyaev, V. D., and Gushchina, E. G. (1971). Reakts. Sposobn. Org. Soedin. $\underline{8}$(1), 219-236.

Ragaini, V., Santi, R., and Carniti, P. (1974). Chim. Ind. $\underline{56}$(10), 687-692.

Rahman, A. and Stillinger, F. H. (1971). J. Chem. Phys. $\underline{55}$, 3336-3359.

Ralston, A. W. and Hoerr, C. W. (1942). J. Org. Chem. $\underline{7}$(6), 546-555.

Ramsay, W. and Young, S. (1886). J. Chem. Soc. $\underline{49}$, 790-812.

Ramsperger, H. C. and Porter, C. W. (1926). J. Am. Chem. Soc. $\underline{48}$, 1267-1273.

Randall, M. and Failey, C. F. (1927). Chem. Rev. $\underline{4}$, 271-286.

Randic, M. (1975). J. Am. Chem. Soc. $\underline{97}$(23), 6609-6615.

Rao, A. K. (1977). Chem. Eng. $\underline{84}$(10), 143-147.

Rao, I. R. and Majumdar, C. K. (1975). Proc. Nucl. Phys. Solid State Phys. Symp. $\underline{18C}$, 318-320.

Rao, M. B. (1975). Indian J. Technol. $\underline{13}$(12), 571-572.

Rao, T. S., Barve, M. S., and Gandhe, B. R. (1974). J. Univ. Poona Sci. Technol. $\underline{46}$, 39-41.

Raoult, F. M. (1888). Z. phys. Chem. $\underline{2}$, 353-375.

Rasumovskii, V. V. and Friedenberg, A. E. (1949). Zh. Obshch. Khim. $\underline{19}$, 92-94.

Ratajczak, H. and Orville-Thomas, W. J., Eds. (1980). Molecular Interactions, Vol. 1. Wiley-Interscience, New York. 415 pp.

Rathmann, D., Bauer, J., and Thompson, P. A. (1978). A Table of Miscellaneous Thermodynamic Properties for Various Substances, with Emphasis on the Critical Properties. Max-Plank Institut für Strömungsforschung, ISSN-0436-1199. Göttingen, West Germany, June. 77 pp. (CA 92:29472r.)

Ratouis, M. and Dode, M. (1965). Bull. Soc. Chim. Fr., No. 5, 3318-3322.

Rauws, A. G., Olling, M., and Wibowo, A. E. (1973). J. Pharm. Pharmacol. $\underline{25}$, 718-722.

Rewat, B. S. and Gulati, I. B. (1977). J. Appl. Chem. Biotechnol. $\underline{27}$(9), 459-464.

Raynes, W. T. (1969). J. Chem. Phys. $\underline{51}$, 3138-3140.

Rebagay, T. and De Luca, P. (1976). J. Pharm. Sci. $\underline{65}$(11), 1645-1648.

Rebert, C. J. and Hayworth, K. E. (1967). AIChE J. $\underline{13}$(1), 118-121.

Rebert, C. J. and Kay, W. B. (1959). AIChE J. $\underline{5}$(3), 285-289.

Redlich, O. (1963). J. Phys. Chem. $\underline{67}$, 496.

Redlich, O. (1976). Thermodynamics: Fundamentals, Applications. Elsevier, Amsterdam. 277 pp.

Redlich, O. and Kister, A. T. (1948). Ind. Eng. Chem. $\underline{40}$(2), 341-348.

Redlich, O. and Kwong, J. N. S. (1949). Chem. Rev. $\underline{44}$(1), 233-244.

Redlich, O., Kister, A. T., and Turnquist, C. E. (1952). Chem. Eng. Prog. Symp. Ser. $\underline{48}$(2), 49-61.

Reed, C. D. and McKetta, J. J. (1959). J. Chem. Eng. Data $\underline{4}$(4), 294-295.

Reed, T. M. (1970). Fed. Proc., Fed. Am. Soc. Exp. Biol. $\underline{29}$(5), 1708-1713.

Reed, T. M. and Gubbins, K. E. (1973). Applied Statistical Mechanics. McGraw-Hill, New York. 506 pp.

Reichardt, C. (1973). Lösungsmittel-Effecte in der organischen Chemie. Verlag Chemie, Weinheim, West Germany. 180 pp.

Reichardt, C. and Dimroth, K. (1968). Fortschr. Chem. Forsch. $\underline{11}$, 1-73.

Reid, D. S., Quickenden, M. A. J., and Franks, F. (1969). Nature $\underline{224}$(5226), 1293-1294.

Reid, R. C., Prausnitz, J. M., and Sherwood, T. K. (1977). The Properties of Gases and Liquids (3rd ed.). McGraw-Hill, New York. 688 pp.

Reilly, J., Kelly, D. F., and O'Connor, M. (1941). J. Chem. Soc., 275-278.

Reinders, W. and De Minjer, C. H. (1947). Recl. Trav. Chim. $\underline{66}$, 573-604.

Reineck, A. E. and Lin, K. F. (1968). J. Paint Technol. $\underline{40}$(527), 611-616.

Reisler, E., Eisenberg, H., and Minton, A. (1972). J. Chem. Soc. Faraday Trans. II $\underline{68}$(6), 1001-1015.

Reisman, A. (1970). Phase Equilibria. Basic Principles, Applications, Experimental Techniques. Academic, New York. 541 pp.

Reiss, H., Frisch, H. L., and Lebowitz, J. L. (1959). J. Chem. Phys. $\underline{31}$(2), 369-380.

Reiss, H., Frisch, H. L., Helfand, E., and Lebowitz, J. L. (1960). J. Chem. Phys. $\underline{32}$(1), 119-124.

Renon, H. (1966). Thermodynamic Properties of Nonideal Liquid Mixtures (Ph. D. thesis). Univ. of California, Berkeley.

Renon, H. and Prausnitz, J. M. (1968). AIChE J. $\underline{14}$, 135-144.

Renon, H., Eckert, C. A., and Prausnitz, J. M. (1967). Ind. Eng. Chem. Fundam. $\underline{6}$(1), 52-58.

Renon, H., Asselineau, L., Cohen, G., and Raimbault, C. (1971). Calcul sur ordinateur des equilibres liquide-vapeur et liquide-liquide. Technip, Paris.

Renuncio, J. A. R., Breedveld, G. J. F., and Prausnitz, J. M. (1977). J. Phys. Chem. $\underline{81}$(4), 324.

Rex, A. (1906). Z. phys. Chem. $\underline{55}$, 355-370.

Reynolds, C. A. and Harris, S. M. J. (1969). Anal. Chem. $\underline{41}$(2), 348-349.

Reynolds, J. A., Gilbert, D. B. and Tanford, C. (1974). Proc. Natl. Acad. Sci. U.S.A. $\underline{71}$(8), 2925-2927.

Rheineck, A. E. and Lin, K. F. (1968). J. Paint Technol. 40(527), 611-616.

Rhodes, W. W. (1947). Refrig. Eng. 53(5), 412, 456, 458.

Rhône Progil (1972). Flugene - Flugex Packaging. Direction Commerciale des Produits Chimiques, Neuilly-Sur-Seine, France. 20 pp.

Rhône Progil (1973). Les Aerosols. Direction Commerciale des Produits Chimiques, Courbevoie, France. 23 pp.

Ricci, J. E. (1951). The Phase Rule and Heterogeneous Equilibrium. D. Van Nostrand, New York. 505 pp.

Ricci, L. J. (1976). Chem. Eng. 83(10), 5.

Rice, S. A. and Gray, P. (1965). Statistical Mechanics of Simple Liquids. Interscience, New York. 582 pp.

Rich, L. A. (1976). Chem. Eng. 83(22), 9-17.

Richards, B. (1966). Austr. Refrig. Air Cond. Heat. 20(10), 29-31.

Richardson, L. T. and Miller, D. M. (1960). Can. J. Bot. 38, 163-175.

Riddick, J. A. and Bunger, W. B. (1970). Organic Solvents. Physical Properties and Methods of Purification (3rd ed.), Vol. 2. Wiley-Interscience, New York. 603 pp.

Riddick, J. A. and Toops, E. E. (1955). Organic Solvents. Physical Properties and Methods of Purification. Technique of Organic Chemistry, Vol. 7 (A. Weissberger, ed.). Interscience, New York. 552 pp.

Rigby, M. and Prausnitz, J. M. (1968). J. Phys. Chem. 72, 330-334.

Riggs, D. M. and Diefendorf, R. J. (1979). The Solubility of Aromatic Compounds. Ext. Abstr. Program--Bienn. Conf. Carbon 14, 407-408. (CA 91:79621p.)

Risbourg, A. and Liebiert, R. (1967). C. R. Acad. Sci. Ser. C 264, 237-240.

Ritter, H. L. and Simons, J. H. (1945). J. Am. Chem. Soc. 67, 757-762.

Robb, I. D. (1966). Aust. J. Chem. 19, 2281-2284.

Robbins, B. H. (1946). J. Pharmacol. 86, 197-204.

Robeck, G. G., Dostal, K. A., Cohen, J. M., and Kreissl, J. F. (1965). J. Am. Water Works Assoc. 57(2), 181-200.

Robertson, J. H., Cowen, W. F., and Longfield, J. Y. (1980). Chem. Eng. 87(13), 102-119.

Robertson, R. E. (1967). Prog. Phys. Org. Chem. 4, 213-280.

Robinson, D. B. and Ng, H. J. (1975). Hydrocarbon Processing 54(12), 95-96.

Roddy, J. W. and Coleman, C. F. (1968). Talanta 15, 1281-1286.

Rodriguez, L. (1978). Ind. Eng. Chem. Fundam. 17(3), 228-230.

Rogalski, M. and Malanowski, S. (1975). Comparison of Different Methods for Correlation of Vapor-Liquid Equilibria. CHISA, Prague.

Ronc, M. and Ratcliff, G. A. (1971). Can. J. Chem. Eng. 49(6), 825-830.

Rook, J. J. (1974). Water Treat. Exam. 23(2), 234-243.

Rosenbaum, C. K. and Walton, J. H. (1930). J. Am. Chem. Soc. 52, 3568-3573.

Rosenzweig, M. D. (1975). Chem. Eng. 82(24), 1124-1126.

Ross, R. G., Andersson, P., and Bäckström, G. (1977). High Temp. High Press. 9(1), 87-96.

Rosseinsky, D. R. (1977). J. Phys. Chem. 81(16), 1578-1579.

Rossky, P. J. and Friedman, H. L. (1980). J. Phys. Chem. 84, 587-589.

Rotariu, G. J., Fraga, D. W., and Hildebrand, J. H. (1952). J. Am. Chem. Soc. 74, 5783.

Roux, G., Perron, G., and Desnoyers, J. E. (1978). Can. J. Chem. 56(22), 2808-2814.

Rowland, F. S. (1974). New Sci. 64(926), 717-720.

Rowland, F. S. and Molina, M. J. (1975). Rev. Geophys. Space Phys. 13(1), 1-35.

Rowland, F. S. and Molina, M. J. (1976). J. Phys. Chem. 80(19), 2041-2056.

Rowland, F. S., Molina, M. J., and Chou, C. C. (1975). Nature 258(5537), 775-776.

Rowlinson, J. S. (1949). Trans. Faraday Soc. 45, 974-984.

Rowlinson, J. S. (1959). Liquid and Liquid Mixtures (2nd ed.). Butterworths, London. 360 pp.

Rowlinson, J. S. (1970). The Structure of Liquids. In Essay in Chemistry (J. N. Bradley, et al., Eds), Vol. 1. Academic, London. p. 1-24.

Rozen, A. M. (1969a). Russ. J. Phys. Chem. 43(1), 88-91.

Rozen, A. M. (1969b). Russ. J. Phys. Chem. 43(1), 92-94.

Rübelt, C. (1969). Mineralölspurenbestimmung in Boden- und Wasserproben. Die Aussage-kraft verschiedener analytischer Methoden (Doctoral dissertation). Univ. des Saarlandes, Saarbrücken, West Germany. 142 pp.

Rusanov, A. I. (1972). Russ. J. Phys. Chem. 46(3), 428-430.

Ryckman, D. W., Irvin, J. W., and Young, R. H. F. (1967). J. Water Pollut. Control Fed. 39(3), 458-469.

Rytting, J. H., Huston, L. P., and Higuchi, T. (1978). J. Pharm. Sci. 67(5), 615-618.

Sabinin, V. E., Kiya-Oglu, N. V., and Gorishnina, V. P. (1970). J. Appl. Chem. USSR 43(8), 1788-1790.

Sadove, M. S. and Wellace, V. E. (1962). Halothane. Blackwell Scientific, Oxford.

Sadovnikova, L. V., Belobokova, N. V., and Aleksandrova, M. V. (1972). Sb. Nauchn. Tr. Ivanov Energ. Inst., No. 14, 221-224.

Saffiotti, U. and Cooper, J. (1976). Chem. Eng. News 54(25), 17-19.

Sahar, A., Apelblat, A., and Michaeli, I. (1975). Israel J. Chem. 13(3), 253-256.

Sahgal, A., La, H. M., and Hayduk, W. (1978). Can. J. Chem. Eng. 56(3), 354-357.

Sahli, B. P., Gager, H., and Richard, A. J. (1976). J. Chem. Thermodyn. 8(2), 179-188.

Saidman, L. J., Eger, E. I., Munson, E. S., and Severinghous, J. W. (1976). Anesthes-iology 27(3), 180-184.

Saito, S. and Tanaka, Y. (1965). Bull. Soc. Sea Water Sci. Jap. 19, 23-33.

Salama, C. and Goring, D. A. J. (1966). J. Phys. Chem. 70(12), 3838-3841.

Salem, L. (1961). Proc. Roy. Soc. Lond. Ser. A 264, 379-391.

Salem, R. R. (1979). Russ. J. Phys. Chem. 53(5), 760-762.

Salkowski, E. (1920). Biochem. Z. 107 191-201.

Salsburg, Z. W., Ed. (1968). Theory of Solutions. Gordon and Breach, New York. 302 pp.

Salsburg, Z. W. and Kirkwood, J. G. (1953). J. Chem. Phys. 21(12), 2169-2177.

Samaha, M. W. (1979). A Free Energy Partitioning Analysis of Solubility and Partition Coefficient Data. (Ph. D. thesis). Univ. of Wisconsin, Madison. 176 pp.

Samoilov, O. Ya. (1946). Zh. Fiz. Khim. 20, 1411-1414.

Samoilov, O. Ya. (1965). Structure of Aqueous Electrolyte Solutions and the Hydration of Ions. Consultants Bureau, New York. 185 pp.

Sandell, D. J. and Johnson, C. A. (1967). Direct Freezing – Wash Separation Processes. Proc. 1st Int. Symp. Water Desalination, October 3-9, 1965, Vol. 3, p. 625-646.

Sanders, P. A. (1979). Handbook of Aerosol Technology (2nd ed.). Van Nostrand Reinhold, New York. 526 pp.

Santoleri, J. J. (1973). Chem. Eng. Prog. 69(1), 68-74.

Saracco, G. and Spaccamela Marchetti, E. (1958). Ann. Chim. (Rome) 48(12), 1357--394.

Sarkisov, G. N., Dashevsky, V. G., and Malenkov, G. G. (1974). 27(5), 1249-1269.

Sarma, T. S. and Ahluwalia, J. C. (1973). Chem. Soc. Rev. 2(2), 203-232.

Saylor, J. H. and Battino, R. (1958). J. Am. Chem. Soc. 62, 1334-1337.

Saylor, J. H., Whitten, A. I., Claiborne, I., and Gross, P. M. (1952). J. Am. Chem. Soc. 74, 1778-1781.

Scatchard, G. (1976). Equilibrium in Solutions. Surface and Colloid Chemistry. Harvard University Press, Cambridge, Mass. 306 pp.

Schäfer, H. (1976). Angew. Chem. Int. 15(12), 713-727.

Schatzberg, P. (1963). J. Phys. Chem. 67, 776-779.

Scheflan, L. and Jacobs, M. B. (1953). The Handbook of Solvents. D. Van Nostrand, New York. 728 pp.

Scheller, W. A. (1965). Ind. Eng. Chem. Fundam. 4(4), 459-462.

Scheraga, H. (1979). Acc. Chem. Res. 12(1), 7-14.

Schiemann, G. (1951). Die Organischen Fluorverbindungen. Dietrich Steinkopff, Darmstadt, West Germany.

Schneider, G. M. (1978). Chemical Thermodynamics (M. L. Glashan, Ed.). Vol. 2, Chem. Soc., London. p. 105-146.

Scholander, P. F. (1947). J. Biol. Chem. 167, 235-250.

Scholte, Th. G. (1949). Physica 15(5/6), 437-458.

Schönfeld, P. and Seibt, H. (1976). Z. Chem. 16(12), 497-498.

Schoor, W. P. (1974). Theoretical Model and Solubility Characteristics of Aroclor 1254 in Water: Problems Associated with Low-Solubility Compounds in Aquatic Toxicity Tests. Natl. Environ. Res. Cent. PB-240 550, Corvallis, Ore. September. 38 pp.

Schoor, W. P. (1975). Water Res. 9(11), 937-944.

Schröder, W. (1973). Chem.-Ing.-Tech. 45(9/10), 603-608.

Schuberth, H. and Kränke, P. (1970). Z. phys. Chem. 245(1/2), 49-67.

Schultz, T. R. (1979). Trichlorofluoromethane as a Ground-Water Tracer for Finite-State Model (Ph. D. thesis). Univ. of Arizona, Tucson. 213 pp.

Schupp, R. L. and Mecke, R. (1948). Z. Elektrochem. 52, 54-60.

Schuster, P. (1973). Z. Chem. 13, 41-55.

Schuster, P., Zundel, G., and Sandorfy, C., Eds. (1976). The Hydrogen Bond. Recent Developments in Theory and Experiments. North-Holland, Amsterdam. 1549 pp.

Schuster, P., Jakubetz, W., Marius, W., and Rice, S. A. (1975). Structure of Liquids. In Topics in Current Chemistry, Vol. 60. Springer-Verlag, Berlin. 205 pp.

Schwarz, F. P. (1977). J. Chem. Eng. Data 22(3), 273-277.

Schwarz, F. P. (1980). Anal. Chem. 52(1), 10-15.

Schwarz, F. P. and Miller, J. (1980). Anal. Chem. 52(13), 2162-2164.

Schwarz, F. P. and Wasik, S. P. (1977). J. Chem. Eng. Data 22(3), 270-273.

Schwarzenbach, R. P., Molnar-Kubica, E., Gieger, W., and Wakeham, S. G. (1979). Environ. Sci. Technol. 13(11), 1367-1373.

Sconce, J. S., Ed. (1962). Chlorine: Its Manufacture, Properties, and Uses. Reinhold, New York. 901 pp.

Scott, J. M. W. (1970). Can. J. Chem. 48, 3307-3318.

Scott, R. L. (1948). J. Am. Chem. Soc. 70, 4090-4093.

Scott, R. L. (1958). J. Phys. Chem. 62, 136-145.

Scotti-Foglieni, L. (1930a). C. R. Seances Soc. Biol. Paris 105, 959-960.

Scotti-Foglieni, L. (1930b). C. R. Seances Soc. Biol. Paris 105, 961-964.

Scotti-Foglieni, L. (1931a). C. R. Seances Soc. Biol. Paris 106, 222-224.

Scotti-Foglieni, L. (1931b). C. R. Seances Soc. Biol. Paris 106, 224-226.

Scotti-Foglieni, L. (1931c). C. R. Seances Soc. Biol. Paris 106, 226-229.

Scotti-Foglieni, L. (1931d). C. R. Seances Soc. Biol. Paris 108, 1203-1205.

Scribner, J. D. (1976). Chem. Eng. News 54(17), 66-67.

Searle, C. E., Ed. (1976). Chemical Carcinogens., Am. Soc. Monogr. No. 173. Washington, D. C. 788 pp.

Sebastiani, E. and Lacquaniti, L. (1967). Chem. Eng. Sci. 22, 1155-1162.

Secher, O. (1971). Physical and Chemical Data on Anaesthetics. Universitetsforlaget. Aarhus, Denmark. 96 pp.

Sedivec, V. and Flek, J. (1976). Handbook of Analysis of Organic Solvents. Ellis Horwood, Chichester, England. 455 pp.

Seidell, A. (1964). Solubilities of Organic Compounds (3rd ed.), Vol. 2. Van Nostrand, New York. 926 pp.

Seidell, A. (1953). Solubilities of Inorganic and Metal Organic Compounds (3rd ed.). Vol. 1. Van Nostrand, New York. 1698 pp.

Seidell, A. and Linke, W. F. (1952). Solubilities of Inorganic and Organic Compounds (Suppl. to 3rd ed.). D. Van Nostrand, New York. 1254 pp.

Seiler, H. K. (1968). The Apparent and Partial Molal Volume of Water in Organic Solvents (Ph. D. thesis). Univ. of Connecticut, Storrs, Conn. 70 pp.

Selenka, F. and Bauer, U. (1978). Detection of Readily Volatile Organochloride Compounds in Water. Org. Verunreinig. Umwelt: Erkennen, Bewerten. p. 242-255. (CA
91:62457s.)

Selinger, B. (1979). Educ. Chem. 16(2), 125-128.

Sellers, P. (1971). Acta Chem. Scand. 25(6), 2295-2301.

Selyuzhitskii, G. V. (1963). Gig. Sanit. 28(12), 9-14.

Selyuzhitskii, G. V. (1967). Prom. Zagryaz. Vodoemov, No. 8, 112-128.

Semenchenko, V. K. and Efuni, V. V. (1974). Russ. J. Phys. Chem. 48(5), 745-746.

Semerikova, I. A., Mironova, N. I., and Sukhotin, A. M. (1976). Soviet Chem. Ind.
8(10), 780-781.

Severance, W. A. N., Akell, R. B., and Fitzjohn, J. L. (1963). Ind. Eng. Chem. Fundam.
2(3), 246-247.

Seymour, K. M., Carmichael, R. H., Carter, J., at al. (1977). Ind. Eng. Chem. Fundam.
16(2), 200-207.

Shanmugasundaram, V. and Thiyagarajan, P. (1980). Indian J. Pure Appl. Phys. 18(1),
36-39.

Sharma, B. K. (1976). Indian J. Pure Appl. Phys. 14(11), 939-942.

Sharma, B. K. (1977). Indian J. Pure Appl. Phys. 15(9), 633-639.

Sharov, V. G. (1975). V. Sb. Dostizh. Nauk-Neftekhim. Proiz-Vam. p. 138-141.

Sharpe, A. G. (1972). The Physical Properties of the Carbon-Fluorine Bond. In Carbon-
Fluorine Compounds, A Ciba Foundation Symposium. Elsevier, Amsterdam. p. 33-54.

Shell Chemical Corporation (1948). Data Sheet Trichloropropane. Houston, Tex.

Shell Chemical Corporation (1949). Allyl Chloride and Other Allyl Halides. New York.

Shenkin, Ya. S., Rozen, A. M., and Gartman, V. L. (1979). Russ. J. Phys. Chem. 53(5),
742-743.

Shepherd, H. R., Ed. (1961). Aerosols: Science and Technology. Interscience, New
York. 548 pp.

Sherwood, M., Rowland, S., Gribbin, J., and Jones, A. (1975). New Sci. 68(969), 7-18.

Shields, R. R. (1976a). J. Electrochem. Soc. 123(8), 254C.

Shields, R. R. (1976b). Solubility of Water in Perfluorohexane as a Function of
Temperature and Humidity. Personal communication, December 9.

Shimanouchi, T. (1970). The Molecular Force Fields. In Physical Chemistry: An Advanced
Treatise. Vol. 4, Molecular Properties (D. Henderson, Ed.). Academic, New York.
p. 233-306.

Shinoda, K. (1977). J. Phys. Chem. 81(13), 1300-1302.

Shinoda, K. (1978). Principles of Solution and Solubility. Dekker, New York. 222 pp.

Shinoda, K. and Fujihira, M. (1968). Bull. Chem. Soc. Jap. 41(11), 2612-2615.

Shostakovskii, M. F. and Druzhinin, I. G. (1942). Zh. Obshch. Khim. 12, 42-47.

Shreiber, E. P. (1976). Russ. J. Phys. Chem. 50(6), 973-977.

Silcock, H., Ed. (1979). Solubilities of Inorganic and Organic Compounds, Vol. 3. Pergamon, Oxford. 3330 pp.

Silverman, P. (1974). New Sci. 61(883), 255-256.

Simmonds, P. G., Kerrin, S. L., Lovelock, J. E., and Shair, F. H. (1974). Atmos. Environ. 8, 209-216.

Simonov, V. D., Pogulyai, V. E., and Shamsutdinov, T. M. (1970). Russ. J. Phys. Chem. 44(12), 1755-1757.

Simonov, V. D., Shamsutdinov, T. M., Pogulyai, V. E., and Popova, L. N. (1974). Russ. J. Phys. Chem. 48(11), 1573-1575.

Simons, J. H., Ed. (1950-1965). Fluorine Chemistry, Academic, New York. Vol. 1 (1950), 615 pp.; Vol. 2 (1954), 565 pp.; Vol. 3 (1963), 240 pp.; Vol. 4 (1965), 786 pp.

Singh, H. B., Fowler, D. P., and Peyton, T. O. (1976). Science 192(4245), 1231-1234.

Singh, S., Murthy, A. S. N., and Rao, C. N. R. (1966). Trans. Faraday Soc. 62, 1056-1066.

Sitting, M. (1969). Water Pollution Control and Solid Wastes Disposal. Chem. Proc. Rev. No. 32. Noyes Development Corp., New York.

Skau, E. L. and Boucher, R. E. (1954). J. Phys. Chem. 58(6), 460-468.

Skjöld-Jørgensen, S., Kolbe, B., Gmehling, J., and Rasmussen, P. (1979). Ind. Eng. Chem. Process Des. Dev. 18(4), 714-722.

Skjöld-Jørgensen, S., Rasmussen, P., and Fredenslund, A. A. (1980). Chem. Eng. Sci. 35, 2389-2403.

Skolnik, H. (1948). Ind. Eng. Chem. 40(3), 442-450.

Skripka, V. G. (1976). Tr. Vses. Neftegazov. Nauchno-Issled Inst. 61, 139-151.

Skripka, V. G. (1979a). Chem. Technol. Fuels Oils 15(1/2), 88-90.

Skripka, V. G. (1979b). Russ. J. Phys. Chem. 53(6), 795-797.

Skripka, V. G. and Boksha, O. A. (1976). Tr. Vses. Naucho. Issled. Inst. Perereb Nefti (Nedra)., No. 61, 96-110.

Slade, R. E. (1945). Chem. Ind. 40, 314-319.

Slesser, C. and Schram, S. R., Eds. (1951). Preparation, Properties, and Technology of Fluorine and Organic Fluoro Compounds. McGraw-Hill, New York.

Slobodin, Ya. M., Baranovich, Z. N., and Bogdanova, L. P. (1964). Zavodsk. Lab. 30 (8), 1203.

Slough, R. J. (1973). Ternary Vapor-Liquid Equilibria Predicted from Binary System Data UF_6-HF, UF-F114, and HF-F114 (M..Sc. thesis). Ohio Univ., Ohio. April 2. 73 pp. (Goodyear Atomic Corp. Rep. GAT-T-2040).

Small, P. A. (1953). J. Appl. Chem. 3(2), 71-80.

Smith, B. L. (1975). Relation of the Dielectric Constant and the Refractive Index to Thermodynamic Properties. In Experimental Thermodynamics, Vol. 2 (B. Le Neidre and B. Vodar, Eds.). Butterworths, London. p. 579-606 and 689-690.

Smith, E. L. (1932). J. Phys. Chem. 36, 1401-1418.

Smith, F. A. (1973). Chem.-Tech., June. 422-429.

Smith, J. W. (1955). Electric Dipole Moments. Butterworths, London. 370 pp.

Smyth, C. P. (1931). Dielectric Constant and Molecular Structure. Am. Chem. Soc.
    Monogr. Ser., Chemical Catalog Co., New York. 214 pp.

Smyth, C. P. (1955). Dielectric Behavior and Structure. McGraw-Hill, New York.

Smyth, C. P. (1972). Determination of Dipole Moments. In Techniques of Chemistry,
    Vol. 1, Physical Methods of Chemistry, Part 4 (A. Weissberger, Ed.). Wiley-
    Interscience, New York. p. 397-429.

Sneer, R. A. (1973). Acc. Chem. Res. 6(1), 47-53.

Snyder, L. (1979). Chem. Tech. 9(12), 750-754.

Snyder, L. (1980). Chem. Tech. 10(3), 188-193.

Sobodka, H. and Kahn, J. (1931). J. Am. Chem. Soc. 53, 2935-2938.

Sokolov, N. D. and Tschulanovskii, V. M., Eds. (1964). Vodorondnaya Svyaz. Nauka,
    Moscow.

Sokolova, S. P. and Pereverzev, A. N. (1977). Zh. Fiz. Khim. 51(5), 1267-1268.

Sørensen, J. M. and Arlt, W. (1979). Liquid-Liquid Equilibrium Data Collection.
    Deutsche Gesellschaft für Chemisches Apparatewesen, Chemistry Data Series Vol. 5,
    No. 1, 660 pp.

Sørensen, J. M., Magnussen, T., Rasmussen, P., and Fredenslund, A. A. (1979). Fluid
    Phase Equil. 2(4), 297-309.

Spaccamela Marchetti, E. and Saracco, G. (1958). Ann. Chim. (Rome), 48, 1371-1394.

Spalthoff, W. and Franck, E. U. (1957). Z. Elektrochem. 61(8), 993-1000.

Speedy, R. J. (1977). J. Chem. Soc. Faraday Trans. II 73(5), 714-721.

Squire, P. W. and Caines, C. M. (1905). Pharm. J. 74, 720, 784-786.

Srebrenik, S. and Cohen, S. (1976). J. Phys. Chem. 80(9), 996-999.

Srinivasan, K. R. and Kay, R. L. (1974). J. Chem. Phys. 60(9), 3645-3649.

Stacey, M., Tatlow, J. C., and Sharpe, A. G. (1963). Advances in Fluorine Chemistry,
    Vol. 3, Butterworths, London. 281 pp.

Stadnik, A. M. and El'tekov, Yu. A. (1975). Russ. J. Phys. Chem. 49(3), 452-453.

Stakhanova, M. S., Mikulin, G. I., Karapet'yants, M. Kh., Vlasenko, K. K., and
    Bazlova, I. V. (1970). Fizicheskaya Khimiya Rastvorov. Izd. Nauka, Symp., Moscow.

Stamford Research Institute (1965). Literature References to Physical Properties of
    Chlorocarbons. New York.

Staveley, L. A. K., Ed. (1971). The Characterisation of Chemical Purity of Organic
    Compounds. Butterworths, London.

Staveley, L. A. K. and Taylor, P. F. (1956). J. Chem. Soc., p. 200-209.

Stavely, L. A. K., Jeffes, J. H. E., and Moy, J. A. E. (1943). Trans. Faraday Soc.
    39(1), 5-13.

Staverman, A. J. (1941a). Recl. Trav. Chim. 60, 827-835.

Staverman, A. J. (1941b). Recl. Trav. Chim. 60, 836-841.

Stawiszynski, A. and Zawadzki, J. (1979). Powloki Ochr. 7(1), 26-34.

Steere, N. V., Ed. (1975). J. Chem. Educ. 52(9), A419-A425 and A468-A470.

Stepakoff, G. L. and Modica, A. P. (1972). The Solubility and Hydrolysis of Freon 114 in Fresh and Saline Water. Am. Chem. Soc., Dic. Water, Air, Waste Chem., Gen. Pap. 12, No. 1, p. 124-129.

Stepakoff, G. L. and Modica, A. P. (1973). Desalination 12, 85-105.

Stephen, C. R. and Little, D. M. (1961). Halothane. Williams & Wilkins, Baltimone, Md.

Stephen, H. and Stephen, T. (1963). Solubilities of Inorganic and Organic Compounds, Vols. 1 and 2. Pergamon, Oxford.

Sterbacek, Z., Biskup, B., and Tausk, P. (1979). Calculation of Properties Using Corresponding-State Methods. Elsevier, Amsterdam, 308 pp.

Steward, A., Allott, R. P., Cowles, A. L., and Mapleson, W. W. (1973-1974). Br. J. Anaesth. 45, 282-293 and 46(4), 310.

Stiel, L. I. and Thodos, G. (1962). J. Chem. Eng. Data 7(2), 234-236.

Stillinger, F. H. (1973). J. Solution Chem. 2(2/3), 141-158.

Stillinger, F. H. (1975). Theory and Molecular Models of Water. In Non-Simple Liquids (I. Prigogine and S. A. Rice, Eds.). Adv. Chem. Phys. Vol. 31. Wiley, New York. p. 1-102.

Stillinger, F. H. and Rahman, A. (1972). J. Chem. Phys. 57, 1281-1292.

Stockmayer, W. H. (Chairman) (1978). National Needs for Critically Evaluated Physical and Chemical Data. Natl. Acad. Sci., Washington, D. C., 69 pp.

Stokes, R. H. and Marsh, K. N. (1976). J. Chem. Thermodyn. 8(8), 709-723.

Stolyarenko, T. E., Stolyarenko, G. S., and Il'in, K. G. (1976). Tr. Novocherk. Politekh. Inst., No. 322, 110-116.

Storvick, T. S. and Sandler, S. I., Eds. (1977). Phase Equilibria and Fluid Properties in the Chemical Industry. Estimation and Correlation., Am. Chem. Soc. Symp. Ser. No. 60. Washington, D. C., 537 pp.

Strakhov, A. N., Krestov, G. A., and Abrosimo, V. K. (1975). Zh. Fiz. Khim. 49(6), 1583-1584.

Straub, J. and Scheffler, K., Eds. (1980). Water and Steam. Their Properties and Current Industrial Applications. Proc. 9th Int. Conf. on the Properties of Steam. Technische Univ. of München, Sept. 10-14 (1979). Pergamon, Oxford. 684 pp.

Streitwieser, A. (1963). Prog. Phys. Org. Chem. 1, 1-30.

Stumm, W., Ed. (1967). Equilibrium Concepts in Natural Water Systems. Am. Chem. Soc., Adv. Chem. Ser. No. 67. Washington, D. C.

Stumm, W. and Morgan, J. J. (1970). Analytical Chemistry. Wiley, New York. 583 pp.

Stupin, D. Yu. and Tevikov, V. N. (1976). Zh. Prikl. Khim. 49(10), 2337-2339.

Stupin, D. Yu., Tevikov, V. N., and Seleznev, A. P. (1978). Zh. Prikl. Khim. 51(3), 589-592.

Su, C.-W. and Goldberg, E. D. (1973). Nature 245, 27.

Sugden, T. M. and West, T. F., Eds. (1980). Chlorofluorocarbons in the Environment: The Aerosol Contraversy. Ellis Horwood, Chicester, England. 183 pp.

Suppan, P. (1968). J. Chem. Soc. A p. 3125-3133.

Surovy, J. and Dojcansky, J. (1976). Chem. Zvesti 30(5), 655-657.

Sutton, C. and Calder, J. A. (1974). Environ. Sci. Technol. 8(7), 654-657.

Sutton, C. and Calder, J. A. (1975). J. Chem. Eng. Data 20(3), 320-322.

Suzuki, K., Taniguchi, Y., Ishigami, T., and Tsuchiya, M. (1974). The Effect of Pressure on the Solubility of Naphthalene in Water and Aqueous Tetraalkylammonium Salt Solutions. Proc. 4th Int. Conf. on High Pressure. Kyoto, Japan. p. 615-618.

Svehla, R. A. (1962). Estimated Viscosities and Thermal Conductivities of Gases at High Temperatures. Natl. Aeronautics and Space Admin., NASA TR R-132. Washington, D. C., 120 pp.

Svetlanov, E. B., Velichko, S. M., Levinskii, M. I., Treger, Yu. A., and Flid, R. M. (1971). Russ. J. Phys. Chem. 45(4), 488-490.

Svoboda, V., Vesely, F., Holub, R., and Pick, J. (1977). Coll. Czech. Chem. Commun. 42(3), 943-951.

Swain, C. G. and Bader, R. F. W. (1960). Tetrahedron 10, 182-199.

Swain, C. G. and Thornton, E. R. (1962). J. Am. Chem. Soc. 84, 822-826.

Swarts, F. (1933). Bull. Soc. Chim. Belg. 42, 114-118.

Swern, D. (1957). Inclusion Compounds. In Kirk-Othmer Encyclopedia of Chemical Technology (1st suppl.) Interscience, New York. p. 429-448.

Swinnerton, J. W. and Lamontagne, R. E. (1974). Environ. Sci. Technol. 8(7), 657-663.

Swinton, F. L. (1976). Mixtures Containing a Fluorocarbon. In Chemical Thermodynamics, Vol. 2 (M. L. McGlashan, Ed.). Chemical Society, London. p. 147-173.

Symons, M. C. R. (1972). Nature 239, 257-259.

Symons, M. C. R. (1977). Structural Guide to Solvents and Solvation. Solvents--The Neglected Parameter, 2nd Solvents Symp., UMIST, Manchester. p. 10-23.

Szczepaniec-Cieciak, E., Dobrowska, B., and Lagan, J. M. (1977). Cryogenics 17(11), 621-627.

Tabor, D. (1970). Gases, Liquids, and Solids. Penguin, Harmondsworth, England. 290 pp.

Tager, A. A. and Kolmakova, L. K. (1980). Vysokomol. Soedin. 22A(3), 483-496.

Tamir, A. (1981a). Chem. Eng. Sci. 36(1), 37-46.

Tamir, A. (1981b). Fluid Phase Equil. 5(3/4), 199-206.

Tammann, G. (1926). Z. anorg. allg. Chem. 158, 17-24.

Tammann, G. and Krige, G. J. R. (1925). Z. anorg. allg. Chem. 146, 179-195.

Tanaka, M. (1971). Purification Coefficients and Extraction Constant of Non-Charged Species as Considered from the Regular Solution Theory. In Solvent Extraction, Proc. Int. Solvent Extraction Conf. ISEC 71, Paper 66, The Hague, April 19-23. Chem. Soc., London. p. 18-24 and D2-D4.

Tanford, C. (1973-1980). The Hydrophobic Effect: Formation of Miscelles and Biological
    Membranes. Wiley-Interscience, New York. 1st ed. (1973), 200 pp.; 2nd ed. (1980),
    234 pp.

Tanford, C. (1979). Interfacial Free Energy and the Hydrophobic Effect. Proc. Natl.
    Acad. Sci. 76(9), 4175-4176.

Tardiff, R. G. and Deinzer, M. (1973). Toxicity of Organic Compounds in Drinking
    Water. 15th Water Quality Conf., U-Ill, Urbana-Champaign, Ill. p. 23-37.

Tarrant, P., Ed. (1969). Fluorine Chemistry Reviews, Vol. 3. Dekker, New York. 156 pp.

Tassios, D. and Van Winkle, M. (1967). J. Chem. Eng. Data 12(4), 555-561.

Taylor, M. D. (1951). J. Am. Chem. Soc. 73, 315-317.

Teague, R. K. and Pings, C. J. (1968). J. Chem. Phys. 48(11), 4973-4984.

Tee, L. S., Gotoh, S., and Stewart, W. E. (1966). Ind. Eng. Chem. 5(3), 356-367.

Temperley, H. N. V. and Trevena, D. H. (1978). Liquids and Their Properties. A mol-
    ecular and Macroscopic Treatise with Applications. Ellis Horwood, Chichister,
    England. 274 pp.

Temperley, H. N. V., Rowlinson, J. S., and Rushbrooke, G. S. (1968). Physics of
    Simple Liquids. North-Holland, Amsterdam. 713 pp.

Teresawa, S., Itsuki, H., and Arakawa, S. (1975). J. Phys. Chem. 79(22), 2345-2351.

Tess, R. W., Ed. (1973). Solvents Theory and Practice. Adv. Chem. Ser. No. 124. Am.
    Chem. Soc., Washington, D. C. 227 pp.

Thermodynamic Research Center (1975). Selected Data on Mixtures, Series A, No, 1.
    A&M Univ., College Station, Tex. 97 pp.

Thomas, L. H. (1948a). J. Chem. Soc., p. 1345-1349.

Thomas, L. H. (1948b). J. Chem. Soc., p. 1349-1354.

Thomas, L. H. (1960). J. Chem. Soc., p. 4906-4914.

Thomas, L. H. (1963). J. Chem. Soc., p. 1995-2002.

Thomas, L. H. (1966). Trans. Faraday Soc. 62, 328-335.

Tiepel, E. W., and Gubbins, K. E. (1972). Can. J. Chem. Eng. 50(3), 361-365.

Tikhonova, N. K., Rogoniva, I. V., Goloborodkin, S. I., and Mozzhukhin, A. S. (1976a).
    J. Appl. Chem. USSR 49(2), 345-348.

Tikhonova, N. K., Rogovina, I. V., Goloborodkin, S. I., and Mozzhukhin, A. S. (1976b).
    Zh. Prikl. Khim. 49(2), 342-346.

Timimi, B. A. (1974). Electrochim. Acta 19, 149-158.

Timmermans, J. (1960). The Physico-Chemical Constants of Binary Systems in Concent-
    Rated Solutions, Vol. 4. Interscience, New York. 1331 pp.

Timmermans, J. and Lewin, J. (1953). Disc. Faraday Soc,, No. 15, 195-202.

Tokunaga, S., Manabe, M., and Koda, M. (1980). Niihama Kogyo Koto Semmon Gakko Kiyo
    Rikogaku Hen. 16, 96-101. (CA 93:156587c.)

Tokura, N. (1970). Kagaku To Kogyo (Osaka) 44(1), 64-73.

Tomanovskaya, V. F. and Kolotova, B. E. (1970). Freon: Properties and Use. Izdatel'stvo
    "Khimiya" Leningradskoe Otdelenie, Leningrad. 182 pp.

Tompkins, F. C., Chairman (1978). Structure and Motion in Molecular Liquids. Faraday
    Disc. Chem. Soc., No. 66, p. 1-313.

Tranchant, J. (1968). Bull. Soc. Chim. Fr. May, 2216-2220.

Treadwell, W. D. and Köhle, A. (1926). Helv. Chim. Acta 9, 681-691.

Treger, Yu. A., Flid, R. M., and Spektor, S. S. (1964). Russ. J. Phys. Chem. 38(2),
    253-255.

Tremaine, P. and Robinson, M. G. (1973). Can. J. Chem. 51, 1497-1503.

Treybal, R. E. (1963). Liquid Extraction (2nd ed.). McGraw-Hill, New York.

Troitskaya, V. S. and Tyulin, V. I. (1976). Vestn. Mosk. Univ., Khim. 17(1), 26-32.

Tsiklis, D. S. (1968). Handbook of Techniques in High-Pressure Research and En-
    gineering. Plenum, New York. 504 pp.

Tsiklis, D. S. and Svetlova, G. M. (1958). Zh. Fiz. Khim. 32, 1476-1480.

Tsonopoulos, C. (1970). Properties of Dilute Aqueous Solutions of Organic Solutes
    (Ph. D. thesis). Univ. of California, Berkeley, Calif. 187 pp.

Tsonopoulos, C. and Prausnitz, J. M. (1971). Ind. Eng. Chem. Fundam. 10(4), 593-600.

Tsykalo, A. L., Kontsov, M. M., Doroshenko, Zh. F., Kartseva, N. I., and Bagmet, A.
    D. (1975). High Temp. 13(6), 1193-1197.

Tugarinova, V. N., Miklashevskii, V. E., Alekseeva, N. P., and Yakovleva, G. P. (1962).
    Sanit. Okhr. Vodoemov Zagryaz. Prom. Stochnymi Vodami, No. 5, 285-307.

Tugarinova, V. N., Miklashevskii, V. E., Yakovleva, G. P., Rakhmanina, N. L., and
    Kogan, G. Z. (1965). Sanit. Okhr. Vodoemov Zagryaz. Prom. Stochnymi Vodami, No.
    7, 41-55.

Tyn, M. T. and Calus, W. F. (1975). Processing 21(4), 16-17.

Tyuzyo, K. (1957). Bull. Chem. Soc. Jap. 30, 851-856.

Udovenko, V. V. and Aleksandrova, L. I. (1960). Zh. Fiz. Khim. 34(6), 1366.

Udovenko, V. V. and Fatkulina, L. G. (1952a). Zh. Fiz. Khim. 26(6), 892-897.

Udovenko, V. V. and Fatkulina, L. G. (1952b). Zh. Fiz. Khim. 26(10), 1438-1447.

Ufnalski, W. (1978). Pol. J. Chem. 52(12), 2443-2453. (CA 91:182264y.)

Uhling, H. H. (1937). J. Phys. Chem. 41, 1215-1225.

Ulbrich, R. (1974). Z. Naturforsch. 29a, 338-341.

Ullmann Encyklopädie der technischen Chemie (1951-1972). Urban and Schwarzenberg,
    München. 19 Vols.

Union Carbide Chemical Co. (1959). Physical Properties of Synthetic Organic Chem-
    icals. Carbon Prod. Div., New York.

Union Carbide Chemicals Co. (1962). Physical Properties of Synthetic Organic Chem-
    icals. Chemical Dept. Rep. F-6136P. New York. 25 pp.

Union Carbide Corporation (1973). Chemicals and Plastics Physical Properties. (1973-
    1974 ed.). Chemicals and Plastics, Rep. F-44086, 2/73-30M. New York. 78 pp.

Urban, M. and Hobza, P. (1975a). Theor. Chim. Acta 36, 207-214.

Urban, M. and Hobza, P. (1975b). Theor. Chim. Acta 36, 215-220.

Utschick, H., Rupp, M., Kapitza, H., and Schubert, M. (1976). Chem. Tech. 28(6), 357-358.

Valentiner, S. (1927). Z. Phys. 42, 253-264.

Vallaud, A., Raymond, V., and Salmon, P. (1957). Les Solvants Chlores et L'Hygiene Industrielle. Inst. Natl. Securite pour le Prevention des Accidents du Travail et des Maladies Professionelles, Paris. 356 pp.

Valvani, S. C., Yalkowsky, S. H., and Amidon, G. L. (1976). J. Phys. Chem. 80(8), 829-835.

Van Arkel, A. E. (1946). Disc. Faraday Soc. 42B, 81-84.

Van Arkel, A. E. and Vles, S. E. (1936). Recl. Trav. Chim. 55, 407-411.

Vand, V. and Senior, W. A. (1965). J. Chem. Phys. 43, 1869-1884.

Van der Waals, J. D. (1873). Over de continuiteit von den has- en vloeistoftoestand. (Dissertation). Leiden, Holland.

Van der Waals, J. H. and Platteeux, J. C. (1957). Adv. Chem. Phys. 2, 1-57.

Vanderzee, C. E. and Rodenburg, W. Wm. (1970). J. Chem. Thermodyn. 2, 461-478.

Van Horn, L. D. (1966). A Study of Low Temperature Vapor-Liquid Equilibria of Light Hydrocarbons in Hydrocarbon Solvents: Chromatographic Measurement of Equilibrium Constants. Development of a Correlation for Such Systems (Ph. D. thesis). Rice Univ., Huston, Tex. 162 pp.

Van Laar, J. J. (1910). Z. phys. Chem. 72, 723-751.

Van Ness, H. C. (1964). Classical Thermodynamics of Non-Electrolyte Solutions. Pergamon, Oxford. 166 pp.

Van Slyke, D. D. and Neill, J. M. (1924). J. Biol. Chem. 61, 523-573.

Varshavskaya, S. P. (1967). Nauchn. Tr. Aspir. Ordinatorov 1-i Mosk. Med. Inst.

Varshavskaya, S. P. (1968). Gig. Sanit. 33(10), 15-21.

Vaughan, W. E. (1969). Dig. Lit. Dielectr. 33, 21-83.

Vaughan, W. E., Smyth, C. P., and Powles, J. G. (1972). Determination of Dielectric Constant and Loss. In Techniques of Chemistry, Vol. 1, Physical Methods of Chemistry, Part 4 (A. Weissberger, Ed.). Wiley-Interscience, New York. p. 351- 395.

Vavruch, I. (1978a). J. Colloid Interface Sci. 63(3), 600-601.

Vavruch, I. (1978b). J. Phys. Chem. 82(25), 2752-2753.

Verschueren, K. (1977). Handbook of Environmental Data on Organic Chemicals. Van Nostrand Reinhold, New York. 659 pp.

Vesala, A. (1973). Thermodynamics of Transfer of Electrolytes from Light to Heavy Water (Ph. D. dissertation). Univ. of Turku, Turku, Finland. 81 pp.

Vesala, A. (1974). Acta Chem. Scand. Ser. A 28(8), 839-845.

Veselovskii, P. F. (1975). Russ. J. Phys. Chem. 49(10), 1505-1507.

Vetere, A. (1977). Can. J. Chem. Eng. 55(1), 70-77.

Villard, P. (1890). C. R. Acad. Sci. 111, 302-305.

Vinogradov, S. N. and Linnell, R. H. (1971). Hydrogen Bonding. Van Nostrand Reinhold, New York. 314 pp.

Vlahakis, J. G., Tsao, T.-C., Richard, H. A., Barduhn, A. J., and Stern, S. A. (1969). Second Report on Hydrate Reaction Kinetics and Ice-Crystal Growth Rates. Res. Dev. Prog. Rep. No. 497. U. S. Dept. of the Interior, Washington, D. C., December. 44 pp.

Volokhonovich, I. E., Nosov, E. F., and Zorina, L. B. (1966). Russ. J. Phys. Chem. 40(1), 146-148.

Von Dahl, W. V. and Cole, R. H. (1966). J. Chem. Phys. 45(5), 1849-1850.

Von Stackelberg, M. (1949). Naturwissenschaften 21(12), 327-333, 359-362.

Von Stackelberg, M. (1954a). Z. Elektrochem. 58(2), 99-104.

Von Stackelberg, M. (1954b). Z. Elektrochem. 58(2), 104-109.

Von Stackelberg, M. (1954c). Z. Elektrochem. 58(3), 162-164.

Von Stackelberg, M. (1956). Recl. Trav. Chim. 75, 902-905.

Von Stackelberg, M. and Meinhold, W. (1954). Z. Elektrochem. 58(1), 40-45.

Von Stackelberg, M. and Meuthen, B. (1958). Z. Elektrochem. 62(2), 130-131.

Von Stackelberg, M. and Müller, H. R. (1954). Z. Elektrochem. 58(1), 25-39.

Voronovich, A. N., Lilich, L. S., and Khripum, M. K. (1969). Zh. Theor. Eksp. Khim. 5(5), 714-715.

Vorozcov, N. N. and Kobelev, V. A. (1938). Zh. Obshch. Khim. 8(12), 1106-1119.

Vukalovitch, M. P. (1967). Thermodynamic Properties of Water and Steem. Izdatel'stvo "Mashinostroyenye," Moscow. 80 pp.

Vuks, M. F. (1969). Russ. J. Phys. Chem. 43(9), 1239-1241.

Wada, G. (1979). Gendai Kagaku 101, 48-55.

Wakahayashi, T., Oki, S., Omori, T., and Suzuki, N. (1964). J. Inorg. Nucl. Chem. 26, 2255-2264.

Wall, T. T. and Hornig, D. F. (1965). J. Chem. Phys. 43, 2079-2087.

Wallnofer, P. R., Koniger, M., and Hutzinger, O. (1973). Analabs. Res. Notes 13(3), 14-16.

Walraevens, R., Trouillet, P., and Devos, A. (1974). Int. J. Chem. Kinet. 6, 777-786.

Walsh, F. and Mitchell, R. (1974). Nature 249(5458), 673-674.

Wang, D. I.-J. and Cheung, H. (1962). Chem. Eng. Prog. 58(11), 74-75.

Ward, R. B. (1976). Nature 261(5561), 540.

Washburn, E. W., Ed. (1928). International Critical Tables of Numerical Data, Physics, Chemistry, and Technology, Vol. 3. McGraw-Hill, New York.

Washburn, E. W., Ed. (1929). International Critical Tables of Numerical Data, Physics, Chemistry, and Technology, Vol. 4. McGraw-Hill, New York.

Wasik, S. P. and Brown, R. L. (1973). Determination of Halocarbon Solubility in Sea Water and the Analysis of Hydrocarbons in Water-Extracts. Proc. Joint Conf. Prev. Contr. Oil Spills., p. 223-227.

Watanabe, H. and Okamoto, J. (1978). Solubility of Mixed Monomers of Tetrafluoroethylene and Propylene in Water and Latex. Japan Atomic Energy Research Institute Rep. M-7548. Tokyo. May. 18 pp.

Watts, R. O. (1971). Rev. Pure Appl. Chem. 21, 167-188.

Watts, R. O. and McGee, I. J. (1977). Liquid State Chemical Physics. Wiley, New York. 350 pp.

Wauchope, R. D. (1970). The Solubility of Aromatic Hydrocarbons in Water (Ph. D. thesis). North Carolina State Univ., Raleigh, N. C., 61 pp.

Wauchope, R. D. and Getzen, F. W. (1972). J. Chem. Eng. Data 17(1), 38-41.

Wauchope, R. D. and Haque, R. (1972). Can. J. Chem. 50(2), 133-138.

Weast, R. C., Ed. (1976). Handbook of Chemistry and Physics (57th ed.). Chemical Rubber Co., Cleveland, Ohio.

Wei, J. (1977). AIChE J. Editorial. 23(3).

Weil, L., Dure, G., and Quentin, K.-E. (1974). Z. Wasser Abwasser Forsch. 7(6), 169-175.

Weinberg, V. L. and Yen, T. F. (1980). Fuel 59(5), 287-289.

Weiss, R. F. (1970). Deep-Sea Res. 17(4), 721-735.

Weiss, R. F. (1971). J. Chem. Eng. Data 16(2), 235-241.

Weissberger, A. (1955). Organic Solvents. Physical Properties and Methods of Purification. In Technique of Organic Chemistry, Vol. 7, Interscience, New York. 552 pp.

Weissberger, A., Ed. (1963). Physical Methods of Organic Chemistry (3rd ed.). Techniques of Organic Chemistry, Vol. 1, Part 1. Interscience, New York.

Weissberger, A. and Rossiter, B. W., Eds. (1972). Physical Methods of Chemistry. Techniques of Chemistry, Vol. 1, Part 4. Determination of Mass, Transport, and Electrical-Magnetic Properties. Wiley-Interscience, New York. 561 pp.

Weltner, W. (1955). J. Am. Chem. Soc. 77, 3941-3950.

Wen, W.-Y. and Hung, J. H. (1970). J. Phys. Chem. 74(1), 170-180.

Wen, W.-Y. and Muccitelli, J. A. (1979). J. Solution Chem. 8(3), 225-246.

Wentorf, R. H., Buehler, R. J., Hirschfelder, J. O., and Curtiss, C. F. (1950). J. Chem. Phys. 18(11), 1484-1500.

Wenzel, H. and Rupp, W. (1978). Chem. Eng. Sci. 33, 683-687.

Weres, O. and Rice, S. A. (1972). J. Am. Chem. Soc. 94, 8983-9002.

Wesson, L. G. (1948). Tables of Electric Dipole Moments. Technology Press, Massachusetts Institute of Technology, Cambridge, Mass., August. 90 pp.

Wheatland, A. B. and Smith, L. J. (1955). J. Appl. Chem. 5, 144-148.

Whelan, E. M. (1981). Nature 290(5803), 183.

Wichterle, I., Linek, J., and Hala, E. (1973-1978). Vapor-Liquid Equilibrium Data Bibliography. Suppl. 1 (1976): Suppl. 2 (1978). Elsevier, Amsterdam.

Wiese, C. S. and Griffin, D. A. (1978). Bull. Environ. Contam. Toxicol. 19(4), 403-411.

Wiley, M. A. (1976). Chem. Tech. $\underline{6}$(2), 134-141.

Wilhelm, E. and Battino, R. (1971). J. Chem. Thermodyn. $\underline{3}$, 379-392.

Wilhelm, E. and Battino, R. (1972). J. Chem. Phys. $\underline{56}$(1), 563-566.

Wilhelm, E. and Battino, R. (1973). Chem. Rev. $\underline{73}$(1), 1-9.

Wilhelm, E., Battino, R., and Wilcock, R. J. (1977). Chem. Rev. $\underline{77}$(2), 219-262.

Wilkniss, P. E., Lamontagne, R. A., Larson, R. E., Swinnerton, J. W., Dickson, C. R., and Thompson, T. (1973). Nat. Phys. Sci. $\underline{245}$(142), 45-47.

Williamson, A. G. (1966). An Introduction to Non-Electrolyte Solutions. Wiley, New York.

Williamson, A. T. (1944). Trans. Faraday Soc. $\underline{40}$, 421-436.

Wilson, G. M. (1964). J. Am. Chem. Soc. $\underline{86}$, 127-136.

Wilson, G. M. (1974). Infinite Dilution Activity Coefficients Estimation of One Binary Component from the Other. In Thermodynamics--Data and Correlations. AIChE Symp. Ser. Vol. 70, No. 140, p. 120-129.

Wilson, G. M. and Deal, C. H. (1962). Ind. Eng. Chem. Fundam. $\underline{1}$(1), 20-23.

Wilson, T. W. (1959). A Study of Equilibria Involved Between Darex Off-Gases and Solutions. A Literature Survey of the Solubility of Ar, Cl_2, HCl, NO, N_2, N_2O_3, NO_2, and O_2 in Water, Nitric Acid, Hydrochloric Acid, and Aqua Regia. Eng. Expt. Stat. Rep. AECU-4319. Georgia Institute of Technology, Atlanta. June 1. 61 pp.

Wing, J. (1956). Studies of the Solubility of Water in Aromatic Solvents Using Tritium as a Tracer (Ph. D. thesis). Purdue Univ., Lafayette, Ind. February. 116 pp.

Wing, J. and Johnston, W. H. (1957). J. Am. Chem. Soc. $\underline{79}$, 864-865.

Winsor, P. A. (1954). Solvent Properties of Amphiphilic Compounds. Butterworths, London. 204 pp.

Winterstein, H. and Hirschberg, E. (1927). Biochem. Z. $\underline{186}$, 172-177.

Wisniak, J. and Tamir, A. (1978). Mixing and Excess Thermodynamic Properties. A Literature Source Book. Elsevier, Amsterdam. **935** pp.

Wisniak, J. and Tamir, A. (1980). Liquid-Liquid Equilibrium and Extraction. A Literature Source Book. Physical Sciences Data 7. Elsevier, Amsterdam. 1252 pp.

Wittstruck, T. A., Brey, W. S., Buswell, A. M., and Rodebush, W. H. (1961). J. Chem. Eng. Data $\underline{6}$(3), 343-346.

Wofsy, S. C., McElroy, M. B., and Sze, N. D. (1975). Science $\underline{187}$(4176), 535-537.

Wohl, A. (1914). Z. phys. Chem. $\underline{87}$, 1-39.

Wolf, H. (1976). Vapour Pressure Studies on Hydrogen and Deuterium Bonding. In the Hydrogen Bond, Vol. 3 (P. Schuster, G. Zundel, and C. Sandorfy, Eds.). North-Holland, Amsterdam. p. 1225-1260.

Wolf, K. L. (1937). Trans. Faraday Soc. $\underline{33}$, 179-190.

Woodcock, L. V. (1971). Nature Phys. Sci. $\underline{232}$, 63-64.

Woodhead-Galloway, J. (1972). New Sci. $\underline{56}$(820), 399-403.

World Health Organization (1971). International Standards for Drinking-Water (3rd ed.). Genova. 72 pp.

Wright, W. H. and Schaffer, J. M. (1932). Am. J. Hyg. 16(2), 325-428.

Wyman, J. (1936). J. Chem. Soc. 58, 1482-1486.

Yaacobi, M. and Ben-Naim, A. (1974). J. Phys. Chem. 78(2), 175-178.

Yalkowsky, S. H. (1979). Ind. Eng. Chem. Fundam. 18(2), 108-111.

Yalkowsky, S. H. and Valvani, S. C. (1979). J. Chem. Eng. Data 24(2), 127-129.

Yalkowsky, S. H. and Valvani, S. C. (1980). J. Pharm. Sci. 69(8), 912-922.

Yalkowsky, S. H. and Zografi, G. (1972). J. Pharm. Sci. 61(5), 793-795.

Yalkowsky, S. H., Flynn, G. L., and Amidon, G. L. (1972). J. Pharm. Sci. 61(6), 983-984.

Yalkowsky, S. H., Flynn, G. L., and Slunick, T. G. (1972). J. Pharm. Sci. 61(6), 852-857.

Yalkowsky, S. H., Orr, R. J., and Valvani, S. C. (1979a). Ind. Eng. Chem. Fundam. 18(4), 351-353.

Yalkowsky, S. H., Valvani, S. C., and Amidon, G. L. (1976). J. Pharm. Sci. 65(10), 1488-1494.

Yalkowsky, S. H., Amidon, G. L., Zografi, G., and Flynn, G. L. (1975). J. Pharm. Sci. 64(1), 48-52.

Yen, L. C. and McKetta, J. J. (1962). AIChE J. 8(4), 501-507.

Yeo, K. O. (1973). Thermodynamic Studies of the Effects of Solvents on Molecular Complex Formation Equilibria: Orientation of Water Around Nonpolar Solutes in Aqueous Solutions (Ph. D. thesis). Univ. of Oklahoma, Norman, Okla. 117 pp.

Yokozeki, A. and Bauer, S. H. (1975). Top. Curr. Chem., No. 53, 71-119.

Young, D. R. and Heesen, T. C. (1974). Inputs of Chlorinated Hydrocarbons. Southern California Coastal Water Res. Proj. Chem. Prog. Annu. Rep. El Segundo, Calif. p. 97-99.

Younglove, B. A. (1972). J. Res. Natl. Bur. Std. 76A(1), 37-40.

Yukhnevich, G. V. and Karyakin, A. V. (1964). Dokl. Akad. Nauk SSSR 156, 556-559.

Zahradnik, R., Hobza, P., and Slanina, Z. (1975). Coll. Czech. Chem. Commun. 40(3), 799-808.

Zahradnik, R., Hobza, P., and Slanina, Z. (1976). Calculation of Henry Constants and Partition Coefficients Using Quantum Chemical Approach. Quantitative Structure-Activity Relationship (M. Tichy, Ed.). Birkhäuser Verlag, Basel, Switzerland, p. 217-230.

Zakharov, E. K. (1967). Russ. J. Phys. Chem. 41(2), 162-164.

Zakuskina, L. M., Martynov, Yu. M., and Oshin, L. A. (1975). Soviet Chem. Ind. 7(1), 831-832.

Zampachova, L. (1962). Chem. Prum. 12, 130-138.

Zarkarian, J. A., Anderson, F. E., Boyd, J. A., and Prausnitz, J. M. (1979). Ind. Eng. Chem. Process Des. Dev. 18(4), 657-661.

Zeininger, H. (1975). Löslichkeiten von Frigen 12 und Frigen 11 im Meerwasser. Hoechst Aktiengesellsch. 5957-III.1. Frankfurt. July 14. 6 pp.

Zellhoefer, G. F. (1939). Ind. Eng. Chem. $\underline{29}$(5), 548-551.

Zellhoefer, G. F. and Copley, M. J. (1938). J. Am. Chem. Soc. $\underline{60}$, 1343-1345.

Zellhoefer, G. F., Copley, M. J., and Marvel, C. S. (1938). J. Am. Chem. Soc. $\underline{60}$, 1337-1343.

Zhukoborskii, S. L. (1973). Kholod. Tekh., No. 11, 36-41.

Zielinski, A. Z. (1959). Chem. Stosowana $\underline{3}$, 377-384.

Zimm, B. H. (1950). J. Phys. Chem. $\underline{54}$(9), 1306-1317.

Zimmerman, H. K. (1952). Chem. Rev. $\underline{51}$(1), 25-65.

Zimmermann, H. (1970). Chem. Unserer Zeit $\underline{4}$, 69-79.

Ziolkowsky, B. (1970). Chem. Tech. Ind. $\underline{66}$(22), 801-806.

Zipp, A. P. (1973). Effects of Solvent and Pressure on the Spectra of Some Chemical and Biological Systems. (Ph. D. thesis). Univ. of Princeton, Princeton, N. J., 184 pp.

Zogorski, J. S., Allgeier, G. D., and Mullins, R. L. (1978). Removal of Chloroform from Drinking Water. Office of Water Res. Technol., Nat. Tech. Inf. Service, PB-285819, Washington, D. C. June. 86 pp.

Zundel, G. (1972). Hydration and Intermolecular Interaction. Academic, New York.

Zürcher, F. and Giger, W. (1976). Vom Wasser $\underline{47}$, 37-55.

Zwanzig, R. (1954). J. Chem. Phys. $\underline{22}$, 1420-1426.

Subject Index